Lecture Notes in Computer Science　　7651

Commenced Publication in 1973
Founding and Former Series Editors:
Gerhard Goos, Juris Hartmanis, and Jan van Leeuwen

X. Sean Wang   Isabel Cruz   Alex Delis
Guangyan Huang (Eds.)

Web Information Systems Engineering - WISE 2012

13th International Conference
Paphos, Cyprus, November 28-30, 2012
Proceedings

 Springer

Volume Editors

X. Sean Wang
Fudan University, Shanghai, China
E-mail: xywangcs@fudan.edu.cn

Isabel Cruz
The University of Illinois at Chicago, IL, USA
E-mail: isabelcfcruz@gmail.com

Alex Delis
University of Athens, Greece
E-mail: alex.delis@gmail.com

Guangyan Huang
Victoria University, Melbourne, VIC, Australia
E-mail: guangyan.huang@vu.edu.au

ISSN 0302-9743   e-ISSN 1611-3349
ISBN 978-3-642-35062-7   e-ISBN 978-3-642-35063-4
DOI 10.1007/978-3-642-35063-4
Springer Heidelberg Dordrecht London New York

Library of Congress Control Number: 2012951834

CR Subject Classification (1998): H.4-5, H.3.3-5, J.1, D.2, H.5, C.2

LNCS Sublibrary: SL 3 – Information Systems and Application, incl. Internet/Web
and HCI

Typesetting: Camera-ready by author, data conversion by Scientific Publishing Services, Chennai, India

Printed on acid-free paper

Springer is part of Springer Science+Business Media (www.springer.com)

Preface

Welcome to the proceedings of the 13th International Conference on Web Information Systems Engineering (WISE 2012) held in Paphos, Cyprus, in November 2012. WISE aims to provide an international forum for researchers, professionals, and industrial practitioners to share their knowledge in the rapidly growing area of Web technologies, methodologies, and applications. Previous WISE conferences were held in Hong Kong, China (2000), Kyoto, Japan (2001), Singapore (2002), Rome, Italy (2003), Brisbane, Australia (2004), New York, USA (2005), Wuhan, China (2006), Nancy, France (2007), Auckland, New Zealand (2008), Poznan, Poland (2009), Hong Kong, China (2010), and Sydney, Australia (2011).

WISE 2012 presented two eminent keynote speakers, Azer Bestavros of Boston University in Boston, MA, USA, and Ricardo Baeza-Yates, VP of Research for EMEA & LatAm at Yahoo! Research in Barcelona, Spain. We received 194 submissions of which 44 were selected as full papers with an acceptance rate of 23%; and another 13 were accepted as short contributions. The program also featured nine demonstration and nine "challenge" papers.

We wish to take this opportunity to thank the "Challenge" Co-chairs: Xiaofang Zhou and Weining Qian; Demo Co-chairs: Georgia Kapitsaki and Gustavo Rossi; Workshop Co-chairs: Armin Haller and Zhisheng Huang; Publication Chair: Guangyan Huang; Local Arrangements Chair: Petros Stratis; Publicity Co-chairs: Demetris Zeinalipour and Jing Yang; and Webmasters: Christos Mettouris and Zhi Qiao.

We would especially like to thank the Program Committee members and the external reviewers for a rigorous and robust reviewing process. Each paper received an average of 3.75 reviews (anonymous to the authors), and in total, we received 723 reviews. We are also grateful to University of Cyprus, the International WISE Society, and the Cyprus Tourist Organization for generously supporting our conference.

October 2012

Yanchun Zhang
George Angelos Papadopoulos
X. Sean Wang
Isabel Cruz
Alex Delis

Preface

Organization

General Co-chairs

Yanchun Zhang   Victoria University, Australia
George Angelos Papadopoulos   University of Cyprus, Cyprus

Program Co-chairs

X. Sean Wang   Fudan University, China
Isabel Cruz   The University of Illinois at Chicago, USA
Alex Delis   University of Athens, Greece

Demo Co-chairs

Georgia Kapitsaki   University of Cyprus, Cyprus
Gustavo Rossi   Universidad Nacional de La Plata, Argentina

WISE Challenge Program/Track Co-chairs

Xiaofang Zhou   University of Queensland, Australia
Weining Qian   East China Normal University, China

Workshop Co-chairs

Armin Haller   CSIRO, Australia
Zhisheng Huang   Vrije University of Amsterdam,
     The Netherlands

Publication Chair

Guangyan Huang   Victoria University, Australia

Local Arrangements Chair

Petros Stratis   Easyconferences Ltd.

Publicity Co-chairs

Demetris Zeinalipour   University of Cyprus, Cyprus
Jing Yang   Graduate University, Chinese Academy of
     Science, China

Program Committee

Karl Aberer	EPFL, Switzerland
Ashraf Aboulnaga	University of Waterloo, Canada
Marco Aiello	University of Groningen, The Netherlands
Huseyin Akcan	Izmir University of Economics, Turkey
Panayiotis Andreou	University of Cyprus, Cyprus
Amitabha Bagchi	Indian Institute of Technology Delhi, India
Wolf-Tilo Balke	University of Hannover, Germany
Luciano Baresi	Politecnico di Milano, Italy
Ladjel Bellatreche	LISI/ENSMA Poitiers University, France
Salima Benbernou	Université Paris Descartes, France
Elisa Bertino	Purdue University, USA
Azer Bestavros	Boston University, USA
Sourav Bhowmick	NTU, Singapore
Walter Binder	University of Lugano, Switzerland
Athman Bouguettaya	RMIT University, Australia
Shawn Bowers	Gonzaga University, USA
Wojciech Cellary	Poznan University of Economics, Poland
Fei Chen	HP Labs, USA
Zhiyuan Chen	University of Maryland Baltimore County, USA
Zhongqiang Chen	Yahoo! Inc., USA
Soon Ae Chun	City University of New York, USA
Oscar Corcho	Universidad Politécnica de Madrid, Spain
Philippe Cudré-Mauroux	University of Fribourg, Switzerland
Bin Cui	Peking University, China
Prasad Deshpande	IBM Research, India
Gill Dobbie	University of Auckland, New Zealand
Guozhu Dong	Wright State University, USA
Schahram Dustdar	The Vienna University of Technology, Austria
Rik Eshuis	Eindhoven Univ. of Technology, The Netherlands
Joao Eduardo Ferreira	University of Sao Paulo, Brazil
Dimitrios Georgakopoulos	CSIRO ICT Centre, Australia
Armin Haller	CSIRO, Australia
Michael Hausenblas	DERI, Ireland
Katja Hose	Max-Planck-Institut für Informatik, Germany
Zhisheng Huang	Vrije University Amsterdam, The Netherlands
Yoshiharu Ishikawa	Nagoya University, Japan
Ibrahim Kamel	University of Sharjah, UAE
Verena Kantere	Cyprus University of Technology, Cyprus
Dimka Karastoyanova	University of Stuttgart, Germany
Helen Karatza	Aristotle University of Thessaloniki, Greece

Ingmar Weber	Yahoo! Research, Spain
Josiane Xavier Parreira	National University of Ireland, Galway, Ireland
Hao Yan	LinkedIn, USA
Jian Yang	Macquarie University, Australia
Peter Z. Yeh	Accenture Technology Labs, USA
Qi Yu	Rochester Institute of Technology, USA
Pingpeng Yuan	Huazhong University of Science and Technology, China
Vladimir Zadorozhny	University of Pittsburgh, USA
Xiangliang Zhang	King Abdullah University of Science and Technology, Saudi Arabia
Gong X. Zhang	Oracle, USA
Xiaofang Zhou	University of Queensland, Australia

External Reviewers

Ahn, Tuan	Ait Sadoune, Idir	Al Aghbari, Zaher
Alam, Shafiq	Allahbakhsh, Mohammad	Aly, Ahmed
Araujo, Luciano V.	Baravalle, Andres	Baryannis, George
Becker, Karin	Beier, Felix	Belk, Marios
Benslimane, Djamal	Catasta, Michele	Cherniak, Andrii
Christoforou, Christoforos	Cordeiro, Daniel	Cuzzocrea, Alfredo
Degeler, Viktoriya	Dey, Atreyee	Difallah, Djellel Eddine
Dragut, Eduard	Dufour, Bruno	Emerencia, Ando
Fang, Qiong	García-Silva, Andrés	Georgiadis, Dimosthenis
Georgievski, Ilche	Germanakos, Panagiotis	Guo, Tian
Hasan, Omar	Hs, Bruhathi	Jain, Prateek
Jankowski, Jacek	Jiang, Di	Jiang, Wei
Jiang, Yu	Kaji, Nobuhiro	Kaldeli, Eirini
Karataev, Evgeny	Khalil, Ibrahim	Khazankin, Roman
Kim, Doo Soon	Konsolaki, Konstantina	Kritikos, Kyriakos
Laskaris, Alexandros	Lazovik, Alexander	Lee, Wookey
Li, Xian	Liarou, Erietta	Lu, Xuesong
Ma, Mike	Mahanti, Anirban	Marconi, Annapaola
Mathew, Sujith Samuel	Moon, Yang-Sae	Møller, Anders
Nguyen Ngoc, Chan	Nizamic, Faris	Nobari, Sadegh
Noor, Talal	Olteanu, Alexandra	Pagani, Giuliano Andrea
Palanisamy, Balaji	Pelechrinis, Konstantinos	Peternier, Achille
Poggi, Nicolas	Pont, Ana	Prodan, Radu
Prokofyev, Roman	Qahtan, Abdulhakim	Qin, Yongrui
Rahman, Samsur	Ramanath, Maya	Ranu, Sayan
Ranvier, Jean-Eudes	Rico, Mariano	Ryan, Caspar
Sarimbekov, Aibek	Sehic, Sanjin	Sharma, Divyasheel
Shebaro, Bilal	Singla, Parag	Song, Andy
Strauch, Steve	Sun, Haiyang	Taher, Yehia

Takai, Osvaldo K.
Tsakalozos, Konstantinos
Vosinakis, Spyros
Wieland, Matthias
Yan, Da
Ye, Henry
Zeginis, Chrysostomos

Thayasivam, Uthaya
Vasquez, Reymonrod
Vukojevic, Karolina
Wijaya, Tri Kurniawan
Yang, Zhenglu
Yongchareon, Sira
Zor, Sema

Tripathy, Rudra M.
Vingralek, Radek
Wang, Mingxue
Wylot, Marcin
Yao, Lina
Yoshinaga, Naoki

Table of Contents

Full Papers

Short Papers

Challenge Papers

Demo Papers

Improving the Performance
of Pipelined Query Processing with Skipping

Simon Jonassen and Svein Erik Bratsberg

Norwegian University of Science and Technology, Trondheim, Norway
{simonj,sveinbra}@idi.ntnu.no

Abstract. Web search engines need to provide high throughput and short query latency. Recent results show that pipelined query processing over a term-wise partitioned inverted index may have superior throughput. However, the query processing latency and scalability with respect to the collections size are the main challenges associated with this method. In this paper, we evaluate the effect of inverted index skipping on the performance of pipelined query processing. Further, we introduce a novel idea of using Max-Score pruning within pipelined query processing and a new term assignment heuristic, partitioning by Max-Score. Our current results indicate a significant improvement over the state-of-the-art approach and lead to several further optimizations, which include dynamic load balancing, intra-query concurrent processing and a hybrid combination between pipelined and non-pipelined execution.

1 Introduction

Two fundamental index partitioning methods, *term-wise* and *document-wise*, have been extensively compared during the last 20 years. The decision about which method gives the best performance relies on a number of system aspects such as query processing model, collection size, pruning techniques, query load, network and disk-access characteristics [3]. Rather than asking whether term-wise partitioning is generally more efficient than document-wise or not, we look at *pipelined query processing* [10] over a term-wise partitioned index and try to resolve the greatest challenges of the method. The advantage of term-wise partitioning is that a query containing l query terms involves only l processing nodes in the worst case. At the same time, it has to process only l posting lists in total. Whether l is the number of lexicon lookups or the total number of disk-accesses depends on the implementation. Pipelined query processing consists of routing a query bundle through the nodes responsible for any of the terms appearing in a particular query, modifying the accumulator set at each of these nodes and finally extracting the results at the last node. Compared to a traditional query processing approach, where each of the nodes retrieves the posting lists and the query processing itself is done solely by a ranker node, this method improves the workload distribution across the nodes and the overall performance [10].

In a recent book, Büttcher et al. [1] enlist three problems of pipelined query processing: poor scalability with increasing collections size, poor load balancing

X.S. Wang et al. (Eds.): WISE 2012, LNCS 7651, pp. 1–15, 2012.

and limited intra-query concurrency due to term-at-a-time/node-at-a-time processing. In this work, we mainly address the first problem, but also show how our solution can be used to solve the other two.

Our contributions in this work are as follows. First, we explore the idea of combining inverted index skipping and pipelined query processing for a term-wise partitioned inverted index. Herein, we present an efficient framework and a skipping optimization to the state-of-the-art approach. Second, we present a novel combination of pipelined query processing, Max-Score pruning and document-at-a-time sub-query processing. Third, we present an alternative posting-list assignment strategy, which improves the efficiency of the Max-Score method. Fourth, we evaluate our methods with a real implementation using the TREC GOV2 document collection, two large query sets and 9 processing nodes and suggest a number of further optimizations.

The remainder of this paper is organized as follows. We briefly review the related work in Section 2 and describe our framework in Section 3. We present our query processing optimizations in Sections 4 and 5 and our posting list assignment optimization in Section 6. We summarize our experiments in Section 7. In Section 8, we finally conclude the paper and suggest further improvements.

2   Related Work

Pipelined query processing was originally presented by Moffat et al. [10] as a method that retains the advantage of term-wise partitioning and additionally reduces the overhead at the ranker node. However, due to a high load imbalance, the method was shown to be less efficient and less scalable than document-wise partitioning. In the following work by Moffat et al. [9] and Webber [13], the load balancing was improved by assigning posting lists in a fill-smallest fashion according to the workload (posting list size × past query frequency) associated with each term. The authors suggested also to multi-replicate the posting lists with the highest workloads and allow the broker to chose the replica dependent on the load. Additionally, Webber [13] has presented a strategy where each node may ask other nodes to report their current load in order to choose the next node in the route. The results reported by Moffat et al. [9] showed a significant improvement over the original (term-wise) approach, but not the document-wise approach. Under extended evaluation with a reduced main memory size performed by Webber [13] pipelined query processing outperformed document-wise partitioning in terms of maximum throughput. However, document-wise partitioning demonstrated shorter latency at low multiprogramming levels.

In a recent publication [4], we have addressed the problem of the increased query latency due to a strict node-at-a-time execution and presented a semi-pipelined approach, which combines parallel disk access and decompression and pipelined evaluation. Additionally, we suggested to switch between semi- and non-pipelined processing dependent on the estimated network cost and an alternative query routing strategy. While the results of this work indicate an ultimate trade-off between the two methods, it has several important limitations.

First, similar to the original description [13], it requires to fetch and decompress posting lists completely. This minimizes the number of disk-accesses, but leaves a large memory footprint and endangers scalability of the method. Second, while the original approach fetches only one of a query's posting lists at a time, the semi-pipelined approach fetches all of them and keeps them in main memory until the query has been processed up-to and including this node. This improves the query latency, but increases the memory usage even further. Third, the compression methods used by the index may be considered outdated.

In the current work, we follow an alternative direction and try to improve pipelined query processing by means of skipping optimizations. In a recent work [5], we have investigated these techniques for a non-distributed index. In particular, we have presented a self-skipping inverted index designed specifically for NewPFoR [14]. Further, we have presented a complete description of document-at-a-time Max-Score. The Max-Score heuristic was originally presented by Turtle and Flood [12] without specifying enough details for skipping itself and later a very different description was given by Strohman et al. [11]. Our algorithm combines the advantage of both previous descriptions. Finally, we have presented a skipping version of the space-limited pruning algorithm by Lester et al. [7], which is the method used in the original description of pipelined query processing.

As a part of the current work, in Section 8, we outline several further extensions. One of these, intra-query concurrent pipelined processing, has already been evaluated in our recent work [6]. The preliminary baseline of [6] is an early version of the Max-Score optimization we are about to present, $\text{MSD}_s^\star$, and the results of [6] are therefore directly applicable to this method.

In contrast to the previous work, we use skipping to resolve the limitations of distributed, pipelined query processing. Herein, we present several skipping optimizations and a new term assignment strategy. In contrast to the previously presented assignment optimizations [2,8,15], our strategy does not try to assign co-occurring terms to the same node or to do load balancing, but rather to maximize the pruning efficiency. Additionally, it opens a possibility for dynamic load balancing with low repartitioning overhead and hybrid query processing. Finally, different from [4,5,6], our experiments use two query logs with very different characteristics, and a varied collection size.

3   Preliminaries

For a given textual query q, we look at the problem of finding the k best documents according to a similarity score $sim(d,q) = \sum_{t \in q} s(t,d)$, where s is a function, such as Okapi BM25, estimating the relevance of a document d to a term t. For this reason, we look at document-ordered inverted lists. Each list I_t stores an ordered sequence of document IDs where the term t appears and an associated number of occurrences of the term in the document $f_{t,d}$.

Our search engine runs on $n+1$ processing nodes, one of which works as a query broker and the remaining n work as query processing nodes. For each of the indexed collection terms we build an inverted list and assign it to one of the processing nodes. By default, we use a hash-based term assignment strategy. Additionally to the inverted index, each query processing node maintains a small lexicon (storing term IDs, inverted file pointers, document and collection frequencies), a small replica of the document dictionary (storing the number of tokens contained in each document) and both partition and collections statistics. The query broker node stores a full document dictionary (document IDs, names and lengths), lexicon (tokens, term IDs, collection and document frequencies, IDs of the nodes a term is assigned to) and collection statistics. At runtime, the inverted index resides on disk and the remaining structures are in main memory.

The broker is responsible for receiving queries, doing all necessary preprocessing and issuing query bundles. The broker uses its lexicon to look-up query terms and to partition the query into several sub-queries consisting of the terms assigned to the same node. Further, it calculates a query route (i.e., a particular order to visit the query nodes) and creates a query bundle. The bundle contains term IDs, query frequencies and an initially empty document-ordered accumulator set. Accumulators represent document IDs and partial scores. Finally, the broker sends the query bundle to the first node in the route.

When a node receives a bundle, it decompresses its content and starts a new query processing task. First, it uses its own lexicon to find the placement of the posting list. In the next step, the node processes its own posting data with respect to the received accumulators and creates a new accumulator set. Then, it updates the query bundle with the new accumulator set and transfers it to the next node. Alternatively, the last node in the route extracts the top-k results, sorts them by descending score and returns them to the broker.

4   Improving Pipelined Query Processing with Skipping

In this section, we describe the optimization to the state-of-the-art query processing approach. With this approach, the terms within each query are ordered by increasing collection frequency F_t and the query itself is routed by increasing minimum F_t. Once a bundle is received by a node, the node extracts the accumulators and initiates posting list iterators. We use Algorithm 1 to describe the following processing step performed by the query processor. The space-limited pruning method itself has been originally presented by Lester et al. [7] and the skipping optimization for a non-distributed index has already been presented in our previous work [5]. Therefore, the goal of this section is to describe the improvements to pipelined query processing. However, for the sake of intelligibility we also explain the most important details derived from the prior work [5,7].

In the following, i_t denotes the iterator of the posting list I_t, which provides methods to get the document ID, frequency, score and position of the current posting, advance to the next posting or to the first posting having $d \geq d'$ (both $next()$ and $skipTo(d')$ return $false$ if the end has been reached), or reset to the beginning. Further, f_t denotes the document frequency of t (i.e., number of documents) and F_t - its collection frequency (i.e., number of occurrences), and $|A|$ - the current size of the accumulator set A.

Algorithm 1. processBundle(b_q) with skip-optimized space-limited pruning

---

Data: iterators $\{i_t\}$ and lexicon entries $\{l_t\}$ for $t \in q_j$ sorted by ascending F_t, accumulator set A, bundle attribute v, system parameters avg_dl, use_opt, L, $h_{\max}$

1 **foreach** $t \in q_j$ **do**
2      $A' \leftarrow A$, $A \leftarrow \emptyset$, $skipmode \leftarrow false$, $p \leftarrow \lceil f_t/L \rceil$;
3   **if** $f_t < L$ **then**
4         $p \leftarrow f_t + 1$, $h \leftarrow 0$;
5   **else if** $v = 0$ **then**
6   set h to the maximum of the first p frequencies retrieved from i_t,
         $v \leftarrow s(l_t, h, avg\_dl)$, $i_t.reset()$;
7   **else**
8   **if** $use\_opt = true$ **and** $s(l_t, h_{\max}, avg\_dl) < v$ **then** $skipmode \leftarrow true$;
9   **else** find $h \in [1, h_{\max}]$ s.t. $s(l_t, h, avg\_dl) \geq v$;
10   **if** $skipmode = false$ **then**
11        $s \leftarrow \max(1, \lfloor (h+1)/2 \rfloor)$, $size_0 \leftarrow |A'|$; merge A' and candidates from i_t into A:
         calculate $i_t.s()$ only when $i_t.d() \in A'$ or $i_t.f() \geq h$, prune candidates having $s < v$;
         Each time $i_t.pos() = p$: $pred \leftarrow |A| + |A'| + (f_t - p) \times (|A| + |A'| - size_0)/p$, **if**
         $pred > 1.2 \times L$ **then** $h \leftarrow h + s$ **else if** $pred < L/1.2$ **then** $h \leftarrow \max(0, h - s)$ **endif**,
         $v \leftarrow s(l_t, h, avg\_dl)$, $s \leftarrow \lfloor (s+1)/2 \rfloor$, $p \leftarrow 2 \times p$;
12   **else**
13   **foreach** accumulator $\langle d', s \rangle \in A'$ **do**
14   **if** $i_t.d() < d'$ **then if** $i_t.skipTo(d') = false$ **then** add remaining accumulators
           s.t. $s \geq v$ to A, proceed to the next term (line 1);
15   **if** $i_t.d() = d'$ **then** $s \leftarrow s + i_t.s()$;
16   **if** $s \geq v$ **then** add $\langle d', s \rangle$ to A;

17 **if** it is the last node in the route **then** use a min-heap to find the k-best candidates from A, sort and return them to the broker **else** update b_q with A and v and send it to the next node;

The idea behind the pruning method (lines 1-7, 9, 11) is to restrict the accumulator set to a target size L. As the algorithm shows, the posting lists of a particular sub-query $q_j \subseteq q$ are evaluated term-at-a-time. As long the document frequency f_t of the current term is below L, each posting is scored and merged with the existing accumulator set. Otherwise, the algorithm estimates a frequency threshold h and a score threshold v, just as suggested by Lester et al. For this reason, h is set to the maximum frequency among the first $p = \lceil f_t/L \rceil$ postings. The rationale here is that, if one out of p postings will pass the frequency filter $f_{t,d} \geq h$, the total number of such postings will be L. Further, in order to prune existing accumulators, v is calculated from h. Differently from Lester et al., our score computation uses also the length of a document. For this reason we calculate the threshold score using l_t, h and the average document length avg_dl. Now, the algorithm is able to prune the existing accumulators having score $s < v$ and avoid scoring postings having $f_{t,d} < h$ and not matching in among the existing accumulators. Each time p postings of I_t has been processed, it predicts the size of the resulting accumulator set. Then, it either increases or decreases the frequency threshold, updates the score threshold and finally cools-down the threshold variation. Finally, if the thresholds have already been defined, it uses the previously computed v to find the corresponding value h. Similar to the implementation in Zettair[1], we try only the values between 1 and a system-specific maximum value $h_{\max}$. Additionally, we apply binary search to reduce the number of score computations.

---

[1] http://www.seg.rmit.edu.au/zettair/

Our optimization (lines 8, 10, 12-16) suggests to switch the query processing into a conjunctive skip-mode when $s(l_t, h_{\max}, avg\_dl)$ is below the previously computed v. In this mode, a posting is scored only when there is already an existing accumulator with the same document ID. For the first sub-query, b_q is initiated with $v = 0$. After processing a sub-query q_j, b_q is updated with the current A and v and forwarded to the next node. This means that each query starts in the normal, disjunctive mode. When the optimization constraint holds, it switches into the conjunctive mode. However, if the next posting list is shorter than L, the processing will switch back to the normal mode, but it will proceed to prune the accumulators having $s < v$.

The benefit of our optimization depends also on the inverted index implementation. We apply the layout described in our previous work [5]. With the basic index, each posting list is divided into groups of 128 entries, which are stored as two chunks containing 128 document ID deltas and 128 frequencies, both compressed with NewPFoR. To support skipping, we build a hierarchy of skip-pointers, which consist of an end-document ID and an end-offset pointer to a chunk level below. Further, we calculate deltas and compress these in chunks of 128 entries using NewPFoR. Next, we prefix-traverse the logical tree while writing to disk in order to minimize the size of skip-pointers and optimize reading. At the processing time, each posting list iterator maintains one chunk from each skip-level decompressed in main memory and applies buffering while reading from disk. Different from the previous work, we decompress frequency chunks only when at least one of the corresponding frequencies has been requested.

With the basic index, the cost of a skip is proportional to the number of blocks (I/O) and chunks (decompression) between the two positions. With the self-skipping index, the operation is done climbing the logical hierarchy up (using already decompressed data) and down (reading and decompressing new data when necessary). Therefore, the upper bound cost of the operation in the number of decompressed chunks and read blocks is $O(\log(D))$.

5   Max-Score Optimization of Pipelined Query Processing

The main drawback of the query processing methods discussed in the previous section lies in the unsafe pruning strategy. Additionally, these techniques are limited to term-at-a-time and node-at-a-time query processing. For this reason, we suggest an alternative query processing approach employing document-at-a-time processing of sub-queries, and later we show how this new method can be extended to provide intra-query parallelism. In order to guarantee safe pruning, we look at the Max-Score heuristic [5,11,12]. To give a better explanation, first we describe the idea for a non-distributed scenario, $q_j = q$, then how it can be applied to pipelined query processing, and finally, present the algorithm.

At indexing time, we pre-compute an upper-bound score $\hat{s}_t$ for each posting list I_t (i.e., the maximum score that can be achieved by any posting). At query processing time, we order $q_j = \{t_1, \ldots, t_l\}$ by decreasing $\hat{s}_t$ and use a_i to denote the maximum score of terms $\{t_i, \ldots, t_l\}$, i.e., $a_i = \sum \hat{s}_t$ s.t $t \in \{t_i, \ldots, t_l\}$. Further, q_j

is processed document-at-a-time and each iteration of the algorithm selects a new candidate, accumulates its score and eventually inserts it into a k-entry min-heap. The idea behind Max-Score is to prune the candidates that cannot enter the heap. As terms are always processed in order t_1 to t_l, they can be viewed as two subsets $\{t_1, \ldots, t_r\}$ (required) and $\{t_{r+1}, \ldots, t_l\}$ (optional), where r is the smallest integer such that $a_r \geq \tilde{s}$ and $\tilde{s}$ is the current k-th best score. It is easy to see that candidates that do not match any of the required terms cannot enter the heap. Therefore, the candidate selection can be based only on the required terms, which also have shorter posting lists. Once a candidate is selected, the terms are evaluated in order and the optional term iterators are advanced with a skip. Finally, at any point, a partially scored candidate can be pruned if its partial score plus the maximum remaining contribution is below the score of the current k-th best candidate, i.e., $s + a_i < \tilde{s}$. The description is so far similar to [5].

Now we explain how to apply these ideas to pipelined query processing. At query processing time, the broker fetches maximum scores along with term ID and location information and includes them in the query bundle. Therefore, when b_q arrives at a particular node, the information about the maximum scores of the terms in the current sub-query, $\hat{s}_t$ s.t. $t \in q_j$, and the maximum contribution of the remaining sub-queries, $\tilde{a} = \sum \hat{s}_t$ s.t. $t \in q_i$, $q_i \subseteq q$ and $i > j$, are available to the query processor. b_q itself is routed by decreasing maximum $\hat{s}_t$ among the sub-queries. Each query processor treats the received accumulator set just as a posting list iterator with $\hat{s}_t$ set to highest score within the set, and processes the sub-query document-at-a-time. Therefore, the required subset is defined by $a_r \geq \tilde{s} - \tilde{a}$, any candidate can be pruned whenever $s + a_i < \tilde{s} - \tilde{a}$ holds, and finally, the candidates with partial scores $s \geq \tilde{s} - \tilde{a}$ have to be transferred to the next node as a modified accumulator set.

We use Algorithm 2 to describe the final query processing approach. First, the algorithm prepares for query processing, calculates a values and defines the required set (lines 1-4). As long as the required set is non-empty, the iterators are processed document-at-a-time (lines 5-13). Each iteration advances the recently used iterators of the required set, selects a new candidate (lines 6-7) and accumulates its score (lines 8-11). If an iterator reaches the end of the posting list, it is removed from the iterator set and the a values of remaining iterators and r are updated (lines 6 and 10). If a candidate succeeds to reach a score higher than the pruning threshold ($s \geq \tilde{s} - \tilde{a}$), it is inserted in the accumulator set (line 12). Potentially, it may also be inserted into the candidate heap (line 13). In this case, $\tilde{s}$ may also be updated. When a sub-query is fully processed, if this is the last node in the route, the candidate heap has to be sorted and returned to the broker as the final result set (lines 14-15). Otherwise, non-pruned accumulators have to be transferred to the next node. In this case, prior to the final transfer, an extra pass through the accumulator set removes candidates having $s < \tilde{s} - \tilde{a}$, which are false positives due to a monotonically increasing pruning threshold. In order to facilitate pruning, $\tilde{s}$ is initiated with the value received from the previous node, or 0 for the first node. In practice, the last node in the route does not have to store non-pruned accumulators, but only the candidate heap.

Algorithm 2. processBundle(b_q) with DAAT Max-Score optimization

---

Data: iterators $\{i_1, \ldots, i_l\}$ and maximum scores $\{\hat{s}_1, \ldots, \hat{s}_l\}$ sorted by descending $\hat{s}$, accumulator set A, bundle attributes $\tilde{s}$ and $\tilde{a}$

1 **if** $A \neq \emptyset$ **then** $l \leftarrow l+1$, **for** $x \leftarrow l$ to 1 **do** $i_x \leftarrow i_{x-1}, \hat{s}_x \leftarrow \hat{s}_{x-1}$ **endfor**, set i_1 to be A's iterator and $\hat{s}_1$ to be A's maximum score;

2 $a_l \leftarrow \hat{s}_l$, **for** $x \leftarrow l-1$ to 1 **do** $a_x \leftarrow a_{x+1}+s_x$ **endfor**, $A' \leftarrow \emptyset$, $minHeap \leftarrow \emptyset$;

3 $r \leftarrow l$, **while** $r > 0$ and $a_r < \tilde{s}-\tilde{a}$ **do** $r \leftarrow r-1$;

4 $d' \leftarrow -1$;

5 **while** $r > 0$ **do**

6   advance iterators having $i_{x \leq r}.d() = d'$, if $i_x.next() = false$: close i_x, update i, s and a sets, decrement l, recompute r (similar to line 3, break if $r=0$);

7      $d' \leftarrow \min(i_{x \leq r}.d())$, $s \leftarrow 0$;

8   **for** $x \leftarrow 1$ to l **do**

9   **if** $s+a_x < \tilde{s}-\tilde{a}$ **then** break and proceed to selection of the next candidate (line 5);

10   **if** $x > r$ and $i_x.d() < d'$ **then** advance i_x to d', if $i_x.skipTo(d') = false$: close i_x, update i, s, a, l and r (break if $r=0$) and proceed to the next iterator (line 8);

11   **if** $i_x.d() = d'$ **then** $s \leftarrow s+i_x.s()$;

12   **if** $s \geq \tilde{s}-\tilde{a}$ **then** add $\langle d', s \rangle$ to A';

13   **if** $s > minHeap.minScore$ **then** add $\langle d', s \rangle$ to $minHeap$, $\tilde{s} \leftarrow \max(\tilde{s}, s)$;

14 **if** it is the last node in the route **then**

15   retrieve candidates from $minHeap$, sort and return them to the broker;

16 **else** $A' \leftarrow \{\langle d, s \rangle \in A'$ s.t. $s \geq \tilde{s}-\tilde{a}\}$, update b_q with A and $\tilde{s}$ and send it to the next node;

---

6 Max-Score Optimization of Term Assignment

The pruning performance of the Max-Score optimization can be limited when long posting lists appear early in the pipeline. Therefore, we suggest to assign posting lists by decreasing $\hat{s}_t$, such that the first node gets posting lists with highest $\hat{s}_t$ and the last node gets posting lists with lowest $\hat{s}_t$. For simplicity, we use equally sized partitions. Since $\hat{s}_t$ increases as f_t decreases (because most of the similarity functions, including BM25, use the inverse document frequency), this strategy implies that the first node now stores only short posting lists and the last node stores a mix of long and short posting lists, and the nodes in between store posting lists with short-to-moderate lengths.

As we show in the next section, this technique significantly improves the performance, but struggles with a high load imbalance. Beyond the experiments presented in the next section, we have tried several load balancing approaches similar to Moffat et al. [9], such as estimating the workload associated with each term t in a past query log Q', $L_t = |I_t| \times f_{t,Q'}$, where $f_{t,Q'}$ is the frequency of t in Q' and $|I_t|$ is the size of I_t, and splitting the index so the accumulated past load would be balanced across the nodes. However, since the load estimator does not take skipping into account, it overestimates the load of long posting lists and assigns nearly half of the index to the first node. As a result, the load imbalance gets only worse. Therefore, we leave load balancing as an important direction for further work and outline a possible solution in Section 8.

7 Experimental Results

For our experiments, we index the 426GB TREC GOV2 corpus. With stemming and stop-word removal applied, it contains 15.4 mil. unique terms, 25.2 mil.

Table 1. Impact of the query processing method on the precision and recall of the Adhoc Retrieaval Topics 701-850

Method	MAP	P@10	Recall
Full/MSD	.153903	.530872	.274597
LT1M	.153746	.530872	.274262
LT100K	.152376	.528859	.270955†
LT50K	.152378	.530872	.271049
LT25K	.150602†	.528188	.267361†
LT10K	.145877‡	.516107†	.256386$^\sharp$
AND	.155073	.531544	.268126

Table 2. Maximum and average document frequency and sample covariance between the document and query frequency distributions in the evaluated query sets

Query set	$\max(f_t)$	$\mathrm{avg}(f_t)$	$\mathrm{cov}(f_t, f_{t,Q})$
A04-06	11256870	997968	149393
E05	11256870	223091	2266801
E06	11256870	290747	5999742

documents, 4.7 bil. pointers and 16.3 bil. tokens. With 8 index partitions, the resulting distributed index is 9.3GB in total, while a corresponding monolithic index is 7.6GB. Most of the overhead (1.54GB) comes from a short replicated version of the document dictionary. Skipping pointers increase the size by additional 87.1MB and the resulting index contains 279 647 posting lists with one skip level, 15 201 with two and 377 with three levels.

We run our experiments on a 9 node cluster. Each node has two 2.0GHz Quad-Core CPUs, 8GB memory and a SATA disk. The nodes are interconnected with a Gigabit network. Our framework[2] is implemented in Java. It uses Java NIO and Netty for efficient disk access and network transfer. For disk access, we use 16KB buffers and the default GNU/Linux OS caching policy (hence, we reset the disk caches before each run). Further, queries are preprocessed in the same way as the document collection and evaluated using the Okapi BM25 model.

We evaluate the following query processing methods. Full/non-pruned evaluation (Full), space-limited pruning described in Section 4 (LT denotes the state-of-the-art method, and SLT denotes the skip-optimized version), Max-Score optimized evaluation described in Section 5 (MSD), and finally an evaluation with intersection semantics and document-at-a-time sub-query processing (AND). We use a subscript N (e.g., SLT$_N$) to denote an execution on a non-optimized index and S on a self-skipping index. To limit the number of experiments, the maximum number of top-results k is fixed at 100. For LT we vary the accumulator set target size L, hence LT1M corresponds to $L = 1\,000\,000$ and LT10K to $L = 10\,000$. Finally, we fix the $h_{\max}$ used by LT/SLT (see Sec. 4) at 2000, which we find to be suitable for our index.

In order to evaluate the impact of the query processing optimizations on the retrieval performance, we use the TREC Terabyte Track Adhoc Retrieval Topics and Relevance Judgments 701-850 from 2004, 2005 and 2006. We use documents with relevance judgments 1 and 2 as a ground truth and consider MAP, precision at 10 results (P@10) and recall at k results as retrieval performance indicators. Table 1 shows the averages over the whole query set. We focus on result degradation with the space-limited pruning (LT) compared to a full evaluation (Full),

---

[2] https://github.com/s-j/laika

which is the retrieval performance baseline. Beyond the average results, we apply a paired t-test at the query level to check the degree of significance. We use † to mark the significance at 0.05 level, ‡ at 0.01 and ♯ at 0.001.

Table 1 shows how the retrieval performance of LT degrades with decreasing L. Degradation becomes significant at lower values of L. The evaluation measures of LT50K and LT100K are different, but without statistical significance when compared to each other. Beyond the presented data, the results obtained with Full and LT10M are identical, and the results obtained with SLT are identical to LT for $L \geq 10\ 000$. From these results, we consider $L \geq 50\ 000$ as a suitable choice with respect to the precision and recall with $k = 100$ (while $L \leq 25\ 000$ is not) and keep LT100K, LT50K and LT25K for further experiments.

Our observations confirm that the results obtained by Full and MSD are identical. Furthermore, we observe a higher precision (MAP and P@10) with AND compared to Full, although the difference is statistically insignificant. A closer look has shown that AND performs better for topics 701-751 and 751-800 and worse for topics 801-850 (no significance). However, with $k = 1000$, Full has both a higher MAP (0.271470 versus 0.255441, no significance) and a higher recall (0.662522 versus 0.596165, significance at 0.01).

In order to evaluate the algorithmic performance, we use two subsets of TREC Terabyte Track Efficiency Topics from 2005 (E05) and 2006 (E06). Both subsets contain 20 000 queries that match at least one indexed term, where the first 5 000 queries are used for warm-up and the next 15 000 to measure the actual performance. To simulate the effect of a result cache, neither of the sets contains duplicated queries.

We observe that E06, which contains government-specific queries, implies higher query processing load than E05, which contains Web-specific queries. Therefore, we consider these two sets as a better (E05) and a worse case (E06) scenarios. We use Table 2 to illustrate the difference between the query sets, including the Adhoc topics marked as A04-06. As the table shows, the term with highest document frequency (i.e., posting list length) is the same in all three of the sets (which is the term 'state'), but the average document frequency and the sample covariance between the document frequency and the query set frequency are significantly different. A04-06 is a very short query set, with a flat query frequency distribution and missing a long-tail distribution among the document frequencies. This explains a high average document frequency and a low covariance. Finally, the results for E05 and E06 illustrate the load difference between these two sets. E06 contains terms with both longer posting lists and a higher correlation between the query and document frequencies.

Figure 1 illustrates the average latency per query (milliseconds) and overall throughput (queries per second) with varied multiprogramming levels. Points on each plot correspond to 1, 8, 16, 24, 32, 48, 56 and 64 concurrent queries (cq). Each run was repeated twice (with the disk cache being reset each time) and the average value was reported. We report results for all methods except Full, which was too slow – even when the multiprogramming level set to 1, a query took on average 474ms for E05, and 1121ms for E06. Therefore, we consider LT_N

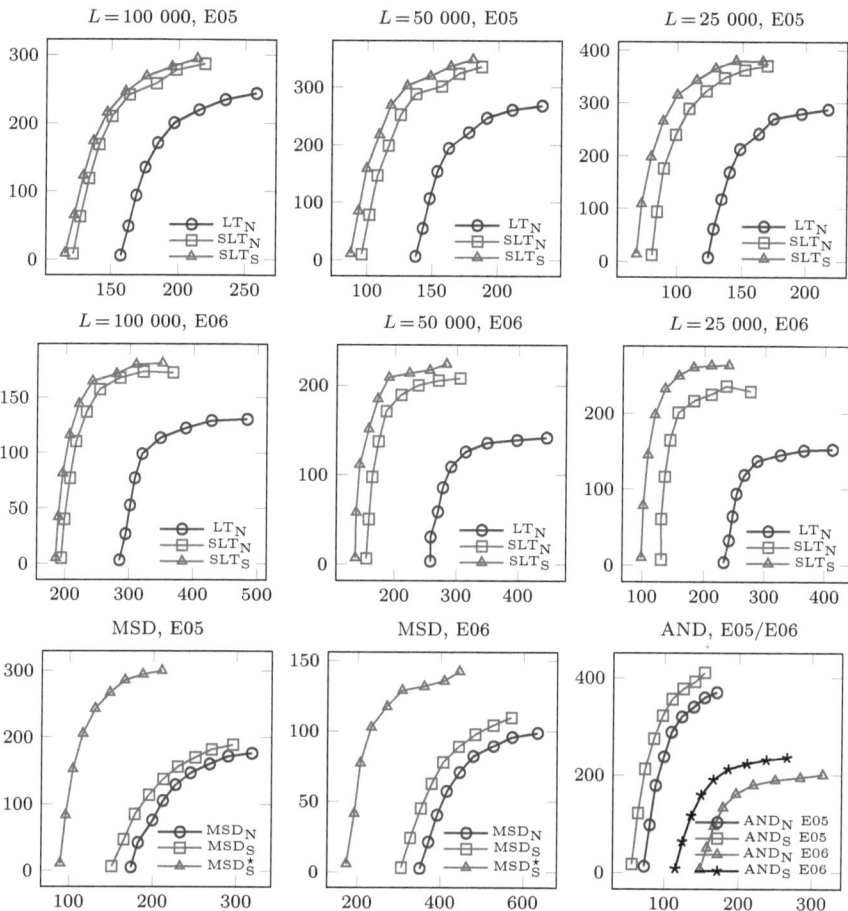

Fig. 1. Throughput (y-axis, qps) and latency (ms) with varied multiprogramming

as the time performance baseline. Our results show that for both E05 and E06, the skipping optimization to LT (SLT_N) significantly improves the performance and the improvement increases with smaller L. Skipping support in the inverted index (SLT_S) provides a further improvement. While the index optimization is not as significant as the algorithmic optimization, we observe that (for E06) the improvement increases as L decreases. We explain these results using Figure 2, which illustrates the number of blocks read, chunks decompressed, unique document IDs evaluated, scores computed and accumulators sent and received by each node, normalized by the evaluation set query count. The figure shows both the average (across the nodes) and the maximum (one node) counts. As the results show, the improvement from $Full_N$ to LT_N lies in reduced score computation and network transfer, which improve as L decreases. SLT_N further improves the number of candidates been considered (Doc.IDs) and SLT_S improves the amount of data been decompressed (Chunks). However, even with $L = 25\,000$ there is no

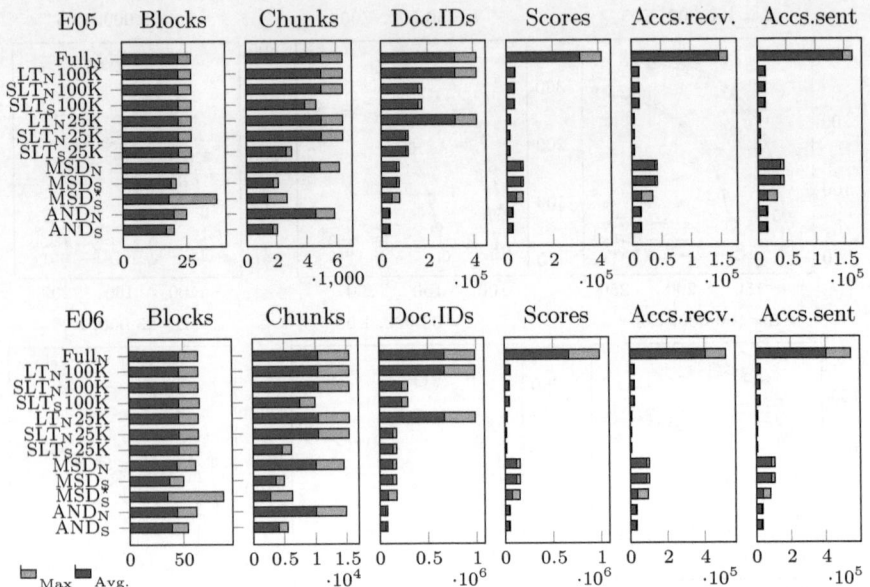

Fig. 2. Maximum and average number of processed entities per node

additional savings in data read from disk (Blocks), which can be explained by a relatively large block size (16KB).

As we show in Figure 1, the Max-Score optimization (MSD_N) gains a modest improvement from the self-skipping index (MSD_S) and a significant improvement from the further term-assignment optimization ($MSD_N^\star$). For E05, $MSD_N^\star$ performs as good as $SLT_S 100K$, but for E06 it struggles with increasing query latency when compared to $SLT_S 100K$. However, having in mind that MSD is equivalent to a full (non-pruned) evaluation, it shows a very good performance. As Figure 2 shows, the Max-Score optimizations significantly improve the total amount of read, decompressed and evaluated data. While these methods increase the number of score computations and the number of transferred accumulators compared to the LT optimizations, they show a significant improvement compared to Full. Finally, our results show that the main challenge of $MSD_N^\star$ is an increased load imbalance, which is the ratio between the maximum and the average counts. This issue should be investigated in the future.

As illustrated in Figures 1 and 2, skipping support in the index improves the performance of AND by reducing the amount of read, decompressed and evaluated data. For E06 it improves the latency at 1cq by 22% and the throughput at 64cq by 18%, for E05 it improves the latency at 1cq by 29% and the throughput at 64cq by 11%. Overall, AND performs better than $SLT_S 25K$ on E05 and slightly worse on E06. As having a better result quality than LT25K on A701-850, we consider AND as a good alternative to the space-limited pruning with a low accumulator set target size (L).

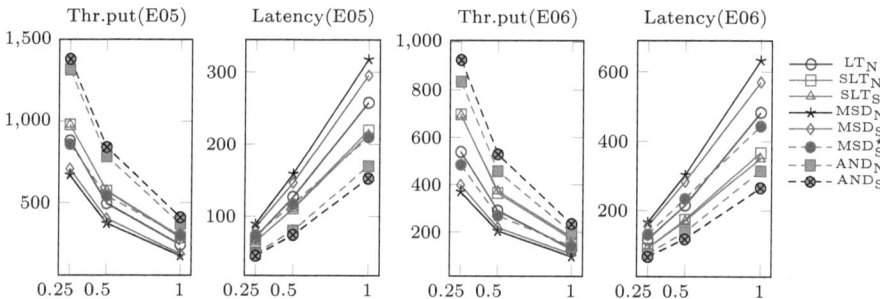

Fig. 3. Throughput (qps) and latency (ms) with varied collection size (x-axis)

Finally, we address performance linearity in Figure 3. In these results, we keep the multiprogramming level at 64cq and vary the collection size to $1/2$ and $1/4$. For LT/SLT we use $L = 100\,000$ scaled with the collection size. The results show that with increasing index size, the methods converge in the absolute throughput and diverge in the latency. In our opinion, the best behavior is given by SLT, MSD and AND. However, our results do not guarantee the performance for a collection larger than the GOV2. This should be addressed in future.

8   Conclusions and Further Work

We have presented and evaluated several skipping optimizations to pipelined query processing. For SLT_N and SLT_S our results indicate a significant improvement over the baseline approach. We also came up with a pruning approach ($MSD_S^\star$) that provides a result quality equivalent to a non-pruned evaluation, while having a considerably good performance. Further, we have observed that processing queries with conjunctive semantics (AND) provides good retrieval performance and efficient query processing. Although the state-of-the-art approach considers disjunctive (OR) queries, in future, we would like to take a closer look at AND queries. Finally, based on our current results, we outline three techniques that can further improve the performance of $MSD_S^\star$:

Dynamic Load Balancing. The load balancing of $MSD_S^\star$ can be improved by gradually moving the posting lists with the highest or the lowest $\hat{s}$ values to one of the neighbouring nodes. Compared to the previously presented fill-smallest and graph-partitioning techniques, this approach will reduce the network volume at repartitioning and can be done dynamically. In order to avoid moving data back-and-forth, we can further replicate the bordering posting lists and fine-tune partitions at the lexicon level, without actual repartitioning.

Hybrid Query Processing. $MSD_S^\star$ tends to place the shortest posting lists (corresponding to rare terms) on the first nodes. Therefore, by transferring these (complete) posting lists to a node corresponding to a later sub-query we can remove decompression, processing and accumulator transfer from the first few nodes (with an additional opportunity for parallelism). The node receiving the posting lists will in this case substitute an accumulator set with a few short posting list. In order to minimize the load on the sending node and the overall network load, the nodes can further cache the received lists.

Intra-query Concurrent Processing. Document-at-a-time processing of sub-queries allows to transfer accumulators to the next node as soon as possible. This feature can be utilized to provide intra-query concurrency and improve the performance at low multiprogramming levels. In [6], we have already evaluated an extension to an earlier version of MSD_8^*. The optimization splits the document ID range into several sub-ranges, called fragments, and does intra-query parallelization at the fragment level, both across the nodes and on the same node. The experiments [6] with a smaller subset of the TREC 2005 Efficiency Topics indicated that this optimization allows to reach a similar peak-throughput at nearly half of the latency. We assume this to be applicable to our current method, however a further evaluation for the 2006 topics is needed.

Acknowledgments. This work was supported by the iAd Centre and funded by the Norwegian University of Science and Technology and the Research Council of Norway.

References

1. Büttcher, S., Clarke, C.L.A., Cormack, G.V.: Information Retrieval: Implementing and Evaluating Search Engines (2010)
2. Cambazoglu, B.B., Aykanat, C.: A term-based inverted index organization for communication-efficient parallel query processing. In: IFIP NPC (2006)
3. Jonassen, S., Bratsberg, S.E.: Impact of the Query Model and System Settings on Performance of Distributed Inverted Indexes. In: NIK (2009)
4. Jonassen, S., Bratsberg, S.E.: A Combined Semi-pipelined Query Processing Architecture for Distributed Full-Text Retrieval. In: Chen, L., Triantafillou, P., Suel, T. (eds.) WISE 2010. LNCS, vol. 6488, pp. 587–601. Springer, Heidelberg (2010)
5. Jonassen, S., Bratsberg, S.E.: Efficient Compressed Inverted Index Skipping for Disjunctive Text-Queries. In: Clough, P., Foley, C., Gurrin, C., Jones, G.J.F., Kraaij, W., Lee, H., Mudoch, V. (eds.) ECIR 2011. LNCS, vol. 6611, pp. 530–542. Springer, Heidelberg (2011)
6. Jonassen, S., Bratsberg, S.E.: Intra-query Concurrent Pipelined Processing for Distributed Full-Text Retrieval. In: Baeza-Yates, R., de Vries, A.P., Zaragoza, H., Cambazoglu, B.B., Murdock, V., Lempel, R., Silvestri, F. (eds.) ECIR 2012. LNCS, vol. 7224, pp. 413–425. Springer, Heidelberg (2012)
7. Lester, N., Moffat, A., Webber, W., Zobel, J.: Space-Limited Ranked Query Evaluation Using Adaptive Pruning. In: Ngu, A.H.H., Kitsuregawa, M., Neuhold, E.J., Chung, J.-Y., Sheng, Q.Z. (eds.) WISE 2005. LNCS, vol. 3806, pp. 470–477. Springer, Heidelberg (2005)
8. Lucchese, C., Orlando, S., Perego, R., Silvestri, F.: Mining query logs to optimize index partitioning in parallel web search engines. In: InfoScale (2007)
9. Moffat, A., Webber, W., Zobel, J.: Load balancing for term-distributed parallel retrieval. In: SIGIR (2006)

10. Moffat, A., Webber, W., Zobel, J., Baeza-Yates, R.: A pipelined architecture for distributed text query evaluation. Inf. Retr. (2007)
11. Strohman, T., Turtle, H., Croft, W.B.: Optimization strategies for complex queries. In: SIGIR (2005)
12. Turtle, H., Flood, J.: Query evaluation: strategies and optimizations. Inf. Proc. and Manag. (1995)
13. Webber, W.: Design and evaluation of a pipelined distributed information retrieval architecture. Master's thesis (2007)
14. Yan, H., Ding, S., Suel, T.: Inverted index compression and query processing with optimized document ordering. In: WWW (2009)
15. Zhang, J., Suel, T.: Optimized inverted list assignment in distributed search engine architectures. In: IPDPS (2007)

Learning to Find Comparable Entities on the Web

Xiaojiang Huang, Xiaojun Wan[*], and Jianguo Xiao

Institute of Computer Science and Technology & The MOE Key Laboratory
of Computational Linguistics, Peking University, Beijing 100871, China
{huangxiaojiang,wanxiaojun,xiaojianguo}@pku.edu.cn

Abstract. Comparison is a popular way for people to discover the commonality
and difference between two entities (e.g. product, person, company, event, etc.).
It would be very useful to automatically provide comparison results for the user.
The prerequisite step of this task is to find comparable entities. In this paper, we
propose a novel Web mining system to address the task of finding comparable
entities for a given single entity. First, the system uses a bootstrapping method to
find candidate entities for the given entity through natural language analysis in
the snippets of search engine results. Then, the system uses set expansion
techniques to find more candidate entities though semi-structured HTML
analysis in the downloaded web pages. Finally, the system uses a supervised
learning method to classify the candidate entities into either comparable or
incomparable by incorporating linguistic, statistical and semantic features.
Experimental results demonstrate that our proposed framework can outperform
the baseline systems.

Keywords: Comparable Entity Finding, Set Expansion, Web Mining.

1   Introduction

Comparison is a popular way for people to understand the commonality and difference
between two or more objects before they make a decision. For example, the users
usually compare different types of cell-phones (e.g. "*Samsung Galaxy*", "*Nokia
Lumia*") on the Web before they decide to buy a particular cell-phone (e.g. "*Apple
iPhone*"). Similarly, users often compare a few similar events (e.g. "*Iraq war*",
"*Kosovo war*") if they want to know more about one particular event (e.g. "*Afghanistan
war*").

It would be very useful to automatically provide comparison results for the users.
The comparison results can be mined from the web pages relevant to the comparable
objects. Given an object, the whole comparison system usually consists of three steps:
comparable objects finding, relevant web pages retrieval, and comparable results
mining. In this paper, we focus on the first step and constrain the objects to be entities
(i.e. persons, companies, products, events, etc.), i.e. the task of finding comparable
entities for a given entities that the user has interest in. We use the entity name to denote
the entity. Note that two entities are comparable only if they are of the same kind.

---

[*] Corresponding author.

X.S. Wang et al. (Eds.): WISE 2012, LNCS 7651, pp. 16–29, 2012.

For example, *"Microsoft Windows"* can be compared with *"Linux"*, but cannot be compared with *"IBM"*.

So far, few previous works have addressed the problem of comparable entities finding. Jain and Pantel [1] proposed a framework for identifying comparable entities on the web. They only make use of natural language patterns to extract entities, and the system suffers from erroneous outputs. Another related task is set expansion of entities. In set expansion, the user issues a query consisting of a small number of seeds (e.g., *"America"*, *"Japan"*) where each seed is a member of some target set. The answer to the query is a list of other probable elements of the target set (e.g., *"Iraq"*, *"China"*, *"Russia"*, etc) [2]. Google Sets[1] and SEAL[2] are two online systems for set expansion. So far these systems work well when two or more seed examples are given, however, their performance are not satisfactory when the users issue only one seed example. The ability of finding comparable entities for only one seed entity not only alleviates the search burden for human users, but also is crucial for utilization by automatic techniques in other NLP and Web Mining tasks. For example, a news recommendation system may recommend related news about a comparable entity of the main entity in a given news report. It is not possible for the recommendation system to provide extra sample entities for expansion if they are not mentioned in the news article.

Inspired by these systems, we formalize the problem of comparable entity finding as the task of set expansion with only a single seed entity as follows:

The user issues a single seed entity q (e.g., "America"), and the task aims to find a list of other probable entity $x_1, ..., x_k$ (e.g., "Iraq", "China", "Russia", etc), where each entity in the list is comparable with the seed entity, i.e. all the entities are of the same kind and share some common attributes.

In this study, we propose a novel system named FCEW using a three-stage approach to address this problem. Given a seed entity (or query entity), we first construct an initial set of candidate comparable entities by using a bootstrapping method to learn text patterns and extract comparable entities from web search results in a dual mode. Based on the initial candidate set, we then employ the set expansion technique to find more candidate comparable entities from the downloaded web pages. The final step is to determine whether each candidate entity in the expanded set is truly comparable, by using a SVM classifier with various linguistic, statistical and semantic features. Our proposed system differs from [1] and [2] in that our system makes full use of linguistic, statistical and semantic clues and machine learning technologies.

We perform experiments on a manually labeled dataset and the experimental results demonstrate that our proposed system can outperform Google Sets and SEAL in the task of finding comparable entities for a given seed entity. The results also demonstrate that the step of entity classification plays an important role in our system.

The rest of this paper is organized as follows: Section 2 introduces the related works. The proposed system is described in Section 3. The experimental results are presented in Section 4. Section 5 discusses some issues about our system. Lastly we conclude this paper in Section 6.

---

[1] It has been shut-down in September, 2011.
[2] http://www.rcwang.com/seal/

2   Related Work

The comparable entity mining task is related to several NLP and Text Mining tasks. In this section, we give a brief review of existing work in related domains, including comparative analysis and set expansion.

2.1   Comparative Analysis

Comparative analysis aims to extract the commonality and difference among objects. It has been widely studied in the linguistic and literature areas, and recently have attracts much attention in the NLP and Web mining communities.

Jain and Pantel propose a system to identify comparable pairs by employing information extraction techniques over a collection of web pages and query logs [1]. The extraction system finds occurrences of seed instances in plain text, learns extraction patterns based on the context between the instances, and then apply the patterns to the text to identify new comparable pairs. Bao et al. use several manually written linguistic patterns to get the pages which may contain information of competitors and extract candidates of competitors [3]. Jindal and Liu propose methods to identify comparative sentences in English text [4], and then extract elements of comparisons, including compared objects in the comparisons [5] by using automatically learned sequential pattern of words in the comparative sentences. Li et al. propose to mine comparable entities from comparative questions using a weakly-supervised bootstrapping method [6]. The above approaches only rely on linguistic and text information. Liu et al. [7] present a system named Compare&Contrast that uses the Web to discover comparable cases for news stories, documents about similar situations but involving distinct entities. The system performs comparable case discovering on the document level.

Zhai et al. [8] propose a cross-collection mixture model to discover comparative themes across all collections. Several models have been proposed to generate comparative summary of comparable topics [9-11]. Sun et al. [12] and Luo et al. [13] propose comparative search systems, which compare two objects by using each object as a query to a search engine and then matching the results. Feldman et al. [14] have examined to extract comparison between products from forum discussions.

2.2   Set Expansion

Set expansion aims to expand a given partial set of entities into a more complete set. Google Sets was a well-known online system that does set expansion using Web search and mining techniques. The method underlying Google Sets is unknown due to commercial privacy, but it most likely performs as described in [15].The method first extracts a list by considering HTML tags, tables, commas or semicolons in web pages, and then ranks all items using the on-topic and off-topic models. SEAL [2] is another online demo system for set expansion of named entities. It automatically constructs page-specific wrappers and then extracts new entities using those wrappers. The weakness of

SEAL lies in that it requires at least two seed entities as input. According to [16], SEAL performs the best with four seed entities. However, in the task of comparable entities finding, it will bring a great burden to the users if the system asks the users to provide more seed entities. Iterative SEAL [16] has been proposed to allow a user to provide a large number of seeds (e.g. ten seeds). We can see that neither SEAL nor Iterative SEAL can address the task of finding comparable entities for a given single entity. Dalvi et al. propose a method to find coordinate-term clusters by merging HTML table columns [17]. Ohshima et al. propose a method for searching coordinate terms using the semantic function of the conjunction "OR" [18]. Distribution similarity [19][20] and topic information [21] are also used in the set expansion task.

Set expansion techniques have been successfully used for improving question answering (QA) when the expected answer is a list of entities belonging to a certain class [22].

3   The Proposed System

In this section, we represent the technical details of our proposed comparable entity extraction system. We first give an overview of the system, including the flowchart of the system and the introduction of the main steps, and then describe each step in details in the following subsections, respectively.

3.1   System Overview

The proposed system is named FCEW (Finding Comparable Entities on the Web). It takes an entity name as the input, and generates a list of entities which are comparable to the given entity as the output. The novelty of FCEW is as follows:

1) FCEW makes full use of natural language text analysis and semi-structured HTML analysis to get candidate comparable objects for a single given entity.

2) FCEW employs a classifier to determine the appropriateness of each candidate comparable entities based on a number of linguistic, statistical and semantic features.

Fig. 1 gives the flowchart of the proposed system, and the example result for each step is attached in the dashed rectangle. Our system consists of three steps: the step of initial candidate set construction aims to extract an initial set of candidate comparable entities from Web search results using only the given single entity; the step of candidate set expansion aims to extract more candidate entities from downloaded web pages by using set expansion techniques; the step of comparable entity classification aims to determine whether the candidate entity is a truly comparable with the given entity, by employing a classifier.

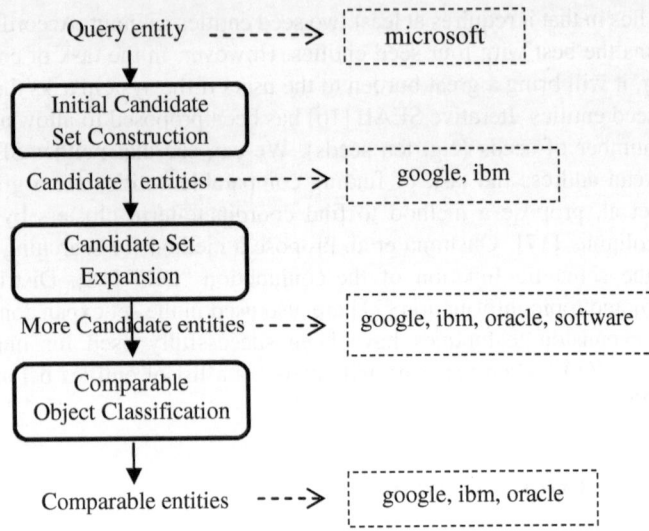

Fig. 1. Flowchart of our system

Note that the proposed system relies on a Web search engine to retrieve search results in each step. In this study, we adopt the widely used Google Search[3], and other search engines are deemed to work well, too.

3.2   Initial Candidate Set Construction

Given a single query entity q, this step aims to find a few candidate comparable entities through natural language text analysis. The initial candidate set is denoted as C_{init}. This step is based on the top 200 result snippets (including titles) returned by Google Search.

Pattern matching is a straightforward method for finding comparable entities for the query entity. However, it is not a trivial task to write all the patterns by hand because of the large number of specific patterns. Instead, we use the bootstrapping method to derive patterns and extract comparable entities in a dual mode. In this study, we define a pattern as follows:

*prefix <**entity1**> middle <**entity2**> suffix*

where the *prefix*, *middle* and *suffix* are strings containing one or more words, or punctuation marks. *<**entity1**>* and *<**entity2**>* are two comparable entities, and usually one is the query entity, and the other is the candidate entity. We assume that the middle string is not null, while the prefix string and the suffix string may be null. For example, the suffix of the pattern "*both <**entity1**> and <**entity2**>*" is null.

Figure 2 illustrates the procedure of the bootstrapping method. Because the bootstrapping method is semi-supervised, it requires either a small number of seed entities or a few seed patterns as input. In this study, we use the following two general patterns as seed patterns for any query entity:

---

[3] http://www.google.com

> *such as <comparable entity> and <query entity>*
> *both <comparable entity> and <query entity>*

```
Input: Query entity q, Pool_pattern = {seed patterns};
Output: candidate comparable entities C_init;

BOOTSTRAPPING ALGORITHM:

1. Candidate entity extraction: extract candidate comparable entities for
q based on patterns in Pool_pattern, then score each extracted entity  and add
top ranked entities into C_init;
2. Pattern derivation: derive new patterns based on the entity pairs between
q and any entity in C_init, then score each derived pattern .and add top ranked
patterns into Pool_pattern.
3. Go to Step 1 and iterate for several times;
```

Fig. 2. Framework of the bootstrapping algorithm

Candidate Entities Extraction. This step aims to extract candidate comparable entities by using pattern matching in the search results. Note that if we only issue the query entity (e.g. "*microsoft*") to Google, it cannot guarantee that the returned search results contain any pattern. Therefore, we formulate a composed query by combining the query entity and each pattern in $Pool_{pattern}$ as follows:

$$\text{"query entity"} + \text{"prefix"} + \text{"middle"} + \text{"suffix"}$$

We issue the query into Google Search and extract the titles and snippets in the top 200 search results. The snippets are segmented into sentences using heuristic rules. A title is considered as a single sentence.

We then apply each pattern in $Pool_{pattern}$ on each sentence to extract candidate comparable entities as follows:

1. For a pattern in which the comparable entity is enveloped by non-null strings, it is easy to extract candidate comparable entitys by pattern matching. For example, we can extract "*ibm*" from sentence "...*both **ibm** and **microsoft**...*" by using the pattern "*both <comparable entity> and <query entity>*", where "*microsoft*" is the query entity.

2. For a pattern in which the comparable entity is not enveloped by non-null strings, it is not easy to decide the cutting point of the entity string.   For example, it is not easy to extract "*ibm*" from "...*both **microsoft** and **ibm** are*" by directly using the pattern "*both <query entity> and <comparable entity>*", where "*microsoft*" is the query entity. We notice that a candidate entity is always a noun or noun phrase, so we to use an in-house NP tagger to extract the noun phrase after "*and*", and then use the noun phrase as the candidate entity. For example, the NP tagger can determine that "*ibm*" is a noun phrase and thus we use "*ibm*" as a candidate comparable entity.

We then assign a score to each distinct entity according to its frequency count. The extracted entities are then ranked according to the scores and the top 10% entities are added into C_{init}.

Pattern Derivation. This step aims to derive new patterns from the search results. For each pair between the query entity and any candidate entity in C_{init}, we perform the following steps to derive new patterns:

1. A composed query consisting of the two entities (e.g *"microsoft"+"ibm"*) is issued into Google Search and we obtain the titles and snippets of the top 200 search results.

2. The sentences containing both the two entities are extracted.

3. The sentences are grouped into a few clusters according to the middle string and the order between the two entities. A cluster with less than five sentences is filtered because it is trivial.

4. Within each cluster, we extract the longest common prefix before the first entity and the longest common suffix after the second entity. The longest common prefix/suffix is defined loosely in this study, i.e. a longest common prefix/suffix should be contained in 80% sentences in the cluster. The prefix or suffix may be null. The pattern is then composed of the prefix string, middle string, suffix string and the entity order.

The patterns are more or less trustworthy and we assign a score to each pattern to indicate the quality of the pattern. And we calculate the precision of each pattern as the pattern score as in [23]. Each pattern is applied to extract candidate entities and the precision is the ratio between the number of pattern presence with the correct entity and the total number of pattern presence. Here, we considered the entities in C_{init} as correct entities. Finally, the top 10% patterns are added into the pattern pool.

3.3   Candidate Set Expansion

This step aims to expand the initial candidate set C_{init} into a more complete set C_{expand}. We note that the initial candidate set is constructed based only on text analysis of titles and snippets in the search results. However, the full web pages are semi-structured and there are many clues in the web pages for extracting comparable entities. For example, a list of items or entities are usually embedded with HTML tags such as "<UL>" and "</UL>", or "<TD>" and "</TD>". We believe that items or entities within the same class will appear in similar formatting structures on the same web page. Thus, the characteristic of semi-structured web pages can be exploited to find more comparable entities for a few seed entities.

Wang and Cohen [2] have proposed the SEAL extractor for set expansion of named entities using the Web. The extractor consists of a wrapper generator and a ranker. The wrapper generator automatically constructs page-dependent wrappers from a few seed entities. A wrapper is defined by two character strings to specify the left-context and right-context necessary for an entity to be extracted from a page. These strings are maximally-long contexts that bracket at least one occurrence of every seed entity on a page. The ranker ranks entities based on a graph walk model.

In this study, we employ the same wrapper construction algorithm to learn wrappers from each web page containing seed entities. Because the algorithm requires multiple entities as input and in this study, we use three entities as seed entities. The seed entities include the query entity and two randomly selected entities in the initial candidate set.

Then we employ the SEAL wrapper construction algorithm [2] to learn wrappers. Based on the wrappers, we can extract candidate entities from the web pages.

In the experiments, we perform the above wrapper construction and entity extraction for three times based on different sets of seed entities, consisting of two random entities in C_{init} and the query entity.

Finally, we rank the extracted entities by their total frequency counts of being extracted by any wrapper from any web page. Note that this simple ranking method can achieve comparable performance to the graph walk based ranker [2]. The entities with frequency count larger than 10 are added and the expanded candidate set is denoted as C_{expand}.

3.4   Comparable Entity Classification

The previous two semi-supervised steps may bring noisy entities into the expanded candidate set, and this step aims to distinguish the truly comparable entities from the noisy entities by using a supervised binary classifier. Formally, given the query entity q and each entity x_i in the expanded set, the task aims to classify the entity pair into either "true" or "false", where "true" indicates that the x_i is comparable with q, and "false" indicates that x_i cannot be compared with q. In this study, we adopt the widely-used SVMLight tool [24] for entity classification.

Given an entity pair $<q, x_i>$, where q is the query entity and x_i is the candidate entity, we define and use various linguistic, statistical and semantic features for classification. The linguistic and statistical features are derived from the search results returned by Google Search, and the semantic features are derived from Wikipedia[4].

The linguistic and statistical features are listed as follows:

1. **Previous Words.** We issue each entity into Google Search and obtain the top 200 search results. We extract the word before each instance of the entity in the search results. The feature value is the proportion of common previous words in the search results for the two entities. We use the Jaccard measure to compute the proportion value as follows:

$$Jcccard(A, B) = \frac{|A \cap B|}{|A \cup B|}$$

where A and B are the extracted word set for the two entity, respectively.
2. **Next Words.** Similar to feature 1), the feature value is the proportion of common next words in the search results for the two entities. The proportion is also computed by using the Jaccard measure.
3. **Previous POS.** We extract the Part-Of-Speech (POS) tags of the previous words in the search results for each entity.   The feature value is the proportion of common previous POS tags in the search results for the two entities.
4. **Next POS.** Similar to feature 3), the feature value is the proportion of common next POS tags in the search results for the two entities.
5. **Middle Words.** We combine the two entities and obtain the top 200 search results for the composed query. This binary feature indicates whether there exists the word "*and*" or "*or*" between any co-occurrence of the two entities in the search results.

---

[4] http://en.wikipedia.org

6. **Capitalized Words.** We combine the two entities and obtain the top 200 search results for the composed query. The feature value is the proportion of the particular co-occurrences of the two entities in the search results, where both the words for the two entities are capitalized.
7. **Cosine Similarity.** For each entity, we concatenate the top 200 search results into a text after removing the entity string. The feature value is the cosine similarity between the two concatenated texts for the two entities.

The semantic features are listed as follows:

8. **Wikipedia Category.** If both the two entities are defined in Wikipedia, we extract the categories which each entity belongs to and the feature value is the proportion of the common categories for the two entities. Otherwise, the feature value is set to -1.
9. **Wikipedia Link.** If both the two entities are defined in Wikipedia, we extract the in-links and out-links for the page of each entity, and the feature value is the proportion of the common links (including both in-links and out-links) for the two entities. Otherwise, the feature value is set to -1.

In order to train the classifier, we issue five query examples into GoogleSets and obtain the returned item list for each query, respectively. We then manually judge whether each item in the list is comparable with the query item. Lastly, we build a training set consisting of 200 entity pairs and the corresponding tags (true/false).

Based on the above features and the training set, we can build a SVM classifier for comparable entity classification by learning a function f: $\{<q, x_i>| x_i \in C_{expand}\} \rightarrow \{true, false\}$. Finally, the candidate entities being classified as "false" are filtered out.

4   Experiments and Results

4.1   Experimental Setup

Finding comparable entities is a novel task and there is no benchmark dataset for evaluation. Thereafter, we manually collected ten sets of entity names, and the entities in each set belong to the same category (e.g. countries, Chinese provinces, car makers, etc.) and thus they are comparable with each other. The average size of the sets is 59 entities.

Note that each set is not absolutely complete. For example, a province in China can be compared with a state in America, other than another province in China. But we consider the provinces in China are a complete set for system comparison, as using the pooling strategy.

We randomly select one entity in each set and then issue each entity into SEAL, GoolgeSet and our system FCEW, respectively. We collect the item lists returned by each system. We use the online demo[5] of SEAL, which is implemented by the original authors. This demo system allows using only one seed entity and can generate some results, however, the technical details of the implementation is not published. In order to validate the effectiveness of the third step (i.e. comparable entity

---

[5] http://www.rcwang.com/seal/

classification) in FCEW, we also collect the results returned by the second step of set expansion in FCEW, and we denote the system without the third step as BaseFCEW.

Given a list of entities returned by any system, we use the precision (P), recall (R) and F-measure (F) as evaluation metrics:

$$precision = \frac{\left|C_{Candidate} \cap C_{Standard}\right|}{\left|C_{Candidate}\right|}, \quad recall = \frac{\left|C_{Candidate} \cap C_{Standard}\right|}{\left|C_{Standard}\right|}, \quad F-measure = \frac{2 \times precision \times recall}{precision + recall}$$

where $C_{Candidate}$ is the object set returned by any system, and $C_{Standard}$ is the corresponding gold-standard set. The evaluation scores are then averaged across all sets.

4.2   Experimental Results

Table 1 gives the comparison results for SEAL, GoolgSets and our systems with or without the third step (i.e. FCEW and BaseFCEW). Seen from the table, our proposed FCEW outperforms SEAL and GoogleSets over all three metrics, which demonstrates the good effectiveness of our system. We can also see that FCEW improves the overall performance (i.e. 40% F-measure increase over BaseFCEW) of BaseFCEW by raising the precision scores (i.e. 75% precision increase over BaseFCEW), which demonstrates that the third step of comparable entity classification plays an important role in FCEW by filtering out inappropriate candidate entities.

Table 1. System comparison results

Seed Entity	SEAL			GoogleSets			BaseFCEW			FCEW		
	P(%)	R(%)	F(%)	P(%)	R(%)	F(%)	P(%)	R(%)	F(%)	P(%)	R(%)	F(%)
Iraq	38.58	24.49	29.96	26.32	4.59	7.82	55.05	89.80	68.25	92.37	61.22	73.64
Tornado	3.51	16.00	5.76	2.63	4.00	3.17	30.77	16.00	21.05	44.44	16.00	23.53
Buffalo Bills	89.29	78.13	83.33	100.00	100.00	100.00	27.68	96.88	43.06	65.91	90.63	76.32
George Washington	14.58	33.33	20.29	90.00	97.62	93.65	49.30	80.95	61.28	55.36	73.81	63.27
Cassiopeia	20.75	12.50	15.60	92.68	43.18	58.91	59.71	94.32	73.13	86.84	75.00	80.49
Acura	51.85	25.00	33.73	62.22	50.00	55.50	10.36	100.00	18.78	51.09	76.79	61.35
Georgia	5.24	20.00	8.30	82.93	68.00	74.73	12.14	100.00	21.65	71.43	80.00	75.47
Beauty and Beast	18.18	28.57	22.22	45.00	34.69	39.18	62.50	55.10	58.57	82.76	48.98	61.54
Boston Celtics	93.75	50.00	64.22	61.22	100.00	75.95	46.88	100.00	63.83	74.36	96.67	84.06
Jiangsu	41.51	64.71	50.57	60.87	82.35	70.00	31.53	94.12	47.27	53.23	94.12	68.00
Average	37.72	35.27	33.40	62.39	58.44	57.89	38.59	**82.72**	47.69	**67.78**	71.32	**66.77**

Seen from the table, GoogleSets outperforms SEAL and BaseFCEW, which can be explained by that GoogleSets works differently from SEAL and BaseFCEW. GoogleSets conducts the task of set expansion in an offline mode and how many seeds are used for set expansion of a particular class is unknown, i.e., the returned list for a given query entity is not produced based only on the single entity, but it has already been produced by using multiple entities in the same class in earlier days. Though the details of GoogleSets are unknown, the system is very competitive.

Table 2 shows the system results for the example entity "Iraq". The incorrect answers are marked in italic font and red color. In comparison with BaseFCEW, FCEW can successfully classify the noisy candidate entities (i.e."terrorism", "news", "hezbollah" and "baghdad") as "false" and filter them out.

Table 2. System results (top 20) for query "Iraq, (The incorrect entities are marked in italic)

SEAL	GoogleSets	BaseFCEW	FCEW
iraq	iraq	iraq	iraq
bush	*baghdad*	syria	syria
politics	iran	afghanistan	iran
war	kuwait	iran	israel
iran	syria	israel	lebanon
washington	*the iraq*	lebanon	north korea
mondo	*of iraq*	north korea	egypt
afghanistan	*iraqi*	*terrorism*	china
baghdad	israel	egypt	libya
military	egypt	pakistan	russia
usa	libya	*news*	turkey
us	*war*	*hezbollah*	india
terrorism	*in iraq*	us	kuwait
israel	afghanistan	china	jordan
government	saudi arabia	libya	*islam*
media	lebanon	russia	palestine
congress	*saddam hussein*	turkey	yemen
democrats	jordan	india	saudi arabia
president	*the country*	*baghdad*	qatar
cheney	qatar	kuwait	cuba

In order to investigate whether the comparable entity classification step can help improve the performance of GoogleSets, we collect five item lists returned by GoogleSets in response to five random query examples, which are different from the training examples, and we then employ the classifier in FCEW to filter out the noisy items. The comparison results between GoogleSets with and without the classifier are presented in Table 3. We can see that the classifier can much improve the performance of the original GoogleSets by raising the precision scores. The results further demonstrate the great importance of the comparable entity classification step.

Table 3. Comparison results for GoogleSets w/o and w/ Classifier

Query Number	GoogleSets			GoogleSet+Classifier		
	P(%)	*R(%)*	*F(%)*	*P(%)*	*R(%)*	*F(%)*
1	14.58	70.00	24.14	55.56	50.00	52.63
2	17.02	88.89	28.57	72.73	88.89	80.00
3	35.71	13.04	19.11	59.09	11.30	18.98
4	45.45	90.00	60.40	78.57	55.00	64.71
5	60.98	26.32	36.76	72.73	25.26	37.50
Average	34.75	**57.65**	33.80	**67.74**	46.09	**50.76**

5   Discussion

In this section, we first discuss the efficiency issue of our propose system. Currently, our system relies heavily on the commercial Google Search, and our system must iteratively access Google Search for a number of times, and download hundreds of web pages in an online way, which lowers the efficiency of our system. In practical use, we can improve the system efficiency in the following two ways:

1. We can collaborate with some search engine company and integrate the proposed system seamlessly with the search engine, and thus the access of the search engine will be much more efficient. Moreover, we can directly use the cached pages instead of downloading the web pages on the Web.

2. We can use our system to find comparable entities offline, and store the entity sets into a database. When a user issues a seed entity, the system can search the database and return the comparable entities directly. This approach is adopted by GoogleSets to make the system real-time.

We then discuss the cons of our system. The major shortcoming of our system is that it cannot deal with the ambiguous query. For example, if a user issues a single query "*apple*", the system does not know the real requirement of the query, because the query can refer to a kind of fruit or an IT company. The above shortcoming can be addressed by using the following two methods:

1. The algorithm of the system can be improved to automatically disambiguate the query and return different lists of comparable entities to the user. For example, the system can generate two lists of entities for the query "*apple*": one list is about fruit, and the other list is about IT company.

2. The system leaves the burden to the user by allowing the user to input another entity for disambiguation purpose. That is to say, the system allows the user to input one or more entities. In most cases, the user inputs a single entity when the entity is not ambiguous. In some cases, the user can input more than one entity when a single entity is ambiguous, and our system can skip the first step of initial candidate set construction and directly use the given two or more entities as the initial candidate set.

6   Conclusions and Future Work

In this paper, we present a novel system named FCOW to find comparable entities for a given query entity. The system adopts a three-stage approach and the system effectiveness has been validated in the experiments. Our system can outperform SEAL and GoogleSets in the task of finding comparable entities for a single entity. The last step of comparable entity classification has been demonstrated to be very important for FCEW, as well for GoogleSets.

In future work, we will improve the system efficiency by integrating the system with the search engine. We will also improve the system capability to handle ambiguous query by developing new disambiguation techniques.

Acknowledgement. This research was supported by the National Natural Science Foundation of China (NSFC) under Grant No. 61170166, Beijing Nova Program under Grant No.2008B03, and the National High Technology Research and Development Program of China under Grant No. 2012AA011101.

References

1. Jain, A., Pantel, P.: How do they compare? Automatic Identification of Comparable Entities on the Web. In: 18th ACM Conference on Information and Knowledge Management, pp. 1661–1664. ACM, New York (2009)
2. Wang, R.C., Cohen, W.W.: Language-Independent Set Expansion of Named Entities using the Web. In: 7th IEEE International Conference on Data Mining, pp. 342–350. IEEE Press, New York (2007)
3. Bao, S., Li, R., Yu, Y., Cao, Y.: Competitor Mining with the Web. IEEE Transactions on Knowledge and Data Engineering 20, 1297–1310 (2008)
4. Jindal, N., Liu, B.: Identifying Comparative Sentences in Text Documents. In: 29th Annual International ACM SIGIR Conference on Research and Development in Information Retrieval, pp. 244–251. ACM, New York (2006)
5. Jindal, N., Liu, B.: Mining Comparative Sentences and Relations. In: 21st National Conference on Artificial Intelligence, pp. 1331–1336. AAAI Press (2006)
6. Li, S., Lin, C.-Y., Song, Y.-I., Li, Z.: Comparable entity mining from comparative questions. In: 48th Annual Meeting of the Association for Computational Linguistics, pp. 650–658. Association for Computational Linguistics, Stroudsburg (2010)
7. Liu, J., Wagner, E., Birnbaum, L.: Compare & Contrast: Using the Web to Discover Comparable Cases for News Stories. In: 16th International Conference on World Wide Web, pp. 541–550. ACM (2007)
8. Zhai, C., Velivelli, A., Yu, B.: A Cross-Collection Mixture Model for Comparative Text Mining. In: 10th ACM SIGKDD International Conference on Knowledge Discovery and Data Mining, pp. 743–748. ACM (2004)
9. Lerman, K., McDonald, R.: Contrastive Summarization: An Experiment with Consumer Reviews. In: North American Chapter of the Association for Computational Linguistics - Human Language Technologies (NAACL HLT) 2009 Conference, pp. 113–116. Association for Computational Linguistics, Stroudsburg (2009)

10. Wang, D., Zhu, S., Li, T., Gong, Y.: Comparative Document Summarization via Discriminative Sentence Selection. In: 18th ACM Conference on Information and Knowledge Management, pp. 1963–1966. ACM, New York (2009)

11. Kim, H.D., Zhai, C.: Generating Comparative Summaries of Contradictory Opinions in Text. In: 18th ACM Conference on Information and Knowledge Management, pp. 385–394. ACM, New York (2009)

12. Sun, J.-T., Wang, X., Shen, D., Zeng, H.-J., Chen, Z.: CWS: A Comparative Web Search System. In: 15th International Conference on World Wide Web, pp. 467–476. ACM, New York (2006)

13. Luo, G., Tang, C., Tian, Y.-L.: Answering relationship queries on the web. In: 16th International Conference on World Wide Web, pp. 561–570. ACM, New York (2007)

14. Feldman, R., Fresko, M., Goldenberg, J., Netzer, O., Ungar, L.: Extracting Product Comparisons from Discussion Boards. In: 7th IEEE International Conference on Data Mining, pp. 469–474. IEEE Press, New York (2007)

15. Tong, S., Dean, J.: System and methods for automatically creating lists. Patent US7350187 (2008)

16. Wang, R.C., Cohen, W.W.: Iterative Set Expansion of Named Entities using the Web. In: 8th IEEE International Conference on Data Mining, pp. 1091–1096. IEEE Press, New York (2008)

17. Dalvi, B.B., Cohen, W.W., Callan, J.: WebSets: extracting sets of entities from the web using unsupervised information extraction. In: 5th ACM International Conference on Web Search and Data Mining, pp. 243–252. ACM, New York (2012)

18. Ohshima, H., Oyama, S., Tanaka, K.: Searching Coordinate Terms with Their Context from the Web. In: Aberer, K., Peng, Z., Rundensteiner, E.A., Zhang, Y., Li, X. (eds.) WISE 2006. LNCS, vol. 4255, pp. 40–47. Springer, Heidelberg (2006)

19. Pantel, P., Crestan, E., Borkovsky, A., Popescu, A.-M., Vyas, V.: Web-scale distributional similarity and entity set expansion. In: 2009 Conference on Empirical Methods in Natural Language Processing, pp. 938–947. Association for Computational Linguistics, Stroudsburg (2009)

20. Lin, D.: Automatic retrieval and clustering of similar words. In: 36th Annual Meeting of the Association for Computational Linguistics and 17th International Conference on Computational Linguistics, pp. 768–774. Association for Computational Linguistics, Stroudsburg (1998)

21. Sadamitsu, K., Saito, K., Imamura, K., Kikui, G.: Entity set expansion using topic information. In: 49th Annual Meeting of the Association for Computational Linguistics: Human Language Technologies, pp. 726–731. Association for Computational Linguistics, Stroudsburg (2011)

22. Wang, R.C., Schlaefer, N., Cohen, W.W., Nyberg, E.: Automatic set expansion for list question answering. In: 2008 Conference on Empirical Methods in Natural Language Processing, pp. 947–954. Association for Computational Linguistics, Stroudsburg (2008)

23. Ravichandran, D., Hovy, E.: Learning Surface Text Patterns for a Question Answering System. In: 40th Annual Meeting of the Association for Computational Linguistics, pp. 41–47. Association for Computational Linguistics, Stroudsburg (2002)

24. Joachims, T.: Making large-scale support vector machine learning practical. In: Advances in Kernel Methods, pp. 169–184. MIT Press, Boston (1999)

Sentiment Analysis by Augmenting Expectation Maximisation with Lexical Knowledge

Xiuzhen Zhang[1,*], Yun Zhou[1,**],
James Bailey[2], and Kotagiri Ramamohanarao[2]

[1] School of Computer Science & IT, RMIT University, Australia
{xiuzhen.zhang}@rmit.edu.au
[2] Dept. of CIS, The University of Melbourne, Australia
{baileyj,kotagiri}@unimelb.edu.au

Abstract. Sentiment analysis of documents aims to characterise the positive or negative sentiment expressed in documents. It has been formulated as a supervised classification problem, which requires large numbers of labelled documents. Semi-supervised sentiment classification using limited documents or words labelled with sentiment-polarities are approaches to reducing labelling cost for effective learning. Expectation Maximisation (EM) has been widely used in semi-supervised sentiment classification. A prominent problem with existing EM-based approaches is that the objective function of EM may not conform to the intended classification task and thus can result in poor classification performance. In this paper we propose to augment EM with the lexical knowledge of opinion words to mitigate this problem. Extensive experiments on diverse domains show that our lexical EM algorithm achieves significantly higher accuracy than existing standard EM-based semi-supervised learning approaches for sentiment classification, and also significantly outperforms alternative approaches using the lexical knowledge.

Keywords: Sentiment Analysis, Expectation Maximisation, Semi-supervised Learning, Text Classification.

1 Introduction

The Web provides a platform for the public to freely express their opinions, where blogs, product reviews and movie reviews are popularly used forums. Sentiment analysis aims to identify the positive or negative opinions and sentiments expressed in documents (blog posts or reviews) [15]. Machine learning approaches have been widely used for sentiment analysis [16,18,12]. Especially Pang et al [16] cast the sentiment analysis of documents as a supervised text classification problem, where documents are classified as carrying *positive* or *negative* labels.

On the other hand, linguistic studies have mostly focused on identifying words and phrases that express subjectivity, either manually or automatically [7,17,20,21]. Opinion words (also called subjectivity or sentiment words)

---

* Work was done while Zhang was on sabbatical leave at the University of Melbourne.
** Currently at The University of Melbourne. Email: yuzhou@student.unimelb.edu.au

X.S. Wang et al. (Eds.): WISE 2012, LNCS 7651, pp. 30–43, 2012.
© Springer-Verlag Berlin Heidelberg 2012

are words that express prior, out of context positive or negative polarity. For example words like *adore* and *perfect* carry prior positive polarity whereas *abhor* and *insane* carry prior negative polarity. Some publicly available opinion word repositories include SentimentWordNet [1] and the General Inquirer positive and negative word lists [6]. As opinion words have sentiment labels, they are also called labelled words [10] or more generally labelled features [4] from the perspective of classification learning.

Generally supervised learning algorithms for text classification require a large number of labelled documents for effective learning. But labelling documents is costly. Semi-supervised learning approaches [13] that leverage unlabelled documents for effective learning from limited training documents are appealing. Recently semi-supervised learning in the new form of learning from labelled words has also attracted lots of attention from the research community [10,12,18].

Expectation Maximisation (EM) [3] has been employed as the key mechanism for both forms of semi-supervised learning. In [13], EM is combined with Naive Bayes to find classifier parameters that maximises the likelihood of both the labelled and unlabelled documents. In [10], word labels are used to construct fictitious exemplar positive and negative documents and unlabelled documents are "softly" labelled according to their distance to the exemplar documents. EM is then applied to build the classification model. A prominent problem with the standard EM procedure is that it may optimise parameters of a generative model whose objective function does not conform to its intended purpose of classification, which can result in poor classification performance [2,13] .

In this paper we propose to augment the EM algorithm with the lexical knowledge of word labels. To this end, we modify the expectation computation step for the probabilistic document labels in the EM procedure by combining any given document class labels and labels derived heuristically from word labels. The Naive Bayes generative model is applied to re-estimate word distribution (data likelihood) under the adjusted document class expectation. It is typically difficult to incorporate prior knowledge into the EM process. In our approach the latent variables in the generative model have been constrained and directed by the lexical knowledge in a simple yet effective approach. Our approach can reduce the problem of mismatch of posterior distribution between latent variable for EM and the objective class label variable for classification.

We conduct experiments on sentiment classification of real-world datasets including blogs on different topics and movie reviews. Experiments show that our lexical knowledge augmented EM (Lexical EM) approach significantly improves semi-supervised classification based on labelled documents [13] as well as based on labelled words [10], where standard EM is employed. It often achieves better results than a recently proposed linear pooling approach of combining lexical knowledge with machine learning [12]. Especially our lexical knowledge augmented EM approach achieves effective learning when there are few labelled documents for training.

2 Related Work

Opinion mining and sentiment analysis have attracted active research recently. Overview of developments in this area has been described in [15] and [9]. Turney [19] employed lexical knowledge phrases (adjectives followed by nouns, or adverbs followed by verbs) to develop an unsupervised learning algorithm for classifying opinions of reviews. The polarity of phrases is computed by searching the Web to compute its similarity to the positive and negative reference word "excellent" and "poor" respectively, and thus the proposed approach can not be easily generalised to applications like blog sentiment analysis where there are not ready-made reference words.

Pang et al. [16] showed that using lexical information in a naive approach of counting the occurrences of positive and negative opinion words for sentiment classification is not as effective as building text classification models using training examples. It is worth noting that their conclusion is conditioned on that a large number of labelled documents are available for training an accurate classification model. But generally labelling documents is a costly task. The objective of this study is to achieve accurate sentiment classification with few labelled documents. We have shown that making use of the lexical knowledge in a more intelligent way can compensate for the shortage of labelled training documents and can achieve fairly accurate sentiment classification.

Incorporating domain knowledge into standard text classification has been actively studied recently [4,12,18]. In [4] and [18] word labels are used directly to constrain the class distribution computation for documents in discriminative models. In contrast the main purpose of our work is to incorporate prior lexical knowledge into the EM process which is based on a generative model such as Naive Bayes. Rather than modifying the class distribution parameters directly we modify the parameters for generating the classification model. In [12], linear pooling is used to combine the word distribution in classes estimated from the training data and the lexical knowledge. Naive Bayes is then employed for classification. Different from their approach of incorporating lexical knowledge by modifying word distributions in classes, we adjust the expected class distribution for documents making use of both the lexical knowledge, and labelled and unlabelled documents. Our experiments show that our approach often achieves better classification accuracy with fewer labelled documents.

There has been previous work on improving the EM process. Graca et al. [8] proposed a general method that incorporates prior constraints into the EM process, focusing on clustering and the alignment problem for statistical machine translation. In [5] a general semi-supervised learning framework is developed to constrain the posterior distribution of latent variables under a set of feature expectation constraints. The generative model parameters are estimated with a coordinate ascent algorithm. Our approach is consistent with this general framework but more importantly has shown a practical way of incorporating domain knowledge into this general framework for document sentiment classification.

3   The Semi-supervised Learning

In this section we describe two forms of semi-supervised learning, in the context of sentiment classification of documents with a limited number of labelled documents. EM has been employed in both semi-supervised settings.

3.1   Semi-supervised Learning with Labelled Documents

EM is an iterative algorithm for maximum likelihood or maximum a posteriori estimation in problems with incomplete data [3]. For text classification, the data is incomplete in the sense that class labels for documents are missing. In the semi-supervised learning paradigm, both the labelled and unlabelled documents are used to derive a classification model [13]. A Naive Bayes (NB) classifier is firstly trained with the available labelled documents, and it is used to assign probabilistic class labels to unlabelled documents. The EM procedure is then employed to train a new classifier using both the originally labelled and unlabelled documents. The EM process iterates to find the word distribution for classes that maximises the likelihood of all documents.

NB is a probabilistic generative model for data, and it is the base classification model to incorporate unlabelled documents for learning. Each document is generated according to a probability distribution defined by a set of parameters — the word distribution for classes. NB estimates the word distribution for classes using only labelled training documents, and then uses the estimated word distribution to classify new documents — computing the probability for a document in each class and the most likely class is thought to have generated the document.

Consider a collection D of documents for training, where each document is labelled with a class label. For ease of discussion we assume that there are only two class labels, the positive and the negative. Suppose that V is the vocabulary for D. Given a document d and class labels C_j, $j \in \{+, -\}$, and under the assumption of independent word distribution for classes, the probability that each class has generated the document is

$$P(C_j|d) = \frac{\prod_{t \in d} P(t|C_j) * P(C_j)}{P(d)}.$$

For a class C_j, its prior probability is the proportion of documents in the collection that belong to class C_j:

$$P(C_j) = \frac{N(d, C_j)}{|D|}.$$

The term distribution in each class is computed as

$$P(t|C_j) = \frac{N(t \in C_j) + 1}{\sum_{t' \in V} (N(t' \in C_j) + 1)}$$

where $N(t \in C_j)$ and $N(t' \in C_j)$ are respectively the total number of occurrences of t and t' in documents with class label C_j. Note that the demoninator is computed from only terms appearing in class C_j.

When NB is given just a small set of labelled training documents, classification accuracy will suffer since variance in the parameter estimates $P(t|C_j)$, $j \in \{+, -\}$ is high. EM can improve the estimation for $P(t|C_j)$ making use of the unlabelled documents. For the originally labelled documents, $P(C_j|d)$ is already known:

$$P(C_j|d) = \begin{cases} 1 & \text{if } d \in C_j \\ 0 & \text{otherwise.} \end{cases} \tag{1}$$

For each originally unlabelled document d, $P(C_j|d)$, $j \in \{+, -\}$ is estimated using the EM process by iterating the following two steps, until $P(t|C_j)$ and $P(C_j)$ converge.

- The Expectation step (E-step): For each document d and a class C_j, $P(C_j|d)$ is estimated as follows:

$$P(C_j|d) = \frac{P(d|C_j) * P(C_j)}{P(d)} = \frac{\prod_{t \in d} P(t|C_j) * P(C_j)}{P(d)} \tag{2}$$

In the above equation $P(d)$ does not need to be computed. Rather for two classes, $P(C_j|d)$ is normalised by the sum of the numerator for all classes.

- The Maximisation step (M-step): At this step, model parameters term distribution $P(t|C_j)$ and class priors $P(C_j)$ are re-computed. For each word t, $P(t|C_j)$ is re-computed based on $P(C_j|d)$:

$$P(t|C_j) = \frac{\sum_{d \in D} N(t \in d) * P(C_j|d) + 1}{\sum_{t' \in V} \sum_{d \in D} N(t' \in d) * P(C_j|d) + |V|}$$

where $|V|$ is the vocabulary size. Note that Laplace smoothing is applied to avoid zero-probabilities. Class priors $P(C_j)$ are re-computed as follows:

$$P(C_j) = \frac{\sum_{d \in D} P(C_j|d)}{|D|}.$$

3.2   Semi-supervised Learning with Labelled Words

For sentiment classification of documents, some opinion words express prior, out of context sentiment polarities. For example "excellent" is typically associated with the positive polarity whereas "abhor" is associated with the negative polarity. For topic classification of documents, some words are strong indicators for topics. For example, to classify documents into the topics "baseball" versus "hockey", the word "puck" strongly indicates that the document is about "hockey". In text classification words that are strongly associated with class labels are called labelled words. Labelling words are reported to be less costly than labelling documents [4,10].

Another semi-supervised document classification scheme is to use a lexicon of labelled words rather than labelled documents. A typical approach of incorporating labelled words into the learning model is by creating pseudo documents using the labelled words, and a representative piece of work is by Liu et al. [10]. In their approach, given a lexicon of representative set of words for each class, representative documents are constructed containing all representative words for each class. The cosine similarity between each unlabelled document and the representative documents are computed. As a result unlabelled documents are softly assigned the label of the class with the highest similarity. Using the soft labels as a start, the EM process is then employed to iteratively improve the word distribution computation and the class probability computation for documents.

It has been shown in [13] that semi-supervised classification based on the standard EM procedure can significantly improve classification accuracy when there are only limited labelled documents. Note that EM finds the word distribution for classes with locally maximal likelihood given both the labelled and unlabelled document. Note also that the underlying assumption for semi-supervised classification based on the standard EM procedure is that components of the generative model correspond to classes for our intended text classification task. However such assumptions do not always hold in real applications. It has been shown that the model may degrade the classification performance when the model parameters are misspecified [2]. In Section 5, we describe making use of the domain knowledge of word labels to modify the basic EM algorithm, aiming to address the issue of performance degradation due to violated assumptions.

4   The Lexical Knowledge Model

When no labelled data or other domain knowledge are available in a new domain, a simple document sentiment classification model can be built using the lexical knowledge of labelled opinion words. Given a lexicon of positive and negative words, the probability that a document d belongs to the positive class can be computed as

$$P'(C_+|d) = \frac{a}{a + b}$$

where a and b represent respectively the number of occurrences of positive and negative words in the document d. Without any prior information regarding the relative frequency of positive and negative words in a domain, a document with more positive words is likely to express overall positive sentiment, while a document with more negative words expresses overall negative feeling. In particular, a document d is classified as positive if $P(C_+|d) \geq 0.5$, and negative otherwise.

In this study we use the comprehensive and generic opinion lexicon developed by Wilson et al. [21]. Words in the lexicon are categorised as strongly subjective or weakly subjective. Words that are subjective in most contexts are *strongly subjective* and those that may only have certain subjective usages are *weakly subjective*. Subjective words are tagged with their prior polarity. The *positive* and *negative* tags are for positive and negative polarities respectively. The *both* tag is

Table 1. Some words in our opinion lexicon

Positive	acclaim, adore, affirm, befit, catalyst, dear, defer, encourage, fantastic, good, hero, loyalty, marvel, nice, perfect, radiant, sane, thrill, understand, want, yearn, zest
Negative	abash, abhor, accuse, admonish, agonize, beg, chao, defunct, excess, fear, grief, hell, insane, lose, malignant, nightmare, object, penalty, quarrel, racist, scandal, thwart, unfair, virus, yawn, zealous

for words that can express both positive and negative polarities. The *neutral* tag is for words expressing subjectivity but not obvious positive or negative polarity.

We apply some filtering criteria to the original subjectivity lexicon to construct the sentiment lexicon for our research. We first remove the small fraction of words with the *both* or *neutral* tags from the original lexicon. As a result only positive and negative words are kept for further consideration. For the positive and negative words, we further remove the weak subjective words. Our final opinion lexicon consists of 3218 unique words after stemming. In total there are 1067 positive and 2151 negative words in our final sentiment lexicon.

Some opinion words randomly chosen from our opinion lexicon are listed in Table 1. The opinion lexicon we use is a generic lexicon without any specific domain in mind. As a result, depending on how applicable the lexicon is to different domains, performance of the simple lexical knowledge-based classification model can vary in different domains. As will be discussed in Section 6, our experiments show that such a simple lexical knowledge classification model achieves modest classification accuracy for blogs while significantly more accurate classification for movie reviews.

5   Augmenting EM with Lexical Knowledge

We now describe how to modify the standard EM process described in Section 3 to effectively incorporate the lexical knowledge model for more accurate sentiment classification. With standard EM, the labelled documents are used to initialise the parameters for the EM hill climbing process. When the labelled documents are limited, during the iteration the unlabelled documents have significant effect on setting the parameters $p(t|C_j)$ ($t \in V$, $j \in \{+, -\}$). As the EM process is set to maximise the likelihood of labelled as well as unlabelled documents, this standard hill-climbing process may result in maximum data likelihood estimation that leads to latent variable estimation drifting away from our target classification function .

Our main idea of modifying EM is to modify the class distribution for unlabelled documents at the expectation step. Our adjustment is intended to constrain the EM process towards generating document class distributions more consistent with our target classification task, and the adjustment is achieved by incorporating the lexical knowledge model. In particular, we modify the posteriori class distribution for documents as in Equation 3, where $P(C_j|d)$ and

---

Algorithm 1. The Lexical EM algorithm

Input:
   A set of labelled documents D^L and a set of unlabelled documents D^U.
   Documents are defined on vocabulary V, and class labels are $\{+, -\}$.
   An opinion lexicon X.
Output:
   A classification model θ with parameters $P(C_j)$ and $P(t|C_j)$, $t \in V$, $j \in \{+, -\}$.
   $\{// \text{ In the description next } t \in V \text{ and } j \in \{+, -\}.\}$
1: Train a Naive Bayes classifier $\theta^0 = \langle P^0(C_j), P^0(w_i|C_j)\rangle$ from D^L
2: **for** $(k = 1; \theta^k$ improves over $\theta^{k-1}; k++)$ **do**
3:    Compute the weight α^{k-1} in Equation 3 for classifier θ^{k-1}
       $\{// \text{ Lines 4–8: E-step}\}$
4:    Compute $P^{k-1}(C_j|d \in D^L)$ from class labels
5:    **for** each document $d \in D^u$ **do**
6:       Compute $P^{k-1}(C_j|d)$ using classifier θ^{k-1}
          $\{// \text{ Line 7: Equation 3}\}$
7:       Adjust $P^{k-1}(C_j|d)$ by α^{k-1}, θ^{k-1} and the lexical knowledge model using X
8:    **end for**
       $\{// \text{ Line 9: M-step}\}$
9:    Compute classifier $\theta^k = \langle P^k(C_j), P^k(t|C_j)\rangle$ from $P^{k-1}(C_j|d \in D^L \cup D^U)$
10: **end for**
11: Return the final classification model $\theta^k = \langle P^k(C_j), P^k(t|C_j)\rangle$

$P'(C_j|d)$ represent respectively the class probability for documents computed from the generative model and lexical model.

$$P(C_j|d) = \begin{cases} P(C_j|d) & \text{if } d \text{ is a labelled document} \\ \alpha P(C_j|d) + (1 - \alpha)P'(C_j|d) & \text{otherwise.} \end{cases} \quad (3)$$

– If a document d has a label, its class probability $P(C_j|d)$ remains unchanged. Following Equation 1, if d has a positive class label $P(C_+|d) = 1$ and $P(C_-|d) = 0$; otherwise $P(C_+|d) = 0$ and $P(C_-|d) = 1$.
– If a document d is an originally unlabelled document, $P(C_j|d)$ is adjusted in each iteration as a weighted sum of $P(C_j|d)$ computed from the current estimation of $P(t|C_j)$ and $P(C_j)$, and $P'(C_j|d)$ according to the lexical knowledge model (Section 4).

In adjusting the class probability for unlabelled documents, generally weighting factors should be set according to the classification accuracy of each component. α is a normalised weight factor according to the classification accuracies of the NB generative and lexical knowledge models of a current iteration. Our lexical EM algorithm is as shown in Algorithm 1. The algorithm starts with initialisation by a Naive Bayes classifier θ^0 trained on the set of labelled documents D^L (line 1). Parameters for classifier θ^0 include $P^0(C_j)$ and $P^0(t|C_j)$, $t \in V$, $j \in \{+, -\}$. The accuracy of θ^0 is estimated from the labelled documents in D^L. The initial class

Table 2. Experimental data sets

Dataset	Description	#Pos	#Neg
Cartoon Blog	Worldwide opinions to the cartoons depicting the Muslim prophet Muhammad printed in a Danish newspaper.	107	107
McDonald's Blog	Opinions regarding the food at McDonald's restaurants.	41	41
Economic Forum Blog	Opinions on the World Economic forum in Davos, Switzerland.	38	38
Challenger Blog	Opinions about the Challenger space shuttle disaster.	104	81
Bolivia Blog	Documents that show opinions about Bolivia.	41	41
Movie Review	Movie reviews from the Internet Movie Database.	1000	1000

distribution expectation for the unlabelled documents is computed from θ^0 — the unlabelled documents in D^U are assigned conditional labels by applying θ^0, and then adjusted by the accuracy of θ^0 and prior lexical knowledge using Equation 3 (lines 5–8). From the class distribution expectations of both the labelled and unlabelled training documents, new parameters $P^1(C_j)$ and $P^1(t|C_j)$ with the maximal likelihood are computed. The new parameters form a new classifier θ^1. The expectation and maximisation steps are iterated until the parameters for model θ^k are not improving.

6   Experiments

We conduct experiments to examine the performance of our approach to augmenting EM with lexical knowledge. Our experimental data sets are extracted from the TREC Blog06 collection [14,11] that was used in the TREC-2006 and TREC-2007 conferences (Blog Track). NIST organised the relevance assessment for the opinion finding task. Given a target topic, if a blog post or its comments is not only on target, but also contains an explicit expression of opinion or sentiment towards the target, showing some personal attitude of the writer(s) then the document is judged as negatively opinionated, mixed or positively opinionated. On five topics from the Blog06 collection where our opinion lexicon has relatively good coverage, we randomly selected a roughly equal number of positive and negative documents for each topic. Table 2 summarises the five blog datasets used in our experiments, and they cover opinion analysis on a wide range of topics, including political figures, restaurants and political events. Table 2 also includes the movie review dataset that is popularly used in literature [16] for sentiment analysis. While the casual writing style is popular in blogs formal language expressions are mostly used in movie reviews. As will be seen next sentiment classification for blogs is significantly more challenging than that for movie reviews.

Table 3. The average accuracies on each dataset for all models

Dataset	Lexical EM	Linear Pooling	Word Supervsion	Doc Supervision
Cartoon Blog	**56.54**	54.73	51.53	53.05
McDonald's Blog	64.70	**65.09**	54.02	58.17
Economic Forum Blog	**70.17**	66.99	59.94	67.68
Challenger Blog	**58.71**	56.56	52.65	49.33
Bolivia Blog	**64.24**	60.46	53.83	53.65
Movie Review	**82.23**	78.61	61.96	76.27

6.1 Overview of Results

We implemented Lexical EM using the accuracy-based weighting function. We compare our lexical EM algorithm with the two semi-supervised learning algorithms based on standard EM (Section 3). We also compare our lexical EM algorithm against the linear pooling algorithm by Melville et al. [12]. All EM processes are set to finish after 10 iterations when the EM parameters start to converge. Classification accuracy is obtained by 10-fold cross validation experiments, where for each fold only a proportion (5%–70%) of the hold-out documents are used as labelled training data. The final results are the average of 10 runs of cross validation.

The average accuracies for all models using different proportions of labelled documents are shown in Table 3, where Word Supervision and Doc Supervision refer to the semi-supervised learning algorithms using labelled words and labelled documents respectively. Overall Lexical EM outperforms other models and has the highest accuracy in five out of 6 domains. According to the paired Wilcoxon signed rank test, the improvements in accuracy of Lexical EM over Word Supervision and Document Supervision are statistically significant ($p < 0.05$) on all datasets. Lexical EM outperforms Linear Pooling statistically significantly on five out of six datasets while shows similar accuracy on the McDonald's blog.

The learning curves for all models using 5%–70% of labelled documents are plotted in Fig. 1. The figure clearly demonstrates the effectiveness of incorporating prior knowledge into the standard EM process, especially when there are limited labelled documents. When there are only 5% labelled training documents Lexical EM shows substantial performance improvement over Linear Pooling, Document Supervision or Word Supervision ($p < 0.05$ for the paired Wilcoxon signed rank test). The lexical knowledge used in the Lexical EM approach provides a reliable source to boost the estimation of $P(C_j|d)$ for a document d than trying to estimate model parameters from a limited set of labels. On the McDonald's blog, with only 8 (10%) labelled documents Lexical EM achieves 14.31% increment in accuracy over Document Supervision, equivalent to relative improvement of 28.37% in accuracy; with only 4 (5%) labelled documents, Lexical EM achieves 17.36% increment in accuracy over NB using labelled documents, equivalent to relative improvement of 47.17% in accuracy.

Not using any labelled documents, the classification accuracy of Word Supervision using only labelled words does not show improvement with more labelled

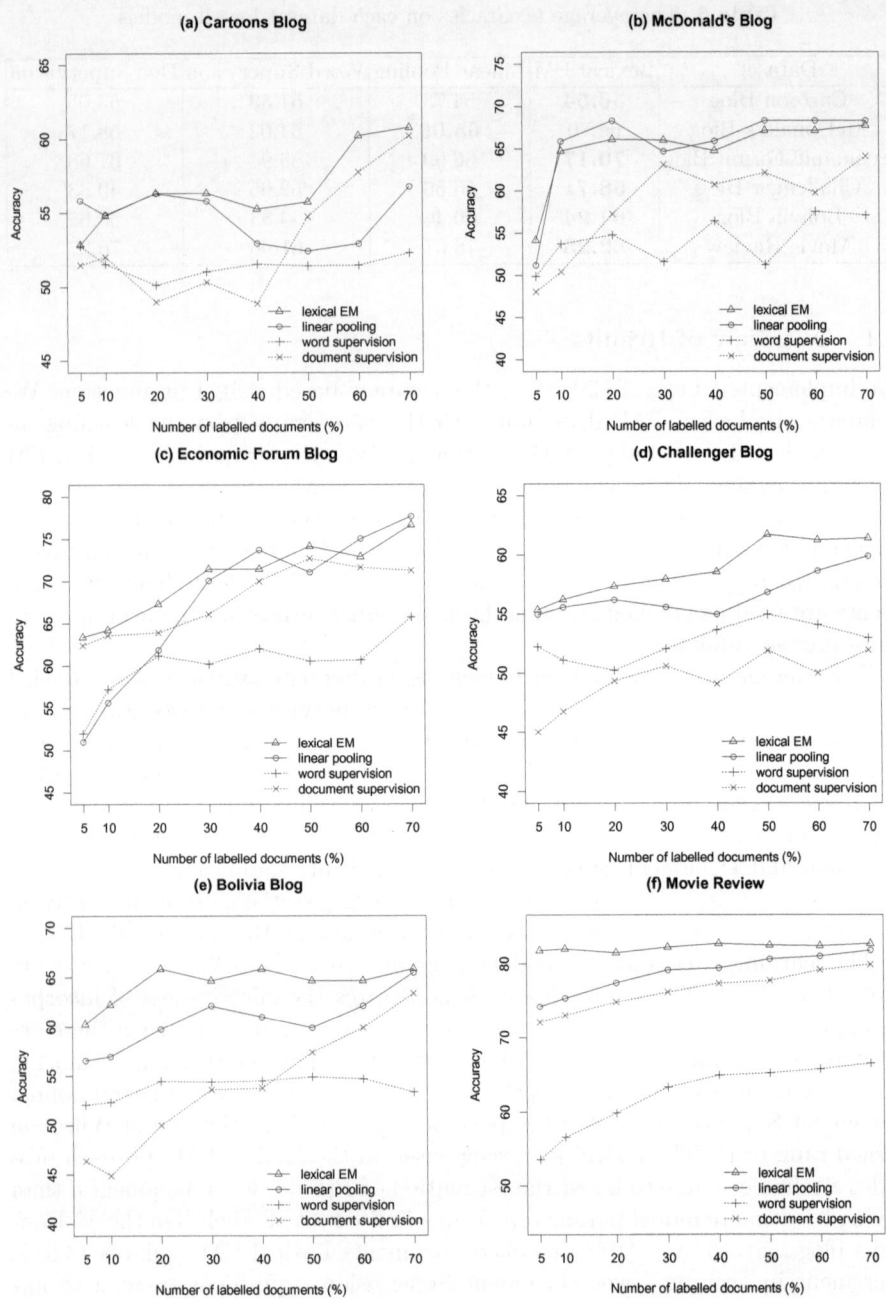

Fig. 1. Learning curves for all models

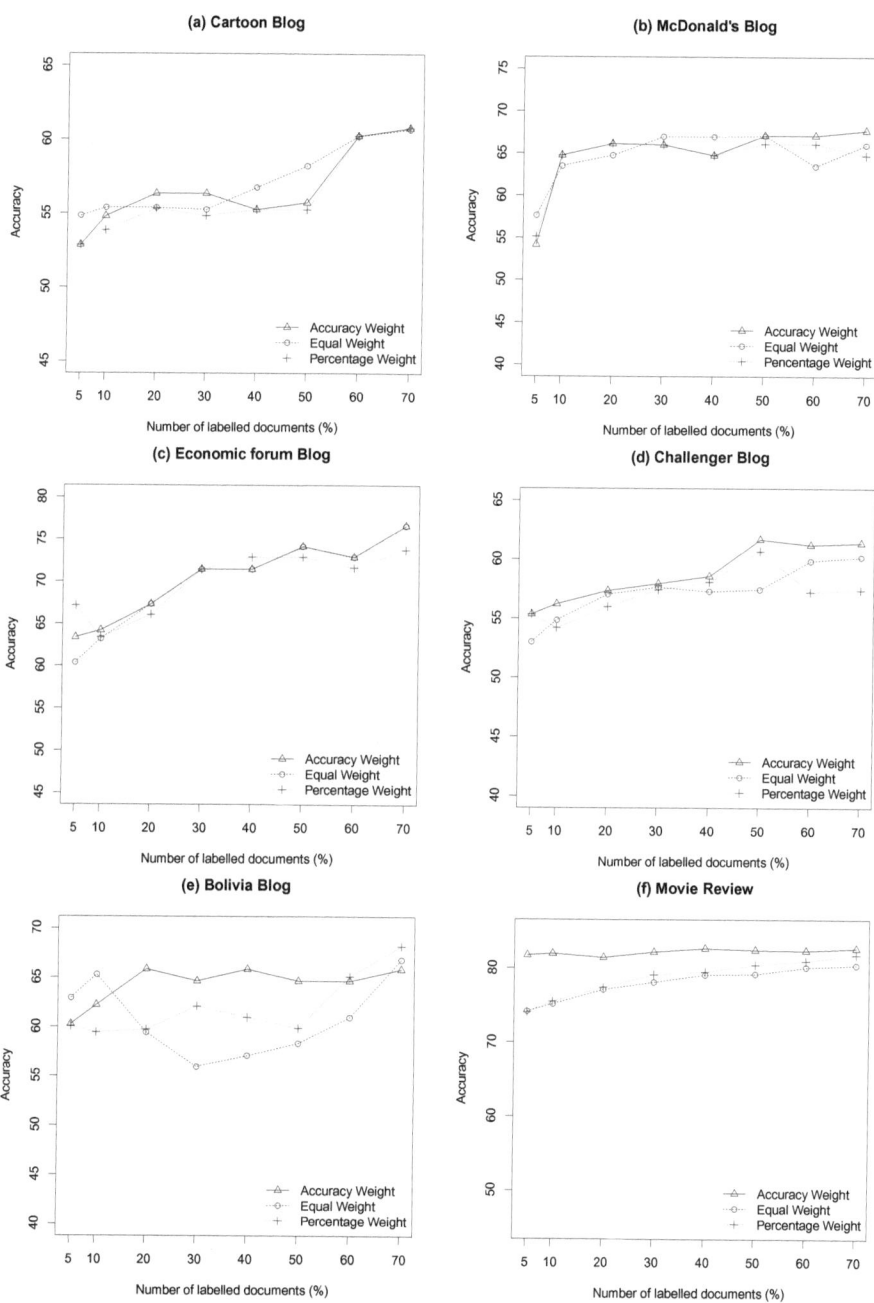

Fig. 2. Performance of Lexical EM with regard to weighting functions

documents. We also observe that Word Supervision very often has the lowest accuracy compared with other models. This result is different from that in [10] where Word Supervision is reported to outperform Document Supervision. Note that however, there are several main differences in our study: First we are using a generic labelled lexicon whereas manually compiled lexicons are used in [10]. Secondly our lexicon has thousands of general labelled words whereas in [10] there are only tens of manually selected words.

We also examined performance of the Lexical Knowledge model described in Section 4. Not surprisingly performance of the Lexical Knowledge model varies considerably on different datasets, from an accuracy of 56.26% on the Cartoon blog to an accuracy of 66.96% on the Economic forum blog. In contrast the Lexical Knowledge model performs surprisingly well on the movie review dataset, with an accuracy of 71.55%. The reason for this modest performance on blogs may be due to the different language style for blogs. The language used in blogs is typically very informal and the vocabulary is far different from the vocabulary for formal writing. As a result our opinion lexicon has poor representation for blogs.

6.2   Analysis of Lexical EM

The only parameter in Lexical EM is the weight α in Equation 3 when the knowledge-based estimation for class distribution is combined with the generative model. Figure 2 plots the accuracies of Lexical EM with different weighting functions for α. We compare our default accuracy-based weighting function against two other weighting functions. "Equal Weight" denotes that $\alpha = 0.5$. "Percentage Weight" denotes that the percentage of labelled training documents is used as the weight for the generative model. The rationale for this strategy is that with an increasing number of labelled documents, the generative model becomes more accurate. The figure shows that Lexical EM is fairly robust with respect to different settings of weight. All three weighting functions lead to the same trend of learning curve – the more available training instances the more accurate is Lexical EM. The Percentage weighting function almost always gives the same accuracy as the accuracy-based weighting function.

7   Conclusions

In this paper we have proposed a simple and effective approach to incorporating the knowledge of word labels into the EM process for document sentiment classification. Experiments show that combining limited domain-specific labelled training documents with general lexical knowledge can achieve significantly better performance than the model derived from only labelled documents. More generally our study strongly suggests that the marriage of limited domain-specific information and domain-independent knowledge is a cost-effective approach to sentiment classification in new domains. In future work we will focus on developing more advanced lexical knowledge models to improve the EM process for document sentiment analysis.

References

1. Baccianella, S., Esuli, A., Sebastiani, F.: Sentiwordnet 3.0: An enhanced lexical resource for sentiment analysis and opinion mining. In: LREC 2008 (2008)
2. Cozman, F., Cohen, I.: Risks of semi-supervised learning: How unlabeled data can degrade performance of generative classifiers. Semi-Supervised Learning 4, 57–72 (2006)
3. Dempster, A., Laird, N., Rubin, D.: Maximum likelihood from incomplete data via the EM algorithm. Journal of the Royal Statistical Society. Series B (Methodological) 39(1), 1–38 (1977)
4. Druck, G., Mann, G., McCallum, A.: Learning from labeled features using generalized expectation criteria. In: SIGIR 2008 (2008)
5. Druck, G., McCallum, A.: High-performance semi-supervised learning using discriminatively constrained generative models. In: ICML 2010, Haifa, Israel (2010)
6. General-Inquirer: The General Inquirer Home Page (2010), http://www.wjh.harvard.edu/~inquirer/ (Online; accessed September 22, 2010)
7. Hatzivassiloglou, V., McKeown, K.R.: Predicting the semantic orientation of adjectives. In: LREC 1997 (1997)
8. Graca, J., Ganchev, K., Taskar, B.: Expectation maximization and posterior constraints. In: Proceedings of NIPS (2007)
9. Liu, B.: Sentiment analysis and subjectivity. In: Handbook of Natural Language Processing, 2nd edn. (2009)
10. Liu, B., Li, X., Lee, W., Yu, P.: Text classification by labelling words. In: AAAI 2004 (2004)
11. Macdonald, C., Ounis, I., Soboroff, I.: Overview of the TREC 2007 blog track. In: TREC 2007 (2007)
12. Melville, P., Gryc, W., Lawrence, R.D.: Sentiment analysis of blogs by combining lexical knowledge with text classification. In: ACM SIGKDD 2009 (2009)
13. Nigam, K., McCallum, A., Thrun, S., Mitchell, T.: Text classification from labeled and unlabeled documents using EM. Machine Learning 39(2), 103–134 (2000)
14. Ounis, I., Rijke, M.D., Macdonald, C., Mishne, G., Soboroff, I.: Overview of the TREC-2006 blog track. In: TREC 2006 (2006)
15. Pang, B., Lee, L.: Opinion mining and sentiment analysis. Foundations and Trends in Information Retrieval 2(1-2), 1–135 (2008)
16. Pang, B., Lee, L., Vaithyanathan, S.: Thumbs up?: sentiment classification using machine learning techniques. In: ACL 2002, pp. 79–86 (2002)
17. Riloff, E., Wiebe, J.: Learning extraction patterns for subjective expressions. In: EMNLP 2003, pp. 105–112 (2003)
18. Sindhwani, V., Melville, P.: Document-word co-regularization for semi-supervised sentiment analysis. In: ICDM 2008, pp. 1025–1030 (2008)
19. Turney, P.D.: Thumbs up or thumbs down?: semantic orientation applied to unsupervised classification of reviews. In: ACL 2002, pp. 417–424 (2002)
20. Wiebe, J., Wilson, T., Cardie, C.: Annotating expressions of opinions and emotions in language. Language Resources and Evaluation 39(2), 165–210 (2005)
21. Wilson, T., Wiebe, J., Hoffmann, P.: Recognizing contextual polarity in phrase-level sentiment analysis. In: EMNLP 2005, pp. 347–354 (2005)

Event-Driven Ontology Updating

Jordy Sangers, Frederik Hogenboom, and Flavius Frasincar

Erasmus University Rotterdam,
PO Box 1738, NL-3000 DR, Rotterdam, The Netherlands
jordysangers@hotmail.com, {fhogenboom,frasincar}@ese.eur.nl

Abstract. Ontologies, as reliable resources in decision making processes, need to be accurate and up-to-date. For this purpose, ontologies have to be maintained regularly. Manual updating is tedious and time consuming, therefore we propose an event-driven automated ontology updating approach. The Ontology Update Language (OUL) and our proposed extensions are inspired by the existing SQL-triggers mechanism and make use of SPARQL and SPARQL/Update statements. We propose different execution models, providing flexibility with respect to the update process. As a proof-of-concept, we implement the language and its execution models in the Hermes News Portal (HNP), an ontology-based news personalization service.

1 Introduction

One of the most important driving factors for information in today's society is news. Every day, millions of people try to keep up-to-date with the latest developments by reading news items. Next to television and newspapers, the World Wide Web has become a good alternative for people to keep track of the state of the world. Developments in the real world – described in news items – influence a variety of activities, ranging from individual daily activities such as buying products to companies' long-term business strategies. Lately, there has been an increasing amount of effort put into automatically processing news data by extracting important information. Applications that make use of this information are plentiful, e.g., automated stock agents that keep track of financial news to exploit extracted knowledge on the stock market, news personalization services that provide users with information that matches user interests, etc.

Traditionally, news is presented as plain text and can be characterized as unstructured data, making it hard for computer systems to interpret it. With the Semantic Web, the World Wide Web Consortium (W3C) provides a framework to add structure to data through the usage of the Web Ontology Language (OWL) [1]. By means of ontologies, domain specific knowledge can be represented by creating concepts and relations between these concepts. The relations are established by defining triples that consist of a subject, a predicate, and an object.

With structured data, information can be easily extracted, and interoperability between computer systems is stimulated. This information, often described using ontologies, is used as an information source that influences the systems'

X.S. Wang et al. (Eds.): WISE 2012, LNCS 7651, pp. 44–57, 2012.

actions. Due to the non-static nature of our society, the information that reflects the real world at any given time has to be updated regularly. Traditional data sources like relational databases have mechanisms for automatic updates. However, a principled way of automatic ontology updating does not yet exist. This forces domain experts to manually update ontologies, which is a tedious, repetitive, error-prone, and time-consuming job.

Numerous applications, like the Hermes News Portal [7] (an ontology-based news personalization service) take advantage of Web news items by exploiting their information through ontology matching. As a classification and querying tool, it is important that the ontology contains up-to-date information. However, such tools often lack an update language for maintaining underlying ontologies and would therefore benefit from an ontology update language. Hence, we propose the use of the event-triggered Ontology Update Language (OUL), where events are defined as phenomena requiring a knowledge base to be updated. In this paper we hypothesize that the language could be extended with techniques from active databases, i.e., with features like prefixes and negation, and also by defining various update execution mechanisms.

2   Related Work

Due to the recent explosion in (meta-)data representation technologies, information can be described in many ways. One way to do this is by making use of relational databases, which store information in tables related to each other. Additionally, the eXtensible Markup Language (XML) [3] can describe the information in a tree-structure, a common way for transportation of information between systems. Last, semantic languages for storing information exist. The Resource Description Framework (RDF) [4] adds meaning to data by using triples and can be serialized in XML. OWL extends RDF with the possibility to express additional constraints and is often used as an ontology representation language.

Common languages for retrieving information from sources are the Structured Query Language (SQL) [5] for relational databases, XPath [6] and XQuery [2] for XML documents, and SPARQL [11] for RDF and OWL. Although SQL is mainly used for querying information from tables, extra functionalities have been added to it, such as the creation, alteration, and removal of tables. These statements can be executed individually, but can also be used in combination with SQL triggers. These triggers react on predefined events based on an Event-Condition-Action model and execute SQL statements either immediately or deferred if a condition is met, hereby creating an automated way of updating relational databases.

Updating XML documents can be realized with XUpdate [9], and updating ontologies is usually done with SPARQL/Update statements [13]. These statements are similar to SPARQL queries, though specifically designed for updating ontologies. Due to the complexity of ontology updating caused by dependencies and physical distributions, we need a principled approach for automatic ontology updating. The Ontology Update Language (OUL) [10] is a blend of active (database) triggers and SPARQL/Update statements, which updates ontologies in an event-driven manner. By defining so-called *changehandlers*, specific ontology change events can be caught and handled individually.

Despite the convenient representation aspects of OUL inspired from active database triggers, the usage of SPARQL and SPARQL/Update, and the implementation of preconditions, the language lacks several key features. First, OUL does not support negation and namespaces. Second, chaining of triggers (changehandlers) is not possible. The changehandlers do not react on actions of other changehandlers. In order to trigger a changehandler, the user has to manually execute an update. Third, there is no differentiation between the order of execution of changehandlers' actions, i.e., there is no distinction between immediate (i.e., once a changehandler is matched, it is executed) and deferred (i.e., the actions are executed all at once after matching the changehandlers and collecting the actions) executions. Fourth, only the first matching changehandler is executed. Hence, when an event occurs, each changehandler is matched against the event; the first changehandler that matches, is handled. Additionally, when updates are triggered and executed, new updates could be triggered, requiring another update cycle. This kind of execution looping is currently not supported.

3   OUL Syntax

Atomic ontology update actions can be executed using SPARQL/Update statements. However, multiple ontology change actions are often required. These actions are hard to express in one single SPARQL/Update statement and can not be edited easily. Therefore, complex ontology updates should be performed as a sequence of atomic SPARQL/Update statements executed in a specific order. The Ontology Update Language (OUL) [10] is based on the automatic update mechanism in active databases: SQL-triggers. Using an Event-Condition-Action model, a list of ontology update actions are performed on event occurrence. This method, using triggers (called changehandlers here), however, does not support a fully automated ontology update process. OUL does feature a dynamic update process using an existing RDF update language, and hence we extend this language in such a way that no human intervention is needed for multiple updates.

OUL makes use of changehandlers that perform SPARQL/Update actions whenever a certain change event (represented as an RDF-graph that is either added to or deleted from the ontology) occurs. If we want to perform particular actions, whenever such triple is added to or deleted from the ontology, we can specify them in a changehandler. Each changehandler has a general form as:

```
CREATE CHANGEHANDLER <name>
FOR <changerequest>
AS
  [ IF <precondition>
    THEN ] <actions>
```

which is analogous to active database triggers. When the *changerequest* matches the change event, a *precondition* on the ontologies is checked. If this precondition is met or if no precondition has been defined, a list of *actions* will be executed. In contrast to active databases, ontology updates do not require SQL statements, but events, conditions, and actions have to be defined using SPARQL and SPARQL/Update statements.

3.1   Requesting Changes

OUL defines two different types of changerequests, i.e., insertion and deletion of information. The add and delete keywords distinguish between the two different types and every changerequest is further defined by a WHERE-clause of a SPARQL SELECT query. The syntax is defined as:

```
<changerequest> ::== add [unique] (<SPARQL>)
                  | delete [unique] (<SPARQL>)
<SPARQL>        ::== WHERE clause of a SPARQL SELECT query
```

When all the triples in the query can be deduced from a change event and the event-type matches the changerequests' type, the changerequest is matched. The set of bindings that are returned from the query can be reused later in the AS-clause of the changehandler definition. A unique property can be used to state whether only one single binding is required. Whenever this property is set, changerequests will not match when their query returns multiple bindings.

3.2   Preconditions

Whenever a changerequest matches, also optional preconditions defined in the changehandler have to be met so that the actions are executed. In contrast to the changerequest, which is used to match the occurring event, the precondition is used to check the current state of the ontology. Three different types of preconditions can be used. First, contains checks whether the ontology contains a set of triples. Second, entails checks whether the ontology entails a set of triples, i.e., using inferencing it can be concluded that the statement is logically entailed by the ontology. Third, entailsChanged checks whether the direct application of the requested change leads to an ontology which entails a set of triples.

Conditions can be combined by and- or or-operators and can be nested as well. Each precondition results in a set of bindings and the and- and or-operators perform join and union operations on the resulting bindings. The syntax for the precondition is defined as follows:

```
<precondition>  ::== contains(<SPARQL>)
                  | entails(<SPARQL>)
                  | entailsChanged(<SPARQL>)
                  | (<precondition>)
                  | <precondition> and <precondition>
                  | <precondition> or <precondition>
<SPARQL>        ::== WHERE clause of a SPARQL SELECT query
```

3.3 Actions

When the changerequest is matched and the precondition is met, a list of actions is executed. Actions make use of the binding information that resulted from matching the changerequest and the precondition. There are four types of actions, i.e., SPARQL/Update queries, feedback actions that give feedback to the user using text containing bounded variables, applyRequest actions that execute the events caught by the changehandler, and last, the for actions that iteratively execute a set of actions with binding information from a for-condition:

```
<actions>         ::== [<action>]|<action><actions>
<action>          ::== <SPARQL update>
                    | for( <precondition> ) <actions> end;
                    | feedback(<text>)
                    | applyRequest
<SPARQL update> ::== a MODIFY action (in SPARQL/Update)
<text>            ::== string (may contain SPARQL variables)
```

3.4 Extensions

In SPARQL it is possible to define prefixes, i.e., labels referring to a namespace. Since in OUL multiple SPARQL WHERE clauses and SPARQL/Update MODIFY clauses may be used, it is necessary to define in each query the used prefixes or to use the full namespaces. The latter provides too much overhead and hence, we propose to define the prefixes for the entire changehandler instead of for every separate SPARQL query.

In OUL, preconditions can be combined by using or- or and-operators. It is, however, not possible to use negation, something that could be desirable whenever ontologies should not contain certain information. Therefore, we implement negation by allowing the usage of an exclamation mark ('!') to denote negation in OUL. The syntax is altered as follows:

```
CREATE CHANGEHANDLER <name>
[<prefixes>]
FOR <changerequest>
AS
  [ IF <precondition>
    THEN ] <actions>

<prefixes>        ::== <prefix>[<prefixes>]
<prefix>          ::== <SPARQL prefix>
<changerequest> ::== add [unique] (<SPARQL>)
                    | delete [unique] (<SPARQL>)
<precondition>   ::== contains(<SPARQL>)
                    | entails(<SPARQL>)
                    | entailsChanged(<SPARQL>)
                    | (<precondition>)
                    | <precondition> and <precondition>
```

```
                       | <precondition> or <precondition>
                       | !<precondition>
<actions>       ::== [<action>]|<action><actions>
<action>        ::== <SPARQL update>
                       | for( <precondition> ) <actions> end;
                       | feedback(<text>)
                       | applyRequest
<SPARQL prefix> ::== PREFIX statement of a SPARQL query
<SPARQL>        ::== WHERE clause of a SPARQL SELECT query
<SPARQL update> ::== a MODIFY action (in SPARQL/Update)
<text>          ::== string (may contain SPARQL variables)
```

4   OUL Execution Models

Updating ontologies in an event-driven manner requires an execution model that controls aspects like selecting the proper changehandlers, executing SPARQL queries, and performing changehandler actions. In [10], the authors provide an execution environment for OUL that allows for ontology updating upon detection of change events in texts. The Ontology Update Manager plays a central role here, as it matches changehandlers based on a changerequest and executes the actions defined in the respective changehandlers. The ontology update specification describes how the ontology can be updated by providing a set of changehandlers.

By default, whenever a change event occurs, all changehandlers defined in the ontology update specification are checked upon their changerequest and precondition to determine if the change event can be handled by a specific changehandler. When a changehandler matches a changerequest with a change event and the precondition is met, the original change event is replaced by the actions defined in the matching changehandler. These actions are then stored and executed all at once later on, i.e., in a deferred manner. In situations where multiple changehandlers match a change event and meet their precondition, only the actions of the first matching changehandler are executed. As OUL does not feature chaining of changehandlers, the execution of the actions cannot trigger other changehandlers, implying that immediate execution would have the same results as deferred execution, when executing only the first matched changehandler.

With respect to the original OUL execution model, we propose several extensions. First, inspired by applications in active databases, we extend OUL by adding support for immediate updating, as opposed to deferred updating. Next, in analogy with active databases where triggers can activate other triggers, we add changehandler chaining. Although this does not ensure termination, it enhances the expressivity of the update language and it enables separation of atomic update operations, thereby enabling modularity. Similarly to active databases triggers, methods for automatic termination evaluation can be developed [12]. Additionally, execution looping is added, which is needed in situations where new updates are required after triggering and executing other updates. Last, we update the OUL execution model in a way that it does not only execute the first matching changehandler, but optionally each matched changehandler.

4.1 Deferred and Immediate Updates

The original (deferred) execution model of OUL comprises three main steps, which are illustrated in Algorithm 1. First, changerequests of all defined changehandlers with respect to the change event are matched and preconditions are verified. Second, actions are collected from the matched changehandlers and SPARQL/Update statements are created. Third and last, the latter statements are applied to the ontology. Note that the method $matchHandlers(\ldots)$ is further specified in Algorithms 3 (first match) and 4 (all matches), and $collectUpdates(\ldots)$ is described in Algorithm 5.

This execution model can be altered in such a way that immediate updating is performed. This implies that during the collection process, update statements are applied immediately to the ontology. Hence, in contrast to deferred updating, we distinguish between two steps, i.e., changehandler matching and update application. Algorithm 2 provides the immediate updating model. Note that the method $matchHandlers(\ldots)$ is further specified in Algorithms 3 (first match) and 4 (all matches), and $applyUpdates(\ldots)$, which applies updates, is described in Algorithm 6.

4.2 First and All Matching Changehandlers

There are two distinct ways of matching changehandlers. The OUL execution model proposed in [10] returns the first changehandler that matches a changerequest and meets its precondition (Algorithm 3). An iterator moves forward through the ontology update specification document until either the end of the document has been reached or a changehandler has been matched to the change event. The matching process returns non-empty binding information which should contain a single binding when a unique keyword is used in the changerequest. For matching preconditions, in case a valid binding is returned, the changehandler is added to the list of matched changehandlers.

However, one could also require multiple changehandlers to be matched. When altering Algorithm 3 by changing the loop conditions, we obtain an execution model that returns all matching changehandlers associated with a change event as given in Algorithm 4. While in Algorithm 3 in line 2 a condition for limiting the list of matched changehandlers is defined, in Algorithm 4, this is removed, making it possible to check all changehandlers defined in the ontology update specification and to add every matching changehandler to the resulting list.

4.3 Chaining Updates

After matching the changehandlers (either the first changehandler encountered, or all changehandlers), their associated update statements have to be collected and applied. This stage depends on the type of execution mechanism. In case deferred execution is applied, all update statements from the matched changehandlers have to be collected before executing them. When immediate execution is used, the statements have to be executed while inspecting them.

Algorithm 1. Deferred ontology updating (updateOntology)

Description: Update ontology with deferred execution of updates
Input: ontology O consisting of axioms,
change event $op(Ax)$ where $op \in \{add, del\}$ and Ax is a set of axioms
Data: $matchedHandlers$ changehandlers that match their changerequest and meet their precondition according to the provided change event,
$updateList$ list of update actions to be applied to the ontology
Output: updated ontology O

1: // Find matched changehandlers
2: $matchedHandlers \leftarrow matchHandlers(O, op(Ax))$
3: // Collect updates from changehandlers
4: $updateList \leftarrow collectUpdates(O, op(Ax), matchedHandlers)$
5: // Apply updates to ontology in deferred way
6: **for all** $update \in updateList$ **do**
7:     apply $update$ to O
8: **end for**
9: **return** O

Algorithm 2. Immediate ontology updating (updateOntology)

Description: Update ontology with immediate execution of updates
Input: ontology O consisting of axioms,
change event $op(Ax)$ where $op \in \{add, del\}$ and Ax is a set of axioms
Data: $matchedHandlers$ changehandlers that match their changerequest and meet their precondition according to the provided change event
Output: updated ontology O

1: // Find matched changehandlers
2: $matchedHandlers \leftarrow matchHandlers(O, op(Ax))$
3: // Apply updates to ontology in an immediate way
4: $O \leftarrow applyUpdates(O, op(Ax), matchedHandlers)$
5: **return** O

Algorithm 3. Returning the first matching changehandler (matchHandlers)

Description: Collect matching changehandlers
Input: ontology O consisting of axioms,
ontology update specification US treated as a list of changehandlers,
change event $op(Ax)$ where $op \in \{add, del\}$ and Ax is a set of axioms.
Data: $handler$ changehandler that is checked for applicability
Output: list of matched changehandlers $matchingHandler$

1: // While not at document's end and no changehandler has been matched
2: **while not** $US.endOfDocument$ **and** $matchingHandlers.count < 1$ **do**
3:     // Take the next changehandler
4:     $handler \leftarrow US.nextChangeHandler$
5:     // Match the changerequest with the change event
6:     $matches \leftarrow SPARQLmatch(handler.changerequest, op(Ax))$
7:     // The bindings form the changerequest should not be empty
8:     **if not** $matches.isEmpty$ **then**
9:         // The number of bindings should be 1 when the unique keyword is used
10:        **if** $(handler.changerequest.unique$ **and** $matches.count == 1)$ **or** **not** $handler.changerequest.unique$ **then**
11:            // Substitute variables in the precondition with changerequest bindings
12:            $instPrecondition \leftarrow substitute(handler.precondition, matches.first)$
13:            // Evaluate the precondition. When this returns any binding, it is met
14:            **if not** $evaluate(instPrecondition, O).isEmpty$ **then**
15:                // Add the changehandler to the list
16:                $matchingHandlers.add(handler)$
17:            **end if**
18:        **end if**
19:    **end if**
20: **end while**
21: // Return the list of matched changehandlers
22: **return** $matchingHandlers$

Algorithm 4. Returning all matching changehandlers (matchHandlers)

Description: Collect matching changehandlers
Input: ontology O consisting of axioms,
ontology update specification US treated as a list of changehandlers,
change event $op(Ax)$ where $op \in \{add, del\}$ and Ax is a set of axioms.
Data: $handler$ changehandler that is checked for applicability
Output: list of matched changehandlers $matchingHandler$

1:   // While not at document's end
2: **while not** $US.endOfDocument$ **do**
3:     // Take the next changehandler
4:     $handler \leftarrow US.nextChangeHandler$
5:     // Match the changehandler with the change event
6:     $matches \leftarrow SPARQLmatch(handler.changerequest, op(Ax))$
7:     // The bindings form the changerequest should not be empty
8:     **if not** $matches.isEmpty$ **then**
9:       // The number of bindings should be 1 when the unique keyword is used
10:       **if** $(handler.changerequest.unique$ **and** $matches.count$ $==$ $1)$ **or** **not** $handler.changerequest.unique$ **then**
11:         // Substitute variables in the precondition with changerequest bindings
12:         $instPrecondition \leftarrow substitute(handler.precondition, matches.first)$
13:         // Evaluate the precondition. When this returns any binding, it is met
14:         **if not** $evaluate(instPrecondition, O).isEmpty$ **then**
15:           // Add the changehandler to the list
16:           $matchingHandlers.add(handler)$
17:         **end if**
18:       **end if**
19:     **end if**
20: **end while**
21: // Return the list of matched changehandlers
22: **return** $matchingHandlers$

Algorithm 5. Update collection from matched changehandlers (collectUpdates)

Description: Collect updates from a list of matched changehandlers using deferred execution
Input: ontology O consisting of axioms,
change event $op(Ax)$ where $op \in \{add, del\}$ and Ax is a set of axioms,
list of changehandlers $matchedHandlers$ that match the change event
Output: list of the update statements $updateList$

1:   // Check whether any changehandler matches the change event
2: **if not** $matchedHandlers.isEmpty()$ **then**
3:     // Loop through all matched changehandlers
4:     **for all** $matchedHandler \in matchedHandlers$ **do**
5:       // Loop through all update statements in the changehandler
6:       **for all** $update \in matchedHandler.updates$ **do**
7:         // Chaining: add the update or the replaced update actions from other
8:         //   changehandlers to the list of update statements
9:         // Find changehandlers that match the update event
10:         $newMatchedHandlers \leftarrow matchHandlers(O, update)$
11:         // Collect updates from changehandlers
12:         $newUpdateList \leftarrow collectUpdates(O, update, newMatchedHandlers)$
13:         // Add the update statements to the list
14:         $updateList.add(newUpdateList)$
15:       **end for**
16:     **end for**
17: **else**
18:     // There is no changehandler matching the change event; therefore, the change
19:     // event itself is added to the list of update statements
20:     $updateList.add(op(Ax))$
21: **end if**
22: // Return the list of update statements
23: **return** $updateList$

The earlier introduced Algorithm 1 defines the execution steps of the deferred execution model. In the first step, update collection, statements are collected from the matched changehandlers. Algorithm 5 explains how this task is performed. First, a check is done to investigate whether any changehandlers match the change event. If this is the case, the update statements from every changehandler in the set of matched changehandlers are collected. If no changehandler matches the change event, the change event itself is applied to the ontology. In lines 7-14, each update statement is treated as a change event, representing the implementation of chaining. This part is similar to Algorithm 1, except for the fact that updates are not applied to the ontology, because this has to happen at the end of the process when using deferred execution. In the end, the algorithm returns a list of update statements that need to be applied to the ontology.

As depicted in Algorithm 6, for immediate ontology updating, no update lists are returned. In contrast to deferred updating, the updates are immediately applied to the ontology. For immediate updating, the algorithm first checks whether any changehandler exists in the list of matched changehandlers. If this is the case, for each update statement in each of the matched changehandlers, a change event is fired as shown in Algorithm 2 using the update as the change event. In this way, we provide a mechanism for chaining. If no changehandler matches the change event, the change event itself is applied to the ontology.

4.4   Looping Updates

Applying updates to the ontology and thereby changing the ontology can trigger new changehandlers to become matched, which can be used for applying additional updates in case the event is handled more than once. Hence, we introduce the possibility to iterate over the changehandlers with the same event and apply updates until there are no matching changehandlers left. In this way, the effect of the update actions can be checked and additional updates can be applied.

Ontology update looping can be implemented by adding a call to the *updateOntology*(...) methods of Algorithms 1 (deferred) and 2 (immediate) at the end of both algorithms, using the same change event and ontology as input. Algorithms 7 and 8 implement looping for deferred and immediate executions, respectively. Before the updated ontology is returned, in both algorithms the *updateOntology*(...) method is called recursively to ensure that additional updates are handled until no updates are available.

5   Implementation

OUL, including the proposed extensions, has been implemented as a standalone software package providing event-driven ontology updates (available at http://people.few.eur.nl/fhogenboom/oulx.html). We used this package in the Hermes News Portal [7], a Java-based news personalization tool implementing the Hermes framework [7]. Hermes uses an ontology for classifying and querying news items. The Hermes domain ontology has to be up-to-date with the

Algorithm 6. Update application from matched changehandlers (applyUpdates)

Description: Apply updates from a list of matched changehandlers using immediate execution
Input: ontology O consisting of axioms,
change event $op(Ax)$ where $op \in \{add, del\}$ and Ax is a set of axioms,
list of changehandlers $matchedHandlers$ that match the change event
Output: updated ontology O

1: // Check whether any changehandler matches the change event
2: **if not** $matchedHandlers.isEmpty()$ **then**
3:     // Loop through all matched changehandlers
4:     **for all** $matchedHandler \in matchedHandlers$ **do**
5:         // Loop through all update statements in the changehandler
6:         **for all** $update \in matchedHandler.updates$ **do**
7:             // Chaining: fire the update as an update event; this way, the update can
8:             //   be handled by appropriate changehandlers
9:             $updateOntology(update)$
10:        **end for**
11:    **end for**
12: **else**
13:    // There is no changehandler matching the change event; therefore, the change
14:    // event itself is applied
15:    apply $op(Ax)$ to O
16: **end if**
17: **return** O

Algorithm 7. Looped deferred ontology updating (updateOntology)

Description: Update ontology with deferred execution of updates and looping
Input: ontology O consisting of axioms,
change event $op(Ax)$ where $op \in \{add, del\}$ and Ax is a set of axioms
Data: $matchedHandlers$ changehandlers that match their changerequest and meet their precondition
according to the provided change event,
$updateList$ list of update actions to be applied to the ontology
Output: updated ontology O

1: // Find matched changehandlers
2: $matchedHandlers \leftarrow matchHandlers(O, op(Ax))$
3: // Collect updates from changehandlers
4: $updateList \leftarrow collectUpdates(O, op(Ax), matchedHandlers)$
5: // Apply updates to ontology in deferred way
6: **for all** $update \in updateList$ **do**
7:     apply $update$ to O
8: **end for**
9: // Execute this algorithm again to check for additional updates
10: **if not** $matchedHandlers.isEmpty()$ **then**
11:    $updateOntology(O, op(Ax))$
12: **end if**
13: **return** O

Algorithm 8. Looped immediate ontology updating (updateOntology)

Description: Update ontology with immediate execution of updates and looping
Input: ontology O consisting of axioms,
change event $op(Ax)$ where $op \in \{add, del\}$ and Ax is a set of axioms
Data: $matchedHandlers$ changehandlers that match their changerequest and meet their precondition
according to the provided change event
Output: updated ontology O

1: // Find matched changehandlers
2: $matchedHandlers \leftarrow matchHandlers(O, op(Ax))$
3: // Apply updates to ontology in an immediate way
4: $O \leftarrow applyUpdates(O, op(Ax), matchedHandlers)$
5: // Execute this algorithm again to check for additional updates
6: **if not** $matchedHandlers.isEmpty()$ **then**
7:     $updateOntology(O, op(Ax))$
8: **end if**
9: **return** O

latest news and hence needs automatic ontology updates. Based on the information extraction plugin for the Hermes News Portal, i.e., Aethalides, information extracted from news items can be used for updating the ontology.

A key aspect of the Hermes News Portal is its financial ontology. For its updates, the ontology is dependent on information extracted from financial news messages, e.g., product releases, CEO appointments, bankruptcies, etc. We implement the OUL update mechanisms and connect them to the information extraction processes of Aethalides. The Aethalides plugin makes use of the Hermes Information Extraction Engine, which is used for matching user-created information extraction rules with text in news items.

To integrate automatic ontology updating, each new news item is processed and information (in the form of events) is extracted using the user-created rules. After validating the extracted information, the ontology is updated using OUL update rules. In our implementation, the execution of the SPARQL WHERE clauses and the SPARQL/Update statements, as well as ontology updating is performed using Jena [8]. In order to work with the latest developments in the Semantic Web, we updated ARQ, the query engine in Jena, to version 2.8.8, which features SPARQL 1.1. The changehandlers can be loaded via a plain text file that contains changehandlers specified in the proposed syntax. Parsing and compiling of changehandlers is performed via a compiler created with JavaCC [14].

6   Evaluation

In order to evaluate the extensions made to OUL, we analyze the characteristics of each proposed execution model. As it is difficult to perform a quantitative analysis and as there are no benchmarks available for OUL, we discuss at a qualitative level the advantages and disadvantages of each execution model. We assume all queries are chained (non-chained queries as originally proposed by OUL are also supported), which provides us with eight execution models:

- Immediate, looped execution of first matching changehandler;
- Immediate, non-looped execution of first matching changehandler;
- Immediate, looped execution of all matching changehandlers;
- Immediate, non-looped execution of all matching changehandlers;
- Deferred, looped execution of first matching changehandler;
- Deferred, non-looped execution of first matching changehandler;
- Deferred, looped execution of all matching changehandlers;
- Deferred, non-looped execution of all matching changehandlers.

Deferred execution of matching changehandlers could lead to erroneous updates, and hence it usually does not make sense to make use of the last four execution models. For example, it could be the case that several changehandlers can originally match, but after executing their corresponding updates in a deferred mode, the updates of the previous matches could be made invalid. Due to the nature of deferred execution, these updates would still be executed. On the other hand, deferred updating could possibly lead to more efficient updates in case multiple changes are to be made to the same entity, as these actions could be merged and transformed into simplified update statements. Additionally, duplicate actions can be merged, eliminating duplicate action executions. So, if deferred updating is used, some caution is required, making sure there are no conflicting dependencies between update actions and change requests.

When comparing models that execute the first matching changehandler with those that execute all changehandlers, one could make the following observations. The latter method is computationally more intensive due to the increased complexity on the execution mechanism. Conversely, updates are more efficient, as in one pass all the matched changehandlers are dealt with, hence eliminating the need for multiple user-triggered iterations.

In case of looped execution models, the advantage is that ontology updates performed during a pass that trigger new changehandlers to be matched are taken into account, hereby improving the efficiency of the ontology updating process, as no separate runs are needed. The looped execution models are on the other hand harder for users to grasp due to the repeated event generation until no changehandler matches the event.

There is a trade-off between easiness of writing update rules and their efficient execution. The all matching and/or looped variants are more efficient due to the automatic execution and possible optimization of their complex actions, while the first matching and/or non-looped counterparts are more intuitive and thus foster easier development of update rules. Also, it should be noted that for chaining there is an increased level of automation, as users do not have to manually trigger updates resulting from earlier updates (as in case of the OUL execution model), as these are automatically handled.

7   Conclusions

The Ontology Update Language (OUL) is based on SQL triggers and focuses on an event-driven ontology update specification. By creating changehandlers, containing an event description, a precondition, and a list of SPARQL/Update statements, update actions are executed when events occur and preconditions are met. OUL features the creation of Event-Condition-Action rules, hereby enabling automatic updates. We identified some drawbacks at language as well as execution model levels, and proposed extensions to address these.

Syntax-wise, in order to facilitate more complex expressions, we extended OUL so that it also supports negation and prefixes. Our main contribution however lies in the extension of OUL's execution mechanism. We incorporated immediate updating, as opposed to deferred updating. Also, we added an internal triggering mechanism for changehandlers called updates chaining, allowing for automatic event triggering based on the matched changehandlers' actions. This contributes to the usability of the language by separating atomic update actions and thus delivering modularity and an increased possibility to reuse changehandlers. Also, we added support for looping for repetitive treatment of an event. Last, it is now also possible to execute all event-related changehandlers, instead of just the first matching handler. The here proposed extensions are viable, provided that technical experts who are accustomed to the update language work together with experts of the knowledge domain.

As future work we would like to evaluate the termination of changehandlers, i.e., which conditions need to be satisfied by a set of changehandlers so that, for any incoming events, the matching changehandlers should always terminate. For this purpose we plan to reuse results from termination of rule-based updates for databases [12]. Alternatively, one could look into developing a principled information extraction language that combines information extraction and ontology updates. For this purpose, we plan to integrate the Hermes Information Extraction Language with OUL, including the here proposed extensions.

References

1. Bechhofer, S., van Harmelen, F., Hendler, J., Horrocks, I., McGuinness, D.L., Patel-Schneider, P.F., Stein, L.A.: OWL Web Ontology Language Reference. W3C Recommendation, February 10 (2004), http://www.w3.org/TR/owl-ref/
2. Boag, S., Chamberlin, D., Fernandez, M.F., Florescu, D., Robie, J., Simeon, J.: XQuery 1.0: An XML Query Language (Second Edition). W3C Recommendation, December 14 (2010), http://www.w3.org/TR/xquery/
3. Bray, T., Paoli, J., Sperberg-McQueen, C., Maler, E., Yergeau, F.: Extensible Markup Language (XML). W3C Recommendation, November 26 (2008), http://www.w3.org/TR/2008/REC-xml-20081126/
4. Brickley, D., Guha, R.: RDF Vocabulary Description Language 1.0: RDF Schema. W3C Recommendation, February 10 (2004), http://www.w3.org/TR/rdf-schema/
5. Chamberlin, D.D., Boyce, R.F.: SEQUEL: A Structured English Query Language. In: Rustin, R. (ed.) 1974 ACM SIGMOD Workshop on Data Description, Access and Control, vol. 1, pp. 249–264. ACM (1974)
6. Clark, J., DeRose, S.: XML Path Language (XPath). W3C Recommendation, November 16 (1999), http://www.w3.org/TR/xpath/
7. Frasincar, F., Borsje, J., Levering, L.: A Semantic Web-Based Approach for Building Personalized News Services. International Journal of E-Business Research 5(3), 35–53 (2009)
8. HP Labs: Jena (2011), http://jena.sourceforge.net/
9. Laux, A., Martin, L.: XUpdate (2000),
   http://xmldb-org.sourceforge.net/xupdate/xupdate-wd.html
10. Lösch, U., Rudolph, S., Vrandečić, D., Studer, R.: Tempus Fugit. In: Aroyo, L., Traverso, P., Ciravegna, F., Cimiano, P., Heath, T., Hyvönen, E., Mizoguchi, R., Oren, E., Sabou, M., Simperl, E. (eds.) ESWC 2009. LNCS, vol. 5554, pp. 278–292. Springer, Heidelberg (2009)
11. Prud'hommeaux, E., Seaborne, A.: SPARQL. W3C Recommendation, January 15 (2008), http://www.w3.org/TR/rdf-sparql-query/
12. Ray, I., Ray, I.: Detecting Termination of Active Database Rules Using Symbolic Model Checking. In: Caplinskas, A., Eder, J. (eds.) ADBIS 2001. LNCS, vol. 2151, pp. 266–279. Springer, Heidelberg (2001)
13. Seaborne, A., Manjunath, G., Bizer, C., Breslin, J., Das, S., Davis, I., Harris, S., Idehen, K., Corby, O., Kjernsmo, K., Nowack, B.: SPARQL Update. W3C Member Submission, July 15 (2008), http://www.w3.org/Submission/SPARQL-Update/
14. Sun Microsystems: JavaCC (2011), http://javacc.java.net/

BiCWS: Mining Cognitive Differences from Bilingual Web Search Results

Xiaojiang Huang, Xiaojun Wan[*], and Jianguo Xiao

Institute of Computer Science and Technology
& The MOE Key Laboratory of Computational Linguistics,
Peking University, Beijing 100871, China
{huangxiaojiang,wanxiaojun,xiaojianguo}@pku.edu.cn

Abstract. In this paper we propose a novel comparative web search system – BiCWS, which can mine cognitive differences from web search results in a multi-language setting. Given a topic represented by two queries (they are the translations of each other) in two languages, the corresponding web search results for the two queries are firstly retrieved by using a general web search engine, and then the bilingual facets for the topic are mined by using a bilingual search results clustering algorithm. The semantics in Wikipedia are leveraged to improve the bilingual clustering performance. After that, the semantic distributions of the search results over the mined facets are visually presented, which can reflect the cognitive differences in the bilingual communities. Experimental results show the effectiveness of our proposed system.

Keywords: Comparative Text Mining, Cross Lingual Text Mining, Information Retrieval.

1 Introduction

With the development of modern technologies, the communication between different cultures and communities is getting more and more convenient and frequent. During the communication, people may assume that others should think similarly as themselves. However, due to the different background, environments, ideologies, religions, etc., the cognition for the same concept or topic in different cultures and communities can be different. For example, in the Chinese culture, the word "dog" usually has pejorative connotation, and means "stooge", "accomplice", "rogue", etc.; while in the western culture, "dog" usually has positive meanings, including "courage" and "loyalty". The ignorant of such cognitive differences sometimes causes misunderstandings and conflicts. Mining the cognitive differences is not only helpful to satisfy the requirement of knowledge discovery, but also helpful to strengthen mutual understanding, and promote cooperation between different civilizations.

As the most important carrier of information, natural language texts record the people's thoughts. The distributions of a topic over semantic aspects in different

---

[*] Corresponding author.

X.S. Wang et al. (Eds.): WISE 2012, LNCS 7651, pp. 58–71, 2012.

languages can reflect the cognitive differences in corresponding communities. For example, the term "长城" (*Great Wall*) occurs in many Chinese articles about "*travel*", "*automobile*", or "*wine*", etc. In comparison, there are only a few English articles talking about the "*automobile*" meaning of "*Great Wall*", because this car brand is not very popular in the Western countries.

In this paper, we propose a novel system to mine cognitive differences in different communities and cultures by comparing the distributions over semantic aspects in texts of different languages. We take the Web as the corpus, which is one of the largest public accessible text collections, and use an existing web search engine to obtain samples of the corpus. Given a topic represented by two queries (they are translations of each other) in two languages, our system retrieves the corresponding search results by using a general web search engine, and then mines bilingual facets for the topic by using a bilingual search results clustering algorithm. After that, it visually presents the semantic distributions of the search results over the mined aspects, which reflects the cognitive differences in the bilingual communities. Experimental results show that our system is able to discover the cognitive aspects in the bilingual search results and visualize the cognitive differences in different languages.

The rest of paper is organized as follows: First, we give an overview of related work in Section 2, and then describe our proposal in details in Section 3. The experimental evaluation is discussed in Section 4. Finally Section 5 concludes our work.

2   Related Work

Comparisons have been researched for a long time in the social science fields, and a number of academic subjects have been founded, such as comparative linguistic [1] and comparative literature [2]. The comparative cultural study is one of the most active fields. The study objects include all sort of culture and cultural products, e.g. languages, disciplines, sociology, psychology, etc. [3].

The comparative mining tasks have also drawn much attention in the information retrieval and natural language processing fields. Different specific subtasks have been studied, including comparative component extraction [4], comparative topic discovery [5] and comparative summarization [6]. Liu et.al proposed an interesting task to compare the semantic distributions of cognitive aspects reflected in the images across multi-language communities [7]. Our study follows the task of cognitive comparison, but focuses on the text domain. The text domain contains much richer semantic information than the image domain, and thus can reflect the differences of cultures more precisely.

Recently, mining conceptual aspects from search results is an active research area [8]. A common kind of methods first groups semantic related result pages by using clustering techniques, and then extracts key phrases for each cluster [9]. Another kind of methods first learns descriptions of aspects, and then assigns each result page to the relevant aspect [10]. So far, most studies focus on mining aspects of a single query. Sun et al. proposed a system to compare the common aspects of two comparable queries by matching search results of the two queries into pairs [11]. The queries in their work are represented in the same language, and refer to different objects, e.g. Canon and Nikon, while our study focuses on compare term translations in different languages.

Inspired by the dream of glitch-free communication with anyone in the world, the cross-language information processing tasks attract the interests of many researchers. In particular, several approaches have been proposed for clustering bilingual documents [12-14]. The basic idea is to map the documents in different languages into a unified feature spaces where the semantic relatedness can be easily estimated. Dagan and Itai proposed an approach for word sense disambiguation (WSD) in one language using statistical data from corpus of another language [15]. Their method is based on syntactic relations between words and requires a large corpus for statistical analysis. In comparison, our system uses the semi-parallel Wikipedia articles to estimate cross-language relations. Khapra et al. proposed a method of using bilingual bootstrapping for bilingual WSD [16]. This method needs annotated data in both languages, while our method is unsupervised, which requires less manual labour and is more suitable for open-domain applications.

3   Proposed Approach

3.1   Overview

In this subsection, we give an overview of our BiCWS system. The aim of BiCWS is to mine the cognitive difference of a topic or concept in different language communities. The topic can be represented by terms in different languages. We ask the user to manually provide the query terms in both languages simultaneously, because machine translations of terms without any contexts are usually not accurate. In spite of a little inconvenience, we do not think it will be a big burden for most users.

Fig. 1 illustrates the procedure of our system. The system requires the user to give two queries q_β and q_θ as input, where q_β and q_θ are translation of each other in two languages L_β and L_θ. Both queries are submitted to a search engine to get the lists of relevant web pages. We use the same global Google Search[1] for both languages in this study, so that the differences in results are not caused by implementation details of different search engines. The underlying cognitive aspects in the results are then mined by using a bilingual clustering algorithm. Both the text-based information and the semantic-based information are used to calculate the semantic relatedness between texts in different languages. The texts are translated into the same language by using an online machine translating service[2], and mapped to a unified external semantic space using the cross language explicit semantic analysis technique. For each aspect, a few salient key phrases are extracted as labels. Finally the distribution of the bilingual search results over mined aspects is visually presented to help the user discover the cognitive differences of the topic in different languages. In this study, we take English and Chinese as example languages. Our methods are also applicable to other pairs of languages, as long as the search engines and Wikipedias in those languages are available.

---

[1] http://www.google.com
[2] http://translate.google.com

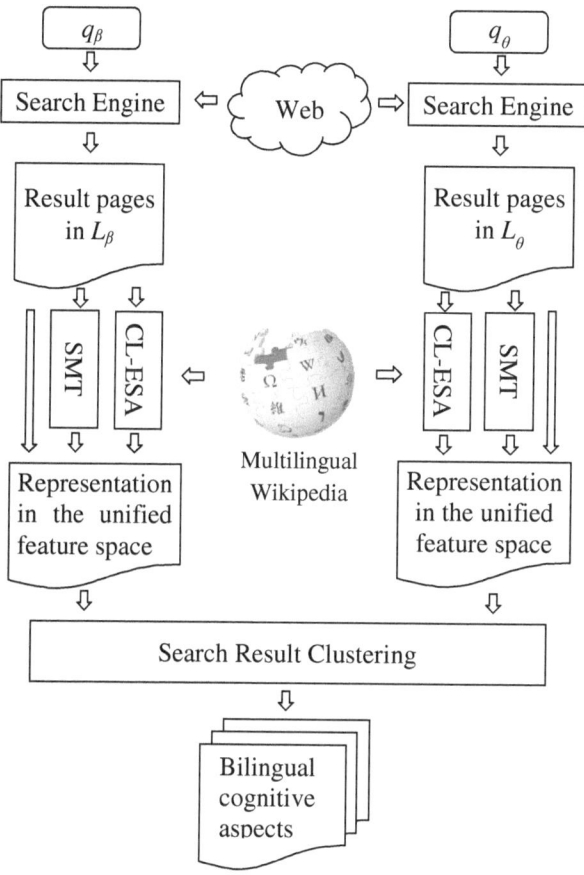

Fig. 1. The procedure of the Bi-CWS system

3.2 Cross-Language Explicit Semantic Analysis

Monolingual Explicit Semantic Analysis

Explicit Semantic Analysis (ESA) attempts to index a document in a high-dimensional space of concepts derived from external categories, e.g. Wikipedia [17]. In the monolingual ESA, this vector space is spanned by a Wikipedia database $W_\delta = \{a_\delta^1,..,a_\delta^n\}$ in language L_δ, where each dimension a_δ^i corresponds to an article in the Wikipedia. The mapping function $\phi_\delta : D \rightarrow \mathbb{R}^{|W_\delta|}$ is defined as follows:

$$\phi_\delta = \left\langle as\left(d, a_\delta^1\right),..., as\left(d, a_\delta^n\right)\right\rangle$$

The function as expresses the association strength between document d and the Wikipedia article a_δ^i. A widely used as function is based on the bag-of-words model:

$$as\big(d,a_\delta^i\big)= \sum_{w_j\in d} tf_d\big(wd_j\big)\cdot tf_{a_\delta^i}\big(wd_j\big)\cdot idf_{W_\delta}\big(wd_j\big)$$

where wd_j is a word in d; $tf_d(wd_j)$ and $tf_{a_\delta^i}(wd_j)$ are the frequencies of wd_j in document d and a_δ^i, respectively; $idf_{W_\delta}(wd_j)$ is the inverse document frequency of wd_j in W_δ, which is defined as follows:

$$idf_{W_\delta}\big(wd_j\big)= \log\frac{|W_\delta|}{\big|\{a\in W_\delta : wd_j\in a\}\big|+1}$$

where $|W_\delta|$ is the number of articles in the Wikipedia, and $|\{a\in W_\delta: wd_j\in a\}|$ is the number of articles which contain the word wd_j.

Cross-Language Explicit Semantic Analysis

The ESA framework has been extended for cross language analysis by transforming the ESA vectors in one language into vectors in another language [18]. This is done by mapping each dimension corresponding to an article a_β^i in Wikipedia W_β of language L_β to the dimension corresponding to the equivalent article a_θ^j in Wikipedia W_θ of language L_θ.

In the Wikipedia, the nearly equivalent or exactly equivalent pages in different languages are linked by inter-language links. For example, some of the inter-language links in the article *"Great Wall of China"* are illustrated as follows:

```
[[af:Groot Muur van Sjina]]
[[ar:الصــين سوق العظيم ]]
[[an:Gran Muralla Chinesa]]
...
[[zh-yue:萬里長城]]
[[bat-smg:Dėdlīsės Kėnėjės robežios]]
[[zh:长城]]
```

In each line the characters before the colon is the code of a language (e.g. the *"af"* stands for *Afrikaans*, and the *"ar"* stands for *Arabic*), and the string after the colon is the title of article in that language (e.g. "长城" links to a page of introduction of the *Great Wall* in Chinese language). The inter-language links can define a function $m_{\theta\to\beta}: W_\theta\to W_\beta$ that maps the Wikipedia articles in language L_θ into articles in another language L_β:

$$m_{\theta\to\beta}\big(a_\theta^i\big)= a_\beta^j,\text{where} <a_\theta^i,a_\beta^j>\in IL_{\theta\to\beta}$$

where $IL_{\theta\to\beta}$ is the collection of all inter-language links from language L_θ to language L_β. Similarly, we can also define the mapping function from L_β to L_θ:

$$m_{\beta\to\theta}\big(a_\beta^i\big)= a_\theta^j,\text{where} <a_\beta^i,a_\theta^j>\in IL_{\beta\to\theta}$$

where $IL_{\beta\rightarrow\theta}$ is the collection of all inter-language links from language L_β to language L_θ.

Given a document d_β in language L_β, CL-ESA allows to index this document with respect to another language L_θ by transforming the vector $\Box_\beta(d_\beta) = <as(d_\beta, a_\beta^1), as(d_\beta, a_\beta^2),...>$ into a corresponding vector in the vector space that is spanned by the Wikipedia articles in the target language. The mapping function is defined as follows:

$$\psi_{\beta\rightarrow\theta}\left(\phi_\beta\left(d_\beta\right)\right) = \left\langle as\left(d_\beta, m_{\theta\rightarrow\beta}\left(a_\theta^1\right)\right), as\left(d_\beta, m_{\theta\rightarrow\beta}\left(a_\theta^2\right)\right),...\right\rangle$$

where a_θ^i are the articles of Wikipedia W_θ. Note that $m_{\theta\rightarrow\beta}(a_\theta^i)$ is an article in L_β, and thus the as function estimates the association strength of two document in the same language.

Based the above settings, it is straightforward to calculate the semantic similarity between documents in two different languages. Given a document d_β in language L_β and a document d_θ in language L_θ, the similarity can be calculated as the cosine value of the CL-ESA vectors, i.e.:

$$sim_\beta\left(d_\beta, d_\theta\right) = cos\left(\phi_\beta\left(d_\beta\right), \psi_{\theta\rightarrow\beta}\left(\phi_\theta\left(d_\theta\right)\right),\right) \text{, or}$$

$$sim_\theta\left(d_\beta, d_\theta\right) = cos\left(\psi_{\beta\rightarrow\theta}\left(\phi_\beta\left(d_\beta\right)\right), \phi_\theta\left(d_\theta\right)\right)$$

Furthermore, if W_β and W_θ are perfectly matched, i.e. for any Wikipedia article in one language, there exists only one equivalent article in the other language, then

$$sim_\beta\left(d_\beta, d_\theta\right) \equiv sim_\theta\left(d_\beta, d_\theta\right)$$

This means that in such case it is arbitrary to choose the target vector space. In other words, for the text in either language, we can represent it in a unified ESA vector space.

In this study, we adapt the WikipediaESA[3] toolkit for cross-language explicit semantic analysis. We use the dumps of Wikipedia in September 2011, which contain more than 3,000,000 articles in English and about 400,000 articles in Chinese. There are about 260,000 language links between English and Chinese articles. The articles which have bidirectional language links are kept as the external categories in cross language explicit semantic analysis.

3.3   Bilingual Search Result Modeling

The core problem of bilingual analysis problem is the estimation of the semantic relatedness between texts in different languages. To calculate the relatedness of result pages, we represent them in a unified feature space, including the text in the original language, the translation in the other language, and the features in the CL-ESA space. Using this representation, the calculation of relatedness between pages in different languages is identical to the calculation of relatedness between pages in the same language.

---

[3] http://www.srcco.de/v/wikipedia-esa

Formally, let q_β and q_θ be a pair of queries about the same topic in language L_β and L_θ respectively. SR_β and SR_θ are the corresponding lists of web search results. For fast processing, we only extract the titles and snippets of result pages given by the search engine, but do not download the whole web pages. For each page $p_\beta^i \in SR_\beta$, we remove q_β from the title and the snippet of p_β^i, and denote the remaining text as $p_\beta^i \backslash q_\beta$. We can represent the text with a vector space model $\mathbf{vt}_\beta^i = <w_\beta^1,...,w_\beta^n>$, where each dimension is the $tf \cdot idf$ weight of a word in language L_β. Meantime, we translate $p_{\beta,i} \backslash q_\beta$ into language L_θ, and represent the translation with another vector space model $\mathbf{vt}_{\beta \to \theta}^i = <w_\theta^1,...w_\theta^m>$, where each dimension is the $tf \cdot idf$ weight of a word in language L_θ. These two vectors represent the text based features for a result page. In addition, we denote the vector model of CL-ESA as $\mathbf{ve}_\beta^i$. The final model for p_β^i is the weighted combination of these three vectors:

$$\mathbf{v}_\beta^i = < ut_\beta^i \cdot \mathbf{vt}_\beta^i,\ ut_\beta^i \cdot \mathbf{vt}_{\beta \to \theta}^i,\ ue_\beta^i \cdot \mathbf{ve}_\beta^i >$$

where ut_β^i and ue_β^i are normalization factors to normalize the above vector to unit length:

$$ut_\beta^i = \frac{\sqrt{\lambda}}{\sqrt{\left|\mathbf{vt}_\beta^i\right|^2 + \left|\mathbf{vt}_{\beta \to \theta}^i\right|^2}}, \qquad ue_\beta^i = \frac{\sqrt{1-\lambda}}{\left|\mathbf{ve}_\theta^i\right|}$$

where the parameter $\lambda \in [0, 1]$ is a factor to tune the impacts of text based features and ESA based features. In the experiments, λ is set to 0.5.

Similarly, a result page $p_\theta^j \in SR_\theta$ can be represented as follows:

$$\mathbf{v}_\theta^j = < ut_\theta^j \cdot \mathbf{vt}_{\theta \to \beta}^j,\ ut_\theta^j \cdot \mathbf{vt}_\theta^j,\ ue_\theta^j \cdot \mathbf{ve}_\theta^j >$$

Note that the first parts of $\mathbf{v}_\beta^i$ and v_θ^j are features in language L_β, the second parts of the two vectors are features in language L_θ, and the third parts are features in the cross-language ESA space. Therefore these three parts together define an unified feature space for search results in both languages.

3.4   Bilingual Search Result Clustering

It is shown that the aspects can be mined from search results by using a clustering algorithm. In this subsection, we describe the approach of using clustering techniques to discover the cognitive aspects in bilingual web search results.

Given two lists of search results SR_β and SR_θ in different languages, there are two strategies to cluster the result pages: the unified strategy and the individual strategy. The unified strategy mixes the search results into a single collection $SR = SR_\beta \cup SR_\theta$, and treats the problem as a traditional web search clustering task. This strategy works because the search results in different languages have been mapped into a unified vector space, as described in the previous subsection. The individual strategy first clusters pages in each language respectively, and then matches the corresponding aspects in the two clustering results.

In this study, we use a clustering algorithm based on the Gaussian mixture model [19] for both strategies. In this model, it is assumed that there are K components (clusters). Each component is a Gaussian distribution $N(\cdot|\mu_k, \Sigma_k)$ parameterized by μ_k, Σ_k. Each data point x is generated by the mixture of these components:

$$p(x\,|\,\pi,\mu,\Sigma)=\sum_{k=1}^{K}\pi_k N(x\,|\,\mu_k,\Sigma_k)$$

where π_k is the possibility that x is generated by component k.

Given a set of data points $X = \{x_i\}$, we assume that they are independent identically distributed. Thus the likelihood function of X is

$$p(X\,|\,\pi,\mu,\Sigma)=\prod_{x\in X}\sum_{k=1}^{K}\pi_k N(x\,|\,\mu_k,\Sigma_k)$$

We can then estimate the parameters π, μ, and Σ by maximize the likelihood using the Expectation Maximization algorithm [20].

For the cluster matching step in the individual strategy, we build a weighted bipartite graph $G = \{V, E\}$, where the vertices include all cluster centroids in each language $V = \{c_\beta^i\}\cup\{c_\theta^j\}$. Each pair of cluster centroids in different languages is linked with an edge, and thus the edge set is $E = \{<c_\beta^i, c_\theta^j>|\forall\ i, j\}$. The weight we_e of an edge $e = <c_\beta^i, c_\theta^j>$ is computed by the cosine similarity of the two cluster centroids c_β^i and c_θ^j. Then the cluster matching can be achieved by finding the maximum total matching weight in the bipartite graph:

$$\max\sum_{e\in E'}we_e, \text{where } E'\subset E : \forall v\in V \rightarrow degree_{E'}(v)\leq 1$$

where $degree_{E'}(v)$ denotes the degree of v in the Graph $G'=\{V, E'\}$.

The maximum weighted matching problem in a bipartite graph has been well studied, and several efficient algorithms have been proposed [21]. In this study, we use the LEDA toolkit[4] to solve this problem.

In practice, we further mix the clusters which contain very few result pages together. These clusters usually reflect noises or unusual aspects, and it is usually annoying for users to see these clusters individually.

3.5   Search Result Cluster Labeling

For better understanding of the mined aspects, we extract key phrases to represent the clusters. The labels come from two different sources. One is the texts and translations of search result pages in the cluster, and the other is the Wikipedia article titles in the corresponding semantic analysis vectors. As a native approach, we evaluate the significance of candidate labels by their weights in the cluster's centroid. For the mixed cluster which is made up by several small clusters, we simply label it as "others /其他".

---

[4] http://www.algorithmic-solutions.com/leda

4   Experiment

4.1   Experiment Setup

For the novelty of this task, there is not public available dataset yet. Thus we collect 10 pairs of queries in English and Chinese, each pair of queries reflecting the similar concept, as illustrated in Table 1. For each query, we submit it to Google, and retrieve the top 100 search result pages. The cognitive aspects of each page are then annotated manually.

Table 1. Bilingual query pairs in the dataset

English	Chinese	English	Chinese
Apple	苹果	Tiger	老虎
Jaguar	美洲虎	Paris	巴黎
Great wall	长城	Dell	戴尔
Jordan	乔丹	Iraq	伊拉克
Saturn	土星	Peking University	北京大学

In the experiment, we use F-measure as the metric to evaluate the systems' discernibility of cognitive aspects. Formally, let $\mathcal{C} = \{ca_1, ..., ca_J\}$ be the set of manually annotated cognitive aspects, where ca_i is a set of pages which belong to the corresponding aspect; let $\Omega = \{\omega_1, ..., \omega_K\}$ be the set of machine generated aspect clusters. The F-Measure of cluster ω_i respect to class ca_j is calculated as:

$$P_{i,j} = \frac{|\omega_i \cap ca_j|}{|\omega_i|}, \quad R_{i,j} = \frac{|\omega_i \cap ca_j|}{|c_j|}, \quad F_{i,j} = \frac{2 \cdot P_{i,j} \cdot R_{i,j}}{P_{i,j} + R_{i,j}}$$

The F-Measure for the overall quality of cluster set Ω is defined by following formula.

$$F = \sum_{j=1}^{J} \frac{|c_i|}{|D|} \cdot \max_{i=1,...K} F_{i,j}$$

where D is the set of all result pages.

4.2   Experiment Results

In the experiment, we compare the performance of several approaches with different data representation models and clustering strategies. Table 2 shows the F-measure values of these systems. The CL-Text model only uses the original titles and snippets, as well as the machine translations to represent the data; the CL-ESA model only uses the vectors spanned on the cross-language Wikipedia categories; and the CL-Text-ESA model integrate both kinds of features together. The unified and individual clustering strategies have been described in the previous section.

Table 2. F-Measure values of systems

	Unified clustering	Individual clustering
CL-Text	0.52	0.50
CL-ESA	0.55	0.51
CL-Text-ESA	**0.62**	0.56

The bilingual search results clustering task is more difficult as compared with the traditional document clustering task, and thus the overall performance values of the systems are not quite high, but are still reasonable. In particular, the CL-ESA model achieves a better performance than the CL-Text model. This demonstrates that the cross-language explicit semantic analysis can discover the underlying aspects within the bilingual texts as well. By integrating the text features and semantic features together, the CL-Text-ESA model surpasses the other models. It is not surprising to find that the individual clustering strategy performs worse than the unified clustering strategy. This is because the individual strategy pays more attention to the language dependent features, while the unified strategy can consider the comparative information between languages in general.

4.3   Case Study

Fig. 2 shows some mined aspects by comparing queries *"Apple"* and "苹果". Due to the space limitations, we are only able to show the top results in each aspect. In both languages, the term can refer to an IT company, or a kind of fruit.

The distributions of bilingual search result pages over the mined aspects are shown in Fig. 3. Different from the annotated result, the system distinguishes *iPhone* from *Apple Inc.* Generally speaking, there are more pages related to *Apple Inc.* than pages related to *fruits.* This reflects a bias of the web users toward the information technologies. It is understandable that there are more Apple Inc. related results in English than in Chinese, since the *Apple Inc.* is a great company lasting for more than three decades in America, but was not quite accepted in China until recently. Meantime, the widespread cooking culture and health culture make Chinese people pay more attention to *fruit apple.*

Fig. 4 shows another example of cognitive differences in English and Chinese. The term *"Jordan"* can refer to a country (*Hashemite Kingdom of Jordan, Al Urdum*) or a person in English, while in Chinese these meanings are represented in different terms. Therefore the semantic of country is missing for the corresponding Chinese term "乔丹". Because *Jordan* is a widely used name in English, there are many search results referring to different persons and organizations, and thus the *"Others"* aspect in English contains many results. In comparison, the semantic distribution in Chinese mainly lies on a few dimensions, because only the world famous *Michael Jordan* and the *shoe brand Jordan* are concerned in China.

中文	英文
Apple Inc. 苹果公司	
Apple Apple 设计并创造了iPod 和iTunes、Mac 便携式和台式电脑、OS X 操作系统以及 革命性的iPhone 和iPad。 **苹果(AAPL.OQ)_财经频道_腾讯网** 苹果公司及附属公司主要从事个人电脑、移动通讯设备和便携式数字音乐播放器的设计、生产和销售，以及各种相关的软件销售、服务和外围设备产品和网络解决方案。 ...	**Apple** Apple designs and creates iPod and iTunes, Mac laptop and desktop computers, the OS X operating system, and the revolutionary iPhone and iPad **Apple (United Kingdom)** Apple Inc. (NASDAQ: AAPL; previously Apple Computer, Inc.) is an American multinational corporation that designs and markets consumer electronics, ...
iPhone	
Apple – iPhone 4 –视屏通话、多任务处理、HD 高清视频以及更多 iPhone 4 GSM 蜂窝电话配备高分辨率显 示器、FaceTime 视频通话、HD 高清摄像、 500 ... **【苹果手机】报价(Apple)苹果手机大全,评测-天极产品库** 苹果(Apple)手机中心为您提供了苹果(Apple)手机的参数、图片、报价、评测、行情、评论及下载,通过我们指定的苹果(Apple)手机经销商选择您的 ...	**All Apple phones** GSMArena.com: Apple GSM cellphones. ... First look · Apple iOS 5 review · Love and hate 2.0 · Apple iPad 2 review · Love it or hate it · Apple iPhone 4 review ... **Apple, Mac, iPhone, iPad, and iPod Reviews, Help, Tips, and News** Up to the minute news, reviews, how-to's and expert opinions on mac computers and software, iPhones, iPads and iPods.
Fruit 水果	
苹果_百度百科 苹果, 落叶乔木, 叶子椭圆形, 花白色带有红晕。果实圆形, 味甜或略酸, 是常见水果, 具有丰富营养成分, 有食疗、辅助治疗功能。苹果原产于欧洲、中亚、西亚和土耳其 ... **苹果_百科_凤凰网** 凤凰网百科>健康百科>食物>水果>仁果类>苹果. 百科分类 ... 不同的地区、不同品种的苹果, 其成熟季节是有差别的。主要品种有国光、元帅、红星、 ..	**Apple - Wikipedia, the free encyclopedia** The apple is the pomaceous fruit of the apple tree, species Malusdomestica in the rose family (Rosaceae). It is one of the most widely cultivated tree fruits, and ... **WHFoods: Apples** Apples belong to the Rose family of plants and are joined in that family by a wide range of very popular foods, including apricots, plums, cherries, peaches, ...
Others 其他	
苹果社区PINGOD(转基因苹果)-楼盘详情-北京搜狐焦点网 苹果社区PINGOD(转基因苹果), 搜狐焦点北京苹果社区PINGOD(转基因苹果)楼盘频道为您介绍位于朝阳附近苹果社区PINGOD(转基因苹果)楼盘最新动态, 包括开盘 ... **上海金苹果学校** 金苹果学校2010年度工作总结 [2010-12-7]; 金苹果学校中学部成功举办2010年秋季...[2010-11-4]	**Walking Off the Big Apple** Walking Off the Big Apple. a strolling guide to New York City .... Enter the site address in your phone, or search for "Walking Off the Big Apple." ... **Apple Cider Vinegar Cures** Apple Cider Vinegar cures – the best NATURAL home remedy for many health issues

Fig. 2. Example of cognitive aspects and corresponding pages of "*Apple*" vs. "苹果"

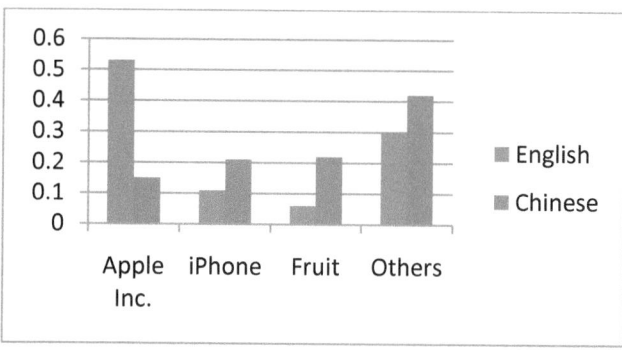

Fig. 3. Semantic distributions of *"Apple"* and "苹果"

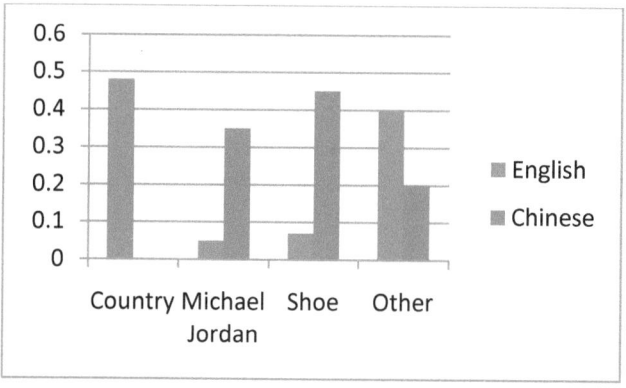

Fig. 4. Semantic distributions of *"Jordan"* and "乔丹"

5   Conclusion

In this paper we propose and study a novel search problem - Bilingual Comparative Web Search. We develop a BiCWS system to help user discover cognitive differences between two different languages from web search results in a multi-language setting. We build a unified feature space for bilingual text relatedness estimation by integrating cross-language explicit semantic analysis and machine translations. The cognitive facets for the user given topic are mined by using a clustering algorithm, and the semantic distributions over the mined facets are visually presented to reflect the cognitive differences in the bilingual communities. The evaluation and case studies show that our system is effective for this task.

The comparative cognition studies can be further extended. In future we plan to extend our system from bilingual comparative mining to multilingual comparative mining. We also plan to make use of visual information for better aspect discovery.

Acknowledgement. This research was supported by the National Natural Science Foundation of China (NSFC) under Grant No. 61170166, Beijing Nova Program under Grant No.2008B03, and the National High Technology Research and Development Program of China under Grant No. 2012AA011101.

Reference

1. Anttila, R.: Historical and Comparative Linguistics (Current Issues in Linguistic Theory). John Benjamins Pub. Co., Amsterdam (1989)
2. Weisstein, U.: Comparative Literature and Literary Theory: Survey and Introduction. Indiana University Press, Bloomington (1974)
3. de Zepetnek, S.: Comparative Central European Culture. Purdue University Press, West Lafayette (2002)
4. Jindal, N., Liu, B.: Mining Comparative Sentences and Relations. In: 21st National Conference on Artificial Intelligence, pp. 1331–1336. AAAI Press, Palo Alto (2006)
5. Zhai, C., Velivelli, A., Yu, B.: A Cross-Collection Mixture Model for Comparative Text Mining. In: 10th ACM SIGKDD International Conference on Knowledge Discovery and Data Mining, pp. 743–748. ACM, New York (2004)
6. Kim, H.D., Zhai, C.: Generating Comparative Summaries of Contradictory Opinions in Text. In: 18th ACM Conference on Information and Knowledge Management, pp. 385–394. ACM, New York (2009)
7. Liu, C., Huang, Q., Jiang, S., Xu, C.: The third eye: mining the visual cognition across multi-language communities. In: 18th International Conference on Multimedia, pp. 431–440. ACM, New York (2010)
8. Carpineto, C., Osiński, S., Romano, G., Weiss, D.: A survey of Web clustering engines. ACM Comput. Surv. 41, 1–38 (2009)
9. Zamir, O., Etzioni, O.: Grouper: a dynamic clustering interface to Web search results. Computer Networks, 1361–1374 (1999)
10. Zeng, H.-J., He, Q.-C., Chen, Z., Ma, W.-Y., Ma, J.: Learning to cluster web search results. In: 27th Annual International ACM SIGIR Conference on Research and Development in Information Retrieval, pp. 210–217. ACM, New York (2004)
11. Sun, J.-T., Wang, X., Shen, D., Zeng, H.-J., Chen, Z.: CWS: A Comparative Web Search System. In: 5th International Conference on World Wide Web, pp. 467–476. ACM, New York (2006)
12. Barrachina, S., Vilar, J.M.: Bilingual clustering using monolingual algorithms. In: 8th International Conference on Theoretical and Methodological Issues in Machine Translation (TMI 1999), pp. 77–87 (1999)
13. Kiran Kumar, N., Santosh, G.S.K., Varma, V.: Multilingual Document Clustering Using Wikipedia as External Knowledge. In: Hanbury, A., Rauber, A., de Vries, A.P. (eds.) IRFC 2011. LNCS, vol. 6653, pp. 108–117. Springer, Heidelberg (2011)
14. Li, B., Gaussier, E., Aizawa, A.: Clustering comparable corpora for bilingual lexicon extraction. In: 49th Annual Meeting of the Association for Computational Linguistics: Human Language Technologies: Short Papers, vol. 2, pp. 473–478. Association for Computational Linguistics, Stroudsburg (2011)
15. Dagan, I., Itai, A.: Word sense disambiguation using a second language monolingual corpus. Comput. Linguist. 20, 563–596 (1994)

16. Khapra, M.M., Joshi, S., Chatterjee, A., Bhattacharyya, P.: Together we can: bilingual bootstrapping for WSD. In: 49th Annual Meeting of the Association for Computational Linguistics: Human Language Technologies, vol. 1, pp. 561–569. Association for Computational Linguistics, Stroudsburg (2011)
17. Gabrilovich, E., Markovitch, S.: Computing semantic relatedness using Wikipedia-based explicit semantic analysis. In: 20th International Joint Conference on Artifical Intelligence, pp. 1606–1611. Morgan Kaufmann Publishers Inc., San Francisco (2007)
18. Potthast, M., Stein, B., Anderka, M.: A Wikipedia-Based Multilingual Retrieval Model. In: Macdonald, C., Ounis, I., Plachouras, V., Ruthven, I., White, R.W. (eds.) ECIR 2008. LNCS, vol. 4956, pp. 522–530. Springer, Heidelberg (2008)
19. Bishop, C.M.: Pattern Recognition and Machine Learning. Springer, Heidelberg (2006)
20. Dempster, A., Laird, N., Rubin, D.: Maximum likelihood from incomplete data via the EM algorithm. Journal of the Royal Statistical Society. Series B (Methodological) 39, 1–38 (1977)
21. Kuhn, H.W.: The Hungarian Method for the assignment problem. Naval Research Logistics Quarterly 2, 83–97 (1955)

Towards a User-Centric Social Approach to Web Services Composition, Execution, and Monitoring

Zakaria Maamar[1], Noura Faci[2], Quan Z. Sheng[3], and Lina Yao[3]

[1] Zayed University, Dubai, U.A.E
[2] Université Lyon 1, Lyon, France
[3] The University of Adelaide, Adelaide, Australia

Abstract. This paper discusses the intertwine of social networks of users and social networks of Web services to compose, execute, and monitor Web services. Each network provides details that permit achieving this intertwine and thus, completing the three operations. A user social-network is used to advise users on the next Web services to select based on their peers' experiences, whereas a Web service social network is used to advise users on the substitutes to select in case a Web service fails, for example. To make the intertwine of these social networks happen, three components are developed: composer, executor, and monitor. The social composer develops composite Web services considering relations between users and the ones between Web services. The social executor assesses the impact of these relations on these composite Web services execution progress. Finally, the social monitor replaces failing Web services to guarantee the execution continuity of these composite Web services. A running example and a prototype illustrate and demonstrate the intertwine of these social networks, respectively.

Keywords: Web service, service composition, social network.

1 Introduction

Over the years, different development waves have shaped the Web. Started as a simple browsing tool to screen Web sites, the Web now is a dynamic and robust platform upon which organizations conduct business and people engage in cross-organization collaborative activities. The latest development wave in the Web triggered by among other things the pressure on organizations to remain agile and Web 2.0 widespread adoption, sheds the light on two major research streams:

- Research on loosely-coupled business applications. The execution of these applications spans several distributed and heterogeneous systems and hence, has to cross organization boundaries transparently. Service-Oriented Architecture (SOA) and its flagship implementation technology known as Web services [15] are a response to the challenges that this type of execution poses on organizations.
- Research on social computing illustrated with the massive deployment of social applications like Facebook and LinkedIn. These applications capitalize on the ability and willingness of users to interact, share, collaborate, and recommend. Users are nowadays referred to as *prosumers*, i.e., providers and consumers at the same

X.S. Wang et al. (Eds.): WISE 2012, LNCS 7651, pp. 72–86, 2012.

time [16,21]. However, the richness and complexity of information in these applications pose challenges on how to capture and structure these information for future use while preserving users' privacy and information sensitivity.

Although the aforementioned research streams are pursued separately, they share a common element, the Web, as an execution platform for cross-organization processes and an exposure means for organizations. It would be tempting to examine why and how both streams can be blended (i.e., interleave their use) together as this might yield interesting results for instance, developing business applications that consider social elements (e.g., users' past experiences) in their operation. However, the success of this blend is subject to addressing questions like how to advise users on the necessary building blocks (e.g., Web services) that they need to use for developing such business applications, what building blocks to select as per these users' needs, how to make sure that conflicts will not raise when separate building blocks are put together, and how to capture the interactions between the building blocks for better use in the future.

Our literature review identified two major but independent research initiatives that look into Web services-based business applications from a *social* perspective. In the first initiative illustrated by [10,20], users are in the center of developing complex, value-added composite Web services. Capitalizing on social networks of users, users are advised on the next step to take based on their social "entourage". In the second initiative [5,7], Web services are in the center of discovering peers when developing complex, value-added composite Web services. Capitalizing on social networks of Web services, Web services are labeled as collaborators, substitutes, or competitors. In this paper, we explore the blend of these two initiatives, i.e., social networks of users and social networks of Web services, to develop a user-centric social approach to Web services composition, execution, and monitoring. The objective is to assess and illustrate the value-added of these networks to the cycle of composition, execution, and monitoring. Composition means making Web services take part in composite Web services based on existing social relations between users and between Web services as well. Execution means triggering Web services with respect to these social relations. Finally, monitoring means keeping Web services on alert in case changes in these social relations happen so that actions are taken.

Section 2 discusses the rationale of interleaving social networks of users and of Web services and suggests a literature review on how these networks permit developing business applications. Section 3 details our user-centric social approach to compose, execute, and monitor Web services. A prototype system implementation is reported in Section 4 before concluding and identifying some future work elements in Section 5.

2   Background

2.1   Rationale of Social Networks Interleaving

As stated in Section 1, interleaving the use of social networks of users and social networks of Web services might yield interesting results for developing Web services-based business applications. We discuss hereafter the motivations of this interleaving and the requirements to satisfy during the interleaving.

On the one hand, social networks of users record interaction experiences of users with Web services over time so that these experiences are captured and shared later with other peers. Assuming that users' feedbacks on these interactions are fair (i.e., unbiased), it becomes possible to advise users on where to look for Web services, how to select Web services, and what to expect out of Web services. Web services' non-functional properties [11] (e.g., response time and execution time) do not include such details, so limited advice is provided. This shows the value-added of social networks of users to the cycle of composing, executing, and monitoring Web services.

On the other hand, social networks of Web services record the situations that Web services come across at run time [5]. These situations known as collaboration, competition, and substitution permit to advise users on which Web services can or like to collaborate with each other, which Web services can be selected over a peer, and which Web services can replace a failing peer. Similarly, Web services' non-functional properties do not include such details, so limited advice is provided. This shows the value-added of social networks of Web services to the cycle of composing, executing, and monitoring Web services. The management of social networks of Web services in terms of creation, access, and maintenance is discussed in [8].

Interleaving the use of these two types of social networks needs to take into account the requirements that are posed on each step of the aforementioned cycle. In particular,

- Requirements on composition refer to Web services discovery and selection, i.e., looking for the necessary Web services while taking into account users' needs, users' social relations, and Web services' social relations.
- Requirements on execution refer to satisfying the requirements posed on compositions at run-time, i.e., assessing the impact of considering users' social relations and Web services' social relations as well on the progress of these compositions.
- Requirements on monitoring refer to the continuity of Web services execution when failures arise, i.e., looking for peers that can substitute for the failing Web services while taking into account the social relations of these failing Web services.

2.2   Literature Review

Our work is at the cross-road of two main research streams: social computing (exemplified with Web 2.0) and service-oriented computing (exemplified with Web services). Existing works either adopt Web services to support social networks of users, or develop social networks of Web services to support users identify collaborator, substitute, and competitor Web services. The combination of both streams is quite new and several research opportunities are still un-taped. To the best of our knowledge, this work is the first attempt to examine such a combination. Existing works adopt either social networks of users or social networks of Web services to build composite Web services. The combination of these networks is totally absent.

In the category of social networks of users, Maaradji et al. propose a social composer (*SoCo*) that advises users on the next actions to take in response to specific events like selecting certain Web services [9]. Xie et al. introduce a framework for semantic service composition based on social networks [20]. Wu et al. rank Web services using non-functional properties and invocation requests at run-time [19]. A Web service's popularity as analyzed by users is the social element used during ranking. Tan et al. apply social networks analysis to mine and analyze a workflow repository, focusing on service usage patterns [18]. Nam Ko et al. discuss the social Web in which a new type of services called "social-networks connect services" help third party develop social applications without having them build social networks [13]. Last but not least, Al-Sharawneh and Williams mix semantic Web, social networks, and recommender systems to assist users in selecting Web services with respect to their functional and non-functional requirements [1]. Besides the "market-leader" concept that refers to the best Web service, Al-Sharawneh and Williams develop two ontologies called "follow-leader" to classify users and "preference" to specify users' preferences.

In the category of social networks of Web services, we cite our research works in [4] and [7]. In the first work, we suggest a method to engineer "social Web services". Questions addressed in this method include what relations exist between Web services, what social networks correspond to these relations, how to build social networks of Web services, and what social behaviors can Web services exhibit. In the second work we use social networks to support Web services discovery. Different social networks permit to describe the situations that Web services encounter for instance collaboration and recommendation. These situations mean that Web services are not isolated components that respond to user requests, only. Contrarily, Web services compete against other similar peers during selection, collaborate with other different peers during composition, and may replace other similar peers during execution despite the competition[1].

3   Proposed Approach

3.1   Overview and Illustration

Our approach to interleave the use of social networks of users and social networks of Web services is built upon *social composer*, *social executor*, and *social monitor* components. The role and duties of each component are described hereafter briefly. Fig. 1 shows these components along with a repository of social networks of users and social networks of Web services, a pool of users who will populate the social networks of users, and a pool of Web services that will populate the future compositions.

The social composer relies on the social networks of users and social networks of Web services to advise users on how to build composite Web services. Examples of advice concern (i) which Web services to include in these compositions [9], (ii) which Web services to check in case some decline to participate in these compositions [6], and (iii) which Web services to select to ensure a better compatibility level of these compositions despite their loose-coupling nature [17].

The social executor does not provide advice on compositions. Instead it assesses the impact of the social composer's advice (when considered) on the execution progress of compositions. The social executor feeds the social composer with details so that the social composer updates the necessary social networks. These details include (i) how the

---

[1] Simultaneous competition and substitution, *a.k.a* coopetition [2], refers to Web services that compete to take part in compositions and also collaborate to support each other during failure.

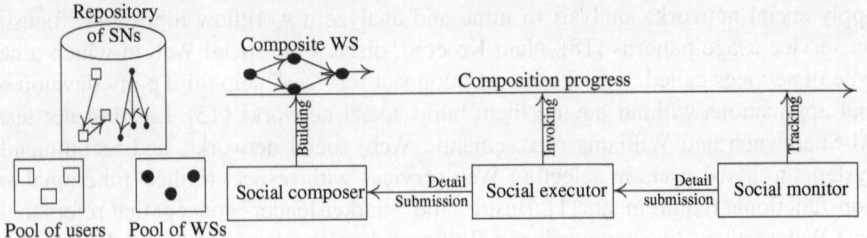

Fig. 1. Composer, executor, and monitor social components in action

Web services that are suggested through the social networks performed and (ii) which Web services that are also suggested did not join the compositions.

The social monitor relies on the social networks of Web services to advise users on which Web services to check in case those that are already taking part in some ongoing compositions fail. The social monitor feeds the social executor with details so that this latter updates the social networks of Web services for the benefit of the social composer. These details include (i) which Web services failed, (ii) which Web services replaced those that failed, (iii) how the replacing Web services performed, and (iv) how the Web services that are already in compositions reacted to the replacing Web services. Out of these details, the social monitor does more than a simple monitoring but puts forward different solutions for the social composer like assessing Web services performance.

To show how the composer, executor, and monitor components support the interleaving of social networks of users and social networks of Web services happens, we suggest the following scenario. Needless to mention the simplifications made in the scenario for the sake of explanation. A businesswoman who has got a 3-day stop over in a city decides to visit some museums among other sightseeing activities. She logs into a Web site and invokes `museumVisitWS` submitting her preferences and constraints. Different cases are listed hereafter to illustrate the role of each component in Fig. 1.

1. Prior to executing `museumVisitWS`, the social composer consults the businesswoman's social networks, finding out that some friends who visited the city before recommend riding taxis at this time of the year due to unexpected heavy rains.
2. To identify a Web service for taxi booking, the social composer consults `museumVisitWS`'s social networks to find out that `museumVisitWS` has frequently and successfully collaborated with `taxiBookingWS`, which is subsequently selected to arrange taxi booking. Another Web service called `translatorServiceWS` is also advised by the social composer as reported in the social networks of `museumVisitWS`, but this time our businesswoman declines the advice since she is familiar with the language spoken in the city.
3. When the selection of Web services is complete, the social executor invokes them while keeping an eye on all the Web services that are added to the composition through the social networks of users and social networks of Web services. The objective is to reflect the performance of these Web services on the different networks.

4. At run time, `museumVisitWS` fails just after `taxiBookingWS` is done. The social monitor takes immediate actions by consulting `museumVisitWS`'s social networks and recommends `museumTourWS` instead, as a substitute.

The aforementioned cases show some of the advantages of the social networks to the cycle of Web services composition, execution, and monitoring. It is for sure that some of these cases can be handled by screening registries, but Web services' previous experiences and users' advice are not captured and hence, overlooked during this screening.

3.2 Laying Down the Foundations

Our user-centric social approach revolves around social networks of users as per Maaradji et al. [9] and social networks of Web services as per Maamar et al. [7]. We interleave these network types as per the requirements listed in Section 2.1.

In Maaradji et al.'s work one type of social networks is developed. It is referred to as recommendation (due to lack of space limitations of recommendation systems like cold start are not discussed). In Maamar et al.'s work three types of social networks are developed. They are referred to as collaboration, competition, and substitution. Fig. 2 illustrates these social networks types. We recall that a social network is a graph that consists of nodes connected to each other through edges. The edges are labeled with elements usually found in people's life like friendship, partnership, and dislike. The edges are sometimes directional, bidirectional, with weight, or a mixture of all of these. The size and shape of a social network vary over time for different reasons. e.g., node deletion that affects outgoing/incoming edges.

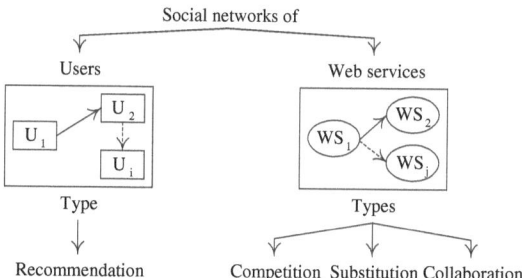

Fig. 2. Social networks of users *versus* social networks of Web services

Social Network of Users. As stated before, recommendation is the sole social network of users that is built to support users develop composite Web services. This network suggests Web services according to the current status of the composition process. The Recommendation Confidence ($\mathcal{RC}$) as discussed in [9] is defined in Equation 1.

$$\mathcal{RC}(ws_k, ws_l) = \sum_{j=1}^{n} \mathcal{NC}_{v_j}(ws_k, ws_l) \times \mathcal{F}it(v_j, ws_l) \times \mathcal{SP}(v_i, v_j) \qquad (1)$$

where:

- $\mathcal{NC}_{vj}(ws_k, ws_l)$ represents how many times user v_j used Web service ws_l following the use of Web service ws_k in compositions.
- $\mathcal{F}it(v_j, s_l)$ quantifies the expertise of user v_j in using Web service ws_l.
- $\mathcal{SP}(v_i, v_j)$ defines v_i's social proximity to v_j in the recommendation network.

Social Networks of Web Services. As stated before, collaboration, competition, and substitution are the social networks of Web services that are built to support the development of composite Web services. They are established based on the functionalities of Web services like `checkWeatherForecast` and `bookAirTicket`. Different techniques assess either the similarity or the complementarity of Web services' functionalities. This is outside this paper's scope and interested readers are referred to [3,12]. In the following, the three types of social networks of Web services are explained:

Competition Social Network. Fig. 3 (a) illustrates a competition social network of Web services. Since this network involves Web services that are similarly functional, they are all in competition against each other and hence, all connected to each other through bidirectional edges.

To evaluate the weight of a competition edge, which we refer to as *Competition Level* ($\mathcal{L}_{Comp}$, Equation 2) between two Web services ws_i and ws_j, we use the Functionality Similarity Level ($\mathcal{L}_{FS}$) to compare their respective functionalities and the No-Functionality Similarity Level ($\mathcal{L}_{NFS}$) to compare their respective non-functional properties (*a.k.a* QoS). We assume that the non-functional properties of Web services are defined with the same taxonomy.

$$\mathcal{L}_{Comp}(ws_i, ws_j) = \mathcal{L}_{FS}(ws_i, ws_j) \times (1 - \mathcal{L}_{NFS}(ws_i, ws_j)) \tag{2}$$

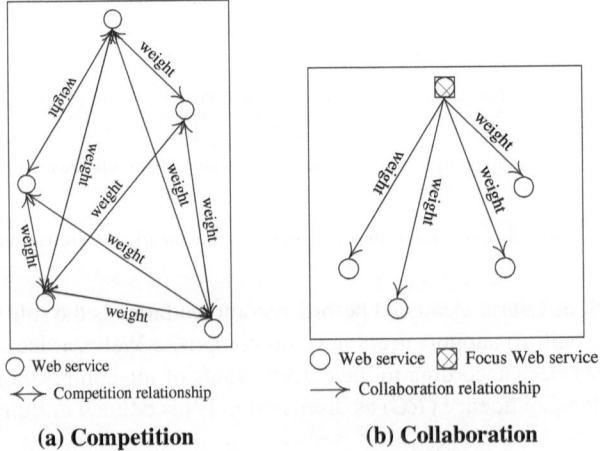

(a) **Competition**   (b) **Collaboration**

Fig. 3. Illustration of social networks of Web services

where:

- $\mathcal{L}_{\mathcal{FS}}(ws_i, ws_j)$ corresponds to the similarity level between the respective functionalities of ws_i and ws_j.
- $\mathcal{L}_{\mathcal{NFS}}(ws_i, ws_j) = \omega_1 \times (|\mathcal{P}(ws_{i,1}) - \mathcal{P}(ws_{j,1})|) + \cdots + \omega_n \times (|\mathcal{P}(ws_{i,n}) - \mathcal{P}(ws_{j,n})|)$ with $\mathcal{P}(ws_{i,k})$ is the value of the kth non-functional property of the ith Web service (assumed to be between 0 and 1), ω_k is a weighting factor representing the importance of a non-functional property, and $\sum_{k=1}^{n} \omega_k = 1$.

As per Equation 2 the more the competition level is close to one, the closer ws_i is to ws_j. As a result, ws_i threatens the competitiveness capacity of ws_j. We recall that only one Web service is selected at a time to complete a task in a composition.

Substitution Social Network. Fig. 3 (a) also illustrates a substitution social network of Web services after changing the edges' name from competition to substitution. Since all the Web services in a substitution social network offer the same functionality, any peer is a potential candidate to replace a failing Web service. To evaluate the weight of a competition edge, which we refer to as *Substitution Level* ($\mathcal{L}_{\mathcal{Sub}}$, Equation 3) between ws_i and ws_j, we use like previously the Functionality Similarity Level ($\mathcal{L}_{\mathcal{FS}}$) and the No-Functionality Similarity Level ($\mathcal{L}_{\mathcal{NFS}}$), in addition to the Reliability Level ($\mathcal{L}_{\mathcal{R}}$) that shows how successful ws_i is when it replaces ws_j.

$$\mathcal{L}_{\mathcal{Sub}}(ws_i, ws_j) = \mathcal{L}_{\mathcal{FS}}(ws_i, ws_j) \times \mathcal{L}_{\mathcal{R}}(ws_i, ws_j) \times (1 - \mathcal{L}_{\mathcal{NFS}}(ws_i, ws_j)) \quad (3)$$

where:

- $\mathcal{L}_{\mathcal{FS}}(ws_i, ws_j)$ and $\mathcal{L}_{\mathcal{NFS}}(ws_i, ws_j)$ are defined in Equation 2.
- $\mathcal{L}_{\mathcal{R}}(ws_i, ws_j) = \frac{\mathcal{SR}(ws_i, ws_j)}{\mathcal{TR}(ws_i, ws_j)}$, with $\mathcal{SR}(ws_i, ws_j)$ as the total number of successful replacements that ws_i made for ws_j (i.e., no failure) and $\mathcal{TR}(ws_i, ws_j)$ as the total number of requests that ws_i received to replace ws_j.

Collaboration Social Network. Fig. 3 (b) illustrates a simple collaboration social network of Web services. It is built when at least one composition of Web services is complete. For navigation purposes, an entry node is required and represented differently from the rest of nodes. We refer to this entry node as "focus" Web service. All the edges that come out of the "focus" Web service are unidirectional pointing towards other peers. To evaluate the weight of a collaboration edge, which we refer to as *Collaboration Level* ($\mathcal{L}_{\mathcal{Col}}$, Equation 4) between ws_i ("focus") and ws_j, we track the number of times that both Web services participated in joint compositions with emphasis on the total number of compositions that ws_i took part in.

$$\mathcal{L}_{\mathcal{Col}}(ws_i, ws_j) = \frac{\mathcal{JC}(ws_i, ws_j)}{\mathcal{TP}(ws_i)} \quad (4)$$

where $\mathcal{JC}(ws_i, ws_j)$ is the total number of participations of ws_i and ws_j in joint compositions and $\mathcal{TP}(ws_i)$ is the total number of participations of ws_i in compositions.

3.3 Social Composer

In Maaradji et al.'s work [9], a social composer advises the user on the next actions to take with respect to the progress of the composition under construction. This progress depends on the Web services selected recently and appended into this composition. The social composer's examples of advice concern the next suitable Web services to select and the necessary data mappings between the pre- and post-Web services. The social composer uses this user's social network to build recommendation strategies for Web services (Fig. 2). More details on advice and data mapping are given in [9].

Relying only on users' recommendations may not be enough to achieve high-quality compositions. Indeed these recommendations do not capture some Web services' characteristics, e.g., (i) refusal (or conditional refusal) to take part in additional compositions as discussed in [6] and (ii) favoritism for some peers over others as discussed in [5]. For each characteristic a social network of Web services provides specific solutions:

Refusal: Since Web services have to maintain a certain QoS level as per the different Service Level Agreements (SLAs) they commit to, it happens that Web services either decline participation requests in forthcoming compositions or delay their participations until further notice. These requests originate from the social composer. In either case a competition social network can be very useful for the social composer. It helps the social composer identify the peers that can be interested in accepting these participation requests. Without a competition social network, a process for discovering Web services has to be launched from scratch, which is time consuming [7]. A competition social network offers direct access to a pool of candidate peers that are competitors to those that decline participation requests. In the businesswoman example, museumVisitWS can decline the social composer's request, which makes the social composer look for a competitor such as museumTourWS using the competition social network of museumVisitWS.

Favoritism: Since Web services are developed separately, semantic and policy compatibility conflicts will for sure arise [14,17]. To minimize the efforts put into addressing these conflicts, a collaboration social network can be useful for the social composer by identifying the peers that participated with the existing Web services in common compositions. In the businesswoman example, museumVisitWS prefers working with taxiBookingWS due to successful previous experiences.

3.4 Social Executor

Although the social executor does not advise users like the social composer does, its role is to assess the impact of the social composer's advice on the progress of compositions, so that the relevant social networks of users and of Web services are updated based on this impact. The various advice are about proposing Web services (i) that can take part in compositions in replacement of those that decline participation invitations in these compositions, and (ii) that can achieve a better compatibility level to compositions by minimizing and avoiding semantic and policy conflicts. In Section 3.3, these two cases are referred to as refusal and favoritism, respectively.

Refusal: On top of the details on the competitiveness level ($\mathcal{L}_{Comp}(ws_i, ws_j)$) that a competition social network carries, this network carries additional details on how Web services responded to the invitations of taking part in compositions since these Web services were not selected initially, i.e., they were less competitive. Therefore, the competitiveness level should be revised ($\mathcal{L}_{RComp}(ws_i, ws_j)$) as per the actions that the social executor performs during a composition progress.

Equation 5 is the revised competitiveness level where ws_i accepts a composition invitation that ws_j declines earlier, α_k is a weighting factor ($\sum \alpha_k = 1$), $\mathcal{L}_{Comp}(ws_i, ws_j)$ is the initial competitiveness level as per Equation 2, $\mathcal{AI}(ws_i, ws_j)$ is the total number of accepted invitations that ws_i received from the social composer following the rejection of ws_j, $\mathcal{TI}(ws_i, ws_j)$ is the total number of invitations that ws_j declined and then, were sent to ws_i, and $\mathcal{P}er(ws_i)$ is the performance of ws_i at run-time using a scale of high, average, and poor. Poor performance means that ws_i "disappointed" the social composer for reasons such as being malicious or inability of delivering the non-functional properties that it posts.

$$\mathcal{L}_{RComp}(ws_i, ws_j) = \alpha_1 \mathcal{L}_{Comp}(ws_i, ws_j) + \alpha_2 \frac{\mathcal{AI}(ws_i, ws_j)}{\mathcal{TI}(ws_i, ws_j)} + \alpha_3 \mathcal{P}er(ws_i)$$

(5)

Favoritism: Like in the refusal case, the social composer needs to update both the recommendation social network of users and the collaboration social network of Web services. We focus on the latter type of social network hereafter, which requires reviewing the collaboration level of Equation 4 into $\mathcal{L}_{RCol}(ws_i, ws_j)$.

Equation 6 is the revised collaboration level where ws_i takes part in a composition as per the collaboration it has with ws_j, α_k is a weighting factor ($\sum \alpha_k = 1$), $\mathcal{L}_{Col}(ws_i, ws_j)$ is the initial collaboration level as per Equation 4, $\mathcal{AC}(ws_i, ws_j)$ is the total number of accepted collaboration requests that ws_i received from the social composer because of the collaboration links it has with ws_j, $\mathcal{TC}(ws_i, ws_j)$ is the total number of collaboration requests that were sent to ws_i because of ws_j, and $\mathcal{P}er(ws_i)$ is the performance of ws_i.

$$\mathcal{L}_{RCol}(ws_i, ws_j) = \alpha_1 \mathcal{L}_{Col}(ws_i, ws_j) + \alpha_2 \frac{\mathcal{AC}(ws_i, ws_j)}{\mathcal{TC}(ws_i)} + \alpha_3 \mathcal{P}er_{ws_i}$$

(6)

3.5 Social Monitor

The social monitor comes into play when a Web service fails so a substitute needs to be found. To this end the social monitor uses the substitution social network of the failing Web service to identify the most appropriate substitute(s) (comparing $\mathcal{L}_{Comp}(ws_i, ws_j)$ to a threshold). Two cases arise: (i) the substitute Web service accepts the request of the social monitor and hence, joins the composition that is put on-hold and now needs to be resumed; and (ii) the substitute Web service rejects the request of the social monitor and hence, another substitute needs to be identified. In either case, the substitution level is updated. Furthermore the collaboration level is updated for the first case.

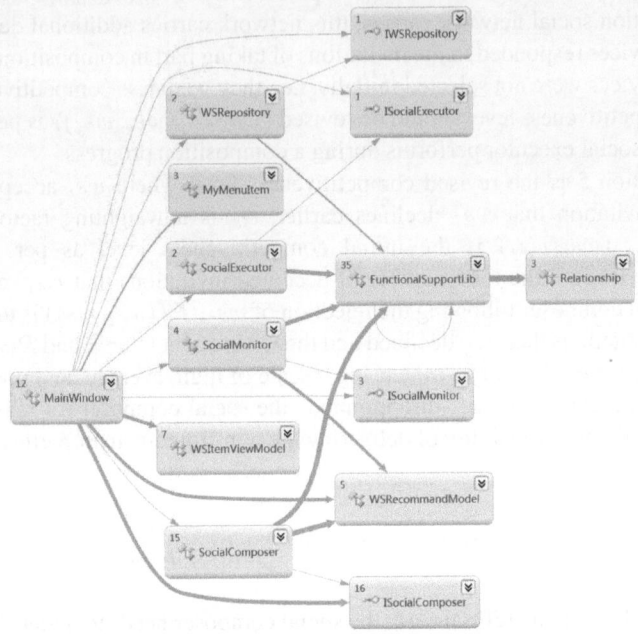

Fig. 4. Main classes for the system prototype

Equation 7 is the revised substitution level where ws_i replaces ws_j, α_k is a weighting factor ($\sum \alpha_k = 1$), $\mathcal{L}_{Sub}(ws_i, ws_j)$ is the initial substitution level as per Equation 3, $\mathcal{AS}(ws_i, ws_j)$ is the total number of accepted substitution requests that ws_i received from the social monitor because of the substitution links it has with ws_j, $\mathcal{TS}(ws_i, ws_j)$ is the total number of substitution requests that were sent to ws_i because of the failure of ws_j, and $\mathcal{P}er(ws_i)$ is the performance of ws_i using a scale of high, average, and poor.

$$\mathcal{L}_{RSub}(ws_i, ws_j) = \alpha_1 \mathcal{L}_{Sub}(ws_i, ws_j) + \alpha_2 \frac{\mathcal{AS}(ws_i, ws_j)}{\mathcal{TS}(ws_i)} + \alpha_3 \mathcal{P}er(ws_i) \quad (7)$$

4   Implementation and Evaluation

In this section, we report a prototype implementation and some preliminary experiments demonstrating the feasibility of developing a user-centric social approach to Web services composition, execution, and monitoring.

The prototype is developed using .Net Framework 4, Windows Presentation Foundation (WPF), and Visual Studio 2010 SP1. The main classes forming the prototype and their dependencies are shown in Fig. 4. The `MainWindow` class offers a visual interface of the prototype to users. The `WSRepository` class manages Web services

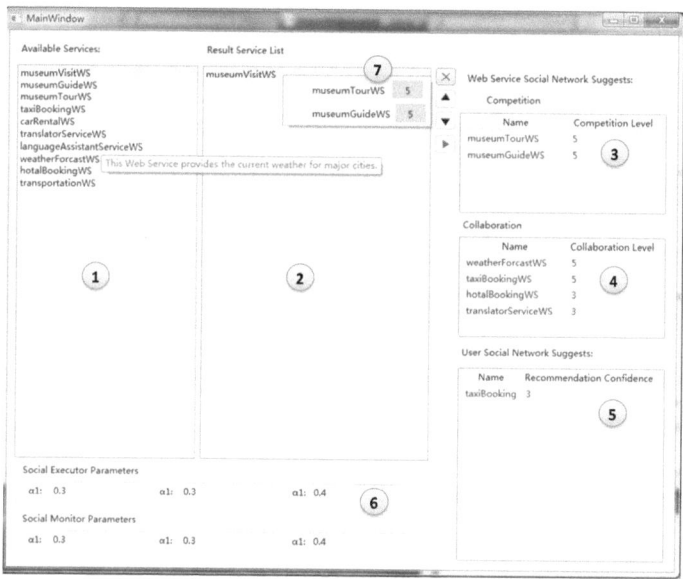

Fig. 5. The system frontend

while the `SocialExecutor`, `SocialMonitor`, and `SocialComposer` achieve the functionalities of the three proposed components in Section 3. Each of these classes is associated with an interface class (e.g., `ISocialExecutor` is the interface class of `SocialExecutor`). The `FunctionalSupportLib` class implements the low level functions of different social networks of users and the ones of Web services. The detailed information of these networks is stored in the `Relationship` class that serves as a data container. Finally, the `MyMenuItem` class specifies how to display customized menu item of the prototype and the `WSItemViewModel` class specifies how to render a Web service item into a listview box.

Fig. 5 shows a snapshot of the prototype system. The left panel (see Part 1) is a container for all available Web services such as `taxiBookingWS`. When double clicking on one of these Web services, the service will be added to the final result list that corresponds to the composite Web service to build (i.e., the middle panel Part 2). By clicking on a Web service in the middle panel (e.g., `museumVisitWS`), the system suggests recommendations through different social networks (see the right panel). For example, `museumTourWS` and `museumGuideWS` are recommended from the competition social network while four others are suggested from the collaboration social network. Each service has a competition or collaboration level value that is calculated by using the formulas developed in Section 3.2. A user can adjust the parameters of the formulas from the bottom panel (see Part 6). At any time, the user can conveniently add Web services from other panels (i.e., Parts 1, 3, 4, 5) to the composition list (i.e., Part 2) by simply double clicking them. If she would like to replace a Web service from Part 2, she can right click on the name of the Web service and a list of similar Web services will pop up (see Part 7), from which a substitute can be chosen.

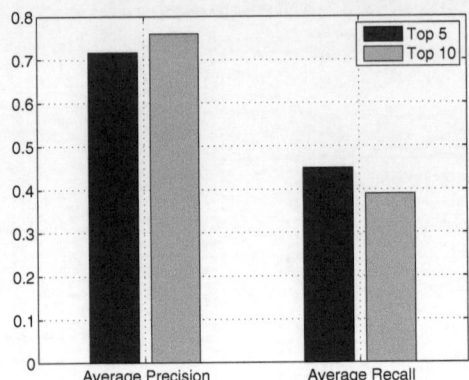

Fig. 6. Precision and recall of the competition social network

Moreover we conducted some preliminary experiments to study the performance of the system. Due to space constraints, we report one of them that studies the recommendation performance of competition social networks in case of refusal during composition. The concepts of *Precision* (P) and *Recall* (R) in information retrieval were used to evaluate the effectiveness of service recommendation. In the experiment, Web services with competition level higher than 7 were considered as valid recommendations. We examined the top N (N=5 and 10) candidates. In Figure 6 we can see that our approach achieves a reasonably good performance (close to 0.8 in precision when N is set to 10). The following was set for the experiment needs. Let p be the number of all Web services that are relevant to the Web service to be substituted, q be the number of relevant Web services recommended, and r be the total number of Web services recommended, then $R = q/p \times 100$ and $P = q/r \times 100$. A high precision value means that there are few false alarms while a high recall value means that there are few false dismissals.

5   Conclusions and Future Work

This paper has discussed the intertwine of social networks of users and social networks of Web services. The former contain details that help users select the necessary Web services when building composite Web services while the latter contain details that permit extending these composite Web services with new Web services or maintaining the operation continuity of these composite Web services in case of failure. This social networks intertwine is taken care by three social components referred to as composer, executor, and monitor. Each performs different operations. For instance the social composer suggests which Web services to check in case some peers decline to participate in composite Web services, the social executor assesses the impact of the social composer's advice on composite Web services execution progress, and last but not least the social monitor feeds the social executor with details so that this latter updates the social networks of Web services for the benefit of the social composer. A prototype has been developed to demonstrate the feasibility of developing a user-centric social approach to Web services composition, execution, and monitoring. Different future research works are identified including adding a social adaptor to the general approach and assessing the overhead of taking into account the different social networks on the progress and performance of composite Web services as well as the quality of the composite Web services that are developed.

References

1. Al-Sharawneh, J., Williams, M.A.: A Social Network Approach in Semantic Web Services Selection using Follow the Leader Behavior. In: Proc. of the 13th Enterprise Distributed Object Computing Conference Workshops (EDOCW 2009), Auckland, New Zealand (2009)
2. Bengtsson, M., Kock, S.: Coopetition in Business Networks to Cooperate and Compete Simultaneously. Industrial Marketing Management 29(5) (2000)
3. Di Martino, B.: Semantic Web Services Discovery based on Structural Ontology Matching. International Journal of Web and Grid Services 5(1) (2009)
4. Maamar, Z., Faci, N., Krug Wives, L., Yahyaoui, H., Hacid, H.: Towards a Method for Engineering Social Web Services. In: Ralyté, J., Mirbel, I., Deneckère, R. (eds.) ME 2011. IFIP AICT, vol. 351, pp. 153–167. Springer, Heidelberg (2011)
5. Maamar, Z., Hacid, H., Hunhs, M.N.: Why Web Services Need Social Networks. IEEE Internet Computing 15(2) (March/April 2011)
6. Maamar, Z., Kouadri Mostéfaoui, S., Yahyaoui, H.: Towards an Agent-based and Context-oriented Approach for Web Services Composition. IEEE Transactions on Knowledge and Data Engineering 17(5) (May 2005)
7. Maamar, Z., Wives, L.K., Badr, Y., Elnaffar, S., Boukadi, K., Faci, N.: LinkedWS: A Novel Web Services Discovery Model Based on the Metaphor of "Social Networks". Simulation Modelling Practice and Theory 19(10) (2011)
8. Maamar et al., Z.: Using Social Networks to Web Services Discovery. IEEE Internet Computing 15(4) (July/August 2011)
9. Maaradji, A., Hacid, H., Daigremont, J., Crespi, N.: Towards a Social Network Based Approach for Services Composition. In: Proceedings of the 2010 IEEE International Conference on Communications (ICC 2010) (2010)
10. Maaradji, A., Hacid, H., Skraba, R., Vakali, A.: Social Web Mashups Full Completion via Frequent Sequence Mining. In: Proceedings of the IEEE 7th World Congress on Services (SERVICES 2011), Washington DC, USA (2011)
11. Menascé, D.A.: QoS Issues in Web Services. IEEE Internet Computing 6(6) (November/December 2002)
12. Min, L., Weiming, S., Qi, H., Junwei, Y.: A Weighted Ontology-based Semantic Similarity Algorithm for Web Services. Expert Systems with Applications 36(10) (December 2009)
13. Nam Ko, M., Cheek, G.P., Shehab, M., Sandhu, R.: Social-Networks Connect Services. IEEE Computer 43(8) (August 2010)
14. Paolucci, M., Kawamura, T., Payne, T.R., Sycara, K.: Semantic Matching of Web Services Capabilities. In: Horrocks, I., Hendler, J. (eds.) ISWC 2002. LNCS, vol. 2342, pp. 333–347. Springer, Heidelberg (2002)
15. Papazoglou, M., Traverso, P., Dustdar, S., Leymann, F.: Service-Oriented Computing: State of the Art and Research Challenges. IEEE Computer 40(11), 38–45 (2007)
16. Pedrinaci, C., Domingue, J.: Toward the Next Wave of Services: Linked Services for the Web Data. Journal of Universal Computer Science 16(13) (2010)
17. Sheng, Q.Z., Yu, J., Maamar, Z., Jiang, W., Li, X.: Compatibility Checking of Heterogeneous Web Service Policies Using VDM++. In: Proc. of the 2009 IEEE Congress on Services, Part I (SERVICES I 2009) (2009)

18. Tan, W., Zhang, J., Foster, I.: Network Analysis of Scientific Workflows: A Gateway to Reuse. IEEE Computer 43(9) (September 2010)
19. Wu, Q., Iyengar, A., Subramanian, R., Rouvellou, I., Silva-Lepe, I., Mikalsen, T.: Combining Quality of Service and Social Information for Ranking Services. In: Baresi, L., Chi, C.-H., Suzuki, J. (eds.) ICSOC-ServiceWave 2009. LNCS, vol. 5900, pp. 561–575. Springer, Heidelberg (2009)
20. Xie, X., Du, B., Zhang, Z.: Semantic Service Composition based on Social Network. In: Proc. of the 17th Intl. World Wide Web Conf. (WWW 2008), Beijing, China (2008)
21. Yu, J., Sheng, Q.Z., Han, J., Wu, Y.: A Semantically Enhanced Service Repository for User-Centric Service Discovery and Management. Data & Knowledge Engineering 72(1) (2012)

On the Use of Optimal Stopping Theory for Improving Cache Consistency

Manos Spanoudakis[1], Dimitris Lorentzos[2], Christos Anagnostopoulos[3], and Stathes Hadjiefthymiades[1]

[1] National and Kapodistrian University of Athens,
Dept. of Informatics & Telecommunications, Athens, Greece
{mspanoud@gmail.com, shadj@di.uoa.gr}
[2] Hellenic Open University, Dept. of Informatics, Patras, Greece
dim.lorentz@gmail.com
[3] Ionian University, Dept. of Informatics, Corfu, Greece
christos@ionio.gr

Abstract. Serving the most updated version of a resource with minimal networking overhead is always a challenge for WWW Caching; especially, for weak consistency algorithms such as the widely adopted Adaptive Time-to-Live (ATTL). We adopt the Optimal Stopping Theory (OST) and, specifically, the Odds-algorithm, to enable the caching server to accurately handle the object refreshing and the stale delivery problem. Simulation results show that the proposed OST-based algorithm outperforms the conventional ATTL.

1   Introduction

Is a cached copy up-to-date? Caching servers try to reply to this question with high accuracy and optimal treatment of the network resources. Ideally, the involvement of the origin server should be minimal or, better, totally inexistent. The caching server faces a dilemma: to reduce the stale object delivery, costly lookups at the origin server are required. We are motivated by the problem associated with *weak* consistency schemes like the Adaptive TTL [1]. In our view, determining the right time t for the update of a certain cached resource i is treated as an Optimal Stopping Time (OST) problem. Therefore, by properly selecting stopping criteria, the problem can be treated quite efficiently. We introduce a control utility function, $U_t(i)$, that is constantly checked against the stopping criteria. The $U_t(i)$ function has two major components: one that reflects the general status of the origin server (e.g., server popularity) at time t and one that deals with the characteristics of the considered resource i, i.e., the resource that is assessed on the refresh need.

Considerable research has been performed w.r.t. problems associated with optimal timing decisions based on OST. The authors in [14] adopt the 'car-parking' OST problem [10] for reducing the dropping or forced termination probability in wireless cellular systems caused by handovers. A distributed opportunistic scheduling scheme for ad-hoc communications based on the maximal 'rate of return' problem is proposed in [15]. The authors in [16] proposed an optimal probabilistic forwarding protocol in delay tolerant networks based on the finite horizon asset-selling problem. The model

X.S. Wang et al. (Eds.): WISE 2012, LNCS 7651, pp. 87–99, 2012.
© Springer-Verlag Berlin Heidelberg 2012

proposed in [17] defines optimal stopping rules for controlling opportunistic spectrum access in radio networks based on maximizing the rate of return. The authors in [18] proposed a mechanism for adaptive optimal time horizon for acknowledgements at the receiving node in ad hoc networks. The model in [19] handles the data delivery problem through on-line search algorithms and the OST 'secretary problem' [21]. Moreover, the model in [20] deals with quality-aware delivery of information in ad hoc networks through the 'discounted asset selling' OST problem. To the best of our knowledge there is no previous approach which adopts OST in a cache consistency mechanism. In the remainder, we report on the basic cache consistency mechanisms.

1.1   Cache Consistency Mechanisms

One of the biggest challenges is to always serve the most updated version of a resource. Since the cache servers keep a local copy of the original resource, they need a mechanism to validate the local copies against the original information at the source. Quite a few architectures have been proposed for this problem also known as *cache consistency*. They can be categorized in two major categories namely *strong* and *weak* consistency. Strong consistency algorithms assure that the information delivered is always accurate and up to date; while with weak consistency there is some tolerance in the delivery of stale data. Specifically:

- In the strong consistency case, the Polling Every-Time is a client-controlled mechanism. For every incoming request, the client treats cached copies as stale and sends an *if-modified-since* request to the Server. This approach results in the exchange of unneeded messages especially in the case of objects that are rarely updated. With Leases, the Server establishes a "contract" with the Client (Cache Server) promising to inform the later about any updates of the leased Object. In order to do so, the Server maintains a Client list for each leased object, and, informs all clients with any update. A client is removed from the list when the leasing contract expires. After the leasing expiration, the client must validate the object on its next access. This validation is also referred to as lease renewal. Hence, in order to maintain strong consistency, the server must invalidate a cached object during its lease, and the client must validate a cached object on its first access after lease expiration. The most important parameter of this mechanism is choosing the duration of the lease. A short duration will increase the number of lease renewal messages and too long may result in maintaining very long client lists on the server side;
- In the weak consistency case, in the approach called Piggyback Server Invalidation (PSI) [1], the origin server assigns version numbers to volumes. This number is incremented when the object is updated. Clients keep track of the volumes to which their cached objects belong. With every if-modified-since request for an object the client includes the version number of the object's volume. With the server response, a list of all objects from this volume that have been updated since the client's version is piggybacked. Subsequently, the client invalidates those objects from the list that exist in its cache. Since the volume information is sent as part of the client's request, the server need not maintain any client lists.

A family of weak consistent algorithms utilizes a metric called Time To Live (TTL). This is a value, which is either set by the origin server (explicit TTL) or is defined implicitly (implicit TTL) [1]. The Cache Server considers an object valid, until the TTL expires. If a request arrives in the Cache Server after the TTL expiration, then it communicates with the origin server to check the object validity. It is quite clear that there is a weak point in this approach. For all requests within the TTL the Cache server "assumes" the object is valid. If there is a change, the cached object will be only updated after the TTL expires. With explicit TTL the value is included in the HTTP response headers. However, since this is not always specified, the Cache Server needs to set it implicitly using heuristic methods. Using a constant value for the TTL is not optimal because a very small value will unnecessarily increase the volume of exchanged messages and a high value may result in the delivery of stale objects [4]. To overcome this weakness, a more efficient mechanism called Adaptive TTL (ATTL) has been introduced. With ATTL, the TTL value is set based on the observation that when an object has not been modified at the origin server for a long time, it is very likely to remain unchanged in the future. In more details TTL is calculated based on the relative age of an object since the last modification time as specified in the "last-modified" HTTP header till the time it was transmitted from the Origin Server to the Cache Server (HTTP "date" header) [1]. Pertinent research has shown that the percentage of stale data can be reduced to less than 5% [5]. Another example is the implementation of Harvest Proxy [6], which calculates the ATTL value as the ½ of the time from the last update of the object.

The structure of the paper is as follows. Section 2 presents the optimal stopping theory and a variant of the odds algorithm, which is adopted by the proposed OST-based algorithm. In Section 3 we introduce the proposed algorithm. Section 4 evaluates our scheme through simulations and Section 5 concludes the paper with ongoing work.

2   Optimal Stopping Theory

The Optimal Stopping Theory (OST) is related to the problem of choosing the best time instance to take the decision of performing a certain action based on sequentially observed random variables in order to minimize the expected risk or cost (or maximize the expected payoff) [8]. Optimal stopping problems are characterized by the availability, at each step of the process, of a control entity (the Decision Maker – DM) that stops the evolution of the process. Hence, at each step the DM observes the current state of a system and decides whether to continue (perhaps at a certain cost) or stop the process and incur a certain loss or discount. The best time, at which the DM stops the process, is the optimal stopping time. There are many problems with optimal solutions derived through OST [2]. The most famous of optimal stopping time problems are the odds algorithm, the secretary problem and the car-parking problem [9], [10]. In a stopping rule problem, the DM observes a sequence $X_1, X_2, \ldots,$ of independent and identically distributed (i.i.d.) random variables, each with $E\{X\} < \infty$. At each step $t = 1, 2, \ldots,$ after observing $x_1, x_2, \ldots, x_t$, the DM may continue and observe x_{t+1}. The optimal stopping rule is to stop at some stage t in order to minimize the expected risk or maximize the expected payoff [11]. An OST problem has a finite horizon if there is a known upper bound on the number of steps at which the DM stops. Such problem is referred to as *finite horizon* OST problem.

2.1   The Odds Algorithm

One class of the finite horizon OST problems, called *last success*, aims at maximize the probability of identifying, in a sequence of sequentially observed independent events, a last specific event. This identification must be done at the time of observation and no recall on preceding observations is allowed. The approach to this class of problems is to observe the sequential events and to stop at the first interesting event after time s, which is defined as the optimal stopping time. Let us denote the finite horizon for this problem as n. The optimal stopping time $s \leq n$ can be calculated using the odds-algorithm [12]. Specifically, consider the $I_1, I_2, \ldots, I_n$ independent indicator functions ($I_t \in \{0, 1\}$) on some probability space with $p_t = E\{I_t\}$. Let also

$$q_t = 1 - p_t, r_t = \frac{p_t}{q_t} \tag{1}$$

that is, r_t presents the odds of the event $\{I_t = 1\}$, i.e., the ratio of the probability of success out of the probability of failure. We may observe the indicators I_t sequentially and may stop on at most one, but only on-line, that is, at the moment of observation. We win if we stop on the last $I_t = 1$ (if any) and lose otherwise (including not stopping at all). The odds-theorem in [12] determines the rule, which maximizes the probability of stopping on the last indicator, which takes the value of unity (if any), i.e., $I_t = 1$. The solution of the odds-algorithm is provided in Figure 1 (left).

Odds Algorithm	**OAU Algorithm**
Input: $n, p_t, t = 1, \ldots, n$	**Input**: $s_d, n, f_t, t = 1, \ldots, n$
Begin	**Begin**
$W_s = r_n + r_{n-1} + \cdots + r_s$	$s = s_d$
$Q_s = q_n q_{n-1} \cdots q_s$	**While** (true)
If $W_1 < 1$ **Then**	$\quad v(s) = I_1 + I_2 + \ldots I_{s_t}$
$\quad s = 1$	$\quad \hat{p}(s, p) = \dfrac{v(s)}{\sum\limits_{t=1}^{s} f_t}$
Else	
$\quad s = \sup\{t: W_t \geq 1\}$	
End-if	**If** $\sum\limits_{t=s+1}^{n} r_t(\hat{p}) < 1$ **Then**
Output: s (odds-optimal stopping time)	$\quad$ **break**
	$\quad$ **Else** $s = s + 1$
	End-if
	End-while
	Output: s (odds-optimal stopping time)

Fig. 1. (left) The odds algorithm, (right) the OAU algorithm

Specifically, the optimal stopping rule to stop on the last $I_t = 1$ is: "stop on the first indicator I_t with $I_t = 1$ and $t \geq s$. If none exists, stop on n and lose". Based on this rule, one wins if the first I_t with $t \geq s$ is the last '1' as indicator value. The stopping time s for this stopping rule is given by:

$$s = \max\left\{1, \sup_t \left\{t \in \{1, \ldots, n\}: \sum_{j=t}^{n} r_j \geq 1\right\}\right\}$$

In other words, s is the maximum time index within horizon n for which the sum of odds r_t from $t = n$ down to s equals or exceeds unity. The optimal win probability (as seen at time/steps 1, 2, ..., s-1) equals to $W_s Q_s$, where $W_s = r_n + r_{n-1} + \cdots + r_s$ and $Q_s = q_n q_{n-1} \cdots q_s$ (see [12] for proof). The odds algorithm is very convenient and allows for many interesting applications, e.g., selection problems for randomly arriving objects, timing problems, buying and selling problems with finite horizon, [8], [9].

2.2   The Odds Algorithm with Sequential Update

In many practical applications, the DM would not know beforehand the values p_t, $1 \leq t \leq n$. The Odds-Algorithm with sequential Updating (OAU) in [13] is a variant of the odds-algorithm, which sequentially estimates the odds for determining the optimal stopping time s. Specifically, the odds r_{t+1}, r_{t+2}, ..., r_n must be estimable from $I_1, I_2, ..., I_t$. This means that the number of unknown parameters on which the p_t, and, thus, the r_t, may depend on n. The OAU algorithm in [13] relies on one unknown parameter, $p \in [0, 1]$. Hence, the p_t is thought of as being deterministic function of one unknown parameter p, i.e.,

$$p_t = pf_t \tag{2}$$

The factor f_t is treated as known quantity. The OAU combines the odds-algorithm with estimating the 'future odds' from preceding observations. Hence, let (f_t), $t = 1, ..., n$ be a sequence of known real non-negative values. Then, $p_t = pf_t$, $q_t = 1 - pf_t$, and $r_t = pf_t/(1-pf_t)$, $p \in [0,1]$, $pf_t \leq 1$. The r_t is the (unknown) odds for $\{I_t = 1\}$. If $I_t(p) = 1$ we denote a success occurrence at time t. Then,

$$E\left\{\sum_{t=1}^{s} I_t(p)\right\} = p \sum_{t=1}^{s} f_t \tag{3}$$

We adopt now the following estimator of p from [13],

$$\hat{p}(s, p) = \frac{\sum_{t=1}^{s} I_t(p)}{\sum_{t=1}^{s} f_t} \tag{4}$$

and

$$v(s) = \sum_{t=1}^{s} I_t(p)$$

indicates the number of successes up to time s. In the case where successes do not occur at the beginning, that is $\hat{p}(s, p)$ is small at the beginning (i.e., no events $\{I_t = 1\}$ during the considered time period), the stopping time s is also small and we could consequently stop too early. This can be dealt with deciding to use some fixed learning samples and never stop on the first $s_d - 1$ values, that is, we start the algorithm at $s = s_d$; $s_d = 1$ corresponds to the OAU with no delay. The adopted OAU algorithm with a delay period s_d is given in Figure 1(right).

We can adopt two different choices for the sequence (f_t). One is $f_t = 1$ for all t. This is reasonable choice for the case when all I_t are i.i.d. Bernoulli random variables. The most frequent choice is $f_t = t^{-1}$. In this case all odds t are different. This is the case of the well-known best-choice problem, the secretary problem [21]. Moreover, the choice of n impacts the estimation of the odds. A relatively small n leads to unreliable odds-estimates.

3   OST-Based Cache Update Algorithm

The proposed OST-base cache update algorithm tries to utilize the odds algorithm with sequential update in order to specify the TTL for each object/resource i. Our scheme uses the utility function $U_t(i)$ on which the stopping time s applies. Such function defines the urgency to update object i. We correlate the value of $U_t(i)$ for object i, with statistics corresponding to all objects from the same origin server. We assume that the caching server hosts objects from the sites–members of the set $X = \{x_1, x_2, ..., x_{|X|}\}$. We denote R_x as the number of requests for the site x and R_X as the total number of requests on the server, i.e., $R_X = \Sigma_{x \in X} R_x$. In addition, S_x denotes the number of hits of objects that belong at site x. The calculation of the $U_t(i)$ value for the object i which belongs to site x depends on the following metrics. The *hit rate* of a site x is defined as:

$$h_x = \frac{S_x}{R_x} \tag{5}$$

The maximum value for h_x is unity when all references for objects of a site x are hits. The *popularity* g_x of site x is defined as:

$$g_x = \frac{R_x}{R_X} \tag{6}$$

The metric $\rho_x(i, t)$ denotes the dependency on the time of existence in the cache of object i and to the average expiration time t_{EXP} for all objects of site x, i.e.,

$$\rho_x(i,t) = \frac{t - t_{LOAD}(i)}{t_{EXP}} \tag{7}$$

The $t_{LOAD}(i)$ refers to the time when the object i has been saved in the Cache Server. The value of $\rho_x(i, t)$ approaches unity when the time the object i remains in cache is close to the average expiration time of the objects belonging to the same site. This implies that the object i has to be refreshed. A relatively low value of $\rho_x(i, t)$ (i.e., $<<$ 1) shows that the object i has been recently saved in Cache, thus, there is no need to refresh. The metric $\xi_x(i, t)$ denotes the dependency on the time till when the object i is considered valid ($t_{EXP}(i)$), that is,

$$\xi_x(i,t)=1-\frac{t_{EXP}(i)-t}{t_{EXP}(i)-t_{LOAD}(i)} \tag{8}$$

The value of $\xi_x(i, t)$ approaches unity when the current time t approaches the expiration time $t_{EXP}(i)$ meaning that it is time to refresh the object. Likewise $\xi_x(i, t)$ tends to zero when the object i has been saved in the Cache. In this case there is no need to refresh. Based on the above metrics in Eq.(5)-Eq.(8), the utility function $U_t(i)$ for object i at time t is:

$$U_t(i)=w_1 h_x + w_2 g_x + w_3 \rho_x(i,t) + w_4 \xi_x(i,t) \tag{9}$$

where $\Sigma_j w_j = 1$ and $w_j \in [0, 1]$, $j = 1, ..., 4$, are the weights of each of the metrics. Assuming that each metric has the same probability affecting the value of $U_t(i)$ at time t, we set $w_j = \frac{1}{4}$ for a site with numerous objects. $U_t(i)$ assumes values in $[0, 1]$. Another option would be to render the utility function calculation adaptive. In this case we could cluster the per-site metrics h_x and g_x and assign a weight w, $w \in [0, 1]$, to their sum. The supplementary weight $(1-w)$ is assigned to the sum of metrics $\rho_x(i, t)$ and $\xi_x(i, t)$, thus,

$$U_t(i)=w(h_x + g_x) + (1-w)(\rho_x(i,t) + \xi_x(i,t))$$

An increase in w would yield either a raise in SH (stale hits) or in SAT/N (success rate) metrics (see Section 4) and be penalized / rewarded accordingly. Such weight w remains unchanged throughout the horizon n. A high w value strong correlation among the site objects. A low w value shows a large heterogeneity between objects. A near zero value reflects the case when there is no need to refresh the object i in Cache. When the $U_t(i)$ value increases then the need for refreshing object i becomes more and more crucial. When $U_t(i)$ reaches unity the Cache server should refresh the object immediately. The question, which arises here, is "when to refresh object i". That is, the proposed scheme observes the $U_t(i)$ values at t ($1 \le t \le n$) and attempts to find an optimal stopping time s within a finite horizon n ($1 \le s \le n$) in which the object i has to be refreshed. If the horizon n has passed and the object i has not been refreshed then the scheme refreshes the object i immediately.

3.1   The Cache Update Algorithm

When the Cache receives a request for a new object i, it stores it locally, similarly to the ATTL case. The proposed OST-based Cache Update algorithm (OCU) derives the object's TTL based on the odds algorithm as shown Section 2.2. Practically, the algorithm's output s is treated as the object's derived (implicit) TTL.

Every time the Cache Server has a hit for object i the timestamp is saved as the *last_access* time for this object and at the same time there is a check whether the object is indeed valid or stale. In regular time intervals of horizon n, the $U_t(i)$ value ($1 \le t \le n$) is calculated based on the above information. When the number of observed values of $U_t(i)$ for object i is adequate, (for a period of s_d) then the OAU is applied. When the OAU decides that there is a possible need to refresh the object i at time t, i.e., the sum of the odds $r_t(\hat{p})$ from $t = s_d + 1$ towards n is less than the unity value for the first time within horizon n, the Cache server needs to *check* if the object has been updated and, then, updates its local copy accordingly. The decision is based on the number of successful observations of the $U_t(i)$ process. Defining success at time t, i.e., $\{I_t = 1\}$ is specific to the problem itself. In our case, an observation $U_t(i)$ is successful when the value of $U_t(i) \geq \theta$, $\theta \in [0, 1]$, otherwise $\{I_t = 0\}$, that is,

$$I_t = \begin{cases} 1, & \text{if } U_t(i) \geq \theta \\ 0, & \text{otherwise} \end{cases} \tag{10}$$

The θ threshold denotes the sensitivity of the algorithm; a low θ value forces the DM to stop the process and update the object i due to a high rate of successes. On the other hand, a high θ value indicates that the DM delays the stopping decision once there is no need for object update due to a low rate of successes. Moreover, we assume that $f_t = t^{-1}$ [21].

4   Performance Evaluation

We develop two prototypes each one implementing the ATTL and OCU algorithms. Since there is no availability of traces containing all the information needed to run the algorithms, we had to rely on synthetic traces with assumptions related to the size of objects, size of Cache Server, and requests generation. In the remainder we present the simulation set up, the introduction of the performance metrics and the performance assessment of the ATTL and OCU algorithms.

4.1   Simulation Setup and Performance Metrics

We generate a trace file by simulating the requests the Cache server needs to serve. In order to achieve this, we assume a set X of seven sites with the following number of objects per site, i.e., $X = \{x_1, x_2, x_3, x_4, x_5, x_6, x_7\}$ with $(|x_1|, |x_2|, |x_3|, |x_4|, |x_5|, |x_6|, |x_7|) = (200, 300, 400, 500, 650, 800, 200)$; $|x_i|$ is the number of objects of site $x_i \in X$. Each request is generated from one of the sites above, based on a random generator following a uniform distribution. The objects of each site are accessed in a way dictated by the Zipfian distribution. The Zipf law defines the probability $P(i)$ of selecting the i-th element (object) from site $x \in X$ of a popularity based sorted list, that is, $P_x(i) = ai^{-z}$ where a is a normalization factor defined such that $\sum_i P_x(i) = 1$ and z the Zipf exponent. Various studies [7] imply that the values of z for WWW Caches belong to [0.8, 1]. From the Zipf law we can determine the index i of the i-th object, i.e., $i = \left(\dfrac{a}{P_x(i)} \right)^{\frac{1}{z}}$

and, thus, we can identify which object of the site x is selected. Moreover, an important parameter for the simulation is time index of the requests. We have adopted the Pareto distribution for directing requests to the Cache Server [22]. Specifically, a random variable T following the Pareto distribution is greater than a value t with probability

$$P(T > t) = \begin{cases} \left(\dfrac{t_{MIN}}{t}\right)^b ,t \geq t_{MIN} \\ 1 \qquad\quad ,t \leq t_{MIN} \end{cases}$$

where t_{MIN} is the lowest boundary of T and b a positive variable. As a result the interarrival time t_{REQ} for requests is defined as

$$t_{REQ} = \frac{t_{MIN}}{P(T > t)^{\frac{1}{b}}}$$

where t_{MIN} is the lowest value of interarrival times. We have, thus, produced a synthetic trace of requests addressed to Cache Server. We have generated 10^4 requests with an average arrival rate $\lambda= 0.0026$ req/min. Finally, we need to define the simulation parameter for the Cache Server storage size M. The size M is defined in storage units (SU) and each object is set to occupy 1 SU. The size of objects is not relevant to our simulation and, thus, the adoption of this approach has no effect on our findings. When the Cache Memory is full the LRU method is used to clean up space. In the simulations, we have that $M \in$ {150, 200, 250, 300, 400} SU. The most important metric for a Cache Server is the *hit ratio*. Since we focus on weak consistency algorithms an interesting metric is: *the number of hits referring to objects, which are stale*. This metric indicates that the Cache Server reports a hit and, thus, serves the local copy, but the actual object is modified at source. We refer to such metric as *Stale Hits* (SH). We also monitor the *number of objects, which are **indeed** Stale After TTL expiration* (SAT) and the *number of objects that are still Valid After TTL expiration* (VAT). We monitor such metrics in order to compare the results with previous studies [5]. Finally, *the number of checks* (N) to verify the object's validity after the TTL expiration is also reported.

4.2   ATTL and OCU Performance Assessment

Every time the Cache Server receives a request for object i, which does not belong in the Cache, it saves a copy locally while serving the object to the client. At the same time it needs to calculate the ATTL for this object. The ATTL calculation -which is also triggered in case of a consistency miss, compulsory miss and slow hit - is performed through the function:

$$ATTL = k \cdot (t_{LOAD}(i) - t_{LM}(i)) \tag{11}$$

where $k = 0.2$ [1], $t_{LOAD}(i)$ is the time the object i was saved in the Cache Server and $t_{LM}(i)$ is the time the object i was last modified at source (last-modified time reference). Table 1 shows the metrics for the ATTL performance for difference M values. In the ATTL, there is a correlation between the number of invalid objects and the Cache memory size M. While the size M decreases, the number of stale objects increases. The SH percentage value is close to 5%, which is the same value calculated in [5].

Table 1. ATTL Simulation Results

M	400	300	250	200	150
Total hits	9548	9479	9410	9293	9078
SH (%)	5.33%	5.328%	5.303%	5.165%	4.527%
SAT (%)	8.35%	7.37%	6.83%	6.28%	4.33%
VAT (%)	91.64%	92.62%	93.16%	93.71%	95.67%
N	467	407	366	318	254

The OCU algorithm is evaluated with the same trace file as in the ATTL algorithm. The success threshold θ assumes values from $\{0.8, 0.75, 0.7, 0.6, 0.5, 0.25\}$ and the finite horizon $n \in \{7, 10, 15\}$ with $s_d = 0.1n$ [3]. Finally, regarding the interval between OCU executions, we run our simulation for $T \in \{1000, 3000, 6000\}$ minutes.

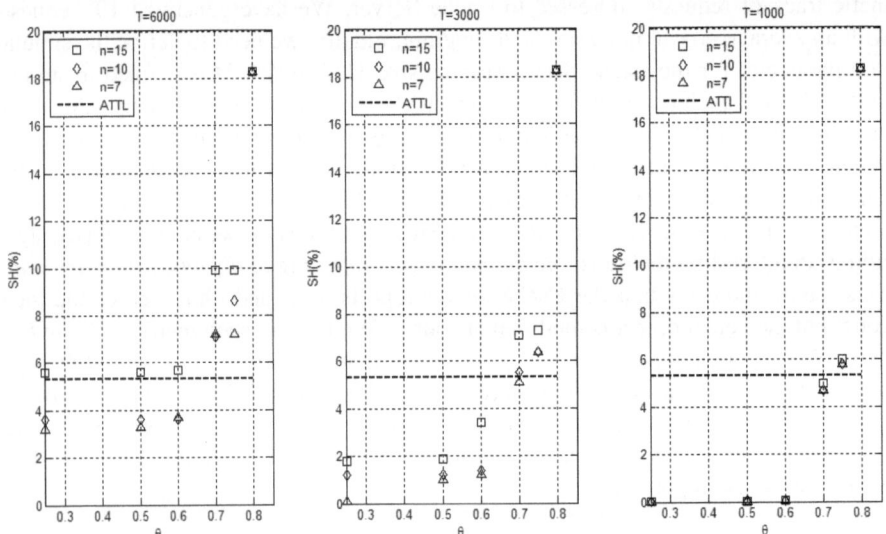

Fig. 2. SH (%) for T = 6000, 3000, and 1000 vs. n and θ.

Figure 2 shows the SH (%) percentage for different T and θ values and the SH percentage of the ATTL for comparison reasons, which is 5.3%. It is worth noting that the SH in the OCU algorithm depends on (i) the execution time of the algorithm (T), (ii) the horizon length n, and, (iii) the success threshold θ. When the T value decreases regardless the other parameters the SH value also decreases. The length n of the horizon also impacts the SH metric. A low n translates to more frequent runs of the OAU algorithm. For instance, for T = 6000, the OAU runs increases from 880 to 1409 when n drops from 15 to 10 while we obtain 35% reduction on the SH value (for $\theta = 0.6$). The θ value impacts the performance of the OCU algorithm since the success threshold indicates the sensitivity of the algorithm. A high θ value implies that the OCU delays the observation of the $U_t(i)$ values, thus, being more self-confident on a decision to check the validity of the object i to the origin server. While decreasing the θ value, more and more observation values are considered successful and thus the sum of odds tends to become less than unity faster.

We can observe from Figure 2 that the lower values of θ results in the delivery of less stale objects; we obtain 44% decrease in the SH value as $\Delta\theta \geq 0.5$ for all n values on average. However this has an implication on the number of executions of the algorithm. Moreover, the OCU algorithm for values of θ less than 0.7 performs better than the ATTL w.r.t. SH; indicatively, we obtain 98% decrease in the SH rate when adopting OCU compared with ATTL for $\theta = 0.6$ and $n = 15$. The difference in SH is bigger when θ value is low values but this has an impact on the number of algorithm executions and, thus, to the number of VAT.

Consider the number of objects (N) whose validity is checked in the origin server in order to verify if they actually need to be refreshed or not. The decision for this check is provided by the OCU algorithm through the OAU decision criteria, i.e., the sum of odds is less than unity and by the ATTL algorithm as TTL expires. The SAT value indicates the number of objects that are actually in need of being update and with VAT the number of objects that were not changed, thus, there was no need to be checked. Figure 3 shows the SAT/N, VAT/N percentages against n and θ values for different T values. We can observe that as θ increases then the ratio SAT/N increases denoting that high θ values result to a low number of invalidation controls however at the expense of high SH values. Figure 4 illustrates the *success rate*, i.e., SAT/N (%)

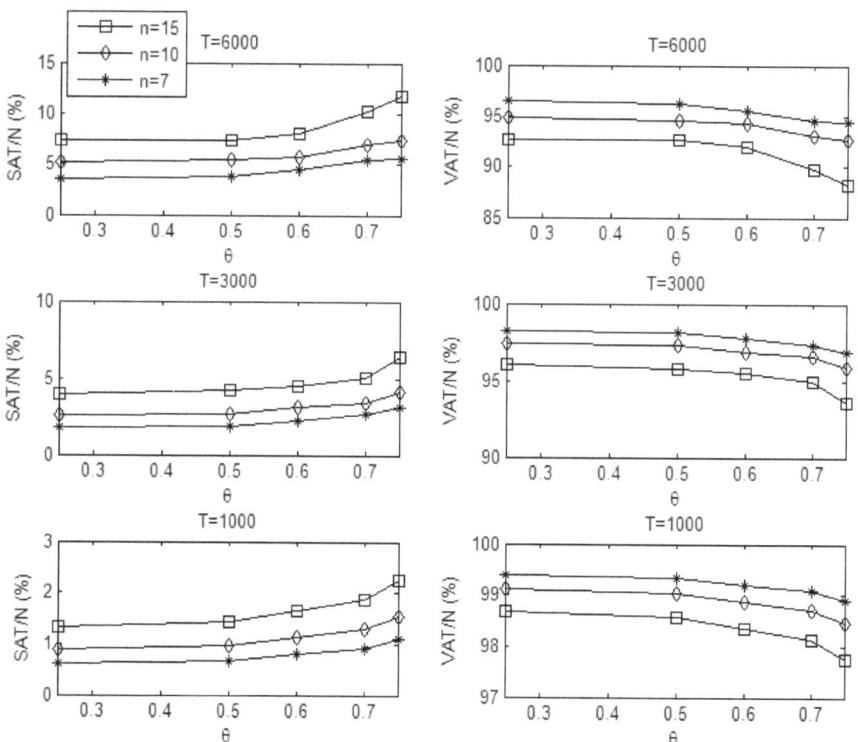

Fig. 3. SAT/N (%) and VAT/N (%) for T = 6000, 3000, and 1000 vs. n and θ

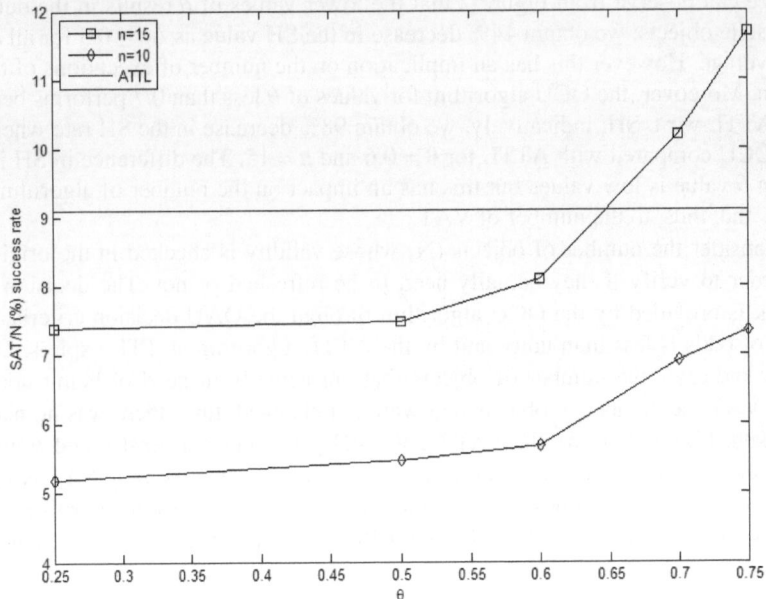

Fig. 4. The SAT/N (%) for OCU and ATTL vs. θ

percentage for OCU (T = 6000) and ATTL algorithm. We can observe that for $\theta > 0.6$ the success rate for OCU is greater than ATTL, which achieves SAT/N = 9.1%, but we obtain lower SH values.

5   Conclusions

In this paper we adopt the Optimal Stopping Theory and, specifically, the finite-horizon odds algorithm with sequential update and propose a novel cache consistency scheme, namely OCU. The OCU scheme is based on the observation of values of a properly structured utility function in order to decide whether a consistency check for a specific object is (indeed) required. The utility function captures different characteristics of the resource, the originating site and the pertinent behavior of the caching server. We conduct simulations in order to evaluate the performance of the proposed scheme w.r.t. the stale hits and the success rate for various parameters. We also compare the OCU performance with the ATTL scheme on the same (synthetic) traces and conclude that we can control the rate of stale hits and the success rate w.r.t. the length of the finite horizon and the success threshold θ of the odds algorithm. The OCU scheme performs better than the ATTL for low θ values. This improvement comes at the expense of a relatively low success rate. We obtain up to 67% reduction on the stale hits rate compared to the ATTL for $\theta = 0.6$ and $n \in \{7, 10, 15\}$. Our ongoing work focuses on the adoption of infinite horizon OST problems for enhancing the OCU algorithm, thus, being independent of the horizon n and the determination of the p_t probabilities of success.

References

1. Rabinovich, M., Spatscheck, O.: Web Caching and Replication. Addison Wesley (2001)
2. Ferguson, T.: Optimal Stopping and Applications. Mathematics Department UCLA, http://www.math.ucla.edu/~tom/Stopping/Contents.html
3. Bruss, F.: A note on the odds theorem of optimal stopping. Ann. Probab. 31(4), 1859–1861 (2003)
4. Lee, J., Whang, K.-Y., Lee, B.S., Chang, J.-W.: An Update-Risk Based Approach to TTL Estimation in Web Caching. In: Proc. 3rd IEEE Intl., Conf. Web Information Systems Engineering (WISE 2002), pp. 21–29 (2002)
5. Gwertzman, J., Seltzer, M.: World-Wide Web Cache Consistency. In: Proc. of the Usenix Technical Conference, ATEC 1996 (January 1996)
6. Chankhunthod, A., Danzig, P.B., Neerdaels, C., Schwartz, M.F., Worrell, K.J.: A Hierarchical internet object cache. In: Proc. of the Annual Conference on USENIX Annual Technical Conference, ATEC 1996 (January 1996)
7. Breslau, L., Cao, P., Fan, L., Phillips, G., Shenker, S.: Web caching and Zipf-like distributions: Evidence and implications. In: Proc. of the IEEE INFOCOM (1), 126–134 (1999
8. Peskir, G., Shiryaev, A.: Optimal Stopping and Free Boundary Problems. ETH Zürich, Birkhauser (2006)
9. Babaioff, M., Dinitz, M., Gupta, A., Immorlica, N., Talwar, K.: Secretary problems: weights and discounts. In: Proc. of 20th ACM-SIAM Symposium on Discrete Algorithms, pp. 1245–1254 (2009)
10. Tamaki, M.: An optimal parking problem. Journal of Applied Probability 19(4), 803–814 (1982)
11. Shiryaev, A.: Optimal Stopping Rules. Springer, New York (1978)
12. Bruss, F.T.: Sum the odds to one and stop. Annals of Probability 28(3), 1384–1391 (2000)
13. Bruss, F.T., Louchard, G.: The Odds-algorithm based on sequential updating and its performance. Advances in Applied Probability 41(1), 131–153 (2009)
14. Poulakis, M., Vassaki, S., Hadjiefthymiades, S.: Proactive radio resource management using optimal stopping theory. In: Proc. of IEEE Intl. Symposium on World of Wireless, Mobile and Multimedia Networks & Workshops, pp. 1–6 (2009)
15. Zheng, D., Ge, W., Zhang, J.: Distributed opportunistic scheduling for ad-hoc communications: an optimal stopping approach. In: Proc. of 8th ACM Intl. Symposium on Mobile Ad Hoc Networking and Computing, pp. 1–10 (2007)
16. Liu, C., Wu, J.: An optimal probabilistic forwarding protocol in delay tolerant net-works. In: Proc. of 10th ACM International Symposium on Mobile Ad Hoc Networking and Computing, pp. 105–114 (2009)
17. Huang, S., Liu, X., Ding, Z.: Opportunistic spectrum access in cognitive radio networks. In: Proc. of the IEEE INFOCOM, pp. 1427–1435 (2008)
18. Chen, J., Gerla, M., Lee, Y.Z., Sanadidi, M.Y.: TCP with delayed ack for wireless networks. Ad Hoc Networks 6(7), 1098–1116 (2008)
19. Anagnostopoulos, C., Hadjiefthymiades, S.: Delay-tolerant delivery of quality information in ad hoc networks. Journal of Parallel and Distributed Computing 71(7), 974–987 (2011)
20. Anagnostopoulos, C., Hadjiefthymiades, S.: Optimal, quality-aware scheduling of data consumption in mobile ad hoc networks. Journal of Parallel and Distributed Computing 72(10), 1269–1279 (2012)
21. Freeman, P.R.: The Secretary Problem and Its Extensions: A Review. International Statistical Review 51(2), 189–206 (1983)
22. Barford, P., Crovella, M.: Generating Representative Web Workloads for Network and Server Performance Evaluation. In: Proc. of the ACM SIGMETRICS, pp. 151–160 (July 1998)

Diversifying User Comments on News Articles

Giorgos Giannopoulos[1,2,*], Ingmar Weber[3], Alejandro Jaimes[3], and Timos Sellis[1,2]

[1] School of ECE, NTU Athens
[2] IMIS Institute, "Athena" Research Center
[3] Yahoo! Research, Barcelona

Abstract. In this paper we present an approach for diversifying user comments on news articles. In our proposed framework, we analyse user comments w.r.t. four different criteria in order to extract the respective diversification dimensions in the form of feature vectors. These criteria involve content similarity, sentiment expressed within comments, article's named entities also found within comments and commenting behavior of the respective users. Then, we apply diversification on comments, utilizing the extracted features vectors. The outcome of this process is a subset of the initial comments that contains heterogeneous comments, representing different aspects of the news article, different sentiments expressed, as well as different user categories, w.r.t. their commenting behavior. We perform a preliminary qualitative analysis showing that the diversity criteria we introduce result in distinctively diverse subsets of comments, as opposed to a baseline of diversifying comments only w.r.t. to their content (textual similarity). We also present a prototype system that implements our diversification framework on news articles comments.

1 Introduction

The last years the concept of social web is growing exponentially. More and more users socialize through facebook, discuss current topics in forums, express their opinions/sentiments through blogs or twitter. The social web has also infiltrated in more traditional aspects of the web, such as news sites. Large corporations, like Yahoo! News[1], allow their users to comment on news articles, facilitating the aggregation and public exposure of a wealth of user contributed information and opinions. Although this feature itself contributes largely to the spread of information and promotes the freedom of expression, data management issues come up due to the large amount of information to be handled.

In our scenario, some news articles can gather tens of thousands of comments, which makes it impossible for interested users to review all of them. However, sometimes, the article's content itself is not enough for a user to form a complete view over a topic. The public opinion is a valuable resource that complements the article and represents

---

* This research has been co-financed by the European Union (European Social Fund - ESF) and Greek national funds through the Operational Program "Education and Lifelong Learning" of the National Strategic Reference Framework (NSRF) - Research Funding Program: Heracleitus II. Investing in knowledge society through the European Social Fund.

[1] http://news.yahoo.com/

X.S. Wang et al. (Eds.): WISE 2012, LNCS 7651, pp. 100–113, 2012.

the "wisdom of the crowds". In this case, the user needs to be able to review a very small amount of as heterogeneous as possible comments, that represent different aspects of the article. On top of that, the user needs to be able, by selecting a specific comment, to review "similar" comments. Another use case scenario regards an archivist that needs to archive web information/resources about a specific topic. In this case too, the archivist should be able to "attach" to the primary resource (news article) complementary information (diverse user comments). This process would help, e.g., a future journalist that is trying to review past events, to gather as much diverse information on a topic as possible.

In this paper, we propose a set of criteria on which we adapt state of the art diversification algorithms in order to obtain a diverse subset of user comments on news articles. To the best of our knowledge, this is the first work handling the specific problem. The rest of the literature involves analysing user comments from several aspects, such as volume, political opinion, etc (see Section 6). On the other hand, diversification is, in most works, handled from the aspect of (mostly unstructured) search results.

The diversification criteria we propose consider the following comment features:

– Textual similarity. This is the baseline diversity criterion that is also used in the rest of the literature to diversify search results. The objective is to obtain comments with diverse content.
– Sentiment. We consider the sentiment of users expressed in the respective comments, w.r.t. the news article content. Sentiment is measured in a 9 grade scale ($[-4, 4]$), expressing negative, neutral, or positive sentiment. The objective is to obtain comments covering the whole range of sentiments.
– Named Entities (NEs). We consider the Named Entities (Persons, Organizations, Locations) found in the news article. Then, for each comment, we examine which of these NEs are referred in its content. Again, the objective is for the selected set of comments to contain as many article's NEs as possible.
– User Co-commenting behavior. We consider, for each user, all the articles she has commented on. Since each comment corresponds to one user, this information applies for comments too. Then, the objective is to select comments, so that their respective users have commented on as many different articles as possible.

After the above features are calculated for each comment, we apply an iterative algorithm that, at each step, compares the current, diversified set of comments with **all** the candidate ones. This comparison gives, for each diversity criterion a separate score. These scores are weighted and integrated into a final diversity score for each comment. However, diversity is not the only objective: although the problem requires that heterogeneous comments are gathered, these comments ought to be relevant to the initial news article. So, the final score of each candidate comment, at each step, is a weighted sum of its relevance score to the news article and its diversity score (distance) to the already selected comments.

We have conducted a preliminary qualitative analysis that demonstrates the effectiveness of our method, as opposed to diversifying comments only on content (textual similarity). We present an intuitive example that shows the importance of adding the previously mentioned criteria into the diversification process. The next step is to conduct a thorough user evaluation study, where users will be asked to evaluate the compared comment sets in terms of novelty, interestingness and topic coverage, concepts closely related to the concept of diversity.

The remaining paper is organized as follows. In Section 2, we discuss some background information on diversification algorithms. In Section 3, we present our method for diversifying news articles comments. In Section 4 we shortly describe the implemented system. In Section 5, we present an example from our preliminary analysis, that demonstrates the effectiveness of the proposed method. Section 6 presents related work and, finally, Section 7 concludes and discusses further work.

2 Background

As analysed in Section 6, there are several works describing diversification algorithms. The main idea, however, for most of them, revolves around the same process. In what follows, we first describe two variations of this process, based on the analysis on dispersion objectives performed in [16] and, then, we demonstrate how we interpret these objectives in our diversification framework.

2.1 Baseline Objectives

The authors in [16] propose 3 diversification objectives which are closely related to the problems of maximum dispersion and facility dispersion [20], [21]. We focus on two of them, *MAXSUM* and *MAXMIN*, that can give us a better intuition about their difference in the diversification optimization process.

The first objective aims at maximizing the sum of the relevance and dissimilarity of the selected set. That is, maximizing the following function f(S):

$$f(S) = (k-1) \sum_{u \in S} w(u) + 2\lambda \sum_{u,v \in S} d(u,v) \tag{1}$$

where S is the set diverse items, $|S| = k$ is the number of diverse items (comments) required, $w(u)$ is the similarity score of item (comment) u to the respective resource (news article), $d(u,v)$ is the diversity score (distance) between items u and v and $\lambda > 0$ is a parameter specifying the trade-off between relevance and similarity.

The second objective aims at maximizing the minimum relevance and dissimilarity of the selected set. That is, maximizing the following function f(S):

$$f(S) = \min_{u \in S} w(u) + \lambda \min_{u,v \in S} d(u,v) \tag{2}$$

2.2 Objectives Interpretation

Consider a set T of candidate comments to be selected into the k-size set S of diverse comments. The candidates are depicted as non-shaded circles in Figure 1. The shaded circles represent already selected comments that constitute the *current* diverse comment

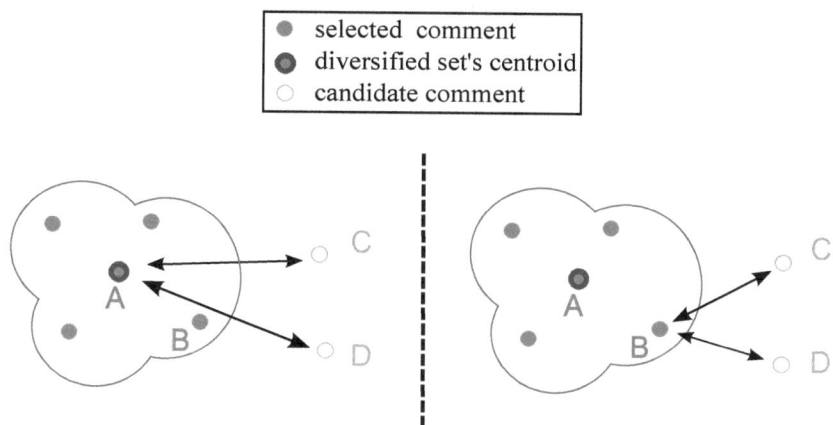

Fig. 1. Diversification Objectives

set. Both objectives described are based on the same iterative process: given the current set of diversified comments, for each candidate comment, produce a diversity score based on a certain score function. Then, select the comment that maximizes this score to insert into the diverse comments set.

The first objective (left figure) consists in selecting the candidate comment that has the maximum diversity distance to the current diverse set's centroid. This way, in every iteration, each candidate is compared to the "average" of the current diverse set. The second objective (right figure) aims at finding the candidate comment that has the maximum distance from the closest comment belonging to the diversified set.

In our example, in the first case, candidates C and D are compared to the centroid of the diversified set, A, in terms of their diversity distance to it. This comparison results to D being the most distant candidate and, thus, the next comment to be inserted into the diverse set. In the second case, candidates C and D are compared to the closest comment of the diversified set, B, in terms of their diversity distance to it. This time, C is the most distant candidate and, thus, the next comment to be inserted into the diverse set. This example demonstrates that, different diversification methods may produce different solutions, that is, different sets of diverse comments. The proper diversification algorithm should be selected by considering the specific problem requirements.

3   News Comments Diversification

In what follows, we describe our method in detail. First, we introduce and analyse the diversification criteria we propose and, then, we describe how they are incorporated into the diversification process.

3.1   Diversification Criteria

As stated in Related Work Section, most works on diversification measure diversity in terms of content, that is textual (dis)similarity between items. Even in works where more complex items are handled, e.g. [18] where items to be diversified are records with attributes, again, the distinct diversification criteria are defined on the textual similarity or matching of distinct attribute values. In this section, we extend the notion of diversity on new dimensions (apart from content) that include sentiment, named entities and user co-commenting.

Sentiment. We consider the sentiment expressed by users through their respective comments. We propose that sentiment (positive, negative, neutral) is a diversification factor, since it expresses users' opinions on the news articles' topics. In this sense, obtaining a set of comments that covers different classes of sentiment and, preferably, in a uniform manner, favors diversity.

We define 9 classes of sentiment within the interval $[-4, 4]$, with -4 denoting very negative sentiment, 4 very positive sentiment and 0 neutral sentiment. Each comment is assigned two different characterizations w.r.t. to the sentiment expressed within it:

- **Maximum/minimum sentiment**. We consider the whole text of the comment. Out of this, we extract the maximum positive sentiment, as well as the minimum negative sentiment value.
- **Average sentiment**. We regard each sentence of the comment separately and extract the respective positive and negative sentiment values. Then, we take the mean average of these values for all the sentences of the comment.

The sentiment extraction process is based on specific words found in the comment's text that express positive/negative sentiment. The above distinction into two types of extracted sentiment is performed in order to capture different facets of the expressed sentiment/opinion. For example, a comment may contain only one sentence that includes a very positive sentiment regarding a specific subtopic (e.g. a specific person) mentioned in the news article. On the other hand, the rest of the comment might be, on the whole, negative towards all the other aspects of the article. With the distinction we propose, we are able to capture these differences in sentiment expression.

After the sentiment extraction process, for each comment, we construct two 9-feature sentiment vectors, one for each type of sentiment extraction, with each feature corresponding to a different sentiment class. Each feature takes a boolean value that denotes whether the specific sentiment class is expressed in the comment.

Named Entities. We consider the Named Entities (NEs) found in the news article's text. These NEs might be Persons, Organizations or Locations. We suggest that NEs are important in terms of diversity, since news articles, most of the times revolve around NEs. Even when an article talks about events or situations, usually one or more Persons or Locations are involved. Given that, it is important for a diversified comment set to cover as many article's NEs as possible.

For each of the aforementioned NE categories, we create distinct NEs vectors, with each vector's features corresponding to the NEs found in the news article. For each comment, its feature values correspond to the frequency of the respective NE within the comment's text. In addition, we consider an aggregative NE vector that contains all NEs, irrespective of category. This results to 4 NEs vectors, that represent, for each comment, the coverage of article's Names Entities.

User Co-commenting Behavior. We consider the whole news article corpus. For each user, we construct a commenting vector, where each feature corresponds to a distinct news article. We suggest that the fact that a user comments a news article implies a relation between the user interests/opinions to the article's topics. Also, the more comments the user has posed to an article, the more closely related to the article's topic she is expected to be. Given that, the objective is to gather a diversified comment set, that corresponds to heterogeneous users. This heterogeneity, in our setting, is measured by the coverage of articles commented by the respective users.

For each comment, each feature value assigned is the commenting frequency of the user for the corresponding article, that is, the number of comments the user has posed on the article. For each user/comment, these feature values are normalized by the total number of comments the user has posed in **all** articles.

Content. Finally, we consider comments' content, which is the baseline diversification criterion, used in most works handling diversification, e.g. in web search results diversification. The importance of comments' content in the diversification process is straightforward. For each comment, we construct its term vector, with each feature corresponding to each distinct term found in the whole articles/comments corpus. Each feature value is computed by normalizing the term's frequency within the comment by the total number of terms the comment contains.

3.2 Diversification Process

The previously described process produces, for each comment, four distinct vector categories, that map the comment to four diversification dimensions: Content, Sentiment, Named Entities and User Co-commenting Behavior. We perform diversification on each of the above dimensions separately, producing, each time, a separate diversity (dissimilarity) score for each comment. Then, these scores are aggregated into a final dissimilarity score that is their weighted sum.

In order to produce a diversity score, we need to define a diversity function that measures the distance between two items. We adopt the widely used cosine similarity score and we define the diversity score of two items, u, v, w.r.t. a specific dimension i, as:

$$d_i(u, v) = 1 - \cos_i(u, v)$$

where $\cos(u, v)$ is normalized in the interval $[0, 1]$.

However, diversity is not the only objective: although the problem requires that heterogeneous comments are gathered, these comments ought to be relevant to the initial news article. So, the final score of each candidate comment, at each step, is a weighted sum of its relevance score to the news article and its diversity score (as described above) to the already selected comments. We define the relevance score of a comment u, w.r.t.

the corresponding news article A, applying the cosine similarity measure on the article's and the comment's term vectors:

$$r(u, A) = \cos(u, A)$$

Depending on the diversification process we follow (as described in Section 2.2) we define two formulas that give the final score for each candidate comment u to be inserted into the set of diverse comments S, w.r.t. a news article A:

$$score_{MAXSUM}(u, A) = (1 - w) \cdot r(u, A) + w \cdot \sum_{i=1}^{4} \lambda_i \cdot d_i(u, C_i)$$

where C_i is the centroid of the current diverse set w.r.t. the diversification dimension i, $w \epsilon [0, 1]$ is the weight of the total diversity score, as opposed to relevance score and $\lambda_i \epsilon [0, 1]$ is the weight of each individual diversity score, with $\sum_{i=1}^{4} \lambda_i = 1$.

$$score_{MAXMIN}(u, A) = (1 - w) \cdot r(u, A) + w \cdot \sum_{i=1}^{4} \lambda_i \cdot d_i(u, \min v_i)$$

where $\min v_i$ is the comment from the current diverse set that has the minimum distance to the closest comment from the candidate ones.

Below, the $MAXSUM$ diversification algorithm is described. The $MAXMIN$ algorithm is omitted since it is straightforward to derive it from the above formulas.

Algorithm 1. Produce diverse set of comments with MAXSUM

Input: Set of candidate comments T, size of diverse set k
Output: Set of diverse comments S
  $S = \emptyset$
  Insert into S the comment u = argmax r(u, A)
  while $|S| < k$ **do**
    Foreach dimension i, compute C_i
    Find the candidate comment u = argmax$score_{MAXSUM}(u, A)$
    Insert u into S
  end while

---

4   System Description

We divide the diversification process in two stages: Offline and online. Offline phase includes downloading the raw news articles and comments data, preprocessing them to extract feature vectors for the diversification dimensions, as well as term vectors for the relevance comparisons and storing them into the system's database. Online phase includes running the diversification algorithms on the extracted feature vectors.

Our system was implemented in Java. We used MySQL to store the preprocessed data. The data used in the specific process were downloaded from NY Times, using

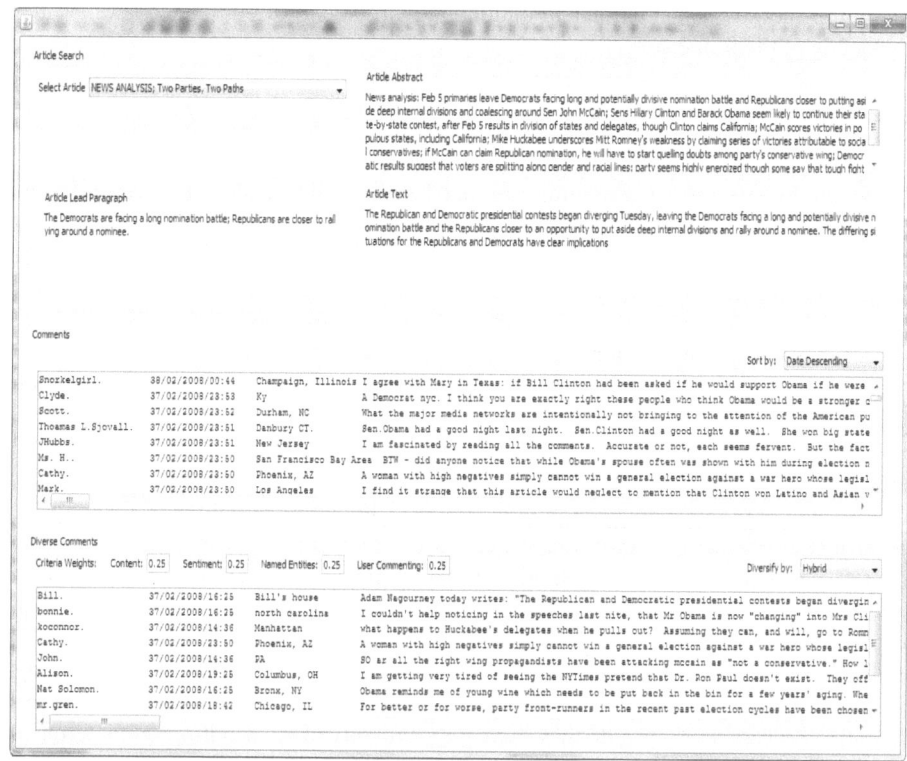

Fig. 2. System Interface

the respective APIs[2] [3]. For sentiment extraction, we usedSentiStrength[4] ([23]) and for Named Entities recognition Stanford Named Entity Recognizer[5] ([22]).

Figure 2 presents a screen of the implemented prototype. Through the upper panel ("Article Search") the user can select a news article and view the available information regarding it (text, abstract, lead paragraph). After an article is selected, its comments appear in the lower panels, sorted depending on the user choices. In the "Comments" panel, all article's comments are presented, either sorted by date, or sorted by their textual relevance to the article. In the "Diverse Comments" panel, a set of top-10 diversified results are presented according to the dimensions weight setup the user has chosen. In the specific example, all diversity dimensions are weighted equally. However, the user can select to diversify only by one dimension (e.g. sentiment) or even set the respective dimension weights on her own.

In the current prototype, we have set as default diversification algorithm the $MAXSUM$ and default total diversity score weigh $w = 0.5$. Of course, these choices

---

[2] http://developer.nytimes.com/docs/read/article_search_api

[3] http://developer.nytimes.com/docs/community_api

[4] http://sentistrength.wlv.ac.uk/

[5] http://nlp.stanford.edu/software/CRF-NER.shtml

require further experimentation and tuning that will be part of our planed user evaluation study of the framework.

5   Evaluation

In this section, we present a preliminary analysis of the different diversification results achieved by our framework, as opposed to diversifying news articles comments only by their content. We note that this is a first cut study on the problem, that can only give us an intuition of the qualitative difference achieved. We are currently planning a thorough user evaluation study, where diversity of comment sets, produced by different approaches (ours, only content, only similarity to the article, etc), will be evaluated by users. In this setting, concepts like coverage, informativeness and novelty, that are closely related to diversity, will be evaluated.

In Table 2 (see last page) we present the abstract of a news article talking about the elections of US president candidates of two US parties, the 11-top diverse comments, diversifying only by content (baseline) and the 11-top diverse comments, diversifying by all the dimensions we propose. We chose 11 comments in each case, because, by default, we choose, for every method, the first comment to be selected, to be the most relevant by content to the news article. So, the first comment is always the same regardless the method used.

From a first look, we can see that the two methods have only 3 out of 10 comments in common. This proves, that, the extra dimensions we propose do matter in the diversification process. However, do they increase the quality of the resulting set, in terms of diversity? Defining the following facets on the specific article's topic, we compare each set's coverage on each of them: critique/analysis/opinion on (a)politicians, (b)media, (c)the election process, (d)voters. Table 1 gives us the coverage, in terms of distinct comments that handle the above issues, for each of the two compared sets. It is obvious that the comments set resulting by our method (Multiple criteria) covers in a more uniform way the several topic facets of the article. Also, most comments in the baseline diversified set involve criticizing or complaining about persons, parties or processes. While criticizing comments exist in our method's comments set, there also exist comments where users perform analysis on the topic and try to give a better insight on the facts (see comments 2 and 8 in Multiple Criteria Diversification of Table 2). The above give us an intuition that our method covers, in general, more aspects of the article's topic, as well selects comments of more heterogeneous objectives (critique, analysis, opinion).

Table 1. Topic facets and their coverage in comments(number of comments handling a facet) for baseline (Content) and our proposed method (Multiple criteria)

Facet	Content	Multiple criteria
Politicians	6	4
Media	1	4
Process	2	4
Voters	3	3

Table 2. Diverse comment sets comparison

Article's Abstract	
	Jared Diamond Op-Ed article holds that biggest global concern is resource consumption rate; notes that consumption rate in North America is 32 times higher than in developing world; says US promise that any country that adopts free-market economy can enjoy 'first-world lifestyle' is cruel hoax; contends that if China's per capita consumption rates rise to US levels, world will run out of resources at even faster rate; says it is futile to tell other countries not to reach for consumption rate that Americans already enjoy; contends that present rate of US consumption is unsustainable; says American consumption is wasteful and contributes little or nothing to quality of life; says that US consumption rates could be lowered if there was political will to tackle problem; drawing (M)
Content Diversification (Baseline)	
1	Adam Nagourney today writes: "The Republican and Democratic presidential contests began diverging Tuesday, leaving the Democrats facing a long...
2	I couldn't help noticing in the speeches last nite, that Mr Obama is now "changing" into Mrs Clinton. He implied he's now going to cover everyone with his healthcare policy program. That and virtually every other content point was Mrs Clinton's idea. What is he going to do, when Mrs Clinton is not around teaching him. Mr Obama (and Mr Kennedy and Ms Winfrey) – please step aside and let the people who know what they are doing, get to work. Talk is cheap and this presidency is not a job for the junior league. Thank you Bonnie Hauser
3	what happens to Huckabee's delegates when he pulls out? Assuming they can, and will, go to Romney, the latter may rightly see this still as a race.
4	A woman with high negatives simply cannot win a general election against a war hero whose legislative record is so indistinguishable from hers, pariticularly on the Iraq war.
5	SO ar all the right wing propagandists have been attacking mccain as "not a conservative." How long before they claim they never really meant it and start spewing lies for him and against the Democrats?
6	I am getting very tired of seeing the NYTimes pretend that Dr. Ron Paul doesn't exist. They offer him almost no coverage – as they know that even bad press is good. The bias is so absolutely blatant; however, educated readers can read between the lines. How come some fair reporting. There are 4 Republican candidates, NY Times; report on all of them equally.
7	Obama reminds me of young wine which needs to be put back in the bin for a few years' aging. When it's more mature, it will probably be far more palatable and substantial.
8	For better or for worse, party front-runners in the recent past election cycles have been chosen more by default and less by way of solid qualification as leaders of vision. While I believe Tuesday's results are in general not all that surprising, the fact that the '08 cycle is unique and ground-breaking in many regards, is refreshing in and of itself. To what extent the much talked about "positive change" can come about when the dust finally settles and the rubber meets the road, of course, remains to be seen. That is the "Crux of the biscuit."
9	What the major media networks are intentionally not bringing to the attention of the American public is that Mike Huckabee's support is only partially derived from evangelical Christians, it's mostly derived from Mike's stated goal of ABOLISHING the INTERNAL REVENUE SERVICE, repealing the Sixteenth Amendment, and instituting the FairTax. This resonates with middle America as demonstrated in yesterday's voting. Many people I spoke with yesterday, who are NOT evangelical Christians by any stretch of the imagination , voted for Huckabee based ENTIRELY on his support of the FairTax, myself included. Get the facts: www.FairTax.org Then get the NYT #1 Bestselling Book by Neal Boortz and US Senator John Linder. WWW.FAIRTAX.ORG
10	I was surprised by the lack of astuteness on the voters' side. They are easily affected by youth and vacant idealism. A smart, effective Hillary Clinton should have been a LANDSLIDE. She has proven herself and her leadership, as well as her caring for all. I find it incredible that people want someone to be PRESIDENT WHO HAS NOT BEEN TRIED OR TESTED. NO WONDER OUR COUNTRY IS IN TROUBLE. ISAPPOINTED IN RESULTS
11	John from Toronto: If Obama had attacked Bush and his policy with the same intensity and vigor with which he campaigns, then I would agree with you. Unfortunately, many Americans feel that he is a lot of bluster and inspiring phrases. These do not necessarily translate well to the Oval Office. Remember Cuomo's soaring speech? Cuomo? Ubi est?
Multiple Criteria Diversification	
1	Adam Nagourney today writes: "The Republican and Democratic presidential contests began diverging Tuesday, leaving the Democrats facing a long...
2	Many pundits are espousing the view that a long and drawn-out nomination battle between Obama and Clinton will turn negative, fracturing a Democratic Party that has found a unifying cause: anti-Bushism. However, there is a far more potentially divisive issue that looms ahead should the race go down to the wire: what is to be done with the polling results from Florida and Michigan? One can easily envision a scenario under which Obama obtains a razor thin majority of delegates heading into the nomination convention, but Clinton supporters claim she should be the party's nominee because Clinton has the greater number of delegates when the votes of Florida and Michigan are taken into account. It doesn't take much imagination to see a court case ensuing, proving once again that the Democratic Party is unequaled when it comes to snatching defeat from the jaws of victory.
3	When is the press going to start referring to the Democrats as being controlled by "Left Wing Liberals" and "non christains". This would appear to be appropriate as the press seems to think the Republican party is composed of "Right Wing Conservatives" and "Evangical Christians".
4	Why don't any of the columnists tell us how many democrats voted in the Tuesday primaries as opposed to republicans? This would give us an idea of the number of democrats and republicans who will be voting in the actual election.
5	Obama should withdraw because he doesn't have strength in the states we democrats need to win the november election. he is winning in red states. he is also winning in states with caucases which it is now clear are not reflective of the true voter sentiments and if he somehow where to get the nomination he would lead the party to defeat because his appeal is limited. it is time Ted Kennedy took him aside and explained reality to him. I think the voters in Mass wanted to send a reality check to Mr.Kennedy. some of us live in the real world and not in the magical kingdom inhabited by Obama,Oprah and the Kennedys
6	what happens to Huckabee's delegates when he pulls out? Assuming they can, and will, go to Romney, the latter may rightly see this still as a race.
7	Excellent!I hope the media and the Obama fans take a good long breath and stop pushing Obama down our throats.Let us make up our minds,without the pundits incessant Hillary bashing.They are treating the primaries the way they treat Paris and Brittney.Please grow up,guys and gal!
8	The results from AK, ID, UT, CO, KS, MO, ND and MN do not seem to fully back up claims by reporters such as your Mr. Adam Nagourney that a?the results suggest that Democrats are fracturing along gender and racial lines as they choose between a black man and a white woman.a I am not aware of large African American populations in most of these states, yet Mr. Obama won them, and in most cases by as much as 60 percent of the vote. Why this continued obsession with gender and race when reporting about the Democrats? I am beginning to think reporters and pundits jump on these a?divisionsa because those are the easiest things to pick out that do not require serious analysis of voteras choices. Give us more insight into how people voted with respect to important issues such as the economy, healthcare, education, and foreign policy. Does it ever occur to you that people may actually select the candidate they vote for because of these issues and not because of their gender or race?
9	I am fascinated by reading all the comments. Accurate or not, each seems fervent. But the fact is that the Republicans are doing what they usually do: unifying; and the Dems are doing what they usually do: squabbling. I'm a Democrat, so I'm not criticizing (business as usual). Oh–and PS: I voted for Hillary. Anybody got a problem with that?
10	For better or for worse, party front-runners in the recent past election cycles have been chosen more by default and less by way of solid qualification as leaders of vision. While I believe Tuesday's results are in general not all that surprising, the fact that the '08 cycle is unique and ground-breaking in many regards, is refreshing in and of itself. To what extent the much talked about "positive change" can come about when the dust finally settles and the rubber meets the road, of course, remains to be seen. That is the "Crux of the biscuit."
11	Obama reminds me of young wine which needs to be put back in the bin for a few years' aging. When it's more mature, it will probably be far more palatable and substantial.

6 Related Work

A stated at the Introduction, to the best of our knowledge, there are no works that can directly be compared with our proposed method. In what follows, we present several approaches that deal with the problems of (a) news comments analysis and (b) search results diversification.

The work in [11] is the closest to ours. The authors present ongoing work on a system regarding online discussion groups. The system first requires that users explicitly state their opinions of specific topics. Then, it exploits this feedback to recommend several opinions, allowing the user to vary the similarity/diversity degree of the recommendations, w.r.t. her own opinions. Apart from the difference in the diversity criteria used, the system described in [11] differs from ours in that it requires explicit, specific feedback from users and, also, it diversifies the recommended opinions w.r.t. each user's personal opinions and not in a global manner.

The authors of [7] propose a news recommendation system in forum-based social media, that exploits user comments to produce news recommendations. The approach aims at building a topic profile, utilizing both the news text and its comments. This profile is then used to retrieve relevant news articles. Similarly, [8] present a method for recommending to users news articles that are likely to be commented by them. The authors propose a hybrid recommendation approach, where they exploit, apart from document content, the co-commenting patterns of users on the respective articles.

In [2] the authors first predict whether a news article is to receive any comments at all and, then, whether it will receive many comments or not. To this end, they apply two separate classification phases. In [1] they try to model and compare commenting distributions from several news sources and, also, predict comment volume by observing a short first period of commenting.

In [4] the authors try to capture commenters sentiment patterns towards political news articles and to predict the political orientation from the sentiments expressed in the comments. The authors apply different learning techniques, depending on whether they predict political orientation for one or more commenters. They also take into account contextual information, such as the vote or links a comment received. In [12] the authors study user comments on political news and evaluate readers' satisfaction on political opinions. In this way, they aim to differentiate between users who seek similar opinions to theirs and users who seek diverse ones.

The authors in [6] study the descriptiveness of comments, i.e. the extent to which comments are similar to the topic they refer to. The authors obtain positive results, in the sense that a sufficient amount of comments can adequately represent the original commented text. In [3] the authors perform a study on users' needs w.r.t. news article comments and conduct a quality analysis on comments posed in the articles of an online newspaper. In [5] an analysis of links, comments and interconnections between blogs is performed. The authors of [9] aim at producing document summaries, utilizing the respective comments. To produce the summaries, they extract sentences from the original document (e.g. blog post), which are biased to keywords extracted from the document's comments. In [10] the authors perform an analysis on blog post comments and their relation to the posts. Specifically, they estimate the overall volume of comments in the blogosphere, analyze the relation between the weblog popularity and commenting patterns in it and measure the contribution of comment content to weblog access.

A thorough review of fundamental works in diversification is given in [19]. [13] describe the maximal marginal relevance method, which attempts to maximize relevance while minimizing similarity to higher ranked documents. To this end, the relevance of search results is calculated using two similarity functions, one measuring the similarity among documents, and the other the similarity between document and query. [14] consider an evaluation metric that penalizes a retrieval model only if it retrieves no relevant results at all. Given that, they propose a method where each result document is selected based on the probability that it is relevant to the previously selected ones.

In [16], the authors introduce a set of diversification axioms and show that it is not possible for a diversification algorithm to satisfy all of them. Also, they propose three diversification objectives. These objectives differ in the level at which the diversity is calculated, e.g. whether it is calculated per separate document or on the average of the currently selected documents. The authors in [15] present a framework for evaluating novelty and diversity. Similarly, [17] propose a greedy diversification algorithm but, also, extend some state of the art IR evaluation measures, so that they can be used in the context of diversification. Finally, [18] present a method for efficient diversification of structured data, where the items to be diversified are not documents, but objects with distinct attributes (i.e. records in a database table).

7   Conclusion

In this paper, we presented a methodology for diversifying user comments on news articles. We introduced comment-specific diversification criteria and extended two state of the art diversification algorithms, so that they can incorporate these criteria. We implemented the above method into a prototype system that works on a publicly available news article dataset. A preliminary evaluation of our approach showed that it is, indeed, meaningful to go beyond textual similarity, when diversifying user comments on news article. Finally, the implemented framework is general enough to be adapted to other (similar) settings, such as forum discussions, tweets, etc.

Our future work lies on enhancing and extending the current diversification criteria, so that they can be integrated to each other and, thus, become more effective for the diversification process. For example, we consider combining the Sentiment and Named Entities criteria, so that we can extract and utilize sentiment values on specific Named Entities found in the news article and commented by users. Also, we plan to perform an extended user study, where users will be asked to evaluate the interestingness, novelty and topic coverage of the diverse comment sets produced by our method, as opposed to baseline comment sets (non-diversified, or diversified only on textual similarity).

Acknowledgements. This research is conducted as part of the EU project ARCOMEM[6] FP7-ICT-270239.

---

[6] http://www.arcomem.eu/

References

1. Tsagkias, M., Weerkamp, W., de Rijke, M.: News Comments:Exploring, Modeling, and Online Prediction. In: Gurrin, C., He, Y., Kazai, G., Kruschwitz, U., Little, S., Roelleke, T., Rüger, S., van Rijsbergen, K. (eds.) ECIR 2010. LNCS, vol. 5993, pp. 191–203. Springer, Heidelberg (2010)
2. Tsagkias, E., Weerkamp, W., de Rijke, M.: Predicting the volume of comments on online news stories. In: Proceedings of the 18th ACM Conference on Information and Knowledge Management (CIKM 2009), pp. 1765–1768 (2009)
3. Diakopoulos, N., Naaman, M.: Towards quality discourse in online news comments. In: Proceedings of the ACM 2011 Conference on Computer Supported Cooperative Work (CSCW 2011), pp. 133–142 (2011)
4. Park, S., Ko, M., Kim, J., Liu, Y., Song, J.: The Politics of Comments: Predicting Political Orientation of News Stories with Commenters Sentiment Patterns. In: Proceedings of the ACM 2011 Conference on Computer Supported Cooperative Work (CSCW 2011), pp. 113–122 (2011)
5. Herring, S.C., Kouper, I., Paolillo, J.C., Scheidt, L.A., Tyworth, M., Welsch, P., Wright, E., Ning, Y.: Conversations in the Blogosphere: An Analysis "From the Bottom Up". In: Proceedings of the 38th Annual Hawaii International Conference on System Sciences (HICSS 2005), p. 107b (2005)
6. Potthast, M.: Measuring the descriptiveness of web comments. In: Proceedings of the 32nd International ACM SIGIR Conference on Research and Development (SIGIR 2009), pp. 724–725 (2009)
7. Li, Q., Wang, J., Chen, Y.P., Lin, Z.: User comments for news recommendation in forum-based social media. Information Sciences: an International Journal 180(24), 4929–4939 (2010)
8. Shmueli, E., Kagian, A., Koren, Y., Lempel, R.: Care to Comment? Recommendations for Commenting on News Stories. In: Proceedings of the 18th International Conference on World Wide Web, WWW 2012 (to appear, 2012)
9. Hu, M., Sun, A., Lim, E.: Comments-oriented document summarization: understanding documents with readers' feedback. In: Proceedings of the 31st Annual International ACM SIGIR Conference on Research and Development in Information Retrieval (SIGIR 2008), pp. 291–298 (2008)
10. Mishne, G.A., Glance, N.: Leave a Reply: An Analysis of Weblog Comments. In: Proceedings of the WWW 2006 Workshop on Weblogging Ecosystem: Aggregation, Analysis and Dynamics, at WWW: The 15th International Conference on World Wide Web (2006)
11. Wong, D., Faridani, S., Bitton, E., Hartmann, B., Goldberg, K.: The diversity donut: enabling participant control over the diversity of recommended responses. In: Proceedings of the 2011 Annual Conference Extended Abstracts on Human Factors in Computing Systems (CHI EA 2011), pp. 1471–1476 (2011)
12. Munson, S.A., Resnick, P.: Presenting diverse political opinions: how and how much. In: Proceedings of the 28th International Conference on Human Factors in Computing Systems (CHI 2010), pp. 1457–1466 (2010)
13. Carbonell, J., Goldstein, J.: The use of MMR, diversity-based reranking for reordering documents and producing summaries. In: Proceedings of the 21st Annual International ACM SIGIR Conference on Research and Development in Information Retrieval (SIGIR 1998), pp. 335–336 (1998)
14. Chen, H., Karger, D.R.: Less is more: probabilistic models for retrieving fewer relevant documents. In: Proceedings of the 29th Annual International ACM SIGIR Conference on Research and Development in Information Retrieval (SIGIR 2006), pp. 429–436 (2006)

15. Clarke, C.L.A., Kolla, M., Cormack, G.V., Vechtomova, O., Ashkan, A., Büttcher, S., MacKinnon, I.: Novelty and diversity in information retrieval evaluation. In: Proceedings of the 31st Annual International ACM SIGIR Conference on Research and Development in Information Retrieval (SIGIR 2008), pp. 659–666 (2008)
16. Gollapudi, S., Sharma, A.: An axiomatic approach for result diversification. In: Proceedings of the 18th International Conference on World Wide Web (WWW 2009), pp. 381–390 (2009)
17. Agrawal, R., Gollapudi, S., Halverson, A., Ieong, S.: Diversifying search results. In: Proceedings of the Second International Conference on Web Search and Web Data Mining (WSDM 2009), pp. 5–14 (2009)
18. Vee, E., Srivastava, U., Shanmugasundaram, J., Bhat, P., Yahia, S.A.: Efficient Computation of Diverse Query Results. In: Proceedings of the 2008 IEEE 24th International Conference on Data Engineering (ICDE 2008), pp. 228–236 (2008)
19. Drosou, M., Pitoura, E.: Search result diversification. ACM SIGMOD Record 39(1), 41–47
20. Hassin, R., Rubinstein, S., Tamir, A.: Approximation algorithms for maximum dispersion. Operations Research Letters 21(3), 133–137 (1997)
21. Ravi, S., Rosenkrantzt, D.J., Tayi, G.K.: Approximation Algorithms for Facility Dispersion. In: Gonzalez, T.F. (ed.) Handbook of Approximation Algorithms and Metaheuristics. Chapman & Hall/CRC (2007)
22. Finkel, J.R., Grenager, T., Manning, C.: Incorporating Non-local Information into Information Extraction Systems by Gibbs Sampling. In: Proceedings of the 43nd Annual Meeting of the Association for Computational Linguistics (ACL 2005), pp. 363–370 (2005)
23. Thelwall, M., Buckley, K., Paltoglou, G., Cai, D., Kappas, A.: Sentiment strength detection in short informal text. Journal of the American Society for Information Science and Technology 61(12), 2544–2558 (2010)

Spelling Suggestion for XML Keyword Search Based on Pairwise Keyword Summaries

Sheng Li and Junhu Wang

School of Information and Communication Technology,
Griffith University, Gold Coast Campus, Australia
sheng.li@griffithuni.edu.au, J.Wang@griffith.edu.au

Abstract. We study the spelling suggestion problem for keyword search over an XML document, which provides users with alternative queries that may better express users' search intention. In order to return the query candidates more efficiently, we calculate the correlations between pairwise keywords, which consider the distribution of keywords and the structures of the XML document. We use the keyword correlations as the summaries of the XML document, which are built off-line. We propose an approach to generating the query candidates, and rank them based on the summaries. Experiments with real datasets verifies the effectiveness and efficiency of our approach.

Keywords: XML, Keyword Search, Spelling Suggestion.

1 Introduction

Keyword search provides a user-friendly information discovery mechanism for people to access XML data without the need of learning a structured query language or studying possibly complex and evolving data schemas [6]. When a user types in a keyword query, there are two processes involved: (1) formulating a keyword search query, and (2) typing the query into systems. However, both steps are susceptible to several kinds of errors [7], e.g., step (2) suffers from typographical errors and incorrect keywords. In these cases, a dirty query is formed, which is likely to return empty or low-quality results, not satisfying users' expectation.

In order to alleviate this problem, *query suggestion (QS)* a.k.a. *query cleaning*, has been proposed, which provides users with alternative queries that better express users' search intention. QS has been studied, and already widely used for text documents.

Most of previous works are designed for Web queries and plaint text data. Besides, most of them rely solely on the query log. As a result, the quality of the suggested queries relies on the quantity and quality of the log, and this approach has biases towards popular queries. Different from previous work, recently [9] and [5,7] tackle the query cleaning problem for relational database and XML data respectively, and they are based on the content of the database, avoiding the biases caused by approaches relying on query logs only.

X.S. Wang et al. (Eds.): WISE 2012, LNCS 7651, pp. 114–127, 2012.

XClean [7] is the first work that focuses on the spelling suggestion problem for XML Keyword Search. It proposes a probabilistic framework that takes the XML structural characteristics into consideration and improves the result quality of suggested queries. Recently, we proposed a new spelling suggestion method [5], which is based on the minimum distance between keyword matching nodes. It adopts nearest keyword search [10] to find the minimum distance between keywords, and rank the candidates mainly based on the minimum distance and spelling error penalty.

The merit of the above methods is that the quality of the suggested queries is improved. However, since they both require to search through the content in the XML dataset after generating variants for each keyword in the given query, the efficiency of the spelling suggestion may be unsatisfactory, especially when the dataset is large and the keywords in the issued query have a large number of variants. For example, in our experiment on the DBLP dataset (experimental configuration is given in Section 4), the runtime for query {gunter, active, objects} is about 2s for XClean (while about 4ms for our work in this paper), which is too slow. In practice, query suggestion should be a lightweight component which suggests query candidates instantly. Thus, how to generate high-quality query candidates more efficiently is an important research issue.

In this paper we propose a method that is based on the summary of the XML data, which summarize the correlation between any two keywords and is generated off-line. When a query is issued, we use the pairwise correlation to estimate the quality of the query. The contribution of our work can be summarized as:

- We design a method to calculate the correlation between keywords in an XML data tree, which considers both the frequencies of keywords and the distance between them.
- We propose an algorithm to generate query candidates and rank them based on the summaries generated off-line, which is far more efficient than previous methods. Meanwhile, it has the similar effectiveness as XClean.

The rest of the paper is organized as follows: we begin with a recall of the previous work in Section 2. Section 3 elaborates on our approach to generating the top-k query candidates. In Section 4 experiments and analysis are presented. Section 5 is the conclusion of the paper.

2   Previous Work

In this section, we will discuss previous work on spelling suggestion and related work.

Spelling suggestion is defined as: given an XML document T and a keyword query Q containing keywords $w_1, w_2, ..., w_q$, we return the top-k candidate queries C_i $(1 \leq i \leq k)$ with the highest score that represent the most likely queries intended by users.

Previous studies, such as [9] and [7], generate the candidate queries in the following two steps:

1. For each keyword w_i in the query Q, a list of variants, denoted as $var_\epsilon(w_i)$, is generated. Each variant is a keyword in the vocabulary of the document, and has no more than ϵ edit errors from w_i.

2. The query candidate space $\mathcal{C}$ is the Cartesian product of the lists, namely $\mathcal{C} = \prod_{i=1}^{q} var_\epsilon(w_i)$.

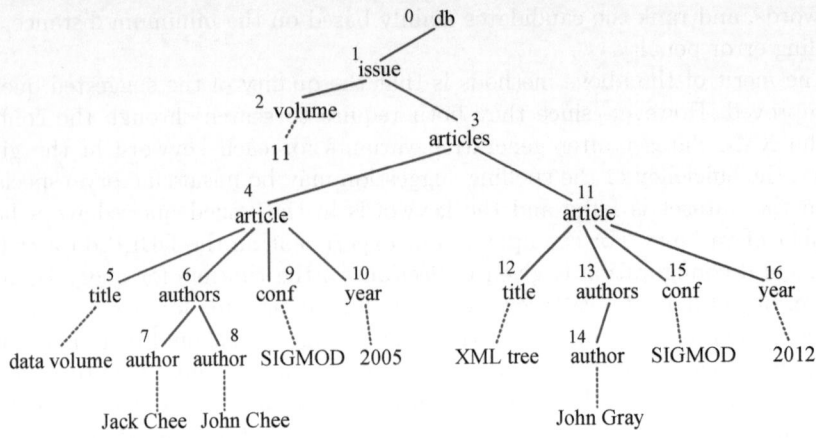

Fig. 1. An example XML tree

For example, consider the XML data tree shown in Fig. 1 and a dirty query $Q = \{xml,\ thee\}$. Suppose $\epsilon = 1$, step 1 generates the variants for keywords in Q based on the vocabulary, $var_1(xml) = \{xml\}$ and $var_1(thee) = \{tree,\ chee\}$. Then step 2 generates the candidate space from the lists of variants, which contains $\{xml,\ tree\}$ and $\{xml,\ chee\}$.

Paper [9] focuses on the query cleaning problem on relational database. It uses a scoring function which combines the spelling error penalty and *term frequency* (tf)/*inverse document frequency* (idf) scores of keywords. The tf/idf has been widely used in the IR field. However, because the structures of XML data carry the relationships between data, applying tf/idf alone on XML data is not sufficient.

XClean [7] proposes a probabilistic framework designed to deal with spelling suggestion problem over XML data. XClean restrains keywords from a query in certain subtrees, which satisfy semantics such as *Search-for Node Type* [1] or SLCA [11]. Each subtree is treated as a virtual document, where *unigram language model* is applied to calculate the *query generation probability* [7] for a query. The probability combined with spelling error penalty is used to rank the candidates. Our previous work [5] adopts the nearest keyword search method [10] to find the minimum distance between the keyword matching nodes which is based on the intuition that the closer the keyword matching nodes are to each other, the more meaningful the query is.

Basically, a query Q is meaningful for a dataset if it has high quality results from the dataset. Let $score(Q)$ be the quality of the query Q for a dataset,

$score(Q)$ can be computed as the total score of all its results. However, this approach could overestimate queries that have numerous results over other queries that have fewer but meaningful results. To be balanced, $score(Q)$ is often evaluated as the total score of its top-k results [8,12], where k is a selected parameter. $score(Q)$ can be computed using the following formula:

$$score(Q) = \sum_{i=1}^{k} score(R_i, Q) \tag{1}$$

where R_i is the i-th top result of Q, and $score(R_i, Q)$ is the relevant score of R_i to Q.

Ideally, we could calculate the $score(C)$ for each query candidate C generated from Q, and then use $score(C)$ combined with the edit distance to rank the candidates. However, when the dataset is huge and there are numerous candidates from Q, the process can be highly time consuming. To avoid querying all the candidates over the original data, summaries can be constructed *off-line* for the dataset and then used to calculate the scores for each candidate C *on-line* [4,8,12]. In this paper, we propose an approach which uses the *Pairwise Keyword Summaries table* (PKS) to estimate the quality of the query candidates.

3   Pairwise Keyword Summaries

In this section, we introduce our spelling suggestion method based on the *pairwise keyword summaries*(PKS).

The main idea of our approach is to generate the PKS *off-line*, when a query Q is issued, we use the PKS to estimate the relevance of the query candidates generated from Q. Our approach for spelling suggestion mainly has the following steps:

1. Generate the PKS *off-line*.
2. When a query $Q = \{w_1, ..., w_q\}$ is issued to the system, we use a version of the FastSS method [2] to generate variants for each keyword w_i within edit distance ϵ, the set of the variants for w_i is denoted as $var_\epsilon(w_i)$.
3. Generate the query candidates from the variants based on the pairwise keyword summaries, and calculate the relevance score of the candidate according to Equation (9).
4. Combine the relevance score with the spelling error penalty to gain the final score of the candidate, and rank the candidates according to the final score.

3.1   Correlation between Pairwise Keywords

In our paper, we adopt the tf/idf in the IR field and use the distance between nodes to represent the relationship between nodes in the XML data tree. Therefore, we first define the distance between two nodes in the XML tree.

Distance between Two Nodes. As observed previously [1,6], an XML document consists of nodes that represent *entities, attributes* and *connection* nodes. XSeek [6] proposed some guidelines to infer those different nodes. In our work, we use the same guidelines and the concept of *entity*, and we refer to the label path of the entity as its *entity type*. For example, in Fig. 1 each *article* is an *entity*, and the path *db/issue/articles/article* is its *entity type*. For simplicity, we will use the label of the node to represent the *entity type* when there is no confusion.

Given *node u* and *node v*, the distance between them is defined as:

$$dist(u, v) = distB(u, v) + distP(u, v) \tag{2}$$

where $distB(u, v)$ stands for the *basic distance* between nodes u and v, which is the number of edges in the path linking u and v; $distP(u, v)$ is the *penalty distance* between u and v, which penalizes nodes in different entities to guarantee that a node within an entity of *entity type E* is unable to be closer to nodes in other entities of the same type than to nodes within the same entity. $distB(u, v)$ can be calculated by finding the *Lowest Common Ancestor (LCA)* of nodes u and v, denoted as $LCA(u, v)$. Suppose that $z = LCA(u, v)$, then

$$distB(u, v) = level(u) + level(v) - 2 \times level(z) \tag{3}$$

where $level(z)$ is the level of z in the tree, which can be easily calculated by comparing the Dewey codes of u and v. The *penalty distance* is defined as

$$distP(u, v) = \begin{cases} depth(E) & \text{if } E \text{ exists} \\ 0 & \text{else} \end{cases} \tag{4}$$

where E is the *type* of the top-level *entities* that exist on both *path $z \to u$* and *path $z \to v$*, where $z = LCA(u, v)$; $depth(E)$, different from $level(u)$, is the depth that the entity type E possesses, which is the maximum number of levels that entities of type E have, e.g., $depth(issue) = 5$ and $depth(article) = 3$.

Example 1. Consider the tree in Fig. 1. For *article* 11, it has the same *basic distance* to nodes 2 and 5, namely, $distB(11, 2) = distB(11, 5) = 3$. However, nodes 11 and 5 are in different *article* entities. After adding the *penalty distance*, $dist(11, 5) = 3 + depth(article) = 6$. $dist(11, 2) = 3$ is smaller, which reflects the fact the nodes 11 and 2 have a closer relationship, compared to nodes 11 and 5.

Keyword Correlation. PKS stores the correlations between pairwise keywords in the XML data. In our paper, each record in the PKS is 3-tuple (w_i, w_j, $corr(w_i, w_j)$), where w_i and w_j are two different keywords that exist in the XML data tree (w_i is smaller than w_j alphabetically to avoid considering both w_i and w_j), and $corr(w_i, w_j)$ represents the keyword correlation between w_i and w_j in the XML data tree.

For example, consider the XML data tree in Fig. 1. The PKS of keywords {chee, sigmod, tree, xml} is given in Table 1. The calculation of the correlation between keywords is explained below.

Table 1. The PKS of keywords {Chee, SIGMOD, tree, XML} , $k = 2$

w_i	w_j	$corr(w_i, w_j)$
chee	sigmod	3.09
chee	tree	1.56
chee	xml	1.56
sigmod	tree	3.29
sigmod	xml	3.29
tree	xml	8.18

Given two nodes u and v which contain two keywords w_i and w_j, the score of w_i and w_j with respect to u and v is defined as:

$$score(w_i, w_j, u, v) = \frac{weight(u, w_i) + weight(v, w_j)}{dist(u, v) + 1} \tag{5}$$

where $dist(u, v)$ is defined in Equation (2) which measures the strength of the relationship between two nodes. The closer the distance between the two nodes, the stronger their relationship is. $weight(u, w_i)$ measures the content relevance of node u with respect to keyword w_i. In order to calculate $weight(u, w_i)$, the tf/idf in the IR field is employed, which measures the content relevance of a document to a keyword query using both tf and idf. To apply the tf/idf to XML data, *term frequency (tf)* of a keyword w_i in node u is the number of occurrences of w_i in node u; instead of idf, *inverse element frequency (ief)* of a keyword w is employed, which is the total number N of elements (nodes) in the XML data tree over the number N_w of elements that contain the keyword w, namely, $ief_w = \frac{N}{N_w}$. Thus, if the frequency of a keyword w is higher in the node u, meanwhile there are few other elements containing the keyword w, the content relevance of the node u w.r.t w is higher. Then the weight of keyword w_i in node u is calculated as:

$$weight(u, w_i) = \log_2 (1 + tf_{w_i}) \log_2 ief_{w_i} \tag{6}$$

For example, consider the nodes matching keywords *SIGMOD* and *XML* over the XML tree in Fig. 1, we have:

$$score(\text{sigmod}, \text{xml}, 9, 12) = \frac{\log_2 \frac{17}{2} + \log_2 \frac{17}{1}}{dist(9, 12) + 1} \approx \frac{3.09 + 4.09}{7 + 1} \approx 0.9$$

$$score(\text{sigmod}, \text{xml}, 15, 12) \approx \frac{3.09 + 4.09}{2 + 1} \approx 2.39$$

Consider two keywords w_i and w_j, and a set $\mathbb{R} = \{\{u_1, v_1\}, ..., \{u_n, v_n\}\}$, where u_t and v_t ($1 \leq t \leq n$) match keywords w_i and w_j respectively. Suppose that $score(w_i, w_j, u_1, v_1) \leq \cdots \leq score(w_i, w_j, u_n, v_n)$, then the keyword correlation between keywords w_i and w_j is measured as the total scores of up to top-k correlations for keywords w_i and w_j in an XML data tree:

$$corr(w_i, w_j) = \begin{cases} \sum_{i=1}^{k} score(w_i, w_j, u_i, v_i) & \text{if } k \leq n \\ \sum_{i=1}^{n} score(w_i, w_j, u_i, v_i) & \text{else} \end{cases} \tag{7}$$

For example, suppose $k = 2$, then we have $corr(\text{sigmod}, \text{xml}) = 0.9 + 2.39 = 3.29$. Similarly, we can calculate the correlation between other keywords, shown in Table 1.

Given a query candidate $C = \{w_1, w_2, ..., w_q\}$, we estimate the relevance of the query C for the database T based on the following equation:

$$CORR\text{-}S(C) = \begin{cases} 0, \text{ if } \exists corr(w_i, w_j) = 0, \text{where } \{w_i, w_j\} \subseteq C, i < j \\ \sum_{\{w_i, w_j\} \subseteq C, i < j} corr(w_i, w_j), \text{ else} \end{cases} \tag{8}$$

Example 2. Consider two queries $C_1 = \{\text{xml, tree, sigmod}\}$ and $C_2 = \{\text{xml, chee, sigmod}\}$ over the XML tree in Fig. 1, and let $k = 2$. We use Equation (8) to estimate the quality of C_1 and C_2. We have $CORR\text{-}S(C_1) = 8.18 + 3.29 + 3.29 = 14.76$ and $CORR\text{-}S(C_2) = 1.56 + 3.29 + 3.09 = 7.94$.

Finally, we combine the spelling error penalty to calculate the score of the query candidate C as follow:

$$SCORE(C) = e^{-ed(Q,C)} \cdot CORR\text{-}S(C) \tag{9}$$

where $ed(Q, C)$ is the edit distance from the query candidate C to the input query Q.

By summarizing the correlations for each pair of keywords and using the correlations to estimate the quality of a query, we can improve the efficiency of the query suggestion process significantly. However, indexing all pairs of keywords at all possible distances can result in extremely large PKS.

Note that not all pairs of keywords in an XML tree are meaningfully related [7,8], especially those keywords that are far away from each other, or even only connected through the root of the data tree. Therefore, to reduce the size of the PKS, a distance threshold d can be used to limit the maximum distance between two keywords. Given two keywords w_i and w_j and a distance threshold d, if for any nodes u and v matching the two keyword respectively, $dist(u, v) \geq d$, then there is no record with respect to the two keywords in the PKS, and $corr(w_i, w_j)$ is defaulted to 0.

Example 3. Consider the correlation between keyword *chee* and *xml* in Fig. 1 when the distance threshold d is set to 7. Because $dist(7, 12) = dist(8, 12) = 5 + 3 = 8 > d$, the two keywords are not meaningfully related. Thus, the correlation between the two keywords are omitted in the PKS, and $corr(\text{chee}, \text{xml}) = 0$ in this case.

3.2   Algorithm

Given an input query $Q = \{w_1, w_2, ..., w_q\}$, instead of considering each query candidate C_i in the candidate space $\mathbb{C}$ separately, we incrementally process the

possible candidates by adding the keywords one by one. First, we process keywords in $var_\epsilon(w_1)$ and $var_\epsilon(w_2)$ to generate all the possible meaningful candidates that are the variants of query $\{w_1, w_2\}$. Suppose w_1' and w_2' are the variants
of w_1 and w_2 respectively, if $corr(w_1', w_2') = 0$, any query containing w_1' and w_2'
would not be meaningful; else $\{w_1', w_2'\}$ is recorded for future computation. Then
we process $var_\epsilon(w_3),...,var_\epsilon(w_q)$ incrementally. The details of our algorithm are
shown in Algorithm 1.

Algorithm 1. Generate top-N query candidates based on summaries

---

Input: Query $Q = \{w_1, ..., w_q\}$; N
Output: a list of suggested queries SQ
 1: generate variants for each keyword in Q
 2: new $cMap$; //create a new hash map to store candidate C and $CORR\text{-}S(C)$
 3: **for** each keyword $w_1' \in var_\epsilon(w_1)$ **do**
 4:    $cMap.put(\{w_1'\}, 0)$;   //initially the score is 0
 5: **end for**
 6: **for** $i = 2 \rightarrow q$ **do**
 7:    new $cMap'$; //create a new hash map
 8:    **for** each candidate $C \in cMap.keySet()$ **do**
 9:      $CORR\text{-}S(C) \leftarrow cMap.get(C)$;
10:      **for** each keyword $w_i' \in var_\epsilon(w_i)$ **do**
11:        $C' = C \cup \{w_i'\}$; //incrementally generate longer candidates
12:        $CORR\text{-}S(C') = CORR\text{-}S(C)$;
13:        $isMeaningful \leftarrow true$;
14:        **for** each keyword $w \in C$ **do**
15:          $corr(w, w_i') \leftarrow getCorrFromPKS(w, w_i')$;
16:          **if** $corr(w, w_i') = 0$ **then**
17:            $isMeaningful \leftarrow false$;
18:            break;
19:          **else**
20:            $CORR\text{-}S(C') = CORR\text{-}S(C') + corr(w, w_i')$;
21:          **end if**
22:        **end for**
23:        **if** $isMeaningful$ is $true$ **then**
24:          $cMap'.put(C', CORR\text{-}S(C'))$;
25:        **end if**
26:      **end for**
27:    **end for**
28:    $cMap = cMap'$;
29: **end for**
30: $SQ = rank(cMap, N)$;

In Algorithm 1, assuming that $w_1, w_2, ..., w_q$ have been arranged in such a
way that $|var_\epsilon(w_1)| \leq \cdots \leq |var_\epsilon(w_q)|$, we first generate the variants for each
keyword in Q (Line 1). We use a hash map $cMap$ to store the query candidates
and the sum of the keyword correlations for each pairwise keywords in the query.
Initially, $cMap$ stores all the queries that only contain one keyword from $var_\epsilon(w_1)$

and the sum of the keyword correlation is 0 (Lines 2-5). Then incrementally, our algorithm processes $var_\epsilon(w_i)$, where i starts from 2 to q (Lines 6-29). For each candidate C from the *key set* of $cMap$ (Line 8), we check the keyword correlation between any keyword w in C and w_i' from PKS (Line 15), where w_i' is any keyword from $var_\epsilon(w_i)$ (Line 10). Let $C' = C \cup \{w_i'\}$, if $\exists w \in C$ such that $corr(w, w_i') = 0$, C' is not a meaningful candidate (Line 17); else, the correlation between any $w \in C$ and w_i' is added up to $CORR\text{-}S(C')$ (Line 20). Besides, the new candidate C' and $CORR\text{-}S(C')$ are stored in the new hash map $cMap'$. After all the candidates in the key set of $cMap$ are processed and before i increases, $cMap$ is replaced by $cMap'$ for the next loop (Line 28). The final query candidates are generated after i reaches q.

Finally, for each candidate C and its $CORR\text{-}S(C)$ stored in $cMap$, we combine the spelling error penalty as shown in Equation (9) to get the final score $SCORE(C)$ as the ranking factor, and return top-k query candidates SQ. These processes are implemented by the $rank(cMap, k)$ function (Line 30).

Example 4. Suppose the distance threshold $d = 7$ and $k = 2$, and the PKS has been generated off-line. Consider an input query $Q = \{$xml, thee, sigmod$\}$. First the variants for each keyword in Q is generated (Here, in order to elaborate the algorithm, we do not arrange the keywords in Q according to the number of variants). Initially, $cMap$ contains $\{$xml$\}$ and the $CORR\text{-}S(\text{xml}) = 0$. Next, we try to merge $\{$xml$\}$ with keywords in $var_\epsilon(thee) = \{$chee, tree$\}$ to generate longer query candidates. Because $corr(\text{xml}, \text{chee}) = 0$, unlike $\{$xml, tree$\}$, candidate $\{$xml, chee$\}$ is not put into $cMap'$. Then $cMap$ is replaced by $cMap'$ to go into next loop for merging keywords in $var_\epsilon(sigmod) = \{$sigmod$\}$. Finally, $cMap$ contains the final query candidate $\{$xml, tree, sigmod$\}$.

3.3   Complexity Analysis

We store the PKS in the structure of binary tree, suppose the vocabulary for the PKS is in V, its size (the number of keywords) is denoted as $|V|$. Given a pair of keywords w_i and w_j, the time complexity to look up the correlation between them (function $getCorrFromPKS$ in Algorithm 1) is $O(\log |V|)$ in the worst case.

Given a query $Q = \{w_1, ..., w_q\}$ which has been arranged in such a way that $|var_\epsilon(w_1)| \leq \cdots \leq |var_\epsilon(w_q)|$. According to Algorithm 1, for each i ($1 < i \leq q$), let $L_j = |var_\epsilon(w_j)|$ be number of variants for w_j, there are maximum $\prod_{j=1}^{i-1} L_j$ number of query candidates in $cMap$ in the worse case. For each candidate C stored in $cMap$, it needs to merge with every keyword w_i' in $var_\epsilon(w_i)$, which costs $O((i - 1) \log |V|)$ for each keyword w_i'. Therefore, the time complexity for our approach in the worse case is:

$$O\left(\sum_{i=2}^{q} \left((i - 1) \prod_{j=1}^{i} L_j\right) \cdot \log |V| \right)$$

4   Experiments

In this section, we present the experimental results and analyze the effectiveness and efficiency of our approach.

4.1   Experimental Setup

All our experiments were carried out on a PC with Intel Core i5 2.50GHz CPU and 4GB RAM. The operating system was Windows 7, and we used the Berkeley DB Java Edition. The algorithms were implemented using Eclipse with JDK 1.6. We used two real world datasets in the experiments:

DBLP is a snapshot of the DBLP database. Its size is 400MB, max depth is 6, and average depth is 5.58;

SigmodRecord is an XML file which contains SIGMOD Records in February 2007. Its size is 467KB, max depth is 6, and average depth is 5.14.

Table 2. Sample Queries

Dataset	Clean	Dirty
DBLP	rose architecture fpga	rose arcitecture fpga
	gunter active objects	guntor artive objects
	harry role processing	garry role procesing
Sigmod	cook database review	cook database preview
	logical modeling arie	logichal modeling arie
	compiler david john	compile davis john

For query cleaning, finding the *ground truth* is difficult [3,7]. We first designed a set of initial clean queries from keywords that occur in the entity nodes, such as *title*, *author*, and *volume*. We then used random edit operations ($\epsilon = 2$) to obtain dirty queries from them. We generated 50 clean queries and 50 dirty queries on each dataset, some query samples are given in Table 2. Like XClean, we employed human assessors to manually identify the *ground truth*. Besides, to ensure enough information is enough information is preserved for the input query, we do not introduce random edit operations to very short keywords whose length is no larger than 4 [7].

We generated the PKS *off-line* first. Theoretically, the quality of the PKS can be affected by the choice of k. However, the effect of k is not very obvious in our experiments. Thus in the following experiments we set $k = 20$. If two keywords are only connected through the root of the XML tree, we assume that the connection between them is not meaningful. The physical space of the PKS for DBLP is about 900MB in our experiment. As we mentioned, we could also customize a distance threshold d to limit the maximum distance between two keywords. However, while constructing the summaries, we set d to a large value, only ignoring the connections between keywords that are connected through the root.

4.2   Algorithms and Measures

We used a re-implementation of XClean and our previous work [5](denoted as minDist) as the baselines of comparison. To measure the effectiveness of algorithms, we used the same measurements that XClean uses:

MRR (Mean Reciprocal Rank) is defined as $\text{MRR} = \frac{1}{|Q|} \cdot \sum_{Q \in \mathcal{Q}} \frac{1}{rank(Q_g)}$, where $\mathcal{Q}$ is a set of queries, and $rank(Q_g)$ is the rank of the ground truth Q_g [7]. The larger the value of MRR is, the better the quality is.

Precision@N is defined as $precision@N = \frac{A}{|Q|}$, where A is the number of queries whose top-N suggestions contain the ground truth, which indicates what percentage of given queries that users will be satisfied if they are presented with at most N suggestions for each query [7]. The higher Precision@N is, the better the quality.

Time We record the running time of the algorithms.

The MRR and precision@N for the three algorithms are shown in Fig. 2, Fig. 3, and Fig. 4; The comparison of time is given in Table 3 and Fig. 5.

4.3   Analysis

Effectiveness. To compare the effectiveness, we first measured the MRR for the three spelling suggestion algorithms on clean and dirty datasets, shown in Fig. 2. As we can see, the three systems achieved a close-to-1 MRR for clean queries on both datasets. For dirty queries, minDist had the best quality of suggested queries out of the three. XClean and our PKS achieved similar MRR on the dirty queries. However, for the dirty queries over DBLP dataset, XClean had higher MRR than PKS.

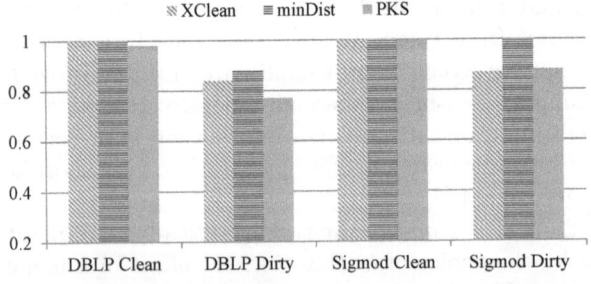

Fig. 2. MRR

Then we measured the Precision@N of the three algorithms to see what percentage of given queries that users will be satisfied if they are presented with at most N suggestions for each query. As we can see in Fig. 3 and Fig. 4, minDist outran the other two systems, and PKS and XClean had the similar quality based on Precision@N.

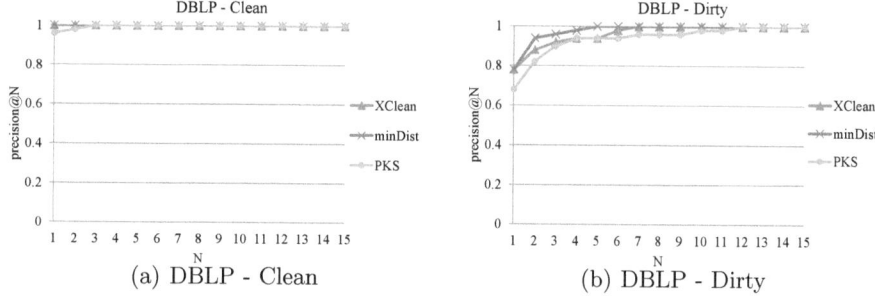

Fig. 3. Precision@N on DBLP with clean and dirty queries

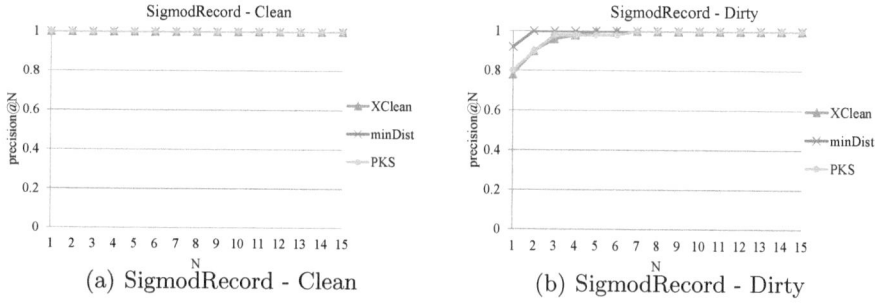

Fig. 4. Precision@N on SigmodRecord with clean and dirty queries

Efficiency. To compare the efficiency, we first compared the average runtime of the three algorithms over the two dataset. With $\epsilon = 2$, the average runtime of them is given in Table 3[1]. As we can see, for both datasets, our PKS outperformed minDist and XClean significantly.

Then, we compared the average runtime of the three algorithms over datasets with different size, shown in Fig. 5. The average runtime of the three algorithms all increased with the size of the datasets. However, our PKS algorithms grew more slowly with the data size. As we analyzed in Section 3.3, the time complexity of PKS depends on the number of variants for each keyword in the given query. As the dataset grows larger, the more number of variants a keyword is more likely to have, which causes more runtime.

To summarize, our PKS in this paper achieves similar effectiveness to XClean. Even though it is slightly less effective than XClean, PKS significantly outperforms XClean and minDist in runtime efficiency.

---

[1] Runtime for XClean varies dramatically to the queries because of the number of variants and the size of inverted list. Due to the different test queries, algorithm implementation and the environment, the runtime for XClean in our experiments is longer than the time reported in [7].

Table 3. Average runtime in milliseconds over 100 queries, $\epsilon = 2$

Dataset	XClean	minDist	PKS
DBLP 400MB	7566.74	320.38	7.35
Sigmod 467KB	1.23	0.91	0.04

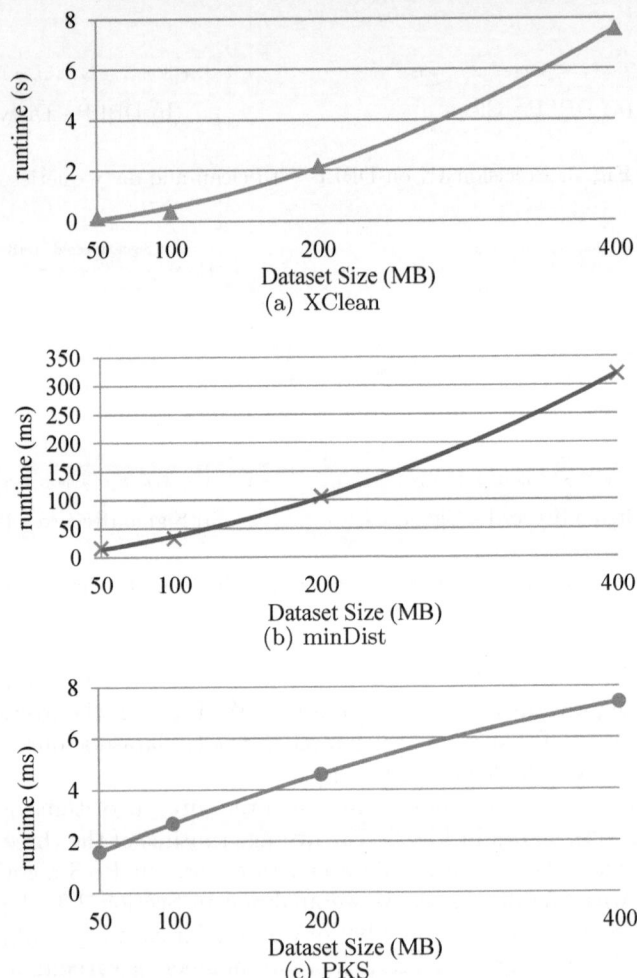

Fig. 5. Average runtime on DBLP with different size, $\epsilon = 2$

5   Conclusion

In this paper, we proposed a spelling suggestion approach for XML data which is based on pairwise keyword summaries. First, our approach generates the *pairwise keyword summaries* which summarize the correlations between pairwise keywords in the vocabulary of the XML data source *off-line*. When a dirty query is issued, we use the pairwise keyword summaries to generate query candidates and estimate the quality of each candidate. Experiments showed that our approach achieved similar effectiveness as XClean, and it is far more efficient compared to previous work. However, it is time-consuming to generate the PKS off-line for large datasets. Thus, how to generate the PKS in a more efficient way is our future work.

Acknowledgement. This work is supported by the Australian Research Council Discovery Grant DP1093404.

References

1. Bao, Z., Ling, T.W., Chen, B., Lu, J.: Effective XML Keyword Search with Relevance Oriented Ranking. In: ICDE, pp. 517–528 (2009)
2. Bocek, T., Hunt, E., Stiller, B.: Fast Similarity Search in Large Dictionaries. Technical Report ifi-2007.02, Department of Informatics, University of Zurich (April 2007)
3. Cucerzan, S., Brill, E.: Spelling Correction as an Iterative Process that Exploits the Collective Knowledge of Web Users. In: EMNLP, pp. 293–300 (2004)
4. Koloniari, G., Pitoura, E.: LCA-based selection for XML document collections. In: WWW, pp. 511–520 (2010)
5. Li, S., Wang, J., Wang, K., Li, J.: A distance-based spelling suggestion method for XML keyword search. In: ER (to appear, 2012)
6. Liu, Z., Walker, J., Chen, Y.: XSeek: A Semantic XML Search Engine Using Keywords. In: VLDB, pp. 1330–1333 (2007)
7. Lu, Y., Wang, W., Li, J., Liu, C.: XClean: Providing valid spelling suggestions for XML keyword queries. In: ICDE, pp. 661–672 (2011)
8. Nguyen, K., Cao, J.: K-Graphs: Selecting Top-k Data Sources for XML Keyword Queries. In: Hameurlain, A., Liddle, S.W., Schewe, K.-D., Zhou, X. (eds.) DEXA 2011, Part I. LNCS, vol. 6860, pp. 425–439. Springer, Heidelberg (2011)
9. Pu, K.Q., Yu, X.: Keyword query cleaning. PVLDB 1(1), 909–920 (2008)
10. Tao, Y., Papadopoulos, S., Sheng, C., Stefanidis, K.: Nearest keyword search in XML documents. In: SIGMOD Conference, pp. 589–600 (2011)
11. Xu, Y., Papakonstantinou, Y.: Efficient Keyword Search for Smallest LCAs in XML Databases. In: SIGMOD Conference, pp. 537–538 (2005)
12. Yu, B., Li, G., Sollins, K.R., Tung, A.K.H.: Effective keyword-based selection of relational databases. In: SIGMOD Conference, pp. 139–150 (2007)

An Extended Compact TVP Index for Finding Top-k Nearest Neighbors over XML Data Tree

Sheng Li and Junhu Wang

School of Information and Communication Technology,
Griffith University, Gold Coast Campus, Australia
sheng.li@griffithuni.edu.au, J.Wang@griffith.edu.au

Abstract. Given a node q and a keyword w, *nearest keyword* (NK) search is to find the nearest w-neighbor of q which carries the keyword w. NK search provides an approach to exploring XML query problems based on the distance between nodes, such as XPath evaluation and XML keyword search. In this paper, we propose an extended compact TVP index (ecTVP) tailored to find top-k nearest w-neighbors of a given node in XML data tree. The ecTVP index is more compact and faster to build. Both theoretical analysis and experimental result show the advantage of the ecTVP index.

Keywords: XML, Nearest Keyword, TVP.

1 Introduction

Nearest Keyword (NK) search [5] provides an approach to exploring XML queries based on the distance between nodes. Define the *distance* between two nodes u and v as $dist(u,v)$, which is the number of edges in the (unique) path linking them in the data tree, an NK query finds the *nearest neighbor* (NN) of a given node q which carries the keyword w. As shown in [5], NK queries can serve as the building bricks to deal with many problems in XML databases, such as *XPath query evaluation* and *group steiner tree (GST)* [3,4] *retrieval*. Besides, NK search can be used for XML keyword search. Previous keyword search semantics, such as *LCA*, *SLCA* [7,8], *ELCA* [2,10], *Search-for Node Type* [1] and so on, are based on the heuristic that far-apart nodes are not as tightly related as nodes that are closer together [6]. However, these semantics only restrain keyword matching nodes within subtrees, without measuring the relationship between the nodes within each subtree. NK provides an approach to finding the keyword matching nodes that are as close to each other as possible.

There are two problems with the approach in [5]. First, for a given node q, there may be several nodes that carry keyword w and are closest to q. However, the approach in [5] returns only one NN. The NN returned may not be the most interesting one to users.

Consider the GST retrieval problem, which is to, given a set of keywords, find a subtree that contains all the keywords in the content of its nodes and has the fewest number of edges. Take the approximate *GST* discussed in [5] on the

X.S. Wang et al. (Eds.): WISE 2012, LNCS 7651, pp. 128–142, 2012.

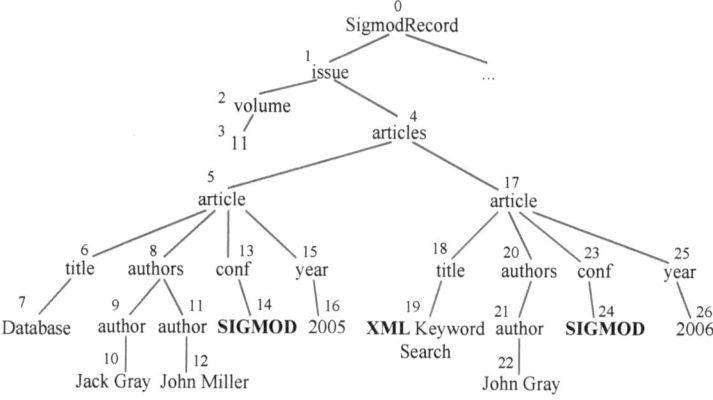

Fig. 1. A sample of SigmodRecord data tree

query Q ={volume, 11, Sigmod, XML} in Fig. 1 for example, node 2 containing *volume* is considered as the given node q in this case. First, NK search finds node 3 containing *11*, then it may find node 14 containing *Sigmod*, finally node 19 carrying *XML* is found. These four nodes {2, 3, 14, 19} act as a vector to determine the approximate *GST*. While a better *GST* (with less edges) that contains node 2, 3, 24 and 19 is missed. Thus, it is insufficient to just return one NN.

Second, [5] uses the *Tree Voronoi Partition (TVP)* index to locate the nearest neighbor. However, the TVP index is not compact. For example, consider the keyword "data" in the DBLP dataset used in our experiments, its occurrences is 38491, while its TVP has 63897 intervals. However, more than 20000 intervals in its TVP index are associated with the same NN, which are removable as we will discuss in Section 3.2 of this paper. If we could make the TVP index more compact, it will cost less physical space, and also it will make the search more efficient.

In this paper, we design a method to construct an *extended compact TVP* (ecTVP) index tailored to find the top-k NNs in XML data. Besides, our ecTVP index is more compact than the TVP index. The contribution of our work can be summarized as:

- We design the improved ecTVP index that can be used to locate the top-k nearest w-neighbors of a given node.
- We design a method to construct the ecTVP index, by constructing a *variant* of Extended Compact Tree (*vECT*). We find the top-k nearest w-neighbors for each node in *vECT* during a bottom-up process and a top-down process while constructing the index, which makes our algorithm more efficient.
- We reduce the redundancy in our ecTVP index, such that it costs less space and query time.

The rest of the paper is organized as follows: we begin with a recall of the previous work [5] in Section 2. Section 3 elaborates on our approach to constructing the improved index. In Section 4 experiments and analysis are presented. Section 5 is the conclusion of the paper.

2   Previous Work

In this section, we recall the work in [5] that we will need in the next section.

Given a node q in an XML tree T and a keyword w, let $U(w)$ be the set of all nodes matching keyword w in T, NK search is defined as:

Definition 1 (Nearest Keyword Search). *Nearest keyword search is to find the nearest w-neighbor of q, denoted as w-NN(q), namely, the node having the minimum distance to q among $U(w)$.*

For example, consider the XML tree in Fig. 2 where each node is assigned an ID that represents the sequence number of the node in the pre-order traversal of the data tree. For node 2, its "Gray"-NN is node 3. For simplicity, we use NN(q) instead of w-NN(q) if there is no confusion. Note that in some cases, for a given node q, there are more than one keyword matching nodes that have the minimum distance to q. In this situation, we assume the w-NN(q) to be the one with the smallest *rank* defined in Definition 2.

Intuitively, given a node q, finding the w-NN(q) can be achieved by comparing the distance between q and each node $u \in U(w)$. Let $N_w = |U(w)|$ be the number of nodes carrying keyword w, this process costs $O(N_w)$. However, this approach is impractical and could be highly time consuming when the frequency of the keyword is high.

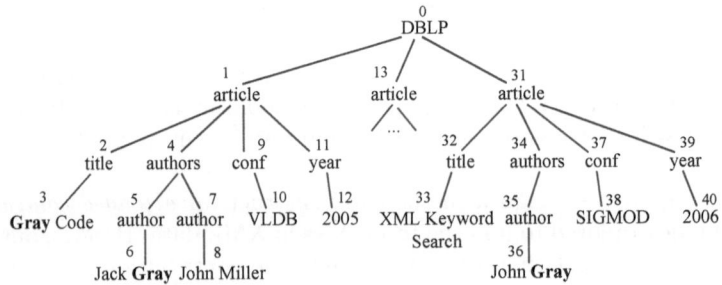

Fig. 2. A sample of DBLP data tree

It is observed [5] that many nodes in the XML data tree share the same w-NN, thus the TVP index is proposed to capture this characteristic through *interval encoding*.

Definition 2 (Interval Encoding). *For each node u of an XML data tree T, define its rank, denoted as $rank(u)$, to be the sequence number of u in the pre-order traversal of T. Node u is associated with an interval $R(u) = [u.start, u.end]$, where u.start is the rank of u, and u.end is the largest rank of the nodes in the subtree rooted at u.*

Definition 3 (TVP of *w*). *The* ranks *of all nodes in an XML tree come from the* rank domain $\mathbb{D}$*, which is $R(root(T))$ where $root(T)$ represents the root of the document T. Domain $\mathbb{D}$ can always be partitioned into a set $\mathcal{I}$ of disjoint intervals such that, for each interval $I \in \mathcal{I}$, the nodes with ranks in I have the same w-NN. The set $\mathcal{I}$ is referred to as a* TVP *of w*.

For example, in Fig. 2, $R(1) = [1, 12]$, $R(4) = [4, 8]$ and the domain $\mathbb{D}$ is $[0, 40]$. The TVP index of keyword "Gray" for the sample tree in Fig. 2 is given in Fig. 3. For example, nodes whose ranks are within interval [4,8] have the same NN, namely node 6.

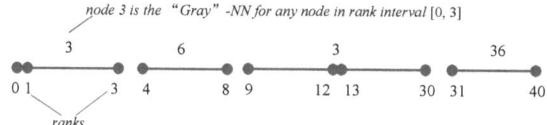

Fig. 3. The TVP of "Gray" for the DBLP sample tree

By storing the TVP index of *w* with a binary tree structure that consumes $O(|\mathcal{I}|)$ space and has query cost $O(\log |\mathcal{I}|)$, the NK search problem can be transformed into finding the interval *I* that covers the *rank* of a given node *q*, and returns the *w*-NN associated with *I*.

TVP index of keyword *w* can be built in $O(N_w \log N_w)$, by constructing the *Compact Tree* (*CT*) and *Extended Compact Tree* (*ECT*) of keyword *w*, denoted as $CT(w)$ and $ECT(w)$ respectively [5]. Fig. 5 shows the examples of the CT and ECT of "Gray" in Fig. 2.

For $CT(w)$, it first includes all the nodes in $U(w)$, which are called the *data* nodes. Then, a non-data node *z* is in $CT(w)$ iff there are at least two child nodes of *z* whose subtrees contain a data node. *z* is called *branching* node. For example, in Fig. 4(a), node 1 is a branching node. The root of the data tree is collected in $CT(w)$, if not present.

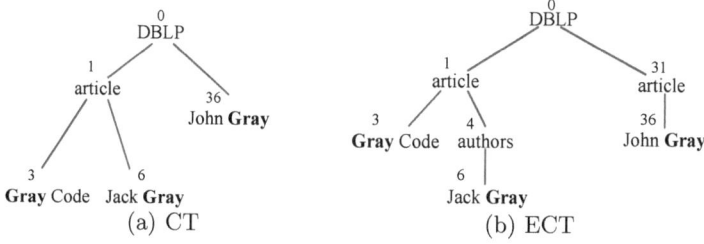

Fig. 4. CT and ECT of "Gray"

After constructing $CT(w)$, $ECT(w)$ is built by adding nodes between edges in $CT(w)$. Given an edge (*u,v*) in $CT(w)$, if *u* and *v* do not have the same NN of *w*, namely $NN(u) \neq NN(v)$, a node must be added in to break the edge (*u*, *v*). Let *z* be the first node on the *u*-to-*v* path in the XML data tree such that

$NN(u) \neq NN(z)$. If $z \neq v$, z is added to $CT(w)$, breaking (u, v) into two edges (u,z) and (z, v). Thus, z is called a *breaking node*. After verifying and updating all the edges of $CT(w)$, the $ECT(w)$ is built. Consider the data tree in Fig. 2 and its CT in Fig. 4(a), edge $(0, 36)$ is broken into two edges $(0, 31)$ and $(31, 36)$ as shown in Fig. 4(b), because $NN(0) \neq NN(36)$ and node 31 is the first node on the 0-to-36 path that does not share the same nearest neighbor with node 0. Similarly, edge $(1, 6)$ is divided by node 4.

When finding the NN of a node in $CT(w)$ during the construction of $ECT(w)$, NK search uses *subtree NK search* [5]. Define the full subtree rooted at q as *subtree*(q) (i.e. the subtree contains q and all nodes and edges under q). *Subtree NK search* is to find the *subtree nearest w-neighbor* of a given node q, which is the keyword matching node within *subtree*(q) having the minimum distance to q. The NN of each node in $ECT(w)$ can be determined after a top-down traversal. To perform subtree NK search, it requires to build an SB-tree [9] structure which costs $O(N_w \log N_w)$, then for each subtree NK query, it costs $O(\log N_w)$. Please refer to [5] for the details of the algorithms for constructing the $CT(w)$ and $ECT(w)$.

Based on $ECT(w)$, the TVP $\mathcal{I}$ of w is generated as shown in Algorithm 1. Once $ECT(w)$ is built, the TVP $\mathcal{I}$ of w can be generated by invoking the Algorithm 2 for every node u in $ECT(w)$. The *idea* is that for a node u in $ECT(w)$, each child v of u means that nodes within the subtree rooted at v may have different NN from $NN(u)$. Consider any node u in $ECT(w)$, and its child nodes $v_1, ..., v_f$ in $ECT(w)$, where $f \geq 0$ ($f = 0$ means node u has no children). For each $1 \leq j \leq f$, Lines 4-6 in Algorithm 2 cut $R(v_j)$ out of $R(u)$, namely removes *subtree*(v_j) from *subtree*(u). Let *subtree*$^{\triangle}(u)$ be the set of nodes in *subtree*(u), but not in *subtree*(v_j) for any j (note that $v_1, ..., v_f$ are the child nodes of u in the $ECT(w)$, not T). All nodes within *subtree*$^{\triangle}(u)$ share the same NN, which is $NN(u)$ [5].

For example, in Fig 4(b), *subtree*$^{\triangle}(4)$ covers nodes 4-5 and nodes 7-8, and *subtree*$^{\triangle}(6)$ covers node 6. That is, $R(4)$ is divided into three intervals, namely interval $[4, 5]$ and $[7, 8]$ associated with $NN(4)$=node 6, and interval $[6, 6]$ associated with $NN(6)$=node 6. Finally, Line 7 in Algorithm 1 merges these three intervals into one interval $[4, 8]$ associated with node 6 as the NN.

Algorithm 1. computeTVP(w)

1: build $CT(w)$
2: build $ECT(w)$ from $CT(w)$
3: $\mathcal{I} = \emptyset$
4: **for** each node u in $ECT(w)$ **do**
5:    voronoiIntv(u) // Algorithm 2
6: **end for**
7: merge consecutive intervals in $\mathcal{I}$ that are associated with the same w-NN
8: return $\mathcal{I}$

It has been proven that for any keyword w appearing in an XML tree T, there is a TVP $\mathcal{I}$ with size less then $8N_w$ [5].

Algorithm 2. voronoiIntv(u)

1: $S = \{R(u)\}$
2: **for** each child node v of u in $ECT(w)$ **do**
3:     $I \leftarrow$ the (only) interval in S covering $R(v)$
4:     break I into intervals I_{left}, $R(v)$, I_{right}
5:     remove I from S
6:     add I_{left} and I_{right} to S
7: **end for**
8: associate the intervals in S with NN(u)
9: add the intervals in S to $\mathcal{I}$

---

3   Extended Compact TVP

In this section, we present the ecTVP index that supports top-k NK search that returns *top-k nearest neighbors* (kNNs) of a given node q, denoted as $kNNs(q)$. If multiple keyword matching nodes have the same distance to q, we rank them by their *rank*s. If a node is the top-i nearest neighbor of q, we denote it as i^{th}NN(q). Given two nodes u and v, $kNNs(u)=kNNs(v)$ iff $\forall i \in [1,k]$ i^{th}NN(u)=i^{th}NN(v).

3.1   Extended TVP

As we can see in Section 2, the TVP index of w can be built by constructing the $ECT(w)$. For each node u in $ECT(w)$, all nodes in $subtree^{\triangle}(u)$ share the same NN as u.

Thus, we propose a *variant* of $ECT(w)$, denoted as $vECT(w)$, such that for each node u in $vECT(w)$, all nodes in $subtree^{\triangle}(u)$ share the same kNNs as u. In order to find the kNNs for each node u in $vECT(w)$, we design a simple but efficient method, which involves a bottom-up process and a top-down process during the construction of the $CT(w)$ and $vECT(w)$, respectively.

vECT. The difference between $vECT$ and ECT is: for an edge (u, v) in $CT(w)$, if u and v do not have the same kNNs, namely $kNNs(u) \neq kNNs(v)$, a *breaking node* z must be added in to break the edge (u, v), where z is the first node on the u-to-v path in the XML data tree such that $kNNs(u) \neq kNNs(z)$.

Proposition 1. *For any node u in* vECT*(w), subtree*$^{\triangle}(u)$ *share the same kNNs as node u.*

The proof of Proposition 1 is similar to that of Lemma 5 in [5]; thus, it is omitted here. Based on Proposition 1, we can build the *extended* TVP index that supports top-k NK search based on $vECT(w)$.

Construction of vECT. The construction of $vECT(w)$ consists of two steps:

1. building $CT(w)$, which is a *bottom-up* process shown in Algorithm 3;
2. building $vECT(w)$, which is a *top-down* process shown in Algorithm 4.

The workflow of building $CT(w)$ and $vECT(w)$ in our work is similar to that in NK search [5]. However, the challenge lies in how to find the top-k nearest neighbors of each node u in $vECT(w)$. Instead of the strategy of *subtree* NK query, we find the kNNs for a node in a *bottom-up* process and a *top-down* process while building $CT(w)$ and $vECT(w)$. For each node u, we maintain an ordered *list* of *nearest neighbors* (NNL) for it, denoted as $u.NNL$, which lists the *nearest neighbors* in ascending order of their distances to node u. $u.NNL$ supports two main functions. The first one is $add(v)$, which means adding a new node v to the right position in $u.NNL$ according to $dist(u, v)$. We denote the size of NNL as $|NNL|$, while adding v, we keep $|NNL| \leq k$, which means once $add(v)$ is performed, if $|NNL| = k$, the $(k+1)^{th} NN$ will be deleted from the list. In addition, $add(v)$ is only performed when needed, namely if v already exists in $u.NNL$ or $dist(u, v)$ can not make v to the top-k NNs of u, v will not be added to $u.NNL$. The second one is $merge(V)$ where V is a set of nodes; for each node $v \in V$, $add(v)$ will be invoked.

Algorithm 3. constructCT(w)

1: $S \leftarrow U(w)$ //S is a list of nodes
2: $r \leftarrow$ the root of the XML data tree
3: **for** each node $u \in S$ **do**
4:     $u.NNL.add(u)$ //initiate each data node's NNL to contain itself
5:     $r.NNL.add(u)$ //compute the kNNs of r
6: **end for**
7: sort the nodes in S in ascending order of ranks
8: **for** each pair of consecutive nodes u, v **do**
9:     $z \leftarrow lca(u, v)$
10:     $z.NNL.merge(u.NNL)$; $z.NNL.merge(v.NNL)$
11:     $S = S.append(z)$ //append node z to the end of S if z is not in S already
12: **end for**
13: **if** r is not in S **then**
14:     put r in S
15: **end if**
16: $computeCTedges(w)$;

Algorithm 4. constructVECT($CT(w)$)

1: sort the edges (u,v) of $CT(w)$ in ascending order of $level(u)$ //v is the child of u in $CT(w)$
2: $E \leftarrow$ the sorted list of edges
3: Suppose r is the root of $CT(w)$, $kNNs(r) \leftarrow r.NNL$
4: **while** $E \neq \emptyset$ **do**
5:     $(u, v) \leftarrow$ the first edge of E; remove it from E
6:     $v.NNL.merge(u.NNL)$
7:     $kNNs(v) \leftarrow v.NNL$
8:     **if** $kNNs(u) = kNNs(v)$ **then**
9:         create edge (u, v) in $vECT(w)$
10:     **else**
11:         $z \leftarrow$ the first node on the u-to-v path such that $kNNs(u) \neq kNNs(z)$, where $kNNs(z) = z.NNL$ after $z.NNL.merge(u.NNL)$ and $z.NNL.merge(v.NNL)$
12:         create edges (u, z) in $vECT(w)$, and add (z, v) (if $z \neq v$) to be the first edge of E
13:     **end if**
14: **end while**

When collecting CT nodes in Algorithm 3 (Lines 1-15), each data node's NNL is set to contain the node itself initially, and compute the kNNs of the root of the XML data (Lines 3-6). Then when a branching node $z = lca(u, v)$ is added, the nodes in both $u.NNL$ and $v.NNL$ are added into $z.NNL$ (Line 10). Note

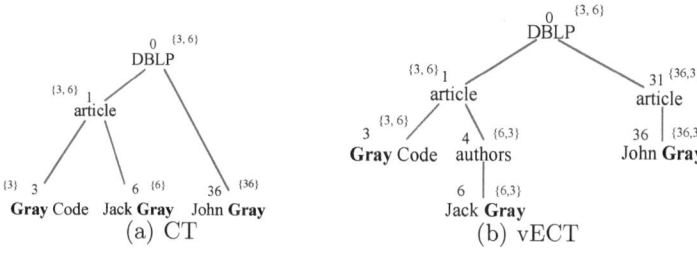

Fig. 5. CT and vECT of "Gray"

that the root of the XML should be put into the S if it is not (Line 13-15). Function $computeCTedges(w)$ (Line 16) is to add edges between the CT nodes, the details of this function can be found in [5].

For example, consider the keyword "Gray" in Fig. 2 and $k = 2$, first we collect the data nodes (3, 6 and 36), and each of them has an NNL that contains the node itself, as shown in the CT in Fig. 5(a). Before *branching* node 1 (the LCA of nodes 3 and 6) is added to CT, the NNL of node 1 is merged from the NNLs of node 3 and node 6, which is $\{3, 6\}$.

After the *bottom-up* process, the CT tree has the following properties:

Proposition 2. *For each node u in* CT(w), $u.NNL$ *contains the* top-k' *subtree nearest neighbors of u, where $k' \leq k$. Let r be the root of the CT tree (also the root of the XML data), $r.NNL$ contains the* top-k' *nearest neighbors of r.*

For the root of the CT tree r, if the keyword occurrence is smaller than k, namely $N_w < k$, then $k' < k$. In our paper, we assume $N_w \geq k$ for the simplicity of presentation. Also, it is easy to prove the following property:

Proposition 3. *Given an edge (u, v) in a CT(w), where v is the child of u, if $u.NNL$ contains the kNNs of u and $v.NNL$ contains the* top-k' *subtree nearest neighbors of v, where $k' \leq k$, then after performing $v.NNL.merge(u.NNL)$, $v.NNL$ contains the kNNs of v.*

Based on the above properties, we can build the $vECT(w)$ in a *top-down* process from the root of the $CT(w)$ whose NNL contains the kNNs, as shown in Algorithm 4. For every edge (u, v), first we merge the nodes in $u.NNL$ to $v.NNL$ (Line 6), so that $v.NNL$ contains the kNNs of v. Then, if $kNNs(u) = kNNs(v)$, we create edge (u, v) in $vECT(w)$ (Lines 8-9); *else*, let z be the first node on the u-to-v path such that $kNNs(u) \neq kNNs(z)$ (Line 11), we create edges (u, z) in $vECT(w)$, and add (z, v) to be the first edge of E (Line 12).

For example, consider the CT in Fig. 5(a) of the data tree in Fig. 2. When dealing with the edge $(1, 6)$, where the kNNs of node 1 is $\{3, 6\}$ stored in the NNL of node 1, first we generate the kNNs of node 6 by merging the NNL of node 1 (Line 6). Then we compare the $kNNs(6)$ with $kNNs(1)$. Because $kNNs(6) \neq kNNs(1)$, we find the first node on the 1-to-6 path, node 4, such that $kNNs(1) \neq kNNs(4)$ (Note that $kNNs(4) = \{6, 3\}$ which is not equal to

$kNNs(1) = \{3, 6\}$), and create the edge $(1, 4)$ in the $vECT$. After verifying every edge, the final $vECT$ of keyword "Gray" for Fig. 2 is given in Fig. 5(b).

3.2 ecTVP

In this section, we present the algorithm to make the TVP index more compact, that is, containing less redundant intervals.

Removable Intervals. Let r be the root of the document T, $r_1, ..., r_n$ be the children of r. If $subtree(r_i)$ does not contain any keyword matching nodes, then nodes within $subtree(r_i)$ share the same NN/kNNs as r. That means in the TVP/ecTVP index, there is an interval that covers $R(r_i)$ and is associated with $NN(r)/kNNs(r)$. For a root node r of a document, such as DBLP, there can be a lot of intervals associated with $NN(r)/kNNs(r)$. If we remove those intervals, when no interval can be found for a given node q, we can assume q's NN/kNNs is $NN(r)/kNNs(r)$. We call those intervals that cover $R(r_i)$ and are associated with $NN(r)/kNNs(r)$ *removable intervals* of r.

More generally, the concept of removable intervals could be applied to a set of nodes S from data tree T, where $\forall r_i, r_j \in S$, $R(r_i)$ is disjoint with $R(r_j)$.

Definition 4 (Extended Compact TVP (ecTVP)).
1. *Given an* extended *TVP index* $\mathcal{I}_s$, *for each* $I \in \mathcal{I}_s$, *nodes within* I *share the* $kNNs$. *If we remove the removable intervals of* r_i *from* $\mathcal{I}_s$, $\forall r_i \in S$, *we call* $\mathcal{I}_s$ *the search intervals.*
2. *For each* $r_i \in S$, $R(r_i)$ *can be expressed as an interval* I_i' *associated with* $kNNs(r_i)$. *Let* $\mathcal{I}_b = \bigcup_{r_i \in S} I_i'$, *we call* $\mathcal{I}_b$ *the backup intervals.*
3. *Let* $\mathcal{I} = \mathcal{I}_s \cup \mathcal{I}_b$, *we call* $\mathcal{I}$ *an ecTVP.*

Given a node q, when there is no interval from $\mathcal{I}_s$ that covers $rank(q)$, if $rank(q) \in R(r_i)$, namely $I_i' \in \mathcal{I}_b$ covers $rank(q)$, then $kNNs(q) = kNNs(r_i)$.

Building ecTVP. In our algorithm, we only consider the situation where intervals that share the kNNs with the root of the document T are removed. To build the ecTVP, we can build the extended TVP first, and then remove the intervals that share the kNNs with the root of T. However, based on the processes of building $CT(w)$ and $vECT(w)$, it is obvious that if $subtree(r_i)$ does not contain any keyword matching node, r_i will exist in neither $CT(w)$ nor $vECT(w)$. Thus, we could build the ecTVP in a simple and efficient way, as shown in Algorithm 5. After building $CT(w)$ and $vECT(w)$ (Lines 1-2), for each node u in $vECT(w)$ and $u \neq root(T)$, we invoke $voronoiIntv(u)$ (Lines 4-9). $voronoiIntv(u)$ is given in Algorithm 2. Note that here in $voronoiIntv(u)$, we associate each interval with kNNs, instead of NN. After merging the consecutive intervals in $\mathcal{I}$ (Line 10), $kNNs(r)$ and $R(r)$ are recorded in $\mathcal{I}_b$ (Line 11). Finally, $\mathcal{I}$ that combined with $\mathcal{I}_b$ is returned as the ecTVP of w.

Consider $k = 2$ for the data tree in Fig. 2, the ecTVP generated by Algorithm 5 is shown in Fig. 6. Suppose the given node is node 13, no interval that covers $rank(13)$ exists in $\mathcal{I}_s$; then in $\mathcal{I}_b$, an interval is found covering $rank(13)$, associated with the kNNs of the root of the document.

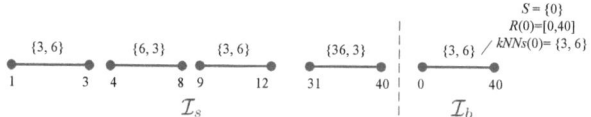

Fig. 6. ecTVP of "Gray" in Fig. 2, $k = 2$

Algorithm 5. computeECTVP(w)

1: build $CT(w)$
2: build $vECT(w)$ from $CT(w)$
3: $\mathcal{I} = \emptyset$ // here, $\mathcal{I}$ is the *search intervals*
4: $r \leftarrow$ the root of the document T
5: **for** each node u in $ECT(w)$ **do**
6:   **if** $u \neq r$ **then**
7:     voronoiIntv(u) //associate each interval with kNNs, instead of NN
8:   **end if**
9: **end for**
10: merge consecutive intervals in $\mathcal{I}$ that are associated with the same kNNs
11: record $kNNs(r)$ and $R(r)$ in $\mathcal{I}_b$
12: **return** $\mathcal{I} = \mathcal{I} \cup \mathcal{I}_b$ //return $\mathcal{I}$ as the ecTVP

3.3 Complexity Analysis

When building $CT(w)$, it costs $O(N_w \log N_w)$ to sort the nodes, and costs $O(k \cdot N_w)$ in the worst case to collect the CT nodes. To build the CT edges, it costs $O(N_w \log N_w + N_w)$. When building $vECT(w)$, for each edge (u, v) in $CT(w)$, it costs $O(k)$ to merge $u.NNL$ to $v.NNL$, $O(k)$ to compare $u.NNL$ with $v.NNL$, and $O(k)$ to merge $u.NNL$ and $v.NNL$ to $z.NNL$. Thus, it costs $O(k \cdot N_w)$ in the worst case (while in [5] it costs $O(N_w \log N_w)$ to build the $ECT(w)$ using subtree NK search). When building ecTVP, it costs $O(N_w)$ to process each edge in $vECT(w)$, and to merge consecutive intervals costs $O(k \cdot N_w)$ in the worst case. To sum up, the overall complexity to construct the ecTVP index is $O(N_w(\log N_w + k))$.

4 Experiments

In this section, we present the experiments, and analyze the compactness and construction time of ecTVP.

4.1 Experimental Setup

All our experiments were carried out on a Laptop with Intel(R) Core(TM) i5-2520 2.50GHz CPU and 4GB RAM. The operating system was Windows 7, and we used the Berkeley DB Java Edition. Algorithms were implemented using Eclipse with JDK 1.6. We used two real word datasets in the experiments: (1) **DBLP** is a snapshot of the DBLP database. Its size is 400MB, max depth is 6, and average depth is 5.58; (2) **SigmodRecord** is an XML file contains SIGMOD Records in February 2007. Its original size is 467KB, we duplicate it to 100MB, max depth is 6, and average depth is 5.14107.

For each dataset, we ran the test on different keywords, and we divided the keywords into two groups according to their frequencies, as shown in Table 1. For example, keyword **w1** is "children", and its frequency in the DBLP dataset, namely N_{w1}, is 500.

Table 1. Sample keywords in the experiments

Dataset	Low Frequency	High Frequency
	w1: children 500	**w6**: design 30259
	w2: central 500	**w7**: data 38491
DBLP	**w3**: electric 507	**w8**: analysis 43070
	w4: children 520	**w9**: computer 48863
	w5: central 522	**w10**: information 52540
	w11: similar 218	**w16**: model 10682
	w12: dimension 218	**w17**: distributed 11990
Sigmod	**w13**: deferred 436	**w18**: research 15042
	w14: type 436	**w19**: relational 17658
	w15: alternative 436	**w20**: query 20462

4.2   Algorithms and Measures

We compared our algorithm with the algorithm that generate TVP index in [5]. To compare the two algorithms, we used the following two measurements:

Size is defined as the number of intervals in the TVP or ecTVP index of a keyword w.

Construction time is the time the algorithm takes to construct the TVP or ecTVP index of a keyword w.

4.3   Analysis

Size. First, we compared the size of TVP and ecTVP when $k = 1$, which means ecTVP only supports top-1 NN search. We also studied how the size of ecTVP changes according to k. Here, we use $|ecTVP(w)|$ and $|TVP(w)|$ to represent the size of $ecTVP(w)$ and $TVP(w)$, respectively.

The size of the TVP and ecTVP index of w1, w2,..., w10 on DBLP dataset is given in Fig. 7. As we can see, $|ecTVP(w)|$ is much smaller than $|TVP(w)|$. Generally, $|TVP(w)|$ and $|ecTVP(w)|$ become larger with the growth of N_w. However, for $|TVP(w)|$, N_w is not the only main factor that influences the size of $TVP(w)$. The distribution of the keyword matching nodes would also have a major impact on the size of $|TVP(w)|$. For example, for w9 and w10, even though N_{w9} and N_{w10} are larger than N_{w8}, because of the distribution of keyword matching nodes, $|TVP(w8)| > |TVP(w9)|$ and $|TVP(w8)| > |TVP(w10)|$. However, there are sufficiently more removable intervals in $TVP(w8)$ than in $TVP(w9)$ and $TVP(w10)$, which causes $|ecTVP(w8)| < |ecTVP(w9)|$ and $|ecTVP(w8)| < |ecTVP(w10)|$.

The size of the TVP and ecTVP index of w1, w2,..., w10 on SigmodRecord dataset is given in Fig. 7 as well. As we can see in Fig. 7(a), for keywords with low frequencies, the compactness of $ecTVP(w)$ on SigmodRecord is similar to that on DBLP. However, for keywords with high frequencies, shown in Fig. 7(b),

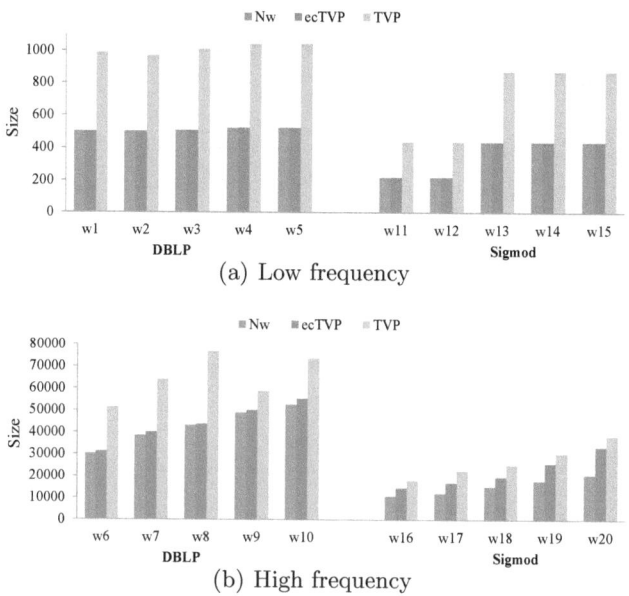

Fig. 7. Comparison of the size between TVP and ecTVP, $k = 1$

the $ecTVP(w)$ on SigmodRecord is less compact compared to that on DBLP. That is because the structure of SigmodRecord is more complicated and the root of Sigmod document is directly connected to less subtrees that contain no keyword matching nodes, which results in less *removable intervals* of the root.

Given a node u and a keyword w, finding its NN/kNNs costs $\log|TVP(w)|$ / $\log|ecTVP(w)|$. As a result, our $ecTVP$ makes the query processing more efficient. We did the experiments on the keywords with high frequencies to compare the average query time, as shown in Fig. 8.

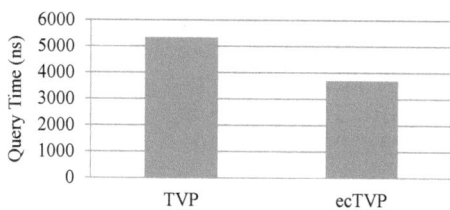

Fig. 8. Average query time for TVP and ecTVP

We tested our algorithm with different values of k. It turns out the size of ecTVP index for keyword w1,...,w20 does not change with k. This situation applies to most keywords in the datasets. However, for some keywords in Sigmod dataset, the size of their ecTVP increases with k. For example, consider keyword "11" and keyword "22", where $N_{11} = 8587$ and $N_{22} = 5450$, the results are

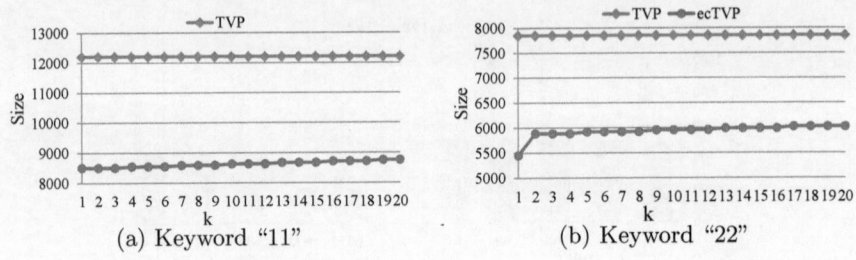

(a) Keyword "11"   (b) Keyword "22"

Fig. 9. Size of ecTVP with different k

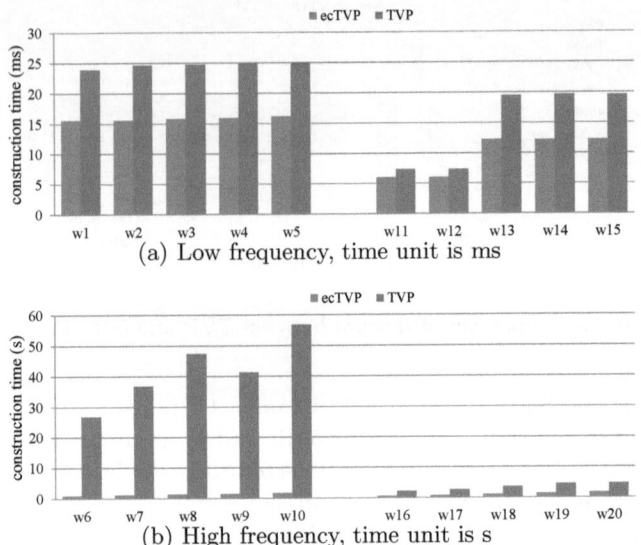

(a) Low frequency, time unit is ms

(b) High frequency, time unit is s

Fig. 10. Comparison of the construction time for TVP and ecTVP, $k = 1$

shown in Fig. 9(a) and Fig. 9(b), respectively. As we can see, the sizes of the ecTVP index for these two keywords generally grow with k slowly. However, compared to TVP, the ecTVPs for these two keywords are still more compact. Even when $k = 100$, $|ecTVP(11)| = 9889$ and $|ecTVP(22)| = 6703$, while $|TVP(11)| = 12208$ and $|TVP(22)| = 7849$.

Construction Time. First, we compared the construction time between TVP and ecTVP index when $k = 1$. Then we tested how the time grows with k.

In Fig. 10, we can see that ecTVP costs less time to build, compared to TVP. Because our algorithm costs less time while constructing vECT and building ecTVP from vECT. For DBLP dataset, the advantage of ecTVP gets more obvious when the frequency of the keyword is high. Because for TVP, it contains a large number of intervals, especially *removable intervals*, which costs more time to generate and merge. For example, for keyword w10 whose frequency is 52540,

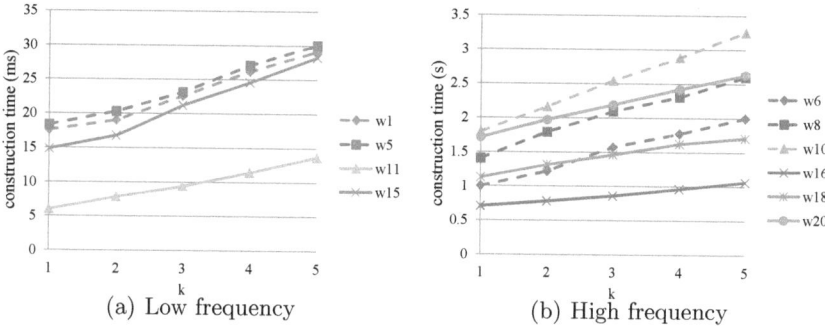

Fig. 11. Construction Time for different k

[5] generates 172787 intervals before merging the consecutive intervals to return the final 73507 intervals in TVP. While for our algorithm, we only generate 55295 intervals before merging them into the final 55294 intervals.

We run experiments on constructing ecTVP with different k to see how construction time varies according to k. To demonstrate the results, we use several keywords from Table 1, shown in Fig. 11. We can see that if N_w is fixed, the construction time is linear to k.

5   Conclusion

In this paper, we proposed the ecTVP index which is used to find the top-k nearest w-neighbors of a given node. In order to build the ecTVP, we proposed an efficient approach to compute the top-k nearest w-neighbors of nodes in vECT in a bottom-up process and a top-down process, such that vECT can be built efficiently. Besides, compared to TVP, our ecTVP is faster to build and more compact. The experiments conducted demonstrated the advantages of ecTVP.

Acknowledgement. This work is supported by the Australian Research Council Discovery Grant DP1093404.

References

1. Bao, Z., Ling, T.W., Chen, B., Lu, J.: Effective XML Keyword Search with Relevance Oriented Ranking. In: ICDE, pp. 517–528 (2009)
2. Guo, L., Shao, F., Botev, C., Shanmugasundaram, J.: XRANK: Ranked Keyword Search over XML Documents. In: SIGMOD Conference, pp. 16–27 (2003)
3. Hristidis, V., Koudas, N., Papakonstantinou, Y., Srivastava, D.: Keyword Proximity Search in XML Trees. IEEE Trans. Knowl. Data Eng. 18(4), 525–539 (2006)
4. Kasneci, G., Ramanath, M., Sozio, M., Suchanek, F.M., Weikum, G.: STAR: Steiner-Tree Approximation in Relationship Graphs. In: ICDE, pp. 868–879 (2009)
5. Tao, Y., Papadopoulos, S., Sheng, C., Stefanidis, K.: Nearest keyword search in XML documents. In: SIGMOD Conference, pp. 589–600 (2011)

6. Termehchy, A., Winslett, M.: Using Structural Information in XML Keyword Search Effectively. ACM Trans. Database Syst. 36(1), 4 (2011)
7. Xu, Y., Papakonstantinou, Y.: Efficient Keyword Search for Smallest LCAs in XML Databases. In: SIGMOD Conference, pp. 537–538 (2005)
8. Xu, Y., Papakonstantinou, Y.: Efficient LCA based keyword search in XML data. In: EDBT, pp. 535–546 (2008)
9. Yang, J., Widom, J.: Incremental computation and maintenance of temporal aggregates. VLDB J. 12(3), 262–283 (2003)
10. Zhou, R., Liu, C., Li, J.: Fast ELCA computation for keyword queries on XML data. In: EDBT, pp. 549–560 (2010)

Architecture-Driven Modeling of Adaptive Collaboration Structures in Large-Scale Social Web Applications

Christoph Dorn and Richard N. Taylor

Institute for Software Research, University of California, Irvine, CA 92697-3455
[cdorn|taylor]@uci.edu

Abstract. Internet-based, large-scale systems provide the technical foundation for massive online collaboration forms such as social networks, crowdsourcing, content sharing, or source code generation. Such systems are typically designed to adapt at the software level to achieve availability and scalability. They, however, remain mostly unaware of the changing requirements of the various ongoing collaborations. As a consequence, cooperative efforts cannot grow and evolve as easily nor efficiently as they need to. An adaptation mechanism needs to become aware of a collaboration's structure and flexibility to consider changing collaboration requirements during system reconfiguration. To this end, this paper presents the human Architecture Description Language (hADL) for describing the envisioned collaboration dynamics. Inspired by software architecture concepts, hADL introduces human components and collaboration connectors for describing the underlying human coordination dependencies. We further outline a methodology for designing collaboration patterns based on a set of fundamental principles that facilitate runtime adaptation. An exemplary model transformation demonstrates hADL's feasibility. It produces the group permission configuration for MediaWiki in reaction to changing collaboration conditions.

Keywords: Design Tools and Techniques, Collaboration Patterns, Adaptation Flexibility.

1   Introduction

The last two decades have witnessed the emergence of numerous web-based, large-scale collaboration tools. Web sites appeared for diverse purposes such as social networking (e.g., Facebook, LinkedIn), collaborative tagging (e.g., Digg), content sharing (e.g., YouTube, Flickr), knowledge creation (e.g., Wikipedia), crowdsourcing (e.g., Amazon Mechanical Turk), or source code production (e.g., GitHub, SourceForge).

Users of such social Web applications typically face one major problem: a rigid, limited set of available collaboration mechanisms in a one-size-fits-all manner. Interaction means such as direct messaging, group chats, discussion boards, task assignments, or shared artifacts remain independent of the collaboration's scale and

X.S. Wang et al. (Eds.): WISE 2012, LNCS 7651, pp. 143–156, 2012.

complexity and thus form a constraint on how large a joint effort can grow, how easily and how efficiently it may evolve. Amazon MTurk, for example, scales the Master/Worker pattern to thousands of users and tasks. The MTurk platform implements a rigid interaction pattern where communication amongst participants is not foreseen. Hence, pattern adaptation for supporting more complex collaborations that require coordination between individual workers is impossible.

We claim that an explicit model of collaboration structures is of uttermost importance for describing a collaboration system's flexibility and subsequently supporting the evolution of collaborative efforts through pattern adaptation.

We take inspiration from software architectures to address this problem for large-scale collaboration systems. A system's software architecture as described in terms of components and connectors has a profound effect on its adaptability, especially scalability [19]. The same holds true for human collaboration (see Sec. 4). Connectors in the form of humans (e.g., forum moderators, secretaries) and software services (e.g., mailing lists, task lists) manage dependencies between collaborators (i.e., human components) when direct interaction amongst all participants is no longer viable. The explicit modeling of humans as components and connectors—a distinction which existing approaches have insufficiently addressed so far (see Sec.3)—draws the focus to the collaboration structure's flexibility and thus facilitates adaptation.

Our contribution in this paper is three-fold. We (i) introduce the human Architecture Description Language (hADL) in Sec. 5, (ii) provide a methodology for defining adaptable collaboration patterns in Sec. 6, and (iii) demonstrate the model's feasibility based on an exemplary model-to-configuration transformation in Sec. 7. We find that not only are components and connectors a very suitable abstraction mechanism for describing collaboration patterns and their adaptation flexibility. As integral part of a human architecture, they also successfully support pattern evolution.

2   Motivating Scenario

Suppose a research project integrates knowledge from the wider research community in the form of Wiki-style articles. After the infrastructure for collecting and managing user contribution goes online, participation remains low but stable. One regular project staff member quality checks changes to existing articles and browses through the content list to check new article entries.

Soon, a report in the media about the research project sparks wide-spread interest with subsequent participation levels soaring. This has significant implications on the quality assurance procedure which has to deal with vandalized or spammed articles. Conflicting opinions amongst contributors of the same article lead to editing wars. A single quality manager is no longer up to the task. Simple replication of her role is one option, but changing the collaboration pattern is potentially more effective. Multiple options exist to handle articles exhibiting high revision rates: (i) updates are checked by an expert — possibly crowdsourced — to decide upon article rollbacks, (ii) contributors vote on changes, or (iii) experts discuss and negotiate changes. Alternatively, articles subject to update wars are temporarily protected or receive a limited write quota. New articles still need no approval to keep participation barriers low but observers now receive notifications about new entries. Depending on the rate of new articles, such monitoring itself may require topic-based subscriptions to ensure that observers receive only notifications relevant to their interests. Planning and subsequently implementing such restructuring requires an explicit model of the underlying collaboration structure and its adaptation flexibility.

3   Related Work

Research efforts that specifically focus on social or collaborative aspects in large-scale systems are still rare. Existing research addresses mainly the general idiosyncrasies of Web 2.0 but remains unaware of specific interaction structures at runtime [21]. Model-driven Web engineering approaches so far focus primarily on software aspects [16] and don't go beyond (user) context-centric adaptations [1]. Requirements elicitation and specification approaches consider collaboration (e.g., CSRML [20]) or adaptation (e.g., [17]) but omit the effects of patterns on adaptation flexibility.

Activity-centric frameworks (e.g., [6,12]) define tasks and their relations for integrating humans and software components [2]. Human-centric workflow systems define business artifacts, their transformations, and interdependencies [8]. The Business Entity Definition Language [13], for example, aggregates access rights, data structure, object state transitions, and events. The human collaboration structure, however, remains implicit. Similar, the Business Process Modeling Language (BPMN [22]) describes human tasks and their dependencies. Recent research efforts on large-scale workflow deployment such as Human-provided Services (HpS) [15], Turkit [9], or CrowdLang [11] differ in their degree of formalizing complex workflows that go beyond simple task assignment in Amazon Mechanical Turk.

Most of these models, tools, and frameworks contain (model) elements similar to hADL (e.g., when specifying human roles and their associated capabilities) but differ in two crucial aspects. First, all these approaches lack an explicit distinction between human components and collaboration connectors. Consequenlty any adaptation knowhow tailored either to coordination or to work execution remains implicit and hidden within each platform. Second, most collaboration platforms focus on a particular collaboration pattern and the associated limited set of adaptation capabilities. Process-centric models, for example, focus only on task execution, no matter whether ad-hoc or rigidly specified. They cannot be applied for describing other patterns such as co-authoring Wiki articles or spreading news on twitter and vice versa.

Extensible software architecture description languages (ADLs e.g., [7,3]) emerged from the need to rigorously define the language's semantics while remaining flexible enough to address the specific needs of a particular domain. Augmenting an existing ADL to describe all details of the human collaboration patterns, however, would be cumbersome as software structure and human interactions reside on different conceptual levels. Nevertheless, ADLs provide fundamental principles that inspire and guide our human architecture description language hADL.

4   The Case for a Human Architecture

The observation that software systems and human collaborations share the same challenges in managing dependencies inspired our concept of a human architecture. Both domains require coordination of (i) shared resources, (ii) producer/consumer relationships, (iii) simultaneity constraints, and (iv) task/subtask relations [10]. An architecture describes how a system addresses these challenges. In the domain of software engineering, following definition of a software architecture fits equally well to collaborative efforts: "A software system's *architecture* is the set of principal design decisions made about the system." [18], p.58.

Components and connectors are the primary building blocks of a software architecture. Components are the loci of computation and data management whereas connectors facilitate and control the interactions between components. Roles such as managers, team leaders, secretaries are rarely described as connectors but they perform a similar task: the coordination of other human (i.e., human components). Highlighting further similarities: in software architecture, architectural styles consist of a set of development context dependent design decisions, constraints, and resulting properties. Collaboration patterns correspondingly describe what combination of human components and coordinators are suitable for a particular joint effort [5,4].

In software architectures, connectors are the key element to system adaptability. For example, connectors allow the dynamic replacement of behavior components in robotic systems without affecting other components. Web proxies are connectors on the Internet that decide which server (component) should process a particular client (component) request. Overloaded or unavailable servers thus become transparent to the client. In the scenario, the article contributors and readers constitute the human components. (Human) quality managers and (software) change monitors implement connector functionality for managing the read and write dependencies amongst the human components. The importance of collaboration connectors grows with the scale and complexity of joint efforts especially in distributed settings where individual collaborators have little opportunity for informal communication.

5   The Human Architecture Description Language

The core human Architecture Description Language (hADL) defines collaborators, their means of interaction through messages, streams, and shared artifacts, and dependencies amongst collaboration objects (Fig. 1). We explain the individual elements based on a hADL model instance (Fig. 2) for the motivating scenario.

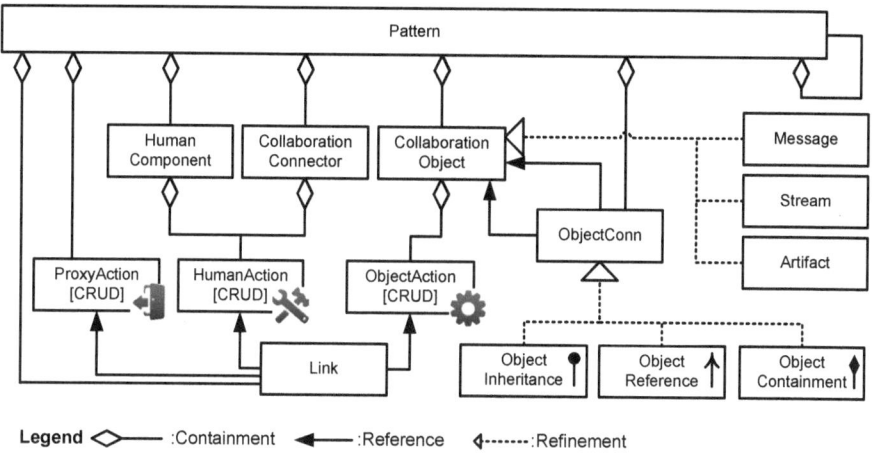

Fig. 1. hADL model (symbols in ObjectConn subtypes and Actions represent the respective visualization in model instances.)

A human architecture describes the configuration of *HumanComponents* and *CollaborationConnectors* to fulfill a particular purpose, for example: carrying out a task, creating a shared artifact, or negotiating a leader. The architecture's purpose determines a suitable collaboration *Pattern*. Typical patterns include Master/Worker, Publish/Subscribe, Shared Artifact, and Peer-to-Peer (e.g., [4]). A HumanComponent has a particular collaboration role that is essential to the completion of the collaborative effort (e.g., Contributor, Reader, Observer in Fig. 2 left and right). A CollaborationConnector provides coordination capabilities to HumanComponents within the pattern's scope (e.g., QualityManager, VandalismDetector, ArticleMonitor in Fig. 2 center). A CollaborationConnector covers the full spectrum from purely human, to software-assisted, to purely software implemented. In the scenario, a quality manager manually approving all edits illustrates a human collaboration connector. In contrast an article monitor notifying users via email about updates exemplifies a software-based collaboration connector.

HumanComponents and CollaborationConnectors are the active collaboration elements in hADL, but they don't specify the means of collaboration. When physically distributed, humans usually communicate through *Messages, Streams,* or shared *Artifacts*. The hADL model considers these three types as *CollaborationObject* variants. A Message is a onetime, immutable object exchanged between a set of collaborators (components and connectors), a typical example is an email. A Stream is a series of messages where sender and receiver maintain a temporary relationship. Two broad types exist: (a) subscriptions characterize a set of independent messages (such as news items in RSS feeds or updates on a user's facebook wall). Alternatively, (b) multimedia streams consist of dependent messages (i.e., frames) that constantly refresh the receiving end (e.g., video chat). A (shared) Artifact is a long-living object that is subject to (si-

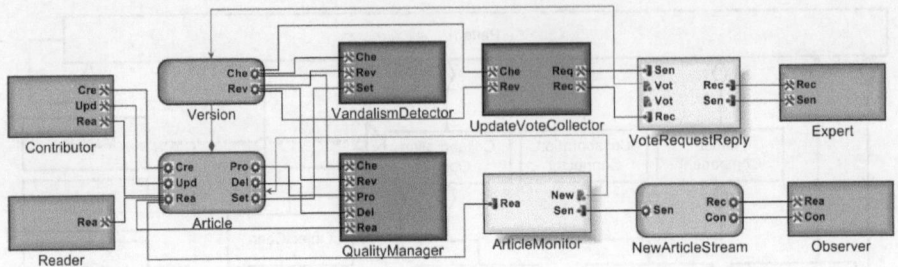

Fig. 2. Scenario hADL model instance: components as light-green shaded boxes, connectors as dark-green shaded boxes, collaboration objects with rounded corners, and substructure patterns with shadow (colors online). Icons represent human, respectively object actions.

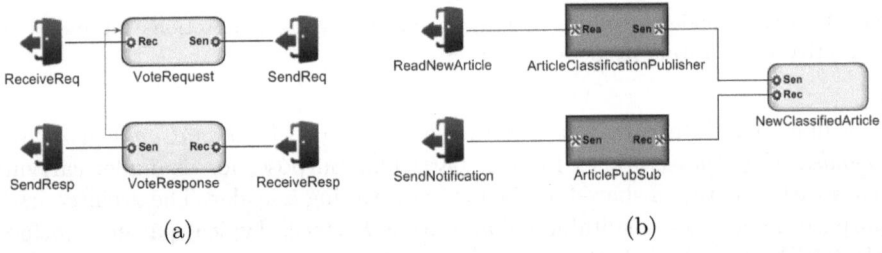

(a)   (b)

Fig. 3. hADL models for (a) Vote Request Reply substructure and (b) Topic-based Article Monitoring substructure.

multaneous) manipulation by multiple collaborators. In the scenario, respective examples are (i) emails sent to Experts to vote on article updates (Fig. 3a), (ii) notifications about new articles (Fig. 3b), and (iii) the articles themselves (Fig. 2). *ObjectConns* describe dependencies amongst CollaborationObjects such as refinement (*ObjectInheritance*), relation (*ObjectReference*), and substructure (*ObjectContainment*). Note that ObjectConns merely highlight such dependencies to improve pattern comprehension but they don't replace data modeling.

The choice of communication means has a profound impact on the collaboration and thus needs to be made explicit. Hence, hADL requires a CollaborationObject between any two or more HumanComponents and/or CollaborationConnectors. This is in contrast to traditional ADLs (e.g., xADL [3] or ACME [7]) where component interfaces link directly to connector interfaces. A rough software architecture interface equivalent in hADL is the *Action*. HumanComponents and CollaborationConnectors exhibit *HumanActions* that specify what access rights a collaborator requires to fulfill its role, whereas a CollaborationObject has *ObjectActions* for defining what rights it grants to particular collaborator. An Action distinguishes between *Create, Read, Update*, and *Delete* (CRUD) privileges. The article Contributor in Fig. 2, for example, exhibits an *Edit* action with Create, Update, and Read rights. Ultimately, CRUD rights need to match when a *Link* connects a HumanAction with an ObjectAction. Multiple Collaborators may connect to the same ObjectAction when they share the same manipulation rights (e.g., several CollaborationConnectors in Fig. 2 connect to the same Article *Read* action).

In some cases, we wish to introduce substructures to hide low-level collaboration details that are irrelevant at the higher-level collaboration scope. In the scenario, a CollaborationConnector monitors new Articles. Whether this connector merely sends an email to all interested Observers or whether observers subscribe to certain article topics is described at a lower level. In the latter case, the substructure defines the appropriate subscription mechanism (Fig. 3b). Pattern substructures are equally well suited to hide complex CollaborationObjects (e.g., tightly coupled request and response messages for voting on article changes, Fig. 3a). In hADL, such substructures are implemented as recursive embedding of Patterns with the use of *ProxyActions*.

6 Designing for Adaptation

Research in software architectures supplies several concepts and tools for designing and analyzing collaboration structures. In our previous work ([5,4]), we applied the BASE framework [19] for studying the adaptation flexibility of various collaboration patterns. Based upon the insights gained in our recent analysis and our experience in architecture-based software adaptation we propose a set of principles that facilitate collaboration adaptation. Specifically, these principles build in part upon an earlier discussion of dynamic software adaptability in the scope of architectural styles [14] and provide best practices when using hADL.

Identifying Adaptable Elements: Collaborative behavior can be modeled at multiple levels of abstraction: from an organization, a department, a team, an individual human, down to a single user's behavior strategies. The finest abstraction level determines the lowest possible level of adaptation. In the presence of modeled, identifiable user behavior, we are able to execute adaptations in the form of recommendations. For example we may suggest switching from "locking an artifact for editing it" to "issuing small but frequent article updates without locking". In contrast, we cannot reconfigure a non-performing team internally but we have to replace it as a whole when the most detailed level merely describes teams. In hADL, pattern substructures allow the simultaneous modeling of multiple abstractions level.

Encapsulating Elements: Collaboration adaptability greatly increases when elements (components, connectors, objects) are easy to replace. Encapsulation describes how tightly an element is woven into its surrounding environment. A worker in the Master/Worker pattern only knows about his personal task copy and about the assignment connector he obtained the task from. This makes him easily replaceable as the assignment connector merely needs to provide a task copy to another worker. In contrast, a group of authors that exchange article drafts directly via email exhibits tight coupling. Removing one author requires considerable effort: notification of all other authors, synchronizing of progress, and ensuring orderly handover of unfinished tasks to the remaining co-authors.

A suitable collaboration pattern in this situation may encourage encapsulation through various mechanisms. For example, replacing direct messages with a shared artifact relieves an individual author from keeping track of involved contributors. Introducing a collaboration connector for continuous integration of individual article sections further limits the coordination dependencies amongst authors. Clearly identified and assigned roles (lead author, data collection, proof reading, figure design, etc) within the group additionally promotes encapsulation. hADL avoids enforcing a particular collaboration pattern. Instead, hADL enables the system designer to flexibly assemble a suitable composition of components, connectors, and collaboration objects.

Just as software architectures suffer from implementations that don't follow the prescribed architectural style at code level, so are informal communication channels jeopardizing the adaptation characteristics of a collaboration pattern. The most adaptive pattern will exhibit potentially catastrophic adaptation consequences when the involved users circumvent the foreseen communication and coordination means and fall back onto multipurpose, pattern external communication channels such as email. The underlying collaboration infrastructure needs discouraging the use of external channels. Strategies are pattern specific, for example, hiding other collaborators, anonymizing collaborators, or providing incentives to communicate within the system.

Controlling Interaction: Fostering encapsulation is one principle that simplifies element replacement. Controlling an element's interactions with its environment is equally important. Coordination dependencies become clear and thus manageable when collaborators utilize explicit interactions. Take as an example a worker producing the input for another worker: transferring the output via precisely specified messages clearly identifies the involved actor roles. Collaboration interdependencies, however, remain largely hidden when such interactions occur via a shared artifact. Connectors are able to provide dedicated support for each interaction type only in the former case. hADL promotes the use of connectors where sensible but does not require them when deemed unnecessary.

Managing State: When replacing a human, we need to address what needs to happen with that user's internal collaboration state. An assignment connector might be waiting for task responses or has unassigned task requests still in his inbox. An article author might be currently working on an unfinished section. Three basic strategies address this challenge: (i) ignore existing state (i.e., work progress) and provide some form of compensation, (ii) provide mechanisms that facilitate the externalization of collaboration state such as shared artifacts or dedicated work progress messages, and (iii) split activities into such fine-grained parts that adaptation may be postponed until completion. hADL encourages the use of collaboration objects to render state explicit but currently lacks support for modeling component or connector internal state.

Making Bindings Malleable: Late binding in collaboration patterns delays addressing of messages until their destination absolutely needs to be determined. In a workflow, for example, the worker carrying out a particular task remains undetermined until shortly before task assignment. In the scenario, experts become part of a voting group just shortly before they are actually needed for deciding on an article update. Shared artifacts yield similar decoupling as contributors need not be known in advance. Patterns with such built-in flexibility allow for adaptation decision just in time. Collaboration objects in hADL constitute specification points for implementation specific addressing mechanisms, thus facilitating just-in-time bindings.

7   Evaluation

In this section, we show that hADL is suitable for capturing flexible collaboration patterns by modeling the MediaWiki platform[1]. Subsequently, we demonstrate runtime dynamic reconfiguration of MediaWiki's underlying collaboration pattern. To this end, we briefly present modeling tool support and then provide the MediaWiki hADL model including its mapping onto explicit and implicit group permissions. The hADL model, introduced model instances, and transformations are available for download at `http://wp.me/P1xPeS-2h`.

We adopted the Generic Modeling Environment[2] (GME) for designing, visualizing, and manipulating the hADL model and model instances. GME provides an automatic model update mechanism that allows for rapid, iterative refinement of the hADL model and model instances. The hADL model, therefore, provides only core elements for describing human collaboration architectures. We outline below how extensions cover domain-specific requirements that are otherwise insufficiently addressed. For most changes of the hADL model, the GME model update mechanism is able to successfully upgrade existing model instances to take advantage of problem-specific extensions.

7.1   Modeling MediaWiki

MediaWiki is the underlying technology platform for Wikipedia (and many other Wikis). Figure 4 visualizes how the project wiki from the scenario might initially be set up. The collaboration objects (Page, TalkPage, WikiPage, ImageOrFile, and Revision) remain the same for all MediaWiki installations as they represent the core MediaWiki collaboration capabilities. The MediaWiki group permissions[3] are a good starting point to define the various actions the collaboration objects make available to human components and collaboration connectors. The permissions, however, are insufficient to grasp the complete collaboration pattern as they include only explicitly defined user rights. Any logged-in user, for example, has access to her WatchList but no corresponding permission exists. We, therefore, add actions (i.e., implicit permissions) that model the streaming of article changes to *ArticleObservers* via the watch list (*WatchListStream*) or notification emails (*NfyEmailStream*). Applying the design methodology from Section 6, we analyze the adaptation flexibility of MediaWiki in general and of this specific instance in particular.

---

[1] `http://www.mediawiki.org/wiki/MediaWiki`
[2] `http://w3.isis.vanderbilt.edu/Projects/gme/`
[3] `http://www.mediawiki.org/wiki/Manual:User_rights_management`

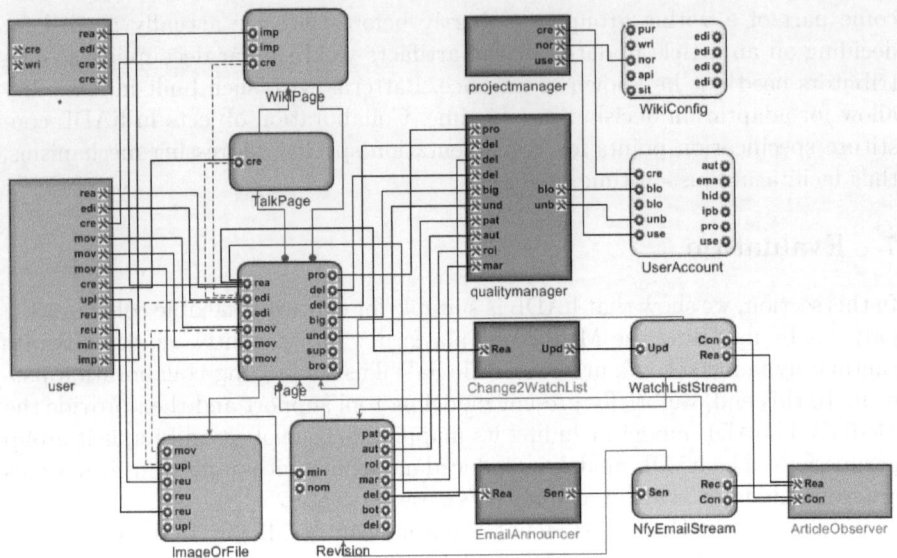

Fig. 4. MediaWiki hADL model for the initial scenario structure

Identifying Adaptable Elements. The smallest, adaptable elements in a MediaWiki installation are individual user and pages (i.e., articles). Structural adaptation actions consist of restructuring user types and (re)assigning users to particular types (i.e., groups). We won't discuss more fine-grained, build-in actions such as blocking a user or protecting a page.

Encapsulating Elements. The individual Wiki authors (component *user*) and readers (component '⋆') exhibit strong encapsulation as all interactions happen via Wiki pages. Discussions on content, structure, etc. are equally restricted to editing of a shared artifact: the respective article TalkPage. *ArticleObservers* receive change notifications without having to rely on authors signaling updates.

Controlling Interaction. For the purpose of writing articles, MediaWiki provides sufficiently precise (inter)actions. Our scenario configuration clearly separates the various components and connectors: authors have edit, move, and upload permissions while *quality managers* have patrol, rollback, revert, delete and protection permissions. There is little to no permission overlap.

Managing State Collaboration state becomes externalized in the form of the Wiki page. A Wiki encourages publishing of frequent and small updates which enables rapid changes in author involvement.

Making Bindings Malleable. Quality managers check (i.e., patrol) article changes by inexperienced and new authors. Which particular quality manager will approve or revert a change, however, is a-priori unknown.

These characteristics and the distinction of human components from collaboration connectors facilitates reconfiguration actions to have minimal effect on active human components. As we will demonstrate next: readers, observers, and authors maintain (largely) the same rights despite considerable pattern evolution.

7.2 Dynamic Structural Adaptation

The scenario highlighted how adding, removing, or replacing users becomes insufficient to address fundamental environmental changes. Figure 5 depicts the evolved MediaWiki structure addressing the needs of the later scenario phase. The adapted structure exhibits new human components and new, reconfigured, or replaced collaboration connectors. Specifically, previous users become *experts*, new users obtain only a limited permission set. The quality managers transfer user blocking privileges to *moderators* and a software-based *editvotecollector* (collaboration connector) contacts *article guardians* for voting on user edits. Instead of receiving emails for all new articles, observers are able to configure topics of interest: the *TopicEmailAnnouncer* replaces the *EmailAnnouncer*.

Planning for reconfigurations is one benefit of modeling MediaWiki with hADL. Another potential use is describing where and how bots as well as extensions provide new functionality. Such additional components and connectors (e.g., TopicEmailAnnouncer) may build upon different collaboration patterns. Here, hADL facilitates the analysis of adaptation implications.

In the case of MediaWiki, hADL goes beyond merely describing the collaboration structure. We developed a model transformation for demonstration purposes that takes the hADL model and generates the group permissions configuration for MediaWiki. Specifically, we export the hADL model as an XML file and

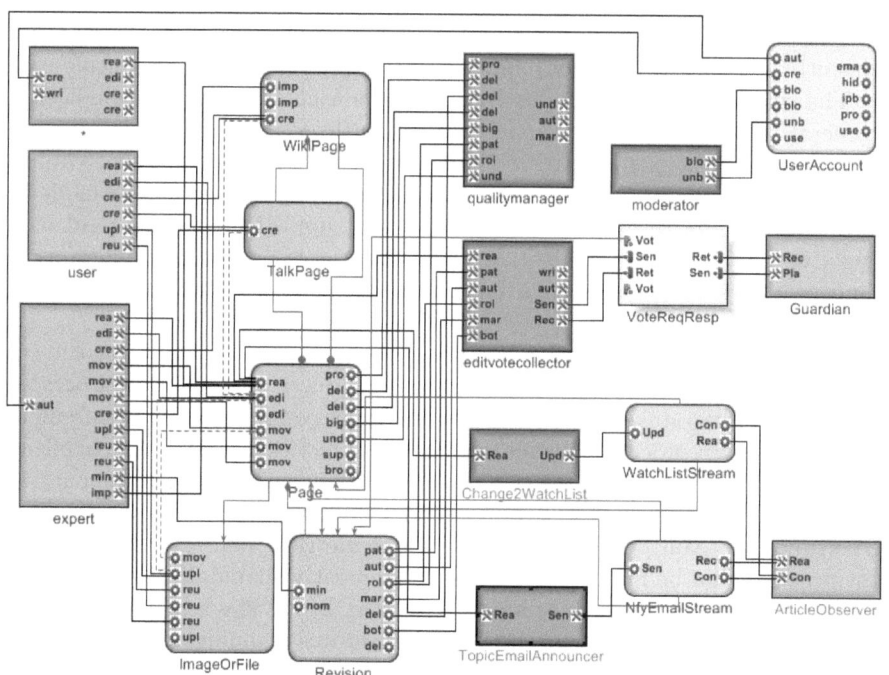

Fig. 5. MediaWiki hADL model for the evolved scenario structure

then process it with the Java Emitter Templates (JET) framework[4]. The transformation interprets every component and connector as a permission group. Each HumanAction becomes an allowed permission when connected to the corresponding ObjectAction, otherwise the permission is denied. Listing 1.1 provides the group permissions for the anonymous user group ('*') in Figure 5. The resulting configuration should not include implicit rights and neither components or connectors that require no *groupPermission* representation (e.g., ArticleObservers, Change2WatchList). To this end, we extend the hADL model with additional properties. The transformation mechanism will thus ignore actions with *isImplicitRight=true* and components and connectors with *isWikiGroup=false*. We also introduce a *Requires* connection in the hADL model (dashed, red lines in Fig. 4 and Fig. 5) for highlighting dependencies between user permissions (e.g., by linking the *move* action to the *edit* action.)

```
1 $wgGroupPermissions['*']['createaccount']   = false;
2 $wgGroupPermissions['*']['read']            = true;
3 $wgGroupPermissions['*']['edit']            = false;
4 $wgGroupPermissions['*']['createpage']      = false;
5 $wgGroupPermissions['*']['createtalk']      = false;
6 $wgGroupPermissions['*']['writeapi']        = false; ...
```

Listing 1.1. GroupPermissions for anonymous MediaWiki users, i.e., '*'

7.3   Discussion

Currently hADL has two main limitations. First, it lacks platform specific models. The evaluation above demonstrates hADL's feasibility but we cannot claim a general purpose tools set for various web platforms. Second, hADL features no integration with existing web modeling methodologies yet. This shortcoming, however, highlights hADL's biggest potential: a recent survey of web modeling approaches emphasizes insufficient support for sophisticated behavioral modeling [16]. Here, hADL would fit in alongside use cases, activity diagrams, or sequence diagrams to enhance current approaches such as WebML, Hera, UWE, or OOWS [16].

Even without such integration, hADL offers considerable benefits at the current stage. An explicit human architecture introduces a collaboration perspective and thus gives stake-holders another means for communicating requirements during the design process. This also enforces a structured approach to explicitly defining adaptation capabilities at the collaboration level. Being implementation independent, hADL provides an opportunity for establishing collaboration patterns tuned to team performance and quality metrics. Thus currently implicit best practises can be made explicit and subsequently shared. When customized to a particular platform such as MediaWiki, hADL provides a high-level view of the collaboration infrastructure. It thereby facilitates planning and documenting the platform's configuration and extensions.

---

[4] http://www.eclipse.org/modeling/m2t/?project=jet#jet

8   Conclusions

We made the case for a human Architecture Description Language for modeling adaptive collaboration structures. Taking inspiration from software architecture, we proposed hADL to specify collaboration patterns in terms of human components, collaboration connectors, and collaboration objects. A set of principles guides the design process to achieve collaboration patterns that facilitate runtime adaptation. Our evaluation successfully demonstrated that hADL supports the dynamic reconfiguration of human components and collaboration connectors at runtime. Nevertheless, even MediaWiki's adaptations capabilities are currently limited to the configuration of group permissions.

Our future work, therefore, will focus on the mapping between the underlying IT infrastructure and collaboration patterns. Ultimately, we aim for techniques that exploit the interdependencies between software elements and collaboration elements for achieving holistic co-adaptation of socio-technical systems. Such work will then also model and exploit diverse relationships between humans such as friendship, rivalry, dis/trust, and organizational hierarchy.

Acknowledgment. This work is supported in part by the National Science Foundation (CCF-0917129, CCF-0820222, and CCF-0808783) and the Austrian Science Fund (FWF) J3068-N23.

References

1. Ceri, S., Daniel, F., Matera, M., Facca, F.M.: Model-driven development of context-aware web applications. ACM Trans. Internet Technol. 7 (February 2007)
2. Chopra, A.K., Paja, E., Giorgini, P.: Sociotechnical Trust: An Architectural Approach. In: Jeusfeld, M., Delcambre, L., Ling, T.-W. (eds.) ER 2011. LNCS, vol. 6998, pp. 104–117. Springer, Heidelberg (2011)
3. Dashofy, E.M., van der Hoek, A., Taylor, R.N.: A comprehensive approach for the development of modular software architecture description languages. ACM Trans. Softw. Eng. Methodol. 14, 199–245 (2005)
4. Dorn, C., Taylor, R.N.: Analyzing runtime adaptability of collaboration patterns. In: International Conference on Collaboration Technologies and Systems (CTS). IEEE Computer Society, Los Alamitos (2012)
5. Dorn, C., Taylor, R.N., Dustdar, S.: Flexible social workflows: Collaborations as human architecture. IEEE Internet Computing 16, 72–77 (2012)
6. Dustdar, S.: Caramba- Process-Aware Collaboration System Supporting Ad hoc and Collaborative Processes in Virtual Teams. Distributed Parallel Databases 15(1), 45–66 (2004)
7. Garlan, D., Monroe, R., Wile, D.: Acme: an architecture description interchange language. In: Proceedings of the 1997 Conference of the Centre for Advanced Studies on Collaborative Research, CASCON 1997, pp. 169–183. IBM Press (1997)
8. Hull, R.: Artifact-centric business process models: Brief survey of research results and challenges. In: Meersman, R., Tari, Z. (eds.) OTM 2008, Part II. LNCS, vol. 5332, pp. 1152–1163. Springer, Heidelberg (2008)
9. Little, G., Chilton, L.B., Miller, R., Goldman, M.: Turkit: Tools for iterative tasks on mechanical turk. In: Human Computation Workshop, HComp 2009 (2009)

10. Malone, T.W., Crowston, K.: The interdisciplinary study of coordination. ACM Comput. Surv. 26, 87–119 (1994)
11. Minder, P., Bernstein, A.: Crowdlang - first steps towards programmable human computers for general computation. In: Proceedings of the 3rd Human Computation Workshop (HCOMP 2011). AAAI Press (January 2011)
12. Moody, P., Gruen, D., Muller, M.J., Tang, J., Moran, T.P.: Business Activity Patterns: A New Model for Collaborative Business Applications (2006)
13. Nandi, P., Koenig, D., Moser, S., Hull, R., Klicnik, V., Claussen, S., Kloppman, M., Vergo, J.: Data4BPM, part 1: Introducing business entities and the business entity definition language (BEDL) (April 2010), http://public.dhe.ibm.com/software/dw/wes/1004_nandi/1004_nandi.pdf
14. Oreizy, P., Medvidovic, N., Taylor, R.N.: Runtime software adaptation: framework, approaches, and styles. In: Companion of the 30th Intl. Conf. on Software Engineering, ICSE Companion 2008, pp. 899–910. ACM, New York (2008)
15. Schall, D.: A human-centric runtime framework for mixed service-oriented systems. Distributed and Parallel Databases 29, 333–360 (2011)
16. Schwinger, W., Retschitzegger, W., Schauerhuber, A., Kappel, G., Wimmer, M., Pröll, B., Castro, C.C., Casteleyn, S., Troyer, O.D., Fraternali, P., et al.: A survey on web modeling approaches for ubiquitous web applications. International Journal of Web Information Systems 4(3), 234–305 (2008)
17. Silva Souza, V.E., Lapouchnian, A., Mylopoulos, J.: System Identification for Adaptive Software Systems: A Requirements Engineering Perspective. In: Jeusfeld, M., Delcambre, L., Ling, T.-W. (eds.) ER 2011. LNCS, vol. 6998, pp. 346–361. Springer, Heidelberg (2011)
18. Taylor, R.N., Medvidovic, N., Dashofy, E.M.: Software Architecture: Foundations, Theory, and Practice. Wiley (2009)
19. Taylor, R.N., Medvidovic, N., Oreizy, P.: Architectural styles for runtime software adaptation. In: WICSA/ECSA, pp. 171–180 (2009)
20. Teruel, M.A., Navarro, E., López-Jaquero, V., Montero, F., González, P.: CSRML: A Goal-Oriented Approach to Model Requirements for Collaborative Systems. In: Jeusfeld, M., Delcambre, L., Ling, T.-W. (eds.) ER 2011. LNCS, vol. 6998, pp. 33–46. Springer, Heidelberg (2011)
21. Wilde, E., Gaedke, M.: Web engineering revisited. In: BCS Int. Acad. Conf. pp. 41–50 (2008)
22. Wohed, P., van der Aalst, W., Dumas, M., ter Hofstede, A., Russell, N.: On the Suitability of BPMN for Business Process Modelling. In: Dustdar, S., Fiadeiro, J.L., Sheth, A.P. (eds.) BPM 2006. LNCS, vol. 4102, pp. 161–176. Springer, Heidelberg (2006)

Modeling Sovereign RFID Data Streams in Collaborative Traceable Networks

Yanbo Wu[1], Quan Z. Sheng[2], Rui Zeng[3], and Jiangang Ma[2]

[1] Beijing Jiaotong Univeresity, Beijing, 100044, China
ybwu@bjtu.edu.cn
[2] The University of Adelaide, SA, 5005, Australia
{michael.sheng,jiangang.ma}@adelaide.edu.au
[3] Yunnan Normal University, Kunming, China
zengruyn@126.com

Abstract. In the emerging environment of the Internet of Things (IoT), through the connection of billions of radio frequency identification (RFID) tags and sensors to the Internet, applications will generate an unprecedented amount of transactions and data that requires novel approaches in RFID data stream processing and management. Unfortunately, it is difficult to maintain a distributed model without a shared directory or structured index. In this paper, we present a fully distributed model for sovereign RFID data streams. This model combines Tilted Time Frame and Histogram to represent the patterns of object flows. It is efficient in space and can be stored in main memory. The model is built on top of an unstructured P2P overlay. To reduce the overhead of distributed data acquisition, we further propose algorithms that use statistically optimistic number of network calls to maintain the model. The scalability and efficiency of the proposed model are demonstrated through an extensive set of experiments.

Keywords: Radio Frequency Identification (RFID), Internet of Things, traceable networks, RFID data streams, scalability.

1 Introduction

Recent advances in wireless sensors, RFID technologies, and Web services have led to the emergence of the "Internet of Things" (IoT), a global network where everyday objects are identifiable, readable, addressable, and controllable via the Internet [11,8]. Such a ubiquitous network offers the capability of integrating the information from both the physical world and the virtual one, which not only affects the way how we live, but also creates tremendous business opportunities such as efficient supply chains, independent living of elderly persons, and improved environmental monitoring.

An important feature in realizing these applications is *traceability*, which is the ability to find the current and historical states of RFID-tagged objects. For example, in a pharmacy supply chain, we want to find where a bottle of problematic medicine comes from. In addition, it is helpful for the managers to understand the patterns of product flows in order to make appropriate decisions on stock management. Enabling traceability is a multiple layer problem and there are still many challenges in realizing effective and efficient tracking and tracing in large networks [12,6,14]. One of them is how

X.S. Wang et al. (Eds.): WISE 2012, LNCS 7651, pp. 157–170, 2012.

to model the movement patterns of objects from sovereign data streams. Through the connection of billions of tags and sensors to the Internet, applications generate an unprecedented amount of transactions and data that requires novel approaches in RFID data stream processing and management [1,11,8]. Many models and techniques have been proposed recently [3,6,11,10]. Unfortunately, most of them require a centralized processing unit that has severe drawbacks such as scalability.

Distributed frameworks such as IBM's Theseos [5] and MOODS [14] require the exact movements to be indexed. As a result, they consume extra storage of indices, and are expensive in bandwidth usage. Instead of maintaining the exact movements of objects, we propose in this paper a probabilistic model that maintains the *object flow patterns*. The object flow pattern is a function of time, which describes the volume/frequency of object movements at a specific time. Our goal is to extract and model the patterns of object flows from high-volume, highly dynamic RFID data streams in autonomous network environments. With these patterns, the efficiency of distributed data processing and mining can be significantly improved. Our contributions are summarized as follows:

- We propose a new model called *Tilted tIme Series of Histograms* (TISH) that combines two techniques, namely *Histogram* and *Tilted Time Frame* [4]. It represents the patterns of the object flow between two nodes using limited memory. Our model suggests the probability that an object comes from a node at a specific time.
- We develop algorithms to maintain the TISH model in a pure Peer-to-Peer (P2P) fashion. To avoid long delays caused by network queries, we further develop an algorithm to choose the most possible neighbors as the target of query rewriting.
- We validate our approach through extensive experiments on large datasets, which indicate the efficiency and scalability of the proposed model and algorithms.

The rest of this paper is organized as follows. We first formally define the problems in Section 2. Sections 3 and 4 introduce the architecture of a distributed RFID system, the details of the TISH model and the algorithms that support this system. Experimental results are reported in Section 5 and related work is discussed in Section 6. Finally, some concluding remarks are given in Section 7.

2 Problem Definition

A traceable network is formed by a set of nodes $\mathcal{N}$: $n_1, n_2, \ldots, n_m$. A node is a place under consideration for tracking and other interactions between objects and the networked system. Objects move along the nodes in the network. When an object o_i arrives at a node n_j at time t_k, it leaves a record of (o_i, n_j, t_k) at n_j. Since RFID readers continuously scan the object, a stream of tuples for the same o_i are generated with increasing timestamps. Without loss of information[1], we simplify the stream at each individual node to a single record of (o_i, n_j, t_s, t_e), where t_s and t_e represents the time when o_i is first and last seen at n_j, respectively. The basic schema for an RFID record is $Record : (Object, Node, Start, End)$. We define an object flow pattern as a function

---

[1] Tracing queries are not sensitive to the internal states of an object at a node.

$f(t_a, t_b, n_i, n_j)$, where t_a and t_b define the range of time under consideration, n_i and n_j are the starting and ending nodes of the flow. The result of the function represents the volume of the flow from n_i to n_j during the period of (t_a, t_b).

Tracking and tracing are two fundamental queries for traceability in traceable networks. Tracking is to find the current or most recent state of an object, while tracing is to find a series of states of the object sorted by time. The following characteristics of RFID data management pose challenges to centralized solutions:

- *Frequent Updates.* Data in traceable networks are generated as streams and update becomes a frequent operation, comparing to traditional database management systems that are optimized for frequent-read scenarios. With a centralized database, frequent updates to the database not only increase the storage cost, but cause high costs of index maintenance.
- *Row Level Security Requirement.* Business applications require high security. In federated systems, the partners want to control their data. For example, a supermarket wants to hide the buying information about the same product from one supplier to another and meanwhile, the suppliers can access the data related to their own products. This requires row level authentication and the authentication overhead for space is high.
- *Archiving.* RFID data, similar to other time series data, is sensitive to time. Recent records are more interesting to the analyzers than the distant ones. For the purpose of storage efficiency, it is often necessary to archive the old data to make room for new data. However, different participants may have different definition on oldness (i.e., when the data should be archived). Furthermore, because the records for different nodes are likely stored in different pages, there may be many fragments in these pages after deletion operations. Rearrangement of records in these pages is very costly.

3   The PeerTrack System

PeerTrack [13] is a federated system focusing on distributed RFID data processing and management. In this system, data is kept where it is collected and each node only governs its own data. In our design, we therefore omit the $Node$ column in the raw schema, i.e., the input to our system at nodes is represented as (o_i, t_s, t_e). In PeerTrack, model establishment and query processing are done in a P2P fashion (see Figure 1).

PeerTrack is built on an unstructured P2P overlay, which has a lower maintenance cost than the structured ones. The difference between PeerTrack and other unstructured P2P systems (e.g., Gnutella [7]) is that neighbor selections are not based on physical measurement. In Gnutella, a query is rewritten to the neighbors with the shortest response time. Contrarily, we select neighbors with the *highest possibility* having the information of the object being queried. This is done by maintaining a *Tilted tIme Series of Histogram* (TISH) model (detailed later) at each node. This model describes the patterns of object flow from/to different neighbors.

The TISH model is updated repeatedly in every event cycle by the *Modeler*. The modeler takes a small random sample out of the large volume of preprocessed records as

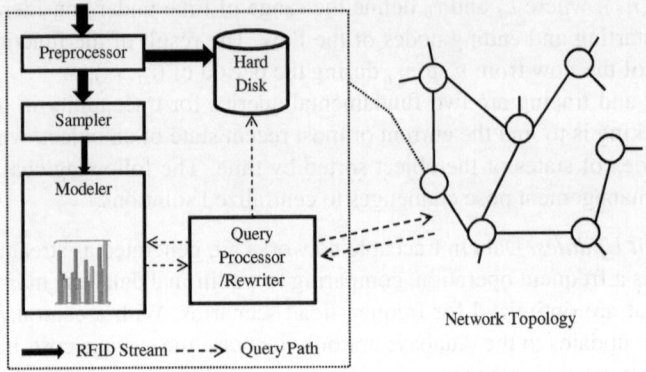

Fig. 1. System Architecture

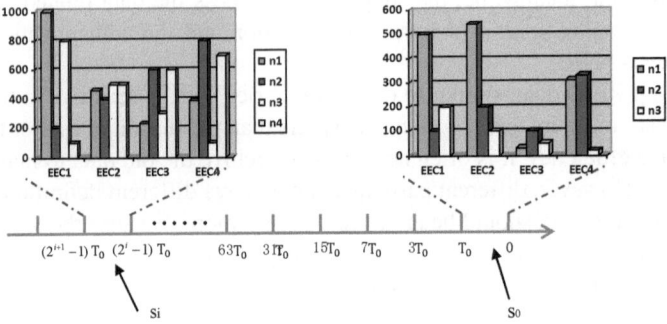

Fig. 2. Illustration of the TISH Model

input, analyzes the sample by querying the neighbors, and refines the TISH model using the result. This sample is generated by the *Sampler*. The *Query Processor/Rewriter* is responsible for answering queries from either local or remote users. It contacts the modeler to find the candidates to rewrite the query if necessary.

This architecture adapts well with the issues discussed in Section 2. We sample the input and use a small portion of incoming data to maintain the model. Thus the system scales well with frequent updates. Each node owns the data physically and fully controls who can access which portion of its data. This model is therefore strictly private. Finally, partners can archive their data whenever they deem appropriate, with flexible strategies.

4   The TISH Model

Our model is based on two important tools in data mining: *Tilted Time Frame* and *Histogram*. Tilted time frame is very useful in data stream analysis, because it gives more details to the recent data than the distant ones. There have been many ways to design a tilted time frame, but the most important ones are : i) *Natural Tilted Time Frame Model*, ii) *Logarithmic Tilted Time Frame Model*, and iii) *Progressive Logarithmic Tilted Time Frame Model*. We will not introduce the details of these models here, interested readers are referred to [4]. In this work, we choose *Logarithmic Tilted Time Frame Model* (LTTF) for its simplicity and flexibility. As illustrated in Figure 2, the right most slot represents the data for the time range of t, while the i_{th} slot S_i represents a range of $2^i t$. The most recent time slot provides the statistics with the finest granularity, while the distant ones are with coarser granularities.

We combine the two techniques and propose a new structure, namely *Tilted tIme Series of Histograms* (TISH). The basic idea of our model is that within each slot, histograms are used to summarize the object flow pattern for each neighbor at the period of time represented by the slot. The height of each bar in the histogram represents the volume of object flow from/to a specific node. Using this model, we can calculate the probability of an object being from/to a specific neighbor at a given time. The symbols used in the following discussion are summarized in Table 1.

Table 1. Symbols

Symbol	Description
w_e	The width of an event cycle
s_i	The ith slot
b_i	The ith neighbor
h_{ij}	The histogram for b_j in slot s_i
f_i	The frequency of object flow for neighbor b_i in an event cycle
w_s	The width of the slots
m	The size of the reservoir sample
n	The number of neighbors
l	The number of slots in the model

We use LTTF to record histories for source and destination nodes, respectively. Time series are split to different *Event Cycles* (also known as the *Sliding Windows*). Each slot in the LTTF represents a certain number of event cycles. The most recent one s_0 maintains the uncompressed event cycles, while the others maintain compressed ones. We name the compressed event cycles as *Exponential Event Cycles (EEC)*, because the ith slot s_i maintains $2^{i-1}(i \geq 1)$ event cycles. For each time slot in the LTTF, we use a histogram to describe the patterns of object flows from/to different neighbors. The x-axis of each histogram represents the EEC. The height of a bar represents the volume of object flow from/to a neighbor, during that cycle. For the histograms of slot $s_i(i > 0)$, the duration of the EEC is $2^{i-1} * w_e$ (width of Event Cycle). The EEC in the slot s_i (the ith EEC) summarizes the object flow patterns for a period of $2^{i-1} * w_e$. The width for slots (i.e., w_s) is a pre-defined constant. It is defined as the maximum number of EECs in each slot. Also we define the *size* of a slot as the current number of EECs, so $s_i.size \leq w_s$. We will discuss how w_e and w_s should be determined later.

Algorithm 1 : Update the Model: *update*

Input: Neighbor set $\mathcal{B} = \{b_1, b_2, \ldots, b_n\}$
　　　Corresponding frequency set $\mathcal{F} = \{f_1, f_2, \ldots, f_n\}$
Output: The refined model $\mathcal{M}$

1: **for** b_i in $\mathcal{B}$
2:　　Gets its histogram $h_{oi} \leftarrow s_0[b_i]$
3:　　**if** h_{oi} is **nil**
4:　　　$H_{oi} \leftarrow$ new array with size w_s, $s_0[b_i] \leftarrow h_{oi}$
5:　　**end if**
6:　　$h_{oi}[s_0.size + 1] = f_i$
7: **end for**
8: $s_0.size \leftarrow s_0.size + 1$
9: **if** s_0 is full
10:　merge($s_1, \mathcal{M}$)
11:　replace s_1 with s_0
12:　clear s_0
13: **end if**

Fig. 3. Algorithm to Update the LTTF Model

The *length* of the model is defined as the number of slots within it. It is easy to see that, a model of length l can store the history of $w_s * w_e * (1 + \sum_{i=2}^{l} 2^{i-2})$. We can get the length of the model which stores the history for the past t time units:

$$l = \log_2 \frac{t}{w_s * w_e} \tag{1}$$

Suppose w_s is 24 and w_e is an hour, we only need $\lceil \log_2 365 \rceil = 9$ slots to store the history of a year. This is efficient in space so that it can be stored in the main memory.

4.1　Update Model

When a new cycle comes, it is sampled and summarized. The synopsis is then added to the most recent slot (s_0). If s_0 is full, it is moved to the slot before it (i.e., s_1), if s_1 is empty. Otherwise, s_1 is summarized and merged into s_2, and then s_0 replaces it. If s_2 is also full, the summarization and merging process repeats until we find a non-full slot, or get to the end of the LTTF. In the latter case, a new slot is appended to the tail.

The algorithm for updating the model is described in Figure 3. First we add the data sampled and summarized from the new event cycle to the slot (line 1–7). If new neighbor joins, a new entry is inserted into the hash table (line 4). If an existing neighbor did not send anything, it is set as zero (this is not shown in the figure). Till now, we have finished processing the new event cycle. If the new event cycle fills the most recent slot s_0, all slots s_i are merged if necessary (the *merge* function in line 10, this is introduced in the next section). Then s_1 is replaced by s_0, and s_0 is cleared to get ready for the coming event cycles.

Algorithm 2 : Merge the Model: *merge*

Input: The slot which is being merged s_i; the model $\mathcal{M}$
Output: The merged model

1: **if** s_{i+1} does not exist in $\mathcal{M}$
2:      $s_{i+1} \leftarrow$ new slot, $\mathcal{M}$.append(s_{i+1})
3: **end if**
4: **if** s_{i+1} is full,   merge($s_{i+1}, \mathcal{M}$)
5: **for** each neighbor b_j in s_i
6:      $h_{i+1,j} \leftarrow s_{i+1}[b_j]$
7:      **if** $h_{i+1,j}$ is *nil*
8:          $s_{i+1}[b_j] \leftarrow h_{i+1,j} \leftarrow$ new array size of w_s
9:      **end if**
10:     move $h_{i+1,j}[1] \sim h_{i+1,j}[w_s/2]$ to $h_{i+1,j}[w_s/2 + 1] \sim h_{i+1,j}[w_s]$
11:     **for** $k \leftarrow 1$ to $w_s/2$
12:         $h_{i+1,j}[k] \leftarrow s_i[b_j][2 * k] + s_i[b_j][2 * k + 1]$
13:     **end for**
14: **end for**

Fig. 4. Algorithm to Merge the LTTF Model

4.2   Merge Logarithmic Tilted Time Frame

When the new event cycle fills the most recent time slot s_0, the second most recent slot s_1 will be merged to the succeeding slots recursively.

The *merge* algorithm, depicted in Figure 4, first checks whether the next slot in the model is full. If yes, it will be merged (line 4). This process is done recursively until either a non-full slot is found or all the existing slots are full. In the latter case, a new slot is created and appended to the model. The slot being merged will be integrated into the first non-full slot (line 5-14). It is compressed by merging two consecutive EECs (line 12). Note that in this algorithm, we assume that the width of slots w_s is even. This assumption does not affect the feasibility and performance of the algorithm.

4.3   RFID Stream Sampling and Summarizing

The model introduced requires the input of the neighbors ($\mathcal{B}$) and the volume of object flow from them ($\mathcal{F}$). However, the object flow is implicit information in the RFID data stream. At node n_i, we only have the local information of the object sent to n_i, such as arrival time and leaving time. To get the information about the object's moving path, it is necessary to query its neighbors. At the worst case, the underlying unstructured P2P overlay is used to locate the object. We need to avoid such case as much as possible. Also because of the large volume of RFID data in the stream, we cannot afford building the model using all the data. Instead, we use a sample of the original data in each event cycle for the stream. In this section, we introduce the sampling and summarizing algorithms to feed the model.

Random sampling can be used as a summarization technique to capture the essential characteristics of data set. The problem of random sampling can be informally defined as "*select a random sample of m records out of a pool of x records*". In data streams,

---

Algorithm 3 : Gather the source node information

Input: A set of sample objects $\mathcal{O} = \{o_1, o_2, \ldots, o_m\}$; a set of neighbors $\mathcal{B} = \{b_1, b_2, \ldots, b_n\}$
      Number of objects in the unsampled event cycle: x

Output: The updated sender LTTF model

1: $\mathcal{P} \leftarrow$ an array of size n
2: **for** each neighbor b_i in $\mathcal{B}$
3:      $\mathcal{P}[i] \leftarrow p_1(b_i)$ // Equation 2
4: $sort(\mathcal{P}, \mathcal{B})$ // sort $\mathcal{B}$ in descending order according to $\mathcal{P}$
5: $\mathcal{F} \leftarrow$ a map from b_i to its frequency
6: **for** each neighbor b_i in $\mathcal{B}$
7:      result set $\mathcal{R} \leftarrow query(b_i, \mathcal{O})$
8:      $\mathcal{F}(b_i) \leftarrow \mathcal{R}.size/m * x$
9:      $\mathcal{O} \leftarrow \mathcal{O} - \mathcal{R}$
10:     **if** $\mathcal{O} = \Phi$,   break
11: **end for**
12: **if** $\mathcal{O} \neq \Phi$
13:     **for** each object o_i in $\mathcal{O}$
14:         $b \leftarrow P2POverlay.locate(o_i)$
15:         **if** b exists in $\mathcal{M}$, $\mathcal{F}(b) \leftarrow \mathcal{F}(b) + x/m$
16:         **else** $\mathcal{F}(b) = x/m$; $\mathcal{B}.append(b)$
17:     **end for**
18: **end if**
19: $update(\mathcal{B}, \mathcal{F})$

Fig. 5. Algorithm to Gather the Source Node Information

the x is often unknown. This makes all algorithms which require to scan the data more than once infeasible. The reservoir sampling algorithm [9] makes only one pass over the data set without knowing its size beforehand. So it is well suited for data streaming sampling in our work. Its output is a uniform sample of the given data set.

After the data within an event cycle is sampled, we need to ask the neighbors to get the source nodes. Instead of flooding the network, the history of object flow is used to find the most possible neighbors who may be the source nodes. With the assumption that the object flow pattern changes granularly, we choose the recent slots in the model to compute the probability of a neighbor being the source node for the objects. Instead of only using the most recent slot, we introduce a zipf-weighted method to compute the probability with several (c) recent slots. This mechanism is introduced to smooth the data flow in case that there are some peak moments for a neighbor. Equation 2 shows how the probability is computed.

$$p_1(b_j) = \frac{1}{\sum_{i=0}^{c} 2^i} * \sum_{i=0}^{c} \left(\frac{1}{2^i} * \frac{\sum_{k=1}^{w_s} h_{i,j}[k]}{\sum_{l=1}^{n} \sum_{k=1}^{w_s} h_{i,l}[k]} \right) \tag{2}$$

The factor $\sum_{i=0}^{c} 2^i$ is for normalization. $\sum_{l=1}^{n} \sum_{k=1}^{w_s} h_{i,l}[k]$ calculates the total number of objects for a slot, while $\sum_{k=1}^{w_s} h_{i,j}[k]$ is the number of objects from b_j. The factor $\frac{1}{2^i}$ assigns weights to different slots. The most recent slot has the highest weight of 1, and the weight decrease exponentially. c is a configurable constant, representing the

---
Algorithm 4 : Trace an Object $trace(o, n)$

Input: The object to trace o; the query initiating node n
Output: A list of nodes that o has been, sorted by time

1: $t_{start} \leftarrow$ select *Start* from Record where Id=o
2: **if** t_{start} is nil, return
3: $s \leftarrow$ the index of slot which covers t_{start}
4: $\mathcal{B} \leftarrow$ the set the neighbors in slot s and adjacent slots
5: $\mathcal{P} \leftarrow$ an array of size $\mathcal{B}$
6: **for** each neighbor b_i in $\mathcal{B}$
7:     $\mathcal{P}[i] \leftarrow p_2(b_i, s)$ // Equation 3
8: **end for**
9: $sort(\mathcal{P}, \mathcal{B})$ // sort $\mathcal{B}$ in descending order according to $\mathcal{P}$ values
10: rewrite the query to all the nodes in $\mathcal{B}$, sequentially
---

Fig. 6. Algorithm to Trace an Object

number of slots under consideration. n is the number of neighbors. In implementation, the sum of frequencies for all nodes in a slot ($\sum_{l=1}^{n} \sum_{k=1}^{w_s} h_{i,l}[k]$) can be calculated and cached. With Equation 2, the neighbors can be sorted by probability p_1. We can first try to contact the neighbor with the highest probability. If there still exists objects without source node after the query, the neighbor ranked the second is queried. The process repeats until we find the source nodes for all objects in the sample, or we have tried all the neighbors. For the latter case, if there are still objects without source node, we will have to rely on the P2P overlay to locate the object.

Figure 5 shows the details of the algorithm to maintain the source node information for the sender model. First the probabilities $\mathcal{P}$ for all the neighbors are calculated and the neighbor list is sorted according to the probabilities (line 1–4). Then we query the neighbors for the list of objects with unknown source, in the order of descendingly sorted probabilities (line 6–11). After querying the neighbors, if there still are objects with unknown source, we have to use underlying P2P overlay to find the sources of the objects (line 12–18). Finally, the neighbors and corresponding frequencies are sent to the data model (line 19). It should be noted that, for implementation, the flooding can be done in parallel. The algorithm to get the recipient model is trivial and it is omitted due to the space constraint.

4.4   Tracing and Tracking Objects

With techniques proposed in this paper, it becomes possible to tracking and tracing objects in a fully distributed, federated environment. The idea is to utilize the history maintained in our model to rewrite the query to the most possible source node. The tracing algorithm is defined in Figure 6. Tracking is almost the same except the direction is reversed to tracing, and instead of all the nodes on the object's moving path, only the last one is retrieved.

The neighbors are sorted (line 9) by the probabilities calculated using Equation 3. Then the query is rewritten to the neighbor with the highest probability (line 10). If it does not return the positive result, the second possible neighbor is queried, and so on.

c_1 and c_2 are two constants which represent how many past/future information to be used in the calculation of probabilities.

$$p_2(b_j, s) = \frac{\sum_{i=min(s-c_1,0)}^{max(s+c_2,l)} \left(\frac{1}{2^{|i-s|}} * \frac{\sum_{k=1}^{w_s} h_{i,j}[k]}{\sum_{l=1}^{n} \sum_{k=1}^{w_s} h_{i,l}[k]} \right)}{\sum_{i=min(s-c_1,0)}^{max(s+c_2,l)} 2^{|i-s|}} \tag{3}$$

s is the slot which covers the time t (Equation 4) when the given object is observed. We can calculate it using the Equation 5.

$$s_0.size + w_s * \sum_{i=1}^{s-1} 2^{i-1} \le t < s_0.size + w_s * \sum_{i=1}^{s} 2^{i-1} \tag{4}$$

$$s = \left\lfloor \log_2 \frac{t - s_0.size}{w_s} + 1 \right\rfloor \tag{5}$$

5   Experimental Evaluation

The proposed techniques have been successfully implemented and are being used in a real-world application for a company tracking their assets. We have conducted an extensive experimental study and this section focuses on reporting some experimental results i) to demonstrate the accuracy of modeling the object flows, ii) to evaluate the performance of the model maintenance, and iii) to prove the scalability of the system built on top of the TISH model by considering various sizes of networks, network topologies and object generation rates.

5.1   Experimental Setup

Experiments were conducted on a Core 2 Quad 2.40GHz machine with 4GB RAM. To study system performance in a large-scale environment, we simulated a network with 1,000 nodes. The object flow network is established randomly by assigning each node a maximum fan-out. We randomly chose a set of 50 nodes as original nodes, which generated objects and put them into the network. All nodes sent objects to neighbors with randomly chosen time-varying patterns from a pre-defined pattern set (Table 3). These patterns vary in the volume of object flow for time t (i.e., $g(t)$) where $\mathbf{V}$ is a constant coefficient and $random(\mathbf{V})$ returns a number which is within $[0, \mathbf{V})$). An RFID object generator has been implemented to generate RFID objects for each pattern. Table 2 shows the default settings.

5.2   The Accuracy of Modeling vs. Time

This experiment evaluated the accuracy of TISH. Because it keeps more information on recent data, we expect that the accuracy drops for the distant data. In this experiment, we selected a random range of time from each time slot in the model, then we calculated the distance between the modeled object distribution among neighbors and the distribution in the pre-set patterns. We define $\mathcal{P}'$ as the real distribution of objects, which is recorded

Table 2. Default Settings

Parameter	Default Value
Number of Nodes	1000
Maximum Fanout	10
Number of Slots	10
Value of V, c and w_s	1000, 2, 10
Size of Sample (m)	50

Table 3. Patterns

Pattern Name	Definition ($g(t)$)
Constant	$V/2$
Random	$random(V)$
Segmentary	V, if $2 * seg \leq t < 2 * seg + 1$
	0, if $2 * seg + 1 \leq t < (2+1) * seg$
	seg is 100 Event Cycles
Sinusoidal	$V * \sin(t)$
Logarithmic	$V * \log(t)$
Hyperbolic	V/t

and calculated using the unsampled, uncompressed data. The distribution modeled by TISH is noted as $\mathcal{P}$. The bias of TISH is defined as the distance between them, which is the distance of the two n-dimension points in an Euclidean space (n is the number of neighbors):

$$P_{bias} = D(\mathcal{P}, \mathcal{P}') = \sqrt{\sum_{i=1}^{n}(\mathcal{P}[i] - \mathcal{P}'[i])^2} \tag{6}$$

We ran the experiment using the default settings in Table 2. During each cycle (called an *epoch*), a random number of objects was sent from one node to another according to the pattern associated with this connection. First, we set up the network using only static/semi-static patterns (Constant and Segmentary , for simplicity, we refer to this setting as *static* setting). We then ran the same experiment with a dynamic setting, in which the time-varying patterns (Sinusoidal, Logarithmic and Hyperbolic) were used.

The result is shown in Figure 7. Firstly, it is clear that the error is higher in the dynamic setting than in the static one. This is reasonable, because in the dynamic setting, patterns constantly change so the bias in the model will increase at the "turning" point of the patterns, such as sinusoidal pattern's highest/lowest points. Secondly, in both settings, the error is higher for the distant records. This is also reasonable because our model gives more details to the recent records while compresses the distant ones. However, the error does not increase significantly for distant records.

5.3   The Performance of Model Maintenance

The main performance bottleneck in our model is the procedure of querying neighbors for new-coming objects. It is also possible that when the new patterns are being established, more network calls are used because the use of the underlying P2P overlay. However, our model is sensitive to this kind of changes and will adapt quickly. In this experiment, we verified this by counting the number of network calls at different time, to see how quickly the model adapts. The system setting is the same as the one in the previous section with all the patterns used. Figure 8 shows the result for this experiment. We can see that during the time of system bootstrap, the network traffic is higher. This is because at that time there was no neighbor, all the objects were found by P2P calls.

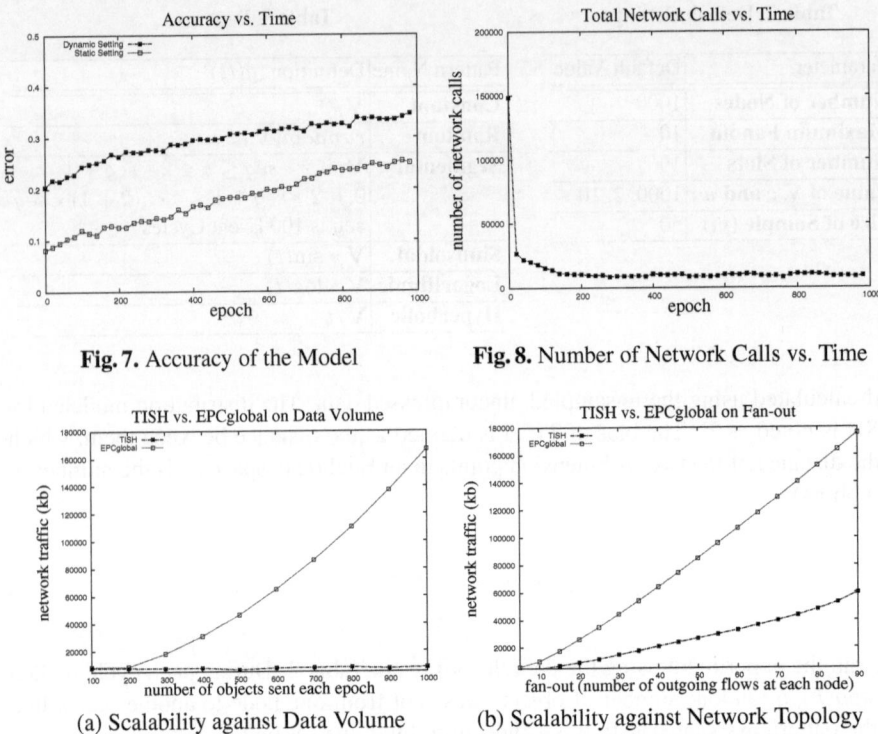

Fig. 7. Accuracy of the Model   **Fig. 8.** Number of Network Calls vs. Time

(a) Scalability against Data Volume   (b) Scalability against Network Topology

Fig. 9. Scalability

However, after the TISH model has been established, only a few network calls are used and the number stays stable.

Another interesting performance evaluation is to see how many network calls are used per query for the algorithm in Figure 5, and what the distribution is. Our finding is that the majority of the queries sent for maintaining the model require only one or two network calls. The average number of network calls in this experiment is only 2.09.

5.4   The Scalability

This experiment examines the scalability of our TISH model against i) volume of object flows and ii) network topology. We exploited the number of network queries used for maintaining the model as the measurement of scalability. We compare our model with the EPCglobal architecture[2] implemented under the same experiment settings. EPCglobal developed a *Discovery Service* (DS) standard which is used to trace individual items [5].

Data Volume. In this test, we examined the impact of the data volume on the performance of the model by comparing the total size of network traffic with EPCglobal

---

[2] http://www.epcglobalinc.org

architecture implementation. We controlled the volume of object flows by varying the value of **V** (Figure 2) from 100 to 1000. The result is shown in Figure 9a. As we can see, the TISH model costs constant network traffic to maintain, meanwhile the cost of EPCglobal architecture raises quickly. This is because the TISH model samples the input and only queries the objects in the sampled set.

Network Topology. In this test, we examined the performance of our model on different network topologies. Using default settings, we varied the maximum fanout to change the connectivity of the network. The maximum fanout varies from 5 to 90. Figure 9b shows that the cost increases for both the TISH model and the EPCglobal architecture when the fanout increases. The larger the number of fanouts is, the more objects are being sent and received at each node. From the figure we can see that the cost of our TISH model increases much slower than the EPCglobal architecture. This is because the TISH model samples the input and chooses the most possible neighbors to query.

6 Related Work

In this section, we briefly survey the modeling techniques for RFID data and the efforts for efficient query processing in large-scale RFID-based collaborative networks. The researchers at Washington University [11] developed an RFID ecosystem that includes an RFID data processing and management system, privacy protection mechanism and a search engine for tracking things. This work mainly focuses on functionality and usability and is limited in a small-scale environment.

In [1], the fundamental problems in RFID data management and query processing are discussed, including challenges in data modeling and tracing. Wang et al. [10] abstract static and dynamic entities including object, reader, location and transaction. Their work models the interactions between these entities as either state or event based relationships. The data model also provides a rule-based data filter engine. In [6], Ng proposes a framework that supports efficient RFID data querying and analysis. As an RFID data warehousing method, the work proposed by Gonzalez et al. in [3] is based on the observation that individual objects tend to move and stay together (i.e., bulky object movements) to compress the data without information loss. This work is further improved in [2] by discovering the *Gateway* nodes which have either high fan-in or high fan-out edges, then the RFID cuboids are created based on this discovery to save more space. These models are all built on a central server and designed for processing data already persisted in DBMS. In [5], Kailing et al. propose a pure distributed RFID data model. Two attributes *sentTo* and *receivedFrom* are associated with each object. The path is a distributed concept, formed by records in correlational nodes. However, this work does not solve the problem on how to acquire these attributes. In [14], Wu et al. introduce a model that indexes the objects in item level in a structured P2P network and algorithms to maintain the model. This model supports item-level and aggregation traceability queries with the cost of indexing spaces.

7 Conclusion

Realizing traceability applications in large-scale, distributed environments such as the emerging Internet of Things presents significant challenges due to their unique characteristics such as large volume of data and sovereignty of the participants. In this paper, we have introduced a distributed model for sovereign RFID data streams in the collaborative traceable networks. We developed distributed algorithms to establish and maintain the patterns of object moving flows. Our proposed model and algorithms are scalable and efficient. Extensive experiments showed the viability, efficiency, and scalability of our approach. Ongoing work includes further performance evaluation of the proposed approach. In this paper, we assume that the preprocessed RFID data is clean and complete. In reality, this is barely true. The noisy, incomplete data introduces *uncertainties* into the model. This is another challenging problem for our future work.

References

1. Chawathe, S.S., Krishnamurthy, V., Ramachandran, S., Sarma, S.: Managing RFID Data. In: Proc. of the 30th Intl. Conf. on Very Large Data Bases (VLDB 2004), Toronto, Canada (2004)
2. Gonzalez, H., Han, J., Cheng, H., Li, X., Klabjan, D., Wu, T.: Modeling Massive RFID Data Sets: A Gateway-Based Movement Graph Approach. IEEE T. on Knowl. and Data Eng. 22, 90–104 (2010)
3. Gonzalez, H., Han, J., Li, X., Klabjan, D.: Warehousing and Analyzing Massive RFID Data Sets. In: Proc. of the 22nd Intl. Conf. on Data Engineering (ICDE 2006), Atlanta, Georgia, USA (2006)
4. Han, J., Kamber, M.: Data Mining: Concepts and Techniques. Elsevier (2006)
5. Kailing, K., Cheung, A., Schönauer, S.: Theseos: A Query Engine for Traceability Across Sovereign, Distributed RFID Databases. In: Proc. of the Intl. Conf. on Data Engineering (ICDE 2007), Istanbul, Turkey (2007)
6. Ng, W.: Developing RFID Database Models for Analysing Moving Tags in Supply Chain Management. In: Jeusfeld, M., Delcambre, L., Ling, T.-W. (eds.) ER 2011. LNCS, vol. 6998, pp. 204–218. Springer, Heidelberg (2011)
7. Ripeanu, M.: Peer-to-Peer Architecture Case Study: Gnutella Network. In: Intl. Conf. on Peer-to-Peer Computing (P2P 2001), Los Alamitos, CA, USA (2001)
8. Sheng, Q.Z., Li, X., Zeadally, S.: Enabling Next-Generation RFID Applications: Solutions and Challenges. IEEE Computer 41(9), 21–28 (2008)
9. Vitter, J.S.: Random Sampling with a Reservoir. T. on Mathematical Software 11(1), 37–57 (1985)
10. Wang, F., Liu, S., Liu, P.: A Temporal RFID Data Model for Querying Physical Objects. Pervasive and Mobile Computing 6(3), 382–397 (2010)
11. Welbourne, E., et al.: Building the Internet of Things Using RFID: The RFID Ecosystem Experience. IEEE Internet Computing 13(3), 48–55 (2009)
12. Wu, Y., Ranasinghe, D.C., Sheng, Q.Z., Zeadally, S., Yu, J.: RFID Enabled Traceability Networks: A Survey. Distributed and Parallel Databases 29(5-6), 397–443 (2011)
13. Wu, Y., Sheng, Q.Z., Ranasinghe, D., Yao, L.: PeerTrack: A Platform for Tracking and Tracing Objects in Large-Scale Traceability Networks. In: Proc. of the 15th Intl. Conf. on Database Technology (EDBT 2012), Berlin, Germany (2012)
14. Wu, Y., Sheng, Q.Z., Ranasinghe, D.C.: Peer-to-Peer Object Tracking in the Internet of Things. In: Proc. of the 40th Intl. Conf. on Parallel Processing (ICPP 2011), Taipei, Taiwan (2011)

Cost-Effective Provisioning
and Scheduling of Deadline-Constrained
Applications in Hybrid Clouds

Rodrigo N. Calheiros and Rajkumar Buyya

Cloud Computing and Distributed Systems (CLOUDS) Laboratory
Department of Computing and Information Systems
The University of Melbourne, Australia
{rnc,rbuyya}@unimelb.edu.au

Abstract. In order to meet distributed application deadlines, Resource
Management Systems (RMSs) have to utilize additional resources from
public Cloud providers when in-house resources cannot cope with the demand of the applications. As a means to enable this feature, called Cloud
Bursting, the RMS has to be able to determine when, how many, and for
how long such resources are required and provision them dynamically.
The RMS has also to determine which tasks will be executed on them and
in which order they will be submitted (scheduling). Current approaches
for dynamic provisioning of Cloud resources operate at a per-job level,
ignoring characteristics of the whole organization workload, which leads
to inefficient utilization of Cloud resources. This paper presents an architecture for coordinated dynamic provisioning and scheduling that is able
to cost-effectively complete applications within their deadlines by considering the whole organization workload at individual tasks level when
making decisions and an accounting mechanism to determine the share
of the cost of utilization of public Cloud resources to be assigned to each
user. Experimental results show that the proposed strategy can reduce
the total utilization of public Cloud services by up to 20% without any
impact in the capacity of meeting application deadlines.

1   Introduction

Advances in Cloud computing made available a virtually infinite amount of resources hosted by public Cloud providers that charge for resource utilization in
a pay-per-use model [1]. Public Cloud infrastructures can be combined with existing in-house resources from organizations in order to accelerate the execution
of their distributed applications. This technique is called *Cloud bursting*, and the
environment comprising such combined resources is termed *Hybrid Cloud*.

When Cloud bursting is applied, the Resource Management System (RMS)
coordinating the access to the resources has to determine when, how many, and
for how long such resources are required and provision them dynamically. The
RMS has also to determine which tasks will be executed on each resource and
in which order (*scheduling*). A common approach to manage such access is to

X.S. Wang et al. (Eds.): WISE 2012, LNCS 7651, pp. 171–184, 2012.

assign an allocation time where a user has exclusive access to a number of resources. More sophisticated resource managers such as Oracle (former Sun) Grid Engine [2] and Aneka [3] operate in a different mode where tasks that compose the application are queued and executed whenever there are free resources in the infrastructure. Priority of tasks are periodically recalculated, what enables enforcement of organization-defined policies about access rights and Quality of Service (QoS) in the form of *deadlines* for application completion.

Even though several research projects focus on each of these steps individually (see Section 2), there is a lack of research in approaches that combine both activities in order to optimize resource utilization, minimize cost during provisioning, decrease execution time of applications, and meet deadlines. Moreover, most approaches for dynamic provisioning operate in a per-job level, and thus they are inefficient because they fail in consider that other tasks could utilize idle cycles of Cloud resources. The latter aspect is especially relevant in the context of typical Infrastructure as a Service (IaaS) providers, which charge users in specific time intervals (typically one hour) even if resources are utilized for just a fraction of the period.

To counter such lack of solutions for cost-effective dynamic provisioning and scheduling in hybrid Clouds, we present a coordinated dynamic provisioning and scheduling approach that is able to cost-effectively complete applications within their deadlines by considering the whole organization workload at individual tasks level when making decisions. The approach also contains an advanced accounting mechanism to determine the share of the cost of utilization of public Cloud resources to be assigned to each user.

The key contributions of this paper are: (i) It proposes an architecture to enable coordinated dynamic provisioning of public Cloud resources and scheduling of deadline-constrained applications; (ii) It proposes a strategy for combined dynamic provisioning and scheduling of tasks; and (iii) It proposes a novel approach for billing users for the utilization of public Cloud resources. Experimental results show that the proposed strategy can reduce the total utilization of public Cloud services by up to 20% without any impact on the capacity of meeting application deadlines.

2   Related Work

The most of the existing scheduling policies for Clusters, Grids [4–10], and hybrid Clouds [11–13] either operate with user specification of allocation slots for utilization of resources or make decisions for a single job without considering jobs already queued. In the latter approach, decisions that optimize one job may cause delays to other jobs or, when Cloud resources are provisioned to complement local resources, may lead to underutilization of the extra resources. In the former approach, users are responsible for ensuring that the job can be executed within the time slot. However, users typically overestimate their jobs' needs, what leads to inefficiencies in the scheduling process. Therefore, we apply a request model where users do not reserve resources during a time interval for job execution. Instead, users submit jobs and specify their deadlines (if any), and the scheduler submits tasks for execution on resources.

The above model is also adopted by the Sun Grid Engine (SGE) [2] and the systems derived from it. Such systems offer a scheduling policy for distributed jobs that allows priority to be assigned to users or groups. It also contains a model of deadline for job execution. However, in such a system deadline is defined in terms of *start time* of the job, whereas our model considers the *completion time* of the job. UniCloud[1] is a software that allows Univa Grid Engine (a derivative from SGE) to provision resources from public Clouds. However, provision of public Cloud resources is manually managed by system administrators.

Lee and Zomaya [14] propose an algorithm for scheduling of Bag of Tasks applications on hybrid Grids and Clouds. This algorithm assigns tasks to Cloud resources only for rescheduling purposes, whereas our approach deploys Cloud resources to meet tight deadlines. Van den Bossche *et al.* [15] propose a heuristic for cost-efficient scheduling of applications in hybrid Clouds. In such work, the application model is similar to the application model addressed by our research. However, their approach makes decision of whether using in-house resources or Cloud resources at job level (i.e., all the tasks that compose the job either run in-house or run on the Cloud) without reutilization of Cloud resources. Our approach, on the other hand, schedules at task level. This has the advantage of enabling a better utilization of Cloud resources by running tasks from other jobs if the billing interval is not over and the job that requested the Cloud resources finished.

Dynamic provisioning of Cloud resources has been explored with different purposes. Vázquez *et al.* [16] present an architecture for dynamic provisioning of Cloud resources to extend the capacity of a Grid in response to events in the RMS. However, the paper does not present any method to determine when the Cloud resources should be deployed or decommissioned. Therefore, the architecture presented in this paper complement such previous work.

Mateescu *et al.* [17] propose a hybrid Cloud environment for HPC applications. Such a system manages requests at single task level. Therefore, deadlines are determined for individual tasks, not for the whole job. It provisions resources from public Clouds to increase probability that tasks start their execution within the start deadline, oppositely to a completion deadline model used in this paper.

Mao *et al.* [18] proposes an auto-scaling mechanism for provisioning resources to jobs in order to meet deadlines. The approach, however, only considers the provisioning problem, while we adopt an integrated provisioning and scheduling mechanism to meet application deadlines.

In our previous work [19], we investigated dynamic provisioning techniques in hybrid Clouds and applied it in the Aneka Cloud platform. However, the approach is applied for individual jobs only and is not integrated with the scheduler; therefore it is not cost-effective in the presence of multiple simultaneous jobs with deadlines.

---

[1] http://www.univa.com/products/unicloud

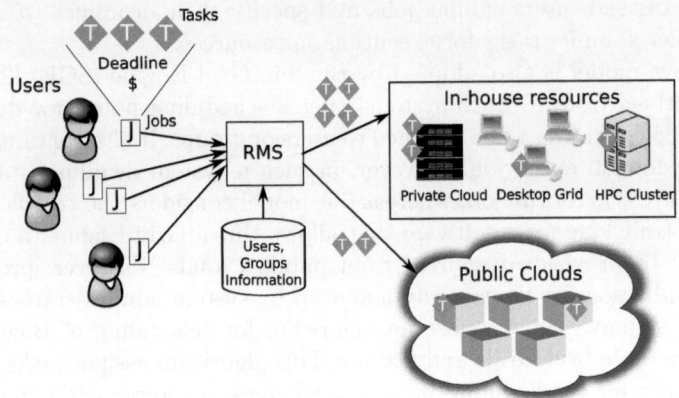

Fig. 1. System and application models assumed in this paper. Jobs are composed of independent tasks, and they can also contain deadline and budget specification. The Resource Management System (RMS) deploys Hybrid Cloud resources to execute tasks and meet deadlines. Decisions are made by the RMS with the support of information about users, groups they belong to, and their access rights.

3   System and Application Models

The system model assumed in this paper is depicted in Figure 1. The central component of the model is the Resource Management System (RMS) that manages a number of local resources (private Cloud). The specific nature of the private Cloud is irrelevant from the system perspective. It may be composed of desktop Grids, a HPC Cluster, or a virtualized data center. Examples of RMSs that follow such a model are Oracle (former Sun) Grid Engine [2] and its derivatives and Manjrasoft Aneka [3].

The RMS has access to one or more public Cloud providers that lease resources in a pay-per use manner. Resources are leased by the RMS via a specific provisioning request sent to the Cloud provider. In such a request, the RMS specifies characteristics of the resources and number of resources required. When Cloud resources are no longer required by the RMS, a decommission request is sent to the public Cloud provider, which releases the resources. Use of Cloud resources is charged in time intervals whose durations are defined by Cloud providers (typically, one hour). Use of a fraction of the time interval incurs in the payment of the whole interval.

The RMS is accessed by users who want to submit loosely-coupled distributed applications in the resources managed by the system. The user request (job) contains (i) description of each task that composes the job, including required files, estimated runtime; and (ii) optional QoS attributes in the form of deadline for job completion and budget to be spent to meet the deadline.

The proposed model does not require that tasks have homogeneous execution time. Therefore, it suits both Bag of Tasks and Parameter Sweep applications.

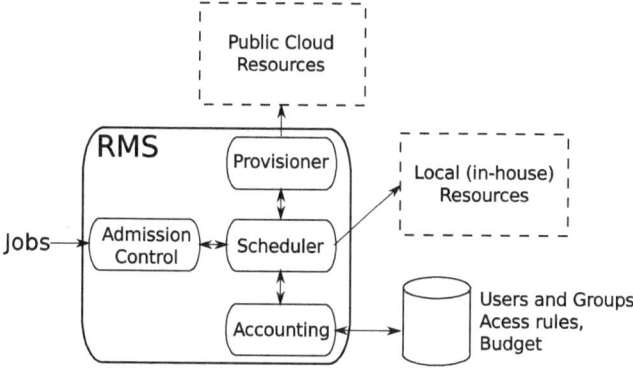

Fig. 2. Proposed Resource Management System architecture for integrated dynamic provisioning and scheduling of applications in hybrid Clouds

Support for dependencies among tasks that constitute a job (which enables support for Workflow applications) is the subject of future work. Furthermore, we assume that files required by tasks are stored in the in-house infrastructure. Therefore, file transferring is required only in the case of Cloud execution.

Tasks from different users compete for resources, and the RMS determines which tasks execute in a given moment and where. However, this has to be done without causing starvation to any job in the waiting queue (i.e., the RMS has to guarantee that each job will eventually complete). Furthermore, the organization can enforce policies about access rights of users and groups, which have to be taken into account by the RMS.

When the RMS detects that one or more jobs are risking missing their deadlines, provisioning policies are applied so that resources are acquired and deployed to speed up such jobs. However, because charge for public Cloud resource utilization is made by a time slot that can be bigger than the runtime of the tasks of the job that required it, the RMS has to apply a reuse policy in order to improve the utilization of the public Cloud resources.

4   Proposed Architecture

Our proposed RMS architecture is depicted in Figure 2. Requests for job execution are received by the *Admission Control* component. Accepted requests are received by the *Scheduler* component. Based on information about job queues, jobs' deadlines, and amount of available resources, requests for extra resources are sent from the Scheduler to the *Provisioner*. The Provisioner is responsible for acquiring resources from public Clouds and making them available to the Scheduler. Finally, the *Accounting* module interacts with the Scheduler to determine whether users have credit and authorization to request and use resources from public Clouds, and also to keep track of utilization of external resources so groups and users can be properly charged for public Cloud utilization.

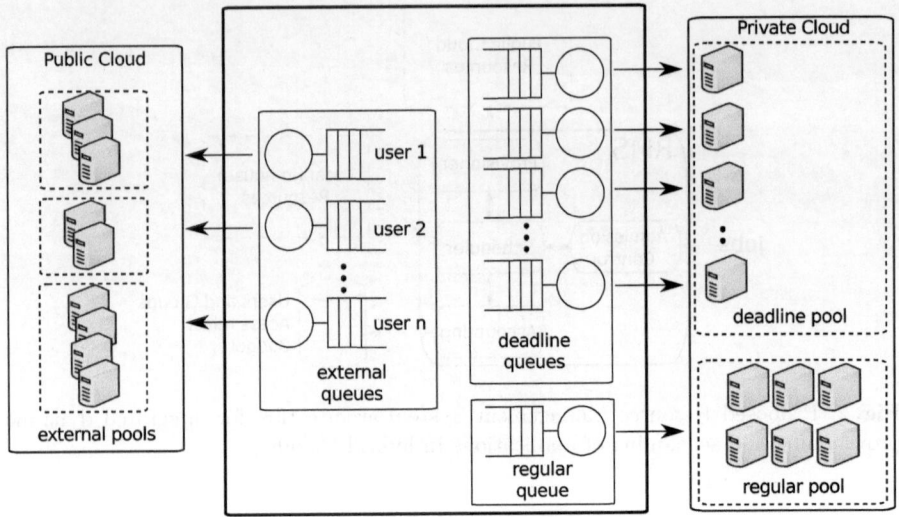

Fig. 3. Organization of resource pools and scheduling queues

The Admission Control accepts all the requests that do not have a deadline constraint. For requests with deadlines, it makes the decision whether the job can be accepted and completed within the deadline or the job must be rejected because it is unfeasible. To determine whether a request can be accepted, the Scheduler module is queried by the Admission Control module. The Scheduler than, considering user estimation, available resources, Cloud resources in use, and user access rights, replies to the Admission Control whether the user has permission and credit to run the job and whether the deadline is feasible or not. The Scheduler's reply is used as the final decision about job acceptance.

4.1   Scheduler

Jobs that are accepted by the Admission Control are received by the Scheduler module, which makes decisions based on a number of factors such as the pool to which the idle resources belongs to and job priority and ownership.

In order to prevent starvation of regular jobs, a minimum amount of resources to be made available for regular tasks can be defined. These resources compose the *regular pool* and its access is coordinated via a *regular queue*. The rest of the local machines belong to the *deadline pool*, whose accesses are coordinated via *deadline queues*. Finally, dynamically provisioned machines belong to *external pools* and are coordinated by *external queues*. Figure 3 depicts the organization of the resource pools and queues in the Scheduler.

Tasks that compose submitted jobs are forwarded either to the *regular* queue or to one of the *deadline* queues (there is one of such queues for each resource that belongs to the deadline pool). They respectively store tasks without deadline-constraints and tasks with such constraints. Tasks on each queue are rearranged every time a new job is received by the Scheduler and every time a task completes. A third set of queues, *external* is also present in the Scheduler. There is one of such queues for each user and it contains tasks that belong to jobs that require dynamic provisioning to complete before the deadline. Tasks on this queue execute preferentially in dynamically provisioned resources, as detailed later in this section.

Algorithm 1 details the procedure for building the regular queue. This procedure runs every time a new job is received and every time a new resource is added to this pool. The total time of each resource used by jobs from a group is summed up to give the total work w_g of group g (Lines 2 to 5). Groups are sorted in ascending order of w_i (Line 6), and each group receives a share of resources N_i that respects the amount of resources assigned to each group defined in the Scheduler (Line 8). The value N_g is the number of tasks from the group g that go to the top of the queue. N_g tasks from the group with the lowest w_i go to the top of the queue, followed by N_h tasks from the group with the second lowest w_i and so on, until all the shares are defined. The rest of the tasks are put in the end of the queue in arrival order (Line 11).

In the case of the deadline pool, whenever a new job is received, tasks are scheduled to different resource queues following a policy such as Round Robin, Worst Fit, Best Fit, and HEFT [20]. We do not apply backfilling techniques to prioritize tasks with closer deadlines because it may motivate users to make late submission of jobs or to overestimate execution time of tasks (both situations that would increase priority of their jobs over others).

Dispatching of tasks for execution depends on the pool that the idle resource belongs to. When a resource from the regular pool becomes idle, the task on top of the regular queue is dispatched for execution in such resource. If the regular queue is empty, the waiting task from deadline queues with the smallest *lag time* (which we define as the difference between the time to the deadline and the estimated execution time) is removed from its queue and dispatched for execution. Finally, if the deadline queue is also empty, the first task on the external queue is dispatched for execution.

When a resource from the deadline pool becomes idle, the next task on its queue is dispatched. If the queue is empty, the task from other queues with the smallest lag time is removed from its original queue and dispatched. If the deadline queue is empty, the first task in the external queue is dispatched or, if the queue is empty, the first task in the regular queue is dispatched.

Whenever a resource from the external queue becomes available, the first task on the external queue that belongs to the user that required the resource is dispatched to the resource. If there is no such a task, a task from the user is sought in the deadline queue. The first task from the user whose estimated execution time is smaller than the time left before the end of the resource's billing period is dispatched. If no task from the user meets this condition, the first task from the user in the regular queue is dispatched.

When the user that requested the resources does not have tasks to execute, the Scheduler applies the same procedure discussed in the previous paragraph

---

Algorithm 1. Regular scheduler queue build up procedure.

Data: res: number of resources in the regular pool.
Data: max_i: maximum number of resources allowed for the group i.
1 empty regular queue;
2 **foreach** *group g_i* **do**
3      $w_i \leftarrow w_i / \sum_j w_j$, the proportional resource utilization by group g_i during
     the current time window;
4   utilizationList $\leftarrow w_i$;
5 **end**
6 sort utilizationList in ascending order of w_i;
7 **foreach** *w_i in utilizationList* **do**
8      $share_i \leftarrow min(max_i, \lceil (1 - w_i) * res \rceil)$;
9   add $share_i$ tasks from group g_i to the regular queue;
10 **end**
11 add remaining tasks in FIFO order to the regular queue;

---

for tasks from the group that required the Cloud resources. If no other task from the same group is found, the procedure is applied for tasks from other groups.

4.2   Provisioner

The Provisioner makes decisions about utilization of public Cloud resources. It calculates the number of extra resources required to execute a job within its deadline and also decides if machines whose billing periods are finishing will be kept for another period or not. The required number of resources is defined at task level: tasks that belong to an accepted job that can run in the deadline pool before the deadline are scheduled locally. Tasks that cannot be completed on time are put in the external queue by the scheduler, and provisioning decision is made based only on such tasks. Currently, the provisioner assumes a single type of VM to be provisioned. This increases the chance of successful allocation of Cloud resources because it enables acquisition of "reserved" or "pre-paid" resources. Most IaaS offer such type of resource, which guarantees that, whenever resources are required, they will be available, as users paid for them upfront or via a premium plus discounted rates for utilization. Alternatively, the provisioner can register multiple providers, and use resources from another provider when the preferable one cannot supply the required resources.

When a virtual machine is reaching the end of its billing period, the Provisioner decides whether the resource should be kept for the next billing period or if it should be decommissioned. This decision is based on the states of external deadline queues. The simplest case is when the external resource is idle or it is running a regular task. It happens when the other queues are empty. In this case, the resource is decommissioned by the Provisioner. A regular task running on the resource is rescheduled in the regular queue. If the provisioned resource is executing a deadline or external task, the resource is kept for the next billing period to avoid risk of missing the job's deadline.

In the case that the resource is no more necessary for the user that originally requested it, and there are still external tasks in the queue, the resource is reassigned for the user that needs the resource (providing it has authorization and credit to use them). In this case, accounting responsibilities for the reassigned resource is also changed, as detailed next.

4.3   Accounting

When Cloud resources are deployed, users and/or groups have to be made accountable for the extra cost incurred by such resources. This is required for reducing the operational costs of the organization. Furthermore, even though accounting is made at user and group level, the system has to apply policies to keep utilization of such external resources as high as possible, so the investment in Cloud resources can be justifiable.

In order to achieve such goals, we propose the *Reassignable Ownership Policy* that operates as follows. Each Cloud resource is associated to an *owner*. Resource ownership is determined by the Provisioner. The resource owner is accountable for any period of idleness of the machine, as well all the period when it was running its tasks on the resource. However, during the period where deadline or external tasks from other users are executed, the corresponding period is assigned to task owners. Moreover, any time a regular task belonging to a user that is not the resource owner is running in Cloud resources, the corresponding period is excluded from the usage period of the owner.

The actual debt the user or group has with the organization corresponds to the fraction of the price per billing period that the user/group was made accountable for. The corresponding fraction of the resource cost is then charged by the accounting module. This enables users to amortize part of the cost related to use Cloud resources and also allows the whole organization to fully utilize Cloud provisioned resources.

5   Performance Evaluation

In this section, we present experiments aiming at evaluating the proposed integrated dynamic provisioning and scheduling technique and its impact in terms of QoS and overall cost of utilization of public Cloud infrastructures.

5.1   Experiment Setup and Workload

Experiments were conducted using the CloudSim toolkit [21] for discrete event simulation. The simulated hybrid Cloud is composed of a local infrastructure managed by a RMS and a public Cloud used for Cloud bursting purposes. The local infrastructure contains 100 virtual machines (VMs). Each machine has 4GB of RAM and a single core processor. The public Cloud accepts requests for up to 100 single core virtual machines from the RMS. Each VM in the public Cloud has the same capacity than the in-house VMs. We assume a negligible latency for communication between the RMS and the in-house infrastructure, and 500 ms latency between the RMS and the public Cloud.

CloudSim applies a "relative" measurement of CPU power, defined as million instructions per second (MIPS), whereas tasks are described in millions of instructions (MIs). Therefore, tasks are defined in terms of how much CPU time is required for its execution assuming no time-sharing of resources. Throughout this section, we refer to this relative time to determine task characteristics.

The RMS is subject to a 24-hours long sequence of job submissions following an adapted version of the BoT workload model proposed by Iosup et al. [22], which was derived from the analysis of utilization traces of seven Grids worldwide. According to this workload model, the interarrival time of a BoT job in peak time follows a Weibull distribution with parameters (4.25, 7.86). However, for this experiment purposes, we assume that during the 24 hours period submission of jobs follows the peak time pattern. Furthermore, we varied the arrival rate of jobs by modifying the parameters of the Weibull distribution. This allowed us to evaluate the system performance subject to different load conditions: we run experiments using two different values for the scale of the distribution (first parameter of the distribution: 4.25 and 8.5) and three different values for the shape (second parameter of the distribution: 7.86, 3.93, 15.72).

The number of tasks of each job request is defined in the workload model as 2^x, where x follows a Weibull distribution with parameters (1.76, 2.11). We assume that tasks that compose a job are homogeneous regarding execution time. The runtime of tasks, as defined by the aforementioned workload model, is 2^x minutes, where x follows a normal distribution with average 2.73 and standard deviation 6.1.

Finally, Iosup's workload model does not contain a description on how to assign deadlines for the each job. Therefore, deadlines for each job were assigned following a method proposed by Garg et al. [23]. Such a method divides jobs in two urgency classes, namely low-urgency and high-urgency jobs. Jobs are assigned to each class uniformly according to a defined share. We evaluated two different shares of high-urgency jobs: 20% and 50%.

Deadlines of jobs on each class vary in the ratio deadline/runtime, as follows. High-urgency jobs have such a rate sampled from a uniform distribution with average 3 and standard deviation 1.4 (i.e, in average the deadline is 3 times of the estimated runtime), whereas low-urgency jobs have such a rate sampled from a uniform distribution with average 8 and standard deviation 3. The obtained value for deadline is counted from the moment the job is submitted for execution to the RMS. Finally, ownership of jobs was assigned to 10 groups, each one with 1 user, following a uniform distribution.

Because the proposed method operates with reservation of resources for composing the regular pool responding to execute regular tasks, and because regular tasks are not subject to QoS metrics, we ignore regular jobs and the regular resource pool for the purpose of these experiments. 24 hours-long workloads with different arrival rates generate according to the above method were submitted for execution in the simulated hybrid Cloud. Experiments for each combination of arrival rates (six different combinations of shape and scale) and urgency (two different rates), which resulted in 12 different scenarios, were repeated 30 times.

For performing the scheduling of the tasks on the deadline queues, we applied the Heterogeneous Earliest Finish Time (HEFT) algorithm [20]. This is a well-known and efficient algorithm for scheduling of applications on heterogeneous environments. Its main advantages are its low complexity and high performance in terms of reducing application execution time.

The HEFT algorithm was used together with two different dynamic provisioning techniques. The first is a job-level provisioning similar to several previous works in the area. When this technique is applied, all the tasks that belong to the new job are removed from the deadline queues and moved to the external queue. Dynamically provisioned resources for the job are decommissioned when the job completes. This technique is labeled as *job-based* in the experiment results. The second technique is the integrated dynamic provisioning and scheduling operating at task-level proposed in this paper: only tasks whose deadline cannot be met with in-house resources are executed in the public Cloud. This technique is applied together with the proposed Reassignable Ownership Policy and is labeled as *Integrated* in the experiment results.

The two strategies of provisioning and scheduling are evaluated with each one of the workloads generated as discussed previously. We report for each combination provisioning-scheduling the amount of jobs whose deadline was missed and the total utilization of public Clouds in terms of number of hours of instances allocated (a metric we call VM-hours).

Finally, it worth noting that, to allow evaluation of a scenario with minimal influence of other types of limitations caused by specific policies, the admission control mechanism was modified to accept all the jobs generated by the workload.

5.2   Results and Discussion

Figure 4 presents results for utilization of public Clouds. The unit used is VM-hours, which we define as the sum of the wall clock time of each dynamically provisioned VM, from its creation to its destruction. Results show that our integrated provisioning approach was able to successfully reduce the utilization of public Cloud resources as a whole. Utilization of public Cloud resources was reduced to up to 20%. The smallest improvement generated by our integrated strategy was 5.2%, and the average reduction in public Cloud utilization was 10.24%. It represents a significant reduction in costs for organizations considering that our experiment simulated 1 day of resources utilization for 10 users. Because typical utilization scenarios are likely to be scaled to a bigger number of users for longer periods of time, the application of our approach can help organizations to significatively reduce their budget of Cloud bursting.

Paired t-tests on the Cloud utilization reported by different policies showed that task-level scheduling and provisioning caused 1% increase in the utilization of local resources (because of tasks that were kept locally instead of being sent to the Cloud). Because the total number of hours of the workload is the same, and the increased local load was smaller, we conclude that the significant reduction

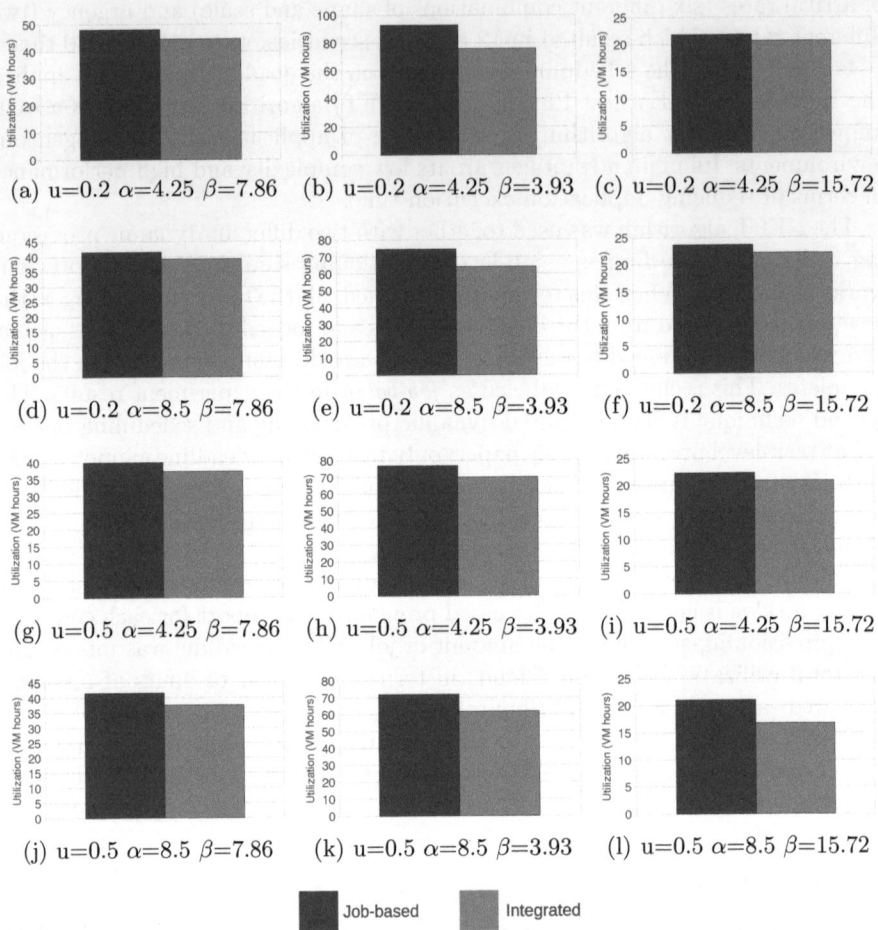

Fig. 4. Public Cloud utilization in VM-hours for a 1-day load workload for different simulation scenarios. Job-based: traditional provisioning techniques applied at job-level. Integrated: our integrated policy with task-level provisioning and reassignment of public resources. α and β denote the scale and shape of the job arrival distribution and u denotes the rate of urgent requests. The scheduling algorithm used in all scenarios is the HEFT algorithm. Note that different plots have different scales.

in the number of Cloud resources is caused by a more effective utilization of such resources. This reduction in Cloud utilization happened without any impact in the capacity of our mechanism in meeting job deadlines. In fact, all the policies were able to meet deadline of 100% of jobs by applying a provisioning strategy. This is an expected effect as, because the system is able to provisioning as many public Cloud resources as necessary, and no unfeasible jobs were submitted, deadlines could always be met with a sufficiently large amount of public Cloud resources.

6   Conclusions and Future Work

Cloud computing transformed the way that distributed applications are executed by making possible to complement in-house resources with pay-per use public Cloud resources. This makes possible for users to define a deadline for application execution and the budget to be spent, if necessary, for the deadline to be met.

In this paper, we presented an architecture that enables Resource Management Systems to support the aforementioned tasks. We describe the architecture, a combined provisioning and scheduling strategy, and an approach for billing users for utilization of Cloud resources that compensates resources reallocated to other users when the deadline application completes before the end of resource billing period. Simulation experiments show that our approach makes an efficient utilization of public Cloud resources and enables deadlines to be met with reduced expenditure with public Cloud resources by organizations.

As future research, we will investigate optimization strategies in order to enable better utilization of multicore resources, when they are available. We will also extend the algorithms to support workflows and other applications where the RMS has to consider dependencies between tasks during the scheduling.

References

1. Buyya, R., Yeo, C.S., Venugopal, S., Broberg, J., Brandic, I.: Cloud computing and emerging IT platforms: Vision, hype, and reality for delivering computing as the 5th utility. Future Generation Computer Systems 25(6), 599–616 (2009)
2. Gentzsch, W.: Sun Grid Engine: towards creating a compute power grid. In: Proceedings of the 1st International Symposium on Cluster Computing and the Grid (CCGrid 2001), Brisbane, Australia, pp. 35–36 (May 2001)
3. Vecchiola, C., Chu, X., Buyya, R.: Aneka: A software platform for .NET-based cloud computing. In: Gentzsch, W., Grandinetti, L., Joubert, G. (eds.) High Speed and Large Scale Scientific Computing, pp. 267–295. IOS Press, Amsterdam (2009)
4. Feitelson, D.G.: Scheduling parallel jobs on clusters. In: Buyya, R. (ed.) High Performance Cluster Computing, vol. 1. Prentice-Hall, Upper Saddle River (1999)
5. Braun, T.D., et al.: A comparison of eleven static heuristics for mapping a class of independent tasks onto heterogeneous distributed computing systems. Journal of Parallel and Distributed Computing 61(6), 810–837 (2001)
6. Silva, D., Cirne, W., Brasileiro, F.: Trading Cycles for Information: Using Replication to Schedule Bag-of-Tasks Applications on Computational Grids. In: Kosch, H., Böszörményi, L., Hellwagner, H. (eds.) Euro-Par 2003. LNCS, vol. 2790, pp. 169–180. Springer, Heidelberg (2003)
7. Cooper, K., et al.: New grid scheduling and rescheduling methods in the GrADS project. In: Proceedings of the 18th International Parallel and Distributed Processing Symposium (IPDPS 2004), Santa Fe, USA (April 2004)
8. Weng, C., Lu, X.: Heuristic scheduling for bag-of-tasks applications in combination with QoS in the computational grid. Future Generation Computer Systems 21(2), 271–280 (2005)
9. Dong, F.: A taxonomy of task scheduling algorithms in the grid. Parallel Processing Letters 17(4), 439–454 (2007)

10. Salehi, M.A., Javadi, B., Buyya, R.: Resource provisioning based on lease pre-emption in InterGrid. In: Proceedings of the 34th Australasian Computer Science Conference (ACSC 2011), Perth, Australia (January 2011)

11. Assunção, M.D., di Costanzo, A., Buyya, R.: Evaluating the cost-benefit of using cloud computing to extend the capacity of clusters. In: Proceedings of the 18th International Symposium on High Performance Distributed Computing (HPDC 2009), Munich, Germany, pp. 141–150 (2009)

12. Salehi, M., Buyya, R.: Adapting Market-Oriented Scheduling Policies for Cloud Computing. In: Hsu, C.-H., Yang, L.T., Park, J.H., Yeo, S.-S. (eds.) ICA3PP 2010, Part I. LNCS, vol. 6081, pp. 351–362. Springer, Heidelberg (2010)

13. Moreno-Vozmediano, R., Montero, R.S., Llorente, I.M.: Multicloud deployment of computing clusters for loosely coupled MTC applications. IEEE Transactions on Parallel and Distributed Systems 22(6), 924–930 (2011)

14. Lee, Y.C., Zomaya, A.: Rescheduling for reliable job completion with the support of clouds. Future Generation Computer Systems 26(8), 1192–1199 (2010)

15. den Bossche, R.V., Vanmechelen, K., Broeckhove, J.: Cost-efficient scheduling heuristics for deadline constrained workloads on hybrid clouds. In: Proceedings of the 3rd IEEE International Conference on Cloud Computing Technology and Science (CloudCom 2011), Athens, Greece, pp. 320–327 (December 2011)

16. Vázquez, C., Huedo, E., Montero, R.S., Llorente, I.M.: Dynamic provision of computing resources from grid infrastructures and cloud providers. In: Proceedings of the Workshops at the Grid and Pervasive Computing Conference, Geneva, Switzerland, pp. 113–120 (May 2009)

17. Mateescu, G., Gentzsch, W., Ribbens, C.J.: Hybrid computing–where HPC meets grid and cloud computing. Future Generation Computer Systems 7(5), 440–453 (2011)

18. Mao, M., Li, J., Humphrey, M.: Cloud auto-scaling with deadline and budget constraints. In: Proceedings of the 11th International Conference on Grid Computing (GRID 2010), Brussels, Belgium, pp. 41–48 (October 2010)

19. Calheiros, R.N., Vecchiola, C., Karunamoorthy, D., Buyya, R.: The Aneka platform and QoS-driven resource provisioning for elastic applications on hybrid clouds. Future Generation Computer Systems 28(6), 861–870 (2012)

20. Topcuoglu, H., Hariri, S., Wu, M.Y.: Performance-effective and low-complexity task scheduling for heterogeneous computing. IEEE Transactions on Parallel and Distributed Systems 13(3), 260–274 (2002)

21. Calheiros, R.N., Ranjan, R., Beloglazov, A., De Rose, C.A.F., Buyya, R.: CloudSim: A toolkit for modeling and simulation of cloud computing environments and evaluation of resource provisioning algorithms. Software: Practice and Experience 41(1), 23–50 (2011)

22. Iosup, A., Sonmez, O., Anoep, S., Epema, D.: The performance of bags-of-tasks in large-scale distributed systems. In: Proceedings of the 17th International Symposium on High Performance Distributed Computing (HPDC 2008), Boston, USA, pp. 97–108 (June 2008)

23. Garg, S.K., Yeo, C.S., Anandasivam, A., Buyya, R.: Environment-conscious scheduling of HPC applications on distributed cloud-oriented data centers. Journal of Parallel and Distributed Computing 71(6), 732–749 (2011)

Good Quality Complementary Information for Multilingual Wikipedia

Yu Suzuki[2], Yuya Fujiwara[1], Yukio Konishi[1], and Akiyo Nadamoto[1]

[1] Konan University, 8-9-1 Okamoto, Higashi-Nada, Kobe, Hyogo 6588501, Japan
[2] Nagoya University, Furo, Chikusa, Nagoya, Aichi 4648601, Japan

Abstract. Many Wikipedia articles lack information, because not all users submit truly complete information to Wikipedia. However, Wikipedia has many language versions that have been developed independently. Therefore, if we supply these complementary information from many language versions, the users must satisfy the amount of information of Wikipedia articles with the complementary information, instead of only one language version of Wikipedia articles. In this study, we specifically examine multilingual Wikipedia and propose a method of extracting good quality complementary information from Wikipedia of other languages. Specifically, we compare Wikipedia articles with less information to those with more information. From Wikipedia articles, which can have the same theme and different languages, we extract different information as complementary information. As described herein, we extract comparison target articles of Wikipedia based on a link graph, because cases exist in which information included in an articles is written in multiple pages of different languages. Furthermore, some low-quality information is extracted as complementary information because Wikipedia articles are written by not only good editors but also bad editors such as vandals. We propose a method to calculate the quality of information based on the editors, and we extract good quality complementary information.

1 Introduction

Many people all over the world use Wikipedia[1], which are accessible throughout the world. The features of Wikipedia is that users create and edit the articles by themselves. Therefore, there are multiple languages of Wikipedia available. For many articles, information is lacking because users can create and edit the information freely. Furthermore, Wikipedia has different levels of value of its information depending on the language version because different users create and edit articles of the respective language versions. Cases exist in which respective articles of the same title and different languages can have sufficient information and scarce information. This seems to be true especially for articles written about culture-related subjects. For example, information of an article about "lawn bowling (Bowling)" is very rich in the English version, but it is poor in the

---

[1] http://en.wikipedia.org/

X.S. Wang et al. (Eds.): WISE 2012, LNCS 7651, pp. 185–198, 2012.

Japanese version because lawn bowling is a popular sport in the U.K., but not in Japan. Almost all articles written in Japanese are produced by Japanese native speakers, few of whom know anything about lawn bowling in any great detail. In contrast, U.K. residents are likely to know about lawn bowling quite well. They can edit an article about lawn bowling with some authority. Consequently, the information of articles differs among respective language versions. We consider that it would be good for users who are browsing Wikipedia, when they browse an article, to have the system complement the information that is lacking in user's browsing article automatically. We specifically examine the multilinguality of Wikipedia and propose a method that complements information of the article which lacks information based on comparing different language articles that have the same title. As described in this paper, we designate an article that a user is browsing and which lacks information as a "browsing article", its Wikipedia as a "browsing Wikipedia", and its language as a "browsing language". We also designate a different language article as a "comparison article", its Wikipedia as a "comparison Wikipedia", and its language as a "comparison language".

When we compare a browsing article with a comparison article, the information granularity differs. The browsing article is only one page, but the comparison article might have multiple pages because these articles are written by intercultural authors. For example, articles about "Sushi" exist in both Japanese and English Wikipedia. In the Japanese version Wikipedia, there are multiple articles such as Inari-sushi, Oshi-sushi, and Chirashi-sushi which are kinds of sushi. However, their English versions are written on the same page under the title of "Sushi" . As that example illustrates, it is necessary to consider the coverage of comparison articles.

Moreover, Wikipedia articles are not always of good quality because anybody can create and edit them. The article becomes low-quality if we complement low-quality information to an article lacking information. Considering the quality of complementary information is necessary. We infer that high-quality Wikipedia information is not deleted by other authors, and therefore that it has a high survival ratio. Furthermore, an author who writes high-quality information tends to write high-quality information consistently. Therefore we also propose a means for calculating the information quality of based on the survival ratio of information and its author. In this time, we regard high-quality information of a Wikipedia article based on featured articles on Wikipedia[2]. As described in this paper, we propose a method that complements high-quality information based on comparing multilingual Wikipedia and author quality. Our proposed complementary information is relevant to the browsing content but it is different (or lack) and high-quality information in the comparison Wikipedia.

Three important technical points are presented as follows.

- Extracting comparison articles from Wikipedia
- Extracting different information
- Calculating Wikipedia information quality.

---

[2] http://en.wikipedia.org/wiki/Wikipedia:Featured_article_candidates

The process used for our proposed system is the following (Fig. 1):

1. A user who browses the Wikipedia article clicks the complement button on the web page, and inputs the comparison language. At this time, the user's browsing article becomes the browsing article. Then its language becomes the browsing language.
2. The system retrieves the comparison articles for which the title of meaning is the same as the user's input query using the language link of Wikipedia.
3. It extracts comparison articles from the comparison Wikipedia using a link graph of Wikipedia.
4. It compares the browsing article with the comparison articles extracted in step 3, and extracts different information. The different information becomes the candidate of complementary information.
5. It calculates the quality of the candidate of complementary information. When the quality degree is greater than threshold, the information becomes complementary information.
6. It browses the browsing article and the complement information related to the web browser.

In our prototype system, we use the English version of Wikipedia as the browsing Wikipedia, and Japanese version of Wikipedia as the comparison Wikipedia. Nevertheless, our system is language-independent. When we use other versions of Wikipedia, we can develop a version for any language. In addition, our target users are people who browse Wikipedia articles.

2   Related Work

Wikipedia Link Structure. Our proposed approach uses link structure analysis techniques. There are many researches which uses Wikipedia link structure. David et al.[8][9] and Michael et al.[12] extract measures of semantic relatedness of terms using the Wikipedia link structure. Nakatani et al.[10] propose a method of ranking search results adaptively by considering a user's comprehension level. They extract technical terms related to a search query by analyzing the link and category structure of Wikipedia. Jaap et al. [6] use Web inlinks for importance research related to the difference between Wikipedia and the Web link structure. Our research differs in that we seek to assess the difference between Wikipedia articles. Nakayama et al. [11] proposed a method to analyze the Wikipedia link structure and produce a thesaurus dictionary. Chen et al.[3] propose a novel approach to construct a domain-specific thesaurus from the Web automatically using link structure information. In contrast, we propose a method to use the term appearance position in the structures of articles. We use the number of interactive links (out-links and in-links) to other articles to calculate relevance.

Comparison of Multilingual Wikipedia. Some studies described in the literature make comparisons between languages using multilingual Wikipedia. Eytan et al.[1] extract the difference between the Wikipedia info-boxes that exist for

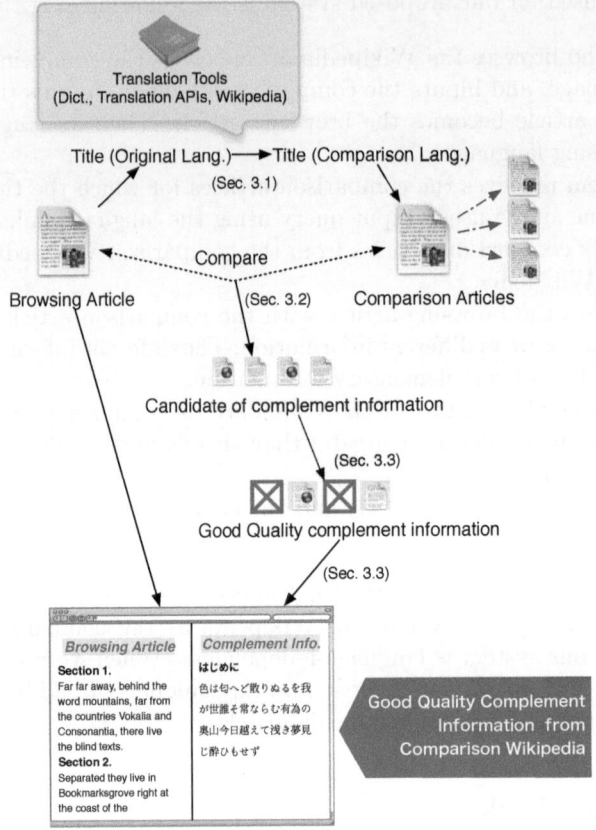

Browsing Article with Good Quality Complement Information

Fig. 1. Overview of our proposed system

other languages. Info-boxes are tables including summary data that exist for
many Wikipedia articles. However, when comparing Wikipedia articles, specifically examining the structure of the table of contents can be superior in terms
of the article contents when the article contents are not known.

Calculating the Quality of Wikipedia. There has been much research related to calculating the degree of quality of products, people, and objects using
reputation-based methods [14,13]. A key concept for evaluating Wikipedia articles is *the peer review process*. Wikipedia is not thought to have a peer review
system because most texts are instantly made and saved, although no one reviews
these texts. However, Stivila et al. [14] mentioned that the open edit system is a
kind of peer review system where editors of the system vote on implicit features
of the texts. In these investigations, many features are extracted from Wikipedia
data in many studies, and they can be divided into two types: explicit and implicit features. Explicit features are the user's decision which are directly input
to the system by users, and implicit features are the user's decision which the

system presumes from their behaviors. In this section, we describe the studies
that have used explicit and implicit features and also describe why we choose to
use implicit features.

Complementary Information. There are some researches which complement
information. Ma et.al. [7] proposed integrating cross-media news content such
as TV programs and Web pages to provide users with complementary information. They extract complementary information from Web pages by using their
proposed topic structure. Their research issue which is complement information
is similar to our issue, however, target of the content is different. Eklou et. al. [4]
proposed the method of information complementation of Wikipedia by the Web.
They extract complementary information from Web pages by using LDA. Their
goal is similar to ours, however the target complement information is different.

3   Extraction of Complementary Information from Articles

3.1   Extraction of Comparison Articles from Comparison Wikipedia

In this subsection, we describe step 3 of the overview in section 1: extraction
of comparison articles from Wikipedia. When we compare the browsing article
with the comparison article on a single particular topic, their information granularity differs between the languages. When a user's browsing topic extends over
multiple articles in the comparison Wikipedia, we extract comparison articles
based on the Wikipedia link graph and our proposed relevance degree.

Link Structure Analysis
First, we create a link graph for a comparison Wikipedia based on the user's
browsing title as described below.

1. The system extracts an article having the same title as the user's input from
   the comparison Wikipedia. We designate the article as "the basic article".
   We regard the basic article as the root node of the link graph.
2. We regard the interactive linked articles which are the subjects of link-out
   and link-in connections with the basic article as important and relevant
   articles to the basic article. The system extracts all interactive linked articles
   from the comparison Wikipedia. It includes the articles as nodes and thereby
   creates the link graph.
3. The system calculates the relevance degree between the root node (basic
   article) and other nodes (interactive linked articles) in the link graph.
4. When the relevance degree is greater than a threshold β value, then the
   system extract the articles. We designate the articles as "relevant articles".

Fig. 2 presents an example of link structure analysis. In the case of Fig. 2, the
basic article is "Sushi" and the threshold is β 0.2. The node "Maguro (Tuna)" is
not an interactive link article. The system deletes the node from the link graph.

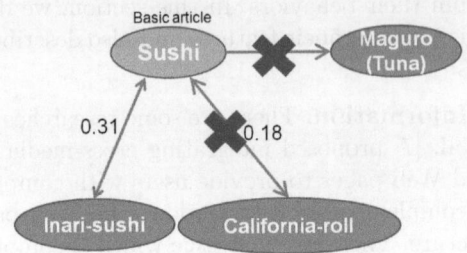

Fig. 2. Analysis of the link graph

The node "California roll" is an interactive link article, but the relevance degree is less than the threshold β. Therefore, the system also deletes the node from the link graph. The relevant article is "Inari-sushi".

When the system extract candidate of complementary information from the comparison articles, a basic article and relevant articles become comparison articles.

Calculating Relevance Degree

The article of an interactive link does not always have a high relation to a basic node. We calculate the relevance degree between a basic article and relevant articles. An important anchor appears many times in a basic article, and appears also in the summary area in Wikipedia. We use that structural feature and calculate the relevance degree between the root node and other node in the link graph as follows described below.

1. The system divides the basic article according to the section which is the structure of the table of contents of the basic article. We designate the divided parts as segments.
2. In the link graph, the system extracts a title from a node that is the interactive linked article. The title becomes the keyword when we extract anchor text from the basic article. We designate the title as the "target title".
3. The system counts the anchor text of the target title in the summary area of the basic article. The number of the anchor is $TFsum_i$. i is the identification number of the interactive link article.
4. It also counts of the anchor in each segment of the basic article. The number of the anchor is TF_{ik}. k is the segment number.
5. It calculates the relevance degree between each segment and the interactive linked article for which the title is the target title. At this time, our hypothesis is that important anchors related to the basic article appear many times in the basic article. They also appear in the summary area. Therefore, we use the following expression to calculate the degree of relevance.

$$R_i = (\alpha \cdot TFsum_i \cdot Ssum_i + \sum_{k=1}^{n}(TF_{ik} \cdot S_{ik}))/ \max(R_{im}) \qquad (1)$$

In the equation presented above, R_i is the relevance degree of each interactive link article i. $Ssum_i$ represents the cosine similarity between summary area of basic article and an interactive link article i. n is the number of segment in the basic article. S_{ik} represents the cosine similarity between segment k of the basic article and an interactive link article i. $\max(R_{\mathrm{im}})$ signifies the maximum value in all R_i.

The system calculates from (2)–(5) for all comparison articles that are interactive linked articles in the link graph. When relevance degree R_i is greater than the threshold β value, it becomes a relevant article.

3.2 Extracting a Candidate of Complementary Information from the Browsing Article and Comparison Articles

After the system extracts comparison articles, it calculates and extracts a candidate of complementary information by comparing the browsing article with each comparison article consisting of a basic article and relevant articles.

Almost all Wikipedia articles are divisible into segments based on the table of contents, which means that the segments are divided semantically. When comparing the similarity of multilingual Wikipedia, we specifically examine the segments of the table of contents of Wikipedia. Particularly, we compare each article based on each segment.

First, we extract nouns from respective articles. Next we translate comparison articles to browsing language using dictionaries. At this time, the translation is insufficient only by the dictionary. We use a translation API such as Google API. Furthermore, we translate proper nouns using Wikipedia's language link. As described in this paper, we are unconcerned about word sense disambiguation. That is left as a subject for future work. We then calculate the similarity between the browsing article and each comparison article using cosine similarity.

$$\cos(x_k, y_l) = \frac{\sum x_i \cdot y_i}{\sqrt{\sum x_i^2 \cdot \sum y_i^2}} \tag{2}$$

In that equation, x_k signifies a segment of the browsing article k; y_l denotes a segment of comparison article l. In addition, x_i represents the term frequency of noun i in a browsing article's segment and y_i stands for the term frequency of noun i in a comparison target article's segment.

If the cosine similarity is less than the threshold γ value, then it is regarded as different information.

3.3 Assessing the Quality of Complementary Information

Finally, we calculate the quality of complement information using survival ratio based approach[2]. Generally, when editors edit articles, they delete any poor quality or misleading texts. As a result, if texts survive beyond multiple edits, many editors who have edited the article have decided that the text should not be deleted or edited. We think that an editor leaving texts unchanged corresponds to someone voting positively, such as giving 5 stars. Using this idea, we can calculate text qualities from survival ratios.

One important policy is that editors do not evaluate themselves. When editors add texts, the texts are not evaluated at that time. We use this policy of self evaluation, because, if we permit editors to evaluate themselves, vandals can easily increase survival ratios of texts they edit, which easily increases qualities of the vandals.

First, we extract all articles from the Wikipedia edit history, and identify which editor edited which texts. Edit history stores the extract title, editor's name, and a snapshot of the article for every version. We extract these data, and store them in a database system. At this time, we identify the editors of the texts using diffs. The texts that editors have added are the texts that differ between the current and previous versions. When a text is not in the previous version but is in the current version, the text must have been written by the editor of the current version. Using this policy, we identify the editor of every text.

Next, we calculate the text quality $\tau(i,e)$ in article i by editor e as follows:

$$\tau(i,e) = \sum_{p(e) \in \overline{P}(e)} \log_2(|p(e)| + 1) \tag{3}$$

where $p(e)$ is a text which is written by e. $\overline{P}(e)$ is a set of texts which is written by e and is not on the version edited by e, because of the policy that editors do not evaluate themselves. $|p(e)|$ is the number of letters in $p(e)$. This equation means the summation of the number of letters on texts that are written by e. We remove the number of letters on the version edited by e himself/herself because of the policy of non-self-evaluation.

Finally, we calculate quality values for each section of complement information using averaged value of $\tau(i,e)$ as follows

$$T = \frac{\sum |p(e)| \cdot \tau(i,e)}{\sum |p(e)|} \tag{4}$$

We calculate T for all sections of all articles of difference information, and normalize these values into $[0,1]$. If the ratio of quality value T is more than the threshold λ, we identify the section as complement information.

4   Prototype System and Experiments

4.1   Prototype System

We developed our prototype system. In our prototype system, the browsing language is English. The comparison language is Japanese. For it, we use Ruby[3] as

---

[3] Ruby http://www.ruby-lang.org/

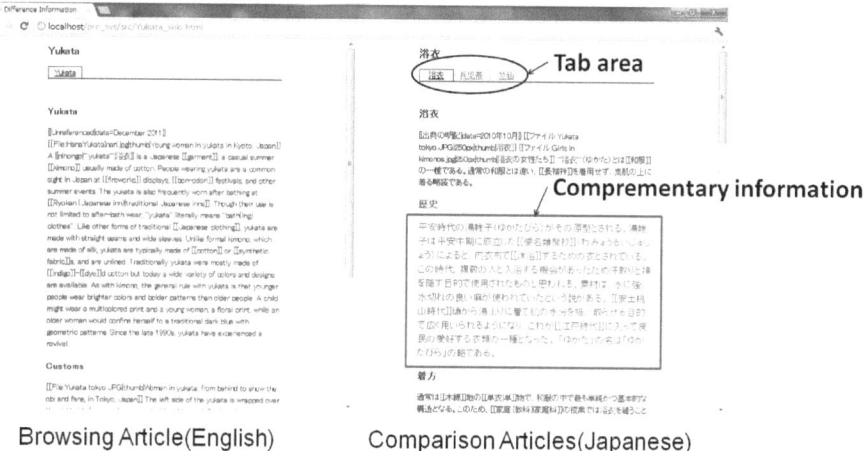

Fig. 3. Output of the prototype system

the programming language, Mecab[4] as the morphological analyzer for Japanese sentences, Tree Tagger[5] as the morphological analyzer for English sentences, and MySQL[6] as the Database server. We use multiple English–Japanese dictionaries. They are GENE95[7], Google Ajax API[8], Microsoft Translator API[9], and Wikipedia.

In our prototype system, the first user browses the browsing article and inputs the comparison language and clicks the complement button. Subsequently, the system outputs English and Japanese Wikipedia articles related to the title of browsing (English) article, with complement (Japanese) information. An overview of our proposed system is presented in Fig. 3. On the left side of Fig. 3 is the English article. The right side of Fig. 3 is the Japanese article. On the right side of Fig. 3, the red frame is a segment that has different information. The upper right side of Fig. 3 is a tab area. When a user clicks a tab, the user can browse complementary information in each comparison article.

4.2   Experimental Evaluation

We conducted experiments of three types to confirm the accuracy of our proposed method using our prototype system. In experiment 1, we confirmed that the system can extract appropriate relevant articles. Also by experiment 1, we confirmed that our proposed extraction of different information methods can

---

[4] Mecab http://mecab.sourceforge.net/

[5] Tree Tagger http://www.ims.uni-stuttgart.de/projekte/corplex/TreeTagger/

[6] Mysql http://www-jp.mysql.com/

[7] GENE95 http://www.namazu.org/~tsuchiya/sdic/data/gene.html

[8] Google Ajax API http://code.google.com/apis/language/

[9] Microsoft Translator API http://www.microsofttranslator.com/dev/

Table 1. Number of collected result pages using all experiments

ID	Query	Number of comparison articles (English)	Number of secs in browsing article (Japanese)	Total number of secs in comparison articles
1	Bowing	2	14	18
2	Rock–paper–scissors	6	23	37
3	Takoyaki	4	1	15
4	My Neighbor Totoro	2	9	25
5	Doraemon	4	9	29
6	Manners	3	1	12
7	Iaido	4	10	24
8	Imagawayaki	3	3	13
9	Ninjya	6	24	28
10	Manzai	4	3	22
11	Yukata	3	1	7

extract appropriate candidates of complementary information. Next in experiment 2, we determined the threshold λ of calculating the quality in section 3. In experiment 3, we confirmed that our proposed calculation of the quality of information method can extract appropriate complementary information.

Dataset. For all experiments, we used 11 queries and correct articles shown in Table1 as the test collection. The targets of our proposed methods are typical titles of the comparison language, such as culture, custom, and so on. The comparison articles of 11 queries are edited by 10 or more users and editing times were 10 times or more because we needed more than both of the conditions when calculating the quality. Three participants gave their opinions. The correct answers are judged by two or three of the participants as correct. We set an appropriate weight $\alpha=3.0$ and threshold $\beta=0.2$ of the relevance degree described in expression (1) in section 3, and threshold $\gamma=0.2$ of cosine similarity in expression (2) in our earlier experiment[5]. We also conducted experiments to assess the availability of our extraction of comparison articles of Wikipedia in the earlier experiment[5]. When we had examination in experiment 3, we used the web pages.

Experiment 1: Availability of Extracting Different Information

In this experiment, we assessed the availability of extracting different information. In our proposed method, our proposed different information is not only ordinary different information but also related information. For example, the browsing article is written about the plot and cast of My Neighbor Totoro, which is a famous Japanese animated movie, if one of the comparison articles is written about Sayama Hill, which is the place where the movie events take place. The geographic information related to Sayama Hill differs, but it is unrelated to the movie. This time, the information does not become different information.

We calculate the precision, recall, and F-measure for each query.

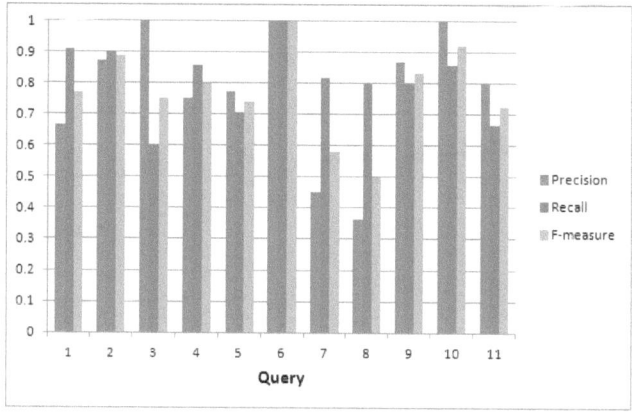

Fig. 4. Results of Experiment 1

Fig. 4 presents the results of experiment 1. The average precision is 0.78, the recall is 0.81, and the F-measure is 0.77. From these improvements, we can extract the information included in comparison articles (Japanese) but not that included in the browsing article (English) using our proposed method. The F-measures of Manner and Manzai indicate good results. A feature of these articles is that there is less information in the browsing article (English) but rich information in the comparison articles (Japanese). Furthermore, the related articles of the comparison articles closely resemble the basic article, which has same title of browsing article. However, the F-measure of Iaido and Imagawayaki are bad results. In this case, the basic article (Japanese) is not rich information, which means that the information of the title that is the same as the browsing content is divided into multiple pages in the comparison language. Our proposed method is not good for information of these types. We must consider another method for such information in the near future.

Experiment 2: Determination of Threshold of Calculating Quality

In this experiment, we set the appropriate the threshold λ of calculating quality of candidate of complementary information by changing the λ. For this experiment, we set the threshold λ from 10% to 100% at 10% intervals. $\lambda = 100\%$ means that the system outputs a browsing article with all comparison articles, and $\lambda = 10\%$ means that the system outputs a browsing article with 10% of good quality comparison articles. We do not set $\lambda = 0\%$ because, this threshold means that the system only outputs a browsing article, do not output any comparison article. Then we calculated the precision, recall, and F-measure for the respective thresholds.

Fig. 5 presents results of experiment 2. The highest F-measure is 20%, which means that the top 20% have high quality. Then we determine the threshold λ as 20%.

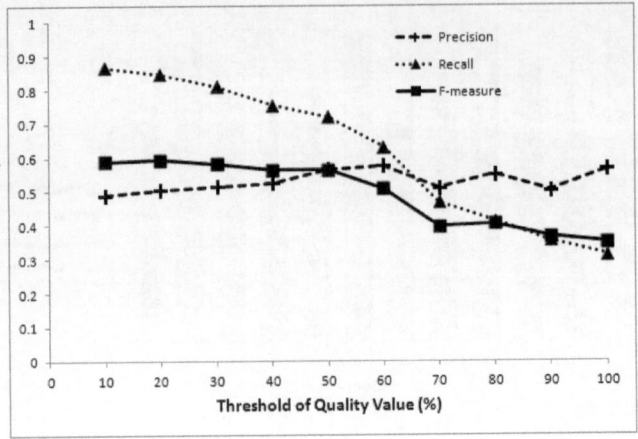

Fig. 5. Results of Experiment 2

Experiment 3: Availability of Extracting High-Quality Complementary Information

We assess the availability of the extracting high-quality complementary information using the result for threshold λ above.

Fig. 6 presents results of the experiment. The average of precision is 0.50, the recall is 0.84, and the F-measure is 0.60. The averages of good results are Bowing, Manners, and Iaido. Their themes are firm. Their sentences tend to be formal. Some examples of good results are that in a case of Rock–paper–scissors, the system extracts the section of how to win Rock–paper–scissors as a candidate of complementary information. After calculating the quality, the system removes the section from the candidate of complementary information

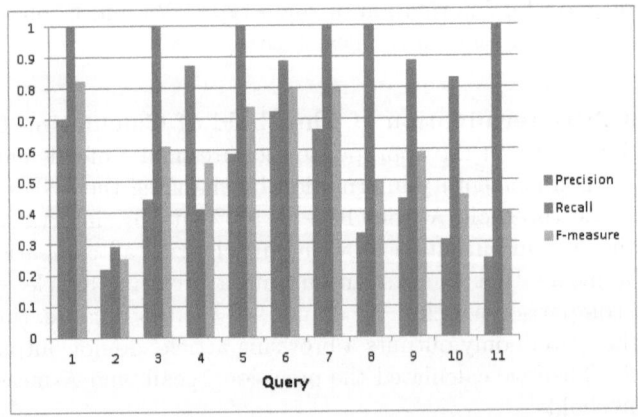

Fig. 6. Results of Experiment 3

because its quality is low. Many theories exist about how to win at Rock–paper–scissors, the participants judged the quality of the section as low. It is a good case. However, an example of bad results is that in the case of Doraemon, a comic character, for which the system extracts the plot section as complementary information. However, the sentences in the section are of low quality. Therefore, it is difficult for ordinary readers to understand the meaning of the sentence.

5   Conclusion

As described in this paper, we proposed a method for extracting high-quality complementary information of Wikipedia by comparing multilingual Wikipedia. Our proposed complementary information is relevant to the browsing content but different (or lack) and high-quality information in the comparison Wikipedia. This paper presented three important technical points, 1) Extracting comparison articles from Wikipedia, 2) Extracting difference information, 3) Assessing Wikipedia information quality.

We conducted an experiment demonstrating the accuracy of our proposed relevance degree by comparing cosine similarity and the accuracy of extracting different information. Results show that our proposed relevance degree is superior to that obtained using existing methods, and that our proposed method is available for extraction of different information from multilingual Wikipedia.

Future work will include the following tasks.

- **Word Sense Disambiguation**
  Many instances of word sense disambiguation exist, but we have ignored such cases in the analyses presented in this paper. We intend to consider word sense disambiguation in later investigations.
- **Presenting Complementary Information**
  As described in this paper, we extracted complementary information by comparing different language versions of Wikipedia. We should also produce and assess a good user interface.
- **Comparing Other Languages**
  As described in this paper, we compared the English Wikipedia with the Japanese Wikipedia in our experiments. However, Wikipedia of many languages exist. In the near future, we intend to compare the Chinese Wikipedia, Korean Wikipedia, and others.

Acknowledgments. This work was partially supported by Research Institute of Konan University, and by Japan Society for the Promotion of Science, Grants-in-Aid for Scientific Research (23700113 and 24500134) .

References

1. Adar, E., Skinner, M., Weld, D.S.: Information arbitrage across multi-lingual wikipedia. In: Proceedings of the Second ACM International Conference on Web Search and Data Mining, WSDM 2009, pp. 94–103. ACM Press, New York (2009)

2. Adler, B., de Alfaro, L.: A content-driven reputation system for the Wikipedia. In: Proceedings of the 16th International Conference on World Wide Web (WWW 2007), pp. 261–270 (2007)

3. Chen, Z., Liu, S., Wenyin, L., Pu, G., Ma, W.Y.: Building a web thesaurus from web link structure. In: Proceedings of the 26th Annual International ACM SIGIR Conference on Research and Development in Informaion Retrieval, pp. 48–55 (2003)

4. Eklou, D., Asano, Y., Yoshikawa, M.: How the web can help wikipedia: a study on information complementation of wikipedia by the web. In: Proceedings of the 6th International Conference on Ubiquitous Information Management and Communication, ICUIMC 2012, pp. 9:1–9:10. ACM, New York (2012), http://doi.acm.org/10.1145/2184751.2184763

5. Fujiwara, Y., Suzuki, Y., Konishi, Y., Nadamoto, A.: Extracting Difference Information from Multilingual Wikipedia. In: Sheng, Q.Z., Wang, G., Jensen, C.S., Xu, G. (eds.) APWeb 2012. LNCS, vol. 7235, pp. 496–503. Springer, Heidelberg (2012)

6. Kamps, J., Koolen, M.: Is wikipedia link structure different? In: Proceedings of the Second ACM International Conference on Web Search and Data Mining, pp. 232–241 (2009)

7. Ma, Q., Nadamoto, A., Tanaka, K.: Complementary information retrieval for cross-media news content. Inf. Syst. 31(7), 659–678 (2006), http://dx.doi.org/10.1016/j.is.2005.12.004

8. Milne, D.: Computing semantic relatedness using wikipedia link structure. In: Proc. of New Zealand Computer Science Research Student Conference, NZCSRSC 2007, CDROM (2007)

9. Milne, D., Medelyan, O., Witten, I.H.: Mining Domain-Specific thesauri from wikipedia: A case study. In: WI 2006: Proceedings of the 2006 IEEE/WIC/ACM International Conference on Web Intelligence, pp. 442–448 (2006)

10. Nakatani, M., Jatowt, A., Tanaka, K.: Adaptive ranking of search results by considering user's comprehension. In: Proceedings of the 4th International Conference on Ubiquitous Information Management and Communication, ICUIMC 2010, CDROM (2010)

11. Nakayama, K., Hara, T., Nishio, S.: Wikipedia Mining for an Association Web Thesaurus Construction. In: Benatallah, B., Casati, F., Georgakopoulos, D., Bartolini, C., Sadiq, W., Godart, C. (eds.) WISE 2007. LNCS, vol. 4831, pp. 322–334. Springer, Heidelberg (2007)

12. Strube, M., Ponzetto, S.P.: WikiRelate! computing semantic relatedness using wikipedia. In: Proceedings of the 21st International conference on Artificial intelligence (AAAI 2006), pp. 1419–1424 (2006)

13. Stvilia, B., Gasser, L., Twidale, M.B., Smith, L.C.: A framework for information quality assessment. Journal of the American Society for Information Science and Technology 58(12), 1720–1733 (2007)

14. Stvilia, B., Twidale, M.B., Smith, L.C., Gasser, L.: Information quality work organization in wikipedia. Journal of the American Society for Information Science and Technology 59(6), 983–1001 (2008)

Generating Tourism Path
from Trajectories and Geo-Photos

Zhixing Zeng, Richong Zhang, Xudong Liu, Xiaohui Guo, and Hailong Sun

School of Computer Science and Engineering, Beihang University,
Beijing, 100191 China
{zengzx,zhangrc,liuxd,guoxh,sunhl}@act.buaa.edu.cn

Abstract. The pervasiveness of GPS devices enables tourists recording
their trajectories and uploading geo-tagged photos. Geo-related data has
emerged as new source for travelers to refer to when making tourism decisions. As the increasing availability of these user-generated experiences
on the social networks, there is a need to automatically discovering useful
patterns for potential travelers. In this paper, we propose a tourism path
by incorporating the trajectories and geo-photos. Specifically, we provide
an algorithm for precisely matching user-uploaded photos to tourism sites
and a density based clustering approach to identify the place of interests
inside tourism sites. We then build a model that adapts the well-known
HITS algorithm to detect interesting points and trajectories with high
utility scores and design an algorithm for efficiently computing rational
routes for visiting tourism sites. Finally, experimental results illustrate
the advantage of the proposed density-based algorithm and confirm the
effectiveness applicability of our tourism path discovering approach.

1   Introduction

Geo-related data, such as geo-photos and geo-spatial trajectories, is growing
exponentially with the pervasive usage of GPS-enabled mobile devices. The pervasiveness of GPS-enabled devices makes it possible for travelers sharing their
geo-location related information in social web communities. Travelers may discover popular visiting places, common visiting sequences and rational visiting
routes from these location based social networks. Especially when tourists are
planning to visit a tourism site, the prior knowledge of the attractiveness of
interesting points and detailed tour paths are needed to be collected to assist
travelers to make visiting plans. However, these patterns often implicitly exist
in various types of data, therefore there is a need for automatically discovering
patterns from the huge spatial data and unifying and matching objects across
social communities.

To solve this problem, existing travel route recommendation approaches [1–3]
simply extract coordinates and timestamps from uploaded photos to generate the
possible visiting paths. However, these approaches are limited to as discreteness
of the geo location of photos, and the extracted path from geo-photos is rough
and only indicates an approximate direction of visiting sequences. In addition,

X.S. Wang et al. (Eds.): WISE 2012, LNCS 7651, pp. 199–212, 2012.

some other researches [4–7] discover temporal-spatial sequences for travelers from collected GPS trajectories. The limitation of these approaches is that places of interest cannot be precisely extracted from raw trajectories and the popularity of the interesting points that trajectories pass is not taken into consideration.

Aware of these limitations, in this paper, we present a more profound approach to discover useful patterns from cross media community and build high quality tourism paths. Particularly, we take the advantages of different communities to mine interesting points. In specific, we present an algorithm for aggregating and mapping geo-photos with geo-spatial trajectory data for tourism sites. We then propose a density based clustering approach to identify the place of interest and estimate the popularity of each place. Furthermore, we build a model based on HITS algorithm to evaluate the quality of trajectories and the popularity of interesting points. Finally, we design a tour path constructing algorithm to generate rational tourism paths under a specified time limitation. The evaluation on Geolife[1] trajectory data and geo-tagged photos from Flickr confirms the efficiency and effectiveness of our proposed approaches.

The rest of the paper is organized as follows. In Sect. 2, we discuss the related works; In Sect. 3, we introduce the approach to discover interesting points and visiting path; Then we evaluate the performances of our proposed algorithms in Sect. 4; Finally in Sect. 5, we conclude the paper.

2   Related Work

Trajectory recommendation researchers have made significant advances in the recent years through trajectory mining [8, 4–6], travelogue mining [9, 10] and photo mining [11, 12, 2, 3].

Some approaches apply clustering algorithm in the photo mapping, such as in [11, 13], authors proposed an approach for clustering photo collections to find popular locations with clustering algorithm. K-means clustering [13] and mean-shift [11] are also been introduced to discover locations and photos. For the same purpose of mapping photos to points with semantic information in [1], textual tags and geo-distance are leveraged to find photos of a given point.

Authors in [3] propose a probabilistic personalized travel recommendation model using knowledge from textural and image features. Moreover, recently, photos are made use to construct visiting sequences with time attribute in order to analyze the pattern of trips in [3, 1, 11]. However, photo sequences can only indicate the order of visited interesting locations but can not provide detailed routes or walking paths between these points. Researches in [6, 7] both recommend travel sequences by mining tourist-generated GPS trajectories. In the paper [6], authors detected the classical travel sequences using location interests and users' travel experiences. An itinerary model concerning elapsed time ratio, stay time ratio, interest density ratio and classical travel sequence ratio was proposed in [7] to recommend itineraries. These two researches convert the

---

[1] http://research.microsoft.com/en-us/projects/geolife/

trajectories to locations sequence and then discard the explicit GPS points sequence. However, these GPS points sequence can show the clear path connecting two locations.

Motivated by these facts, in this paper, we proposed an improved density-based clustering algorithm to mine attractive places in tourism sites. Then we extract travel paths connecting these attractive places. At last, an intelligent tour path recommendation approach is proposed to generate a specific path within time limitation for tourists.

3 Methodology

Our work focuses on analyzing previous user-uploaded trajectories and photos, and discovering high quality visiting paths for tourism sites. In this section, we begin with introducing the definitions used throughout this paper. We then propose the algorithm for finding attractive spots from photos taken by previous travelers. With these discovered attractive spots, we build a graph for each tourism site, define utility functions for trajectories passing these spots and provide an algorithm for composing possible high utility trajectories.

3.1 Preliminary

To build a generic model for solving the tourism site visiting path discovering problem, we formulate definitions which would be used in this paper.

Definition 1 (Trajectory). *As defined in [14, 4], a trajectory, denoted by t, is a sequence of GPS points. Each GPS point, denoted by p, consists of latitude (p.lat), longitude (p.lng) and timestamp (p.time). Formally, a trajectory can be represented by:*

$$t = \{p_0,\ p_1,\ p_2,\ \ldots, p_n\}, \ \forall i \in [0, n]\ p_i.time < p_{i+1}.time$$

As the interest of this paper is on finding movement patterns in a tourism site, we only consider tourism trajectories in the following sections.

Definition 2 (Geo-photo). *A geo-photo, denoted by gp, represents a photo that is geo-tagged and consists of a set of tags, latitude(lat) and longitude(lng) of where this photo is taken. The set of geo-photos in i_{th} tourism site can be represented by GP_i.*

Definition 3 (Interesting Point). *An interesting point, denoted by ip, means an attractive place inside a tourism site.*

In this work, we assume that travelers usually take photos at interesting points, such the interesting points can be extracted from the geo-photos set GP. We denote the set of interest points of i_{th} tourism site as IP_i.

Definition 4 (Tourism Site). *A tourism site consists of a set of interesting points, photos and trajectories. Formally, the ith one can be represented by:*

$$s =< name_i, lat_i, lng_i, IP_i, GP_i, T_i >$$

where T_i denotes the trajectories set in i^{th} tourism site. The coordinate lat_i and lng_i represent roughly location on map. It will be used in mapping photos approach. It can be get from Google Geocoding API[2].

Definition 5 (Trajectory Segment). *A trajectory segment, denoted by ts, is a segment of a trajectory connecting two adjacent interesting points. The trajectory segment connecting the jth and k^{th} interesting points can be denoted by $ts_{j,k}$.*

We note that there may exist several *ts* connecting two same interesting points, we may introduce a superscript x on *ts* (ts^x) to distinguish these trajectory segments.

Definition 6 (Tourism Site Graph). *In this paper, we use a graph $G =< E, V >$ where $E = \{ts_1, ts_2, \cdots, ts_n\}$ and $V = \{ip_1, ip_2, \cdots, ip_m\}$ to formal indicate the interesting points topology in a tourism site. In addition, we denote the entries and exits as sets V_s and V_e standing for entries and exits respectively.*

Definition 7 (Route). *A route, denoted by r, is a trajectory that consists of some interesting points and trajectory segments which can be formally defined as below:*

$$< ip_0 \xrightarrow{ts_{01}^p} ip_1 \xrightarrow{ts_{12}^q} ip_2 \ldots ip_n >, ip_0 \in V_s, ip_n \in V_e$$

where ts_{ij}^p denotes the pth trajectory segment of ones connecting from ip_i to ip_j.

3.2   Photo Mining

Given a set of photos GP and a set of tourism sites S, the first goal of our system is to map photos to their corresponding tourism sites and extract interesting points in each site.

Mapping Photos to Tourism Sites. When mapping photos to tourism sites, by photo annotations or tags, some unexpected photos may be located inside or near a specific tourism site, although, they share the same tags with expected ones. The solution proposed in [11] used both geo and tag information to map photos whose locations are close to a certain tourism site within 100 meters, as well as photos whose tags are matched with the expected ones. However, the size of tourism sites varies significantly. Photos would be mismatched merely by a fixed range.

---

[2] https://developers.google.com/maps/documentation/geocoding/

In this approach, we propose a method which can adapt to different sizes of tourism sites for filtering photos. Given a set of geo-photos GP and a specified tourism site s, we take the photos tagged with $s.name$ as the photo candidate set for s. To adaptively adjust various ranges of tourism sites, we exploit DBSCAN[15], a density-based cluster algorithm, to dynamically and precisely discover photos for each tourism site. We treat a tourism site and its every geo-photo as points to be clustering. After clustering, photo points sharing the same cluster with the tourism site point are mapped to the tourism site s. Empirically, two parameters of DBSCAN, Eps and $MinPts$, are set as $Eps = 25m$ and $MinPts = 2$.

Algorithm 1. Find Neighborhood of Each Point(Geo-photo)

Require: Input: $Points$, K, Eps_D, Eps_S $MinPts$.
    Output: A link array neighborhood
1: **for** $i = 1$ **to** $Points.size$ **do**
2:    **for** $j = 1$ **to** $Points.size$ **do**
3:       **if** $dist(i, j) < kdistance[i]$ **then**
4:          $sortedInsert(knearest[i], j)$;
5:          $kdistance[i] \leftarrow dist(knearest[i][last], i)$;
6:       **end if**
7:    **end for**
8: **end for**
9: **for** $i = 1$ **to** $Points.size$ **do**
10:    **for** $j = 1$ **to** $Points.size$ **do**
11:       **if** $dist(i, j) < Eps_D$ and $similarity[i, j] < Eps_S$ **then**
12:          $neighborhood[i].add(j)$;
13:       **end if**
14:    **end for**
15: **end for**
16: **return** $neighborhood$;

Interesting Points Extraction. To extract interesting points of a tourism site s from mapped photos set GP, we use a hybrid density-based clustering algorithm based on DBSCAN [15] and Shared Nearest Neighbor (SNN) [16]. We find that DBSCAN cannot achieve a good clustering performance when data density is relatively high, while SNN is not effective for distinguishing noise data points as the data density is low. However, photo densities inside a tourism site vary significantly at different area. We take the best of both algorithms to improve the performance of clustering. We let Eps_D and Eps_S denote the original parameter Eps in DBSCAN and SNN respectively. According to the definitions in [17], we redefine the neighborhood of a point p as follows:

$$N_{Eps}(p) = \{q \in P | dist(p, q) \leq Eps_D \text{ and } similarity(p, q) \geq Eps_S\} \quad (1)$$

where $dist(p, q)$ denotes the distance between p and q, and $similarity(p, q)$ denotes the SNN similarity [16] between p and q. After finding neighborhood, the subsequent algorithm are the same as that defined in the DBSCAN algorithm.

Algorithm 1 depicts the process of obtaining the neighborhood of each point that represents a geo-photo here. We let $k - distance$ denote the distance from a specific point to its kth nearest point and use $kdistance[i]$ at line 3 to denote the ith point's k-distance. At line 4, $knearest[i]$ is a K length array to save k nearest points of the ith point in ascending order. In [15], authors define a function $regionQuery(Point, Eps)$ to represent the neighborhood of $Point$ within Eps. In this paper, we redefine this function as $regionQuery(p)$, p denoting the index of a point, which returns $neighborhood[p]$ obtained in Algorithm 1.

Each cluster generated by this process can be treated as an interesting points and the set of generated interesting points compose the vertex set, V, for the tourism graph representation of G.

3.3   Trajectory Processing

As we focus on trajectory inside tourism sites in this study, we first discover tourism trajectories inside each tourism site and further work in this paper is all on these filtered trajectories.

Trajectory Segment Extraction. To discover trajectories inside a tourism site, we need to confirm the boundaries of every tourism site. We then filter out the part of a user-generated trajectory located inside every boundary. As we have discovered the cluster of photos for each tourism site in the previous subsection, the most northeast and most southwest points of these photos can be identified and we build a rectangular area which covers all photos mapped to this site.

It is obvious that we can identify the interesting points sequence of a trajectory t inside a tourism site s by finding the correlation between the discovered interesting points from the photo set GP and the trajectory t in s. Formally, we denote the interesting points sequences by $< ip_1 \rightarrow ip_2 \rightarrow \cdots ip_n >$ when it satisfies following conditions:

1. $\forall ip_i, \exists p_i \in t, \exists gp_j \in GP_i, dist(p_i, gp_j) < \varepsilon$ and $dist(gp_j, c_i) < \sigma$, where GP_i is the photos set clustered to ip_i, c_i is the central geo point of ip_i.
2. $p_1.time < p_2.time < \cdots p_m.time$, namely the interesting point sequence of t is in chronological order.

With the discovered interesting point sequences, it can be easily convert the sequences into several trajectory segments. We denote the set of trajectory segments on site t_i as TS_i and each segment ts including two interesting points it connects, the time cost and the path. The path is a geo point sequence representing the detailed movement pattern between two successive interesting points. The set of trajectory segments TS_i can also be seen as the set of edges E_i in the tourism site graph G_i.

3.4 Discovering Visiting Path

To find high quality visiting path inside a specific tourism site, interesting points that are attractive to tourists should be discovered. In this study, we exploit HITS algorithm[18] to identify the popularity of interesting points. In the context of this work, the authority scores of trajectories can be seen as the typicality of trajectories and the hub scores of interesting points can represent the popularity of interesting points. We let $auth(t)$ denotes the authority score and $hub(ip)$ denotes the hub score. A matrix M representing the correlation between trajectories and interesting points is introduced where rows correspond to the trajectories and columns correspond to the interesting points, and item v_{ij} in M represents whether t_i passes ip_j (passing denoted by 1, otherwise by 0). In this work, we use the following two formulas to calculate the authorities and hubs. First, the authority scores of a trajectory t_i is formulated as:

$$auth_n(t_i) = \frac{\alpha_i \cdot hub_{n-1}(ip)}{\sqrt{(M \cdot hub_{n-1}(ip))^T \cdot (M \cdot hub_{n-1}(ip))}} \tag{2}$$

where $\alpha_i = (v_{i1}, v_{i2}, \cdots, v_{in})$. Then, the hub score of an interesting point ip_i is formulated by:

$$hub_n(ip_j) = \frac{\beta_j \cdot auth_n(t)}{\sqrt{(M^T \cdot auth_n(t))^T \cdot (M^T \cdot auth_n(t))}} \tag{3}$$

where $\beta_j = (v_{1j}, v_{2j}, \cdots, v_{mj})$.

In (2) and (3), $auth(t) = (auth(t_1), auth(t_2), \cdots, auth(t_n))$, $hub(ip) = (hub(ip_1), hub(ip_2), \cdots, hub(ip_m))$.

We iteratively compute the authorities and hubs until algorithm converges. The hub scores and authority scores are initialized as follows:

$$hub_0(ip_i) = \frac{N(ip_i)}{\sum_{ip \in s} N(ip)}$$

$$auth_0(s_i) = 1$$

where $N(ip_i)$ denote the number of photos in the cluster of ip_i, s denotes a specified tourism site.

In practice, time constrains is often an important factor to be considered in path planning approaches. The time cost of a path, in this study, is formulated by:

$$t(r) = \sum_{i=0}^{n-1} t(ts^x_{i,i+1})$$

where $ts^x_{i\,i+1}$ denotes the xth trajectory segment connecting ip_i and ip_{i+1}, and $t(ts^x_{i\,i+1})$ denotes the time cost of $ts^x_{i\,i+1}$.

The utility of the planned path is measure to represent the quality of paths. We let $u(r)$ denote the utility of the path trajectory r:

$$u(r) = \sum_{i=0}^{n} hub(ip_i) \tag{4}$$

where r contains an interesting point sequence $< ip_1 \to ip_2 \to \cdots ip_n >$, $hub(ip_i)$ denotes the hub score of ip_i generated by the HITS algorithm.

Our objective is to find a path r with the maximum utility $u(r)$ and r meets the constraint of $t(r) < \varphi$, where φ is time limitation requested by a tourist.

To improve the efficiency of our algorithm, we prepare a vector τ, which denotes minimum time that each interesting point costs to exit, based on shortest path algorithm Dijkstra [19], i.e. $\tau(ip)$ denotes the minimum time it costs to exit from the interesting point ip. In Algorithm 2, we describe a backtracking algorithm which discovers the best route. We apply τ as a pruning condition to reduce the time complexity at line 5. At the same line, $isPassed(path, l, ip)$ denotes if $path$ contains the trajectory segment from l to ip.

Such, the paths in tourism sites with the best utilities and under the user requested constraints are discovered.

Algorithm 2. The algorithm for mining route with time limitation

---

Require: Input: A graph $G =< E, V >$, a vector τ denotes minimum time each interesting point costs to exit, a sequence $path$ including chosen trajectory segments and interesting points, time limitation t and current interesting point l. Output: A route r.

1: **if** $l \in V_e$ **and** $u(path) > u(best)$ **then**
2:     $best \leftarrow path$
3: **end if**
4: **for all** $ip \in V$ **do**
5:     **if** $\tau(ip) \geq t$ **and not** $isPassed(path, l, ip)$ **then**
6:         **for all** ts connecting from l to ip **and** ts not in $path$ **do**
7:             **if** $t - t(ts) > \tau(ip)$ **then**
8:                 $newPath \leftarrow path$ add ip, ts;
9:                 $RouteMining(G, \tau, newPath, t - t(ts), ip)$;
10:            **end if**
11:        **end for**
12:    **end if**
13: **end for**

---

4 Experiments

In this section, we first present the dataset. Second, we introduce the cluster and path generation evaluation approaches. Last, some discussions of the results are given.

Table 1. The number of photos per considered tourism site

Destination	Summer Palace	Temple of Heaven	Houhai	Forbidden City	Tiananmen Square
Photo Number	4250	1996	1058	3809	727

4.1   Dataset

The tourism trajectories are filtered out from Geolife dataset which is collected by 178 users in a period of over four years and contains 17621 trajectories. As the majority of the data is created in Beijing, we only consider the tourism sites in Beijing and crawled 23,649 geo-tagged photos taken at Beijing from Flickr. The tourism sites are crawled from tourism web sites like TripAdvisor[3]. Their coordinates are queried from Google Geocoding. In Table 1, we present the number of geo-photos of five considered tourism sites on which we will do experiment.

4.2   Clustering Evaluation

In this evaluation, we choose $Eps_D = 30$ and $MinPts = 5$ for DBSCAN, $KN = 9$, $Eps_S = 6$ and $MinPts = 5$ for SNN. These parameters are also used by our proposed Hybrid algorithm. We let σ denote the average distance of all elements in each cluster.

$$\sigma = \frac{1}{n} \sum_{p \in IP} dist(p, c_{ip}), c_{ip} = < \frac{1}{n} \sum_{p \in IP} p.lat, \frac{1}{n} \sum_{p \in IP} p.lng >$$

where IP denotes the photo set of a specific interesting point ip, n is the number of photos and c_{ip} denotes the center coordinate of ip.

The goal of this evaluation is to compare DBSCAN, SNN and Hybrid algorithms with respect to the clustering result. In Fig. 1, we illustrate five greatest σ of each clustering algorithm on three different tourism sites. It can be observed that the proposed Hybrid algorithm outperforms the other two compared density-based clustering algorithms in terms of the intra-cluster average distance.

As finding interesting points is the main purpose of the proposed algorithm. We evaluate the results based on precision and recall. We find ten people who are quite familiar with these five tourism sites. They can tell how many interesting points there are and where they are located in each tourism site. They also check out the clustering results that whether a cluster can represent an interesting point. We let the number of confirmed clustering interesting points divided by the number of all ones be the recall and divided by the number of clusters be the precision. Then we compare the results with respect to F-measure as shown in Fig. 2. We can observe that the Hybrid algorithm outperforms other two.

To visually illustrate the cluster results, we plot the exacted clusters of photos on a map, as shown in Fig. 3. In Fig. 3(b), we show the photo clusters generated

---

[3] http://www.tripadvisor.com

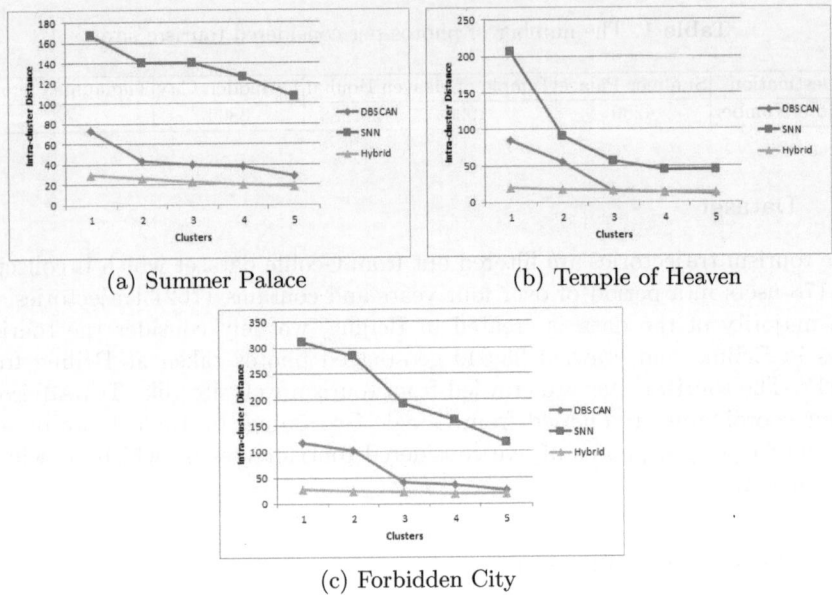

(a) Summer Palace   (b) Temple of Heaven

(c) Forbidden City

Fig. 1. Five the greatest average intra-cluster distance by DBSACN, SNN and Hybrid on three tourism sites

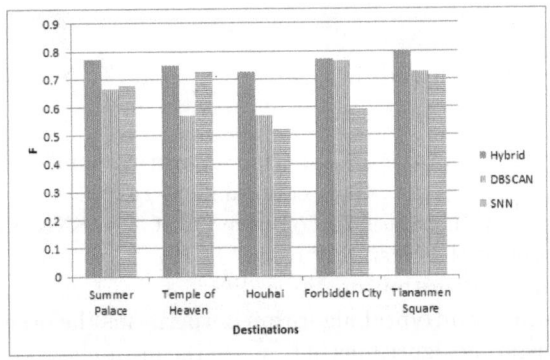

Fig. 2. Comparision of the clustering results on five tourism sites based on F-measure

by DBSCAN. It can be observed that DBSCAN performs well at proper density area, but it regards a high density area, like area A in Fig. 3(b), as a cluster rather than divides it into clusters. So σ of the cluster covering high density area is great. In Fig. 3(c), we show the photo clusters calculated by SNN and it can be seen that it clusters well at the same area. However, as SNN only concerning nearest neighborhoods without distance limit, it mistakenly integrates some low density area as clusters like area B in Fig. 3(c). As shown in Fig. 3(d), Hybrid overcomes these two limitations and achieves a better clustering performance.

In summary, in comparison with other commonly-used density-based algorithms in terms of cluster performances, our proposed hybrid algorithm outperforms them.

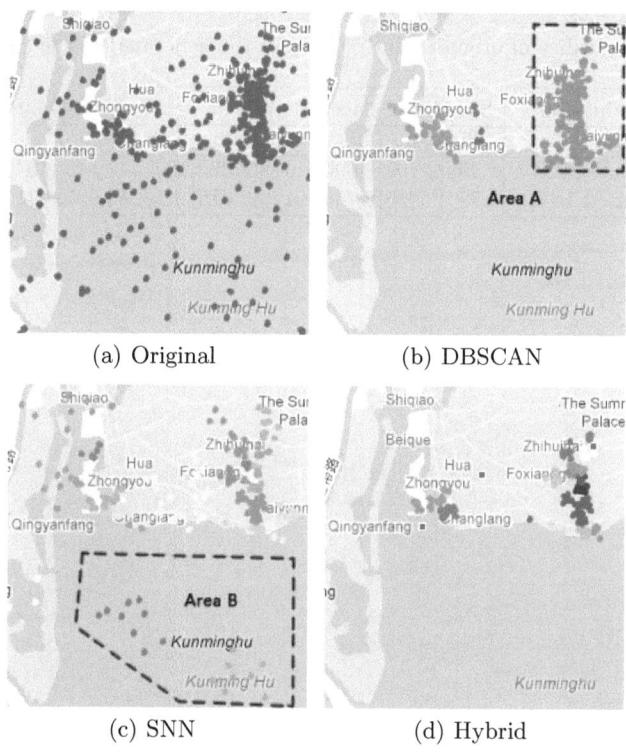

(a) Original   (b) DBSCAN

(c) SNN   (d) Hybrid

Fig. 3. Part of clustering result of three algorithms in the Summer Palace. Dots with a same color are in one cluster.

4.3 Recommended Route Evaluation

In order to evaluate the tour path generation approach, we find three best original tour paths within three different periods of time at three tourism sites, the Summer Palace, the Temple of Heaven and Houhai. We evaluate the utility, as defined in E.q. 4, of the generated tourism path by the hybrid algorithm. Table 2 show the results for the Summer Palace, the Temple of Heaven and Houhai respectively. From the experimental results, we find that the generated tour paths all achieve better utility values. We note that the original tour path utility increase slower than generated. One of the possible reason is because that the longer time the tourists travel for, the more time they will take to rest.

Similar as the previous experiment, we show some typical generated tourism path on the map. Fig. 4(a) and Fig. 4(b) illustrate a 4 hours and a 9 hours generated tour path in the Summer Palace. From these two figures we can observed that in the 9 hours route, our algorithm plans a boating event on the lake as some discovered interesting points are located on the lake. If the tourist only request for a short time period, the representative interesting points take priority over others, as shown in Fig. 4(c) and Fig. 4(d).

Table 2. The utility of original tour paths and the generated in the Summer Palace

Destination	Summer Palace			Temple of Heaven			Houhai		
Time	2h	4h	9h	1h	2h	3h	1h	2h	3h
Original Utility	2.409	2.797	3.000	1.908	2.072	2.615	1.843	2.043	2.384
Generated Utility	3.797	5.915	6.756	2.143	2.615	4.190	4.987	6.415	6.978

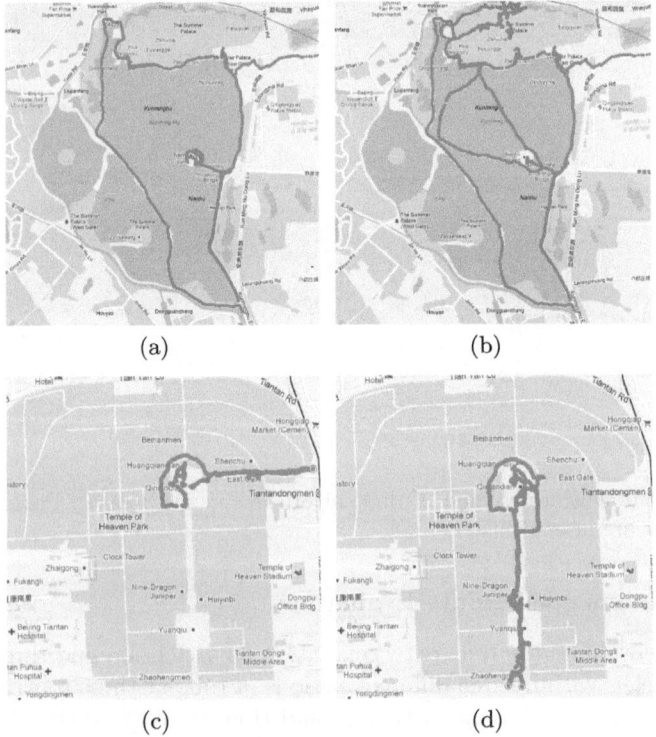

(a)   (b)

(c)   (d)

Fig. 4. Four generated tour routes: (a) is a 4 hours tour route in the Summer Palace; (b) is a 9 hours route in Summer Palace; (c) is a 1 hour tour route in the Temple of Heaven; (d) is a 2 hours route in Temple of Heaven

4.4   Discussion

Traditional tourism route planning algorithms which are simply invoke APIs from other web site that provide map related services. These approaches no longer serve the increasing needs of travelers for different tourism sites. Indeed, nowadays, travelers would like to visit as much interesting points as they can. However, the quality of interesting points is not always available for route planning algorithm. With the help of user-uploaded geo-photos on the social web, this study takes the advantage of the location histories of previous travelers and presents algorithms for discovering attractive interesting points and organizing a satisfactory visiting path for potential travelers.

5 Conclusions

In this paper, geo-tagged photos and user-generated trajectories are collected to extract travel knowledge for discovering tourism site routes. We first maped photos to tourism sites with an algorithm including annotation matching and density-based clustering. After mapping photos, a hybrid algorithm was proposed to discover interesting points in tourism sites. We then proposed a density based clustering approach to identify the places of interests and estimate the popularity of each place and trajectory. Finally we compared our density-based algorithm to two original ones in terms of average intra-cluster distance and F-measure. The experimental results of route generation have shown the intelligence of the proposed approach.

In the future, we are going to collect more geo related contents to discover the interesting points and their hub score more precisely, to build more complex topologies for tourism sites, and to evaluate the performance of the proposed approach on a much larger data set. We would like to take more aspects, e.g. season, tourism site type and travel time, into account to generate more precise paths.

Acknowledgement. This work was supported by China 863 program (No. 2012AA011203), National Natural Science Foundation of China (No. 61103031), Specialized Research Fund for the Doctoral Program of Higher Education (No. 20111102120016), the State Key Lab for Software Development Environment (No. SKLSDE-2012ZX-12), and the Fundamental Research Funds for the Central Universities (No. YWF-12-LXGY-023) and (No. YWF-12-RHRS-016).

References

1. De Choudhury, M., Feldman, M., Amer-Yahia, S., Golbandi, N., Lempel, R., Yu, C.: Automatic construction of travel itineraries using social breadcrumbs. In: Proceedings of the 21st ACM Conference on Hypertext and Hypermedia, pp. 35–44. ACM (2010)
2. Lu, X., Wang, C., Yang, J., Pang, Y., Zhang, L.: Photo2trip: generating travel routes from geo-tagged photos for trip planning. In: Proceedings of the International Conference on Multimedia, pp. 143–152. ACM (2010)
3. Cheng, A., Chen, Y., Huang, Y., Hsu, W., Liao, H.: Personalized travel recommendation by mining people attributes from community-contributed photos. In: Proceedings of the 19th ACM International Conference on Multimedia, pp. 83–92. ACM (2011)

4. Li, Q., Zheng, Y., Xie, X., Chen, Y., Liu, W., Ma, W.: Mining user similarity based on location history. In: Proceedings of the 16th ACM SIGSPATIAL International Conference on Advances in Geographic Information Systems, vol. 34. ACM (2008)
5. Xiao, X., Zheng, Y., Luo, Q., Xie, X.: Finding similar users using category-based location history. In: Proceedings of the 18th SIGSPATIAL International Conference on Advances in Geographic Information Systems, pp. 442–445. ACM (2010)
6. Zheng, Y., Zhang, L., Xie, X., Ma, W.: Mining interesting locations and travel sequences from gps trajectories. In: Proceedings of the 18th International Conference on World Wide Web, pp. 791–800. ACM (2009)
7. Yoon, H., Zheng, Y., Xie, X., Woo, W.: Social itinerary recommendation from user-generated digital trails. Personal and Ubiquitous Computing, 1–16 (2011)
8. Farrahi, K., Gatica-Perez, D.: What did you do today?: discovering daily routines from large-scale mobile data. In: Proceedings of the 16th ACM International Conference on Multimedia, pp. 849–852. ACM (2008)
9. Hao, Q., Cai, R., Wang, C., Xiao, R., Yang, J., Pang, Y., Zhang, L.: Equip tourists with knowledge mined from travelogues. In: Proceedings of the 19th International Conference on World Wide Web, pp. 401–410. ACM (2010)
10. Ji, R., Xie, X., Yao, H., Ma, W.: Mining city landmarks from blogs by graph modeling. In: Proceedings of the 17th ACM International Conference on Multimedia, pp. 105–114. ACM (2009)
11. Crandall, D., Backstrom, L., Huttenlocher, D., Kleinberg, J.: Mapping the world's photos. In: Proceedings of the 18th International Conference on World Wide Web, pp. 761–770. ACM (2009)
12. Arase, Y., Xie, X., Hara, T., Nishio, S.: Mining people's trips from large scale geotagged photos. In: Proceedings of the International Conference on Multimedia, pp. 133–142. ACM (2010)
13. Kennedy, L., Naaman, M.: Generating diverse and representative image search results for landmarks. In: Proceeding of the 17th International Conference on World Wide Web, pp. 297–306. ACM (2008)
14. Monreale, A., Pinelli, F., Trasarti, R., Giannotti, F.: Wherenext: a location predictor on trajectory pattern mining. In: Proceedings of the 15th ACM SIGKDD International Conference on Knowledge Discovery and Data Mining, pp. 637–646. ACM (2009)
15. Ester, M., Kriegel, H., Sander, J., Xu, X.: A density-based algorithm for discovering clusters in large spatial databases with noise. In: Proceedings of the 2nd International Conference on Knowledge Discovery and Data Mining, vol. 1996, pp. 226–231. AAAI Press (1996)
16. Jarvis, R., Patrick, E.: Clustering using a similarity measure based on shared near neighbors. IEEE Transactions on Computers 100(11), 1025–1034 (1973)
17. Popescu, A., Grefenstette, G.: Mining social media to create personalized recommendations for tourist visits. In: Proceedings of the 2nd International Conference on Computing for Geospatial Research & Applications, vol. 37. ACM (2011)
18. Kleinberg, J.: Hubs, authorities, and communities. ACM Computing Surveys (CSUR) 31(4es), 5 (1999)
19. Dijkstra, E.: A note on two problems in connexion with graphs. Numerische Mathematik 1(1), 269–271 (1959)

A Framework and a Language
for On-Line Analytical Processing on Graphs

Seyed-Mehdi-Reza Beheshti[1], Boualem Benatallah[1],
Hamid Reza Motahari-Nezhad[2], and Mohammad Allahbakhsh[1]

[1] University of New South Wales, Sydney, Australia
{sbeheshti,boualem,mallahbakhsh}@cse.unsw.edu.au
[2] HP Labs Palo Alto, CA, USA
hamid.motahari@hp.com

Abstract. Graphs are essential modeling and analytical objects for representing information networks. Existing approaches, in on-line analytical processing on graphs, took the first step by supporting multi-level and multi-dimensional queries on graphs, but they do not provide a semantic-driven framework and a language to support n-dimensional computations, which are frequent in OLAP environments. The major challenge here is how to extend decision support on multidimensional networks considering both data objects and the relationships among them. Moreover, one of the critical deficiencies of graph query languages, e.g. SPARQL, is the lack of support for n-dimensional computations. In this paper, we propose a graph data model, *GOLAP*, for online analytical processing on graphs. This data model enables extending decision support on multidimensional networks considering both data objects and the relationships among them. Moreover, we extend SPARQL to support n-dimensional computations. The approaches presented in this paper have been implemented on top of FPSPARQL, Folder-Path enabled extension of SPARQL, and experimentally validated on synthetic and real-world datasets.

Keywords: Graph OLAP, SPARQL, Query Processing.

1 Introduction

In traditional databases (e.g., relational DBs), data warehouses and OLAP (On-Line Analytical Processing) technologies [1,7] were conceived to support decision making and multidimensional analysis within organizations. To achieve this, a plethora of OLAP algorithms and tools have been proposed for integrating data, extracting relevant knowledge, and fast analysis of shared business information from a multidimensional point of view. Moreover, several approaches have been presented to support the multidimensional design of a data warehouse. Cubes defined as set of partitions, organized to provide a multi-dimensional and multi-level view, where partitions considered as the unit of granularity. Dimensions defined as perspectives used for looking at the data. Furthermore, OLAP operations have been presented for describing computations on cells, i.e. data rows.

X.S. Wang et al. (Eds.): WISE 2012, LNCS 7651, pp. 213–227, 2012.

Unlike traditional data warehouses, where the focus is on storage and retrieval of data items, graph databases are used to represent information-rich, inter-related and multi-typed networks.

Existing approaches [11,12,26,8,30,20,14,10], in on-line analytical processing on graphs, took the first step by supporting multi-dimensional and multi-level queries on graphs, but they do not provide a semantic-driven framework and a language to support n-dimensional computations, which are frequent in OLAP environments. For example, considering a bibliographical network, it would be interesting to analyze the collaboration patterns (e.g., frequency of collaboration, degree of collaboration, mutual impact, and degree of contribution) among authors or analyze the reputation of a book, an author, or a publisher in a specific year. Such operations, requires supporting n-dimensional computations on graphs, providing multiple views at different granularities, and analyzing set of dimensions coming from the attributes of graph entities, e.g., authors, papers and venues, and the relationship among them.

The major challenges here are: (i) how to extend decision support on multidimensional networks considering both data objects and the relationships among them: traditional OLAP technologies cannot recognize patterns among graph entities. Consequently, enabling users to analyze multidimensional graph data may become complex and cumbersome; and (ii) providing multiple views at different granularities is subjective: depends on the perspective of OLAP analysts how to partition graphs and apply further operations to constructed partitions. The unique contributions of the paper are as follows:

– We propose a graph data model, *GOLAP*, for online analytical processing on graphs. This data model enables extending decision support on multi-dimensional networks considering both data objects and the relationships among them. We use the notions of folder and path nodes to support multi-dimensional and multi-level views over large graphs. We redefine OLAP data elements (e.g., dimensions and measures) by considering the relationships among graph entities as first class objects.

– We extend SPARQL to support n-dimensional computations on graphs. The extension supports partitioning graphs (using folder and path nodes) and allows evaluation of OLAP operations on graphs independently for each partition, providing a natural parallelization of execution. We propose two types of OLAP operations: *assignments*, to apply operations on entity attributes, and *functions*, to apply operations on network structures among entities. GOLAP operations support UPDATE and UPSERT [27] semantics. We describe optimizations and execution strategies possible with the proposed extensions.

The remainder of this paper is organized as follows: We present related work in Section 2. Section 3 presents an example scenario. In Section 4 we present a graph data model for online analytical processing on graphs. In Section 5 we propose a query language for applying OLAP operations on graphs. In Section 6 we describe the query engine implementation and evaluation experiments. Finally, we conclude the paper with a prospect on future work in Section 7.

2 Background and Related Work

OLAP (On-Line Analytical Processing) [1,7] is part of the broader category of business intelligence which encompasses *data decision support*, focusing on inter-actively analyzing multidimensional data from multiple perspectives, and *data mining*, focusing on computational complexity problems. There have been a lot of works, discussed in a recent survey [21] and a book [25], dealing with mul-tidimensional modeling methodologies for OLAP systems, introducing OLAP data elements (e.g., dimensions, measures, and cubes) and their applications. They discuss that one fact (the subjects of analysis) and several dimensions to analyze it give rise to what is known as the data cube. Many works in OLAP focused on efficient computation of data cubes [28,25], clustering/partitioning data warehouses [16,25], and querying multidimensional models [3,27].

In recent years, a new stream of work [11,12,26,8,30,20,14,10] has focused on online analytical processing on graphs. Chen and Qu et. al. [8,20] proposed a conceptual framework for data cubes on graphs and classify their framework into informational (dimensions coming from node attributes) and topological (dimen-sions coming from node and edge attributes) OLAP. Tian et. al. [26] proposed operations to produce a summary graph (by grouping nodes) and controlling the resolutions of summaries (by providing the drill-down and roll-up abilities). Zhao et. al. [30] introduced a new data warehousing model, Graph-Cube, that supports OLAP queries on graphs. They considered both attribute aggregation and structure summarization of the networks. Kämpgen et. al. [14] presented a mapping from statistical Linked Data that conforms to the RDF Data Cube vocabulary. Etcheverry et. al. [10] introduced Open-Cube, an RDFS vocabulary for the specification and publication of multidimensional cubes on the Semantic Web. Although these line of works took the first step to put graphs in a rigid multi-dimensional and multi-level framework, none of them provide a semantic-driven framework and a language to support n-dimensional computations on graphs, considering both informational and topological dimensions of graphs.

Another line of related work [20,9,23,2,13,19] focused on mining and query-ing information networks. BiQL [9] focused on the uniform treatment of nodes and edges in the graph and supports queries that return subgraphs. Other works [24,2,19] focused on clustering and classification of networks by study-ing systematically the methods for mining information networks. Some cluster-ing techniques, e.g. [23,24], developed a ranking-based clustering approach that generates interesting results for both clustering and ranking efficiently [11]. Clas-sification techniques [2,13] classify graphs into a certain number of categories by similarity. All these works provide some kind of (network) summaries incorpo-rates OLAP-style functionalities.

SPARQL [18] is an RDF query language, standardized by the World Wide Web Consortium, for semantic web. Initial specifications of SPARQL did not provide any facility for expressing path queries, reachability expressions, or sup-porting closure property on query results [18]. The recent specifications of the SPARQL 1.1 has addressed some of these limitations in a limited manner. How-ever, no systems have been proposed, yet, to support these features. We proposed

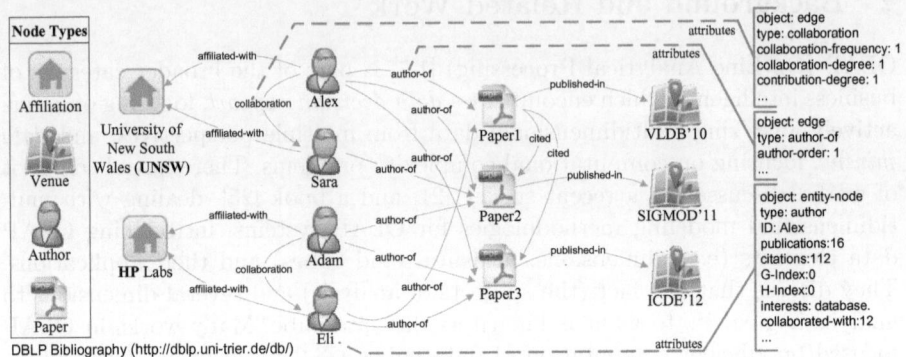

Fig. 1. Motivating Scenario

FPSPARQL [6], a graph query processing engine, which is a Folder-Path enabled extension of SPARQL. FPSPARQL supports primitive graph queries, constructing folder/path nodes, applying further queries to constructed folder/path nodes, and applying external tools and algorithms to graph. In this work, we extend FPSPARQL to support n-dimensional computations on graphs.

3   Example Scenario

DBLP (Digital Bibliography and Library Project) is a computer science bibliography database, listed more than 2 million publications (in May 2012), and tracked most of journals and conference proceedings. We use DBLP dataset to generate graph models, e.g., we enriched nodes typed as 'author' with attributes such as number of publications/citations, G-index, and H-index, and we added edges typed as 'collaboration' between two authors, where collaboration can be defined as a directed link between every two authors having at least one co-authored paper, i.e., to show the properties of the pairwise relation between the two authors such as: frequency of collaboration, degree of collaboration, mutual impact, and degree of contribution. Figure 1 illustrates a simplified DBLP graph. This dataset contains over 1,670,000 nodes (i.e. authors, papers, venues, and affiliations) and over 2,810,000 edges (i.e., author-of, published-in, affiliated-with, cited, and published-in).

4   Representing Analytics over Graphs

In this work, we focus on providing users with an explorative method to analyze multidimensional graph data from multiple perspectives and granularities. We use folder and path nodes (our previous work [6]) to provide network summaries and to support multidimensional and multi-level views over graphs. We present a data model and a query language to facilitate the analysis of online analytical processing on graphs in an explorative manner.

4.1   Data Model

We consider extending decision support on multidimensional networks by introducing a model, i.e. GOLAP. In [6], we proposed a graph data model to represent information networks. We use this data model to represent online analytical processing on attributed graphs. In particular, GOLAP data model supports: (i) uniform representation of nodes and edges; (ii) structured and unstructured entities; and (iii) *folder* and *path* nodes, which can represent a network snapshot, i.e. a subgraph, from multiple perspectives and granularities. We represent entities and relationships as a directed attributed graph $G = (V, E)$ where V is a set of nodes representing entities and folder/path nodes, and E is a set of directed edges, relationships, among nodes.

Entities. An entity is a data object that exists separately and has a unique identity. Entities could be structured or unstructured, and simple or composite. A folder node contains related entities and relationships among them. A path node contains set of related paths, where a *path* is the result of the transitive relationship between two entities. Folder/Path nodes can be defined at several levels of abstractions and can be considered as a network snapshot used to support multidimensional analysis of graphs from multiple perspectives and granularities. Folder/Path nodes may conform to an entity type and can be described by a set of attributes. The set of (related) entities in a folder/path node is the result of a given query that requires grouping graph entities based on set of dimensions coming from the attributes of: (i) graph entities; or (ii) network structures, i.e., patterns among graph entities. The transitive relationship between two entities, in a path, can be codified using *regular expressions* (RE) [6] in which alphabets are the nodes and edges from the graph.

Relationships. A relationship is a directed link between a pair of graph entities, which is associated with a predicate defined on the attributes of nodes that characterizes the relationship. A relationship may conform to a type, can be described by a set of attributes, and can be explicit or implicit.

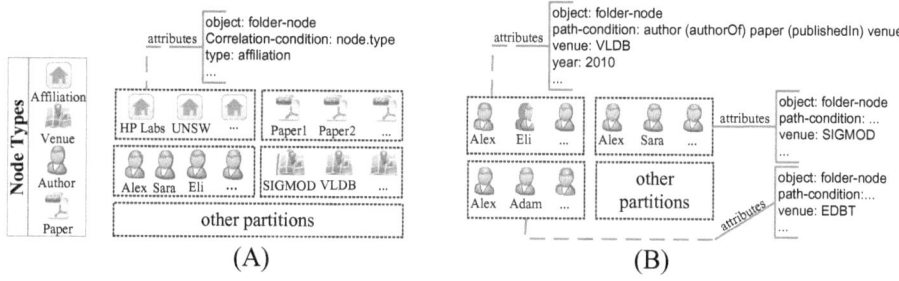

Fig. 2. Examples of folder-partitions: (A) CC-Partitions; and (B) PC-Partitions

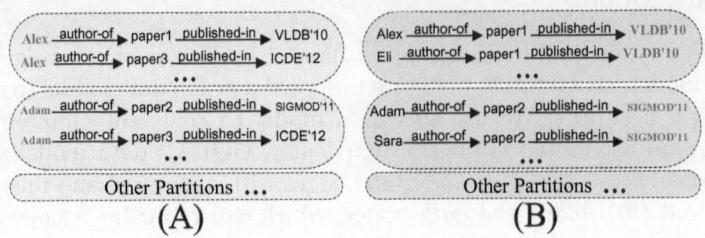

Fig. 3. Examples of path-partitions, e.g., grouping paths by authors(A) and venues(B)

4.2 GOLAP Data Elements

Cubes (Q). A cube enables effective analysis of the graph data from different perspectives and with multiple granularities. We reuse and extend the definition for graph-cube proposed in [30]. In particular, given a multidimensional network N, the graph cube is obtained by restructuring N in all possible aggregations of set of node/edge attributes A, where for each aggregation A' of A, the measure is an aggregate network G' w.r.t. A'. We use partitions to define possible aggregations upon multidimensional networks. In particular, $Q = \{q_1, q_2, ..., q_n\}$ is a set of n cubes, where each q_i is a cube, a placeholder for set of partitions, and can be modeled using folder nodes. A partition can be considered as the unit of granularity, supports multi-dimensional and multi-level views over graphs, and allows evaluation of (OLAP) operations providing a natural parallelization of execution. Three types of partitions are recognized: CC/PC/Path-Partitions.

CC-Partitions. A *Correlation Condition* ψ is a binary predicate defined on the attributes of data objects that allows to identify whether two or more entities (in a given graph) are potentially related [17]. We use correlation condition (CC) to group related entities in a graph, based on set of dimensions coming from the attributes of graph entities. For example, Adam, an OLAP analyst, is interested in partitioning the DBLP graph into a set of related nodes having the same type, and store each partition in a folder node. The correlation condition $\psi(node_x, node_y) : node_x.type = node_y.type$ can be defined over the attribute *type* of two node entities $node_x$ and $node_y$ in the DBLP graph. This predicate is true when $node_x$ and $node_y$ have the same type value and false otherwise. Figure 2-A represents the result of this example.

PC-Partitions. A *Path Condition* (PC) can be used to group related entities in a graph based on set of dimensions coming from the attributes of network structures, patterns, among graph entities. A path condition ϕ is a binary predicate defined on the attributes of a path that allows to identify whether two or more entities (in a given graph) are potentially related through that path. For example, Adam can be interested in partitioning the DBLP graph into a set of related authors satisfying the path condition $\phi(author, venue, RE)$ which defined on the existence of the path codified by the regular expression

RE:[*author*(*authorOf*)*paper*(*publishedIn*)*venue*] between an author and a venue. This predicate is *true* if the path exists, and *false* otherwise. Authors satisfying the path condition, will be added to the related partitions. Folder nodes can be used to represent PC-Partitions. Figure 2-B represents the result of this example.

Path-Partitions. There are cases where OLAP analysts may be interested in partitioning the graph into set of related paths. For example, Adam is interested in partitioning the DBLP graph into a set of related paths having the pattern 'RE: *author*(*authorOf*)*paper*(*publishedIn*)*venue*', to calculate quality metrics for venues by analyzing related papers and/or authors quality metrics. Set of related paths in a path node can be grouped, for example, by authors (Figure 3-A) or by venues (Figure 3-B). Regular expressions can be used to discover paths. Path nodes, a placeholder for a set of related paths, can be used to represent Path-Partitions.

Dimensions (D). Dimensions can be defined as perspectives used for looking at the data. In particular, $D = \{d_1, d_2, ..., d_n\}$ is a set of n dimensions, where each d_i is a dimension name. Each dimension d_i is represented by a set of elements (E) where elements are the nodes and edges of the graph. In particular, $E = \{e_1, e_2, ..., e_m\}$ is a set of m elements, where each e_i is an element name. Each element e_i is represented by a set of attributes (A), where $A = \{a_1, a_2, ..., a_k\}$ is a set of k attributes for element e_i, and each a_i is an attribute name. A dimension d_i can be considered as a query that require grouping graph entities in a certain way. Correlation/Path conditions can be used to define such queries.

Cells (C). A dimension uniquely identify a subgraph within each partition, which we call a *cell*. In particular, $C = \{c_1, c_2, ..., c_n\}$ is a set of n cells, where each c_i is a cell name. In order to identify cells, dimensions may have *levels* used for drilling up/down, where levels enable visiting the general/detailed view of dimensions. For example, it is important to see if the number of publications for a specific author (or group of related authors) are higher in a particular year, or drill down to see if they were higher in a specific month.

Measures (M). Dimensions can be used as an index in order to analyze measures. A measure can be considered as numerical and computational attributes of dimensions' elements. In particular, $M = \{m_1, m_2, ..., m_n\}$ is a set of n measures, where each m_i is a measure name. Measures can be calculated by applying operations to multidimensional graph data, where operations can support upsert/update semantics. Three types of measures can be recognized: (i) *entity attributes*: for example G-index/H-index can be considered as a measure for an author, a node, or collaboration-frequency can be considered as a measure for an edge 'collaboration' between two authors; (ii) *aggregated nodes*: are subgraphs including set of related nodes and relationships among them. For example, in motivating scenario, an OLAP analyst may be interested in the collaborative relationship between researchers affiliated with HP Labs and UNSW; and (iii) *inferred edges*, are new edges which can be added between two nodes in the graph as a result of an OLAP operation on graphs. For example, collaboration between two authors (or group of authors) can be calculated and a weighted edge can be constructed between two authors.

Operations (O). Typical operations on data cubes (e.g., roll-up, drill-down, and slice-and-dice) are supported to explore different multidimensional views and allow interactive querying and analysis of the underlying data. In particular, $O = \{o_1, o_2, ..., o_n\}$ is a set of n operations, where each o_i is an operation name, can be used for describing a computation on cells, and can be ordered based on the dependencies between cells. Operations support UPSERT and UP-DATE semantics. In order to provide a natural parallelization of execution, each operation should be evaluated independently for each partition within a graph cube. Two types of OLAP operations are recognized: (i) *assignments*: are defined to apply operations on entity attributes. Assignments support correlation between the left side (which designates target cells) and right side (which contains expressions involving cells or ranges of cells within the partition), i.e., to simulate the effect of multiple joins and UNIONs using a single access structure; and (ii) *functions*: are defined to apply set of related operations on network structures among entities. Please refer to the extended version of the paper [5], for more details and examples regarding GOLAP data elements.

5   A Query Language for OLAP on Graphs

Online analytical processing on graphs requires dividing objects and relationship among them into partitions, dimensions, and measures. To model that, we extend FPSPARQL by introducing the 'GOLAP' clause. A basic GOLAP query looks like this:

```
Select <measures> Where{
  # <existing parts of a query block>
  GOLAP{
   ?analytic [@CC- | @PC- | @Path-Partition] <identify partitions>.
   ?analytic @dimension <identify dimensions>.
   ?analytic @measure <identify measures>.
   <operation>; <operation>; ... <operation>;  }
}
```

The GOLAP command is evaluated after aggregations but before ORDER BY clause. The variable ?ANALYTIC is defined to specify partitions, dimensions, and measures. We use the '@' symbol for representing attribute edges and distinguishing them from the relationship edges between graph nodes. Attributes '@CC-Partition', '@PC-Partition', and '@Path-Partition' are used to identify CC-, PC-, and Path-partitions. The attribute '@dimension' is used to identify dimensions on entity attributes or network structures among entities. The attribute '@measure' is used to identify expressions computed by GOLAP clause. Finally, set of operations can be defined to identify cells and update/upsert measures. By default, the evaluation of operations occurs in the order of their dependencies. However, there are scenarios in which sequential ordering of evaluation is desired. We provide an explicit processing option, i.e. SEQUENTIAL, for that as in: 'SEQUENTIAL(...<operation>;<operation>;...)'.

Following we will explain FPSPARQL extension for OLAP on graphs through examples. See extended version of the paper [5], for more details and examples.

Example 1 [entity attributes]. Adam is interested in partitioning the DBLP graph into a set of authors having same interests (e.g., cloud and database). He will apply the following operations on constructed partitions: (i) insert a new attribute (i.e., 'contribution-degree') for authors within each partition. For each partition, this attribute will be calculated from division of number of publications for an authors into the sum of publications of all authors in the partition; and (ii) update the value for the 'rank' attribute, a real number, of the authors who has: (a) less than 200 publications to '6'; and (b) more than 200 publications and more than 1000 citations to 50% higher than the authors having same number of publications but less than 1000 citations. Following is the FPSPARQL query for this example.

```
1   Select ?ar,?cd Where{
2     GOLAP{
3       ?analytic  @CC-Partition 'node[type="author"].interest'.
4       ?analytic  @dimension 'publications as ?ap, citations as ?ac'.
5       ?analytic  @measure 'rank as ?ar, contribution-degree as ?cd'.
6       upsert ?cd[*] = ?ap / ?sum_publications;
7       update ?ar[?ap <= 200]= 6;
8       update ?ar[?ap > 200 AND ?ac > 1000] =
9           ?ar[?ap > 200 AND ?ac < 1000]*1.5;
10    } AGGREGATE { (?sum_publications, SUM, {?ap} ) }}
```

In this example, the variable '?analytic' is used to define GOLAP data elements. The attribute '@CC-Partition' partitions the DBLP graph into a set of related nodes. The NODE keyword helps in partitioning the graph by filtering graph nodes through set of nodes attributes (e.g., 'type') and their values in the bracket following by a node attribute, e.g., "node[type='author'].interest" will partition the graph into set of authors having same interest. The attribute '@dimension' is used to identify entity attribute dimensions 'publications' and 'citations'. In order to write operations in a simple way we support aliases (using AS statement). The attribute '@measure' used to identify entity attribute measures 'rank' and 'contribution-degree'. The Keyword UPDATE/UPSERT used to update/upsert a measure in an operation.

Evaluation of operations will apply independently for each partition providing a natural parallelization of execution. An assignment can designate a single object reference (e.g., '[author-id = 2]' author whose ID is 2) or set of objects (e.g., '[publications >= 200]' set of authors having more than 200 publications). Assignments support the correlation between the left side and right side of assignment (lines 8 and 9). To remove unnecessary computations, assignments whose results are not referenced in outer blocks will be removed automatically.

To calculate contribution-degree we use aggregate functions. Using AGGRE-GATE statement (see [4,5] for details about aggregation extension), we calculate sum of publications of all authors in the partition and assign the value to variable '?sum_publications'. Notice that all aggregates are computed before evaluation of operations so they are available for the operations. Then we calculate contribution-degree, '?cd', and add it as a new attribute for all author (line 6).

Example 2 [inferred edges]. Adam is interested in partitioning the DBLP graph into a set of related authors collaborating on (specific) papers. To achieve this, set of dimensions coming from the attributes of authors, papers and the relationship among them should be analyzed. The path-condition ϕ_a(author,paper, RE:'author (authorOf) paper') can be used to partition authors, where each partition will contain authors of a paper. Adam can establish a pairwise attributed edge, i.e. 'collaboration' edge, between authors in each partition. Following is the FPSPARQL query for this example.

```
1 Select ?m,?edge,?n Where{
2   ?path-condition @RE '?author (?authorOf) ?paper'.
3   ?path-condition @groupBy ?paper.
4   ?path-condition @partition-item 'distinct ?author'.
5   # defining regular-expression (@RE) elements
6   ?author @isA entityNode. ?author @type author. ?authorOf @isA edge.
7   ?authorOf @type author-of. ?paper @isA entityNode. ?paper @type paper.
8
9    GOLAP{ ?analytic @PC-Partition ?path-condition.
10     ?analytic @dimension '?m, ?n'. ?analytic @measure '?edge'.
11     function.F1; } }
12
13 functions{
14   F1{ ?m @isA entityNode. ?m @type author. ?m @name ?m_name.
15       ?n @isA entityNode. ?n @type author. ?n @name ?n_name.
16       correlate{ (?m,?n,?edge,FILTER(?m_name <> ?n_name))
17        ?edge @isA edge. ?edge @type 'collaboration'.
18        ?edge @collaboration-frequency '1'.  } } }
```

In this example, '@PC-Partition' (line 9) is used to partition the graph using the path condition ϕ_a. In particular, the variable '?path-condition' defined to identify the regular expression (line 2), group discovered paths by papers (line 3), and add distinct authors who satisfy the regular expression into related partitions (line 4). The second block of codes (lines 5 to 7), defines regular expression elements, e.g., variables '?author', '?authorOf', and '?paper'. Dimensions ('?m' and '?n') and measure ('?edge') are defined in the function block, where the function 'F1' is used to construct an attributed collaboration edge between two authors in partitions. The FUNCTIONS keyword (line 13) is used to define a block of functions. Each function defined by a name (e.g., 'F1') and a block of SPARQL patterns. Defined functions can be called using FUNCTION keyword (line 11) following by function name (e.g., 'function.F1').

In our previous work [6], we introduced CORRELATE statement to establish a directed edge between two nodes in a graph. In function 'F1', variables $?m$ and $?n$ represent authors. The condition in *correlate* statement (i.e., $?m_{name} <> ?n_{name}$) makes sure that only two different authors will be connected. For simplicity reason, in this example, we assign a constant value for collaboration edge attribute (line 18). Please refer to the extended version of the paper [5] for dynamic calculation of edge attributes.

Example 3 [aggregated nodes]. Adam is interested in the collaborative relationship between researchers. He needs to partition the DBLP graph using the path-condition ϕ_b(author,affiliation,RE:'author (affiliated-with) affiliation'), where each partition will contain authors affiliated with specific affiliations. Then he plans to store set of related authors, e.g., an aggregated nodes for authors affiliated with HP or another aggregated node for authors affiliated with UNSW. This way Adam will be able to construct relationships (e.g. 'collaboration' edge) among aggregated nodes in the graph, i.e., to show the collaborative relationship between researchers. Please refer to the extended version of the paper [5] to see the FPSPARQL query and details for this example.

Example 4 [path partitions]. Adam is interested in partitioning the DBLP graph into a set of related paths having the pattern 'RE: author (authorOf) paper (publishedIn) venue', group discovered paths by authors, and store related paths in path nodes. As next step, Adam is interested to: (a) Update the number of publications for each author. This query can be done by counting papers in each partition; (b) Update the number of citations for each author. This can be done by calculating the summation of all papers' citations; and (c) calculate ERA (Excellence in Research for Australia) rank for each author. For example, consider that in ERA ranking, papers published in venues which ranked: (i) 'A*', have 4 points; (ii) 'A', have 1 point; (iii) 'B', have 0 point; and (iv) 'C', have -1 point. In this example, the predicate @PATH-Partition can be used to partition the graph into set of related paths. Please refer to the extended version of the paper [5] to see the FPSPARQL query and details for this example.

Query Optimization. Partitioning of the graph provides an obvious way to parallelize OLAP operations on graph partitions and provide scalability. For efficient processing of partitions, we apply techniques for indexing RDF triples. We use path-based indexing [29] (and cycle elimination) technique to build path indices and partitioning the graph based on patterns among graph entities. From the operation dependencies point of view, the order of evaluation of operations is determined from their dependency graph, where the dependency graph is a directed graph representing dependencies of several operations towards each other. For example, given a several assignments like "i-Index = sum(publications)*c-measure; c-measure = avg(citations)", then 'i-Index' depends on 'c-measure' which should be calculated before 'i-Index'. Moreover, dependency graph can be used to identify the operations that can be *pruned*, e.g., the evaluation of an assignment becomes unnecessary when the cells it updates are not used in the evaluation of other assignments.

FPSPARQL supports *aggregate* and *keyword* search queries [6]. To enhance the capability of the keyword search technique on triple tables, we develop the aggregate keyword search method which finds aggregate groups of entities jointly matching a set of query keywords, i.e. both for folder and path nodes. Moreover, for efficient access to single cells we built a partition level hash access structure. We avoid spilling to disk for evaluating the operations: the partitions will be kept in memory and the operations will evaluated for one partition at a time.

6   Implementation and Experiments

Implementation. The query engine is implemented in Java. Implementation details, including optimization techniques, architecture and graphical representation of the query engine, can be found in [5,6]. We carried out the experiments on two real-world graph *datasets*: (i) DBLP: this dataset was introduced in the example scenario; and (ii) AMZLog: the rating log of Amazon online rating system collected by Leskovec et. al. [15] for analyzing dynamics of viral Marketing.

Evaluation. We report the evaluation results of the query engine extension, GOLAP, in term of performance, scalability and quality of results. All experiments were conducted on a HP PC with a 3.30 Ghz processor, 8 GBytes of memory, and running a 64-bit Linux. To the best of our knowledge, we couldn't find any open source system that support similar functionalities of FPSPARQL to compare with. We report performance in terms of running time in seconds. We applied optimization techniques on DBLP and AMZLog datasets. The optimization took 62 minutes for DBLP and 41 minutes for AMZLog dataset. The performance results for queries in Examples 1 to 4 can be found in [5]. Moreover, we provided 10 CC-, 10 PC-, and 10 Path-Partition queries (each having one operation) for DBLP and AMZLog datasets. Figures 4-A and 4-D show the average execution time for applying the queries to DBLP and AMZLog datasets respectively. We divided each dataset into regular number of graph nodes and ran the experiment for different sizes of graph datasets. We sampled DBLP graph according to venues and AMZlog graph according to products, to guarantee the properties of the sampled graphs. In Figures 4-G, 4-H, and 4-I we compare the performance of queries applied to DBLP dataset (i.e., CC/PC/Path-Partition queries in Figure 4-A) with and without query optimization.

Figures 4-A and 4-D show an almost linear scalability between the response time of queries and the number of nodes in the graph. We increased the number of assignments and functions for 10 queries applied to DBLP dataset. Figures 4-B and 4-E show an almost linear scalability between the average response time of queries and the number of assignments and functions respectively. Figures 4-C and 4-F show the performance of our access structure as a function of available memory for folder partitions (where, we execute a single assignment from the query in Example 1) and path partitions (where, we execute a single function, ERA, from the query in Example 4) respectively. The memory size is expressed as a percentage of the size required to fit the largest partition of data in the hash access structure in physical memory.

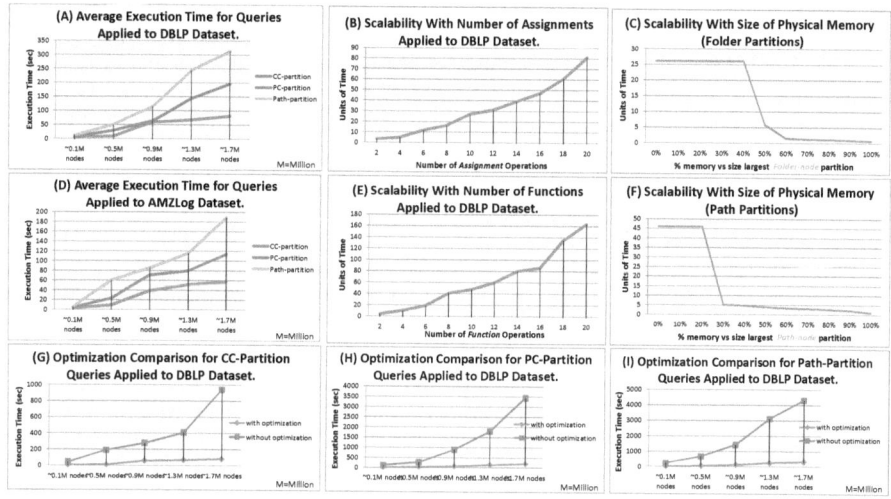

Fig. 4. The query engine extension, GOLAP, evaluation results

The quality of the results is assessed using classical *precision* metric which defined as the percentage of discovered results that are actually interesting. For evaluating the interestingness of the result, we asked domain experts who have the most accurate knowledge about the dataset to construct OLAP queries on graphs. The quality evaluation applied on both DBLP and AMZLog datasets, where 18 queries constructed. As a result, all partitions and query results (e.g., updated/upserted measures) examined by domain experts and all considered relevant. The evaluation shows that the approach is performing well. Please refer to the extended version of the paper [5] for details about evaluation.

7 Conclusion and Future Work

In this paper, we have proposed a graph data model, GOLAP, and a query language for online analytical processing on graphs. We use folder and path nodes to support multi-dimensional and multi-level views over large graphs. We extended FPSPARQL to support n-dimensional computations. We described optimizations and execution strategies possible with the proposed extensions. We have conducted experiments on synthetic and real-world datasets. As future work, we plan to employ interactive graph exploration and visualization techniques (e.g., storytelling systems [22]) to design a visual query interface. Moreover, we are performing an empirical study of computations on very large graphs in three well-studied platform models: relational, data-parallel, and special-purpose in-memory models.

References

1. Abelló, A., Romero, O.: On-line analytical processing. In: Encyclopedia of Database Systems, pp. 1949–1954. Springer US (2009)
2. Aggarwal, C.C., Wang, H. (eds.): Managing and Mining Graph Data. Springer (2010)
3. Balmin, A., Papadimitriou, T., Papakonstantinou, Y.: Hypothetical queries in an olap environment. In: VLDB, pp. 220–231 (2000)
4. Barbieri, D.F., et al.: C-sparql: Sparql for continuous querying. In: WWW, pp. 1061–1062 (2009)
5. Beheshti, S.-M.-R., Benatallah, B., Motahari-Nezhad, H.R., Allahbakhsh, M.: On-line Analytical Processing on Graphs (GOLAP): Model and Query Language. unsw-cse-tr-201214, University of New South Wales (2012)
6. Beheshti, S.-M.-R., Benatallah, B., Motahari-Nezhad, H.R., Sakr, S.: A Query Language for Analyzing Business Processes Execution. In: Rinderle-Ma, S., Toumani, F., Wolf, K. (eds.) BPM 2011. LNCS, vol. 6896, pp. 281–297. Springer, Heidelberg (2011)
7. Chaudhuri, S., Dayal, U.: An overview of data warehousing and OLAP technology. SIGMOD Record 26(1), 65–74 (1997)
8. Chen, C., Yan, X., Zhu, F., Han, J., Yu, P.S.: Graph OLAP: Towards online analytical processing on graphs. In: ICDM, pp. 103–112 (2008)
9. Dries, A., Nijssen, S., De Raedt, L.: A query language for analyzing networks. In: CIKM, pp. 485–494 (2009)
10. Etcheverry, L., Vaisman, A.A.: Enhancing OLAP Analysis with Web Cubes. In: Simperl, E., Cimiano, P., Polleres, A., Corcho, O., Presutti, V. (eds.) ESWC 2012. LNCS, vol. 7295, pp. 469–483. Springer, Heidelberg (2012)
11. Han, J., Sun, Y., Yan, X., Yu, P.S.: Mining knowledge from data: An information network analysis approach. In: ICDE (2012)
12. Han, J., Yan, X., Yu, P.S.: Scalable OLAP and mining of information networks. In: EDBT (2009)
13. Ji, M., Sun, Y., Danilevsky, M., Han, J., Gao, J.: Graph Regularized Transductive Classification on Heterogeneous Information Networks. In: Balcázar, J.L., Bonchi, F., Gionis, A., Sebag, M. (eds.) ECML PKDD 2010, Part I. LNCS, vol. 6321, pp. 570–586. Springer, Heidelberg (2010)
14. Kämpgen, B., Harth, A.: Transforming statistical linked data for use in OLAP systems. In: I-SEMANTICS, pp. 33–40 (2011)
15. Leskovec, J., Adamic, L.A., Huberman, B.A.: The dynamics of viral marketing. TWEB 1(1) (2007)
16. Lima, A.A.B., et al.: Adaptive virtual partitioning for OLAP query processing in a database cluster. JIDM 1(1), 75–88 (2010)
17. Motahari-Nezhad, H.R., et al.: Event correlation for process discovery from web service interaction logs. The VLDB Journal 20(3), 417–444 (2011)
18. Prud'hommeaux, E., Seaborne, A.: Sparql query language for rdf (working draft). Technical report, W3C (March 2007)
19. Qian, T., Yang, Y., Wang, S.: Refining Graph Partitioning for Social Network Clustering. In: Chen, L., Triantafillou, P., Suel, T. (eds.) WISE 2010. LNCS, vol. 6488, pp. 77–90. Springer, Heidelberg (2010)
20. Qu, Q., Zhu, F., Yan, X., Han, J., Yu, P.S., Li, H.: Efficient Topological OLAP on Information Networks. In: Yu, J.X., Kim, M.H., Unland, R. (eds.) DASFAA 2011, Part I. LNCS, vol. 6587, pp. 389–403. Springer, Heidelberg (2011)

21. Romero, O., Abelló, A.: A survey of multidimensional modeling methodologies. IJDWM 5(2), 1–23 (2009)
22. Satish, A., Jain, R., Gupta, A.: Tolkien: an event based storytelling system. Proc. VLDB Endow. 2, 1630–1633 (2009)
23. Sun, Y., et al.: Rankclus: integrating clustering with ranking for heterogeneous information network analysis. In: EDBT, pp. 565–576 (2009)
24. Sun, Y., et al.: Relation strength-aware clustering of heterogeneous information networks with incomplete attributes. PVLDB 5(5), 394–405 (2012)
25. Thomsen, E.: OLAP Solutions: Building Multidimensional Information Systems, 2nd edn. John Wiley & Sons, Inc., New York (2002)
26. Tian, Y., Hankins, R.A., Patel, J.M.: Efficient aggregation for graph summarization. In: SIGMOD Conference, pp. 567–580 (2008)
27. Witkowski, A., et al.: Spreadsheets in RDBMS for OLAP. In: SIGMOD Conference, pp. 52–63 (2003)
28. Xin, D., Shao, Z., Han, J., Liu, H.: C-cubing: Efficient computation of closed cubes by aggregation-based checking. In: ICDE (2006)
29. Yan, X., Yu, P.S., Han, J.: Graph indexing: A frequent structure-based approach. In: SIGMOD Conference, pp. 335–346 (2004)
30. Zhao, P., Li, X., Xin, D., Han, J.: Graph cube: on warehousing and OLAP multidimensional networks. In: SIGMOD 2011, pp. 853–864 (2011)

Scenario-Driven Development
of Context-Aware Adaptive Web Services[*]

Mahmoud Hussein, Jian Yu, Jun Han, and Alan Colman

Faculty of Information and Communication Technologies,
Swinburne University of Technology, Melbourne, Victoria, Australia
{mhussein,jianyu,jhan,acolman}@swin.edu.au

Abstract. *Context-awareness* and *adaptability* are highly desirable features for web services that operate in dynamic environments. In recent years, a number of approaches have been proposed to support the development of such services. However, the requirements elicitation of this kind of services and the synthesis of their design models from the requirements are still major challenges. In this paper, we propose a novel *scenario-driven* approach to developing context-aware adaptive web services. Our approach enables the elicitation of a web service's requirements as two sets of scenarios: functional and adaptation. The *functional scenarios* capture the service's functionality while the *adaptation scenarios* represent the service's adaptation logic to cope with runtime context changes. We also support the *synthesis* of the service's design model from its scenarios, and the automatic transformation from the service's design model to the executable service code. To demonstrate the applicability of our approach, we have used it to develop a context-aware travel guide service.

Keywords: Scenario-driven Development, Context-awareness, Self-adaptation, Requirements Elicitation, Web Services.

1 Introduction

There is an increasing demand for web services that have the ability to *dynamically adapt* their behaviours in response to changes in their operating environments (i.e. their contexts) without explicit user intervention to bring better usability and also effectiveness. In recent years, a number of approaches have been proposed to support the development of such context-aware adaptive services [1-2]. However, two major challenges still remain that hinder easy and effective development of this kind of services. First, to elicit the requirements of a context-aware adaptive service, there is a need for an approach that considers the service's context and its adaptation in response to context changes during the requirements elicitation. Second, to ease a service design and maintain a causal connection between the service requirements and its design model [3], an approach is needed to automatically synthesize the service's design model from its requirements.

[*] This research is partly supported by the Australian's Cooperative Research Centre for Advanced Automotive Technology (www.autocrc.com).

X.S. Wang et al. (Eds.): WISE 2012, LNCS 7651, pp. 228–242, 2012.
© Springer-Verlag Berlin Heidelberg 2012

To tackle the above two challenges, in this paper, we propose a novel *scenario-driven* approach to developing context-aware adaptive services. Our approach elicits a service's requirements as two sets of scenarios [4]: *functional* and *adaptation*. The functional scenarios capture the service functionality while the adaptation scenarios represent the service's runtime adaptation in response to context changes to keep achieving the users' needs. To specify these two types of scenarios, we have extended the UML sequence diagram with three types of scenario's participants: *functional*, *contextual*, and *management*. The functional participants provide the service's functionality while the contextual participants provide context information that is either needed by the service functionality or can trigger the service adaptation. The management participants decide adaptation actions to cope with the context changes. These participants interact with each other through special types of messages (e.g. a management message can only be exchanged between management and functional participants). Our approach also supports the *synthesis* of the service's design model from its scenario's descriptions. Based on the synthesized model, the service engineer may add elements that are related to the service's solution space (e.g. partner services that provide the service's core functionality). A technique to automatically transform the service's design model to the executable service code is also developed. A tool has been developed to support the scenario-driven development of context-aware adaptive services. To demonstrate the applicability of our approach, a case study of developing a context-aware travel guide service is presented.

The remainder of the paper is organized as follows. We start by introducing a motivating scenario in Section 2. In Section 3, we present our approach to eliciting a context-aware adaptive service's requirements, synthesis of the service's design model from its requirements, and generating the executable service code from its design model. The implementation of our approach is discussed in Section 4. Section 5 analyses existing and our approaches. Finally, we conclude the paper in Section 6.

2   Motivating Scenario

Let us consider a travel guide web service that a tourist can use to plan her trip. Based on her preferences (e.g. outdoor attractions) and the weather forecast for that day (e.g. sunny), the travel guide service suggests to her a number of attractions. She selects some of these attractions to visit in her rented car, and then a set of routes are displayed to her. These routes are calculated based on her current location, her attractions list, her driving preferences (e.g. shortest route), and the live traffic information (e.g. congested roads). She selects a suitable route and starts to explore the attractions.

To develop the context-aware travel guide service, two types of requirements need to be elicited: *functional* and *adaptation*. The functional requirements are a set of functions that need to be provided by the service (e.g. route planning). These functions also need to take the context information into account to give the tourist better suggestions. For example, the route planner needs the traffic information to provide better estimations for the routes travel times. The adaptation requirements specify the service's adaptation in response to runtime context changes. For example, the travel guide provider may want to provide the attractions finder service free, while the tourist should pay to use the route planner service. Therefore, while the travel guide service is in operation, the tourist may want to integrate the route planner service, which is not provided free to her initially, into the existing service. To integrate such service, several changes need to be applied into the running service. Firstly, the service needs to be adapted by incorporating the route planning service (i.e. *adding a functional service*). Secondly, to find a suitable route for the tourist, there is also a need to acquire the tourist's driving preferences and the live traffic information and use them in calculating and suggesting the routes (i.e. *adding new context providers*).

3   The Approach

To support the development of context-aware adaptive services, in this section, we describe our approach to eliciting a service's requirements as a set of scenarios (*Step 1*), synthesis of the service's design model from its scenarios' descriptions (*Step 2*), and transforming the service's design model to the executable service code (*Step 3*).

3.1   Requirements Elicitation of Context-Aware Adaptive Web Services

To enable the elicitation of a context-aware adaptive service's requirements (i.e. *Step 1*), we adopt a *scenario*-based approach because its simplicity and intuitive graphical representation facilitate stakeholder involvement in eliciting the requirements [4]. Following such approach, we elicit the service requirements as two sets of scenarios: *functional* and *adaptation*. The functional scenarios capture the service's functionality while the adaptation scenarios represent the service's runtime adaptation to cope with context changes.

To document a service's scenarios, a number of scenario-based notations have been introduced [4]. In this paper, we adopt and extend the widely accepted UML sequence diagram to specify the functional and adaptation scenarios. Before introducing our extensions, we describe below the core elements of the UML sequence diagram.

Following the UML sequence diagram, a scenario consists of *lifelines*, *messages*, and *interaction fragments* (see the top part of Figure 1). The lifelines represent the scenario participants, while the messages capture the interactions between them. To capture the scenario flow, the sequence diagram has a set of interaction fragments such as an interaction use, an occurrence specification, and a combined fragment. First, the *"interaction use"* specifies that a scenario can refer to (or use another scenario). For example, in Figure 2, the scenario "FS1" has a reference to the "SelectRoute" scenario. Second, the *occurrence specification* concerns the intersection point between a participant and a message, defining what events are sent or received by a participant. For example, the intersection point between the "plan a route" message "FM4" and the user participant (see Figure 2) specifies that the user can request the route planning operation from the route planner. Third, the *combined fragment* is used to group a set of interactions and define their relationships (e.g. loop, alterative, etc.). Each combined fragment consists of an *operator* and *operands* (see Figure 1). The operator defines the type of the combined fragment (e.g. alternative) while the operands specify interactions that are grouped by the combined fragment operator. In addition, each operand can have a *constraint* as a Boolean expression that specifies when this interaction can be executed. An example combined fragment is the *alternative fragment*. It is used to specify when an interaction or a set of interactions are enabled or disabled where other interaction(s) are enabled. For example, the interaction "FM4: PlanRoute1" is executed when the user had selected a set of attractions while "FM5: PlanRoute2" is executed otherwise as shown in Figure 2.

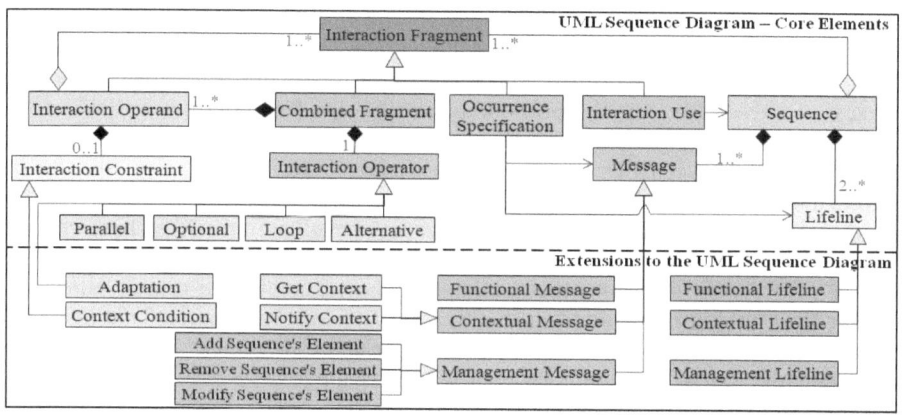

Fig. 1. An extended meta-model of the UML sequence diagram

(1) The Service's Functional Scenarios: To capture the service functionality while taking context information into account as a set of scenarios, we have extended the UML sequence diagram with two types of scenario participants (i.e. *functional* and *contextual* participants) that interact with each other through special types of message as shown in the bottom part of Figure 1.

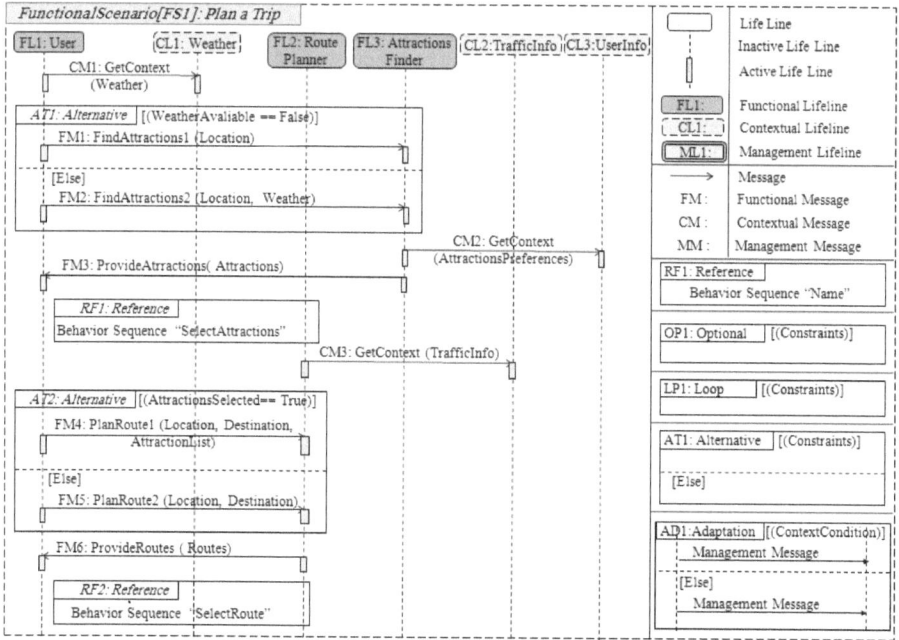

Fig. 2. A scenario shows the tourist interactions with the travel guide service to plan a trip

A. Functional Participants and their Messages: The functional participants are responsible for providing the service's functionality. For example, the route planner participant is responsible for calculating the routes as shown in Figure 2. The messages between the functional participants are either requesting or providing a functional operation, and therefore called "functional messages".

B. Contextual Participants and their Messages: Context information is needed in a context-aware service to better carry out its functionality (e.g. considering the weather condition in suggesting the attractions). We make the *contextual* participants explicit in the functional scenarios to provide the context information (e.g. the traffic information provider in Figure 2). The functional participants can be notified of the context changes or they can request the context information on-the-fly. Thus, we have two types of contextual messages (*notify* and *get* context) that can be exchanged between functional and contextual participants. For example, the route planner can acquire the traffic information from its provider (i.e. CM3: "get context" in Figure 2).

An example functional scenario is "FS1" shown in Figure 2. This scenario represents the tourist's interactions with the travel guide service to find attractions, select a set of attractions for visit, and plan a route to visit these attractions.

In naming the scenario's elements, we give each element an *identifier*, so that it can be referenced by the adaptation scenarios (described below). The identifier has two letters that represent the element type and an auto-generated number to differentiate between elements of the same type in each scenario (see the right part of Figure 2 for abbreviations of each element type and their meanings).

(2) The Service's Adaptation Scenarios: To elicit the adaptation requirements of a service, we propose the concept of *adaptation scenarios*. These scenarios describe the required adaptations to the service's functional scenarios in response to runtime context changes. To define this type of scenarios, we have introduced *management lifelines*, *management messages*, and *adaptation fragments* as shown in the bottom part of Figure 1. The adaptation scenarios also have functional and contextual participants.

A. Functional and Contextual Participants: In an adaptation scenario, functional participants represent the service functionality (or the functional scenarios) that needs to be adapted in response to context changes (e.g. the service's functionality "FL1" in Figure 3). The adaptation scenario also has contextual participants that provide context information that triggers the service adaptation. For example, a change in a user's selected features causes the travel guide service's adaptation. Thus, we have a contextual lifeline "CL2" for making this information available as shown in Figure 3.

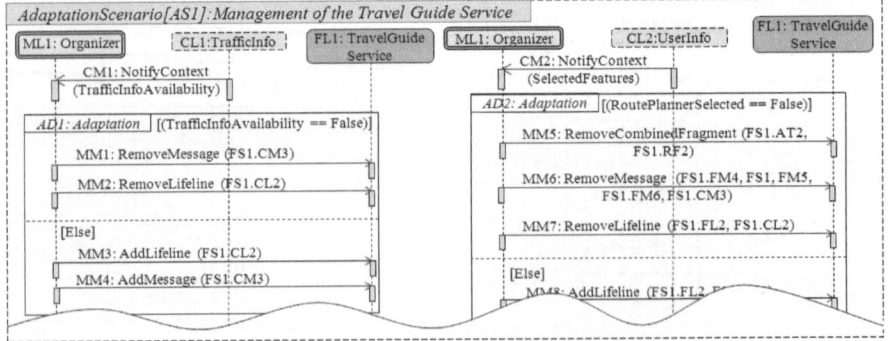

Fig. 3. An adaptation scenario in the travel guide service to cope with context changes

B. Management Participant and its Messages: In an adaptation scenario, a *management* participant is responsible for deciding and performing the service adaptation in response to context changes, e.g. the organizer participant "ML1" in Figure 3. The management participant uses a set of *management* messages to adapt the service, which specify what adaptation actions can be applied to the service's functional scenarios at runtime. Example actions that can be applied to the functional scenario "FS1" are: add the lifeline "CL2", remove the message "CM3", modify combined fragment "AT2" by removing the message "FM4", etc. as shown in Figure 3. In general, we have actions to *add*, *remove*, or *modify* a sequence's (scenario) element (e.g. lifeline, combined fragment, messages, etc.) as shown in Figure 1.

C. Adaptation Fragment: To specify what to adapt in response to context changes, we introduce the *adaptation operator* as shown in Figure 1. This operator groups a set of management messages (i.e. adaptation actions). We also extend the interaction constraint with *context condition* (see Figure 1) to specify a *contextual situation* in which the service needs to adapt itself in (e.g. the user wants to include the route planner service). We call the combined fragment that has the adaptation operator and the context condition as "*adaptation fragment*". For example, the adaptation fragment "AD1" (shown in Figure 3) specifies the required changes to the functional scenario "FS1" (presented in Figure 2) in response to changes in the traffic information status. When the traffic information is not available, the "plan a trip" scenario is adapted by removing the traffic information participant "CL2" and the get context message "CM3". The reverse of these two adaptation actions are performed when the traffic information become available as shown in the ELSE part of "AD1" (see Figure 3).

3.2   Synthesis of the Service's Design Model

In this section, we describe our approach to designing a context-aware adaptive web service and synthesizing the service's design model from its scenarios (i.e. *Step 2*).

A. Designing Context-Aware Adaptive Web Services

To design context-aware adaptive services, we adopt an *organizational* approach because it represents relationships between the service elements explicitly. Therefore, the service elements and their relationships can be clearly captured. It also keeps the service's composition structure alive at runtime, so that it can be easily adapted [5].

A service composition as an *organisation* consists of a set of *dynamic relationships* between its *roles* to maintain the service's viability in a changing environment [5]. The *relationships* (defined as contracts) specify permissible interactions between the service's roles (i.e. what tasks a role can request from others). Thus, they are used to define the "position descriptions" of the service's roles. These position descriptions specify what tasks the service's *roles* should do, while there are *players* (specific partner services) who actually perform the tasks by playing these roles. In addition, to coordinate interactions between the service's roles to carry out composite tasks, a set of workflow processes are specified. These processes define in what order a set of simple tasks are performed to achieve the composite tasks. Furthermore, in response to environment changes, the service manager (organizer) changes the service's roles, the role-player bindings, and the roles' relationships to maintain the service viability.

Following the *organizational approach*, a meta-model for a context-aware adaptive web service is shown in Figure 4. The service's composition consists of two main elements: *functional* and *management* composites. In the following, we describe the basic elements of these two composites. We also introduce a set of graphical notations for the meta-model concepts (see Figure 5) that are more convenient for the service engineer to design such services.

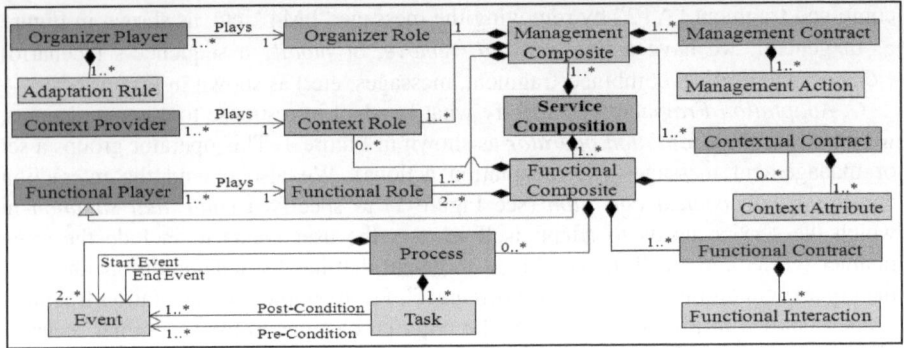

Fig. 4. A meta-model for a context-aware adaptive service composition[1]

The functional composite consists of *functional roles* that interact with each other through *functional contracts* to capture the service's functionality as shown in Figure 4. For example, the route planner role (see Figure 5-A) is responsible for calculating the routes. This role is also connected with the user role through the functional contract "FC2" to make the user able to request the routes calculation operation. In addition, to make the service's functionality context-aware, there is a set of *context roles* bound to context providers to make the context information available. The functional composite also has *contextual contracts* formed between context and functional roles to specify the context information required by the functional roles to continue their operations. For example, the traffic information role bound to its player is responsible for providing the traffic information to the route planner role through the contextual contract "CC3" as shown in Figure 5-A. Thus, the route planner can give better estimation for the routes travel times. Furthermore, to define the service behaviour, a set of *behaviour processes* are specified. In designing these processes, we follow an event-based approach [6], where a process is designed as a set of tasks. Each task has a *pre* (events that enable the execution of a task) and *post* (events that are generated upon a task completion) conditions as a set of events. The process also has two events that specify the process's start and end as shown in Figure 4. We adopt this approach because it represents a process as a set of loosely coupled *tasks* that are related with each other through *events*, so that the process can be easily adapted by changing the pre and post events of its tasks. An example process is "plan a trip"

---

[1] To simplify the Figure, we have removed associations between the service's roles and contracts, and they are only described in the text. In general, a contract is formed between two roles only.

presented in Figure 5-B (to ease the understandability of an event-based process, we visualized the process in the form of an Event-driven Process Chain (EPC) [6]). This process starts by suggesting a set of attractions to the user. She can select some of them for visit, and then a route to visit the selected attractions is suggested.

The management composite specifies how the service composition is adapted in response to context changes to keep achieving the users' needs. Based on the organizational approach, this composite has a number of items. First, the composite has three types of roles: *functional*, *contextual*, and *management*. The functional role represents the functional composite that need to be adapted at runtime (e.g. the travel guide service's composite as shown in Figure 5-A) while the contextual roles represent context information that causes the service's runtime adaptation (e.g. the user information as shown in Figure 5-A). The management (organizer) role bound with its player is used to decide the adaptation actions in response to context changes by a set of adaptation rules as shown in Figure 4. Second, to capture the relationships between the composite roles, two types of contracts are used: *management* and *contextual*. The management contract is formed between a management role and a functional role to capture adaptation actions that can be applied to the functional role, while the contextual contract captures context information that can trigger the service adaptation and it is formed between a management role and a contextual role (e.g. "MC1" and "CC1" shown in the top of Figure 5-A).

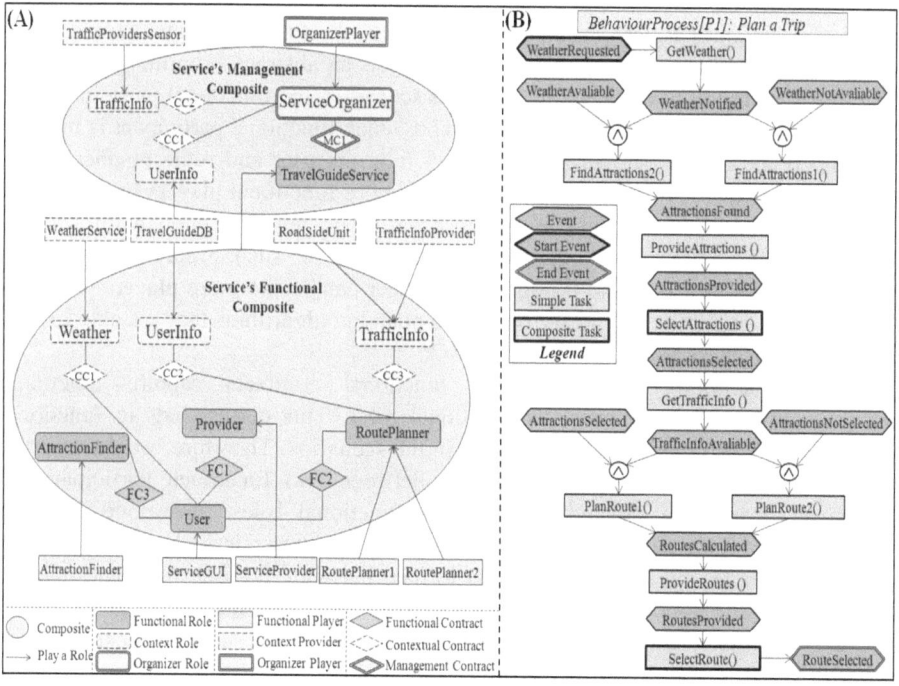

Fig. 5. The travel guide service's design models

B. Synthesis of the Service's Design Model

In this section, we describe the automatic synthesis of a service's design model from its scenarios' descriptions. Table 1 summarises the mappings between the scenarios' elements and the service's design model concepts. In general, the functional scenarios are mapped to the service's *functional* composite, while the service's *management* composite is derived from the adaptation scenarios.

Table 1. The mappings between the service's scenarios and the its design model

Service's Scenarios	Service's Design Model	Service's Scenarios	Service's Design Model
Functional Scenarios	Functional Composite	Adaptation Scenarios	Management Composite
Functional Participant	Functional Role	Contextual Participant	Context Role
Management Participant	Organizer Role	Functional Message	Functional Interaction
Contextual Message	Context Attribute(s)	Management Message	Management Action(s)
Sequence (Scenario)	Process	Message	Task
Sequence's First Message	Process's Start Event	Sequence's Last Message	Process's End Event
Interaction Constraint	Task Pre-Condition	Adaptation Fragment	Adaptation Rule

(1) Synthesis of the Functional Composite: To synthesize the service's functional composite, a number of items are derived from the functional scenarios.

(1.a) Functional Roles and their Players: First, the functional roles represent the service functionality where each role description is an abstract definition of the tasks that this role should provide. In the service's scenarios, the functional participants are involved in providing or requesting tasks. Thus, each functional participant is mapped to a functional role as shown in Figure 5-A (e.g. the user and route planner roles). Second, a functional role needs to have one or more functional players to provide its actual functionality at runtime. But, the scenarios do not give enough information about how many players a synthesized role can have. As such, a player is generated for each synthesised role and the service engineer can define more players later. For example, the engineer can specify two route planning algorithms that are able to play the route planner role as shown in Figure 5-A.

(1.b) Functional Contracts: The functional contracts capture functional *interactions* between the service's functional roles. This corresponds to functional messages between functional participants in the scenarios. Therefore, we traverse the scenarios and if there is any interaction between two functional participants, a functional contract is created between the functional roles correspond to these participants (if the contract does not already exists) as shown in Listing 1 (Lines 2 to 11). For example, the contract "FC2" is formed between the user and route planner roles (see Figure 5-A). Then, for each functional message, a functional interaction is added to the created contract (see the example contract "FC2" shown in Listing 1).

Listing 1. An algorithm (pseudocode) to synthesis the service's functional contracts

```
 1: void SynthesisOfFunctionalContracts (FunctionalComposite FC, FunctionalScenarios FS){
 2:      FOR each Scenario S in FS
 3:          ArrayList <Message> messages= S.getMessages();
 4:          FOR each Message M in messages{
 5:              IF M.getType() is "Functional" THEN
 6:                  FunctionalInteraction FI= new FunctionalInteraction(M);
 7:                  FunctionalContract contract=FC.getContract(M.getRoleA(), M.getRoleB());
 8:                  IF contract is not exists THEN
 9:                      contract = new FunctionalContract(M.getRoleA(), M.getRoleB()) ;
10:                      FC.addFunctionalContract(contract);
11:                  END IF
12:                  contract.addFunctionalInteraction(FI);
13:              END IF
14:          END FOR
15:      END FOR
16: END
//Part of the Synthesized Contract "FC2"
Contract ID              FC2: User_RoutePlanner
Parties:                 Role A: User;      Role B: RoutePlanner;
Interaction Clauses:     i1: {requestRoutes1 (Location, Destination, Attractions), AtoB, void};
                         i2: {requestRoutes2 (Location, Destination), AtoB, void};
```

(1.c) Behaviour Processes: To synthesize the functional composite's behaviour, each scenario is transformed to a process where the first interaction in the scenario defines the start event of the process while the process's end event is the completion of the scenario's last interaction. For example, the event "weather requested" is the start event of the "plan a trip" process while the event "route selected" is the process end event as shown in Figure 5-B. Then, each interaction fragment in the scenario is transformed to a task or a set of tasks based on its type. If an interaction is a message or a sequence reference, it is transformed to a single task. The task pre-condition is its previous interaction's completion and its post-condition is an event that specifies the task completion to enable the execution of other dependent tasks. For example, the "provide attractions" message is transformed to "provide attractions" task that has "attractions found" as pre-condition and "attraction provided" as post condition as shown in Figure 5-B. On the other hand, if an interaction is a combined fragment, a set of tasks correspond to this fragment are generated. For example, the alterative fragment is transformed to two groups of tasks. The first group contains tasks to be executed when the fragment condition is true, while the second group has tasks to be performed when the fragment condition is false. An example of a transformed fragment is "find attraction" fragment shown in Figure 5-B. In this fragment, the operation used to find the attractions depends on the weather information availability where "FindAttractions2" is used when the weather information is available while "FindAttractions1" is used otherwise. Due to space constraint, we described the synthesis of the behaviour processes briefly and more details can be found in [7].

(1.d) Context Roles and their Providers: Similar to mapping the functional participants to functional roles, the contextual participants are mapped to context roles (e.g. weather and TrafficInfo roles shown in Figure 5-A). To make the context information available, there is a need for context providers that monitor the context information. These providers are part of the service's solution space, and then they need to be specified by the service engineer and bound to the synthesized roles. For example, the traffic information role bound with its provider (e.g. road side unit) is responsible for providing the live traffic information.

(1.e) Contextual Contracts: A contextual contract specifies what context information is required by a functional role. In the scenarios, the context information is captured through contextual messages. Thus, we map a set of contextual messages between a contextual participant and a functional participant to a contextual contract. The algorithm to synthesis the contextual contracts is similar to the one we have used to synthesis the functional contracts. Thus, we removed the algorithm details to save the space. An example synthesized contract is "CC3" (see Figure 5-A). This contract specifies that the route planner role needs to know the live traffic information.

(2) Synthesis of the Management Composite: The service's management composite corresponds to the service adaptation scenarios. Therefore, we use these scenarios to derive this composite's elements. In the following, we discuss how to synthesize the composite's management roles, players, and contracts while the other items in this composite (e.g. context and functional roles) can be synthesized as discussed above.

(2.a) Management Roles: A management participant in the adaptation scenarios has the same purpose of a *management role* in the service's management composite. Thus, each management participant is mapped to a management role in the service's design model. For example, the organizer role that corresponds to the organizer participant in Figure 3 is added to the management composite shown in Figure 5-A.

(2.b) Management Players: To decide the service adaptation actions in response to context changes, we model the management player as Event-Condition-Action rules. We adopt the rule-based approach because of its expressiveness and tool support. The events that activate an adaptation rule are usually context changes which the service needs to have a response to. The rule condition specifies the context situation that needs the service reaction(s). The rule actions are a set of adaptation actions to cope with the context changes. In general, the adaptation actions are to *add*, *remove*, or *modify* any service element. For example, to change the service's roles, we have three adaptation actions: add role, remove role, and change role-player binding.

To synthesize a management player, the adaptation fragments in the adaptation scenarios are translated to adaptation rules. Each rule's events are inferred from an adaptation fragment's condition, where the events are changes in context attributes that are included into the fragment's condition. The rule condition is same as the adaptation condition. The rule's adaptation actions are actions that correspond to management messages specified into the adaptation fragment where each message at the scenario level has a corresponding adaptation action at the service's design level. For example, the action *removeLifeline("x")* is transformed to *removeRole("x")*. An example of a synthesized adaptation rule is shown in Listing 2. This rule corresponds to the IF part of the adaptation fragment "AD1" presented in Figure 3.

Listing 2. An adaptation rule to cope with the unavailability of the traffic information

```
Rule "AdaptationRule1":  {
When     ValueChanges (TrafficInfoAvailability);
if       TrafficInfoAvailability == False;
do            RemoveContract("CC3"),    RemoveRole("TrafficInfo"),    RemoveTask
         ("P1","GetTrafficInfo"),   RemoveEvent ("P1", "TrafficInfoAvaliable")};
```

(2.c) Management Contracts: To apply the adaptation rules' actions, the service that needs to be adapted should support the application of the actions. To do so, a management contract between the service role and the organizer role is formed (e.g. the contract "MC1" in Figure 5-A). This contract defines adaptation actions that need to be performed into the service at runtime. Similar to synthesizing the functional contracts, the management contracts can be synthesised [7].

3.3   Realizing Context-Aware Adaptive Web Services

In our previous work, we have proposed the ROAD framework which is an extension to Apache Axis2 to realize adaptive service compositions [8]. To realize context-aware adaptive services using this framework (i.e. *Step 3*), we *transform* the service's design model described in Section 3.2 to a model that is compatible with the ROAD framework. The following are the major transformations (details can be found in [7]).

Firstly, the service's functional and management composites are mapped to ROAD composites in a straightforward manner where the ROAD model also follows an organizational approach. But, in ROAD model, the context information is maintained as a set of facts. Each fact contains one or more context attributes. In our model, each context role position description is a collection of context attributes that describe a context entity. This makes a correspondence between a fact in the ROAD model and a context role in our model. As such, we transform each context role to a fact in the ROAD model. Secondly, to enable the execution of the adaptation rules, we transform them into Drools rules, so that the Drools rule engine can be used for their execution to decide the required adaptation actions while the service is in operation.

4   Implementation

To support the scenario-driven development of context-aware adaptive services, we have developed a tool. This tool enables the service engineer to *represent* a service's functional and adaptation scenarios (see the top part of Figure 6), *synthesis* of the service design from its scenarios automatically as discussed in Section 3.2 (the bottom of Figure 6 is a screenshot from the tool during the synthesis of the service design), and *generate* an executable service model from its design model through transforming the service model to an executable ROAD model as described in Section 3.3.

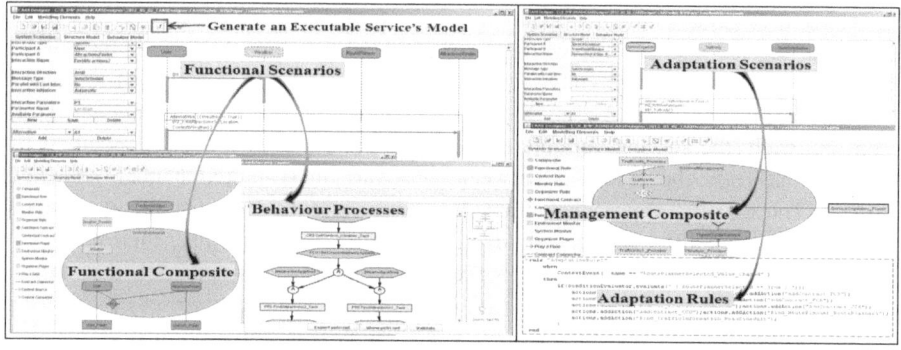

Fig. 6. Tool's screenshots during the development of the travel guide service

Following our approach, we have developed the travel guide service described in Section 2. To make it fully functioning, we have developed a set of players and context providers as web services. An example of the travel guide service's adaptation is shown in Figure 7. In Figure7-A, the service only includes the attraction finder function which is provided to the tourist for free initially. In response to the tourist request to include the route planning function, the service is adapted to include such functionality as shown in Figure 7-B.

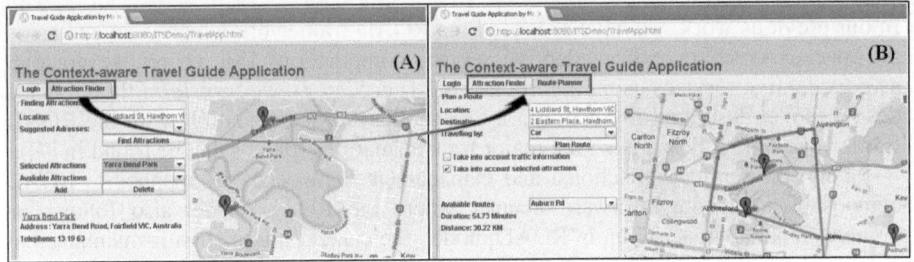

Fig. 7. Evolving the travel guide service to include the route planning function

5   Related Work

A number of approaches have been proposed to support the development of context-aware adaptive services. Below, we discuss the approaches dealing with a service's *requirements elicitation* and *synthesis* of a service's *design model* (i.e. our focus).

Requirements Elicitation: Recently, a number of approaches have been proposed to support the requirements elicitation of context-aware adaptive services. *RELAX* is an approach to capture an adaptive service's requirements by relaxing the service requirements to take into account context changes [9]. They capture the service requirements as SHALL statements in a textual representation. Then, they classify the elicited requirements as *invariant* or *relax-able* requirements. Finally, they introduce RELAX operators (e.g. as early as possible and as close as possible to) to the relax-able requirements, so that they become flexible to cope with runtime context changes. Another approach is proposed by Salifu et al. to capture a service's monitoring and switching requirements [10]. The *monitoring* requirements are context variables that cause the service adaptations, while the *switching* requirements specify the service switching between different states to keep achieving its goals in response to context changes. Dalpiaz et al. have used Troops to capture a service's requirements [11]. In their approach, the service variants are identified and a goal model is specified for each variant. Then, the switching between these variants (captured as goal models) is defined. The existing approaches elicit the service's requirements as a set of goals. However, it is often that the stakeholders find it difficult to articulate their needs [4]. To tackle this difficulty, we have proposed a scenario-based approach that can be used by the requirement engineer to capture the users' needs as two set of scenarios. We also adopted the UML sequence diagram which is widely accepted in practice. Similar to our approach, there are scenario-based approaches for eliciting a service's requirements in general [12]. However, they do not support the elicitation of the service's context and adaptation requirements.

Synthesis of the Service's Design Model: A number of approaches have been proposed to designing and realizing context-aware adaptive services. CEVICHE is a framework that combines complex event processing and business process adaptation to enable runtime changes to composite services [13]. It has a language called SBPL (*Standard Business Process Language*). This language is an extension to the BPEL to make business processes more flexible through defining adaptation points in a process at the design time. *PLASTIC* is a model-driven approach to support a web service life cycle from design to implementation and to runtime execution [14]. It also supports the runtime adaptation of the service based on context or quality of service changes to ensure that the users get the best quality of service. Sheng et al. have proposed a model-driven approach to ease the development of context-aware web services [15]. In this approach, the context information is used by a set of adaptation rules to adapt the service's output parameters (i.e. they do not support adaptation of the service functionality and its context model). In existing approaches, the service designer is responsible for translating the service's requirements to its design model manually. Therefore, this process is error-prone in complex services and the causal connection between the service's requirements and its design is maintain manually which is a complex task. To tackle these challenges, our approach synthesizes the service initial design from its scenarios automatically, so that the causal connection between the service's design and its requirements is maintained and the service's design effort is reduced. Also, there are existing approaches to synthesize a service design from its scenarios automatically (e.g. [4]). But, they only synthesize the service functionality (i.e. no support for synthesizing the service's context and adaptation), and this hinders their adoption to synthesize the design of context-aware adaptive web services.

6   Conclusion and Future Work

In this paper, we have proposed a scenario-driven approach to ease the development of context-aware adaptive services. Compared with existing approaches, our approach has the following key contributions. First, our approach elicits a web service's requirements as two sets of scenarios: functional and adaptation. The *functional scenarios* capture the service's functionality while the *adaptation scenarios* represent the service's adaptation to cope with runtime context changes. Second, we support the automatic synthesis of the service's design model from its scenarios' descriptions. The synthesized model can be also transformed to the service's executable code.

As future work, we will extend our approach to enable a web service's runtime evolution using a scenario-driven approach. In addition, we have applied our approach to the travel guide case study and the results were promising. We will perform more validations to assess our approach's applicability.

References

1. Kakousis, K., Paspallis, N., Papadopoulos, G.A.: A survey of software adaptation in mobile and ubiquitous computing. Enterprise Information Systems 4(4), 355–389 (2010)
2. Kapitsaki, G.M., et al.: Context-aware service engineering: A survey. Journal of Systems and Software 82(8), 1285–1297 (2009)
3. Sawyer, P., et al.: Requirements-Aware Systems: A Research Agenda for RE for Self-adaptive Systems. In: RE 2010, Sydney, NSW, Australia (2010)
4. Uchitel, S., Kramer, J., Magee, J.: Synthesis of Behavioral Models from Scenarios. IEEE Trans. Softw. Eng. 29(2), 99–115 (2003)
5. Colman, A., Han, J.: Adaptive service-oriented systems: an organisational approach. Int. Journal of Computer Systems Science & Engineering 21(4), 235–246 (2006)
6. Mendling, J.: Event-Driven Process Chains (EPC). In: Metrics for Process Models, pp. 17–57. Springer, Heidelberg (2009)
7. Hussein, M., et al.: Scenario-driven Development of Context-aware Adaptive Software Systems Technical Report #C3-516_07, Swinburne University of Technology (2012), http://www.ict.swin.edu.au/personal/mhussein/papers/ C3-516_07.pdf
8. Kapuruge, M., Colman, A., King, J.: ROAD4WS – Extending Apache Axis2 for Adaptive Service Compositions. In: EDOC 2011, Helsinki, Finland (2011)
9. Whittle, J., et al.: RELAX: Incorporating Uncertainty into the Specification of Self-Adaptive Systems. In: RE 2009, Atlanta, Georgia, USA (2009)
10. Salifu, M., Yijun, Y., Nuseibeh, B.: Specifying Monitoring and Switching Problems in Context. In: RE 2007, Delhi, India (2007)
11. Dalpiaz, F., Chopra, A.K., Giorgini, P., Mylopoulos, J.: Adaptation in Open Systems: Giving Interaction Its Rightful Place. In: Parsons, J., Saeki, M., Shoval, P., Woo, C., Wand, Y. (eds.) ER 2010. LNCS, vol. 6412, pp. 31–45. Springer, Heidelberg (2010)
12. Cheng, B.H.C., Atlee, J.M.: Research Directions in Requirements Engineering. In: 2007 Future of Software Engineering, Minneapolis, MN (2007)
13. Hermosillo, G., Seinturier, L., Duchien, L.: Creating Context-Adaptive Business Processes. In: Maglio, P.P., Weske, M., Yang, J., Fantinato, M. (eds.) ICSOC 2010. LNCS, vol. 6470, pp. 228–242. Springer, Heidelberg (2010)
14. Autili, M., Berardinelli, L., Cortellessa, V., Di Marco, A., Di Ruscio, D., Inverardi, P., Tivoli, M.: A Development Process for Self-adapting Service Oriented Applications. In: Krämer, B.J., Lin, K.-J., Narasimhan, P. (eds.) ICSOC 2007. LNCS, vol. 4749, pp. 442–448. Springer, Heidelberg (2007)
15. Sheng, Q.Z., et al.: ContextServ: A platform for rapid and flexible development of context-aware Web services. In: ICSE 2009, Vancouver, BC, Canada (2009)

WebPut: Efficient Web-Based Data Imputation

Zhixu Li[1], Mohamed A. Sharaf[1], Laurianne Sitbon[2],
Shazia Sadiq[1], Marta Indulska[1], and Xiaofang Zhou[1]

[1] The University of Queensland, QLD 4072 Australia
[2] Queensland University of Technology, QLD 4000 Australia
{zhixuli,shazia,zxf}@itee.uq.edu.au, m.sharaf@uq.edu.au,
laurianne.sitbon@qut.edu.au, m.indulska@business.uq.edu.au

Abstract. In this paper, we present WebPut, a prototype system that adopts a novel web-based approach to the data imputation problem. Towards this, Webput utilizes the available information in an incomplete database in conjunction with the data consistency principle. Moreover, WebPut extends effective Information Extraction (IE) methods for the purpose of formulating web search queries that are capable of effectively retrieving missing values with high accuracy. WebPut employs a confidence-based scheme that efficiently leverages our suite of data imputation queries to automatically select the most effective imputation query for each missing value. A greedy iterative algorithm is also proposed to schedule the imputation order of the different missing values in a database, and in turn the issuing of their corresponding imputation queries, for improving the accuracy and efficiency of WebPut. Experiments based on several real-world data collections demonstrate that WebPut outperforms existing approaches.

Keywords: Web-based Data Imputation, WebPut, Incomplete Data.

1 Introduction

Data incompleteness is one of the most pervasive data quality problems especially for web databases [13]. The process of filling in missing attribute values is well-known as *Data Imputation* [5, 16]. Commonly used data imputation approaches assign a missing value to "the most common attribute value", "a special value" [8], or a "closest-fit" value from the most similar context within the data set [6, 11]. Another line of data imputation approaches attempts to predict an estimation for the missing values using models built on the incomplete data set [15, 18, 21]. Such approaches could effectively smooth the influence of missing attribute values on some statistical data analysis. Both approaches, however, typically fall short in replacing the individual missing attribute values, especially when those values are unique within the data set (e.g., the missing Email addresses in Table 1), which is precisely the context we address in this paper.

The premise underlying our work is that most of the missing data in a wide range of online databases is typically available from some external data sources on the world wide web. The effective and efficient retrieval and extraction of this data, however, remains a challenging task, which motivated us to propose our novel *Web-based data imputation (WebPut)* approach.

X.S. Wang et al. (Eds.): WISE 2012, LNCS 7651, pp. 243–256, 2012.
© Springer-Verlag Berlin Heidelberg 2012

Table 1. A Personal Information Table with Missing Data

	Name(N)	Email(E)	Title(T)	University(U)	State(S)
1	Jack Davis	jdavis@mit.edu	Professor	MIT	MA
2	Tom Smith	tomsmith2@cs.cmu.edu		CMU	PA
3	Bill Wilson		Doctor	UIUC	IL
4	Bob Brown	bbrown7@yale.edu	A/Professor	Yale	NY
5	Ama Jones		Ms		CA
6		lank@ucla.edu			

In principle, WebPut could be perceived as a novel Information Extraction (IE) approach. In particular, classical IE tasks typically process web documents for the purpose of recognizing a category of entities (such as *locations*), or relations (such as *company-headquarter pairs*) [1, 4, 19, 17]. WebPut, however, formulates the extraction tasks around the missing attribute values in a database, which greatly challenges the existing IE techniques. Towards this, WebPut utilizes the available information in the incomplete data set in conjunction with the data consistency principle. That is, the values in the same tuple are of the same instance, while the values in the same column are of the same domain. To this end, WebPut extends effective IE methods for the purpose of formulating web search queries that are capable of effectively retrieving the missing values in a database. The main contributions of this paper are summarized as follows:

1. We implement a WebPut prototype system that employs and extends a suite of traditional IE methods for the purpose of formulating effective web-based data imputation queries.
2. We propose a confidence-based scheme that efficiently leverages our suite of data imputation queries so that to automatically select the most effective imputation query for each missing value.
3. We propose a novel greedy iterative algorithm to schedule the imputation order of the different missing values in a database, and in turn the issuing of their corresponding imputation queries, for improving the accuracy and efficiency of WebPut.

Our experimental results on several real-world data collections show that WebPut outperforms all previous approaches and can achieve up to 80% accuracy. In addition, our proposed query scheduling techniques improve the efficiency of WebPut by up to 50%, as compared to the baseline approach.

Roadmap: We discuss both our system model and problem definition in Sec. 2, then present our proposed WebPut approach in Sec. 3. We report on our experimental results in Sec. 4, and then cover related work in Sec. 5. Then we conclude this paper in Sec. 6.

2   Model and Problem Definition

Our Web-based approach for data imputation (*WebPut*) leverages traditional *Information Extraction* (IE) methods together with the capabilities of Web search engines towards the goal of completing missing attribute values in relational tables such as the one shown in Table 1. Towards this, WebPut introduces what we call a *Web-based data imputation query*, which is basically a web search query specially formulated for the purpose of data imputation. Such data imputation query is formally defined as follows:

Definition 1. *For a relational tuple t, a* **data imputation query** $q(X \rightarrow y)$ *is a web search query formulated to utilize the existing values of a certain set of attributes* $\{X\} = \{x_1, x_2, ...\}$ *to retrieve the missing value of a certain attribute y. Hence, in* $q(X \rightarrow y)$, *X is denoted as the* **utilized attributes** *and y is denoted as the* **target attribute**.

In addition to formulating effective data imputation queries, the distribution of complete and missing values throughout the relational table makes the data imputation problem further challenging. Specifically, in a typical incomplete table, each tuple t might contain multiple target attributes: $y_1, y_2, ...$ and for each one of those target attributes there might exist multiple sets of possible utilized attributes: $X_1, X_2, ...$ For instance, in the 5th tuple of Table 1, there are three existing attribute values (Name, Title, State) and two missing values (Email, University). Any combination of the three existing attributes $\{N\}, \{T\}, \{S\}, \{N, T\}, \{N, S\}, \{T, S\}$ or $\{N, T, S\}$ could be utilized in a data imputation query targeting one of the two missing values.

Problem Definition: Given the above, our work presented in this paper addresses and proposes solutions to the following critical questions:

1. Given a pair $< X, y >$, how to **formulate** $q(X \rightarrow y)$ to effectively retrieve the missing value y? (Sec. 3.1)
2. In the presence of multiple sets of possible utilized attributes: $X_1, X_2, ...$ for the same target attribute y, how to **select** which set X is the best to impute y? (Sec. 3.2), and
3. In the presence of multiple target attributes: $y_1, y_2, ...,$ how to **schedule** the imputation order of each y_j? (Sec. 3.3).

3   The WebPut Approach

In this section, we present WebPut, our web-based approach for data imputation. WebPut integrates several novel methods that provide effective and efficient solutions to the challenges listed above.

3.1   Web-Based Data Imputation

WebPut extends popular and effective IE methods for the purpose of formulating effective data imputation queries. In particular, WebPut leverages existing complete tuples together with IE methods to *learn* the different query formulations suitable for the imputation of each missing value in a relational table. Central to that idea is searching for Web documents that contain some of the data in those complete tuples and extracting some *auxiliary information* from those documents to use in future data imputation queries. Given that approach, our previous definition of data imputation query could be further refined as: $q(X, A \rightarrow y)$ where A is the auxiliary information required for query formulation. The particular nature of this auxiliary information depends on the adopted IE method. While our approach described above is general enough to accommodate and extend any relevant IE method towards our goal of Web-based data imputation. In this paper, however, we focus on two particular IE methods, namely: the Pattern based IE method [1] and the Co-occurrence based IE method [12]. Our choice is based on the high effectiveness of these methods as shown in [12]. Investigating other methods remains part of our future work.

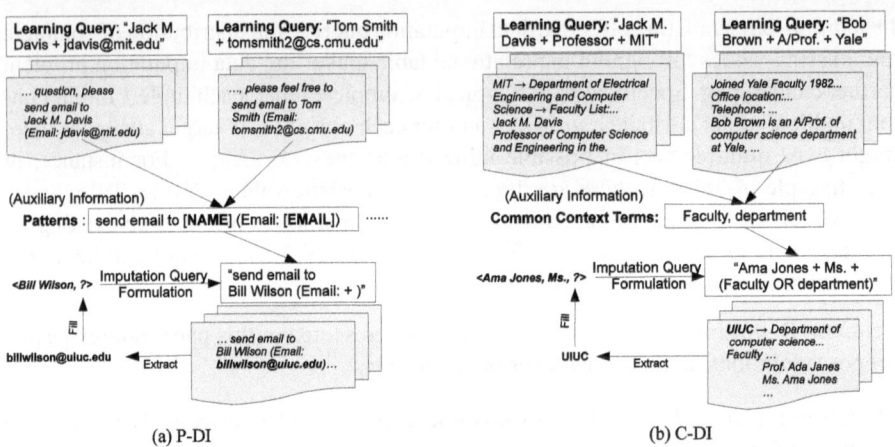

(a) P-DI   (b) C-DI

Fig. 1. Example Learning and Retrieving Process of the Two Retrieval Ways

Pattern Based Data Imputation (P-DI). Our pattern based data imputation extends the classical Pattern based IE method [10, 1], which relies on syntactic patterns to identify instances of a given entity type (such as *University*) or relation type (such as *(University, State)*). Early research in that area focused only on high-quality patterns known as Hearst Patterns (such as *"... such as..."*) to identify new entities or relations [10]. Later, bootstrapping Pattern based IE methods were proposed, in which patterns learned from seed instances are used to find more instances of the same type (such as *Snowball* [1]).

Applying pattern based data imputation in WebPut involves learning and using auxiliary information in the form of patterns, which is accomplished via the following three tasks: (1) identifying all possible utilized and target attributes pairs as $< X_i, y_j >$ based on the existing attribute values and missing values per tuple, (2) learning auxiliary information $A_{i,j}$ for each possible pair $< X_i, y_j >$ in the set of complete tuples, and (3) applying those learned auxiliary information to formulate data imputation queries for incomplete tuples.

For example, as shown in Figure 1(a), to learn auxiliary information (i.e., patterns) based on complete instances such as (*"Jack M. Davis"*, *"jdavis @mit.edu"*) and (*"Tom Smith"*, *"tomsmith2 @cs.cmu.edu"*), we issue a *Learning Query* based on each one of those complete tuples. A learning query is a Web search query that returns a set of documents that are further utilized for pattern extraction. In particular, from the retrieved documents, we may learn patterns corresponding to $< \{Name\}, Email >$ such as: "send email to [NAME] (Email: [EMAIL])" (as shown in Figure 1(a)). Finally, we can easily formulate a data imputation query for each tuple with a missing Email value using the values of Name and the extracted pattern. For example *"send email to Bill wilson (Email:"* + *")"*(The string in a quotation is taken as an unseparated keyword) and extract the missing Email value from the retrieved documents.

Co-Occurrence Based Data Imputation (C-DI). Co-occurrence based data imputation extends the co-occurrence based IE method [12]. In the context of IE, the co-occurrence based method was proposed to circumvent the limitations of the pattern based approach, given that patterns are sometimes too *strict* to capture most of the existing entities or relations on the web. In particular, a Co-occurrence based extraction method learns *common context terms* instead of patterns from seed instances of a given relation [12]. For example, from instances such as (MIT, MA), (CMU, PA), (Yale, NY), we could learn some common context terms for the relation (University, Location) such as "located at", which are mentioned closely and frequently with these instance pairs in some web pages. With these context terms, we expect to find the Location of another university like (UIUC, ?). Different from pattern-based extraction, the co-occurrence based extraction method relies on *Named Entity Recognition(NER)* [14] to extract the entity or relation from the documents.

Employing C-DI in WebPut involves almost the same three steps as the P-DI. There are three minor differences, however, which could be observed in the example in Figure 1(b): (1) C-DI learns common context terms, instead of patterns, as the auxiliary information, (2) The formulated imputation query only requires the retrieved documents to contain the common context terms at any position, (3) From the retrieved documents, we apply NER to identify all Universities, the one with the highest frequency will be taken as the missing value we are looking for.

Multiple Data Imputation Queries: To simplify the presentation, so far we have assumed that all the learning queries for each $< X_i, y_j >$ pair will return the same auxiliary information, being a pattern or a set of context terms, and in turn result in formulating a single data imputation query. For instance, Figure 1(a) shows that all the learning queries return the same pattern *"send email to [Name] (Email: [Email])"*. In reality, however, WebPut could extract multiple auxiliary information for the same $< X_i, y_j >$ pair from the retrieved documents. For instance, the learning queries in Figure 1(a) could have returned other patterns such as: *"contact [Name] through [Email]"* and *"Name: [Name], Email: [Name]."*. Similarly, for each target attribute y_i there might exist multiple sets of possible utilized attributes: $X_1, X_2, ...$ For example, for the {Email} target attribute, {Name} could be a utilized attribute (as shown Figure 1(a)) but also all other combinations of attributes form a valid set of utilized attributes (e.g., {Name, University}, etc.). As such, for each y_j WebPut can formulate a significantly large number of data imputation queries. The decision of which of these queries to use brings additional complexity to our WebPut approach which is addressed in the next section.

3.2   Confidence-Based Selection of Data Imputation Queries

In this section, we propose the use of confidence values as a method for effective and efficient data imputation in the presence of multiple applicable data imputation queries. As discussed in the previous section, it is typically the case that WebPut can formulate a significantly large number of data imputation queries for each target attribute y_j. In particular, for each y_j WebPut formulates a set of imputation queries $\mathbb{Q}_j$ where the number of queries in $\mathbb{Q}_j$ is equal to the number of valid combinations of utilized attributes and auxiliary information that is applicable to the target attribute y_j and each $q(y_j) \in \mathbb{Q}_j$ represents exactly one of those valid combinations.

In a naive implementation, WebPut would fire all queries in $\mathbb{Q}_j$ for the imputation of each y_j and select for imputation the value agreed upon by most of the queries in $\mathbb{Q}_j$ (i.e., high frequency). This approach is inefficient as it incurs a large overhead in terms of the number of issued queries and their corresponding delays. Moreover, applying a voting mechanism based on frequency might jeopardize the accuracy of imputation since frequency does not necessarily translate into high-quality.

In the rest of this section we focus on our proposed confidence-based scheme for effective imputation. In summary, our scheme *ranks* each candidate value returned by a a data imputation query $q \in \mathbb{Q}_j$ according to two factors: 1) the confidence in the data imputation query, and 2) the confidence in the values utilized by that query. Then in turn selects the value with the highest rank.

The first factor (i.e., confidence of data imputation query) is simply based on the query's success in retrieving values similar to those already existing in the complete tuples during the learning phase. Hence, our scheme naturally relies on the concepts of *coverage* and *support*, which are defined as:

Definition 1. *A data imputation query* $q(y_j)$ **covers** *a target attribute* y_j *in tuple t, if and only if: (1) all the utilized attribute values of* $q(y_j)$ *exist in t; and (2)* $q(y_j)$ *could retrieve a value for* y_j.

Definition 2. *A complete tuple t* **supports** *a data imputation query* $q(y_j)$, *if and only if: (1)* $q(y_j)$ *covers t; and (2) the missing value returned by* $q(y_j)$ *equals to the existing value of the target attribute* y_j *in the complete tuple t.*

The second condition in Definition 1 is necessary because sometimes a query might fail to retrieve any value for its target attribute. Moreover, notice that even though the definition of support is based on that of cover, it only applies for complete tuples.

1. Confidence of Data Imputation Queries Given the definitions of both coverage and support, the *confidence* of a data imputation query is estimated as:

$$Conf(q) = \frac{|Support(q)|}{|Cover(q)|} \tag{1}$$

where $|Cover(q)|$ is the number of complete tuples that are covered by q, whereas $|Support(q)|$ is the number of complete tuples that support q, as defined above.

Notice, however, that a data imputation query q is only *valid* (i.e., can be applied to impute missing values), if and only its confidence and cover are higher than predefined thresholds τ and η respectively. This approach is mainly to avoid employing outlier queries (i.e., low cover) or poor-quality queries (i.e., low support). The values of τ and η are system's parameters and their settings are discussed in Section 4.

2. Confidence of Imputed Values The confidence of a new imputed value is directly computed according to the confidence of its corresponding data imputation query q as well as the set of utilized attribute values X. More specifically,

$$Conf(y_{j,t}, q) = Conf(q) \prod_{x_i \in X} Conf(x_{i,t}) \tag{2}$$

where the confidence of each utilized attribute value $x_{i,t}$ is either pre-determined (for existing values) or it is computed (for already imputed values) in the same way. When several data imputation queries are applied to a missing value, and different data imputation queries lead to different values, only the value with the highest confidence will be used to impute the missing value.

3.3   Efficient Scheduling of Data Imputation Queries

WebPut aims to impute each missing value with the highest-confidence value while at the same time minimizing the number of issued data imputation queries. In this section, we propose efficient algorithms for the scheduling of data imputation queries that achieve these goals. As opposed to the naive approach discussed in the previous section (Sec. 3.2), the following algorithms share the principle idea of leveraging the confidence of data imputation queries to achieve high efficiency and effectiveness.

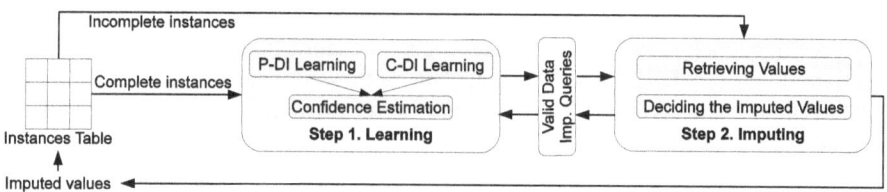

Fig. 2. The Architectures of the Iterative WebPut Algorithm

1. One-Pass Scheduling Algorithm (Baseline). In this baseline algorithm, WebPut imputes all missing values in one pass. First, based on existing complete tuples, it estimates the confidence of all the data imputation queries targeting any incomplete attributes in the given database. Second, it uses those valid data imputation queries to retrieve values for missing values. However, when several queries are available for the same missing value, it selects the one with the highest confidence to be issued under the assumption that it should provide the highest confidence value.

For example, consider the 5th tuple in Table 1 in which there are two missing values (=Email and University) and three existing values (=Name, Title, State). As shown in

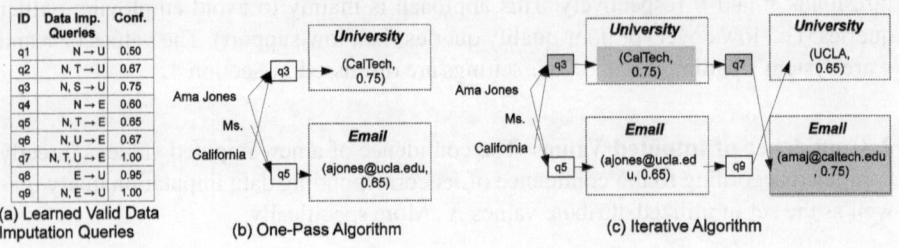

(a) Learned Valid Data Imputation Queries

ID	Data Imp. Queries	Conf.
q1	N → U	0.50
q2	N, T → U	0.67
q3	N, S → U	0.75
q4	N → E	0.60
q5	N, T → E	0.65
q6	N, U → E	0.67
q7	N, T, U → E	1.00
q8	E → U	0.95
q9	N, E → U	1.00

(b) One-Pass Algorithm

(c) Iterative Algorithm

Fig. 3. Imputation Results in an Example Tuple with Two Different Algorithms

Figure 3(a), data imputation queries q_1, q_2, q_3 can be applied to the University missing value, among which q_3 has the highest confidence. Meanwhile q_4 and q_5 can be applied to the Email missing value, among which q_5 has the highest confidence. By applying both of the high-confidence queries (i.e., q_3 and q_5), we assign "CalTech" with 0.75 confidence for the University missing value; and "ajones@ucla.edu" with 0.65 confidence for the Email missing value as shown in Figure 3(b). While both values have a reasonable confidence, they are inconsistent with each other: people from one University are unlikely to use an email belonging to another university. This example shows that one-pass algorithm suffers from one major drawback: the imputed value of one imputed missing data is not further utilized to impute values of other missing data. Consequently, it may lose the chance to retrieve a better value with a higher confidence for some missing values.

2. Iterative Scheduling Algorithm. To address the drawbacks of the one-pass algorithm described above, we introduce an iterative scheduling algorithm. In each iteration of this algorithm, we impute the missing values, similar to the one-pass algorithm. However, after each iteration, the imputed values are stored in an *intermediate* instance database (Figure 2). In the next iteration, some of those imputed values are further utilized to impute already imputed ones if they are of high confidence and they support high-confidence queries.

To further illustrate the iterative algorithm, let's consider again the 5th tuple in Table 1. After the 1st iteration (Figure 3(c)), we fill the two missing missing values (Email and University missing values) with two imputed values, which are similar to those provided by the one-pass algorithm. In the 2nd iteration, however, two more queries (=q_8 and q_9) that target the University missing value become available as they can utilize the Email value imputed in the 1st iteration. Similarly, queries q_6 and q_7 targeting the Email missing value become available as they can utilize the University value imputed in the 1st iteration. According to our query rankings, we apply q_7 and q_9 in this iteration, and we get "amaj@caltech.edu" with 0.75 confidence for the Email missing value; and "UCLA" with 0.65 confidence for the University missing value. Since the confidence of the new value "amaj@caltech.edu" is higher than the value "ajones@ucla.edu" obtained in the 1st iteration, we replace the imputed value of Email with the new one. However, note that our schedule prevents cycles, that is, the value of missing value y_1 that is imputed by utilizing the value of missing value y_2 cannot be utilized in imputing values for missing value y_2. Thus, we could stop the iterative algorithm at this moment. Now we have more consistent imputed values for the two missing values.

3. Greedy Iterative Scheduling Algorithm. In the above example, four data imputation queries (q_3, q_5, q_7, q_9) were applied, while only two (q_3 and q_7) provided the highest-confidence values for the two missing values. Motivated by this observation, we present a greedy shceduling algorithm, which identified "effective" data imputation queries that provide the highest-confidence values for missing values apriori, and in turn, minimizes the number of issued queries. For an incomplete tuple, all identified data imputation queries are processed in a particular order so they form a *greedy schedule* for each incomplete tuple.

The optimal schedule is *acyclic* and associates all missing values with the minimum number of data imputation queries, as shown in the highlighted path in Figure 3(c).

To generate that optimal schedule, in each iteration we only select one data imputation query that is estimated to provide the highest-confidence attribute value for a not imputed missing value. As long as the data imputation query could provide a value for that missing data, we impute the missing data with the already imputed values. However, if no value is provided, WebPut drops that query and re-selects a new data imputation query from the remaining unused imputing queries based on the same rule. Finally, when all missing values are imputed or there are no more unprocessed data imputation queries, the algorithm stops. The optimal schedule for the 5th tuple in Table 1 is highlighted (shaded boxes) in Figure 3(c).

The greedy version of the iterative algorithm always provides the same imputed values as the pure iterative algorithm described earlier. However, in the best case scenario, the greedy algorithm requires only one data imputation query to apply per missing value, which is more efficient than the iterative algorithm.

4 Experimental Evaluation

We have implemented a WebPut prototype in Java which uses Google API to answer data imputation queries. We have experimented with three real-world data sets:

- *Multilingual Disney Cartoon Table (Disney):* This table contains names of 51 classical disney cartoons in 8 different languages collected from Wikipedia.
- *Personal Information Table (PersonInfo):* This is a 2.5k-tuples, 5-attributes table, which contains contact information for academics including name, email, title, university and country. This information is collected from 20 different universities in the USA, UK and Australia.
- *DBLP Publication Table (DBLP):* This is a 10k-tuples, 5-attributes table. Each tuple contains information about a published paper, including its title, first author, conference name, year and venue. All papers are randomly selected from DBLP.

The three data sets above are complete relational tables. To generate incomplete tables for our experiments, we remove attribute values at random positions from the complete table, while making sure that at least one key attribute value will be kept in each tuple.

(For the Disney dataset, all attributes are key attributes. For the PersonInfo dataset, the name and email are key attributes. For the DBLP dataset, the paper title is the only key attribute.) Each reported result is the average of 5 evaluations, that is, for each missing value percentage (1%, 5%, 10%, 20%, 30%, 40%, 50%, 60%), 5 incomplete tables will be generated with 5 random seeds, and the experimental results we present are the average results based on the 5 generated incomplete tables. We then impute these generated incomplete tables using WebPut and evaluate the performance of our solutions by using the original complete table as the ground truth.

4.1   WebPut vs. Related Approaches

We compare the accuracy of WebPut against two state-of-the-art data imputation approaches: (1) *Close-fit*: a substitute-based approach which finds a "close-fit" value from a similar context using association rules [20]; (2) *Mixing*: a model-based approach where a mixture-kernel based iterative estimator is advocated to impute mixed-attribute datasets [23]. The accuracy of an algorithm is the percentage of missing values filled with correct values among all missing values in the table.

As shown in Figure 4, WebPut (using the greedy iterative scheduling algorithm) reaches a much higher accuracy than both of the other two algorithms for the 3 data sets at all the missing ratios. For the 1st data set both *Close-fit* and *Mixing* can not impute any missing values at all, since all values in this dataset are unique and thus missing values are unlikely to be imputed either from a similar context within the dataset or through a established model built on the dataset. For the same reason, the two methods impute no more than 50% of missing values on the other two datasets.

The performance of WebPut is determined by whether the missing values could be retrieved from the web based on existing values. For the 1st data set, since only the English name of a Disney movie is usually mentioned together with names in other languages, once the English name is missing, it is pretty hard to retrieve missing names in other languages through existing names in other languages. With the iterative algorithm, we may impute the missing English name first, and then identify other names through the imputed English name. But once the English name is incorrect, all the other names that imputed based on the English names are also incorrect. Therefore, the accuracy on the 1st data set is not high. For the other two data sets, as long as a key value of a tuple exists, such as the name in PersonInfo, or the paper title in DBLP, it is not difficult to impute the missing values based on the key values. As a result, WebPut retrieves approximately 70-80% of missing values for the two data sets.

4.2   P-DI vs. C-DI

We measure the effectiveness of WebPut when the data imputation queries are formulated based on: 1) only P-DI; 2) only C-DI; and 3) a suite of P-DI and C-DI formulated queries (as described in Section 3.1). In this experiment, for a query to be valid it must be supported by at least 3 complete tuples and its confidence must be more than 0.3. For P-DI, the length of Prefix and Suffix in each pattern must be at least 5 characters long. Finally, when applying a pattern to extract the value of an entity, that candidate value

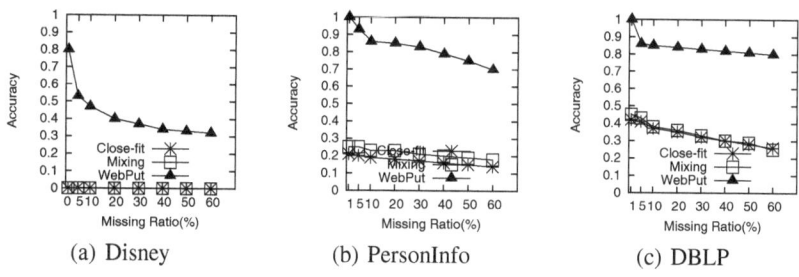

Fig. 4. Comparing the Accuracy of WebPut against Previous Approaches

must be within 300 characters. For C-DI, we adopt all the settings from [12] to get the best expansion terms for each query.

As shown in Figure 5, the accuracy of using both P-DI and C-DI queries in WebPut is always higher than that of using either one alone. For instance, P-DI can only fill up to 20% of the missing values due to the strictness of pattern matching. While C-DI can fill much more missing values than P-DI due to its flexibility, the combination of both still achieves the highest accuracy.

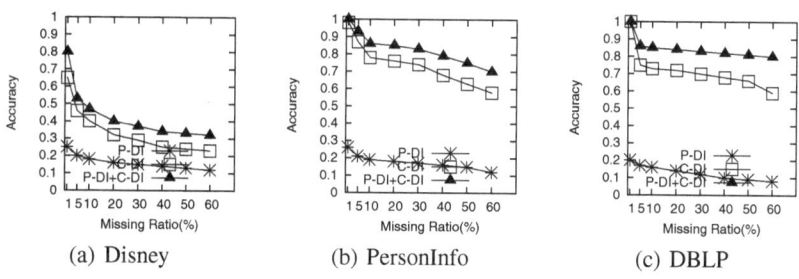

Fig. 5. Comparing the Accuracy of the different Query Formulation Methods

4.3 Evaluation of the Scheduling Algorithms

In this experiment, we first evaluate the accuracy of three WebPut algorithms proposed in Section 3. As shown in Figure 6(a)(b)(c), when the missing ratio is low (<5%), the accuracy of the three algorithms is almost the same. As the missing ratio increases from 5% to around 50%, the accuracy of *One-Pass* drops faster than *Iterative* for all the three data sets. When the missing ratio becomes more than 50%, the accuracy of *Iterative* is about 20% higher than that of *One-Pass*. The figure also shows that the accuracy of *Greedy Iterative (GreedyIter)* is always the same as that of *Iterative* as discussed in Section 3.3.

In Figure 6(d)(e)(f) we compare the costs of the 3 algorithms measured in terms of the number of data imputation queries issued by WebPut. The figure shows that the imputation costs of both *GreedyIter* and *One-Pass* is approximately proportional to the missing ratio, since both of them need only one query to impute each missing value.

The cost of *Iterative*, however, increases much faster than *GreedyIter* and *One-Pass*. For instance, when the missing ratio is larger than 50%, the cost of *Iterative* is about 2 times that of *GreedyIter*. From the two figures, it is clear that *GreedyIter* reaches the best accuracy at the minimum cost as intended.

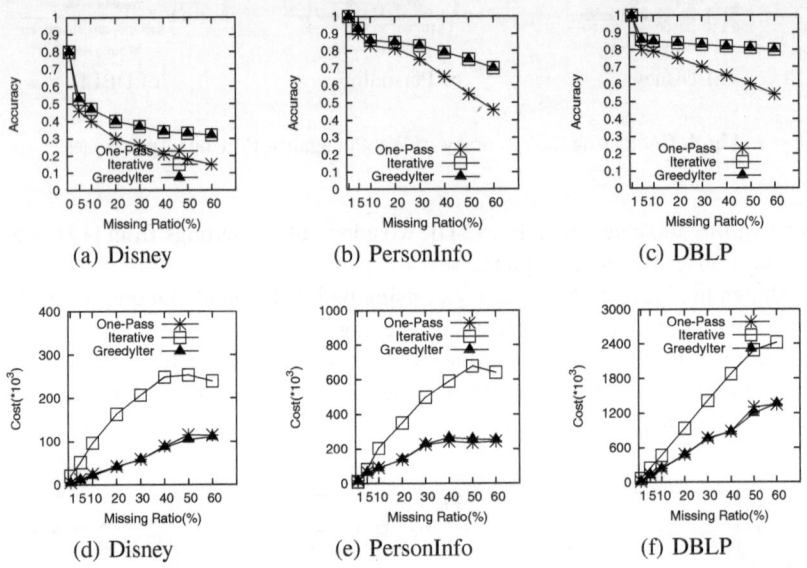

Fig. 6. Comparing the Accuracy and Cost of the Scheduling Algorithms

5   Related Work

Data imputation aims at providing estimations for missing attribute values by reasoning from observed data on a data set [3]. Existing data imputation approaches can be divided into three categories: (1) substitute-based data imputation; (2) model-based data imputation; (3) external resource based data imputation.

The substitute-data imputation approaches arbitrarily find a substitute value for the missing one from the same data set. Nine approaches have been introduced and compared in [8], such as selecting the "most common attribute value" as the missing value, or "assigning all possible values of the attribute restricted to the given concept", or "treating missing attribute values as special values" and so on. Later, k-Nearest Neighbor [7] and association rules [20] were also used to find a "close-fit" value from a similar context. The substitute-based data imputation approaches attempt to smooth the influence of missing values on statistical data analysis results by replacing them with suitable substitutes, but are unlikely to find the right missing attribute values for missing values, especially when the missing value is a unique one within the data set.

The model-based data imputation approaches build a prediction model based on the data set, and then estimate a value for a missing value based on the model. Among the proposed approaches, some [2, 18] were developed for continuous attributes only, while others [15, 22] were designed to deal with discrete attributes. There are also some approaches [21, 23] targeting imputing mixed attributes, which either take the discrete attributes as continuous ones, or smooth the mixed regressors. The model-based data imputation approaches have advantage on estimating a vey close value for the missing one, which could greatly smooth the influence brought by incomplete data. However, close estimations can not replace the missed original values, especially the values of discrete attributes.

The third type of data imputation approaches aim at finding missing attribute values from external resources. The WebPut belongs to this category. This type of data imputation approach do not find substitutes or estimations for the missing ones, but the missing values themselves from external resources. There were some effects on augmenting a table with very few example rows by constructing new rows from unstructured lists on the web [9]. Although table augmentation does not have the same purpose with data imputation, it is also extracting data from external sources and put them into local table.

6   Conclusions and Future Work

In this paper, we present WebPut - an approach to impute missing data from external data sources on the world wide web. In that sense, WebPut could be perceived as a novel Information Extraction (IE) approach, which formulates the extraction tasks around the missing attribute values in a database. Our experimental results based on several real-world data collections demonstrate that WebPut can effectively retrieve a large percentage (approximately 70-80%) of correct missing values in an incomplete table, outperforming previous approaches.

An underlying assumption of the work presented in this paper is that all existing attribute values are faultless (i.e., clean data), meanwhile we leave the problem of data imputation in the presence of incorrect and dirty data as part of our future work. We are also working on a hybrid approach that combines and integrates our web-based approach with previous model-based data imputation methods.

Acknowledgments. The authors would like to thank the anonymous reviewers for their valuable comments and suggestions. This work was supported in part by Australian Research Council grant DP110102777.

References

[1] Agichtein, E., Gravano, L.: Snowball: Extracting relations from large plain-text collections. In: ACM DL (2000)

[2] Barnard, J., Rubin, D.: Small-sample degrees of freedom with multiple imputation. Biometrika 86(4), 948–955 (1999)

[3] Batista, G., Monard, M.: An analysis of four missing data treatment methods for supervised learning. Applied Artificial Intelligence 17(5-6), 519–533 (2003)

[4] Brin, S.: Extracting patterns and relations from the world wide web. In: The World Wide Web and Databases, pp. 172–183 (1999)

[5] Dempster, A., Laird, N., Rubin, D.: Maximum likelihood from incomplete data via the em algorithm. Journal of the Royal Statistical Society. Series B (Methodological), 1–38 (1977)

[6] Grzymala-Busse, J.: Three approaches to missing attribute values: A rough set perspective. In: Data Mining: Foundations and Practice, pp. 139–152 (2008)

[7] Grzymala-Busse, J., Grzymala-Busse, W., Goodwin, L.: Coping with missing attribute values based on closest fit in preterm birth data: A rough set approach. Computational Intelligence 17(3), 425–434 (2001)

[8] Grzymala-Busse, J.W., Hu, M.: A Comparison of Several Approaches to Missing Attribute Values in Data Mining. In: Ziarko, W.P., Yao, Y. (eds.) RSCTC 2000. LNCS (LNAI), vol. 2005, pp. 378–385. Springer, Heidelberg (2001)

[9] Gupta, R., Sarawagi, S.: Answering table augmentation queries from unstructured lists on the web. PVLDB 2(1), 289–300 (2009)

[10] Hearst, M.: Automatic acquisition of hyponyms from large text corpora. In: COLING (1992)

[11] Li, J., Cercone, N.: Assigning missing attribute values based on rough sets theory. IEEE Granular Computing, 607–610 (2006)

[12] Li, Z., Sitbon, L., Zhou, X.: Learning-based Relevance Feedback for Web-based Relation Completion. In: CIKM (2011)

[13] Loshin, D.: The Data Quality Business Case: Projecting Return on Investment. Informatica (2006)

[14] Mikheev, A., Moens, M., Grover, C.: Named entity recognition without gazetteers. In: EACL (1999)

[15] Quinlan, J.R.: C4. 5: programs for machine learning. Morgan Kaufmann (1993)

[16] Ramoni, M., Sebastiani, P.: Robust learning with missing data. Machine Learning 45(2), 147–170 (2001)

[17] Shi, S., Zhang, H., Yuan, X., Wen, J.R.: Corpus-based semantic class mining: distributional vs. pattern-based approaches. In: COLING (2010)

[18] Wang, Q., Rao, J.: Empirical likelihood-based inference under imputation for missing response data. The Annals of Statistics 30(3), 896–924 (2002)

[19] Wang, R.C., Cohen, W.W.: Automatic set instance extraction using the web. In: ACL/AFNLP (2009)

[20] Wu, C., Wun, C., Chou, H.: Using association rules for completing missing data. In: HIS (2004)

[21] Zhang, S.: Parimputation: From imputation and null-imputation to partially imputation. IEEE Intelligent Informatics Bulletin 9(1), 32–38 (2008)

[22] Zhang, S.: Shell-neighbor method and its application in missing data imputation. Applied Intelligence 35(1), 123–133 (2011)

[23] Zhu, X., Zhang, S., Jin, Z., Zhang, Z., Xu, Z.: Missing value estimation for mixed-attribute data sets. IEEE TKDE 23(1), 110–121 (2011)

Representing Service-Relationships as First Class Entities in Service Orchestrations

Malinda Kapuruge, Jun Han, and Alan Colman

Faculty of Information and Communication Technologies
Swinburne University of Technology, Melbourne, Australia
{mkapuruge,jhan,acolman}@swin.edu.au

Abstract. Service orchestration approaches are widely used to composing multiple business services (partner services) into a business process to achieve a particular business objective. The business relationships captured in such a service orchestration are primarily those between the partner services and the business process itself. This however results in tight-coupling between processes and partner services and inadequate capturing of relationships between partner services that participate in an orchestration. These limitations create problems concerning the stability and runtime adaptability of a service orchestration. To address these limitations, we propose in this paper an approach that represents the *service-relationships* as first-class entities in service orchestrations during design-time and runtime. It provides the required stability and improves the runtime adaptability for service orchestrations amidst changing business requirements. A novel process enactment platform supporting the approach has been implemented by further extending the Apache Axis2 Web service engine.

Keywords: Service Orchestration, Service-relationships, Adaptability, Stability.

1   Introduction

Service Oriented Computing (SOC) has gained popularity as a computing paradigm because of its ability to easily compose distributed services in a loosely coupled manner. These distributed services are orchestrated to support and enable business operations according to well-defined business processes, known as service orchestrations.

Service orchestration approaches combine the benefits offered by both the SOC and Business Process Management (BPM) disciplines [1]. They complement each other and exhibit a fine *fit* in the enterprise architecture. This fit is mainly due to the fundamental and yet complementary differences between them. SOC is primarily *IT-driven* and used to specify loosely coupled and reusable distributed systems, enabling business capabilities across organisational boundaries. In contrast, BPM is primarily *business-driven* and used to capture a way to achieve a business objective of an organisation by aligning with IT [2]. For a service orchestration approach to truly support business-IT alignment, the nature and characteristics of the real-world business need to be appropriately and timely captured in its design and enactment.

X.S. Wang et al. (Eds.): WISE 2012, LNCS 7651, pp. 257–270, 2012.
© Springer-Verlag Berlin Heidelberg 2012

The current approaches for service orchestrations attempt to achieve the business-IT alignment primarily by modelling a service orchestration as a process that wires a set of partner services together. The relationships between the process and its partner services are well represented, e.g., through *partnerLinks* in WS-BPEL [3]. However, the *relationships between the partner services* (*service-relationships*), which exist in real-world business environments, are not explicitly captured in the service orchestration. An explicit representation of service-relationships in a service orchestration allows reflecting the underlying partner services and their relationships; and thereby the ability to define business processes based on the representation. The tight coupling of business process and the partner service interactions in existing approaches create problems concerning the *stability* and the *adaptability* of the service orchestration as the means to provide IT support for the business.

In this paper we propose a *meta*-model, associated *language* and *enactment platform* to address these challenges by explicitly representing the service-relationships in a service orchestration. These service-relationships collectively form an abstract organisational structure providing the required *stability* to define and enact business processes. In addition, the defined service-relationships and the coordination logic (*Organisational Behaviour*) of the organisation can be dynamically changed at runtime providing the required *adaptability* for the service orchestration. Such adaptability allows the timely capture of changing business relationships and adjustment to the organisational behaviour reflecting the changing relationships.

The rest of the paper is organised as follows. In Section 2, we use a motivational business example to analyse the problem. In Section 3 we examine the related work. Based on the understanding gained, Section 4 presents our approach and how it is capable of solving the problem. The implementation details are discussed in Section 5. Finally, conclusive remarks are presented in Section 6.

2   Problem Analysis

In this section, we first present a simplified business example to analyse the problem and motivate the research. Then, we explain *what* the service-relationships are and *why* they need to be explicitly represented in a service orchestration providing the required adaptability and stability.

2.1   Business Example

RoSAS is a business organisation that provides Roadside Assistance Services to motorists. RoSAS aggregates a number of third party services as part of its roadside assistance business. This includes Garage/Auto-repair services (GR) to repair cars, Tow-Truck (TT) services to tow cars, and Case-Officers (CO) to handle the assistance requests and payment claims. The consumers of RoSAS are motorists (MM) who have subscribed to the RoSAS membership.

RoSAS is inherently a collaborative environment. During the runtime, service-providers interact with each other to serve consumer requests. RoSAS needs to define the allowed interactions and coordinate the service invocations in an automated manner. For this, RoSAS expects to use the service orchestration technology due to its capability of composing and coordinating distributed applications and services.

The third party service providers expose their business offerings as Web services as shown in Fig. 1. For example, Garage and Tow-Truck chains may provide Web service endpoints for their legacy systems to accept the repair and towing requests. Moreover, RoSAS may hire Case-Officers (e.g., Human-Resource/Call-Centre Services) and provide necessary software systems for handling assistance requests remotely. In addition, RoSAS exposes its own unified roadside assistance service as a Web service endpoint for motorists to submit assistance requests via client applications installed in their mobile phone or their In-Car Emergency-Alert-Systems.

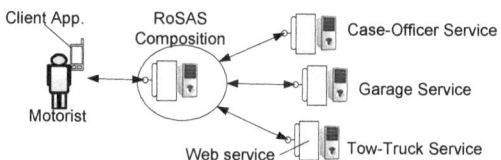

Fig. 1. RoSAS business model realised using Web services

2.2   Service-Relationships

The third party services need to be integrated and interact with each other as defined by RoSAS. RoSAS has imposed certain obligations on them according to its business model. These obligations need to be realised and maintained in the IT-model. Any partner service bound to the RoSAS composition should abide by these obligations.

The interactions and obligations between two partner services collectively reflect and represent the *service-relationship* between the two services within the RoSAS business model. For example, for every 5th completed repair, the CO is to make a bonus payment to the Garage (GR) service. If the car is towed to an assigned repair station beyond a distance of 100 km, e.g., due to the Garage failing to assign a nearby repair station, the Garage is to make a *reward* payment to the Tow-Truck. On the other hand, if the towing is delayed or the car is towed to a wrong repair station, the Tow-Truck must make a *Penalty Payment* to the Garage.

Service-relationships should not be misunderstood as *Service Level Agreements* (SLAs). Service-relationships and SLAs are orthogonal to each other and serve different purposes (Fig. 2). An SLA is an agreement between two business parties and usually specifies the quality properties of service delivery. Obviously RoSAS as the service aggregator maintains such SLAs with its partner services, e.g., with *Fast-Tow.com*. In contrast, a service-relationship is an internal representation maintained by the service aggregator, and defines the relationship between two such partner services *in the context of the aggregation*, e.g., between the Tow-Truck and Garage services. The service-relationships need to be represented in the design of the service orchestration and accessible to the enactment platform, so that the allowable interactions and obligations between the services can be clearly defined and managed.

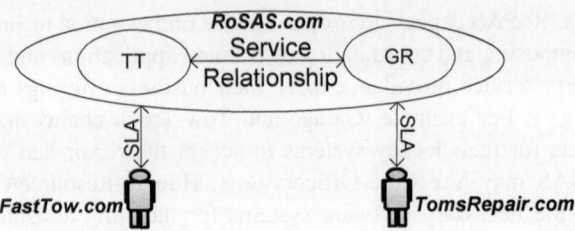

Fig. 2. Service-relationship vs SLA

2.3   Adaptability and Stability

The business requirements of RoSAS change over time and the IT realisation should follow and adapt to these changes. New business regulations imposed by RoSAS as a business organisation, needs to be captured in its IT realisation (i.e., the defined service orchestration). For example, RoSAS may wish to remove the currently imposed *Penalty-Payment* for delayed towing after having failed to attract suitable Tow-Truck services after few months of operation. Furthermore, to maintain an advantage over its competitors, there may be frequent changes of this nature in the business environment, demanding corresponding changes in its IT realisation. In a competitive business environment, a system re-deployment/re-start is not a viable solution. Therefore, the service orchestration approach used by RoSAS needs to provide the runtime adaptability allowing these dynamic changes in the business model and service orchestration to be realised with no or minimal impact on its runtime operation.

During the runtime the bound partner services may also change for both fault-handling and optimisation purposes. For example, if the currently bound Garage service quits the business or becomes unavailable for technical reasons, a new garage chain needs to be bound. If the currently bound Tow-Truck chain frequently delays the towing and better alternatives are available, it needs to be replaced. If the service orchestration is directly ground upon the concrete services in terms of its control flow and data flow, such changes in partner services frequently call for changes in the core orchestration process, challenging its stability. Therefore it is necessary to decouple the process and the partner services via a suitable abstract representation. This abstract representation should provide a stable basis for defining business processes irrespective of the availability or changes of the partner services.

In summary, the following three requirements need to be satisfied by a service orchestration approach to achieve the required stability and adaptability.

— *(Req1) Representation of service-relationships*: The service-relationships need to be captured in the service orchestration as the basis for enacting processes.
— *(Req2) Adaptability*: The core orchestration process, service-relationships and partner services should be dynamically modifiable to meet the changing business requirements.
— *(Req3) Stability*: A stable basis should be provided to define core business processes in terms of partner services in a decoupled manner.

3   Related Work

A prevalent choice to implement the above business example is to use WS-BPEL [3]. However, WS-BPEL provides little support for adaptability. In this regard, there have been efforts to improve its adaptability [4-7]. For example, the static-proxy-based [7] and dynamic-proxy-based [4] solutions are proposed by Ezenwoye et al. to handle the failures in partner services of a BPEL process. Yet, they do not provide the capability of handling changes in the core service orchestration process.

The above limitation has been addressed by approaches integrating changeable business rules. For example, Rosenberg et al.[5] proposes to integrate/map rules *before/after* a BPEL activity via a *Rule Interceptor Service*. By exploiting the benefits of Aspect-Oriented Programming (AOP), Charfi et al.[6] integrates business rules with a WS-BPEL process by treating business rules as aspects to be woven into the process. MoDAR [8] combines the benefits of both AOP and model-driven-development to integrate business rules into processes. Several patterns on how business rules can be integrated with business processes has also been introduced by Graml et al.[9]. These approaches have immensely improved the adaptability of a service orchestration primarily due to the adaptability offered by business rules. But they distinguish a fixed part and a volatile part of a service orchestration. The fixed part is modelled as a process (e.g., in BPEL) and the volatile part is modelled as Business Rules. However, in real world business scenarios such an assumption of distinguishing fixed and volatile parts can be difficult as the business evolves leading to major process changes where the assumption may no longer hold.

In addition, these approaches do not sufficiently capture the service-relationships in a service orchestration. Consequently, the mutual obligations and interactions between partner services cannot be explicitly represented or therefore managed in the modelling or enactment of the service orchestration.

In the *Enterprise Service Bus* (ESB) architecture, connections are used to mediate the messages [10]. However, ESBs neither capture nor maintain the mutual obligations among partner services. Therefore ESB-based service orchestration solutions (e.g., via integration with a BPEL orchestrator) cannot provide the required level of abstraction to model a stable and adaptable service composition.

The *Service Component Architecture* (SCA) [11] provides an abstract representation of a service composition based on the abstraction provided by the *components*. *Components* in SCA are implementation independent modules that can be assembled to create larger composite applications. However, the connections (bindings) among the components are *not* considered as first class entities of the composite design (while components are treated as first-class entities). The connections neither provide the abstraction nor modularity to specify the interactions among services (as represented by components), because they are viewed as implementation specific communication protocols established at the binding time. This limitation in SCA makes it difficult to capture the mutual interactions and obligations (service-relationships) among partner services at a required level of abstraction. Such limitations inherently challenge the stability and the adaptability of a service composition.

Table 1. Summary of evaluation of related work

Approach/Architecture	[7]	[6]	[5]	[8]	[9]	[4]	[11]	[10]
Req1 (Service-relationships)	-	-	-	-	-	-	-	-
Req2 (Stability)	~	-	-	-	-	~	~	~
Req3 (Adaptability)	~	+	+	+	+	~	-	-

+ Supported, - Not Supported, ~ Limited Support

Table 1 summarises the existing approaches' support for the requirements of having stable and adaptable service orchestrations as presented in Section 2. The limitations in the existing approaches pose the need for a service orchestration approach that can explicitly capture the service-relationships as first-class entities of a service composition. These service-relationships need to be *adaptable,* supporting changes in the business model. Moreover, the relationships need to collectively provide the required level of abstraction over the interactions of the underlying partner services, providing a *stable* basis for defining and enacting business processes.

4   The Approach

In order to address the aforementioned challenges, we propose a novel approach (*Serendip*) that envisions a service composition as an *organisation*, where the relationships among services (*service-relationships*) are explicitly represented as first class entities. A service-relationship captures the interactions and mutual obligations between two partner services (represented by *Roles*) in a declarative manner.

A collection of such *service-relationships* and *roles* define an abstract *Organisational Structure*. This organisational structure provides a stable basis for defining one or more business processes. As such, it decouples the actual participating services from the processes in the service orchestration. During runtime, concrete services (*Players*) can be bound to the roles. Bound players are obliged to perform the *Tasks* assigned to roles according to the defined service-relationships.

These service-relationships are adaptable at runtime. That is, new interactions and obligations can be dynamically added and existing ones can be modified or removed without requiring a system restart. Similar to service-relationships, the *Business Processes* that coordinate the Tasks are also defined in a declarative and adaptable manner to support the changes in the service-relationships.

A high-level design of RoSAS, modelled as such an adaptable organisation, is given in Fig. 3. We represent a service-relationship as a *Contract*. In the example, there are four contracts (CO_MM, CO_TT, CO_GR and GR_TT) between four roles (CO, MM, GR and TT). During runtime, concrete services (e.g., *TomsRepair.com*) may be bound to or unbound from these roles. Yet, the composite structure remains, because the interactions and mutual obligations between the partners of the composition are specified independent of the bound players. Moreover, a role does not define its interactions or obligations of its own. Instead, the contracts that a role has with other roles define its obligations and interactions. For example, the interactions and obligations of role TT are defined by its adjoining contracts i.e., CO_TT and GR_TT.

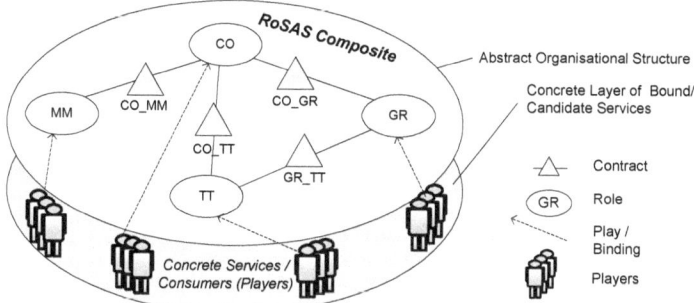

Fig. 3. RoSAS business model designed as a ROAD composite

4.1   Explicit Representation of Service-Relationships

A contract defines a service-relationship. It identifies the two participants (*Roles*) and specifies *Interaction Terms* (defining the allowed interactions), *Facts* (representing contract state), and *Business Rules* (evaluating interactions and enforcing obligations).

The listing in Fig. 4 shows a description of the contract CO_GR between the roles CO and GR. It specifies two facts: *RepairCounter* records the number of repairs done; *ContractGrade* maintains the current grade (e.g., Peak/Low) of the contract. Interaction Terms (ITerm) define how the roles bound by the contract should interact. The messages exchanged between CO and GR must conform to these ITerms. For example, the first ITerm (*iOrderRepair*) allows a CO to order repairs from the GR. It defines the parameters of interaction (String:repairInfo,   int:caseId) and the interaction direction (AtoB: from CO to GR).

```
Contract CO_GR{
    A is CO, B is GR;  //The two roles bound by the contract.
    Fact RepairCounter(int:counter) ;
    Fact ContractGrade(String:grade);
    ITerm iOrderRepair(String:repairInfo, int:caseId) from AtoB;
    ITerm iRepairNotify(String:notifInfo, int:caseId) from BtoA;
    ITerm iRepairPay (int:amount, String:invoiceId, int:caseId)from AtoB;
    RuleFile "CO_GR.drl";
}
```

Fig. 4. (Listing) Contract CO_GR between two roles CO and GR

A *RuleFile* may contain a number of business rules that evaluate interactions and enforce obligations. To do so, these rules use the messages that flow across the contract (as allowed by ITerms) and the facts that are maintained within the contract. For example, according to the RoSAS requirements, for every 5th repair done, the CO is obliged to make a bonus payment to the Garage. To support this requirement, contract CO_GR maintains the fact *RepairCounter* and specify a rule, *If (counter>=5) then a bonus needs to be paid and counter is reset. Or else the counter needs to be incre-*

mented. This rule is captured using two[1] Drools [12] rules as shown in the listing of Fig. 5. Here, the operationName of the message (msg) is same as the identifier of corresponding ITerm. Based on the runtime evaluation, *events* (e.g., *ePayRepairBonus* or *ePayRepairNormal*) can be triggered via special method *triggerEvent()*. These events will cause the corresponding tasks to be carried out .

```
declare RepairCounter
   counter : int
end
rule "paySomeBonus"
    when      $msg:MessageRecieved(operationName=="iRepairNotify")
              $ctr:RepairCounter(counter>=5)
    then      $msg.triggerEvent("ePayRepairBonus"); //Pay With Bonus
              $ctr.setCounter(0);   //Reset Counter
end
rule "incrementCounter"
     when     $msg:MessageRecieved(operationName=="iRepairNotify")
              $ctr:RepairCounter(counter<5)
    then $msg.triggerEvent("ePayRepairNormal"); //Pay Normal
              $ctr.setCounter($ctr.getCounter()+1); //Increment Counter
end
```

Fig. 5. (Listing) Sample rules of contract CO_GR to evaluate interaction *iRepairNotify*

4.2   Service-Relationship-Driven Process Coordination

Tasks are defined in roles. Players perform tasks and they interact with each other via roles according to the contracts. Fig. 6 shows a sample task *tRepair* defined in the role GR. The interaction message *iOrderRepair* is used to perform task and as a result creates a new message as part of interaction *iRepairNotify*, as defined in contract CO_GR. Such tasks defined in roles need to be *coordinated*. In other words it should be possible to specify *when* to invoke the bound players, e.g., *TomsRepair.com*.

```
Role GR playedBy  TomsRepairPlayerBinding {
   Task tRepair{
      UsingMsgs CO_GR.iOrderRepair;  ResultingMsgs CO_GR.iRepairNotify; }
}
```

Fig. 6. (Listing) Sample task defined in role GR

We decouple the coordination logic and the contractual interactions via *events*. An event is a passive element that marks a situation. Events[2] are used to define the

---

[1] Drools rules are *condition-action* rules rather than *if-then-else*. The 'else' part need to be captured in a separate rule. Hence, two rules.

[2] Events (situation) usually associate with a process instance. Further discussion on such association is beyond the scope of this paper.

dependencies among the tasks of the organisation. For example, the triggered event (Fig. 5) *ePayRepairBonus* [3] initiates the task *tPayRepairBonus* defined in the role CO.

To capture task dependencies in a declarative manner we define coordination rules in *behaviour units*. An organisation may have multiple behaviour units grouping the related tasks together, e.g., coordinating tasks for towing and for repairing. A description of the repairing behaviour unit (*bRepairing*) is given in the listing of Fig. 7.

As shown, the dependencies among tasks are captured via events (or pattern of events). For example, *GR.tRepair* (Task *tRepair* of GR) is initiated (InitOn) when event *eRepairReqd* is triggered. When the task *tRpeiar* completes, the event *eRepair-Done* and (AND =*) either of the events (XOR=^) *ePayRepairNormal and ePayRe-pairBonus* are triggered. These events act as triggers for consequent tasks, e.g., *tPayRepair* and *tPayRepairBonus*.

```
Behavior bRepairing{
   TaskRef CO.tOrderRepair{
      InitOn "eTowDone";    Triggers "eRepairReqd"; }
   TaskRef GR.tRepair{
      InitOn "eRepairReqd";
      Triggers "eRpairDone *(ePayRepairNormal^ePayRepairBonus)"; }
   TaskRef CO.tPayRepair{
      InitOn "ePayRepairNormal";    Triggers "eRepairPaid";    }
   TaskRef CO.tPayRepairBonus{
      InitOn "ePayRepairBonus";    Triggers "eRepairBonusPaid"; }
}
```

Fig. 7. (Listing) Behaviour Unit bReparing specifies the coordination logic

To elaborate how the runtime works, let us use the scenario in Fig. 8. When CO sends (marked by #1) the repair request (*iOrderRepair*), the contract triggers (#2) event *eRepairReqd*. Consequently, task *GR.tRpeiar* become executable and the role GR invokes (#3) the Web service exposed by its bound player (e.g., *TomsRepair.com*). When received, the response is routed (#4) to CO (*iRepairNotify*). This interaction is evaluated at the contract and more events are triggered (#5) depending on the contract state and the rules. For example, if the value of the fact *RepairCounter* maintained in the contract is equal to 5, then the event *ePayRepairBonus* is triggered, instead of *ePayRepairNormal*. Depending on the triggered events, more tasks become doable and are executed, and the cycle continues.

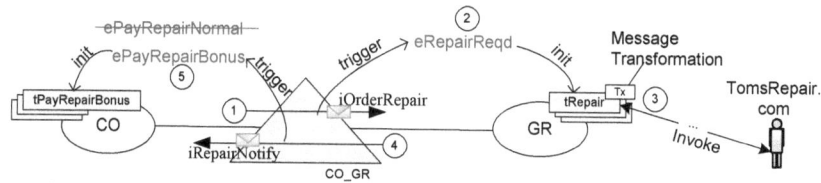

Fig. 8. Events triggered by contracts are used to initiate tasks defined in roles

---

[3] We use prefixes, *e,t,i,b* to name *events, tasks, interactions* and *behaviours* respectively.

A Process Definition in the organisation is a lightweight description that refers to the relevant behaviour units (via BehaviorRef). It also defines the conditions-of-start (CoS) and conditions-of-termination (CoT) based on the events as shown in Fig. 9, which are used by the enactment engine to start and terminate a process instance. In fact, this decoupling of process and behaviour is introduced to re-use behaviours to capture commonalities and allow variations as between multiple processes (described in our previous work [13]).

```
ProcessDefinition roadsideAssistanceProcess {
    CoS "eAsstReqd";
    CoT "(eTowPaid * eRepairPaid * eMMNotified)";
    BehaviorRef bClaim; BehaviorRef bTowing; BehaviorRef bRepairing;
}
```

Fig. 9. (Listing) A process definition refers to behaviour units

Overall, both the coordination-logic (in behaviour units) and the message-evaluation-logic (in contracts) are specified in a declarative manner. The events decouple the interactions (over contracts) from the coordination.

The triggered events (ε) are determined by evaluating the interactions (ι) against facts (φ) and rules (ρ) of a contract. The next set of executable tasks (τ) is determined based on defined behaviours (β) and triggered events (ε), i.e.,

$$\varepsilon = f_{contract\text{-}evaluation} (\varphi, \rho, \iota)$$
$$\tau = f_{coordination} (\varepsilon, \beta)$$

Hence, the changes in contracts (service-relationships) determine the changes in the set of triggered events, which in turn decide changes in the next set of executable tasks. This makes service-relationships not only being represented explicitly in a service orchestration, but also *influential* in determining the path of process execution. The Serendip meta-model underpinning the work of this paper is given in Fig. 10.

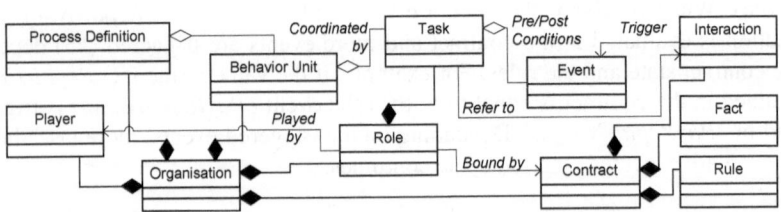

Fig. 10. Serendip Meta-Model

4.3   Providing Stability

The organisation structure formed by the contracts and roles provides the required abstraction to decouple the concrete services from the business processes as shown in Fig. 11. For example, the organisation structure models the RoSAS environment and acts as a stable basis to define and enact business processes. A role represents a functional requirement/position that should be filled by a concrete service. The interactions and mutual obligations that capture the service-relationships are defined *between the roles* rather than *between concrete services*. Although there are other approaches [4, 7] that decouple the core process from the services, in our approach we provide a complete representation of underlying service layer.

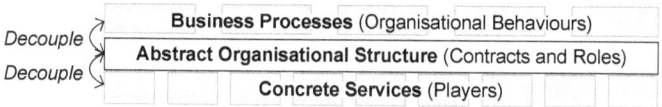

Fig. 11. Abstract organisational structure decouples Process from concrete services

Having such an abstract but explicit organisational structure is important for stability because during the lifetime of a service composition many services need to be bound and later be replaced. Yet, the service composition that represents the business needs to remain. The business processes are relatively independent of who the players are, the way the players interact and what type of message formats they use. For example, due to its frequent unavailability *TomsRepair* may be replaced with *mobileRepair* who uses asynchronous communication protocol and use a message format different from *TomsRepair*. These changes do not impact on the definition and behaviour of the core business processes because of the decoupling provided by the organisational structure.

On the other hand the coordination defined by the core organisational behaviours can also change, transparently to the bound partner services. For example, new variants of existing behaviours and multiple processes can be defined to suit the new or different business requirements of RoSAS on top of the abstract organisational structure [13]. Again, the stability provided by the abstract organisational structure supports such transparent changes in the coordination.

4.4   Supporting Adaptability

The adaptability in a Serendip service orchestration is supported from two perspectives within the scope of this paper[4]: (1) Adaptations in service-relationships, and (2) Adaptations in organisational behaviour.

To support *adaptations in service-relationships*, all the aspects of the contracts i.e., *Facts, Rules* and *Interactions* are defined and implemented as loosely-coupled and adaptable entities. These entities can be *added, removed* or *modified* during the runtime. All such adaptations in service-relationships can influence the path of execution of a service orchestration. For example,

- The fact *RepairCounter* of contract CO_GR is updated via rules in an automated manner (rule-based fact derivation). Alternatively, the value of the fact *ContractGrade* can be manually changed from *Low* to *Peak* by an administrator.

---

[4] New roles and contracts can also be added dynamically. However, such details are beyond the scope of this paper.

- A new rule can be dynamically inserted into contract CO_GR to specify that bonus payment is allowed upon 5th repair, only when the *ContractGrade* is *Peak*. i.e., *(RepairCounter.counter >=5) && (ContractGrade.grade==Peak)*.
- A new interaction term *iGetRepairUpdate* can be defined in contract CO_GR for CO to receive an update from GR.

Some adaptations in service-relationships may demand adaptations *in organisational behaviour*. For example, suppose that RoSAS requires imposing a *penalty* payable to CO by GR if the repair has been delayed. To support this requirement, an event *eRepairDelayed* can be triggered by interpreting the interaction *iRepairNotify*, a new task *CO.tChargePenalty* can be inserted into CO with the pre-condition *eRepairDelayed*, and the new dependency for this task can be added dynamically to behaviour unit *bRepairing*. Note that in contrast to [5, 6, 8, 9] we do not use rules to distinguish a volatile part and a fixed part, and new tasks and task dependencies can be dynamically added to the *behaviour units (and processes)*.

5   Implementation

The Serendip framework consists of two parts, *Deployment Platform* and *Core*. The *Deployment Platform* has been implemented by further extending the Apache Axis2 Web service engine [14]. This makes the service orchestration compatible with the existing Web services standards such as WSDL and SOAP as supported by Axis2. We have extended the Deployment and Message-Handling aspects of Axis2, yet without requiring any modifications to its code base, but through an additional layer. The *Deployment extension* ensures that Serendip orchestrations are deployed in Axis2 (in addition to default Axis2 services). The *Message-Handling extension* ensures that the messages are routed to/from the service orchestration by extending the Axis2 message routing mechanism [14]. The *Deployment Platform* reacts to the changes in the *Core* accordingly by changing service interfaces and handling messages [15].

The *Core* consists of an *event-driven enactment engine* and an *adaptation management engine*, by further extending the ROAD adaptive middleware framework [16]. The *Core* is independent of the Web service technology. The enactment engine maintains a runtime representation of the organisation (i.e., contracts, roles, behaviours, process definitions and process instances). It records and reasons about the triggered events, and coordinates the tasks of roles as discussed above.

The adaptation engine supports a *script-based*[5] mechanism to reconfigure the organisation and change the processes. An adaptation script can be issued to the adaptation engine to modify the service-relationships and behaviours of the organisation. The sample script given below updates the fact *ContractGrade* in contract *CO_GR* with a new value *"Peak"*.

```
updateFactOfContract ctId="CO_GR" factId="ContractGrade" prop="grade"
val="Peak";
```

---

[5] http://www.ict.swin.edu.au/personal/mkapuruge/files/
TR-Serendip-AdaptScripts.pdf

One of the important aspects concerning the adaptability described in this paper is the implementation-design of contracts. The framework uses *Drools Stateful-Knowledge-Session* [12] to maintain all the facts and rules of a contract. The session is started using the *fireUntilHalt()* method during instantiation so that it runs continuously. The session is fed with new facts and rules at runtime to update the contract to support changes in service-relationships as shown in **Fig. 12**. The existing facts and rules also can be modified or removed from the session at runtime.

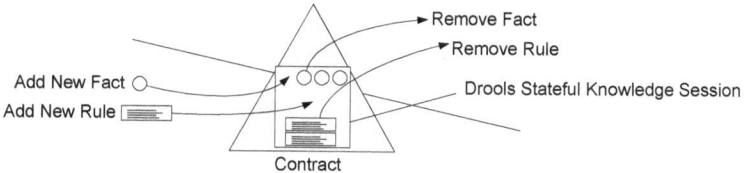

Fig. 12. A Drools Stateful-Knowledge-Session is used to maintain rules and facts of a contract

Tool support is provided to model, enact, monitor and adapt *Serendip* service-orchestrations. A screenshot of the *Serendip-Modelling-Tool* (an Eclipse plugin) is given in **Fig. 13**. The modelling tool generates the deployable descriptor and Drools rule templates. These rule templates can be edited via *Drools Rule IDE[6]* for Eclipse.

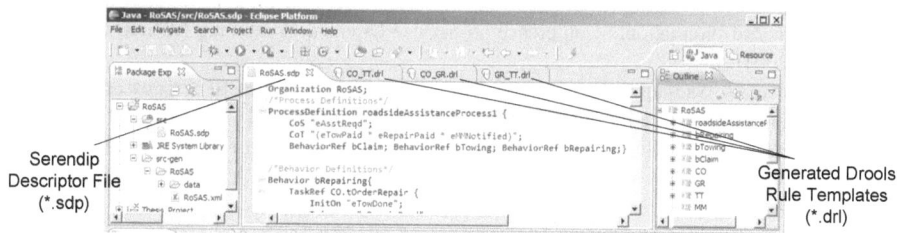

Fig. 13. A screenshot of Serendip-Modelling-Tool (an Eclipse plugin)

6   Conclusion

In this paper we have presented the *Serendip* approach to service orchestration that supports the *explicit representation of service-relationships* in modelling and enacting service orchestrations. Instead of a process-centric viewpoint, *Serendip* takes an organisational viewpoint to design a service orchestration. The explicitly defined service-relationships form an abstract organisational structure that provides the *stability* required to define business processes independent of underlying concrete services.

The service-relationships are declaratively captured as *Contracts* between *Roles*. During the runtime the *facts*, *interactions* and *rules* maintained in contracts can be added, modified or removed according to the changing service-relationships.

---

[6] http://www.jboss.org/drools/downloads

The coordination logic is also defined declaratively and is changeable to support the changes in service-relationships. Both the coordination and the contract evaluation logic can be modified without requiring a system restart. This provides better *adaptability* to support unforseen business requirements.

As the future work, we are expecting to evaluate the scalability of the approach and the performance overhead of the implemented framework.

References

1. Cummins, F.: BPM Meets SOA. In: Vom Brocke, J., Rosemann, M. (eds.) Handbook on Business Process Management1, pp. 461–479. Springer, Heidelberg (2010)
2. Weske, M.: Business Process Management: Concepts, Languages, Architectures. Springer (2010)
3. OASIS, Web Services Business Process Execution Language Version 2.0, http://docs.oasis-open.org/wsbpel/2.0/wsbpel-v2.0.html
4. Ezenwoye, O., Sadjadi, S.M.: RobustBPEL2: Transparent Autonomization in Business Processes through Dynamic Proxies. In: International Symposium on Autonomous Decentralized Systems (ISADS), pp. 17–24 (2007)
5. Rosenberg, F., Dustdar, S.: Business rules integration in BPEL - a service-oriented approach. In: 7th IEEE International Conference on E-Commerce Technology (CEC), pp. 476–479 (2005)
6. Charfi, A., Mezini, M.: Hybrid web service composition: business processes meet business rules. In: 2nd International Conference on Service Oriented Computing (ICSOC), pp. 30–38. ACM (2004)
7. Ezenwoye, O., Sadjadi, S.M.: Enabling Robustness in Existing BPEL Processes. In: 8th International Conference on Enterprise Information Systems (ICEIS), pp. 95–102 (2006)
8. Yu, J., Sheng, Q.Z., Swee, J.K.Y.: Model-Driven Development of Adaptive Service-Based Systems with Aspects and Rules. In: Chen, L., Triantafillou, P., Suel, T. (eds.) WISE 2010. LNCS, vol. 6488, pp. 548–563. Springer, Heidelberg (2010)
9. Graml, T., Bracht, R., Spies, M.: Patterns of business rules to enable agile business processes. In: IEEE International Conference on Enterprise Distributed Object Computing (EDOC), pp. 385–402 (2008)
10. Chappell, D.A.: Enterprise Service Bus. O'Reilly (2004)
11. OSOA, SCA Service Component Architecture: Assembly Model Specification, http://docs.oasis-open.org/opencsa/sca-assembly/sca-assembly-1.1-spec.html
12. Amador, L.: Drools Developer's Cookbook. Packt Publishing (2012)
13. Kapuruge, M., Colman, A., Han, J.: Achieving Multi-tenanted Business Processes in SaaS Applications. In: Bouguettaya, A., Hauswirth, M., Liu, L. (eds.) WISE 2011. LNCS, vol. 6997, pp. 143–157. Springer, Heidelberg (2011)
14. Jayasinghe, D.: Quickstart Apache Axis2. Packt Publishing (2008)
15. Kapuruge, M., Colman, A., King, J.: ROAD4WS – Extending Apache Axis2 for Adaptive Service Compositions. In: IEEE International Conference on Enterprise Distributed Object Computing (EDOC), pp. 183–192. IEEE Press (2011)
16. Colman, A.: Role-Oriented Adaptive Design. Ph.D. dissertation, Swinburne University of Technology, Melbourne (2007)

Assembling the Optimal Sentiment Classifiers

Yuming Lin, Xiaoling Wang, Jingwei Zhang, and Aoying Zhou

Institute of Massive Computing, East China Normal University,
200062 Shanghai, China
{ymlinbh,gtzhjw}@gmail.com, {xlwang,ayzhou}@sei.ecnu.edu.cn

Abstract. Sentiment classification aims to classify documents according to their overall sentiment orientation, which plays an important role in many web applications, such as electronic commerce. Machine learning is an effective method for such tasks. In general, a classifier is determined by a feature type, a weighting function and a classification algorithm for a given training set. Thus, users are required to predetermine which ones should be applied, that is a troublesome problem for them, because each classifier always achieves different performance for different domains. To deal with this problem, we develop a three phase framework based on assembling multiple classifiers. In order to choose the optimal combination of classifiers, we propose a criterion for estimating the quality of the combination based on sentiment classification accuracy and diversity of the results generated by these classifiers. Moreover, we study the effect of the number of classifiers selected experimentally. With our solution, users can achieve a good performance without making a choice among plentiful combinations of different classifiers. We perform extensive experiments to demonstrate the effectiveness of our solution for different domains.

Keywords: sentiment classification, multiple features, classifier combination, ensemble learning.

1 Introduction

With the rapid development of Web applications such as microblogging, online comment website and blog, more and more users prefer to contribute contents for sharing their opinions. These user-generated data are sentiment-rich, and are valuable to many applications like recommendation system, business and government intelligence. Therefor, automatic identification of user sentiment included in such contents becomes an urgent demand. Sentiment classification is a task of classifying the documents according to their overall sentiment orientation, which has seen a great deal of attention in recent years [1–5].

Machine learning is an effective method for sentiment classification [6]. To construct a good classifier for such a task, users need to predetermine three basic factors: feature type, feature weighting function and classification method. The unigram, bigram and the mixture of unigram and bigram, etc. are the feature types used commonly in sentiment classification. Classification methods contain

X.S. Wang et al. (Eds.): WISE 2012, LNCS 7651, pp. 271–283, 2012.

NaiveBayes, maximum entropy and SVM etc. Moreover, feature weights are determined according to frequency, presence or not, Δ *tfidf* [1], *tf*MI* [7] and so on. Unfortunately, there is no classifier that always performs optimally in all domains due to its inherent properties. Thus, it is difficult for users to predetermine which classifier should be applied due to lots of alternatives, especially for those who are not familiar with the analyzed domains. Table 1 shows the sentiment classification accuracy using SVM with different feature types and weighting functions on three product reviews. The feature types and weighting functions are listed in the first row of Table 1, the left part of underline stands for feature type and the right for weighting function. For example, the 'uni+bi_fre' means the feature type is the mixture of unigram and bigram, and frequency is used to weight the features. We can observe that the optimal combination of feature type and weighting function with SVM is different for different domains, namely 'uni+bi_fre' is the best for book domain, 'uni+bi_pre' and 'uni_pre' achieve the best performance in electronics domain and kitchen appliance domain respectively in our dataset. Notably, we consider only the SVM, other effective classification methods like NaiveBayes and maximum are not considered in Table 1, because SVM is usually superior to the others in sentiment classification [6]. Therefore, it brings the pressing need for a solution, by which sentiment classification tasks are performed well without user's predefinition on the combination of feature type, weighting function and classification method.

Table 1. The classification accuracy (%) on different feature sets with SVM

	uni_fre	bi_fre	uni+bi_fre	uni_pre	bi_pre	uni+bi_pre	uni_Δ*tfidf*
Book	79.90	75.75	**80.45**	78.25	75.55	79.50	79.85
Electronics	83.80	79.90	83.85	83.10	80.45	**85.15**	84.40
Kitchen appliance	84.95	80.40	86.65	**87.00**	80.55	86.50	86.20

In this paper, we focus on three main problems: (1) How to utilize the existing feature types, weighting functions and classification methods to perform a sentiment classification task as well as possible? (2) How to select a set of classifiers to maximize the classification performance? (3) How many classifiers should be included into this set?

We devised a three phase framework (shown in Figure 1) to track these problems. In the evaluating phase, we proposed a criterion for assessing the quality of a combination of classifiers (also called base classifiers) based on the classification accuracy and diversity of these classifiers, which will be discussed in detail in Section 4. Then the optimal combination will be chosen for our classification task. In the training phase, we use a part of training set to train the base classifiers. The rest training samples are applied to train the meta-level classifier with *stacking* technique in the assembling phase.

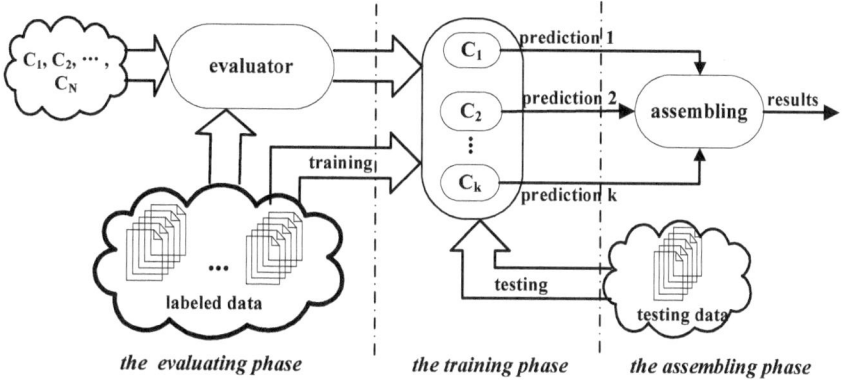

Fig. 1. A three phase framework for sentiment classification

Our main contributions are as follows:

1. Introducing a general framework for sentiment classification, in which users are not required to predetermine the feature type, weighting function and the classification method but can achieve good classification performance.
2. Proposing a criterion to assess the quality of a set of base classifiers based on the classification accuracy and diversity, the latter is quantified by the *Fleiss' kappa* value[20].
3. Studying the effect of the number of total base classifiers on overall classification performance experimentally.
4. Conducting extensive experiments for different domains on a real dataset, the experimental results show our solution outperforms the traditional single classifier methods significantly.

The remainder of this paper is structured as follows. The related work is presented in Section 2. Section 3 states the problems. Our solution is presented in detail in Section 4. Experimental results and discussion are reported in Section 5, and Section 6 concludes our work.

2 Related Work

In this section, we briefly present the related work on sentiment classification based on three aspects (feature type, weighting function and classification method).

Feature Type. In the bag-of-words model, N-gram is an very important feature type. Pang et al. compared different feature types (unigram, bigram, POS, adjectives, and the mixture of unigram and bigram, etc.) according to weighting functions and learning methods in movie reviews [6], they reported the SVM with unigram and presence achieved the highest accuracy (82.9%). [8] expanded the features with frequent word sub-sequences and dependency sub-trees, which improved the classification performance. Emoticons are often used to express individual sentiment, thus they are treated as a type of feature for analyzing the sentiment of contents on social website like Twitter[1][9, 10]. [11] divided features into parts-of-speech (POS) features and the others, in which POS features stood for the features capturing statistics about parts-of-speech of words. [14] made a comparative study on five feature selections including term selection based on document frequency, information gain, mutual information, χ^2 test and term strength, they reported that the χ^2 test was the most effective.

Weighting Function. Term frequency and present or not are used generally in sentiment classification [6–8]. The classic *tfidf* schema was compared with its variants for sentiment analysis [12], and the authors emphasized that expressing sample vectors with emotional information via supervised methods was helpful for predicting sentiment polarity. [1] proposed a new feature wighting function, $\Delta tfidf$, for sentiment polarity classification, by which the importance of discriminative terms could be identified and boosted. In [7], we applied mutual information to identify the term's sentiment polarity and evaluated its contribution by term frequency, by which we achieved a good classification performance. Moreover, features can be weighted with the some external resources, such as WordNet[13]. But it does not always perform well by this way, since the sentiment of word is domain-aware.

Classification Method. SVM is one of the most frequently used learning methods in sentiment classification, which was compared with the NaiveByes and the maximum entropy on movie reviews in [6], the experimental results showed the SVM outperformed the others. [15] applied Markov random walk model to a relatedness graph for estimating the polarity of a word. Kamps et al. constructed a network based on WordNet synonyms, they used the shortest path starting from a given word to the words 'good' and 'bad' to identify the word's sentiment polarity [16]. Turney identified word polarity by measuring its point-wise mutual information (PMI) with some seeds like 'excellent' and 'poor' in unsupervised sentiment classification. To get the PMI value, he used search engine 'near' operator to search instances where the given word was close to the seed words in the returned pages. [3] applied hierarchical learning process with a defined sentiment ontology tree to label the sentiments in product reviews. In [17], the user's social relationships (follow relationship and @-convention) were used to analyze the user's sentiment based on the principle of *homophily* without considering the context information. [18] applied an extend structural correspondence learning (SCL) for a cross-domain sentiment classification problem, in which a key step was to select the pivot features to link the source and target domains.

3   Problem Statement

For a traditional supervised classification problem, we seek a classification function C to map a sample S expressed as a feature vector $\mathbf{x}$ to the corresponding

---

[1] www.Twitter.com

label y. The classifier is trained on a finite labeled sample set $\{(S_i, y_i)|i = 1, ..., N\}$, and the object is to minimize the expected error, i.e.,

$$C^* = \underset{C \in H}{argmin} \sum_{i=1}^{N} loss(C(S_i, y_i)) \tag{1}$$

where $loss$ is a predefined loss function and H is a set of classification functions called the hypothesis. In sentiment classification, the vector $\mathbf{x}$ is constructed based on feature type and weighting function. The label y stands for the sentiment class. We focus on sentiment polarity classification in this paper, which means to detect whether a document is positive or negative.

The traditional methods work with the predetermined feature type, weighting function and learning method, which are uncertain in our setting, i.e., we do not know how to set the three basic factors in advance. By using the assembling classification method, we design a combination method with M classifiers, which will be used to predict the label of a sample separately, then the ultimate label of this sample is determined based on these predictions.

We define some symbols for narrating conveniently. Given a set of feature types $\{f_1, ..., f_u\}$, a set of weighting functions $\{w_1, ..., w_m\}$, and a set of classification methods $\{c_1, ..., c_n\}$, we denote $\mathbf{x}_{(r,j)}$ as the vector constructed by the feature type f_r and the weighting function w_j, and the $C_{(k,r,j)}$ as the generated classifier using classification method c_k trained on training vectors expressed with $\mathbf{x}_{(r,j)}$. Thus, the triples $\Phi = (k, r, j)$ indicates a classifier uniquely. Let $com_{(C_{\Phi_1}, ..., C_{\Phi_M})}$ to be a combination of M base classifiers, $\Phi_i \in \{(k, r, j)|k = 1..n, r = 1..u, j = 1..m\}$. Our target is to find the prior combination of classifiers for the classification tasks:

$$com^* = \underset{com \in COM_{all}}{argmin} \sum_{i=1}^{N} loss(\Gamma(com_{(C_{\Phi_1}, ..., C_{\Phi_M})}(S_i, y_i))) \tag{2}$$

where COM_{all} is the set of all possible combinations with M classifiers, Γ is a fusion function which maps the M predicted results of sample S_i generated by M base classifiers to an ultimate label.

4   Proposed Solution

In this section, we present our solution for assembling multiple base classifiers for sentiment classification tasks. We focus on the problem of how to choose the prior combination of base classifiers firstly, then we explore the fusion function Γ for assembling the predictions made by base classifiers.

4.1   Evaluating the Quality of a Classifier Combination

Given u feature types, m weighting functions and n classification methods, we can determine umn classifiers. If we choose a combination with k classifiers, there will be $\binom{umn}{k}$ possible options. It is a thorny problem for users to make a choice without an effective evaluating criterion.

Classification accuracy is one of the most important indicators for estimating the classifier performance. Thus, our criterion treats the accuracy as a key composition. On the other hand, the diversity of classification results is another important factor for improving the ultimate accuracy after assembling. For example, if we select 5 classifiers with the same classification results, the assembling accuracy would not exceed that of the single one. Then we want to select the combination, for which the predictions made by each classifier are disagreement as high as possible while ensuring high average accuracy. We call this disagreement diversity of base classifiers.

The Fleiss' kappa [20] is a statistical measure for assessing the agreement among several classifiers when assigning labels to a number of samples. Let N be the total number of samples, k the number of classifiers, and n the number of labels. The $n_{i,j}$ refers to the number of classifiers which assign sample S_i to the label y_j. Then the kappa, κ, is defined as following:

$$\kappa = \frac{p_o - p_e}{1 - p_e} \tag{3}$$

where p_o is the overall proportion of observed agreement, p_e is the overall proportion of chance-excepted agreement:

$$p_o = \frac{1}{Nk(k-1)} \left(\sum_{i=1}^{N} \sum_{j=1}^{n} n_{ij}^2 - Nk \right) \tag{4}$$

$$p_e = \frac{1}{(Nk)^2} \sum_{j=1}^{n} \left(\sum_{i=1}^{N} n_{ij} \right)^2 \tag{5}$$

The greater κ is, higher agreement the classifiers are. If the classifiers are in full agreement then $\kappa = 1$. If there is no agreement among the classifiers then $\kappa = 0$. In this paper we focus on disagreement, then $1 - \kappa$ is applied to estimate the diversity of classifiers. Now we evaluate the quality of a combination of classifiers based on the average accuracy and diversity:

$$Q_{com(C_{\Phi_1}, \dots, C_{\Phi_M})} = \alpha \frac{\sum_{i=1}^{M} accuracy_{C_{\Phi_i}}}{M} + (1-\alpha)(1 - \kappa_{com(C_{\Phi_1}, \dots, C_{\Phi_M})}) \tag{6}$$

where $accuracy_{C_{\Phi_i}}$ is the accuracy of the ith base classifier, α ($0 \leqslant \alpha \leqslant 1$) is used to weight the average accuracy and diversity. Notably, α should be assigned a relatively great value ($\alpha = 0.8$ in our experiments), since the accuracy is more important. The α is stable for different domains when the best performance is achieved in our dataset, which will be discussed further in Section 5.

As discussed earlier, the combination of base classifiers with great average accuracy and diversity is a good choice. We analyze that whether the accuracy implicates the diversity, namely, whether accuracy improvement leads to diversity increasement inevitably. If it does, the diversity should be removed from

Formula (6). Example 1 shows there is nothing deterministic about accuracy improvement leading to higher diversity, namely, higher $1 - \kappa$.

Example 1. Assuming 3 classifiers (C_1, C_2, C_3) assign 10 samples $(S_1, ..., S_{10})$ with 2 labels (0, 1) independently, the predictions are shown in Table 2. The true label of each sample is shown in the last row of Table 2. The accuracy of C_1 is 80%, C_2 70% and C_3 80%, then the average accuracy is 76.67% and the $1 - \kappa$ is about 0.34 according to the Formula (3),(4) and (5). We make some modifications in Table 2: $C_2(S_6) = 1$, $C_2(S_8) = 0$ and $C_3(S_7) = 1$. Now the average accuracy increases to 80%, but the $1 - \kappa$ decreases to 0.33.

Table 2. The predictions of three classifiers

	S_1	S_2	S_3	S_4	S_5	S_6	S_7	S_8	S_9	S_{10}
C_1	0	0	1	1	0	1	1	1	1	1
C_2	1	0	0	0	0	0	0	1	1	1
C_3	0	0	0	0	0	0	0	1	1	1
true label	0	0	0	0	0	1	1	1	1	1

Likewise, we can verify that there is nothing deterministic about the reducing of average accuracy leading to lower diversity. Therefor, both of them should be included into our estimation criterion at the same time.

4.2   Assembling Base Classifiers

Now we can select the optimal combination of base classifiers according to the Formula (6). These base classifiers are used to predict the label of a sample, and there will be multiple predictions. Then, we will devise a fusion function Γ to generate the ultimate label of the sample based on these predictions.

For assembling multiple classifiers, *stacking* is a well-known technique and has been shown to be very effective [19, 21]. Moreover, we compared the weighted voting with *stacking* for sentiment classification in [5], the later achieves better performance. Therefore, we apply *stacking* technique on assembling multiple base classifiers in this paper.

The *stacking* consists of the base-level training phase and the meta-level training phase. In the base-level training phase, the training set is divided into two disjoint subsets, $Part_1$ and $Part_2$, of different sizes, assuming $|Part_1| > |Part_2|$, the base classifiers $C_{\Phi_1}, ..., C_{\Phi_M}$ are trained on the $Part_1$ respectively. Then, the samples in $Part_2$ are regarded as the testing samples, and are input into the base classifier to predict their labels. For example, for $S_i \in Part_2$, its predictions are combined whit its true label y_i to form a meta-level training sample $(C_{\Phi_1}(S_i), ..., C_{\Phi_M}(S_i), y_i)$. Then, we get $|Part_2|$ meta-level training samples. To construct more meta-level training samples, we can apply the k-fold cross validation process on the original training set. In the meta-level training phase, a meta-level classifier C_{meat} is trained with meta-level training samples. For a testing sample T_j, it's label $y^*_{T_j}$ is generated with following formula:

$$y^*_{T_j} = C_{meta}(C_{\Phi_1}(T_j), ..., C_{\Phi_M}(T_j)) \tag{7}$$

4.3   Algorithm

Our solution is implemented by algorithm 1. Firstly, we determine the set COM_{all} of all possible base classifiers, each classifier is identified by a feature type, a weighting function and a classification method uniquely. The COM_{all} is constructed by selecting different combinations with M base classifiers (line 2). The combination of base classifiers with the highest quality is selected based on the Formula (6) (line 3). Line 2 and 3 correspond to the evaluating phase in Figure 1. The selected base classifiers are trained on a part of training set, and the other training samples are predicted by the trained base classifiers (line 4, 5 and 6), while these three steps compose the training phase. Then, the predictions are used to train the meta-level classifier, i.e., the fusion function Γ in Formula (2) (line 7 and 8). For a testing sample, we apply the base classifiers to predict its labels, these predicted labels are treated as the input of meta-level classifier, which will generate the ultimate label of the testing sample (line 10 and 11). The assembling phase in Figure 1 consist of line 7-11.

Algorithm 1. Assembling Multiple Classifiers for Sentiment Classification

---

Input: the training set $\mathbf{S} = \{(S_1, y_{S_1}), ..., (S_N, y_{S_N})\}$,
        the testing set $\mathbf{T} = \{T_1, ..., T_V\}$,
        the number of base classifiers M,
        the set of all possible base classifiers $\mathbf{C} = \{C_{\Phi_1}, ..., C_{\Phi_n}\}$
Output: the ultimate predicted result set $\mathbf{Y} = \{y^*_1, ..., y^*_V\}$

1:   $S_{meta} = \varnothing, \mathbf{Y} = \varnothing$;
2:   generate the set COM_{all} of combinations with M base classifiers in $\mathbf{C}$;
3:   select the combination $com_{\Phi_1, ..., \Phi_M}$ with the greatest value Q from COM_{all}
     according to the Formula (6);
4:   divide $\mathbf{S}$ into two subsets $Part_1$ and $Part_2$ ($|Part_1| > |Part_2|$);
5:   for i=1 to M do
6:      train C_{Φ_i} on $Part_1$;
7:      construct the prediction set $P_{C_{\Phi_i}}$ for the samples in $Part_2$ with C_{Φ_i};
8:      $S_{meta} = S_{meta} \cup P_{C_{\Phi_i}}$;
9:   train meta-level classifier C_{meta} on S_{meta};
10: for j=1 to V do
11:     $\mathbf{Y} = \mathbf{Y} \cup \{C_{meta}(C_{\Phi_1}(T_j), ..., C_{\Phi_M}(T_j))\}$;
12: return $\mathbf{Y}$;

---

5   Experiments

In this section, we conduct a serial of experiments to demonstrate the effectiveness of our solution for different domains in a real-world dataset.

5.1   Experimental Setting

We carry out our experiments on the labeled product reviews from three do-mains[2]: books (**B**), kitchen appliances (**K**) and electronics (**E**). The reviews marked with 4 or 5 stars are labeled with a positive label, those with 1 or 2 stars are labeled with a negative label, the rest is omitted. Each product domain contains 1000 positive reviews and 1000 negative ones.

All punctuations are removed but the stop words retained. The terms occur-ring less than 5 times are filtered. Furthermore, we omit the negatory words and append the tag "not_" to the words following the negatory word in a sen-tence. For instance, the sentence "It doesn't work smoothly." would be altered to become "It not_work not_smoothly.".

We focus on the following feature types commonly used in sentiment classifi-cation: unigram, bigram and the mixture of unigram and bigram. Four feature weighting functions are considered in our experiments: frequency, present or not, $\Delta tfidf$ and $tf * MI$. For the classification methods of base classifiers, we ap-ply the NaiveByes[3], maximum entropy and SVM[4]. For $\Delta tfidf$ and $tf * MI$, we consider only the unigram for simplicity. Then, we have 24 base classifiers, and five of them compose a combination. The maximum entropy classifier is used as meta-level classifier[5]. The classification accuracy is used to evaluate the effec-tiveness of different methods, we treat the methods with highest accuracy for different domains in Table 1 as our baseline. The 4-fold cross validation proce-dure is used on the training set to construct the meta-level training set. The 5-fold cross validation is used to estimate the overall classification performance.

5.2   Results and Discussion

The first experiment concerns at the effect of our solution. The best single clas-sifier refers to the classifier that achieves the best performance in a domain. For instance, the accuracy of SVM with frequency and the mixture of unigram and bigram in book domain is 80.45%, which is the best among the comparison items in Table 1. The comparison results are shown in Figure 2. We set the α to 0.8 in this experiment. Our method outperforms the best single classifier significantly, this means we are required to determine only what feature types, weighting functions and classification methods can be used without determining which one would be applied, but do not decrease the performance at the same time. On the other hand, the lower accuracy achieved by single classifier for a domain, the more improvement on accuracy would be achieved with our method. For example, the accuracy of our method is increased over 5% for book domain, but 2.9% for electronics domain and 2.6% for kitchen appliance domain.

---

[2] These reviews are collected from Amazon and reorganize by Blizer et al., and avail-able at `http://www.cs.jhu.edu/~mdredze/datasets/sentiment/`

[3] Implemented by WEKA. Available at `http://www.weka.net.nz`

[4] Implemented by LIBSVM with a linear kernel function. Available at `http://www.csie.ntu.edu.tw/cjlin/libsvm/`

[5] We also use NaiveByes and SVM as meta-classifier independently, but their perfor-mances are poorer than the *stacking* with maximum entropy in our dataset.

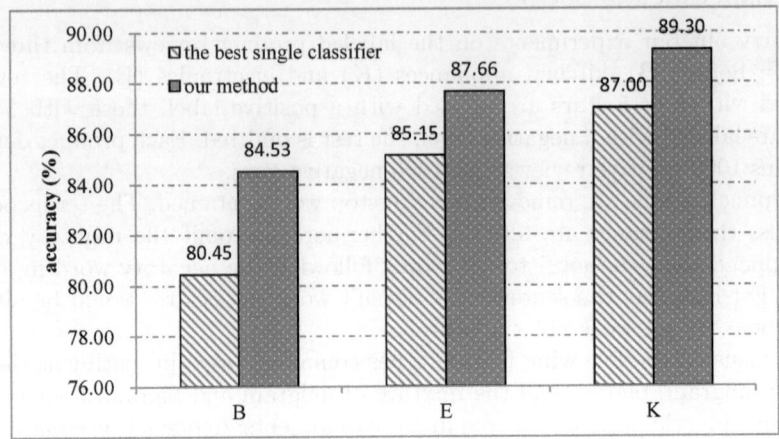

Fig. 2. Comparisons on accuracies of single classifier and our method

We estimate the quality of a combination of base classifiers with two factors: accuracy and diversity. Now we compare the effects with different α values in the coming experiment, by which we can adjust the weight of each factor. In other words, the higher value the α is set, more important the average accuracy of base classifiers is. Otherwise, the diversity is more important. As shown in Figure 3, with the improvement of α, the accuracies do not decrease until $\alpha = 0.8$ for all domains. Thus, the average accuracy of base classifiers plays a key role in our solution, and the diversity makes an indispensable complementarity, because the performance is decreased when the α is set to 0.9 or 1. Consequently, the α is stable for different domains in our experiments, that is a good property.

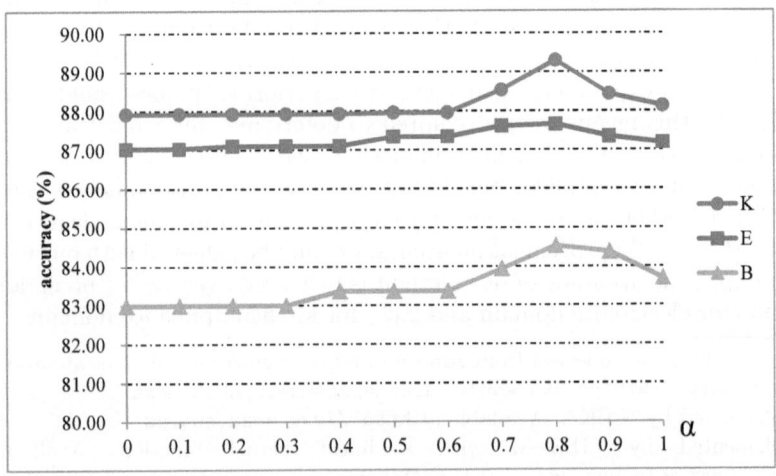

Fig. 3. Varying the value of α

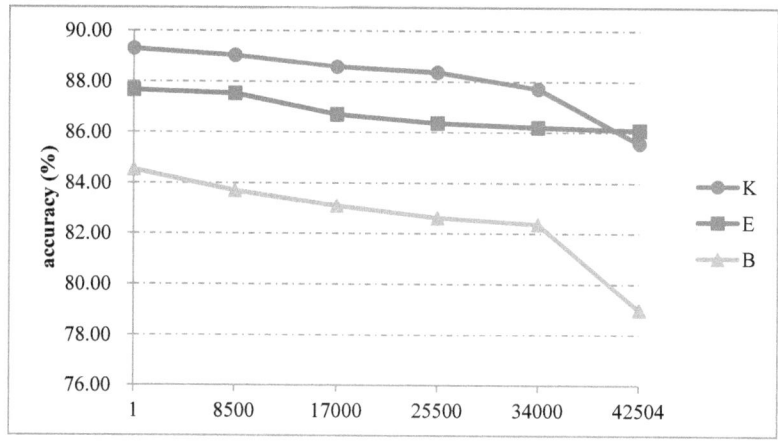

Fig. 4. The effect of selection criterion

The third experiment focus on the effect of the selection criterion. As discussed as before, we now have 24 base classifiers, and will choose 5 of them as a combination, then there are 42,504 possible combinations. We compute the estimation value Q for each combination with Formula (6), and sort these values by descending order. Six combinations of base classifiers are picked out in turn, which locate at the 1th, 8500th, 17000th, 25500th, 34000th and 42504th location of the sorted results respectively. These six points divide the sorted results five approximative equal parts. Figure 4 shows that the accuracy is descending with the declining of the combination quality. Consequently, our selection criterion is effective for capturing the combination quality of base classifiers.

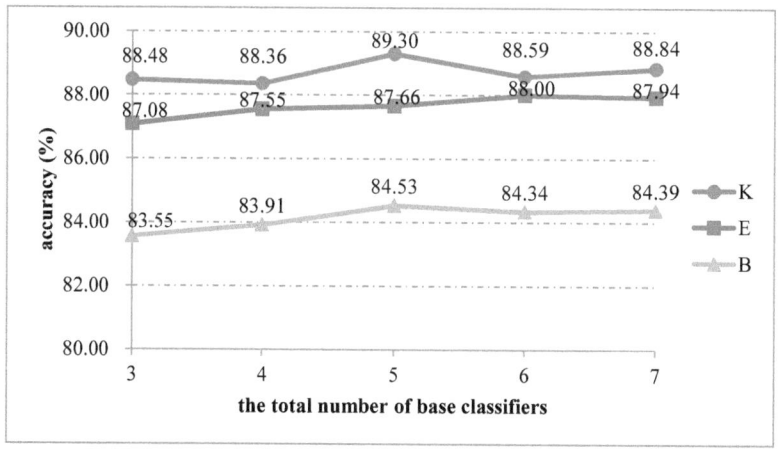

Fig. 5. Varying the total number of base classifiers

The last experiment considers the problem of how many base classifiers should be used in our framework. We change the number of base classifiers with 3, 4, 5, 6 and 7 respectively, and compare their classification accuracy for each domain. The effects of the base classifiers number are shown in Figure 5. A key observation is that the number of base classifiers does not affect the performance too much, namely, the accuracy difference between two cases in each domain is less than 1% in our dataset. Therefore, the number of base classifiers is easy to determine, which is another good property of our solution.

6   Conclusions

Sentiment classification has seen a great deal of attention in recent years. In general, it is difficult for users to determine which combination of feature type, weighting function and classification method should be applied for a domain, especially for the users who are not familiar with the analyzed domains. In this paper we propose a three phase framework to assemble multiple base classifiers for improving the classification accuracy. To select the optimal set of base classifiers, we devise an estimation criterion based on the average accuracy and diversity of the base classifiers. The only parameter in our estimation formula is stable for different domains. Moreover, we find that the number of base classifiers does not affect the performance of our solution too much by studying experimentally. At last, a series of experiments are carried out for different domains in a real-world dataset to demonstrate the effectiveness of our solution.

Acknowledgements. This work was supported by the National Major Projects on Science and Technology under grant number 2010ZX01042-002-003-004, 973 project(No. 2010CB328106), NSFC grant (No.61170085 and 61033007), Program for New Century Excellent Talents in China (No.NCET-10-0388) and Supported by Innovation Program of Shanghai Municipal Education Commission(No. 10ZZ33).

References

1. Martineau, J., Finin, T., Joshi, A., Patel, S.: Improve binary classification on text problems using differential word features. In: CIKM 2009, pp. 2019–2024 (2009)
2. Dasgupta, S., Ng, V.: Mine the easy, classify the hard: a semi-supervised approach to automatic sentiment classification. In: Proc. of the 47th ACL and the 4th IJC-NLP of the AFNLP, pp. 701–709 (2009)
3. Wei, W., Gulla, J.A.: Sentiment Learning on product reviews via sentiment ontology tree. In: Proc. of the 48th ACL, pp. 404–413 (2010)
4. Tan, C., Lee, L., Tang, J., Jiang, L., Zhou, M., Li, P.: User-level sentiment analysis incorporating social networks. In: KDD 2011, pp. 1397–1405 (2011)
5. Lin, Y., Zhang, J., Wang, X., Zhou, A.: Sentiment classification via integrating multiple feature presentation. In: WWW 2012, pp. 569–570 (2012)
6. Pang, B., Lee, L., Vaithyanathan, S.: Thumbs up? Sentiment Classification Using Maching Learning Technique. In: Proc. of the 7th EMNLP, pp. 79–86 (2002)

7. Lin, Y., Zhang, J., Wang, X., Zhou, A.: An information theoretic approach to sentiment polarity classification. In: WebQuarity 2012, pp. 35–40 (2012)
8. Matsumoto, S., Takamura, H., Okumura, M.: Sentiment Classification Using Word Sub-sequences and Dependency Sub-trees. In: Ho, T.-B., Cheung, D., Liu, H. (eds.) PAKDD 2005. LNCS (LNAI), vol. 3518, pp. 301–311. Springer, Heidelberg (2005)
9. Alm, C.O., Roth, D., Sproat, R.: Emotions from text: machine learning for text-based emotion prediction. In: HLT/EMNLP, pp. 579–586 (2005)
10. Alec Go, Richa Bhayani, Lei Huang: Twitter sentiment classification using distant supervison. Technical report, Stanford (2009)
11. Agarwal, A., Xie, B., Vovsha, I., Rambow, O., Passonneau, R.: Sentiment analysis of Twotter data. In: LSM, pp. 30–38 (2011)
12. Paltoglou, G., Thelwall, M.: A study of informationretrieval weigheing schemes for sentiment analysis. In: Proc. of the 48th ACL, pp. 1386–1395 (2010)
13. Miller, G.A., Beckwith, R., Fellbaum, C., Gross, D., Miller, K.: Introduction to wordnet: An on-line lexical database. International Journal of Lexicography 3(4), 235–312 (1990)
14. Yang, Y., Pedersen, J.O.: A comparative study on feature selection in text categorization. In: Proc. 14th ICML, pp. 412–420 (1997)
15. Hassan, A., Radev, D.: Identify text polarity using random walks. In: Proc. of the 48th ACL, pp. 395–403 (2010)
16. Kamps, J., Marx, M., Mokken, R.J., Rijke, M.D.: Using wordnet to measure semantic orientations of adjectives. In: Proc. of the 4th LREC, pp. 1115–1118 (2004)
17. Tan, C., Lee, L., Tang, J., Jiang, L., Zhou, M., Li, P.: User-level sentiment analysis incorporating social networks. In: KDD 2011, pp. 1397–1405 (2011)
18. Blitzer, J., Dredze, M., Pereira, F.: Biographies, bollywood, boo-boxes and blenders: Domain adaptation for sentiment classification. In: Proc. of the 45th ACL, pp. 440–447 (2007)
19. Džeroski, S., Ženko, B.: Is Combining Classifiers with Stacking Better than Selecting the Best One? Machine Learning 54(3), 255–273 (2004)
20. Fleiss, J.L., Levin, B.: Statistical Methods for Rates and Proportions, 3rd edn. Wiley, New York (2003)
21. Ženko, B., Todorovski, L., Džeroski, S.: A comparison of stacking with MDTs to bagging, boosting, and other stacking methods. In: ICDM 2001, pp. 669–670 (2001)

Panoramic Image Search by Similarity and Adjacency for Similar Landscape Discovery

Meng Zhao, Hiroaki Ohshima, and Katsumi Tanaka

Department of Social Informatics, Graduate School of Informatics, Kyoto University,
Yoshida-honmachi, Sakyo, Kyoto 606-8501, Japan
{zhao,ohshima,tanaka}@dl.kuis.kyoto-u.ac.jp

Abstract. In this paper, we propose a new image search method, called "panoramic image search", and show its application to similar landscape discovery. In order to perform the "panoramic image search", we introduce an image ranking method called PanoramaRank: a combination of image similarity and image adjacency, where image similarity is the retrieval score obtained from the classic vocabulary tree based image retrieval framework, and image adjacency is computed using a RANSAC verified SURF matching process. Our proposing notion means to search for images physically surrounded to given query image(s). A landscape is a view of an area comprising several geographical features, having a common and meaningful atmosphere. We believe a collection of images is necessary for describing a landscape. Besides, images in this collection have to be roughly similar and roughly adjacent to each other directly or indirectly. In order to discover similar landscapes, (1)find images describing the same landscape as user-selected query image(s) by employing PanoramaRank. (2)Similar images taken in different locations are retrieved, of which belong to the same location are treated as an insufficient representation of a similar landscape to the original one. (3)PanoramaRank is applied once more to find a whole landscape for each location separately. (4)Based on several comparison criteria, landscape similarity ranking has been worked out. Moreover, images of landscapes similar to a given landscape image, especially those not presented in results based on the individual pair-wised measure, can be found. Experimental results and evaluation are also presented.

Keywords: landscape, image similarity, image adjacency.

1 Introduction

In recent years, the increasing development of the Internet and its related technologies, has led to the appearance of a variety of multimedia content formats on the Internet, such as images, audio, and video. Various search technologies are often used to retrieve desired contents. To improve user satisfaction in multimeadia retrieval, content-based image retrieval has been studied extensively and incorporated into retrieval frame. Currently, it is possible to search for images based on their contents, irrespective of features such as their color, brightness,

X.S. Wang et al. (Eds.): WISE 2012, LNCS 7651, pp. 284–297, 2012.

and texture. As a result, some recent studies have focused on the development of search systems that try to detect images similar to the query image. An example is TinEye [1], a Web search service, which returns images that are very similar to a query image and there is no need to enter a keyword.

Among a variety of search intents, we believe that attempts to identify locations similar to a location, to which a user is familiar, are common. For example, users may imagine a scene in which there is a garden around the Kyokochi pond , which reflects Shari-den of the Golden Pavilion temple on its surface. The task is to find locations with scenes that are similar to those in users' imaginations. In this case, the desired results should be different from the location in users' mind. It is straightforward to search by a keyword that indicates a location in the user's imagination, or an image that represents the scene imagined by the user, but it is still difficult to realize this goal.

Even if each photo is attached with a geo-tag, it is still challenging. Because it is only useful for short-range views, but for intermediate or distant views, the spot where a photo is taken may be distant from the place described in the photo. For example, we can take photos of Mt. Fuji in Tokyo, Chiba prefecture, or Saitama prefecture. However, the geo-tags attached to these photos are obviously different from where Mt. Fuji locates. Besides, it is hard to find boundaries of landscapes and photos are likely to appear somewhat different even belonging to the same landscape.

In this paper, we propose a novel kind of image search that given an image or a few images of a location, images which is atmosphere-similar to query image(s), especially, those of other locations, will be returned. In general, atmosphere-similar images in a location compose a representation of a "landscape". Notice there is no need for images to be visually similar to each other. As images in the same landscape always have some overlaps with each other, we incorporate adjacency (See details in Sec.3.2) into our scheme. We propose an image ranking method called PanoramaRank: a combination of image similarity and image adjacency to discover a certain landscape in a location. In Section 2, we describe a simple survey and give our definition of "landscape". In Section 3, the basic idea and the problems are addressed. Section 4 introduces some related works. Section 5 explains our proposed method PanoramaRank and Section 6 presents how to discover similar landscapes. Our evaluation experiments are addressed in Section 7. Finally, conclusions and possible directions for future work are stated in Section 8.

2   Brief Introduction to "Landscape"

2.1   Landscape

The word "landscape" is widely used in many fields, such as geography, urban engineering, social technology, and architecture. In 2004, the Landscape Act was enacted in Japan, but there is no precise legal definition of what constitutes a

---

[1] http://www.tineye.com/

landscape. According to our survey, there are two main arguments. The main difference between them is whether or not the components that constitute a landscape should contain human elements.

In [7], Nakamura et al. determined that the concept of a landscape can be summarized in the following five points. First, different objects such as buildings or structures exist simultaneously and are associated with each other. Second, a certain space has a specific form. Third, there is a hierarchy where the size of the space is concerned. Fourth, the landscape is a type or a model. Finally, a landscape will change over time. For example, a bridge cannot by itself be referred to as a bridge landscape, but it is an element of a bridge landscape. It must therefore be considered in relation to other elements such as the river, traffic network and community.

On the other hand, in architecture, landscape can be divided into its scenery, which is viewed by people, and feeling, which is felt by people viewing the landscape. Furthermore, a scene may be subdivided into locality, integration, and the extent of exposure to the public. A feeling is subdivided into diversity, lifestyle and participation. In particular, even without seeing an actual landscape, visualization is possible from past experiences and information obtained. In this way, images conjured by persons are referred to as landscape images (different with the below-mentioned "landscape images"). Considering the types of appearance pertaining to landscapes, even if the same objects are seen, the impression on the viewer may be different because of the distance (whether it is a short-range view, a distant view, or an intermediate view). Likewise, our impression of a landscape would vary according to the season, or time at which it was seen.

In this paper, we exclude factors that refer to feelings, i.e. psychological factors, which include but are not limited to differences that result from personal experiences, background, education, and religion. We prefer to define "landscape" as an area that produces the same response from most people. Moreover, since a landscape is a specific area, an image set, a set of images of a particular area, is necessary to integrally and factually represent a landscape.

2.2 Landscape Images

As mentioned in Section 2.1, due to changes in season, or time, there may be diverse landscapes at the same location. Furthermore, what we refer to as an area is not an exact geographical area. For example, the Golden Pavilion Temple does not refer to the golden reliquary hall, or the entire area belonging to the temple. It is actually a subarea of an exact geographical area, or sometimes even includes an extended surrounding area. Fig.1 is an example of a landscape of the Golden Pavilion Temple, mainly describing the environment of the golden reliquary hall and the surrounding Kyokochi pond. Fig.2 reflects the Japanese style in a quiet and calm atmosphere. Although the area changes little in Fig.3 compared to that in Fig.1, it is a different landscape because of the seasonal change. We consider these three landscapes at the same location as distinct ones. From these three examples, we observe that

Fig. 1. A typical landscape of the Golden Pavilion Temple

Fig. 2. An atypical landscape of the Golden Pavilion Temple

(a) A whole space can be divided into several distinct landscapes, as each landscape image set introduces a somewhat different impression.
(b) Even if the area represented in an image does not change much, it can become another landscape as time progresses.
(c) These three distinct landscapes, together with others not listed here, generate an upper-level landscape of the Golden Pavilion Temple, which means there exists a hierarchy related to the word "landscape".

3 Basic Idea

Similar landscapes are geographically separated but can invoke similar impressions in most people. A simple example is provided in Fig.4. Thus, the problem described in this paper is as follows:

– **Input:** At least one image, which partly indicates the desired landscape
– **Output:** Ranked landscape image sets that represent different landscapes similar to the given landscape

Fig. 3. A winter landscape of the Golden Pavilion Temple

(a) A landscape of Ryoan-ji

(b) A landscape of Tenryu-ji

Fig. 4. Both Fig.4(a) and Fig.4(b) are typical representations of a traditional Japanese dry landscape garden. From this point of view, we consider landscapes described by those two image sets to be similar.

Based on our hypothesis that a specific area usually cannot be represented by just a single image, we consider the following two distinct relations between images.

3.1 Image Similarity

This problem is as follows, given an image as our input, other images that are similar to the input image are expected to be flagged based on their similarity to the input image. Suppose there is an image dataset I_n as follows, $I_n=\{img_1, img_2, ..., img_i, ..., img_n\}$, where n is the total number of images in this dataset. When any one image img_i in the dataset is given as an input, the expected output is similarities between other images in the dataset and the input image. The functional form representing this problem is as follows,

$$Sim(img_i, img_j)$$

where img_j is any other image in the dataset, $1 \leq i \leq n$, $1 \leq j \leq n$. It is desired that this function returns a real number, which has a high value when the content of img_j is highly similar to that of img_i, and vice versa.

3.2 Image Adjacency

In this section, we will explain the concept of image adjacency in details and describe how it is applied to image search. Fig.5(a)-5(c) is the left, middle, right part of the clock tower, respectively. Note that there is no overlap between Fig.5(a) and Fig.5(c). However, there are common features in the photo pairs of Figs.5(a) and 5(b), and Figs.5(b) and 5(c). For example, both Figs.5(a) and 5(b) contain the top of the clock tower and a small part of the camphor tree. Therefore, these two photos are referred to as adjacent images. The same applies to Figs.5(b) and 5(c). Notice that we don't aim to recognize each object, but only consider overlap between two images.

(a) Left   (b) Middle   (c) Right

Fig. 5. Three photos taken in front of the main gate of Kyoto University

The problem is, then similar to that of image similarity. Given an image as our input, it is expected that other images adjacent to this image will be highlighted in terms of adjacency with the input image. When any one image img_i in the dataset is given as an input, the expected output is adjacencies between other images in the dataset and the input image. The functional form representing this problem is as follows,

$$Adj(img_i, img_j)$$

$1 \leq i \leq n, 1 \leq j \leq n$. A high value will be returned when img_j is highly adjacent to img_i, in other words, when it is most likely that img_j and img_i can generate a panorama. And vice versa.

We combine image similarity and image adjacency together to find images roughly similar and roughly adjacent to query image(s) no matter whether directly or indirectly.

4   Related Work

The use of a set of local interest points for image matching can be traced back to 30 years ago. Recently, there have been extensive studies into methods used to find words, in the same manner as those used in natural language processing and information retrieval, (referred to as visual words). This has been done in order to generate a similar process for retrieval using visual words to search for content-similar images based on Term Frequency Inverse Document Frequency (TF-IDF). Descriptors extracted from the local region are applied to a clustering algorithm, such as k-means, to create clusters that refer to visual words. In a variety of interest point detectors and feature descriptors, the Speeded Up Robust Features (SURF) proposed by Bay et al. in 2006 [1] is a novel scale- and rotation-invariant detector-descriptor. They apply integral image for image convolutions, and a Hessian matrix-based method for the detector (Fast-Hessian detector), and calculate the Haar-wavelet to identify the orientation of the interest points for invariance to rotation. Finally, 64 dimensional descriptors are determined.

Nister and Stewenius [8] propose a scheme, which is robust to background clutter and occlusion. They use Scale-Invariant Feature Transform (SIFT)[6], a feature descriptor that inspired SURF, for extraction from local regions. They then perform k-means clustering on extracted local region descriptors to hierarchically generate a vocabulary tree. However, k represents the number of children for each node of the tree, as opposed to the final number of clusters. As a result, for the purpose of retrieving each descriptor, it is required to traverse the tree and at each level, the descriptor is compared to each current cluster center and the closest one is chosen to continue propagating downwards.

Regarding the overlap among images, we determined that it resembles the construction of panoramas, in which detection of regions that can be connected is required. Brown and Lowe[2] describe an approach based on extracting invariant local image features to select matching images. They extract SIFT features from all of the images and determine k nearest-neighbors for each feature using a kd-tree. Then they detect m images(they use $m = 6$) as candidate matching images, which have the maximum number of feature matches to the current image. Next, they use RANSAC, an iterative method to estimate parameters of a mathematical model, to find geometrically consistent feature matches, with the purpose of determining the homography between pairs of images. To verify image matches, they employ a probabilistic model. Up until this step, the connected parts of the panorama are confirmed.

What we called adjacency is essentially a degree of spatially verified inliers, which denotes how much overlap there is between two images. Although it has been used in re-ranking of retrieved images in [3][5][10][11], in which it is called spatial verification, their objective is to retrieve a specific object, while we aim to search the surroundings of a specific object rather than the object itself. Especially, in [5], an approach for modeling landmark sites is stated, which seems a similar job as what we are trying to do. However, note "landmark" is a particular part of a building, while "landscape" is a whole environment, containing "landmark" and also its surroundings. Object-based image retrieval, which devotes to associate a set of images of the same object, has been extensively explored. However, as far as know, visual surrounding search has not been extensively studied yet.

5   PanoramaRank

Here we explain our graph-based image ranking method PanoramaRank in detail. Roughly speaking, we employ PanoramaRank to discover a landscape in a location. Image similarity and image adjacency is combined in our method to find images both similar and adjacent to chosen image(s).

5.1   Calculating Image Similarity

In this section, we apply existing methods to calculate image similarity.

Search through a Vocabulary Tree: In [8], a descriptor vector is retrieved to propagate down the hierarchical vocabulary tree, in which each node is assigned a weight, so that we get the searching image query vector. Then, the relevance score assigned to the database image can be defined based on the normalized difference between the searching image query vector img_i and the database image vector img_j, indicated by $NisterScore(img_i, img_j)$.

Color: We compare images by employing color coherence vectors(CCV), a histogram-based method to compare images that incorporate spatial information, which was proposed by Pass et al. in 1996 [9]. For each color, the number of coherent pixels relative to the number of incoherent pixels is stored. The method used to compare the difference between two images is based on the differences between coherent pixels for each discretized color in both images as well as the differences between incoherent pixels. We use $CCVScore(img_i, img_j)$ to indicate the difference estimated by this method.

Our image similarity between two images img_i and img_j is considered as

$$Sim(img_i, img_j) =$$

$$w_1 NisterScore(img_i, img_j) + w_2 CCVScore(img_i, img_j)$$

where w_1 and w_2 are weight factors of different scores, and $w_1 + w_2 = 1$.

5.2 Calculating Image Adjacency

In this section, we apply the panorama construction algorithm [2] to find corresponding pixels that can generate a panorama using two images.

Given two images img_i and img_j, first we extract SURF descriptors from each image, separately. Then we find correlation points in these two images. Finally, we find geometrically consistent feature matches using RANSAC to solve for the homography between these two images. We treat the inner-most points as the corresponding points between these two images. Here we use N_{cp} to refer to the number of corresponding points in query image img_i and database image img_j, and N_{img_i} to the number of extracted SURF descriptors in the query image, N_{img_j} to the number of extracted SURF descriptors in any database image img_j, respectively. Then image adjacency is defined as

$$Adj(img_i, img_j) = \frac{N_{cp}}{(N_{img_i} + N_{img_j})/2}$$

5.3 Similarity/Adjacency Graph and PanoramaRank

In [4], VisualRank, which is an inferred visual similarity graph-based ranking model for image-ranking problems, was introduced. In this model, edge weights have also been considered when estimating the score associated with a vertex in the graph. The random walk algorithm is employed to rank images based on the visual hyperlinks among the images. It is assumed that if a user is viewing an image, other related(similar) images may also arouse the user's interest. Similar to PageRank algorithm, if image u is visually hyperlinked to image v, such hyperlink is treated as a vote of confidence, which means it is possible that the user will go from viewing u to viewing v. As a result, images related (similar) to the query image will have many other images pointing to them and will therefore be viewed often.

Here, let $G_S = (V, E_S)$ be an undirected graph with a set of vertices V and a set of edges E_S, where E_S is a subset of $V \times V$, $E_S = \{e = (u, v) | u$ is similar to v$\}$. Likewise, let $G_A = (V, E_A)$ be an undirected graph with a set of vertices V and a set of edges E_A, where E_A is a subset of $V \times V$, $E_A = \{e = (u, v) | u$ is adjacent to v$\}$. Then the final similarity/adjacency graph(SA graph for short below) for ranking is defined as $G = G_S \cup G_A$, where $E_S \cap E_A \neq \phi$.

In our case, we apply the above-mentioned random walk algorithm to the defined SA graph G. Thus, after iterative calculation, images roughly similar and roughly adjacent to query image(s) will be deemed important, since those are viewed as images composing the same landscape according to our hypothsis. Given n images, our proposed PanoramaRank (PR) is defined as follows:

$$PR = dS^* \times PR + (1 - d)p$$

where p_i is the initial value of V_i, and we refer to vertex V_i of an image img_i. d is a damping factor, and we set it to 0.8 in the following experiments empirically.

$$p_i = \begin{cases} \frac{1}{|SI|}, img_i \in SI \\ 0, img_i \notin SI \end{cases}$$

where SI is the image set whose element is selected by the user, and S^* is the column normalized adjacency matrix S, but here $S_{u,v}$ denotes the combination of visual similarity and adjacency between image u and v. Since both similarity and adjacency should be considered, the geometric mean and harmonic mean are employed to weigh edges between images in G.

$$S_{u,v} = Sim(u, v)^\alpha * Adj(u, v)^{1-\alpha}$$

or

$$S_{u,v} = \frac{1}{\frac{\alpha}{Sim(u,v)} + \frac{1-\alpha}{Adj(u,v)}}$$

where α is the weight factor.

6   Discovering Similar "Landscapes"

We assume each image attached with a tag that indicates its geographic location. This tag can be the name of a certain location, such as "Tokyo Tower", or a geo-tag containing both longitude and latitude information. According to this information, images are separated into different sets.

When at least one image is selected, our PanoramaRank will be applied limited in the set to which selected image(s) belong. Given an example in Fig.6. An image of Location A is chosen as the query image q. Therefore, PanoramaRank is employed in the image set of Location A. Only images whose score is above the

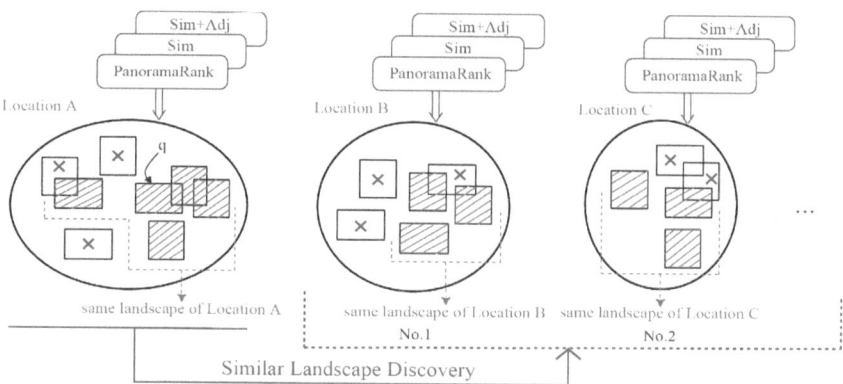

Fig. 6. Illustration of our approach

threshold (empirically set to 0.075 in the following experiments), are preserved and referred to as a supplement that represents a certain landscape together with the selected one(s). This step is also considered as query formulation, as user intent is about searching for a certain landscape but it is difficult to point out all images described that landscape. In a word, our proposed method is to discover a certain landscape in a location according to the given image example(s). In the above-mentioned example, after PanoramaRank is employed, some images, with a cross, are discarded. The rest ones, with some diagonals, are considered to form a representation of a landscape of Location A.

Meanwhile, similar images taken in different locations are retrieved, which is only based on image similarity mentioned in Sec.5.1 and graph-based ranking algorithm is not employed. For the top K (K=20) images, ones belonging to the same location are treated as an insufficient representation of a similar landscape to the original one. Images with the same geographic tag in every distinct location are regarded as the given image examples. Then PanoramaRank is applied in every set found before to find a whole landscape for every distinct location. Similar to what happened in the image set of Location A, PanoramaRank is then applied in the image set of Location B, C and so on. As a result, some images with a cross are also discarded and landscapes of Location B, C (shown in Fig.6) turn to be similar landscapes of that of Location A. For each image in the final result set, similarity between each one in the original landscape is calculated. The maximum (used in experiments), or minimum, or average similarity score is referrd to as its final score. We believe this score can reflect landscape similarity between images. As a result, a ranked image list will be returned. Moreover, each image in every landscape is compared with each one in the original landscape. The maximum (used in experiments), or minimum, or average similarity score is treated as landscape similarity between different landscapes. Therefore, a ranked landscape list will also be returned. In our example, the landscape of Location B is the most similar landscape, compared to the original landscape of Location A. The landscape of Location C is the second-most similar one.

7 Experiments and Evaluations

We created a small dataset having 6 categories, with 180 photos in total, shown in Fig.9. The performance of our proposed method is evaluated by

- comparing the resulting landscape image ranking list to the list obtained by three evaluators
- examining how well the expected landscape image sets are ranked at the top.

7.1 Similar "Landscape Image" Search

In detail, our evaluation is divided into two steps. In the first instance, ranking results, referred to the listed-up landscape images, are estimated in comparision with the original selected query image(s). Especially, here we only chose one image of a certain location as the query so that all we need to do is to evaluate ranking results from the very begining to the end according to their relations with the query image. Relation here indicates whether this pair of images looks landscape-similar or not. Since it will fairly vary based on diverse personality, three evaluators are employed for estimation with the purpose of reducing bias. Besides, five-step evaluation is applied, then the average score for each image is calculated. As a result, we got the ideal ranked list for each query.

We find after combining image similarity and adjacency, performance will be improved as the ranked list is closer to the human judgment, above-mentioned ideal ranked list. Fig.7 shows our PanoramaRank performance for each query under harmonic mean of image similarity and adjacency, evaluated by nDCG. The final score of an image is the maximum similarity score among the similarity scores between this image and each one in the original landscape. When $\alpha = 0$, it means search only based on image adjacency, likewise, when $\alpha = 1$, it means search only based on image similarity. Queries such as the image of the Eikan-do temple, shown in Fig.9, are good examples to support the effectiveness of our proposed method. When only applying image similarity-based search ($\alpha = 1$), the nDCG score is 0.84, and only applying image adjacency-based search ($\alpha = 0$), the nDCG score is down to 0.75. However, after the combination of image similarity and adjacency, we can get a higher nDCG score, which is 0.91 when $\alpha = 0.3$ or $\alpha = 0.5$. From this, we can point out that it has been significantly improved with the purpose of discovering similar landscape images compared to search based on either image similarity or image adjacency. On the other hand, queries such as the image of Jiuzhaigou didn't go well. No matter adjacency is incorporated or not, there is little change in performance. The reason is that there are not so many images adjacent to each other in the image set of Jiuzhaigou as those in others. Thus, after taking harmonic mean of similarity and adjacency, the weight for each edge will turn smaller instead, which leads that discovered landscape images vary little so that final results hardly vary.

7.2 Similar "Landscape" Discovery

In the second step, evaluation is not based on a single image, but rather an intergral landscape image set. In Fig.9, every location in the same category is

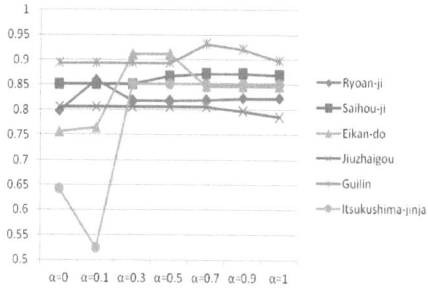 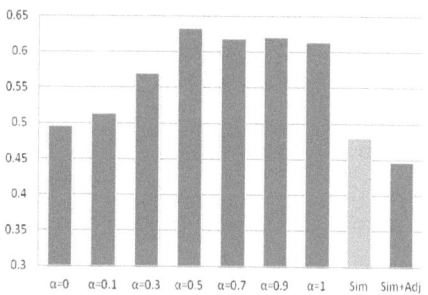

Fig. 7. nDCG scores for each query when α is set to different values, using harmonic mean of image similarity and adjacency

Fig. 8. MAP scores when α is set to different values shown above, using harmonic mean of image similarity and adjacency

a candidate similar landscape of the rest, such as given an image of the rock garden in the Ryoan-ji temple, images about similar rock garden in the Daisen-in or Manshu-in are deemed to be similar landscape images. Thus, it can be said for rock garden landscape, the Daisen-in and Manshu-in are similar to the Ryoan-ji temple. We prepared two baseline methods as follows.

- **Sim:** Instead of applying our proposed PanoramaRank, only image similarity with the query image(s) is taken into account to find images belonging to the same landscape. Pay attention that graph-based ranking method is not employed here, which is the essential difference with the case when $\alpha = 1$. See Fig.6, Sim is employed to discover a landscape in a certain location, for example, in the image set of Location A, B, C and so forth.
- **Sim+Adj:** Instead of applying our proposed PanoramaRank, images belonging to the same landscape are found according to harmonic mean of image similarity and image adjacency ($\alpha = 0.5$) with the query image(s), shown in Fig.6.

Harmonic mean is taken and the performance is shown in Fig.8. We find that when $\alpha = 0.5$, 0.02 has been improved compared to search only based on image similarity in MAP, and more than 0.1 has been improved compared to search only based on image adjacency, and also ourperforms two baseline methods, whose MAP score is 0.48 and 0.44, respectively. This strongly illustrates that it is useful to incorporate image adjacency to discover similar landscapes.

8   Conclusions and Future Works

In this paper, we proposed a method that can be used to find a landscape that is similar to the one whose image is given.

To discover a certain landscape, we applied a biased-VisualRank method, called PanoramaRank. In addition, we incorporated image similarity and image

Fig. 9. All images in our dataset. Actually, we assume six categories: traditional Japanese dry landscape garden, moss temple, maple leaves, karst landform, hills and waters, and torii in sea, respectively. For each category, we chose three different locations, with each containing ten images about that location. The leftmost in each row shows the exact location of the rest of the images in the same row. Meanwhile, category information is also shown at the top of the three locations belonging to the same category. Images with red frames are used as inputted query images in our experiments.

adjacency to weigh each edge between two vertices in the SA graph constructed for ranking. Experiments are presented to show the effectiveness of our proposed method.

Our future works will be as follows. First, we will incorporate other image features into our scheme to improve the performance of our proposed method, and then employ a machine learning method to determine weight factors in related formulas. Second, for tags that are attached to photos, we will explore the relationships that exist among similar landscape image sets. Finally, we will extend the categories and generate a larger dataset to objectively evaluate our proposed method.

Acknowledgements. This work was supported in part by the following projects: Grants-in-Aid for Scientific Research (Nos. 24240013, 24680008) from MEXT of Japan.

References

1. Bay, H., Ess, A., Tuytelaars, T., Gool, L.V.: Surf: Speeded up robust features. Computer Vision and Image Understanding (CVIU) 110(3), 346–359 (2008)
2. Brown, M., Lowe, D.G.: Recognising panoramas. In: Proceedings of Ninth IEEE International Conference on Computer Vision, vol. 2, pp. 1218–1225 (2003)
3. Chum, O., Philbin, J., Sivic, J., Isard, M., Zisserman, A.: Total recall: Automatic query expansion with a generative feature model for object retrieval. In: Proceedings of the 11th International Conference on Computer Vision, ICCV 2007 (2007)
4. Jing, Y., Baluja, S.: Visualrank: Applying pagerank to large-scale image search. IEEE Transactions on Pattern Analysis and Machine Intelligence 30(11), 1877–1890 (2008)
5. Li, X., Wu, C., Zach, C., Lazebnik, S., Frahm, J.-M.: Modeling and Recognition of Landmark Image Collections Using Iconic Scene Graphs. In: Forsyth, D., Torr, P., Zisserman, A. (eds.) ECCV 2008, Part I. LNCS, vol. 5302, pp. 427–440. Springer, Heidelberg (2008)
6. Lowe, D.G.: Distinctive image features from scale-invariant keypoints. International Journal of Computer Vision 60(2), 91–110 (2004)
7. Nakamura, K., Ishii, H., Tezuka, A.: Region and Landscape. Kokin Shoin, Tokyo (1991)
8. Nister, D., Stewenius, H.: Scalable recognition with a vocabulary tree. In: Proceedings of the IEEE Computer Society Conference on Computer Vision and Pattern Recognition, pp. 2161–2168 (2006)
9. Pass, G., Miler, R.Z.J.: Comparing images using color coherence vectors. In: Proceedings of the Fourth ACM International Conference on Multimedia, pp. 65–73 (1996)
10. Philbin, J., Chum, O., Isard, M., Sivic, J., Zisserman, A.: Object retrieval with large vocabularies and fast spatial matching. In: Proceedings of the IEEE Conference on Computer Vision and Pattern Recognition, CVPR 2007 (2007)
11. Philbin, J., Zisserman, A.: Object mining using a matching graph on very large image collections. In: Proceedings of the 2008 Sixth Indian Conference on Computer Vision, Graphics, and Image Processing (ICVGIP 2008), pp. 738–745 (2008)

A Framework for Distributed Managing Uncertain Data in RFID Traceability Networks

Jiangang Ma[1], Quan Z. Sheng[1], Damith Ranasinghe[1],
Jen Min Chuah[1], and Yanbo Wu[2]

[1] School of Computer Science, The University of Adelaide, Australia
jiangang.ma@adelaide.edu.au
[2] Beijing Jiaotong Univeresity, Beijing, 100044, China
ybwu@bjtu.edu.cn

Abstract. The ability to track and trace individual items, especially through large-scale and distributed networks, is the key to realizing many important business applications such as supply chain management, asset tracking, and counterfeit detection. Networked RFID (radio frequency identification), which uses the Internet to connect otherwise isolated RFID systems and software, is an emerging technology to support traceability applications. Despite its promising benefits, there remains many challenges to be overcome before these benefits can be realized. One significant challenge centers around dealing with *uncertainty* of raw RFID data. In this paper, we propose a novel framework to effectively manage the uncertainty of RFID data in large scale traceability networks. The framework consists of a global object tracking model and a local RFID data cleaning model. In particular, we propose a Markov-based model for tracking objects globally and a particle filter based approach for processing noisy, low-level RFID data locally. Our implementation validates the proposed approach and the experimental results show its effectiveness.

Keywords: RFID, Internet of Things, Uncertainty, Traceability Networks.

1 Introduction

Radio Frequency Identification (RFID) is a wireless communication technology that is useful for identifying objects. RFID uses radio-frequency waves to transfer identifying information between tagged objects and readers without line of sight, thus enabling automatic identification [1]. The ability to track and trace individual items—especially through large-scale and distributed networks—is the key to realizing many important business applications such as supply chain management, asset tracking, and counterfeit detection [2–5].

One of the important technological advances that targets large-scale traceability (e.g., nation-wide supply chain management across companies) is the so-called "Networked RFID" [6]. The basic idea behind Networked RFID is to use the Internet to connect otherwise isolated RFID systems and software. With Networked RFID, traceability applications analyze automatically recorded identification events to discover the current location of an individual item. They can also retrieve the historical information, such as previous locations, time of travel between locations, and time spent in storage.

X.S. Wang et al. (Eds.): WISE 2012, LNCS 7651, pp. 298–311, 2012.

Such technological advances will revolutionize our ability to monitor the world around us, allowing critical decisions and required interventions to be made in a timely fashion.

While RFID provides promising benefits in many applications, there remains significant challenges to be overcome before these benefits can be realized. Central to these challenges is the *uncertainty* of the data collected by the underlying RFID networks. Due to the sensitivity of sensing to the orientation of reading, interference, malfunction of reading components, and many environmental factors, RFID data are typically incomplete, imprecise, and even misleading [7, 8]. Obviously, when such data streams are used directly in monitoring and tracking applications (e.g., product recall), the quality of the applications can be a significant concern.

The inherent uncertainty of the raw RFID data makes it impossible to be used directly in high-level applications. Instead, sophisticated approach needs to be developed to be able to support uncertainty as a first-class citizen in RFID applications. In this paper, we design and implement a novel framework for uncertainty management in large scale RFID traceability networks, geared towards efficiently and accurately supporting traceability applications. In particular, we develop techniques that cope with ambiguous and imprecise RFID data by transforming low-level RFID readings into probabilistic events. In a nutshell, our key contributions are as the following:

- We introduce a data model for globally tracking moving objects. The proposed approach is based on a Markov-based process to infer objects' actual location according to probabilistic RFID observations.
- We propose a sampling-based inference technique to capture the uncertainty of RFID raw data. The technique also produces the probability distribution of RFID objects from dynamic and noisy low-level RFID data.
- We validate the proposed techniques in prototype implementation and the experimental results show the effectiveness of the proposed techniques.

The remainder of this paper is organized as follows. In Section 2, we discuss some background information related to managing uncertain data in RFID traceability networks. In Section 3, we introduce a data model for tracking moving objects and in Section 4, we discuss local RFID data management. In Section 5, we present our experiments to show the effectiveness and efficiency of our approach. Section 6 is dedicated to the related work and Section 7 concludes the paper and discusses some future research directions.

2   Preliminaries

In this section, we first describe a scenario of recalling problematic drugs in a supply chain application. We then propose an overall architecture and show how to efficiently manage, query and analyze uncertain information obtained from RFID data sources.

2.1   An Example of Uncertainty in Object Tracking

We first use a simple example to illustrate locating and recalling problematic products in a supply chain network, as shown in Figure 1. Suppose a particular type of drug to be distributed from its manufacturer `Brisbane Plant` in Brisbane to its destination

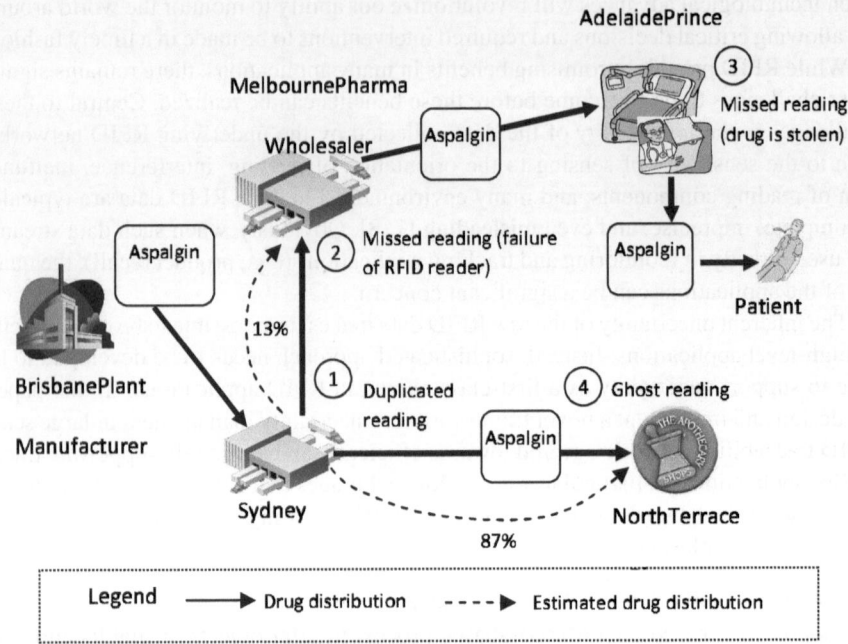

Fig. 1. An example of uncertainty in an RFID-enabled supply chain

`Royal Adelaide Hospital` in Adelaide. On the path of distribution of the drug, RFID readers are positioned at different nodes like `Melbourne Wholesaler` and `Sydney Distributor` to detect the event when the drugs pass by. Further suppose that problematic drugs are detected at the `Royal Adelaide Hospital` in Adelaide. In this case, the source in Brisbane of problematic drugs needs to be identified. Having tracked the source, the system may need to find similar problematic drugs that may be distributed in other places (e.g., warehouses and pharmacies).

Unfortunately, due to the sensitivity of sensing to the orientation of readings, interference, malfunction of reading components, and many environmental factors, raw RFID data are highly noisy [7, 8]. For example, there could be duplicated reads, missed readings (either due to malfunction of an RFID reader, or due to the objects being stolen or misplaced), and even "ghost" reads, meaning that a tag "captured" by an RFID reader does not exist or is not within the reader's detecting field. Obviously, when such data streams are used directly in monitoring and tracking applications (e.g., product recall), the quality of the applications can be a significant concern.

2.2   Overview of the Framework

In this section, we provide a high level overview of the proposed traceability framework. The framework involves two main tracking components: a global object tracking component and a local RFID data management component.

- The global object tracking component (GOTC), as shown in Figure 2, is used to support tracking and tracing the objects moving from one organization to the other. Our traceable network includes three main entities: sensor nodes, receptors and objects. First, a node is equipped with battery-powered devices that have basic sensing modules and communication interfaces. In particular, each node governs a number of receptors (e.g., RFID readers) whose signals can cover and monitor some areas. These nodes are deployed and dispersed in different sites for monitoring and tracking objects. For example, in a supply chain network, a node may be a distribution center, or a retail store and RFID readers are deployed at the fixed locations such as the entrances of a warehouse. Second, an object is any monitored entry attached with RFID tags such as items and goods, etc. Very often, objects move across the nodes confined within this traceability network. We will use a Markov-based process to model and infer objects actual locations based on probabilistic observations, which will be detailed in the next section.
- Local RFID Data Management (LRDM), as shown in Figure 3, represents the internal structure and functionalities of each node in a traceability network. LRDM component is executed at individual senor node. To capture uncertainty of raw RFID data, we employ a sampling-based inference technique called particle filtering [9] to compute the probability distribution of objects from dynamic and noisy low-level RFID data. The generated probabilistic RFID data is then processed in order to extract the high-level events that can be used by applications. In addition, the processed data is stored in databases in terms of `records` according to a data model developed in the framework, readily for querying and mining.

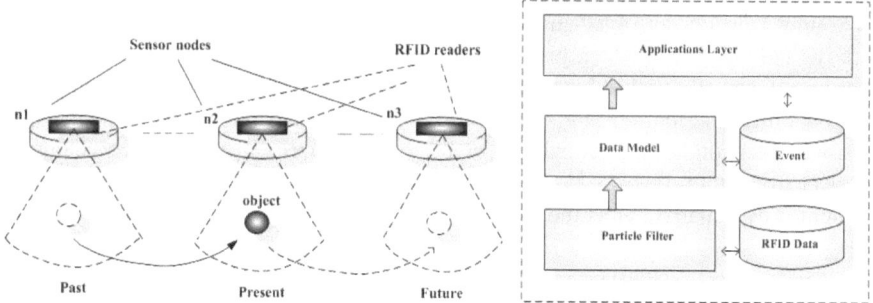

Fig. 2. Global object tracking   **Fig. 3.** Local data management

3   A Global Data Model for Tracking Moving Objects

In this section, we introduce a data model for globally tracking moving objects. The proposed approach is based on the Markov-based process to infer objects' actual location according to probabilistic RFID data observations.

3.1   Location Information Flows vs. Observation Information Flows

We start with the basic description of tracking approach in traceable networks, shown in Figure 2. When tracking and tracing objects that move along a certain route, we are interested in two types of information flows: *the location information flows* (LIF) and *the observation information flows* (OIF). Physically, one object moves along a certain route (e.g., a pallet moves in a supply chain network) and such movement of objects causes the change of object's physical locations. We call this LIF. On the other hand, once the objects are captured by RFID readers, the movement of objects generates digitalized information, we call this OIF.

However, due to different factors such as inaccurate objects identifications and missed readings caused by RFID readers, the observed information (OIF) may not exactly reflect the actual location of the object (LIF), which generates uncertain information. In addition, LIF are often unobservable state variables due to the uncertainty, so we refer LIF as hidden state variables. On the other hand, OIF are observable evidence variables (e.g., observed by RFID readers) [10]. Thus, in this paper, one of our aims is to model and reason objects' actual locations according to OIF. We deal with this problem using a Markov-based probability approach.

3.2   A Markov-Based Model for Global Object Tracking

Formally, we consider a distributed traceability network, comprising a collection of n nodes. The LIF for the nodes are represented as $\mathcal{X}=\{x_1, x_2, ..., x_n\}$, which are often unobservable due to the uncertainty. $\mathcal{Y}=\{y_1, y_2, ..., y_n\}$ denotes a set of OIFs that are generated by RFID readers. We use $\mathcal{X}_t$ to denote unobservable state variable at time t and and $\mathcal{Y}_t$ for observable evidence variable at time t.

We now consider an object moving across a traceability network as a first-order Markov random process. At a given point of time, although an object can be at any location of a set of possible locations $\mathcal{X}$, the first-order Markov process assumes that the probability distribution at $t + 1$ is entirely determined by the current state at t. Thus, we use a conditional probability $p(\mathcal{X}_t|\mathcal{X}_{t-1})$ to represent the location change of an object from state $\mathcal{X}_{t-1}$ to $\mathcal{X}_t$. The transition probability $p(\mathcal{X}_t|\mathcal{X}_{t-1})$ can be further represented by a matrix $\mathcal{M}$ as the following:

$$M_L = \begin{bmatrix} m_{11} & \cdots & m_{1n} \\ \vdots & & \vdots \\ m_{n1} & \cdots & m_{nn} \end{bmatrix} = \begin{bmatrix} p(x_1|x_1) & \cdots & p(x_n|x_1) \\ \vdots & p(x_j|x_i) & \vdots \\ p(x_1|x_n) & \cdots & p(x_n|x_n) \end{bmatrix}$$

where the entry $m_{ij} = p(x_j|x_i)$ represents the transition probability by which state x_i changes to state x_j. Based on the above transition model, an object should be in a certain location after state x_i, so $\sum_{j\neq i} m_{ij} = \sum_{j\neq i} p(x_j|x_i) = 1$ should hold.

Similarly, we can further model the relationships between location flows $\mathcal{X}$ and observation flows $\mathcal{Y}$. Intuitively, the observable variables $\mathcal{Y}$ are often affected by actual state variables $\mathcal{X}$. In other words, state variables $\mathcal{X}$ cause the observable variables $\mathcal{Y}$ to take different values. Formally, let $\mathcal{Y}$ be a set of the observable variables taking values from RFID readers. Based on Markov assumption, we can use a conditional distribution to describe how state variables $\mathcal{X}$ affect the observable variables $\mathcal{Y}$. This model is represented by a matrix $\mathcal{N}$:

$$N = \begin{bmatrix} n_{11} & \cdots & n_{1m} \\ \vdots & & \vdots \\ n_{n1} & \cdots & n_{mm} \end{bmatrix} = \begin{bmatrix} p(y_1|x_1) & \cdots & p(y_m|x_1) \\ \vdots & p(y_j|x_i) & \vdots \\ p(y_1|x_m) & \cdots & p(y_m|x_m) \end{bmatrix}$$

where the entry $n_{ij} = p(y_j|x_i)$ represents the probability that system will demonstrate observation value y_j given that the object is in location x_i. Based on the above transition model, an object should be in some location after state x_i, so $\sum_{j \neq i} m_{ij} = \sum_{j \neq i} p(x_j|x_i) = 1$ should hold. An example of transition and observation model for an RFID traceability network is shown in Figure 4.

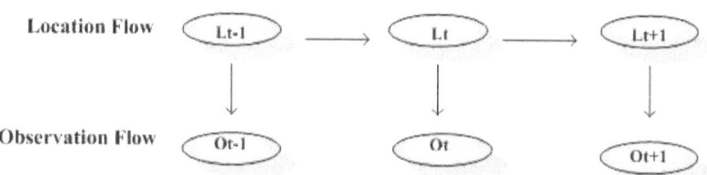

Fig. 4. Transition model for RFID traceable network

4   Local RFID Data Management in Traceable Networks

In this section, we describe our local RFID data management in traceable networks. We first describe a probabilistic model of uncertain data. We then present a sampling-based inference technique to capture uncertainty of RFID raw data.

4.1   A Probabilistic Model for Capturing Uncertain Data

In RFID-based tracking applications, raw data often exhibit uncertainty and impreciseness. For example, the products may be stolen, counterfeited, damaged, and missplaced. Thus, data produced by sensors (e.g., RFID readers) are often noisy. These uncertainties can be further classified into four categories: *false negative*, *true negative*, *false positive*, and *duplicate readings*. To capture such uncertainty in data, we use a continuous random variable x and describe data's uncertainty by using probability density functions $pdf(x)$.

Definition 1. *(Uncertain Object) An uncertain object o is defined to be a point x within a region $\mathcal{R}_u$ covered by RFID readers. The positions of the object o follow a probability density function pdf(x), denoting the likelihood of the object's position,and the condition holds:* $\int_{x \in \mathcal{R}_u} pdf(x)dx = 1$.

We further use Bayes rule to infer an object's location. As mentioned above, objects' locations in RFID applications are regarded as hidden variables x and observation variables y are the readings produced by RFID readers. Thus, our aim is to compute the joint probability distribution $p(x, y)$ over both hidden and observed variables. In addition, given an observed values y, this joint model induces the computation of a conditional distribution $p(x|y)$, which can be used to predict the objects' locations. More precisely, based on Bayes' rule, the conditional distribution $p(x|y)$ can be represented as:

$$f(x|y) = \frac{f(y|x)f(x)}{\int f(y|x)f(x)dx}$$

Suppose we have n observations $\mathcal{Y}^n = (\mathcal{Y}_1, \mathcal{Y}_2, ..., \mathcal{Y}_n)$ of RFID data at each time-step, we replace $f(y|x)$ with:

$$f(y_1, y_2, ..., y_n|x) = \prod_{i=1}^{n} f(y_i) = \Gamma(x)$$

Thus, the current true state is defined by the posterior density over the random variable x conditioned on all RFID data:

$$f(x|y^n) = \frac{f(y^n|x)f(x)}{\int f(y^n)f(x)dx} = \frac{\Gamma(x)f(x)}{h_n}$$

where

$$h_n = \int \Gamma(x)f(x)dx$$

is the normalizing constant. In this case, one location variable in $\mathcal{X} = (x_1, ..., x_k)$ (e.g., x_1) is determined by computing the following marginal posterior density:

$$f(x_1|\mathcal{Y}^n) = \int \int \cdots \int f(x_1, ..., x_k|\mathcal{Y}^n)dx_2 \ldots dx_k$$

The challenge here is to perform accurate inference for a large number of RFID data because the cost of computing such posterior densities grows exponentially over time. In order to deal with this issue, we next introduce a sampling-based inference called particle filtering to approximate the target conditional distribution.

4.2 Processing Uncertain RFID Data Based on Particle Filtering

The main idea of the particle filter [9, 11] is to approximately represent the posterior probability by a set of limited random state samples (also called particles), extracted from this posterior probability. One of the main advantages for particle filtering is that such approximation of original conditional probability is able to achieve the expected solutions with only a limited number of samples. It can also represent a broader range of distributions than common probabilistic distributions like Gaussian Distribution.

Intuitively, the main task of a tracking application in RFID is to estimate states (e.g., the location of an object), based on the available evidence states (e.g., RFID readings). The estimating inference includes two main steps: *prediction* and *importance sampling*. The prediction stage is to construct the set of candidate particles, while the update stage is to construct the set of qualified particles from candidate particles. These qualified particles will approximate the original posterior *pdf*.

Algorithm 1. State Update

Input: raw RFID readings, $\mathbf{X} = \{x^i | i = 1...I\}$ **Output**: estimation of an object's location

1. **for** $\forall x^i \in \mathbf{X}$ **do**
2. $x^i\_location + = motion\_model.$
3. $x^i\_location + = diffuse\_rate * random\_number.$
4. **end for**
5. $prob \leftarrow meausre\_probability(x^i\_location).$
6. calculate normalization_value
7. **for** $\forall x^i \in \mathbf{X}$ **do**
8. $x^i\_prob \ / = normalization\_value.$
9. $calculate x^i\_prob.cumulative\_prob$
10. **end for**
11. **for** $\forall x^i \in \mathbf{X}$ **do**
12. $r \leftarrow random\_num\_coffie.$
13. $p \leftarrow first x^i\_prob.cumulative\_prob >= r$
14. $ressampled\_x^i \leftarrow p$
15. **end for**

Algorithm 2. Sampling Process

---

1. **if** object reading is missing
2.   **for** $\forall x^i \in \mathbf{X}$ **do**
3. $find\_nearest\_reader(x^i).$
4. $x^i \leftarrow position\_of\_nearest\_sensor$
5. $d \leftarrow x^i\_distance\_from\_reader$
6.   **end for**
7. **else**
8.   **for** $\forall x^i \in \mathbf{X}$ **do**
9. $dis \leftarrow compute\_distance(x^i - reader).$
10.   **end for**
11. **end if**

The input to the particle filter algorithm are a set of samples (states) of an object, $\mathcal{X}_{t-1} = \{x_{t-1}^1, x_{t-1}^2, ..., x_{t-1}^m\}$, and a set of current observation y_t. At the first step of constructing set $\mathcal{X}_t^c$ of candidate particles, the algorithm processes each particle in $\mathcal{X}_{t-1}$. At the end of this step, the set $\mathcal{X}_t^c$ of candidate particles includes a set of weighted particles $\mathcal{X}_t^c = \langle x_t^i, w_t^i \rangle$, where w_t^i is the weight of a particle x_t^i and $0 \leq i \leq m$. At the second step, the algorithm selects the particles from $\mathcal{X}_t^c$, and puts the particles that have higher weights as qualified particles into the set $\mathcal{X}_t^q$ of qualified particles . At the end of this step, the set $\mathcal{X}_t^q$ includes all qualified particles. Finally, these qualified particles in the set $\mathcal{X}_t^q$ approximate the original posterior probabilistic distribution pdf. The particle filtering algorithm is a recursive process that includes constructing set $\mathcal{X}_t^c$ of candidate particles, and the set $\mathcal{X}_t^q$ of qualified particles.

Prediction. This step updates the particles to reflect the new states of the observed objects (see Algorithm 1). A new state x_t at time t will be generated based on the particle x_{t-1} and the observation y_t. In our implementation, we define the difference between the location of readers and object's previous location as the motion model. The motion model describes how the states of an object change. Thus, in the prediction stage, we can update each of the particles based on the results from the motion model.

Algorithm 3. Managing Missing Readings

---

1. read input $\mathbf{X} = \{x^i | i = 1...I\}$
2. **do**
3.   get readings
4.   **if** reading_time = current_time
5.   **if** *object* ≠ *exist*
6.   *initialize_object.*
7.   **else**
8.   *get_object*
9.   **end if**
10.   *calculate_motion_model*
11.   *update_state_of_object*
12.   **else**
13. $\mathbf{M} = \{m^i | i = 1...H\} \backslash * check\_missing\_readings$
14.   **for** $\forall m^i \in \mathbf{M}$ **do**
15.   *simulate_uncertainty.*
16.   *update_object_state.*
17.   **end for**
18.   **end if**
19. **while** ! more readings

Sampling. This step involves taking samples for transition distribution $p(x_t | y_t, x_{t-1})$. In our case, the probability of each particle is calculated using a standard normal distribution (Gaussian) $f(x) = \frac{1}{\sqrt{2\pi\sigma^2}} e^{-\frac{(x-\mu)^2}{2\sigma^2}}$, where μ is mean, and σ^2 variance. Then for each particle x_t^i, we compute its weight $w_t^i = p(y_t | x_t^i)$. For example, if an observed object is within the range that the RFID reader covers, the particles produced by the object are assigned higher weights because these particles have a higher likelihood of representing the correctly estimated locations. At the end of this step, the set $\mathcal{X}_t^c$ of candidate particles includes a set of weighted particles $\mathcal{X}_t^c = \langle x_t^i, w_t^i \rangle$. An algorithm of sampling process is shown in Algorithm 2.

Resampling. This step converts the set $\mathcal{X}_t^c$ of candidate particles into the set $\mathcal{X}_t^q$ of qualified particles. However, only those particles in set $\mathcal{X}_t^c$ that have bigger values of weights are selected and transformed into $\mathcal{X}_t^q$.

We also design an algorithm (see Algorithm 3) to deal with missed readings in order to further improve the effectiveness of our approach. When the readings are missed, the location of an object indicates uncertainty. To simulate the uncertainty of an object's locations due to missed readings, we set the locations of some particles randomly, and estimate the locations of the remaining particles according to $p(X_t|X_{t-1})$.

4.3   A Data Store Model

We develop a new schema to store cleaned RFID data for supporting querying and mining. Our model incorporates RFID data's uncertain information, including time, tagID, location, and probability. RFID records form a probabilistic database **PDB**:

Definition 2. *Each record* $r \in$ **PDB** *is represented by a tuple* $r = \langle t^r, d^r, l^r, p^r \rangle$, *where* t^r, d^r, l^r p^r *denotes the 'time', 'tagID', 'location', 'probability'.*

For instance, one instance of record (2:05pm, EPC0001, Sydney-D, 0.57) indicates that at 2:05pm, the object with ID of EPC0001 was located at Sydney Distributor (i.e., Sydney-D) with the probability of 0.57. This model also stores the probability distribution over an RFID object's location at a given time. For example, there might be another tuple (2:05pm, EPC0001, Melbourne-W, 0.43), indicating that there was also a 0.43 probability that the object with ID of EPC0001 was at Melbourne Wholesaler (i.e., Melbourne-W). An example of RFID data records is shown in Table 1.

Table 1. RFID data records

Raw Data Records	(t_1, r_1, l_1), (t_1, r_2, l_1), (t_1, r_3, l_2), (t_2, r_2, l_1)
	(t_2, r_1, l_3), (t_3, r_1, l_2), (t_4, r_4, l_1), (t_4, r_3, l_3)
	(t_2, r_3, l_4), (t_5, r_3, l_4), (t_5, r_1, l_3), (t_5, r_4, l_5)
Probabilistic RFID Data Records	$(t_1, r_2, l_1, 0.1)$, $(t_1, r_3, l_2, 0.4)$, $(t_2, r_2, l_1, 0.5)$
	$(t_2, r_1, l_3, 0.4)$, $(t_3, r_1, l_2, 0.2)$, $(t_4, r_4, l_1, 0.3)$
	$(t_2, r_3, l_4, 0.3)$, $(t_5, r_3, l_4, 0.4)$, $(t_5, r_1, l_3, 0.6)$

5   Experiments

To validate the proposed approach, we conducted two groups of experimental studies. The first group tested the effectiveness of the basic functionality of the system, and the second group explored the accuracy of the particle filtering-based approach.

We use NetBeans to develop two Java applications to support experiments. The first Java application is an RFID data generator that produces a series of simulated RFID readings. The generated RFID data are then used as the input of the particle filtering system. The probabilistic outputs generated from the particle filtering system are fed into the second Java application. This Java application translates probabilistic outputs into a SQL file for supporting updating and querying the database.

To test the basic functionality of the system, we have designed a scenario where two pallets travel through different stages within a warehouse. The missed readings for the first pallet occur at 3.02pm shown in Figure 5. The testing results are represented by four three-dimensional graphs, as shown in Figure 6, each representing the first pallet's predicted location at a given time.

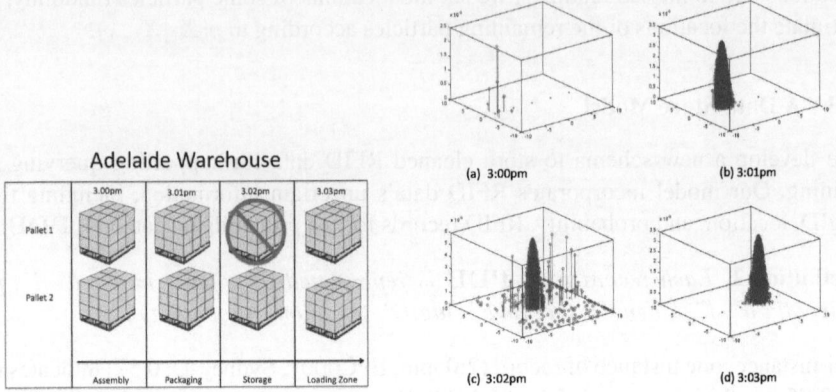

Fig. 5. Missed reading in warehouse scenario   **Fig. 6.** The result of testing particle filtering system

In these graphs, the X axis and Z axis of the visualization represent the X and Y coordinates of the warehouse, while the Y axis represents the probability values. The red dots signify the different RFID readers and their positions within the warehouse. In addition, each blue circle represents a particle of the object observed and the probability of that particle. This experimental results show that the proposed system is able to predict the locations of moving objects. For example, Figure 6 shows that the particles are correctly predicted to be around the third reader with a probability of 0.489, although some readings are missed at 3.02pm.

5.1   Experimental Results

To show the accuracy of our tracking system, we carried out two experimental studies. The first experiment explores the accuracy of particle filtering system whilst the second group demonstrates the effect of optimization on the runtime.

Accuracy of Particle Filtering System. In this experiment, we studied the effect of missed readings on the accuracy of location estimation. For each category of missed readings, our data generator generated an input file containing data with a certain percentage of missed readings. Each input data file includes 100 records, and these files were used to examine the average accuracy. Figure 7 shows the experimental results. Overall, our system is able to perform object location estimation with an accuracy of 76%. It indicates a good property of our system even if some readings are missing.

Effect of Optimization on Runtime. We have also examined the effect of optimization on the runtime speed. For each experiment with a set of objects, we executed each version of the program five times to determine the average runtime. Our experiments show that our approach can reduce the numbers of objects to be processed. This also has an impact on the runtime of the particle filter algorithm. From Figure 8, we can see that when the number of objects exceed 1,000, the time used in our approach is much

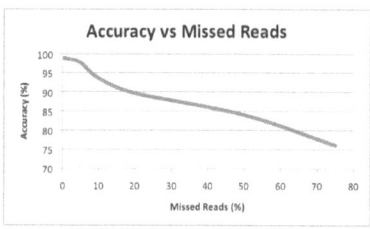

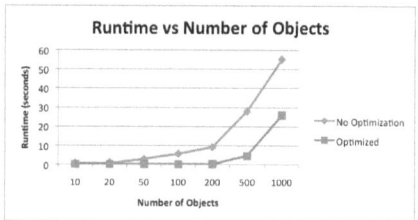

Fig. 7. Accuracy vs percentage of missed reads   **Fig. 8.** Runtime speedup

less than the time used without our approach. Therefore, together with the ability to perform object location estimation in reduced time, we can conclude that our system is suitable for large-scale practical applications.

6 Related Work

Managing uncertain data is an important research topic in RFID-based applications. There are a number of researches focusing on RFID networks [12–14], uncertain data management[7, 15, 16], data model [17, 18] and probabilistic databases [4, 19–21]. However, there still lacks of effective framework to support uncertain data management in a distributed traceability network.

EPCglobal[7] provides rich standards to support RFID-based tracking networks. Currently, EPCglobal Architecture Framework (EAF) is widely regarded as one of the most well-known RFID network architectures in industry [22]. In addition, works in [13, 23] propose generic approaches for tracing and tracking of objects in large-scale and distributed environments such as the Internet of Things. In particular, by analyzing a wide range of traceability applications, these works provide models for moving objects in discrete spaces, which are mainly built on top of the DHT (Distributed Hash Table) based overlay network. Furthermore, the proposed frameworks separate functionalities into three isolated modules: identity, capture, and exchange.

The work related to uncertain data management can be found in [7, 15, 16]. The work by Diao et al. [16] proposes a probabilistic model and approach to transform raw data streams into tuple streams with quantified uncertainty. This approach approximates object location estimation based on data streams processing, making their research relevant to our approach. However, their techniques cannot be directly used in our work because their work focuses on general estimations for objects tracking and not deal with the location estimation with the large-volume of data in distributed applications such as supply chain networks. To understand the containment relationships of objects for location estimating, the work in [15] presents a SPIRE framework, which relies on packaging level information extracted from tagIDs of objects to form time-varying graphs and depict the relationships of inter-objects' containments. Based on the graphs, estimating locations of an object can be done by inferring the edges and the nodes in the graph, so that a probabilistic distribution over all possible states for each node could be built.

In the area of particle filtering in RFID-based applications, the Cascadia project [17] applies a particle filter approach to infer the probability distributions over an object's location. In particular, each particle is associated with a location. Upon receiving a raw RFID event, the particles are weighted based on the consistency of their coordinates with the raw RFID event. However, since this project targets applications from a single administrative domain, it is not required to process data in a large-scale and widely distributed environments.

7   Conclusion

Effectively managing uncertain data in RFID traceability networks still remains a challenge. In this paper, we have studied the main problems related to tracking the movements of objects and particularly focused on processing ambiguous and imprecise RFID data. We have designed and implemented a novel framework, which improves the existing techniques for tracking the movement of objects and cleaning RFID raw data. Within the framework, we use a Markov-based process to infer objects' actual location according to probabilistic RFID data observations. In addition, we propose a sampling-based inference technique called particle filtering to capture uncertainty of RFID raw data, and to compute the probability distribution of RFID objects from dynamic and noisy low-level RFID data. Our probabilistic reasoning approach enables efficient and accurate support for traceability applications. The experimental results have shown that our approach is capable in object estimation with high accuracy and scalable in large-scale applications.

Our further work include conducting more experiments to further study the system performance. We also plan to develop a model for probabilistic RFID event generation. This will particularly consider the issue of Unordered Event Stream (UnES) where the order of arriving RFID events might not match with the order of the occurrence of the events in the real world (e.g., due to network routing delay).

References

1. Nath, B., Reynolds, F., Want, R.: Rfid technology and applications. IEEE Pervasive Computing 5, 22–24 (2006)
2. Wu, Y., Ranasinghe, D.C., Sheng, Q.Z., Zeadally, S., Yu, J.: RFID Enabled Traceability Networks: A Survey. Distributed and Parallel Databases 29, 397–443 (2011)
3. Franklin, M., Jeffery, S., Krishnamurthy, S., Reiss, F., Rizvi, S., Wu, E., Cooper, O., Edakkunni, A., Hong, W.: Design considerations for high fan-in systems: The hifi approach. In: Proceedings of the 2nd Biennial Conference on Innovative Data Systems Research, CIDR 2005 (2005)
4. Gonzalez, H., Han, J., Li, X., Klabjan, D.: Warehousing and analyzing massive rfid data sets. In: Proceedings of the 22nd International Conference on Data Engineering, ICDE 2006, pp. 83–83. IEEE (2006)
5. Ilic, A., Andersen, T., Michahelles, F.: Increasing supply-chain visibility with rule-based rfid data analysis. IEEE Internet Computing 13, 31–38 (2009)
6. Roussos, G., Duri, S., Thompson, C.: Rfid meets the internet. IEEE Internet Computing 13, 11–13 (2009)

7. Jeffery, S., Franklin, M., Garofalakis, M.: An adaptive rfid middleware for supporting meta-physical data independence. The VLDB Journal 17, 265–289 (2008)
8. Sheng, Q., Li, X., Zeadally, S.: Enabling next-generation rfid applications: Solutions and challenges. Computer 41, 21–28 (2008)
9. Fox, V., Hightower, J., Liao, L., Schulz, D., Borriello, G.: Bayesian filtering for location estimation. IEEE Pervasive Computing 2(3), 24–33 (2003)
10. Russell, S.: Artificial intelligence: A modern approach, December 30 (2002)
11. Ng, B., Peshkin, L., Pfeffer, A.: Factored particles for scalable monitoring. In: Proceedings of the Eighteenth Conference on Uncertainty in Artificial Intelligence, pp. 370–377 (2002)
12. EPCGLOBAL, http://www.epcglobal.com
13. Wu, Y., Sheng, Q., Ranasinghe, D.: P2p object tracking in the internet of things. In: Proceedings of International Conference on Parallel Processing (ICPP 2011), pp. 502–511. IEEE (2011)
14. Cambridge University: Serial-level inventory tracking model. Bridge WP03, Cambridge University, BT Research (2007)
15. Nie, Y., Cocci, R., Cao, Z., Diao, Y., Shenoy, P.: Spire: Efficient data inference and compression over rfid streams. IEEE Transactions on Knowledge and Data Engineering, 141–155 (2012)
16. Diao, Y., Li, B., Liu, A., Peng, L., Sutton, C., Tran, T., Zink, M.: Capturing data uncertainty in high-volume stream processing. In: Proceedings of the 4th Biennial Conference on Innovative Data Systems Research, CIDR 2009 (2009)
17. Welbourne, E., Khoussainova, N., Letchner, J., Li, Y., Balazinska, M., Borriello, G., Suciu, D.: Cascadia: A system for specifying, detecting, and managing rfid events. In: Proceedings of the 6th International Conference on Mobile Systems, Applications, and Services (MobiSys 2008), New York, USA, pp. 281–294 (2008)
18. Wang, F., Liu, P.: Temporal management of rfid data. In: Proceedings of International Conference on Very Large Databases (VLDB 2005), Norway, pp. 1128–1139 (2005)
19. Soliman, M., Ilyas, I., Chen-Chuan Chang, K.: Top-k query processing in uncertain databases. In: Proceedings of the 23rd International Conference on Data Engineering (ICDE 2007), pp. 896–905. IEEE (2007)
20. Zhang, Y., Lin, X., Zhu, G., Zhang, W., Lin, Q.: Efficient rank based knn query processing over uncertain data. In: Proceedings of the 26th International Conference on Data Engineering (ICDE 2010), pp. 28–39. IEEE (2010)
21. Cormode, G., Garofalakis, M.: Sketching probabilistic data streams. In: Proceedings of the 2007 ACM SIGMOD International Conference on Management of Data, pp. 281–292. ACM (2007)
22. EPCglobal: EPCglobal Specifications, http://www.epcglobalinc.org/standards/specs
23. Mo, J., Sheng, Q., Li, X., Zeadally, S.: Rfid infrastructure design: a case study of two australian rfid projects. IEEE Internet Computing 13, 14–21 (2009)

Controlled Knowledge Base Enrichment
from Web Documents

Yassine Mrabet[1], Nacéra Bennacer[2], and Nathalie Pernelle[1]

[1] LRI, Université Paris-sud, PCRI, bât. 690, 91405 Orsay, France
{yassine.mrabet,nathalie.pernelle}@lri.fr
[2] Supélec, E3S 3 rue Joliot Curie, 91192 GIF-SUR-YVETTE
nacera.bennacer@supelec.fr

Abstract. The Linked Open Data initiative brought more and more RDF data sources to be published on the Web. However, these data sources contain relatively little information compared to the documents available on the surface Web. Many annotation tools have been proposed in the last decade for the automatic construction and enrichment of knowledge bases. But, while noticeable advances are achieved for the extraction of concept instances, the extraction of semantic relations remains a challenging task when the structures and the vocabularies of the target documents are heterogeneous. In this paper, we propose a novel approach, called REISA, which allows to enrich RDF/OWL knowledge bases with semantic relations using semistructured documents annotated with concept instances. REISA produces weighted relation instances without exploiting lexico-syntactic or structure regularities in the documents. Neighbor domain entities in the annotated documents are used to generate the first sets of candidate relations according to the domain and range axioms defined in a domain ontology. The construction of these candidate sets relies on automated semantic controls performed with (i) the existing knowledge bases and (ii) the (inverse) functionality of the target relations. The weighting of the selected relation candidates is performed according to the neighborhood distance between the annotated domain entities in the document. Experiments on two real web datasets show that (i) REISA allows to extract semantic relationships with interesting precision values reaching 76,5% and that (ii) the weighting method is effective for ranking the relation candidates according to their precision.

Keywords: Knowledge Base Enrichment, RDF(S)/OWL, domain ontologies, semantic annotation, semi-structured documents.

1 Introduction

Many semantic Web projects aim to publish RDF data on the Web and to link them with existing RDF knowledge bases. Such initiatives allow semantic search engines to query different RDF knowledge bases and to perform reasoning tasks on their data. However, these data sources contain much less information

X.S. Wang et al. (Eds.): WISE 2012, LNCS 7651, pp. 312–325, 2012.

than the HTML documents available on the surface Web. Semantic annotation tools allow to associate ontology-based metadata to documents and/or document parts. Therefore, their automation is a key factor in order to exploit the contents of HTML documents at Web scale. Diverse techniques could be used and combined to annotate Web documents, ranging from machine learning and natural language processing to knowledge management.

Many research works focused on the extraction and annotation of named entities or domain terms [10,2]. Such approaches can exploit lexico-syntactic patterns representing regular expressions appearing in the documents' texts [14,11,5] or (onto-)lexical resources describing sets of named entities and domain terms [11,15]. Some of these approaches consist in the identification of the existing concept instances that may be referred by document entities. These concept instances are described in populated ontologies (or RDF knowledge bases). For example, SOFIE allows to identify references to a concept instance i by comparing the document context of the reference with the labels of concept instances that are linked to i in the knowledge base [14].

Another aspect of the enrichment of RDF knowledge bases is the production of semantic relation instances. Some document-annotation approaches tackled this task with lexico-syntactic patterns defined manually or learned from available examples in document corpora [14,1,12]. PLATO allows to construct partonomic relations using LOD entities and to validate these candidates using generated linguistic patterns that are tested on the Web [8]. Other semantic relation extraction approaches target document parts that have regular structures such as HTML tables [9,7,4] or specific structures such as Wikipedia infoboxes [13,3]. The main shortcoming of such methods occurs when the documents have heterogeneous structures and vocabularies. In such cases, using lexico-syntactic patterns or regular structures won't allow to extract a major part of the semantic relations that are described in the documents. Thus, complementary approaches are needed in order to discover these unreachable relations with regularity-independent methods.

In this paper, we present the REISA approach which allows to enrich RDF-OWL knowledge bases with property instances using HTML documents annotated with concept instances. REISA does not use structural or lexico-syntactic patterns. It exploits the structural closeness between document entities referring to concept instances (or litteral values). Doing so produces a huge number of property instances. REISA controls and filters these candidates at two main levels. First, REISA exploits the existing knowledge bases and ontology axioms to avoid the selection of wrong or redundant property candidates. Second, REISA associates confidence values to the candidates according to the closeness of the references in the document. The uncertainty of the produced property instances is represented using weighted RDF named graphs. REISA has been evaluated on two corpora using knowledge bases from the Linked Open Data project[1] and HTML documents describing call-for-papers and geographic entities. This paper is structured as follows: in section 2, we present the semantic integration model

---

[1] http://linkeddata.org

(SIM) that we defined to represent both knowledge bases and annotation bases. In section 3, We present the REISA approach. In section 4, we present the results obtained in our first experiments. The final section is a discussion of the main contributions and results.

2   Semantic Integration Model

We define the Semantic Integration Model (SIM) that represents both knowledge bases and annotation bases (cf. figure 1). More precisely, the SIM model includes a homogeneous representation of (i) domain entities (i.e. concept and relation instances), (ii) document parts and (iii) annotation links between both elements.

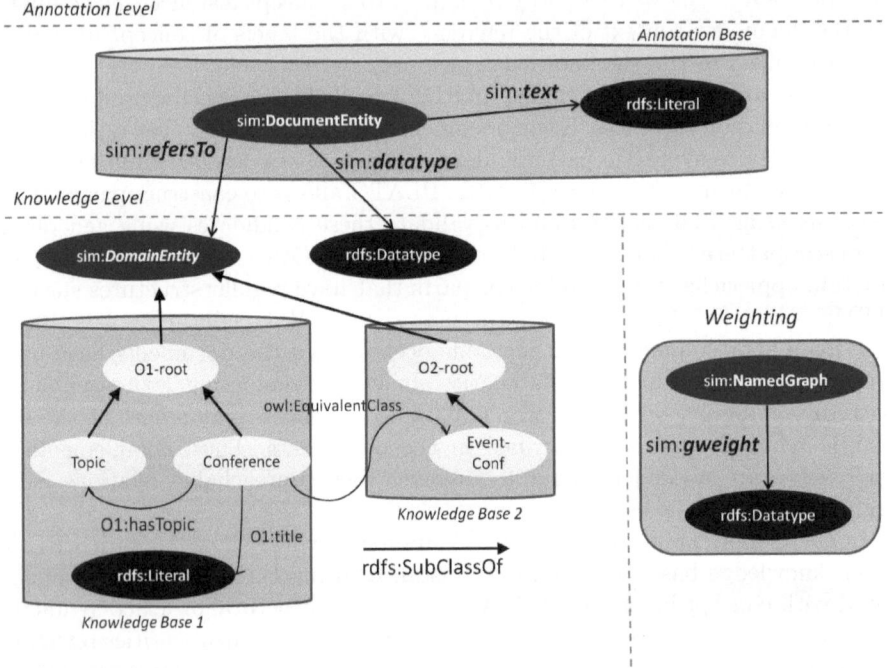

Fig. 1. Semantic Integration Model (SIM)

Knowledge Base. A knowledge base KB can be defined by the couple (O, T_O) where O is an OWL ontology defined by the quadruplet (C_O, R_O, A_O, X_O). C_O is the set of concepts, R_O is the set of properties defined between concepts (called relations), A_O is the set of properties defined between concepts and literals (called attributes). X_O is the set of axioms which define, for example, subsumption links between concepts or between properties, properties' domains and ranges, (inverse) functionality of the defined properties, equivalence or subsumption links issued from ontology alignment processes. The second element of a knowledge base KB, is the set of facts T_O that describes the instances of the ontology's concepts and properties.

In the SIM model, the concepts of C_O are domain concepts represented by the abstract class *DomainEntity*. Several knowledge bases $KB_1....KB_n$ can be represented.

Knowledge Base Weighting. We use RDF named graphs in order to associate a confidence measure to fact sets from the knowledge bases. Named graphs are RDF graphs identified by a URI[2]. The weight of a named graph represents the weight of the RDF triples it contains. This weight can be manually associated by domain experts or by automatic weighting methods. The property *sim:gWeight* allows to represent this weight as a real number between 0 and 1. A domain expert assigns a weight value of 1 to validated knowledge bases and inferior weight values to knowledge bases constructed automatically by (un)supervised annotation tools. The facts that are produced by our enrichment approach are weighted automatically using the closeness of the annotated entities in the document (see section 4.2).

Annotation Base. The Annotation Base AB is a set of document entities and annotation links generated by document annotators between the document entities and the knowledge base. In the SIM model, document entities are represented as instances of the *DocumentEntity* class. They are parts of XHTML documents that are semantically annotated. Thus, document entities are defined by a URI. Their textual content is represented by the *sim:text* attribute. In our approach, document entities may be (i) document nodes defined in the DOM tree of the document (in that case, their textual content is the concatenation of all their child text nodes) or (ii) a text window which may correspond to a named entity or to word sequences.

Annotated document entities are linked to concept instances of the knowledge base by the *sim:refersTo* property. The notation *refersTo(e,i)* means that a document entity e refers to a domain concept instance i. The links between document entities and datatypes (*rdfs:datatype*) are represented by the *sim:datatype* property. *DataType(e,t)* means that the textual content of a document entity e is a potential value of the datatype t[3]. Figure 2 shows an extract of KIM knowledge base and the annotation base corresponding to a document extract.

3   The REISA Approach

REISA consists in three main modules (cf. figure 3): (1) the integration module which allows to represent the KBs and the AB according to the SIM model, (2) the enrichment module which uses the KBs, the AB and the documents structure in order to produce new property instances and (3) the interrogation

---

[2] http://www.w3.org/2004/03/trix/
[3] XML schema datatypes are instances of the *rdfs:datatype* class

Extract of a semistructured document
\<div\> ... \<p\> **Laos** traces its history to the kingdom of Lan Xang ... took over **Vientiane** with ... along the Annamite mountains in **Vietnam**. \<span\> ... tools discovered in northern **Laos** attest ... communities along the **Mekong** River ... \</span\> \</p\> ... \<p\> Following the military defeat of Japan ... the Viet Minh occupied **Hanoi** and proclaimed a provisional government, which asserted national independence ...\</p\> ... \</div\>

Extract from the WKB (KIM) knowledge base
@prefix graphs: \<http://lri.fr/reisa/graphs/\> graphs:knowledgebase = { kimkb:Laos.0 **rdf:type** onto:Country kimkb:Laos.0 **onto:partOf** kimkb:Continent.2 kimkb:Continent.2 **rdf:type** onto:Continent kimkb:Continent.2 **rdfs:label** "Asia" kimkb:Vientiane.0 **rdf:type** onto:City kimkb:Vientiane.0 **onto:capital** kimkb:Laos.0 }

Extract from the Annotation base of the document
graphs:annotationsbase = { corpus:doc0/html/body/div/p[3]/a.0 **rdf:type** sim:DocumentEntity corpus:doc0/html/body/div/p[3]/a.0 **sim:refersTo** kimkb:Vietnam.0 corpus:doc0/html/body/div/p[3]/a.0 **sim:text** "Vietnam" corpus:doc0/html/body/div/p[3].12 **sim:refersTo** kimkb:Laos.0 corpus:doc0/html/body/div/p[3].12 **sim:text** "Laos" corpus:doc0/html/body/div/p[3].20 **sim:refersTo** kimkb:Mekong.0 corpus:doc0/html/body/div/p[3].20 **sim:text** "Mkong" corpus:doc0/html/body/div/p[3]/a[2].0 **sim:refersTo** kimkb:Hanoi.0 corpus:doc0/html/body/div/p[3]/a[2].0 **sim:text** "Hanoi" } graphs:knowledgebase sim:gweight 1 graphs:annotationsbase sim:gweight 0.9

Fig. 2. Extracts from the WKB knowledge base, a semistructured document and its associated annotation base

module which allows to answer user queries from the KBs and the enrichment base in a transparent manner and ranks the answers according to the weights of returned facts.

3.1 Construction of the Enrichment Base

In order to extract new property instances, we exploit the following hypotheses: (1) the document structure can indicate a part of their semantics and (2) annotated document entities which are closer to each other in the documents are more likely to be semantically related. More precisely, the structural closeness of annotated document entities is used to propose new domain properties between concept instances (or between concept instances and literals). The candidate properties are weighted by a confidence measure computed according to

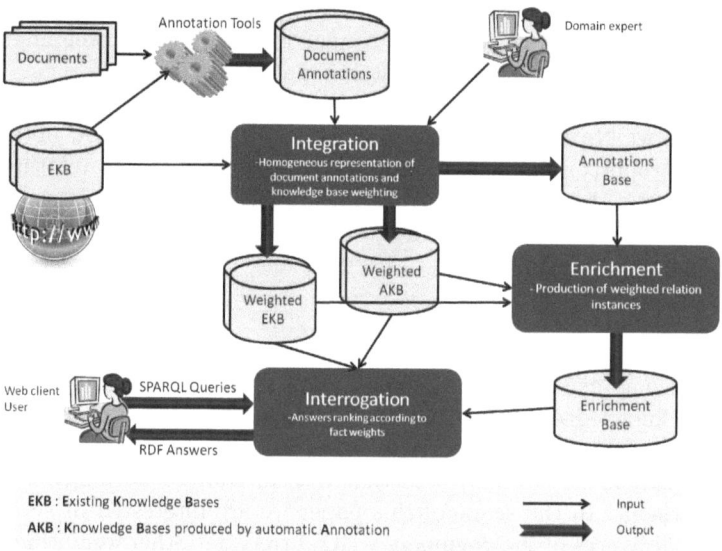

Fig. 3. REISA modules

the distance between the annotated document entities. The distance between two document entities is the shortest distance between the deepest document nodes containing them in the DOM tree of the document. We note here that all font-related tags are deleted from the documents before the distance calculation. In a second step, the domain ontology and the available property instances in the KBs are used to control and filter the candidate instances. In this section, we present the enrichment process which is based on the notion of semantic neighbors.

Semantic Neighbors. Two concept instances i_1 and i_2 are said to be semantic neighbors at a distance d for a property P, noted $V_P^d(i_1, i_2)$, if two document entities e_1 and e_2 referring to i_1 and i_2 are at a distance d from each other in the document, and the instances i_1 and i_2 belong to concepts that are respectively domain and range of P :

$$refersTo(e_1, i_1) \wedge refersTo(e_2, i_2) \wedge distance(e_1, e_2) = d \wedge$$
$$type(i_1, C_1) \wedge type(i_2, C_2) \wedge domain(P, C_1) \wedge range(P, C_2) \rightarrow V_P^d(i_1, i_2)$$

In a similar manner, a concept instance i and the textual content l of a document entity are said to be semantic neighbors at a distance d for a property P if the litteral associated to this text and a concept of the domain instance are respectively range and domain of the property P.

$$refersTo(e_1, i) \wedge distance(e_1, e_2) = d \wedge datatype(e_2, T) \wedge type(i, C)$$
$$\wedge domain(P, C) \wedge range(P, T) \wedge text(e_2, l) \rightarrow V_P^d(i, l)$$

The enrichment base is constructed using the knowledge bases. We note G^p the named graph having a weight value p. Named graphs having a weight value of 1 are considered as reference graphs containing a valid set of facts.

Neighborhood Named Graph. Neighborhood named graphs, noted $G^{w(d)}$, contain the property instances produced by enrichment from annotated document entities at a distance d in the document. We define the weight of a neighborhood named graph, $w(d)$, such that it is inversely proportional to the distance d. It is computed as follows, for $\alpha \leq 1$ a constant:

$$w(d) = \frac{\alpha}{d+1} \tag{1}$$

The value of α is fixed by the domain expert who associated weight values to the existing knowledge bases. This allows (i) to keep some coherence between all the weight values and (ii) to put a maximum limit to the weight of the candidate relations at distance 0 so that it is inferior to he weight of existing (valid) knowledge. In the scope of this paper we are interested in studying the relevance of distances in the document's DOM tree, but other weighting methods could also be considered (e.g. number of words between the two entities). A property instance $P(i_1,i_2)$, between two concept instances i_1 and i_2, is added to the graph $G^{w(d)}$ if:

- $V_P^d\ (i_1,i_2)$
  i_1 and i_2 are semantic neighbors at distance d for the relation P.

- $\neg \exists\ P(i_1,i_2) \in G^p$ s.t. $p > w(d)$
  The fact does not already exist in graphs having a better weight.

- if P is functional $\neg \exists\ z$ s.t. $P(i_1, z) \in G^1$
  If the property is declared as functional in the ontology, and i_1 already have a matching instance z for the property P in a valid KB, then one of these two facts holds: (1) i_2 and z refer to the same world entity (should be linked with *owl:sameAs*) and, accordingly, $P(i_1,i_2)$ does not bring new knowledge or (2) $P(i_1,i_2)$ is a wrong fact.

- if P is inverse functional then $\neg \exists\ z$ s.t. $P(z,i_2) \in G^1$
  Similar reasoning.

A property instance $P(i,l)$ between a concept instance i and a literal l is added to the graph $G^{w(d)}$ if:

- $V_P^d\ (i,l)$
- $\neg \exists\ P(i,l) \in G^p$ s.t. $p > w(d)$
- if P is functional then $\neg \exists\ m$ s.t. $P(i,m) \in G^1$

A property instance $P(x,y)$ between a concept instance x and an instance or a literal y is added to the graph $G^{w(d)}$ if:

- $\neg\exists\ P(x,y) \in G^p$ s.t. $p > w(d)$
- if P is functional $\neg\exists\ z$ s.t. $P(x,z) \in G^1$
- if P is inverse functional then $\neg\exists\ z$ tq. $P(z,y) \in G^1$
- $\exists\ P'$ s.t. $subPropertyOf(P',P)$ and $P'(x,y) \in G^{w(d)}$

  if a subproperty of P links x and y in $G^{w(d)}$, $P(x,y)$ is added to $G^{w(d)}$. Property instances are spread with the same weight according to the *rdfs:subPropertyOf* links. The probability of belonging to a subset $A \subset B$ is inferior or equal to the probability of belonging to B ([16]).

The enrichment process consists in constructing the weighted property instances and their corresponding neighborhood named graphs until a distance threshold is reached. Shortest neighboring distances are used first. Figure 4 shows an example of neighborhood graphs generated by the enrichment module according to the extracts presented in figure 2. Here, the weights were computed with $\alpha=0.9$ and a distance threshold $=2$ (structural distances 0, 1 and 2).

```
@prefix graphs: <http://lri.fr/reisa/graphs/>
graphs:candidates.distance.0 {
kimkb:Mekong.0 dbpedia:country kimkb:Laos.0
}
graphs:candidates.distance.1 {
kimkb:Mekong.0 dbpedia:country kimkb:Vietnam.0
}
graphs:candidates.distance.2 {
kimkb:Hanoi.0 kim:capital kimkb:Vietnam.0
}
graphs:candidates.distance.0 sim:gweight 0.9
graphs:candidates.distance.1 sim:gweight 0.45
graphs:candidates.distance.2 sim:gweight 0.3
```

Fig. 4. Example of three neighboring graphs constructed by enrichment

In this example, REISA generated three candidate property instances. For example, the property *kim:capital* is asserted between "Hanoi" and "Vietnam" which are at distance 2 in the document but not between "Vientiane" and "Vietnam" which are at distance 0. This is due to the fact that the *kim:capital* property is functional, which allowed to avoid proposing the city "Vientiane" as capital of "Vietnam", since the city is already known as the capital of another country (*Laos*) in the knowledge base (see knowledge base extract in figure 2).

4   Evaluation

In this section, we present the two experimentations performed to evaluate the REISA approach. The aim of these experiments is to evaluate the precision of the facts produced by our enrichment method according to (i) their weight (or neighboring distance) and (ii) the impact of the semantic controls exploiting the *KBs*.

4.1   First Experimentation

Knowledge Base. We constructed a reference knowledge base with a weight value of 1, noted KB_1, by combining extracts from the DBLP-RDF knowledge base[4] which describes bibliographic references and extracts from the WKB knowledge base of KIM[5] related to cities, countries and locations. DBLP-RDF uses the *SWRC* and *FOAF* ontologies. *WKB* uses the *PROTON* ontology. In our experiments we used the union of these ontologies[6]. In order to represent the domain links between the conferences and their locations, we defined the property *reisa:hasLocation* and its functional sub-properties *reisa:hasCountry* and *reisa:hasCity*. We considered the following instances in KB_1:

- Proceedings of events held between 2003 and 2007 in DBLP RDF (5608 instances of *swrc:Proceedings* and their descriptions).
- Cities, countries and locations described in KIM *WKB* and their descriptions (12484 instances of *kim:Location*, 3093 instances of *kim:City*, 502 instances of *kim:Country*).
- 5608 instances of *reisa:Event* constructed automaticaly from the *swrc:Proceedings* instances and described by a date (the proceeding date), a title (the title of the corresponding series) and potentially a city and a country, when the information is available in the DBLP title of the conference.

Annotation Base. We extracted 511 HTML documents from 57 call-for-paper web sites[7]. In these documents, we focused on retrieving the locations of the conferences as this information is not always available in KB_1. The documents were annotated with the KIM platform in order to extract references to cities, countries and other locations. We extracted references to events and their dates using lexico-syntactic patterns that we defined manually. This annotation allowed to extract 1429 document entities referring to scientific events, 840 entities referring to cities and 1618 entities referring to countries. In the 1429 document entities referring to events, 348 refer to instances from KB_1 and 1081 refer to new instances created in the annotation step. These new instances are collected in a second knowledge base KB_2 which have an inferior weight value, fixed to 0.9.

Enrichment. REISA was applied using the knowledge bases KB_1 and KB_2 and the annotated corpus of 511 documents. REISA produced 187 instances of the properties *reisa:hasCountry*, *reisa:hasCity* and *reisa:hasLocation*, with a neighborhood distance threshold set to 2 and the weighting constant α set to 0.9.

Results. In the following, we give the precision of the relation instances in the enrichment base for the functional properties *reisa:hasCountry* and *reisa:hasCity* and the property *reisa:hasLocation* (i.e. the number of correct facts divided by the number of retrieved facts). The correctness of the retrieved locations were

---

[4] http://thedatahub.org/dataset/fu-berlin-dblp
[5] http://www.ontotext.com/kim/semantic-annotation
[6] http://www.lri.fr/~mrabet/reisa-onto.rdf
[7] http://www.lri.fr/~mrabet/reisa-dataset.zip

verified manually in the HTML pages describing the events. We defined different configurations (cf. table 1) to evaluate the precision of the generated facts
according to the controls made with the KBs. Table 2 presents the precision
(in %) and the number of correct facts retrieved (per property, distance and
configuration).

Table 1. Configurations for the construction of the enrichment base

V	Semantic neighbors only
VR	V + redunduncy elimination
VRF (REISA)	VR + functionality-based control

Table 2. Precision (nb of correct facts) per property, distance and test

	hasCity			hasCountry		
	d=0	d=1	d=2	d=0	d=1	d=2
V	45,5 (10)	38,2 (13)	28,1(16)	72,0(18)	44,0(22)	31,4(32)
VR	36,8 (7)	35,7 (10)	21,7 (10)	58,8 (10)	33,3 (14)	22,2 (20)
VRF (REISA)	46,7 (7)	34,5 (10)	23,8 (10)	76,5 (10)	77,8 (14)	39,2 (20)

	hasLocation		
	d=0	d=1	d=2
V	60,4 (29)	42,8 (36)	30,8 (49)
VR	47,4 (18)	35,2 (25)	22,6 (31)
VRF (REISA)	47,4 (18)	35,2 (25)	22,6 (31)

The quality of the new facts varies according to the neighborhood distance. At
a distance 0, almost one fact on two is correct for the *hasCity* relation and
76.5% of the facts are correct for the *hasCountry* relation. Precision decreases
significantly when the neighboring distance increases. However, an exception is to
be noted for the *hasCountry* relation, where the precision increases from 76,5%
to 77,8% at distance 1. This is due to the fact that (i) new facts are discovered
at distance 1 and (ii) the functionality-based control filters out more wrong facts
at distance 1 than distance 0. The results also show that the functionality-based
control has an important impact on the facts precision. For example, for the
property *hasCountry*, this control increases the precision from 58,8% to 76,5%
at distance 0. Table 3 presents the percentage of new facts (not existing in KB_1
and KB_2) produced by the REISA approach among the number of retrieved facts
(i.e. facts of the enrichment base). The proportion of new facts varies between
55 and 70% of the *hasCity*, *hasCountry* and *hasLocation* instances generated
by enrichment at distance $d = 0$.

4.2   Second Experimentation

In a second experimentation on the geographic domain, we used Wikipedia articles and DBpedia as a reference knowledge base. We constructed the KB automatically with a SPARQL query submitted to DBpedia and considered all the 27

Table 3. % of new facts among the retrieved facts (VRF/V)

hasCity			hasCountry			hasLocation		
d=0	d=1	d=2	d=0	d=1	d=2	d=0	d=1	d=2
70,0%	76,9%	62,5%	55,6%	63,6%	62,5%	62,1%	69,4%	63,3%

documents corresponding to mountain ranges of France. We used Wikipedia surface forms, in a similar approach to [6], in order to annotate the documents with concept instances. A surface form is the text of hyperlinks to other Wikipedia documents. As each wikipedia page is associated to one concept instance from DBpedia, a *sim:refersTo* link is asserted between the surface form and the DBpedia instance associated to the document pointed by the hyperlink.

In this experimentation, we targeted the 16 relations from the DBpedia ontology related to the concepts *NaturalPlace*, *PopulatedPlace* and their subclasses. These relations consist mainly in the relations expressed in the ontology between the concepts (i) *BodyOfWater* and *City/Country* (e.g. source country, country, mouthPlace) (ii) *BodyOfWater* and *BodyOfWater* (e.g. branchOf, hasJunctionWith) and (iii) *City* and *Country* (e.g. capital, country). REISA was applied on this experimental setting with a distance threshold set to 10 and a weight constant α set to 0.9. With this distance threshold, REISA produced relation instances for 6 relations with 3 relations having an insufficient number of candidates for an effective evaluation. The three remaining relations are (i) *mouthCountry* and *sourceCountry* which indicate the mouth and source countries of a river and (ii) *BodyOfWater/Country* which indicates the country of a body of water. A total of 429 relation instances was produced by REISA for these three relations.

Like in the first experiment, the precision of candidate relation instances decreased when the distance threshold increased. The overall precision of all relations is 71,42% at distance 2 (no relation instances were produced with distances 0 and 1) and decreased progressively to 49,54% at distance 10. With a distance threshold of 2, constructing the enrichment base with the semantic neighboring only (configuration *V*) allowed to reach 81,8% precision for the *BodyOfWater/Country*, 33% precision for the *sourceCountry* and 44,4% precision for the *mouthCountry*. Constructing the enrichment base with the elimination of redundancies (configuration *VR*) deletes the correct relation instances that are already available in the knowledge base, leading to an overall precision of 43,47% in the remaining (new) facts. With the *VRF* configuration (REISA), the functionality-based control allowed to increase this precision to 50%.

5   Discussion

In a general observation, these first results show the interest of exploiting knowledge bases in controlling the enrichment process. In the first experimentation, the overall precision of new facts increased from 47,29% to 53,03% for the relations *hasCountry*, *hasCity* and *hasLocation*. In the second experimentation,

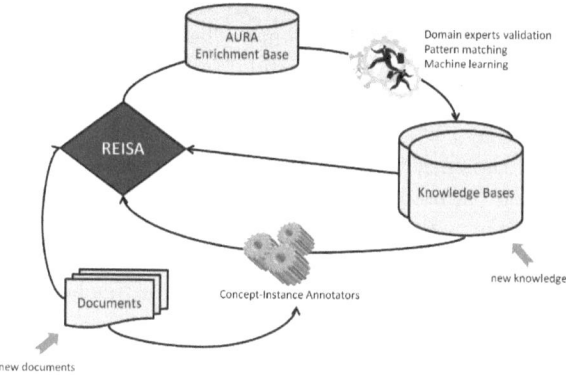

Fig. 5. Knowledge Acquisition Loop

User query	Rewrited query

SELECT ?c ?t WHERE {
 ?c rdf:type kim:Country
 ?t rdf:type kim:City
 ?t kim:capital ?c
 }
	SELECT ?c ?t WHERE {
	 GRAPH ?g1 { ?c rdf:type kim:Country }
	 GRAPH ?g2 { ?c rdf:type kim:City }
	 GRAPH ?g3 { ?t kim:capital ?c }
	 ?g1 sim:gweight ?p1
	 ?g2 sim:gweight ?p2
	 ?g3 sim:gweight ?p3
	 } ORDER BY avg(?p1,?p2,?p3)

Fig. 6. Example of SPARQL query rewriting

the overall precision of new facts increased from 43,4% to 50% for the relations *sourceCountry*, *mouthCountry* and *BodyOfWater/Country*. These semantic controls can have a bigger impact on the precision values if the used annotation tools discover more reference links between document entities and existing knowledge bases. These results could also be further enhanced if domain-level rules are applied as we plan to do in the near future (e.g. the city of an event must be part of the country of the same event). These first precision values are interesting since they were obtained without using document-level regularities. However, if the domain ontology defines several possible relations between the same couple of concepts, the current approach is expected to be less effective. A solution to this semantic heterogeneity issue may be to include it in the weighting function. For example, if two concept instances are semantic neighbors for several properties P_1, .., P_n, the number of possible relations should be taken into account to reduce the weight values.

The aim of this work is not to develop an end-to-end or domain-specific extraction approach (in which case better precision values are needed) but rather to propose a novel knowledge acquisition strategy. In the scope of this strategy, the REISA approach is complementary to "classic" pattern-based and machine-learning approaches and can be combined with them. A first combination consists in the knowledge acquisition loop as described in figure 5. In this loop, the

REISA approach produces new facts that are controlled by the existing knowledge bases using semistructured documents. Classical pattern-based or machine learning approaches or domain experts validate the enrichment bases constructed by REISA and adds the validated facts to the existing knowledge bases. These added facts are in their turn used to strengthen the semantic controls and enhance the quality of the enrichment base produced by REISA. Also, all validated facts can be exploited "immediately" to filter the candidate relations, in contrast to resolution systems (e.g. [14]) which have to be re-launched from scratch if new knowledge is available. Another combination can also be to use REISA to filter the relation candidates extracted by pattern-based or machine learning methods according to semantic and knoweldge-based criteria.

The interrogation module of REISA can also be used to perform direct querying of the enrichment base and the existing knowledge bases in a transparent manner. This module performs a query rewriting to use the named graphs weights in the answers ranking. For example, figure 6 shows a query rewriting that uses the average function to aggregate the weights of the returned facts.

6   Conclusion

In this paper, we presented the REISA approach for the controlled enrichment of RDF knowledge bases with property instances from semi-structured documents. This approach is complementary to "classical" approaches that uses lexico-syntactic patterns or automatic classifiers. Weighted property instances are generated using semistructured documents annotated with concept instances and existing knowledge bases without requiring any lexico-syntactic or structural regularity. The quality of the exploited knowledge bases is taken into account through a confidence measure associated to fact sets according to their provenance or to their construction method. We used RDF named graphs to associate weight values to fact sets. The REISA approach could be exploited to populate (semi-)automatically RDF knowledge bases by validating candidate facts with domain experts or pattern-based methods. The new validated facts can then be used to enhance the automatic controls performed by REISA. We plan now to test the effectiveness of REISA on other domains (e.g. biomedical, cinema) and to use domain-level rules in the selection/filtering of discovered properties.

References

1. Aussenac-Gilles, N., Jacques, M.-P.: Designing and Evaluating Patterns for Ontology Enrichment from Texts. In: Staab, S., Svátek, V. (eds.) EKAW 2006. LNCS (LNAI), vol. 4248, pp. 158–165. Springer, Heidelberg (2006)
2. Bikel, D.M., Miller, S., Schwartz, R., Weischedel, R.: Nymble. In: Proceedings of the Fifth Conference on Applied Natural Language Processing, Morristown, NJ, USA, pp. 194–201. Association for Computational Linguistics (March 1997)
3. Bizer, C., Lehmann, J., Kobilarov, G., Auer, S., Becker, C., Cyganiak, R., Hellmann, S.: Dbpedia – a crystallization point for the web of data. Web Semantics: Science, Services and Agents on the World Wide Web 7, 154–165 (2009)

4. Buitelaar, P., Siegel, M.: Ontology-based information extraction with soba. In: Proc. of the International Conference on Language Resources and Evaluation (LREC), pp. 2321–2324 (2006)
5. Cimiano, P., Ladwig, G., Staab, S.: Gimme'the context: Context driven automatic semantic annotation with c-pankow. In: WWW Conference (2005)
6. Gerber, D., Ngonga Ngomo, A.-C.: Bootstrapping the linked data web. In: 1st Workshop on Web Scale Knowledge Extraction, International Semantic Web Conference (1), vol. 7031. LNCS. Springer (2011)
7. Hignette, G., Buche, P., Dibie-Barthélemy, J., Haemmerlé, O.: Fuzzy Annotation of Web Data Tables Driven by a Domain Ontology. In: Aroyo, L., Traverso, P., Ciravegna, F., Cimiano, P., Heath, T., Hyvönen, E., Mizoguchi, R., Oren, E., Sabou, M., Simperl, E. (eds.) ESWC 2009. LNCS, vol. 5554, pp. 638–653. Springer, Heidelberg (2009)
8. Jain, P., Hitzler, P., Verma, K., Yeh, P.Z., Sheth, A.P.: Moving beyond sameas with plato: partonomy detection for linked data. In: Proceedings of the 23rd ACM Conference on Hypertext and Social Media, HT 2012, pp. 33–42. ACM, New York (2012)
9. Limaye, G., Sarawagi, S., Chakrabarti, S.: Annotating and searching web tables using entities, types and relationships. Proc. VLDB Endow. 3, 1338–1347 (2010)
10. Nadeau, D., Sekine, S.: A survey of named entity recognition and classification. Lingvisticae Investigationes, 3–26 (2007)
11. Popov, B., Kiryakov, A., Kirilov, A., Manov, D., Ognyanoff, D., Goranov, M.: Kim - semantic annotation platform. Journal of Natural Language Engineering 10(3), 375–392 (2004)
12. Suchanek, F.M., Ifrim, G., Weikum, G.: Combining linguistic and statistical analysis to extract relations from web documents. In: Proceedings of the 12th ACM SIGKDD International Conference on Knowledge Discovery and Data Mining, KDD 2006, New York, USA, pp. 712–717 (August 2006)
13. Suchanek, F.M., Kasneci, G., Weikum, G.: Yago: A large ontology from wikipedia and wordnet. J. Web Sem. 6(3), 203–217 (2008)
14. Suchanek, F.M., Sozio, M., Weikum, G.: Sofie: A self-organizing framework for information extraction. In: WWW Conference (2009)
15. Thiam, M., Bennacer, N., Pernelle, N., Lô, M.: Incremental Ontology-Based Extraction and Alignment in Semi-structured Documents. In: Bhowmick, S.S., Küng, J., Wagner, R. (eds.) DEXA 2009. LNCS, vol. 5690, pp. 611–618. Springer, Heidelberg (2009)
16. Zadeh, L.A.: Fuzzy sets. Information and Control, 338–353 (1965)

The Impact of Conceptualization on Text Classification

Shereen Albitar, Sébastien Fournier, and Bernard Espinasse

Université d'Aix marseille, LSIS, av. escadrille Normandie Niemen, 13397, Marseille, France
{shereen.albitar,sebastien.fournier,bernard.espinasse}@lsis.org

Abstract. Aiming at more efficient search on the Internet, it seems adequate to deploy classification techniques using semantic resources restricting this search to the user's domain of interest. In this work, we try to assess the impact of integrating semantic knowledge on text classification. This integration can be realized in different ways. The one we choose in this paper is the conceptualization. We examine the impact of the different conceptualization strategies on text classification using three traditional text classification methods: Rocchio, Support Vector Machines (SVMs) and Naïve Bayes (NB). We restrain our experimentation to the biomedical domain so conceptualization is applied on OHSUMED corpus, mapping terms in text to their corresponding concepts in UMLS Metathesaurus in order to take their meaning into consideration during text classification. Rocchio, SVMs, and NB are tested using different conceptualization strategies in order to evaluate their effect on classification. Preliminary results demonstrate promising improvements.

Keywords: Text Classification, Semantic classification, Information retrieval, Rocchio, SVMs, NB, Similarity measures, conceptualization.

1   Introduction

Nowadays and due to the explosive increase in published information on Internet, existing search engines seem to be unable to respond efficiently to user requests. This is often related to the traditional keyword-based indexing techniques neglecting search domain [1]. Aiming at more efficient and less time expensive search, it seems adequate to involve classification techniques in order to analyze the contents of search engines' answers, applying thorough filtering and ranking. Concerning information published on internet in Web pages, preprocessing treatments and content extraction are necessary to prepare this information for classification [2].

Text classification is currently a challenging research topic, particularly in areas such as information retrieval, recommendation, personalization, user profiles etc. Generally, text classification methods use syntactical and statistical models for text document representation. This applies to the most popular text classification methods such as: Naïve Bayes Classifier (NB), Support Vector Machines (SVMs), Rocchio, and so forth. Making a comparative study on the different traditional classification methods is out of the scope of this paper, for detailed comparisons please refer to [3, 4].

X.S. Wang et al. (Eds.): WISE 2012, LNCS 7651, pp. 326–339, 2012.
© Springer-Verlag Berlin Heidelberg 2012

These models suffer the lack of sense in resulting representations ignoring all semantics that reside in the original text that can help in text classification. However Vector-based (binary or TF/IDF) representations used by preceding methods permit semantic integration or "Conceptualization" that enriches document representation model using background knowledge bases [5, 6].

In this work, according to experiments, we try to estimate the impact of different text conceptualization strategies on traditional methods. These experiments are realized particularly in the biomedical domain on the OHSUMED corpus, using domain specific knowledge base UMLS. Many works have already tried to improve biomedical text representation for better classification using UMLS [7, 8] or MESH [9, 10]. To the best of our knowledge, few works investigated in details the gain of integrating these semantic resources in biomedical text classification process and more particularly, the impact of conceptualization.

Next section presents how semantic resources (thesaurus or ontologies) can be taken into account during text classification, through text conceptualization task, leading to a "semantic" text classification. In third section, we apply Rocchio (with different similarity measures) SVMs and NB methods to the original and the conceptualized OHSUMED corpus. This conceptualization is realized using UMLS (Unified Medical Language System) Metathesaurus and MetaMap tool. The forth section presents some preliminary results using these methods. Conceptualization effects on the performance of the studied methods, according to different conceptualization strategies are discussed. Then we expose some related works. Finally, we conclude with an assessment of our work, followed by different research perspectives.

2   Text Conceptualization Task

In order to overcome the previous limitations of traditional text representation models, semantic resources such as thesaurus or ontologies, can be used to replace term-based representation by a concept-based one. Thus, text classification using conceptualized vectors is so called "Semantic Classification". This section presents text conceptualization task, introducing different possible conceptualization and disambiguation strategies.

Conceptualization is the process of moving from terms literally occurring in treated text to their semantically corresponding concepts or senses in semantic resources that might permit better classification results. As an example of semantic resources that might be used for conceptualization: Wordnet, Wikipedia and other domain specific resources usually called domain ontologies such as UMLS thesaurus in the medical domain. In general, text conceptualization is realized in two steps:

- Analyze text in order to find candidate terms for term to concept mapping.
- Search for corresponding concepts related to candidate terms, and then the integration of these concepts in text producing the final conceptualized text.

2.1   Text Conceptualization Strategies

Three different strategies can be used for text conceptualization:

- *Adding Concepts*: in this strategy the original text is extended and corresponding concepts are added.
- *Partial Conceptualization*: in this strategy terms are substituted by corresponding concepts. Terms having no related concepts are reserved in the text.
- *Complete Conceptualization*: similarly to Partial Conceptualization, in this strategy terms are substituted by concepts whereas remaining terms are eliminated from the final text.

The second strategy seems to be the most appropriate one as it removes no term before replacing it with a related concept so no original feature is removed from the text (compared to the third one), and no extra feature is added (compared to the first one) resulting in minimized efficiency effects. Yet, the classification method has to be adapted to hybrid (concepts + terms) representation.

2.2   Disambiguation Strategies

While searching polysemic term corresponding concepts in semantic resources, multiple matches are detected and introducing some ambiguities in final document representation. For example: the term "Book" signifies in English a book and also a reservation (Ticket, accommodations, etc.). Three strategies for disambiguation can be used:

- *All*: this strategy accepts all candidate concepts as matches for the considered term.
- *First*: this strategy accepts the most frequently used concept among candidates using language statistics.
- *Context*: this strategy accepts the candidate concepts having the most similar semantic context compared to the term's context in the document.

The first strategy, despite being the simplest, is the least reliable as it accepts all candidate concepts without choosing a specific sense of the term. In cases where a term is used in the document signifying its rarely used sense, the second strategy gives bad decision. Despite its complexity, the last strategy seems to be more accurate as concept context can be derived from semantic resources using: concepts definition, its descriptive terms or from text corpus.

3   Platform for Conceptualized Text Classification

This section presents an experimental platform for assessing the impact of different conceptualization strategies on text classification, using three traditional text classification methods: Rocchio, SVMs and NB. This platform is illustrated in figure 1. First we present the components constituting this platform: the biomedical knowledge base

UMLS (Unified Medical Language System), and the MetaMap tool that deploys UMLS in order to realize text to concept mapping or conceptualization on the OHSUMED corpus. Conceptualized text is then transformed into vectors of terms and/or concepts depending on the used conceptualization strategy.

During Training phase the training corpus is prepared for training the classifier resulting in a classification model whereas during the Classification phase the test corpus is prepared in order to attribute classes to its document by the classifier using the learned classification model.

This architecture is modular and generic so different components can be modified and even replaced. In this work we use three different traditional classification algorithms, Rocchio, NB and SVMs, to realize Training and Classification phases.

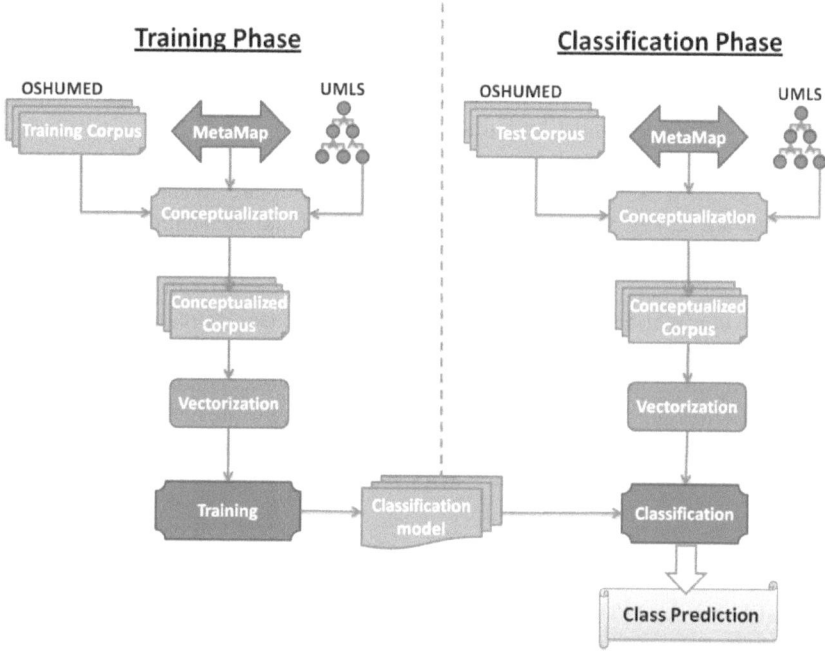

Fig. 1. The architecture of a platform for conceptualized text classification using MetaMap and UMLS

3.1 Unified Medical Language System (UMLS)

The Unified Medical Language System (UMLS) [11] was developed in order to model the language of biomedicine and health. UMLS' knowledge sources enhance the development of information systems in the biomedical domain.

The UMLS knowledge base consists of three main resources: the Metathesaurus, the Semantic Network and the SPECIALIST Lexicon. The Metathesaurus is a multilingual vocabulary database of biomedical concepts, their names, their attributes and the relations among them. This database organizes concepts of the various source vocabularies (like Mesh, SNOMED-CT, etc.) according to their senses grouping common concepts together. Concepts and relations among them are assigned at least one type from the Semantic Network. Indeed, the Semantic Network provides a higher level of abstraction through concept and relation categorization in inter-related Types constituting a network of 133 semantic types and 54 relationships. The SPECIALIST lexicon contains a large variety of general as well as medical terms and words.

3.2   MetaMap Tool

In addition to the UMLS knowledge resources, many tools are developed and provided by NLM in order to facilitate deploying these sources for medical information system developers. In this work we are interested particularly in MetaMap [12]. The major goal of MetaMap developers was to improve biomedical text retrieval using UMLS Metathesaurus. Indeed, MetaMap can discover the links between biomedical text and the knowledge in the Metathesaurus.

This mapping is the result of a rigorous linguistic analysis of each phrase of the text: First the text is tokenized and phrase boundaries are identified, then part of-speech-tags are added. Second, the Specialist lexicon and the shallow parser are used to analyze syntactically these phrases. Finally, different candidates are identified in the Metathesaurus and then final mappings combining these candidates are evaluated resulting in confidence scores for each mapping. In cases were ambiguities are detected, MetaMap keeps the most semantically similar mappings to the surrounding text following the context strategy (section 3.1). Text conceptualization using UMLS as a semantic resource, if compared to other more generic semantic resources such as WordNet, is more relevant in the medical domain. This allows the mapping between text and related specific concepts in UMLS.

3.3   OHSUMED Corpus

OHSUMED corpus [13] is composed of abstracts of biomedical articles of the year 1991 retrieved from the MEDLINE database indexed using MeSh (Medical Subject Headings) [14]. The first 20000 documents of this database were selected and categorized using 23 sub-concepts of the Mesh concept "Disease".

The corpus is divided in Training and Test sets, so experimentations are realized in two phases: Training and Test. In this work, we restricted this corpus to the five most frequent classes [15]. Training is realized on the corpus and so five class centroïds are calculated for each of the classes listed in Table 1.

Table 1. OHSUMED Corpus

Category	Description
C04	Neoplasms
C23	Pathological Conditions, Signs and
C06	Digestive System Diseases
C14	Cardiovascular Diseases
C20	Immune System Diseases

4   Experiments and Results

Using the previously presented platform, we have performed experiments in order to evaluate the effect of conceptualization on the classification process using three traditional classification methods: SVMs, Rocchio and NB.

In this section we present first the text conceptualization task performed on the OHSUMED corpus according to UMLS Metathesaurus, according to different strategies, and using the MetaMap tool. Then we present classification results obtained with each of the tree traditional classification methods (Rocchio, SVMs and NB) for each of these conceptualization strategies. Finally we analyze and discuss these results obtained.

4.1   Text Conceptualization Task

During the conceptualization task, different strategies can be implemented as previously described (adding concepts, partial conceptualization and complete conceptualization). Futhermore, according to MetaMap text-to-concept matching results, we can choose two complementary strategies:

- *Best concept strategy.* The best concept among several candidate concepts matched to the text. This depends on a matching score computed by MetaMap [12].
- *All concepts strategy.* All candidate concepts are kept.

Candidates resulting from matching have many properties. In this work we choose to use the concept name or the concept ID. In fact, during the tokenization step, the concept ID is considered as a single token so it stays intact. Concept names, being sometimes compound words, are cut during tokenization when applied on a text conceptualized using concept name strategy.

In this work, conceptualization is done using all combinations of the different strategies (12 combinations).

4.2   Classification Task

The use of MetaMap enables mapping text to UMLS concepts. Conceptualization is realized as a first step before vectorization. In our experiments, we tested all methods on the OHSUMED corpus before and after conceptualization that is realized according to twelve different conceptualization strategies.

Three traditional classification methods we used in experiments:

- Rocchio, using three different similarity measures: Cosine, Jaccard, KullbackLeibler [16].
- SVMs, using the library LIBSVM [17].
- NB, using the platform Weka [18].

In all experiments, methods are evaluated through Holdout Validation. F1-Measure [19] is considered as the criterion for performance comparison. This measure is given though the following equations:

$$\text{Precision} = \frac{t_p}{t_p + f_p} \tag{9}$$

$$\text{Recall} = \frac{t_p}{t_p + f_n} \tag{10}$$

$$\text{F1} - \text{Measure} = \frac{2 * \text{Precision} * \text{Recall}}{\text{Precision} + \text{Recall}} \tag{11}$$

Considering a specific category C:

t_p: The number of correctly classified documents in C.

t_n: The number of correctly classified documents in other categories.

f_p: The number of documents classified in C while they belong to other categories.

f_n: The number of documents classified in other categories while they belong to C

4.3   Results Analysis

Table 2 shows the F1-measure obtained applying Rocchio (with three different similarity measures: Cosine, Jaccard and KullBack), SVMs and NB to conceptualized OHSUMED corpus, for each of five categories {C04, C06, C14, C20, C23}. First we test the five classification methods on *original* OHSUMED corpus without conceptualization. Then methods are tested using each of the twelve different conceptualization strategies that are introduced previously. The two last columns present Micro and Macro averaged F-measure obtained for each pair of classification methods and conceptualization strategies. In micro-averaging, F-measure is computed globally over all category documents, whereas in macro-averaging it is equal to the average of locally calculated F-measure for each class.

Table 2. Results of applying Rocchio to conceptualized OHSUMED

Classifier	Strategy	Method	C04		C06		C14		C20		C23		Macro		Micro	
		Original	69.67%		55.61%		70.00%		57.33%		52.13%		60.95%		61.14%	
Rocchio with Cosine	AddConcept	AllConcepts	67.10%	-3.69%	51.94%	-6.61%	67.68%	-3.12%	53.20%	-7.19%	49.71%	-4.64%	57.92%	-4.96%	58.58%	-4.19%
		AllConceptsID	68.87%	1.15%	53.68%	-3.48%	69.45%	-0.78%	57.79%	+0.80%	52.22%	+0.18%	60.40%	-0.89%	60.85%	-0.47%
		BestConcept	69.71%	+0.05%	55.74%	+0.22%	70.41%	+0.59%	57.07%	-0.45%	52.68%	+1.06%	61.12%	+0.28%	61.38%	+0.39%
		BestConceptID	69.50%	-0.24%	55.17%	-0.80%	69.71%	-0.42%	58.60%	+2.22%	53.35%	+2.35%	61.27%	+0.52%	61.51%	+0.60%
	Partial	AllConcepts	66.26%	-4.89%	50.23%	9.68%	66.62%	-4.83%	52.78%	-7.93%	49.10%	-5.80%	57.00%	6.48%	57.75%	-5.55%
		AllConceptsID	68.44%	-1.77%	52.89%	-4.89%	68.97%	-1.48%	57.70%	+0.65%	51.90%	-0.44%	59.98%	-1.59%	60.47%	-1.10%
		BestConcept	69.29%	0.55%	54.36%	-2.26%	69.33%	-0.96%	55.90%	-2.49%	53.37%	+2.38%	60.45%	-0.82%	60.98%	-0.26%
		BestConceptID	62.89%	-9.74%	45.84%	-17.57%	61.94%	-11.52%	58.25%	+1.62%	53.38%	+2.41%	56.46%	7.36%	57.27%	-6.34%
	Complete	AllConcepts	66.29%	-4.86%	50.03%	-10.03%	66.52%	-4.97%	52.92%	-7.68%	49.05%	-5.90%	56.96%	-6.54%	57.71%	-5.60%
		AllConceptsID	68.44%	1.77%	52.68%	-5.27%	68.99%	-1.44%	57.63%	+0.54%	51.82%	-0.58%	59.91%	1.70%	60.42%	-1.18%
		BestConcept	69.33%	0.49%	54.38%	-2.21%	69.36%	-0.91%	55.99%	-2.32%	53.38%	+2.40%	60.49%	-0.75%	61.01%	-0.21%
		BestConceptID	62.73%	-9.96%	47.17%	-15.18%	62.18%	-11.17%	57.19%	-0.25%	53.05%	+1.77%	56.46%	-7.36%	57.15%	-6.52%
		Original	70.30%		48.81%		67.99%		55.23%		7.82%		50.03%		52.74%	
Rocchio with Jaccard	AddConcept	AllConcepts	67.34%	-4.22%	47.58%	-2.51%	64.68%	4.87%	51.34%	-7.06%	10.99%	+40.63%	48.39%	-3.29%	51.01%	-3.28%
		AllConceptsID	69.32%	-1.40%	48.65%	-0.32%	67.32%	-0.99%	53.73%	-2.72%	11.30%	+44.58%	50.06%	+0.07%	52.48%	-0.49%
		BestConcept	70.62%	+0.45%	50.69%	+3.85%	67.52%	-0.69%	55.58%	+0.64%	8.40%	+7.52%	50.56%	+1.06%	53.20%	+0.88%
		BestConceptID	69.99%	-0.44%	48.26%	-1.12%	67.18%	-1.19%	56.63%	+2.52%	5.84%	-25.27%	49.58%	-0.90%	52.34%	-0.76%
	Partial	AllConcepts	66.54%	-5.36%	46.47%	-4.79%	63.74%	-6.25%	50.68%	-8.24%	12.33%	+57.82%	47.95%	-4.15%	50.50%	-4.25%
		AllConceptsID	68.62%	-2.39%	48.03%	1.59%	67.04%	-1.40%	53.29%	-3.52%	12.33%	+57.74%	49.86%	-0.34%	52.16%	-1.09%
		BestConcept	69.29%	-1.45%	50.90%	+4.28%	66.86%	-1.66%	54.86%	-0.68%	8.98%	+14.89%	50.18%	+0.29%	52.75%	+0.01%
		BestConceptID	60.62%	-13.78%	40.30%	-17.45%	61.47%	-9.58%	56.54%	+2.36%	3.63%	-53.58%	44.51%	-11.03%	46.48%	-11.87%
	Complete	AllConcepts	66.58%	-5.30%	46.45%	-4.84%	63.72%	-6.28%	50.63%	-8.34%	12.25%	+56.79%	47.92%	-4.21%	50.46%	-4.31%
		AllConceptsID	68.49%	-2.58%	48.24%	-1.16%	66.93%	-1.55%	53.35%	-3.40%	12.56%	+60.77%	49.92%	-0.23%	52.18%	-1.06%
		BestConcept	69.28%	-1.46%	51.27%	+5.05%	66.65%	-1.97%	54.69%	-0.97%	8.98%	+14.94%	50.17%	+0.29%	52.75%	+0.03%
		BestConceptID	60.79%	-13.53%	41.23%	-15.53%	60.84%	-10.51%	56.73%	+2.72%	3.80%	-51.34%	44.68%	-10.69%	46.70%	-11.44%
		Original	69.56%		54.19%		68.92%		55.52%		18.97%		53.43%		55.01%	
Rocchio with Kullback	AddConcept	AllConcepts	69.15%	-0.58%	53.41%	-1.45%	69.65%	+1.06%	54.88%	-1.14%	27.91%	+47.15%	55.00%	+2.94%	56.07%	+1.92%
		AllConceptsID	68.94%	-0.88%	53.80%	-0.72%	69.13%	+0.30%	55.00%	-0.94%	25.86%	+36.35%	54.55%	+2.09%	55.62%	+1.11%
		BestConcept	69.93%	-0.45%	54.49%	+0.56%	68.79%	-0.19%	56.46%	+1.71%	20.63%	+8.79%	54.06%	+1.19%	55.55%	+0.99%
		BestConceptID	69.52%	-0.05%	53.12%	-1.98%	68.51%	-0.60%	55.86%	+0.62%	16.83%	-11.29%	52.77%	-1.24%	54.61%	-0.73%
	Partial	AllConcepts	68.82%	1.06%	52.96%	-2.27%	69.13%	+0.30%	54.47%	-1.88%	29.50%	+55.50%	54.98%	+2.90%	55.92%	+1.60%
		AllConceptsID	68.48%	-1.54%	53.00%	-2.20%	68.84%	-0.11%	53.72%	-2.88%	26.68%	+40.68%	54.19%	+1.41%	55.22%	+0.38%
		BestConcept	68.97%	-0.84%	53.44%	-1.38%	68.11%	-1.18%	56.24%	+1.31%	19.02%	+0.27%	53.16%	-0.51%	54.85%	-0.29%
		BestConceptID	64.12%	-7.82%	45.92%	-15.26%	64.59%	-6.29%	54.90%	-1.11%	10.70%	-43.57%	48.05%	-10.08%	50.14%	-8.84%
	Complete	AllConcepts	68.65%	-1.30%	53.14%	-1.91%	69.03%	+0.15%	54.43%	-1.95%	29.64%	+56.27%	54.98%	+2.90%	55.87%	+1.57%
		AllConceptsID	68.31%	-1.80%	53.13%	-1.95%	68.89%	-0.05%	53.80%	-3.09%	27.30%	+43.96%	54.29%	+1.60%	55.28%	+0.49%
		BestConcept	68.94%	0.89%	53.91%	-0.52%	68.21%	-1.03%	56.26%	+1.34%	20.75%	+9.39%	53.61%	+0.34%	55.19%	+0.32%
		BestConceptID	64.25%	7.64%	46.32%	-14.53%	64.91%	-5.82%	54.39%	-2.02%	12.13%	-36.04%	48.40%	-9.42%	50.46%	-8.26%
		Original	71.20%		43.00%		68.60%		64.90%		58.10%		61.16%		62.50%	
SVMs	AddConcept	AllConcepts	69.70%	-2.11%	47.70%	+10.93%	66.30%	-3.35%	64.50%	-0.62%	54.90%	-5.51%	60.62%	-0.88%	61.00%	-2.40%
		AllConceptsID	71.40%	+0.28%	53.70%	+24.88%	69.20%	+0.87%	68.70%	+5.86%	57.60%	-0.86%	64.12%	+4.84%	64.00%	+2.40%
		BestConcept	70.70%	-0.70%	47.90%	+11.40%	67.40%	-1.75%	65.30%	+0.62%	56.70%	-2.41%	61.60%	+0.71%	62.20%	-0.48%
		BestConceptID	70.30%	1.26%	46.90%	+9.07%	67.70%	-1.31%	65.50%	+0.92%	58.30%	+0.34%	61.76%	+0.95%	62.70%	+0.32%
	Partial	AllConcepts	70.30%	-1.26%	48.50%	+12.79%	67.90%	-1.02%	64.90%	+0.00%	56.10%	-3.44%	61.54%	+0.62%	62.00%	-0.80%
		AllConceptsID	71.40%	+0.28%	51.70%	+20.23%	69.40%	+1.17%	67.60%	+4.16%	58.10%	+0.00%	63.64%	+4.05%	63.90%	+2.24%
		BestConcept	70.00%	-1.69%	40.30%	-6.28%	65.20%	-4.96%	60.10%	-7.40%	57.00%	-1.89%	58.52%	-4.32%	60.30%	-3.52%
		BestConceptID	66.90%	-6.04%	14.10%	-67.21%	57.10%	16.76%	54.20%	16.49%	58.30%	+0.34%	50.12%	-18.05%	55.00%	-12.00%
	Complete	AllConcepts	70.30%	-1.26%	48.50%	+12.79%	67.90%	-1.02%	64.50%	-0.62%	56.10%	-3.44%	61.46%	+0.49%	62.00%	-0.80%
		AllConceptsID	71.60%	+0.56%	52.00%	+20.93%	69.40%	+1.17%	67.50%	+4.01%	58.20%	+0.17%	63.74%	+4.22%	64.00%	+2.40%
		BestConcept	70.00%	-1.69%	41.20%	-4.19%	65.70%	-4.23%	60.70%	-6.47%	57.40%	-1.20%	59.00%	-3.53%	60.70%	-2.88%
		BestConceptID	66.60%	-6.46%	13.90%	-67.67%	57.10%	16.76%	54.10%	16.64%	58.40%	+0.52%	50.02%	-18.21%	55.00%	-12.00%
		Original	64.80%		46.80%		65.70%		51.30%		29.30%		51.58%		49.40%	
NB	AddConcept	AllConcepts	65.30%	+0.77%	50.50%	+7.91%	67.50%	+2.74%	53.70%	+4.68%	29.80%	+1.71%	53.36%	+3.45%	50.70%	+2.63%
		AllConceptsID	64.00%	-1.23%	52.50%	+12.18%	67.00%	+1.98%	54.60%	+6.43%	29.80%	+1.71%	53.58%	+3.88%	50.60%	+2.43%
		BestConcept	65.00%	+0.31%	48.20%	+2.99%	66.10%	+0.61%	54.10%	+5.46%	28.60%	-2.39%	52.40%	+1.59%	49.70%	+0.61%
		BestConceptID	65.70%	+1.39%	50.20%	+7.26%	66.10%	+0.61%	56.10%	+9.36%	32.70%	+11.60%	54.16%	+5.00%	51.70%	+4.66%
	Partial	AllConcepts	64.90%	+0.15%	49.60%	+5.98%	67.40%	+2.59%	53.70%	+4.68%	29.40%	+0.34%	53.00%	+2.75%	50.30%	+1.82%
		AllConceptsID	63.20%	-2.47%	51.30%	+9.62%	66.90%	+1.83%	53.20%	+3.70%	29.70%	+1.37%	52.86%	+2.48%	50.10%	+1.42%
		BestConcept	63.00%	-2.78%	45.80%	-2.14%	64.30%	-2.13%	51.40%	+0.19%	25.70%	-12.29%	50.04%	-2.99%	47.30%	-4.25%
		BestConceptID	57.50%	-11.27%	38.80%	-17.09%	60.70%	-7.61%	45.80%	-10.72%	26.50%	-9.56%	45.86%	-11.09%	44.30%	-10.32%
	Complete	AllConcepts	64.60%	-0.31%	49.60%	+5.98%	67.40%	+2.59%	52.90%	+3.12%	28.00%	-4.44%	52.50%	+1.78%	49.70%	+0.61%
		AllConceptsID	63.30%	-2.31%	51.10%	+9.19%	67.70%	+3.04%	53.00%	+3.31%	29.40%	+0.34%	52.90%	+2.56%	50.10%	+1.42%
		BestConcept	63.50%	-2.01%	45.80%	-2.14%	65.70%	+0.00%	50.90%	-0.78%	25.00%	-14.68%	50.18%	-2.71%	47.40%	-4.05%
		BestConceptID	59.00%	-8.95%	39.70%	-15.17%	62.60%	-4.72%	47.20%	7.99%	25.80%	-11.95%	46.86%	-9.15%	45.00%	-8.91%

As illustrated in the table, in most cases the original system outperforms the new system integrating conceptualization phase. In fact this applies to approximately 60% of results evaluated using the micro-average F-measure. However, after a thorough look into the results, it seems clear that the system using the similarity measure of KullbackLeibler and NB shows some amelioration. Indeed, results using Micro averaged F-measure are improved in two thirds of the cases after conceptualization.

Considering each classes independently in the results, we can observe that conceptualization improves the outcome in about 58% of cases for the class "C23". All methods, except for SVMs, when tested on the original corpus, showed the worst performance in treating this class. This difference is significant according to the McNemar [20] test on classification results considering C23's documents before and after conceptualization ($\rho \ll 0.01$). Moreover, we observe a significant improvement of the outcome in about 66% of cases for the class "C06" when using SVMs.

Concerning conceptualization strategies, one of the 12 conceptualization strategies tested in these experimentations seems to provide an improvement in most cases. Indeed, Add Best Concept strategy outperforms others as it retrieves the best candidate concept provided by MetaMap as the mapping result. Best concept names rather than IDs are used in text conceptualization. However this strategy does not present a significant improvement over the original system without conceptualization. Indeed, the gain is in most cases around 1% or less. The highest increases are obtained when using conceptualization strategies taking into account all the candidates found by MetaMap and not only the best. Thus, in some cases this increase exceeds 60%. Furthermore, largest increase in the Micro-average F-Measure values are attained using the strategy of adding Concepts applied to the NB. Previous improvements seem to be significant according to McNemar [20] test having ($\rho \ll 0.01$).

Concerning Rocchio method used with different similarity measures, the least improvement of conceptualization can be observed for the method using the similarity measure Cosine. In fact, among all Micro-averaged F-measure values for this method, only one surpasses its value when compared to system results using Cosine without conceptualization. Rocchio using Jaccard similarity measure outperforms the original system in conceptualizing text using the best concept name according to three different strategies: (AddConcepts, Partial and Complete) conceptualization. Considering Rocchio with Kullback, the improvement is obtained in the case of class "C23" with all strategies (Partial, Addconcept, or Complete). Moreover, only the strategy Add Bestconcept improves the original method in four of five cases.

Concerning SVMs is improved in most cases when using *AllConceptID* regardless the strategy (AddConcept, Partial or Complete).

Considering NB, an improvement is observed in almost all cases when the strategy is *AddConcept*. This improvement occurs in approximately 58% of cases.

4.4   Discussion

According to the results presented in the preceding section, here we list some remarks. First of all, in most case, lowest results are observed when terms are replaced by IDs of their corresponding concepts in the UMLS. This performance degradation might be principally related to replacing all terms corresponding to a concept by its ID; only the IDs of concepts can participate in vectorization. Terms that are shared among concept with different IDs are excluded from vectors even if they had a high importance.

Second, when the system performance has a good F1-measure value (i.e. exceeds 60%), no significant effect can be observed for the integration of the conceptualization task into the system. In fact, as the same similarity measures are used for both cases with/without conceptualization, results' amelioration was limited.

Third, when the system performance using a specific similarity measure has a low F1-measure value, as it the case for the class "C23" or "C06" in the case of SVMs, introducing conceptualization can significantly improve this value with a maximum gain reaching (60%) in some cases. Indeed, the class "C23" is very large compared to others and so enriching class representation by semantics might result in a better identification of this class and also in better results.

Fourth, the best conceptualization strategy is Addconcept adding the Best concept among mapped candidates into the text. In fact, best mappings retrieved by MetaMap are added into text in order to enrich it with semantics avoiding any information loss.

Finally, even if the results are still preliminary, it seems useful to introduce semantic enrichments to classification methods in order to ameliorate their predictions. However, these improvements are relatively dependent on the behavior of the method and also on used corpus and its class distribution. Consequently, it seems necessary to experimentally define the conditions under which the introduction of semantics can improve classification. Moreover, the exploitation of semantic resources was limited in this work. For example, it ignores all relations (like Subsumption and Transversal relations) among concepts used in the conceptualization task. Thus, it seems adequate to deploy these relations in the classification process.

5   Related Work

Text classification is a challenging task due to the sparse and high dimensional feature space. Moreover, the complex nature of semantics residing in text makes text classification so difficult due to ambiguities that are tricky to resolve. Most of these difficulties are related to the widely used text representation model (VSM) that is unable to extract useful and meaningful features from text.

Considering Kernel based classifiers like SVMs, many works define kernel function on the knowledge base hierarchy producing semantic kernels. Séaghdha [21] proposes multiple semantic kernel functions on WordNet and proves that SVMs work better with semantic kernels.

Many works proposed new extensions to the traditional VSM in order to overcome its limitations. Numerous weighting schemes for the traditional VSM are proposed in [22], all aiming at optimizing feature weights, which might ameliorate text classification. Moreover, other works demonstrate some improvements by aide of new feature extraction methods. In order to overcome the VSM's limitation that is related to composed words, authors in [23] propose a Bag of Phrases (BOP) model instead of the traditional Bag of Words (BOW) taking frequently occurring N-gram phrases into account during feature extraction. According to tests on KNN, Decision Trees, SVMs and NB, the proposed BOP outperforms the original BOW. Despite the improvements demonstrated in this work, few of VSM limitations are treated.

In order to take into account ambiguities and polysemous words beside the previous limitations, background knowledge bases are frequently used. Indeed, thesaurus and domain specific ontologies can help in determining the accurate semantics residing in text. According to these approaches, the original BOW is transformed into Bag of Concepts (BOC) [24] so resulting vectors are constituted of text-related concepts. These concepts can be discovered in the original text by means of background knowledge. It is also important to incorporate related concepts to those already adopted through conceptualization step. This deployment enriches document representation and might improve classification results.

Authors in [24] present a new representation model using concepts extracted from background knowledge. AdaBoost algorithm is tested on three different corpora in order to support the approach. During the experimentations on the corpus Reuters-21578, Wordnet is used as the background knowledge whereas MESH ontology is used with OHSUMED dataset. Different strategies of sense disambiguation were used. Moreover, the superconcepts of specific concepts discovered in text are also integrated into the vector of concept representing text documents. These superconcepts are searched up to a maximal distance into the ontology. Deploying superconcept in document representation model is called generalization which helps in improving results especially with general purpose ontologies like WordNet. Nevertheless, this strategy does not seem to be adequate when dealing with domain specific knowledge bases like MESH.

Bai, Wang [25] align three general purpose ontologies: WordNet, OpenCyc and SUMO in their system and use the resulting knowledge base. The traditional BOW model is then replaced by BOC through new ontological indexing of text documents by means of this knowledge base. The context strategy is used in order to resolve ambiguities. Text classification using SVMS is realized on three corpora: Reuters-21578, OHSUMED and 20Newsgroups. Significant improvements are demonstrated especially with OHSUMED data set. Authors conclude that integrated concepts are helpful for special domain datasets.

Guisse, Khelif [26] propose also a BOC approach with generalization using weight propagating algorithm in order to attribute appropriate weights to superconcepts in the domain specific ontology. This work demonstrated significant improvement in patent classification.

New representation models using parts of ontology hierarchy are also proposed in the literature. These parts constitute semantic trees [27] or forests [28] where each concept is assigned an importance score. Using the semantic hierarchy of WordNet, significant improvement is demonstrated in classifying Yahoo! document [27]. Concept Forests [28], that are constituted of parts of WordNet, help in improving classification results when tested on Reuters-21578.

According to the literature, Authors seem to disagree in assessing the importance of using semantic information in classification [29]. Nevertheless, it seems to be a promising approach taking into consideration the particular context of the classification task [6].

6   Conclusion and Perspectives

Due to the explosive growth of published data, many search engines demonstrate poor performance that cannot meet the needs of users. This leads to a challenging need for efficient filtering, ranking and classification techniques. This paper concerns text classification that is currently a challenging research topic, particularly in domains as information retrieval, recommendation, personalization, user profiles etc.

Generally, text classification uses syntactical and statistical models for text document representation as it is the case for traditional methods as Naïve Bayes Classifier (NB), Support Vector Machines (SVMs), Rocchio. Representations resulting from these models suffer the lack of sense ignoring all semantics that reside in the original text that can help in text classification. Vector-based representation used in these methods permits the integration of semantics through "Conceptualization" that enriches document representation.

This work, depending on experiments, tries to estimate the impact of different text conceptualization strategies on traditional classification methods. These experiments are realized in the biomedical domain and particularly on the OHSUMED corpus, using domain specific knowledge base UMLS. Three traditional classification methods are chosen for these experiments: Rocchio (using different similarity measures), SVMs and NB. Indeed, the results of our experiments show, in some cases, considerable performance improvements. We also observed that these improvements are related to many factors: the classification method, the corpus, the class and the conceptualization strategy.

Finally, it seems useful to integrate semantic knowledge into text representation in order to ameliorate text classification. Nevertheless, experimental analysis is necessary to determine if this integration is appropriate and in which conditions it might be applied. A combination of the original and the enriched representation model can be used in order to combine their advantages and overcome their limits. In addition, other useful information contained in semantic resources such as relations can also be deployed in decision making leading to semantic classification that can improve search in restricted domains.

References

1. Asirvatham, A.P., Ravi, K.K.: Web page classification based on document structure (2001)
2. Ferreira, R., et al.: Improving News Web Page Classification Through Content Extraction. In: IADIS International Conference WWW/Internet 2011 (2011)
3. Sebastiani, F.: Machine learning in automated text categorization. ACM Computer. Survey 34(1), 1–47 (2002)
4. Joachims, T.: Text Categorization with Support Vector Machines: Learning with Many Relevant Features. In: Nédellec, C., Rouveirol, C. (eds.) ECML 1998. LNCS, vol. 1398, pp. 137–142. Springer, Heidelberg (1998)
5. Hotho, A., Staab, S., Stumme, G.: Text clustering based on background knowledge (2003)
6. Ferretti, E., Errecalde, M., Rosso, P.: Does Semantic Information Help in the Text Categorization Task? Journal of Intelligent Systems 17, 91–107 (2008)
7. Garla, V.N., Brandt, C.: Ontology-guided feature engineering for clinical text classification. J. Biomed. Inform. (in press)
8. Yetisgen-Yildiz, M., Pratt, W.: The effect of feature representation on MEDLINE document classification. In: AMIA Annu. Symp., pp. 849–853 (2005)

9. Zhang, X., Jing, L., Hu, X., Ng, M., Zhou, X.: A Comparative Study of Ontology Based Term Similarity Measures on PubMed Document Clustering. In: Kotagiri, R., Radha Krishna, P., Mohania, M., Nantajeewarawat, E. (eds.) DASFAA 2007. LNCS, vol. 4443, pp. 115–126. Springer, Heidelberg (2007)

10. Camous, F., Blott, S., Smeaton, A.F.: Ontology-Based MEDLINE Document Classification. In: Hochreiter, S., Wagner, R. (eds.) BIRD 2007. LNCS (LNBI), vol. 4414, pp. 439–452. Springer, Heidelberg (2007)

11. Unified Medical Language System (UMLS®), http://www.nlm.nih.gov/research/umls/

12. Aronson, A.R., Lang, F.M.: An overview of MetaMap: historical perspective and recent advances. J. Am Med. Inform. Assoc. 17(3), 229–236 (2010)

13. Hersh, W., et al.: OHSUMED: an interactive retrieval evaluation and new large test collection for research. In: 17th Annual International ACM SIGIR Conference on Research and Development in Information Retrieval, pp. 192–201. Springer-Verlag New York, Inc., Dublin (1994)

14. Medical Subject Headings (MeSH®), http://www.nlm.nih.gov/pubs/factsheets/mesh.html

15. Yi, K., Beheshti, J.: A hidden Markov model-based text classification of medical documents. J. Inf. Sci 35(1), 67–81 (2009)

16. Huang, A.: Similarity measures for text document clustering. In: Sixth New Zealand Computer Science Research Student Conference, Christchurch, New Zealand, pp. 49–56 (2008)

17. Chang, C.-C., Lin, C.-J.: LIBSVM: A library for support vector machines. ACM Trans. Intell. Syst. Technol. 2(3), 1–27 (2011)

18. Hall, M., et al.: The WEKA data mining software: an update. SIGKDD Explor. Newsl. 11(1), 10–18 (2009)

19. Sokolova, M., Lapalme, G.: A systematic analysis of performance measures for classification tasks. Information Processing Management 45(4), 427–437 (2009)

20. Dietterich, T.G.: Approximate statistical tests for comparing supervised classification learning algorithms. Neural Comput. 10(7), 1895–1923 (1998)

21. Séaghdha, D.O.: Semantic classification with WordNet kernels. In: Proceedings of Human Language Technologies: The 2009 Annual Conference of the North American Chapter of the Association for Computational Linguistics, Companion Volume: Short Papers, pp. 237–240. Association for Computational Linguistics, Boulder (2009)

22. Lan, M., et al.: Supervised and Traditional Term Weighting Methods for Automatic Text Categorization. IEEE Trans. Pattern Anal. Mach. Intell. 31(4), 721–735 (2009)

23. Li, Z., Li, P., Wei, W., Liu, H., He, J., Liu, T., Du, X.: AutoPCS: A Phrase-Based Text Categorization System for Similar Texts. In: Li, Q., Feng, L., Pei, J., Wang, S.X., Zhou, X., Zhu, Q.-M., et al. (eds.) APWeb/WAIM 2009. LNCS, vol. 5446, pp. 369–380. Springer, Heidelberg (2009)

24. Bloehdorn, S., Hotho, A.: Boosting for Text Classification with Semantic Features. In: Mobasher, B., Nasraoui, O., Liu, B., Masand, B. (eds.) WebKDD 2004. LNCS (LNAI), vol. 3932, pp. 149–166. Springer, Heidelberg (2006)

25. Bai, R., Wang, X., Liao, J.: Using an Integrated Ontology Database to Categorize Web Pages. In: Kim, T.-H., Adeli, H. (eds.) AST/UCMA/ISA/ACN 2010. LNCS, vol. 6059, pp. 300–309. Springer, Heidelberg (2010)

26. Guisse, A., Khelif, K., Collard, M.: PatClust: une plateforme pour la classification sémantique des brevets. In: Conférence d'Ingénierie des connaissances, Hammamet, Tunisie (2009)

27. Peng, X., Choi, B.: Document classifications based on word semantic hierarchies. In: International Conference on Artificial Intelligence and Applications (AIA 2005), pp. 362–367 (2005)
28. Wang, J.Z., Taylor, W.: Concept Forest: A New Ontology-assisted Text Document Similarity Measurement Method. In: Proceedings of the IEEE/WIC/ACM International Conference on Web Intelligence 2007, pp. 395–401. IEEE Computer Society (2006)
29. Stein, B., Eissen, S.M.Z., Potthast, M.: Syntax versus semantics: Analysis of enriched vector space models. In: Third International Workshop on Text-Based Information Retrieval (TIR 2006). University of Trento, Italy (2006)

Efficient Execution of Web Navigation Sequences

José Losada, Juan Raposo, Alberto Pan, and Paula Montoto

Information and Communications Technology Department, University of A Coruña
Facultad de Informática, Campus de Elviña, s/n, 15071, A Coruña, Spain
{jlosada,jrs,apan,pmontoto}@udc.es

Abstract. Web automation applications are widely used for different purposes such as B2B integration and automated testing of web applications. Most current systems build the automatic web navigation component by using the APIs of conventional browsers. While this approach has its advantages, it suffers performance problems for intensive web automation tasks which require real time responses and/or a high degree of parallelism. In this paper, we outline a set of techniques to build a web navigation component able to efficiently execute web navigation sequences. These techniques detect what elements and scripts of the pages accessed during the navigation sequence are needed for the correct execution of the sequence (and, therefore, must be loaded and executed), and what parts of the pages can be discarded. The tests executed with real web sources show that the optimized navigation sequences run significantly faster and consume significantly less resources.

Keywords: Web Automation, Navigation Sequence, Optimization, Efficient Execution.

1   Introduction

Most today's web sources do not provide suitable interfaces for software programs. That is why a growing interest has arisen in so-called web automation applications that are able to automatically navigate through websites simulating the behavior of a human user. For example, a flight meta-search application can use web automation to automatically search flights in the websites of different airlines or travel agencies. Web automation applications are widely used for different purposes such as B2B integration, web mashups, automated testing of web applications, Internet meta-search or technology and business watch.

A crucial part of web automation technologies is the ability to execute automatic web navigation sequences. An automatic web navigation sequence consists in a sequence of steps representing the actions to be performed by a human user over a web browser to reach a target web page. Figure 1 illustrates an example of a web navigation sequence to access to the content of the first message in the Inbox folder of a Gmail account.

This work is focused in improving the performance of the execution of automatic web navigation sequences. The approach followed by most of the current web automation systems [5] [8] [9] [11] [12] consists in using the APIs of conventional web

X.S. Wang et al. (Eds.): WISE 2012, LNCS 7651, pp. 340–353, 2012.

browsers to automate them. This approach does not require to develop a custom navigation component, and guarantees that the accessed web pages will behave the same as when they are accessed by a regular user.

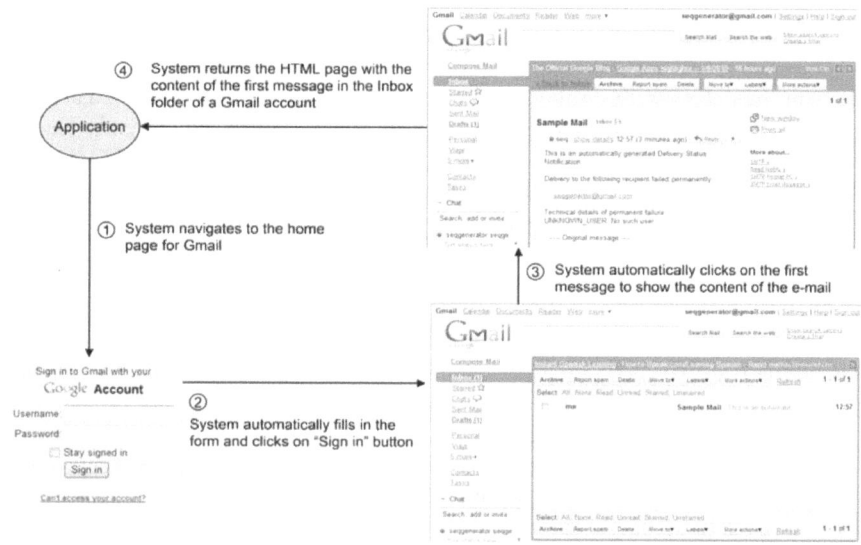

Fig. 1. Navigation Sequence Example

While this approach is adequate to some web automation applications, it presents performance problems for intensive web automation tasks which require real time responses and/or to execute a significant number of navigation sequences in parallel. This is because commercial web browsers are designed to be client-side applications and, therefore, they consume a significant amount of resources, both memory and CPU. In this work we address this problem by using a custom browser specially built for web automation tasks. This browser is able to improve the response times and save a significant amount of resources (memory and CPU). We present a set of techniques and algorithms to automatically optimize the navigation sequences, detecting which parts of the accessed pages can be discarded (not loaded), and which of the automatic events that are fired each time a new page is loaded can be omitted (not fired) without affecting to the correct execution of the navigation sequence.

There exist other systems which use the approach of creating custom browsers to execute web navigation sequences [4] [6]. Since they are not oriented to be used by humans, they can avoid some of the tasks of conventional browsers (e.g. rendering). Nevertheless, they work like conventional browsers when loading and building the internal representation of the web pages. Since this is the most important part in terms of the use of computational resources, their performance enhancements are much smaller than the ones achieved with our approach.

The rest of the paper is organized as follows. Section 2 briefly describes the models our approach relies on. Section 3 presents an overview of the solution. Section 4 explains the designed techniques in detail. Section 5 describes the experimental evaluation of the approach. Section 6 discusses related work. Finally, section 7 summarizes our conclusions and future work.

2   Background

The main model we rely on is the Document Object Model (DOM) [3]. This model describes how browsers internally represent the HTML web page currently loaded in the browser and how they respond to user-performed actions on it. An HTML page is modelled as a tree, where each HTML element is represented by an appropriate type of node. An important type of nodes are the script nodes, used to place and execute a script code within the document (typically written in a script language such as JavaScript). The script nodes can contain the code directly or can reference an external file containing it. Those scripts are processed when the page is loaded and they can contain element declarations (e.g. a function or a variable used from other nodes).

In addition, every node in the tree can receive events produced (directly or indirectly) by the user actions. Event types exist for actions such as clicking on an element (*click*), moving the mouse cursor over it (*mouseover*), or to indicate that a new page has just been loaded (*load*), to name but a few. Each node can register a set of listeners for different types of events. Each event is dispatched following a path from the root of the tree to the target node (capture phase) and then from the target node to the root of the tree (bubbling phase), and it can be handled locally at the target node or at any target's ancestor in the tree (at the capture or bubbling phase). An event listener executes arbitrary script code, which normally calls a function declared in script nodes. The scripting code has the entire page DOM tree accessible and can perform actions such as modifying existing nodes, removing them, creating new ones or even launching new events.

In addition to the events caused by the user actions on the page, there are also some events that are automatically generated by the browser when a new page is loaded. The most typical example is the *load* event, which is fired by the browser over the *body* element of the HTML page when the page has just been loaded. We will name these events as "automatic events".

3   Overview

This section presents an overview of our proposal.

The input for the automatic web navigation component is a navigation sequence specification. In most systems, this specification is created by example: the user performs the desired sequence manually and her actions are recorded by some plugin in the browser. The exact format used to specify navigation sequences is different in each web automation system but all of them basically consist in a list of events which must be generated on certain elements of the website pages. Between executing one event and the next, it is needed to wait for the effects of the previous event to take place (e.g. wait for a new page to be loaded in the browser). See [7] for a discussion of the different approaches for recording and executing web navigation sequences.

The basic idea of our approach consists in detecting which parts of the accessed pages can be discarded (not loaded) and which events can be omitted (not fired) without affecting the execution of the desired navigation sequence. Our approach works in two phases:

- In the optimization phase the navigation sequence is executed once, and, in the meantime, the navigation component automatically calculates which nodes of the HTML DOM [3] tree of each loaded page are needed to execute the sequence, and which ones can be discarded. Then, it stores some information to be able to detect those nodes in subsequent sequence executions (the information to identify the nodes should be resilient to small changes in the page). At the same time, the navigation component calculates which of the automatic events fired each time a page is loaded are necessary to execute the sequence.
- In the execution phase the navigation component executes the sequence using the optimization information previously calculated. When each page is loaded, a reduced HTML DOM tree is built, containing only the relevant nodes needed to execute the sequence, and only the necessary automatic events are fired.

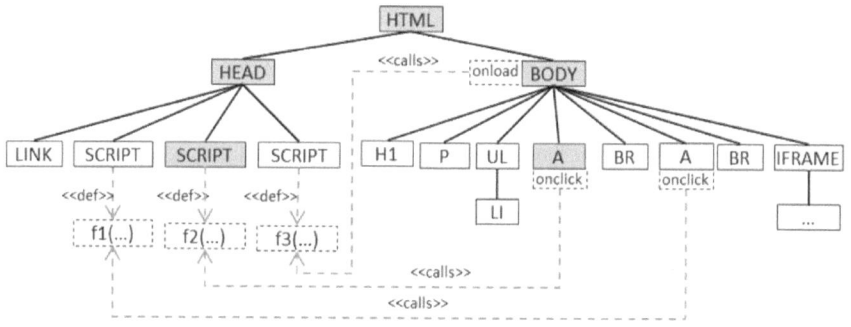

Fig. 2. DOM tree of an example page

Figure 2 shows the DOM tree of a simple example page. We use boxes to represent the nodes of the tree, and continuous lines to represent its parent-child relationship. Event listeners are represented as dashed boxes adjacent to the corresponding tree node (*onclick*, *onload*). Arrows with dashed lines are used to indicate the functions defined in script nodes (*<def>*), and the functions defined in script nodes which are invoked from event listeners (*<calls>*). Suppose that the only action specified by the navigation sequence for this page is executing a *click* on the first A node. When the *click* event is produced, the *onclick* event listener is executed, and the function *f2* performs a navigation to the desired page (e.g. *window.location* = *'http://acme.com';*).

The shaded nodes are those that are needed to simulate the *click* action and properly perform the navigation to the next page (we call them *relevant* nodes). In this case, the relevant nodes are: the A node which is the target of the *click* event, the *SCRIPT* node which defines the *f2* function executed by the *onclick* event listener, and their respective ancestors (the exact rules to compute the relevant nodes will be described later).

The rest of the nodes can be discarded (not loaded) without any problem (we call these ones *irrelevant* nodes). Besides, the automatic *load* event does not need to be fired when the page is loaded, since the code of the *onload* listener is not needed for the execution of the sequence.

This will produce significant performance and resource usage improvements:

- We will save memory, since much less nodes need to be represented.
- We will save CPU and execution time since unneeded scripts are not executed. For instance, in this case, the script nodes not shaded do not need to be executed.
- We will save bandwidth and execution time because unneeded navigations are not performed. For instance, in this case, the navigations specified by the *LINK* and *IFRAME* nodes will not be performed.

The main problem we need to address is how to calculate what we call *node dependencies*. For instance, in this example the *SCRIPT* node which defines *f2* is a dependency of the *A* node when the *click* event is fired on it. Notice that in the DOM model, scripts are "black boxes" and, therefore, these dependencies cannot be inferred directly. By using a custom browser, where we have full control over the script execution engine, we have a way to uncover these hidden dependencies.

Also notice that dependencies can get much more complex than in this example. For example, in the previous figure, a *click* on an anchor may produce the execution of a script that requires another script in a different node in the DOM tree to be executed previously. Another difficult example would be that the *load* event listener of the *BODY* node could generate content dynamically, including the *A* node that invokes the script that will lead us to the next page. It could even happen that the script required another script contained in an iframe and, therefore, the iframe would need to be loaded too. We will see how to deal with these problems in the next section.

4   Proposed Techniques

In this section we begin stating some definitions and properties which will help us to model all the possible dependencies between the DOM tree nodes we are interested in (section 4.1). After that, we describe the techniques used during the optimization phase of our approach, (section 4.2). Then, we briefly explain the method used to generate expressions to identify the irrelevant nodes at the execution phase (section 4.3). Finally we outline the operation at the execution phase (section 4.4).

4.1   Node Dependencies

Definition 1: We say that there exists a dependency between two nodes *n1* and *n2* when the node *n2* is necessary for the correct execution of the node *n1*. We say that the node *n2* is a dependency of the node *n1* and denote it as *n1*→ *n2*. The following rules define this type of dependencies:

- If the script code of a node *s1* uses an element (e.g. a function or a variable) declared in a script node *s2*, then *s1*→ *s2*. Rationale: to be able to execute the script code of the node *s1* the node *s2* must be executed previously.
- If the script code of a node *s* uses a node *n*, then *s*→ *n*. Rationale: to be able to execute the script code of the node *s*, the node *n* must be loaded previously. For instance, if *s* obtains a reference to an *anchor* node (e.g. using the JavaScript function *document.getElementById*) and navigates to the URL specified by its *href* attribute, then it will not be possible to execute *s* unless the *anchor* node is loaded.
- If the script code of a node *s* makes a modification in a node *n*, then *n*→ *s*. Rationale: the action performed by *s* may be needed to allow *n* to be used later. For instance, if *s* modifies the *action* attribute of a *form* node to set the target URL, then it will not be possible to submit the form unless *s* is executed previously.

Definition 2: We say that there exists a dependency conditioned to the event *e* being fired over the node *n*, between two nodes *n1* and *n2*, when the node *n2* is necessary for the correct execution of the node *n1*, when the event *e* is fired over the node *n*. We denote this as $n1 \rightarrow^{e|n} n2$. Analogous rules to the ones explained before define this type of dependencies, which, in this case, involve nodes containing event listeners:

- If the script code of an event listener *l* for the event *e* in the node *n* uses an element (e.g. a function or a variable) declared in a script node *s*, then $n \rightarrow^{e|n} s$. Rationale: if the event *e* is fired over the node *n*, then the event listener *l* is executed, and it requires the script node *s* to be executed previously.
- If the script code of an event listener *l* for the event *e* in the node *n1* uses a node *n2*, then $n1 \rightarrow^{e|n1} n2$. Rationale: if the event *e* is fired over *n1*, then the event listener *l* is executed and the node *n2* must be loaded previously.
- If the script code of an event listener *l* for the event *e* in the node *n1* makes a modification in a node *n2*, then $n2 \rightarrow^{e|n1} n1$. Rationale: the action performed by *l* may be needed to allow *n2* to be used later. For instance, if *l* modifies the *action* attribute of a *form* node to set the target URL, then it will not be possible to submit the *form* unless *l* is executed previously. Since *l* will only be executed when the event *e* is fired over *n1*, then *n1* is needed.

Observe that the following transitivity properties apply to node dependencies (we will explain them through examples).

Property 1: If *n1*→ *n2* and *n2*→ *n3* then *n1*→ *n3*.
The example of Figure 3.a shows a fragment of the DOM tree of a page where the script code of the node *SCRIPT1* invokes a function *f1* which is defined in the node *SCRIPT2* (*SCRIPT1*→ *SCRIPT2*), and the code of function *f1* calls a function *f2* which is defined in the node *SCRIPT3* (*SCRIPT2*→ *SCRIPT3*). For the correct execution of the script code of the node *SCRIPT1*, both the second and the third *SCRIPT* nodes are necessary, so both are dependencies of it (*SCRIPT1*→ *SCRIPT3*).

Property 2: If $n1 \rightarrow^{e|n} n2$ and *n2*→ *n3* then $n1 \rightarrow^{e|n} n3$.
The example of Figure 3.b shows a fragment of a page DOM tree where the *click* event listener of the node *A* calls a function *f1* which is defined in the *SCRIPT* node

$(A\rightarrow^{click|A} SCRIPT)$, and the code of the function *f1* uses the *src* attribute of the *IMG* node $(SCRIPT\rightarrow IMG)$. For the correct processing of the *A* node when the *click* event is fired over it, both the *SCRIPT* and *IMG* nodes are necessary, so both are dependencies of it $(A\rightarrow^{click|A} IMG)$.

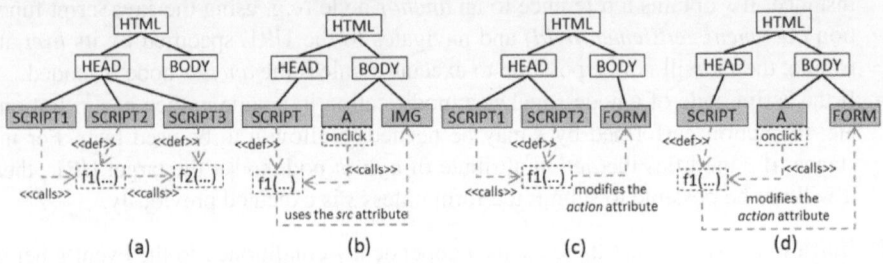

Fig. 3. Transitivity Dependency Examples

Property 3: If $n1\rightarrow n2$, and $n3\rightarrow n2$ because $n2$ is a script node which makes a modification in $n3$, then $n3\rightarrow n1$.

The example of Figure 3.c shows a fragment of a page DOM tree where the script code of the node *SCRIPT1* invokes a function *f1* which is defined in the node *SCRIPT2* $(SCRIPT1\rightarrow SCRIPT2)$, and the code of the function *f1* modifies the *action* attribute of the *FORM* node $(FORM\rightarrow SCRIPT2)$. For the correct processing of the *FORM* node (for example to correctly submitting it), we need to ensure that *f1* is both defined (and, therefore, we need *SCRIPT2*) and executed (and, therefore, we need *SCRIPT1*). That is why both are dependencies of it $(FORM\rightarrow SCRIPT1)$.

Property 4: If $n1\rightarrow^{eln} n2$ and $n3\rightarrow n2$ because $n2$ is a script node which makes a modification in $n3$, then $n3\rightarrow^{eln} n1$.

The example of Figure 3.d shows a fragment of a page DOM tree where the *click* event listener of the *A* node calls a function *f1* which is defined in the *SCRIPT* node $(A\rightarrow^{click|A} SCRIPT)$, and the code of the function *f1* modifies the *action* attribute of the *FORM* node $(FORM\rightarrow SCRIPT)$. For the correct processing of the *FORM* node (e.g. to correctly submitting it), when the *click* event is fired over the *A* node, both the *SCRIPT* and *A* nodes are necessary, so both are dependencies of it $(FORM\rightarrow^{click|A} A)$.

4.2   Calculating the Relevant Nodes and Automatic Events

The main goal of the optimization phase is finding the set of relevant nodes for the navigation sequence. During this phase, the browser works in a similar manner to a conventional browser: the full page is loaded, generating the DOM tree, downloading all external elements (e.g. style sheets, script files) and executing all the script nodes defined in the page. Also, all the automatic events (recall section 2 for the definition of automatic events) are automatically fired by the browser when each new page is completely loaded (e.g. the *load* event is fired over the *body* element). After that, the browser will reproduce the desired navigation sequence by firing the necessary events on the adequate elements to emulate the user interaction with the page (e.g. clicking on elements, firing mouse events, etc.), until a navigation to a new page is started.

During all this process, the browser interacts with the script execution engine (we use Mozilla Rhino) to detect the node dependencies, according to the rules defined in the previous section. For instance, when a *script* node is executed, the browser interacts with the scripting engine to monitor what functions are called during its execution. Then, according to the first rule of Definition 1, the nodes defining those functions are marked as dependencies of the *script* node. Similarly, if the code of the *script* node creates or modifies another node, then, according to rule 3 of Definition 1, the *script* node will be a dependency of the changed node.

In a similar way, when an event (be it automatic or generated by the navigation sequence) is fired, the browser monitors which other nodes are used during the execution of the listeners associated to the event, which other events are generated and which nodes are modified by the execution of the event listeners. The appropriate dependencies according to the rules of Definition 2 will be generated.

Once the dependencies have been computed, the set of relevant nodes is built according to the following rules:

1. The nodes which are directly used in the target navigation sequence are relevant. For instance, if one step in the sequence is generating the *click* event on a *A* node, then that *A* node is relevant

2. If a node n is relevant, all its ancestors are relevant. Note, that the ancestors could be needed because of the capture and bubbling phases of the event dispatching model of the DOM trees (see section 2).

3. By definition, if a node $n1$ is relevant and $n1 \rightarrow n2$ then $n2$ is relevant (all its dependencies are relevant too).

4. By definition, if a node $n1$ is relevant, $n1 \rightarrow^{eln} n2$, and the event e was fired over the node n, then $n2$ is relevant (all its dependencies conditioned to the event e being fired over the node n are relevant too, if the event e was fired over n).

5. Some special rules apply to *form*-related nodes, to be able to properly submit forms: (a) if a *form* node is relevant, all the nodes corresponding to *input* and *select* elements contained in the *form* are relevant, (b) if an *input* or *select* node is relevant, the *form* node containing it is relevant, (c) if a *select* node is relevant, all its child *option* nodes are relevant.

6. A small set of nodes corresponding to some special element types are always considered relevant because they are needed to properly process other nodes of the page DOM tree. For instance, the *base* element sets the *base URL*, which means that the URLs specified by other elements are relative to it.

From the set of relevant nodes, we can easily calculate the set of irrelevant nodes which will be removed at the execution phase. First, all the DOM tree nodes not contained in the set of relevant nodes are added to the set of irrelevant nodes. Then, all the irrelevant nodes which have an ancestor also contained in the set are removed from it. The resulting set contains only the root nodes of the sub-trees whose descendants are all irrelevant. We call them irrelevant sub-trees.

Finally, to determine which of the automatic events are necessary for the correct execution of the sequence, the system checks, for each automatic event, if any of the relevant nodes has any dependency derived from it (i.e. it checks if a relevant node has been affected by the listeners executed as result of firing the event). If that is the case, the event is added to the list of automatic events that should be fired at execution time when the current page is loaded.

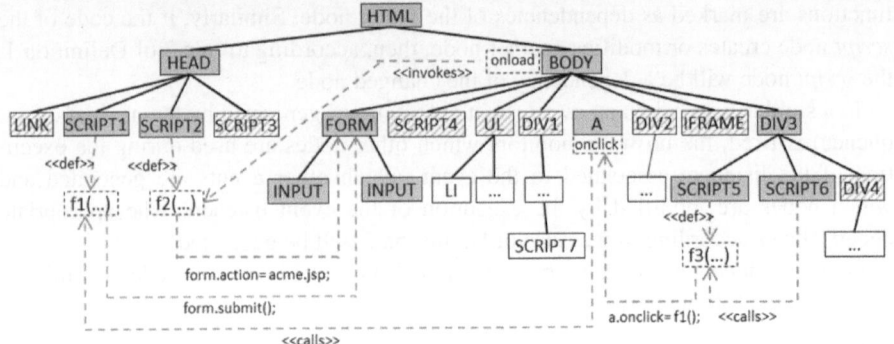

Fig. 4. Example

Let's see an example. Figure 4 shows a fragment of the DOM tree of a page. Suppose the target sequence specifies that the navigation component should execute a *click* over the *A* node. The relevant nodes for this interaction are shaded in the figure. Let's see how they are computed:

- According to rule 1, the node *A* is relevant (since it is the target of the action).
- According to rule 2, all the *A* ancestors are relevant: *BODY* and *HTML*.
- According to rule 3, all *A* dependencies are relevant: *SCRIPT5* and *SCRIPT6* (and its ancestors: *DIV3*). In this case they are needed because they execute script code which modifies the *click* event handler of the node *A* when the page is loaded.
  - The function *f3* (defined in *SCRIPT5*) modifies the *click* event handler of the node *A*, so *A →SCRIPT5*.
  - *SCRIPT6*, which is executed when the page is loaded, invokes the function *f3*, so *SCRIPT6→ SCRIPT5*, and due to the transitivity rules explained in section 2, *A→ SCRIPT6*.
- According to rule 4, all *A* dependencies conditioned to the event *click* being fired over *A* are relevant too: *SCRIPT1* and *FORM* (and all its ancestors: *HEAD*). They are needed because the event handler of the node *A* invokes a function defined in *SCRIPT1* which submits the *form*.
  - The *onclick* event handler of the node *A* invokes the function *f1* defined in *SCRIPT1*, so *A→$^{click|A}$ SCRIPT1*.
  - The function *f1* uses the node *FORM*, so *SCRIPT1→FORM*, and due to the transitivity rules explained in section 2, *A→$^{click|A}$ FORM*.

- According to rule 5, if a *form* node is relevant, all the *input* nodes contained in the form are relevant: *INPUT1* and *INPUT2*. To properly submit the form all its input fields are necessary.
- According to rule 3, all *FORM* dependencies are relevant: *SCRIPT2* and *BODY* (and all its ancestors, already included in the set of relevant nodes). They are needed because the *load* event handler of the node *BODY* invokes a function defined in *SCRIPT2* which modifies the *action* attribute of the *form*.
  - The *onload* event listener of the node *BODY* invokes the function *f2* defined in *SCRIPT2*, so $BODY \to^{load\backslash body} SCRIPT2$.
  - The function *f2* (defined in *SCRIPT2*) modifies the *action* attribute of the node *FORM*, so $FORM \to SCRIPT2$, and due to the transitivity rules explained in section 2, $FORM \to^{load\backslash body} BODY$.

The nodes which are stripped in Figure 4 are those which are identified as the roots of the irrelevant sub-trees, which can be discarded in the following executions.

The automatic event *load*, which is fired over the *BODY*, must be added to the list of necessary automatic events, because the *FORM*, which is a relevant node, has a dependency derived from it ($FORM \to^{load\backslash body} BODY$). Note that, to properly submit the form, the *load* event listener of the *body* element (*onload*) must have been executed, because it invokes *f2* which sets the action of the form.

4.3   Identifying the Irrelevant Subtrees at Execution Phase

Once the root nodes of the irrelevant sub-trees have been calculated, we need to generate expressions to be able to identify them at the execution phase. There are two requirements for this process. On one hand, the generated expressions should be resilient to small changes in the page because in real web sites there are usually small differences between the DOM tree of the same page loaded at different moments (e.g. new advertisement banners can appear or different data records can be shown). On the other hand, the process of testing if a node matches an expression should be very efficient, because, at the execution phase the browser should check if each node matches with any of those expressions.

To uniquely identify a node in the DOM tree we use an XPath-like [13] expression. For our purposes, we need to ensure that the generated expression identifies a single node, but is not too specific to be affected by the aforementioned small changes in the pages. For this, we use an enhanced version of the algorithm explained in [7], which is not described in detail here due to space constraints.

4.4   Execution Phase

The general functioning of the navigation component at this phase is the following one: before loading each page, it checks if it has optimization information regarding relevant nodes associated to that page, that is, a set of expressions to identify the root nodes of the irrelevant sub-trees. That information is used to build a reduced version of the HTML DOM tree, containing only the relevant nodes. Then it checks if it has optimization information related to automatic events that should be fired in that page. If that is the case, only the appropriate events are fired.

The process of checking if a node is the root of an irrelevant sub-tree should be very efficient because it is executed for all the elements present in the page to decide if they must be added to the HTML DOM tree or not. That is why we do not use a conventional Xpath matching algorithm. Instead, we leverage on the fact that the XPath-like expressions we generate use a strict subset of Xpath and always verify certain restrictions. This allow us to use a much faster algorithm for those particular expressions. The algorithm is not described due to space constraints.

5   Evaluation

To evaluate the validity of our approach we implemented a custom browser. This browser was fully developed in Java using open-source libraries including Apache Commons-Httpclient to handle HTTP requests, Neko HTML parser to build DOM structures, and Mozilla Rhino as JavaScript engine.

This section explains the set of experiments that we have performed. We selected a set of websites of different domains included in the top 500 sites on the web according to Alexa [1]. In each website we recorded a navigation sequence representative of its main function (e.g. a product search in an e-commerce website). Every sequence executes events to fill and submit forms, to navigate through hyperlinks and, in some cases, to display content collected with AJAX requests.

In the first experiment, we compared the resources consumed by our custom browser when it uses its optimization capabilities, with the resources consumed in a normal execution (which emulates the behavior of the commercial browsers, loading the accessed pages entirely and firing all the automatic events). We ran a first execution of the navigation sequence to collect the optimization information. Then, we ran one normal execution without using the optimization information, and another one using it. Table 1 shows the metrics measured for each of the two executions in the selected websites (each cell shows the result of the normal execution followed by the result of the optimized one). To prevent the problem of small variations in web pages when they are accessed in different moments, each sequence was executed 5 times and the result is the average of the 5 executions.

Measuring the resources used in all the navigation sequences, the optimized executions only require the 18% of the nodes. Discarding those nodes, the browser also avoids unnecessary downloads and the execution of unnecessary scripts, so the memory and CPU usage, is highly minimized. The optimized executions only execute the 27% of the scripts and download the 60% of the HTML documents, the 34% of the external scripts and CSS stylesheets, and execute the 36% of the AJAX requests.

In the second experiment we compared the execution time of our custom browser using and without using its optimization capabilities, with the execution time of other representative navigation components. We used a navigation component based on another custom browser, in this case, we chose HtmlUnit [4] because it is an open source project and also supports JavaScript and CSS, and a navigation component using the APIs of a commercial web browser, in this case Microsoft Internet Explorer 9. Table 2 shows the average execution time of 20 consecutive executions of each of our test navigation sequence, discarding those that don't fit in the range of the standard deviation. The table also shows the percentage of time in comparison with the execution time of our custom browser using its optimization capabilities.

The execution time of the custom browser using its optimization capabilities always got better results. Compared with the executions without optimization, the execution time varies from 125% in the worst case to the 472% in the best case. Calculating the average of the percentages, and discarding those that do not fit in range of the average ± standard deviation (the standard deviation is 38%), the execution time of the custom browser without optimization is 2.04 times slower (204%) than the execution time with optimization. The median value of the executions indicates that the custom browser without optimization is 2.08 times slower. With the same calculations, the custom browser based on HtmlUnit is 3.15 times slower (315%) than the custom browser with optimization and the median of the executions is 3.13. In the case of the navigation component based on Microsoft Internet Explorer, the average execution time is 5.03 times slower (503%) than the execution time of the custom browser with optimization and the median of the executions is 4.87.

Table 1. Metrics comparing normal and optimized executions in some websites

	HTML DOM Nodes created	Scripts Executed	Frames and Windows	HTML pages Downloaded	External objects Downloaded	AJAX Requests
Alexa	1377 / 91	46 / 17	1 / 1	2 / 2	24 / 13	0 / 0
Amazon	8604 / 4496	223 / 77	23 / 3	22 / 5	43 / 7	5 / 2
Apple	2900 / 395	51 / 15	1 / 1	3 / 3	14 / 10	1 / 1
Barnes	8814 / 6021	73 / 41	3 / 1	6 / 3	45 / 26	0 / 0
Bloomberg	5880 / 397	204 / 46	4 / 4	5 / 5	82 / 17	1 / 0
CNET	3650 / 104	110 / 45	5 / 4	7 / 6	53 / 26	0 / 0
CNN	4167 / 97	138 / 14	9 / 1	12 / 3	46 / 7	2 / 0
EBay	4051 / 1134	80 / 51	3 / 1	6 / 4	25 / 19	0 / 0
Flickr	1519 / 98	140 / 20	1 / 1	4 / 4	97 / 8	0 / 1
GoogleNews	6408 / 102	45 / 8	1 / 1	4 / 4	6 / 1	0 / 0
Imdb	4250 / 820	213 / 47	36 / 1	23 / 4	65 / 14	7 / 4
Linkedin	2025 / 152	53 / 6	3 / 1	5 / 3	17 / 4	3 / 0
Reference	2260 / 285	112 / 20	6 / 1	7 / 2	28 / 3	0 / 0
Reuters	2446 / 277	230 / 37	6 / 2	7 / 3	143 / 31	4 / 1
Softonic	3928 / 241	89 / 22	9 / 1	11 / 3	30 / 17	0 / 0
Spiegel	3517 / 302	89 / 19	6 / 1	7 / 4	20 / 13	0 / 0
StackOverflow	4486 / 322	43 / 13	1 / 1	3 / 3	20 / 5	0 / 0
Taringa	2668 / 1251	178 / 59	9 / 1	10 / 5	58 / 26	0 / 0
TheGuardian	3924 / 332	236 / 84	4 / 2	4 / 3	82 / 35	0 / 0
TripAdvisor	9949 / 202	372 / 39	1 / 1	5 / 5	24 / 9	0 / 0
W3CSchools	2078 / 42	66 / 20	4 / 1	4 / 3	31 / 11	0 / 0
Walmart	1627 / 124	57 / 8	2 / 1	1 / 1	14 / 3	3 / 0
Wikipedia	5246 / 152	49 / 24	1 / 1	4 / 4	34 / 21	3 / 3
Wordpress	694 / 96	68 / 20	1 / 1	2 / 2	29 / 8	1 / 0
WSJournal	6416 / 1125	85 / 65	28 / 24	49 / 46	44 / 36	3 / 0
Yahoo	2755 / 954	162 / 78	7 / 2	9 / 4	31 / 17	0 / 0
Yelp	1655 / 24	47 / 2	5 / 1	2 / 2	7 / 0	0 / 0
TOTAL	107294 / 19636 (18%)	3259 / 897 (27%)	180 / 61 (33%)	224 / 136 (60%)	1112 / 387 (34%)	33 / 12 (36%)

Table 2. Average execution times in milliseconds

	Custom browser with optimization	Custom browser without otimization	HtmlUnit	Internet Explorer
Alexa	3431	5152 (150%)	6292 (183%)	14678 (427%)
Amazon	6488	14552 (224%)	15718 (242%)	23441 (361%)
Apple	2502	4425 (176%)	8339 (333%)	16350 (653%)
Barnes	11593	16405 (141%)	20513 (176%)	33622 (290%)
Bloomberg	3879	18335 (472%)	20750 (534%)	25196 (649%)
CNET	5844	11587 (198%)	23907 (409%)	28500 (487%)
CNN	3355	15083 (449%)	26190 (780%)	22149 (660%)
EBay	6332	8448 (133%)	11583 (182%)	20388 (321%)
Flickr	8662	22572 (260%)	17879 (206%)	22962 (265%)
GoogleNews	2497	4931 (197%)	11614 (465%)	22651 (907%)
Imdb	8314	20468 (246%)	19049 (229%)	25130 (302%)
Linkedin	2901	5942 (204%)	6760 (233%)	16102 (555%)
Reference	3555	11640 (327%)	22893 (643%)	19514 (548%)
Reuters	3647	13393 (367%)	14828 (406%)	19004 (521%)
Softonic	4450	9372 (210%)	19797 (444%)	38279 (860%)
Spiegel	2492	6540 (262%)	7812 (313%)	21022 (843%)
StackOverflow	3684	7551 (204%)	16186 (439%)	17350 (470%)
Taringa	10239	21339 (208%)	21075 (205%)	28025 (273%)
TheGuardian	5631	13043 (231%)	19358 (343%)	25104 (445%)
TripAdvisor	6102	9047 (148%)	15139 (248%)	24994 (409%)
W3CSchools	3142	6746 (214%)	13205 (420%)	20541 (653%)
Walmart	1350	4594 (340%)	7276 (538%)	12280 (909%)
Wikipedia	4850	7791 (160%)	10054 (207%)	19505 (402%)
Wordpress	2081	4655 (223%)	5031 (241%)	13890 (667%)
WSJournal	13896	17415 (125%)	18629 (134%)	20941 (150%)
Yahoo	7815	13755 (176%)	21501 (275%)	25359 (324%)
Yelp	2344	4354 (185%)	7692 (328%)	18875 (805%)

6   Related Work

Currently, web automation applications are widely used for different purposes. The approach followed by most of the current web automation systems, like Smart Bookmarks [5], Wargo [8], QEngine [9], Sahi [11], Selenium [12], and Montoto et al. [7] consists in using the APIs of conventional web browsers to automate them.

This approach has two important advantages: it does not require to develop a new browser (which is costly), and it is guaranteed that the page will behave in the same way as when a human user access the page with her browser. Nevertheless, it presents performance problems for intensive web automation tasks which require real time responses and/or to execute a significant number of navigation sequences in parallel. This is because commercial web browsers are designed to be client-side applications and, therefore, they consume a significant amount of resources.

Other systems use the approach of creating simplified custom browsers specially built for the task. WebVCR [2] and WebMacros [10] rely on simple HTTP clients that lack the ability to execute complex scripting code or to support AJAX requests. Our custom browser supports all those complexities.

HtmlUnit [4] and Kapow [6] use their own custom browser with support for many JavaScript and AJAX functionalities. They are more efficient than commercial web browsers, because they are not oriented to be used by humans and can avoid some tasks (e.g. rendering). Nevertheless, HtmlUnit works like conventional browsers when loading and building the internal representation of the web pages. The last versions of Kapow are not downloadable, but to the best of our knowledge it also works like conventional browsers regarding this issue. Since this is the most important part in terms of the use of computational resources, their performance enhancements are much smaller than the ones achieved with our approach.

7   Conclusions

In this paper, we have presented a novel set of techniques and algorithms to efficiently execute web navigation sequences. Our approach is based on executing the navigation sequence once, to automatically collect information about the elements of the loaded pages that are irrelevant for that navigation sequence. Then, that information is used in the next executions of the sequence, to load only the required elements and fire only the required events. According to our experiments these techniques are very effective: smaller DOM tree nodes are built, unneeded scripts are not executed and unneeded navigations are not performed. This way, the techniques allow to save bandwidth, memory and CPU usage, and to execute the navigation sequences faster.

Acknowledgments. This research was partially supported by the Spanish Ministry of Science and Innovation under projects TIN2009-14203 and TIN2010-09988-E, and the European Commission under project FP7-SEC-2007-01 Proposal N° 218223.

References

1. Alexa, The Web Infomration Company, http://www.alexa.com
2. Anupam, V., Freire, J., Kumar, B., Lieuwen, D.: Automating web navigation with the WebVCR. Computer Networks 33(1-6), 503–517 (2000)
3. Document Object Model (DOM), http://www.w3.org/DOM/
4. HtmlUnit, http://htmlunit.sourceforge.net/
5. Hupp, D., Miller, R.C.: Smart Bookmarks: automatic retroactive macro recording on the web. In: Proceedings of the 20th Annual ACM Symposium on User Interface Software and Technology, pp. 81–90. ACM, New York (2007)
6. Kapow, http://www.openkapow.com
7. Montoto, P., Pan, A., Raposo, J., Bellas, F., López, J.: Automated browsing in AJAX websites. Data Knowl. Eng. 70(3), 269–283 (2011)
8. Pan, A., Raposo, J., Álvarez, M., Hidalgo, J., Viña, A.: Semi automatic wrapper-generation for commercial web sources. In: IFIP WG8.1 Working Conference on Engineering Information Systems in the Internet Context, pp. 265–283. Kluwer, B.V. Deventer (2002)
9. QEngine, http://www.adventnet.com/products/qengine/index.html
10. Safonov, A., Konstan, J., Carlis, J.: Beyond Hard-to-Reach Pages: Interactive, Parametric Web Macros. In: 7th Conference on Human Factors & the Web, Madison (2001)
11. Sahi, http://sahi.co.in/w/
12. Selenium, http://seleniumhq.org/
13. XML Path Language (XPath), http://www.w3.org/TR/xpath

Today's Top "RESTful" Services
and Why They Are Not RESTful

Dominik Renzel, Patrick Schlebusch, and Ralf Klamma

Advanced Community Information Systems (ACIS),
RWTH Aachen University
Informatik 5, Ahornstr. 55, 52056 Aachen, Germany
{renzel,schlebu,klamma}@dbis.rwth-aachen.de
http://dbis.rwth-aachen.de

Abstract. Since Fielding's seminal contribution on the REST architecture style in 2000, the so-called class of RESTful services has taken off to challenge previously existing Web services. Several books have since then emerged, providing a set of valuable guidelines and design principles for the development of truly RESTful services. However, today's most popular "RESTful" services adopt only few of these guidelines, resulting in overburdening developers integrating multiple services in mashup applications. In this paper we present an in-depth analysis for the top 20 RESTful services listed on programmableweb.com against 17 RESTful service design criteria found in literature. Results provide evidence that hardly any of the services claiming to be RESTful is truly RESTful, probably due to the lack of rigidness and ease-of-use of currently available decision criteria. To improve the situation, we provide recommendations for various stakeholder groups.

Keywords: RESTful Web Services, Mashup Applications, Compliance Analysis, Design Criteria.

1 Introduction

REST [4], as described by Fielding, is an abstract concept, which supposedly led to the choice of the more general term *architecture style* as a description for this type of concept. As such, it is not a standard and hence does not mandate the strict adherence to any methods or protocols. Its application to Web services is thus subject to many different (mis-)understandings and interpretations. Although the original concept of REST is decoupled from concrete technologies, the use of well-adopted technologies such as the Hypertext Transfer Protocol [5] conveying important aspects of REST making the Web successful has evolved to the de-facto standard. RESTful service implementations today exhibit a great deal of heterogeneity. Some are adhering much closer to the style described by Fielding, whereas many implementations confuse the concept with just being another way to access existing RPC-style services. The result are often hybrid services, which are still designed as RPC-style services, but directly accessible via HTTP. HTTP provides

X.S. Wang et al. (Eds.): WISE 2012, LNCS 7651, pp. 354–367, 2012.

a lot of different functionality, some of which is adopted and used frequently by RESTful services, while other concepts are mostly ignored. In order to get a clear picture of the concepts and technologies actually adopted, we conducted a manual in-depth analysis of the most popular RESTful services used today with the result that most of them are not truly RESTful. Thus, although in theory the orchestration of RESTful services should be rather seamless, reality shows that developers have to follow different rules involving different representation schemes, etc., resulting in extensive efforts for the integration of such services to complete applications. Furthermore, there exists a certain danger that without any further measures from the side of research as well as development tool and major service providers the situation is progressively worsened. Therefore, we provide recommendations how to improve the situation.

The rest of this paper is structured as follows. First, we discuss related research with regard to RESTful services in Section 2. We then present our analysis methodology in Section 3 and results in Section 4, showing discrepancies between what is called RESTful and what actually is RESTful. Section 5 concludes the paper with a set of recommendations on how to solve current issues.

2 Related Work

The application of the REST architecture style to Web services has been described in detail in literature [14,15]. Richardson and Ruby also provide an overview over techniques for realizing RESTful services, ranging from microformats to the application of (XHTML) forms in representations. Although these techniques fit well in the concepts of REST and provide guidelines and practical examples for realizing RESTful services, it remains open whether these techniques actually gained acceptance among service developers. Over the last years, RESTful services have been subject to a variety of research questions.

Adamczyk et al. [1] and Hausenblas [7] discuss both theoretical and practical aspects of RESTful services. Their discussion of services in practice is based on an analysis of existing implementations, similar to our work. Liskin et al. [8] conducted a study on 22 randomly picked services examining four criteria in particular interesting for their work. All authors commonly note that many RESTful service implementations neglect certain REST principles. However, the results are presented as selected examples only or examine services with a narrower focus and thus fail to provide the comprehensive overview and empirical evidence we provide in this paper.

With rising popularity, RESTful services and *Resource Oriented Architecture (ROA)* have become an often considered alternative to the classical *Service Oriented Architecture (SOA)*, mainly centered around SOAP services and the related WS-* standards. This led to debates over the superiority of either of the two alternatives and generated research interest in comparing both approaches.

Pautasso et al. [11] analyzed RESTful and SOA Web Services presenting a quantitative comparison in terms of design decisions and alternatives. They conclude to have provided evidence for the perceived simplicity of ROA, which they found to reduce the number of design decisions to be taken. However, while SOA is more complex and requires more decisions, it also provides more alternatives to be considered. Thus, the authors recommend RESTful services for ad-hoc integration over the Web and WS-* Services for enterprise application integration scenarios with longer lifespan and advanced QoS requirements.

Pautasso et al. [10] analyzed loose coupling in different kinds of Web services as a fundamental concept in the REST architecture style and in distributed systems in general. They adressed loose coupling by defining coupling for Web services as a multidimensional metric with respect to 12 different facets (e.g. identification, interaction, evolution). After considering RESTful services, general RPC-style services over HTTP and WS-* services in particular, their findings state that RESTful services provide the possibility of loose coupling in all facets, while both RPC-style and WS-* services showed tight coupling in several of the twelve considered facets.

Lucchi et al. [9] provide insights into REST principles and the necessary design work for ROAs. The authors go on to describe a case study of application of the ROA principles to the spacial information infrastructure project Inspire, discussing necessary work to be done in order to switch from a SOA implementation to ROA as well as possible benefits. They conclude their analysis by stating that considering the effort and the projects time schedule, a ROA implementation would not be feasible and would provide no benefits in their case.

Discussions comparing SOA and ROA are still ongoing. However, there exist significant differences in terms of interpretation and adherence to REST principles in the domain of RESTful services. The contribution of this paper is to provide empirical evidence on the current situation, namely that most services claiming to follow RESTful design are in reality only adhering to very few of the guidelines found in literature. In the following, we present our analysis of the current RESTful service landscape on the example of some of the top used services of today.

3   Analysis Methodology

As a first step, we pre-selected the 25 most popular RESTful Web Services from http://programmableweb.com, a popular platform for Web services and mashups resp. applications using these services. The selection was based on the number of registered mashups using the services as of May 2011. Of the 25 pre-selected services, 20 remained after sorting out outdated or discontinued services and duplicates.

Table 1 lists the services selected for further analysis. This selection amounts to 53% of the mashups listed on programmableweb.com in which RESTful Web Services are used.

Each of the services was examined with respect to a catalog of 17 criteria based on the REST architecture style, the relevant features of the HTTP protocol as described in the previous section of this chapter and characteristics of RESTful services described by Richardson and Ruby [14]. The primary source of information about all criteria for each service was the provided service documentation. It should be noted that due to the lack of common structure in

Table 1. List of the 20 most popular RESTful services selected for analysis (taken from programmableweb.com)

API	Description	Category
Twitter	Microblogging service	Social
Flickr	Photo sharing service	Photos
Amazon Product Advertising API	Online retailer	Shopping
Facebook (Graph API)	Social networking service	Social
Twilio	Telephony service	Telephony
eBay	Online auction marketplace	Shopping
Last.fm	Online radio service	Music
del.icio.us	Social bookmarking	Bookmarks
Yahoo Map Service	Mapping services	Mapping
Google Custom Search API	Web search components	Search
Yahoo PlaceFinder	Geocoding services	Mapping
GeoNames	Geographic name and postal code lookup	Reference
Amazon S3	Online storage services	Storage
Box.net	Online file storage	Storage
Google Chart API	Chart creation service	Other
Amazon EC2	Elastic Compute Cloud virtual hosting	Internet
Digg	Community driven news links and ratings	News
Google Content/Search API for Shopping	Platform for structure and semi-structured data	Shopping
Wikipedia (MediaWiki API)	Online collaborative encyclopedia	Reference
Yahoo Local Search	Local search service	Search

service documentation, automatic data extraction was impossible and was thus done manually. In the following, we briefly discuss each survey criterion in detail, before we present a summary of results.

4   RESTful Services – State of the Art and Best Practices

In this section we discuss the results of our analysis for each of the 17 criteria considered. It should be noted that criteria appear in rank order of particular relevance for RESTful service design. While criteria 1 - 12 are considered highly relevant, criteria 13 - 17 are related, but not highly relevant. However, we decided to not drop these criteria for the sake of completeness. It should be noted that most of the criteria used in our analysis are not completely decisive on their own meaning that one cannot say a service is good or bad according to the properties given for that criterion. Sometimes we do not take any valuation, but only use the properties for descriptive purposes, making a point for other criteria. However, they at least serve as indicators for further manual analysis. A graphical overview over the results of each examined criterion is presented in Figure 1. Table 2 provides an overview of all services including their individual properties for each criterion. To the best of our knowledge, we assigned colors to each table cell, attempting to provide a classification in good (green), still ok (lightgreen), slightly bad (lightred) and bad (red). However, given the previous comment on the decisive quality of the criteria, sometimes assigning a color was not completely easy and/or doable at all. This shows that there is still need for more rigid criteria if a service is truly RESTful or not.

1. **Availability of Formal Description.** This criterion examines the use of formal descriptions for RESTful services. Formal descriptions allow for easier automated service access and are often used for automated service composition [13,12]. Of the examined services, 15 (75%) provide no formal descriptions, while 4 (20%) provide WSDL [3] descriptions. A single service (5%) at least provides an XML schema of its XML responses. Clearly, RESTful services so far rely on informal documentation, which further contributes to the heterogeneity to be handled in order to use them. The lack of formal descriptions often requires developers to rely on out-of-band agreements specified in the informal documentation of services resulting in high efforts on part of service clients in case a service changes (see also Criterion 5). In conjunction with Criterion 2 the availability of links in representations pointing to formal descriptions of the services are desirable, but more than often not given.

2. **Links in Representations.** Interlinked hypermedia representations can provide clients with possible follow-up actions to a service request and thus can realize navigational help, similar to Web site navigation. This constitutes a core principle of the REST architecture style described by Fielding with the term *Hypermedia As The Engine Of Application State (HATEOAS)*. This criterion examines the adoption of HATEOAS by RESTful services. Of all examined services, only four (20%) provide links to other resources of the service in their representations. The lack of this principle implicates the need for out-of-band agreements for the transitions of the application state, thus coupling clients much tighter with the service than otherwise necessary.

3. **Forms in Representations.** The concept of forms is closely related to the previously described links in representations. As suggested by [14], forms as a commonly used input concept for regular websites can be applied to RESTful services as well. We broaden the concept presented by Richardson and do not only consider XHTML forms, but also links in the form of URI-templates. These URI-templates contain placeholders for variable parts of a resource identifier and thus can provide service clients with a formal description of how to construct requests to different resource types. However, of the 20 examined services, only one (5%) includes URI-templates in its representations, while no service makes use of XHTML forms as proposed in [14].

4. **Number of Resource Types.** This criterion relates to the resource design of RESTful services. RESTful services may expose a potentially unlimited number of resources. However, we can distinguish groups of exposed resources that share the same functionality or semantics. We refer to these groups as resource types. The question how many resource types a service exposes is useful for determining whether it is designed as a purely RESTful service or following an RPC-style approach. An often found phenomenon is

to disguise a conceptually RPC-style service as RESTful, usually by exposing a single resource, to which all service requests are sent and that provides varying functionality, depending on the request parameters. The characteristic distinguishing these services from truly RESTfully designed services that expose only a single resource type is the fact that they overload its functionality. Of the examined services, eight (40%) expose multiple resource types. Five services (25%) expose only one resource type without overloading its functionality. A total of seven services (35%) expose a single overloaded resource. This shows the widespread use of the term *RESTful* for services that are essentially designed as RPC-style services or violate important REST principles. It should be noted that this phenomenon is also related to overloading the use of HTTP methods (cf. Criterion 9).

5. **Versioned service endpoint.** Service endpoints usually constitute a single entry point for accessing all methods of an RPC service. We lift this concept to fit RESTful services by considering a URL-prefix shared by all exposed resources of a service as the service endpoint. The fact that at the time of our analysis, some of the preselected services were already discontinued and/or replaced indicates the development activity around many RESTful services. In order to not break compatibility with each release, some services provide versioned access. While this allows service providers to maintain multiple versions of the same service accessible to clients, we argue that exposing the version in the URI violates the concept of resource addressability. Of the examined services, only 11 services (55%) provide a versioned service endpoint URL.

6. **Scoping Information.** In the context of service request information, Richardson and Ruby [14] distinguish between *method information* determining the action a service should perform, and *scoping information* determining the data the service should operate on. Technical possibilities of expressing scoping information are numerous, leading to a choice issue for service implementation. However, for the sake of addressability, scoping information should be encoded in resource URIs. Hierarchical information is best suited as part of the URI path, while additional scoping information can be encoded in the form of request parameters in the query portion of a URI. This criterion analyzes the source of the scoping data for the different services. 12 services (60%) encode scoping information in the URI path, fitting well into the concept of URIs. Four services (20%) use the URI path as well as parameters and three services solely use request parameters for scoping information. Finally, one service (5 %) uses header fields for scoping information. During this analysis it became apparent that the distinction between what constitutes scoping information and what constitutes a parameter to the service request can be hard and is thus hardly automizable.

7. **Parameter Sources.** This criterion examines the different sources used for providing request parameters. 17 of the examined services (85%) make use of URI-encoded request parameters only. Two services (10%) also take some parameters from form-encoded *POST* requests, while a single implementation (5%) only takes form-encoded requests. One explanation for this

significant tendency towards URI-encoded parameters is ease of use from the developer and experienced user point of view. However, we argue that using form-encoded parameters inherently expresses richer parameter semantics. Furthermore, simultaneous presentation of information and controls is desirable such that the information becomes the affordance through which any agent - either being human or machine - obtains choices and selects actions.

8. **Meaningful HTTP Status Codes.** With this criterion we examine the adoption of HTTP status codes in communication with service clients. By sending out reasonable HTTP status codes in error cases, services can make use of a well-defined vocabulary for communicating problems to their clients. Of the examined services, 13 (65%) make meaningful use of HTTP status codes. The remaining services (35%) send out the status of responses in unstandardized form in the response body, thus increasing complexity of error-handling on client-side.

9. **HTTP Methods Used.** A central part of the REST architecture style is the uniform interface, usually realized with HTTP methods. This criterion examines which of the available HTTP methods are used by RESTful service implementations. While the *GET* method is used by almost all of the analyzed services and more than half of the services use the *POST* method, there are only five services which make use of the *PUT* or *DELETE* methods. While some services may just not require these methods, use of only a single HTTP method can also point to an overloading of functionality, a possible indicator for an RPC-style service disguised as RESTful service (cf. Criterion 4).

10. **HTTP Method Override.** Although more methods have been specified in the standard for a long time, some old browsers do not support any methods except for *GET* and *POST*. In order to be able to realize RESTful services to their full potential despite this restriction, some services allow to override the method passed in a request by other means. This criterion intends to analyze which mechanisms are used for this purpose. Of the five examined services that make use of HTTP methods beyond *GET* and *POST*, three (15%) provide a method override through a request parameter. One service (5%) realizes a method override by a custom header field.

11. **Representation Formats.** With this criterion we examine the different types of representation formats commonly produced by RESTful Services. We found that 13 (65%) of the 20 services support multiple representation formats. The most popular ones are XML, produced by 17 (85%) of the services, and JSON, produced by 12 (60%) services. Six services (30%) produce their data in a serialized PHP format, comparable to JSON. ATOM and HTML representations are produced by two services (10%) each. During the analysis we encountered another eight different representation formats produced by services, ranging from plain text and CSV to the image format PNG or specific XML formats such as SOAP. It should be noted that the formats consumed by the services are strongly related to Criterion 7 (Parameter Sources) - most services let users pass information as URL-encoded parameters, while only some of them use form-encoded POST data.

12. **Representation Format Selection.** For this criterion we examine the mechanisms for selecting a specific representation format in case a service offers multiple formats. The HTTP protocol supports such a selection through its content negotiation mechanism, allowing clients to specify the media types they understand and define a preference for certain representations. Following the same ease-of-use argument as given in Criterion 7, a simple alternative is to allow the specification of the desired format in the form of a request parameter. Finally, some people suggest to encode it in the URI in analogy to a filetype suffix. However, this practice ignores the separation of resources from their representations and can lead to a number of problems for example when URIs are shared. As mentioned in the discussion of the previous criterion, seven services offer only a single representation format and thus do not require a selection mechanism. The most common method used is to pass the desired format as a parameter to the request. Two services (10%) encode the representation format as a URI suffix, similar to the suffixes common in many filesystems. Only one service (5%) uses standard HTTP content-negotiation. This service additionally offers the specification of a parameter. One service (5%)offers different formats as completely different resource types, thus encoding the format in the URI. It is unclear why the use of standard HTTP features is so commonly avoided. A possible explanation is that clients may not always have full control over the request headers, in which case the use of request parameters provides a more simple alternative.

13. **User Authentication.** Most services require user authentication at least for parts of the offered functionality. E.g. some information might be publicly available, while changing user-specific settings usually requires some form of user authentication. HTTP provides a general mechanism, including specific authentication protocols to be used. Other standards such as the OAuth protocol [6] have emerged. Only four of the selected services do not require any kind of authentication. Seven services only allow authentication via proprietary mechanisms. The OAuth protocol is supported by five of the services (25%). Only four of the services (20%) allow authentication via standard HTTP methods. It is notable that although there exist a number of open standards in this area, their adoption among the examined services is rather low with more than a third of the services only supporting proprietary methods. Adopting well-established standards can significantly lower the effort necessary for using a service because reusable implementations exist for several programming languages.

14. **Secure Connections.** This criterion examines the support for SSL-secured connections provided by the services. The use of secured connections is generally recommended for any request containing sensitive data. Of the examined services, 14 services (70%) support usage of a secure connection, six of which (30%) require secure connections throughout all communications. The remaining six services (30%) do not support secure connections at all.

15. **Accessiblity via other protocols.** With this criterion we examine the parallel use of different protocols for the same services. Most of the examined

Table 2. Analysis Results Overview

	Twitter	Flickr	Amazon Product Advertising API	Facebook Graph API	eBay	Last.fm	del.icio.us	Twilio	Yahoo Map Service	Google Custom Search API
1. Formal Description	-	-	WSDL	-	WSDL	-	-	-	-	-
2. Links in Representations	no	no	?	yes	no	no	no	yes	no	yes
3. Forms in Representations	no	no	no	no	no	no	no	?	no	yes, URI template
4. Resource Types	many	1 (overloaded)	1 (overloaded)	many	1 (overloaded)	1 (overloaded)	many	many	1	1
5. Versioned End-Point	yes	no	no	no	no	yes	yes	yes	yes	yes
6. Scoping Information	URI Path	Parameters	Parameters	URI Path	HTTP Headers	Parameters	URI Path	URI Path	URI Path	URI Path
7. Parameters	Query Parameters	Query Parameters	Query Parameters	Query Parameters	Form-encoded	Query Parameters / Form-encoded	Query Parameters	Query Parameters / Form-encoded	Query Parameters	Query Parameters
8. Using HTTP Status	yes	yes	yes	no	no	no	no	yes	yes	yes
9. HTTP Methods	GET, PUT, POST, DELETE	GET, POST	GET	GET, POST, DELETE	POST	GET, POST	GET	GET, POST, PUT, DELETE	GET	GET
10. Method Workaround	Query Parameter	-	-	Query Parameter	-	-	-	Query Parameter	-	-
11. Formats	XML, JSON, RSS, ATOM	XML, JSON, PHP, SOAP, XMLRPC	XML, HTML	JSON	XML	XML, JSON	XML	XML, JSON, CSV, HTML	XML, PHP	JSON, ATOM
12. Format Selection	URI-Suffix	Query Parameter	Query Parameter	-	-	Query Parameter	-	URI-Suffix	Query Parameter	Query Parameter
13. Authentication	HTTP, OAuth	Proprietary, OAuth	Proprietary	OAuth	Proprietary	Proprietary	HTTP, OAuth	HTTP	-	-
14. HTTPS Support	yes	yes	yes	yes	yes, mandatory	yes	yes, mandatory	yes, mandatory	no	yes
15. Other Protocols	SOAP	SOAP	SOAP	Proprietary	SOAP	XML-RPC	-	Proprietary	-	-
16. Registration	API-Key	API-Key	API-Key	API-Key	Dev-ID, API-Key (App-Id)	API-Key	-	Account	API-Key (AppId)	API Key
17. Limitations	Rate Limit	-	Rate Limit	-	-	Rate Limit	Rate Limit	-	Rate Limit	Rate Limit

Table 3. Analysis Results Overview (continued)

	Yahoo PlaceFinder	GeoNames	Amazon S3 API	Box.net	Google Chart API	Amazon EC2 API	Digg API	Google Content/Search API	MediaWiki API (Wikipedia)	Yahoo Local Search
1. Formal Description	-	-	WSDL	-	-	WSDL	-	-	-	Response XML Schema
2. Links in Representations	no	no	no	no	no	no	no	yes	no	no
3. Forms in Representations	no	no	no	no	no	no	no	no	no	no
4. Resource Types	1	many	many	few (overloaded)	1	1 (overloaded)	many	many	1 (overloaded)	1
5. Versioned End-Point	yes	no	no	yes	no	no	yes	yes	no	yes
6. Scoping Information	URI Path/Parameters	URI Path	URI Path	Parameters/URI Path	URI Path	Parameters	URI Path	URI Path	URI Path/Parameters	URI Path
7. Parameters	Query Parameters	Query Parameters	Query Parameters / Form-encoded	Query Parameters	Query Parameters	Query Parameters	Query Parameters	Query Parameters	Query Parameters	Query Parameters
8. Using HTTP Status	no	no	yes	yes	yes	yes	yes	yes	yes	yes
9. HTTP Methods	GET	GET	GET, POST, PUT, DELETE	GET, POST	GET, POST	GET, POST	GET	GET, POST, PUT, DELETE	GET, POST	GET
10. Method Workaround	-	-	no	-	-	-	-	Header Field	-	-
11. Formats	XML, JSON, PHP	XML, JSON	XML	XML	PNG	XML	XML, JSON, Javascript, PHP	XML, JSON	XML, JSON, PHP, WDDX, YAML, TXT	XML, JSON, PHP
12. Format Selection	Query Parameter	URI	-	-	-	-	HTTP Content Negotiation, Query Parameter	Query Parameter	Query Parameter	Query Parameter
13. Authentication	-	-	HTTP	proprietary	-	proprietary	OAuth	OAuth and more	proprietary	-
14. HTTPS Support	no	no	yes	yes, mandatory	yes, mandatory	yes	no	yes, mandatory	no	no
15. Other Protocols	-	-	SOAP	SOAP, XML-RPC	-	SOAP	-	-	-	-
16. Registration	API-Key	Account	Account	API-Key	-	Account, Developer-Key	-	API-Key	depends on Wiki	API-Key
17. Limitations	Rate Limit	Rate Limit	-	-	-	implicit	implicit	-	Rate Limit	Rate Limit

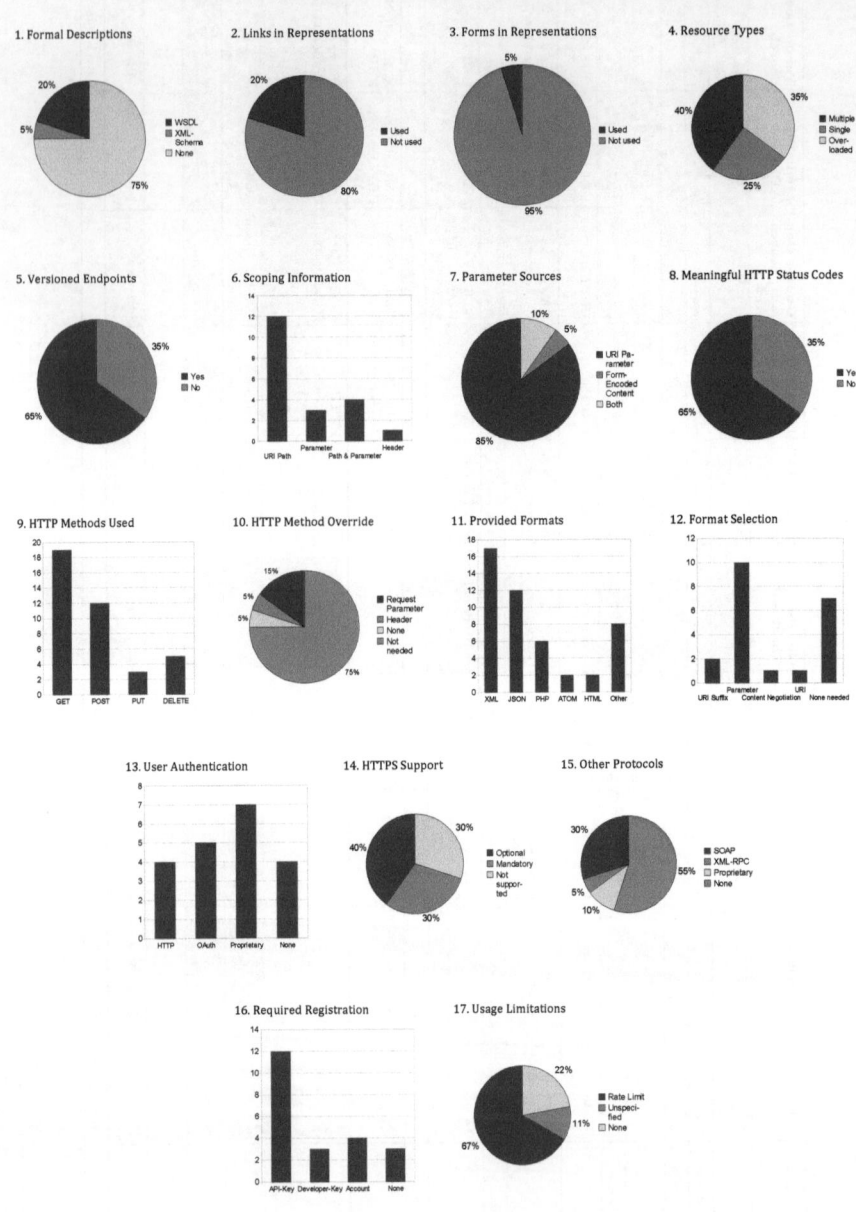

Fig. 1. Analysis Results Overview

services do not provide support for any other protocol than HTTP. Six of the services (30%) support SOAP as well. It is worth noting that five of these six services cannot be considered truly RESTful, but rather hybrid in terms of borrowing only few RESTful design concepts, while mainly being RPC-style. The rather outdated XML-RPC protocol is supported by two services (10%), while another two services (10%) provide proprietary access protocols, based on markup languages. Examining the services providing RPC-style access, it becomes apparent that most of them use the term REST to describe an easy way to access the functionality via HTTP-requests, but do not apply REST concepts to service design.

16. **Type of Registration Required.** Usually, even services offering free-of-charge usage require some form of registration of users, developers or applications making use of the service. This criterion examines the different types of registration required and their technical consequences on service usage. The most widely used concept is the API-Key, an identifier that can be obtained from the service provider and may be unique to each application using the service or to an application developer. A total of 12 (60%) services requires the usage of API-Keys. Four services (20%) require user accounts, and three services (15%) require developers to register for a developer-key. Only three (15%) of the examined services are usable without any kind of registration.

17. **Usage Limitations** Although most of the selected services were free-of-charge at least for non-commercial use, most of them impose certain usage limitations. Only three of the examined services (15%) do not express any usage limits, another service (5%) provides only a vague, not explicitly specified limit. All 16 other services (80%) allow only a certain amount of requests during a fixed timeframe. Often, a higher allowed request rate can be purchased from the service provider. This wide distribution of explicit usage limits stresses the need for efficient handling of service requests, which can be supported by a RESTful architecture through the use of caching features.

5   Conclusion

Our analysis clearly shows the existing differences in terms of adoption and interpretation of RESTful concepts among service implementations. While RESTful Web Services often claim to be a much simpler alternative to RPC-style services, this claim is contrasted by the reality of today's implementations of popular services. Obviously, the term *RESTful* has taken on a life of its own, expressing careful application of design decisions that made the Web popular on one side and being a buzzword for just about any service that is accessible via HTTP URIs on the other. While in [10] the authors show the potential of a RESTful approach for realizing loose-couping in Web Services, this is usually not achieved in practice due to the neglect of RESTful principles, whether it is the result of deliberate business decisions or ongoing misconceptions about REST. Either way, the result is often to the disadvantage of the users of RESTful service APIs, i.e. developers and providers of mashup Web applications as well as their end-users. Already now voices of API users become louder complaining about the rather informal documentation, increased burden with much more work than needed to combine services, and finally frequent changes in service APIs rendering their applications unusable[1]. It still takes a lot of effort for developers to fully grasp the rather abstract concept of REST. Some of the successful concepts (e.g. to use HTTP methods as uniform interface) already found broad uptake. However, many other concepts contributing to the evolution of a coherent ecosystem of interoperable services (e.g. linking between resources, HATEOAS) are neglected. The result is a certain form of "REST smattering". Although literature exists, it seems to be the duty of research and providers of RESTful service development tools to further streamline and make the rather abstract concepts even more understandable, convincing, and easy to apply in practice. Documentation often looks compact and is usable for smaller projects, but more complex projects raise questions not answerable anymore with rather informal documentation. Thus, we argue that increased efforts in formalization and documentation are required in particular. Desirable and helpful tools for developers supporting REST guidelines, profiles, automated checks of a service being RESTful would become possible at least partially. Recent work by Zuzak and Schreier [16] and Liskin et al.[8] leads to this direction. Major service providers act as role model for their customers. Smaller businesses interested in providing new services based on existing ones rather tend to learn from previously successful services as good examples. Thus, if major providers do not adhere to the overarching concepts, then followers will even more blur these concepts. Furthermore, service providers should come up with more formal definitions of representations, thus strengthening the interoperability between services and facilitating their seamless orchestration (cf. [2]).

However, we also see a bright future for RESTful services. From the side of technology providers (e.g. browser vendors, providers of development tools and libraries) the constant progress in widely implementing full-featured HTTP support as well as RESTful service development tools will help to reduce complexity in the context of realizing mashups based on RESTful services. Finally, the time factor will also contribute to a better future situation. Still, the wide uptake of many old browsers not fully supporting HTTP is one factor forcing service providers into deliberately dropping RESTful principles in favor of wide adoption of their services. Such old browsers will become increasingly obsolete and replaced with modern browsers offering completely new modes of operation and interaction with services (e.g. mobile devices incl. sensors). All these factors will contribute to make it easier for companies to find arguments for dropping backwards compatibility requirements and become compliant with REST principles after all.

Acknowledgments. The research leading to these results has received funding from the European Community's Seventh Framework Programme (FP7/2007-2013) under grant agreement no 231396 (ROLE project).

---

[1] http://blog.yourtrove.com/2011/08/11/api-integration-pain-survey-results/

References

1. Adamczyk, P., Smith, P.H., Johnson, R.E., Hafiz, M.: REST and Web Services: In Theory and In Practice. In: Wilde, E., Pautasso, C. (eds.) REST: From Research to Practice, pp. 35–57. Springer, Heidelberg (2011)
2. Alarcon, R., Wilde, E.: Linking Data from RESTful Services. In: Bizer, C., Heath, T., Berners-Lee, T., Hausenblas, M. (eds.) Proceedings of the WWW 2010 Workshop on Linked Data on the Web (LDOW 2010). CEUR Workshop Proceedings (2010) ISSN 1613-0073, http://ceur-ws.org/Vol-628/
3. Christensen, E., Curbera, F., Meredith, G., Weerawarana, S.: Web Service Definition Language (WSDL). Tech. rep., W3C (March 2001), http://www.w3.org/TR/wsdl, (last access: June 2012)
4. Fielding, R.T.: Architectural Styles and the Design of Network-based Software Architectures. Ph.D. thesis, University of California, Irvine, California, USA (2000)
5. Fielding, R.T., Gettys, J., Mogul, J., Frystyk, H., Masinter, L., Leach, P., Berners-Lee, T.: RFC 2616, Hypertext Transfer Protocol – HTTP/1.1. Tech. rep., The Internet Society (1999), http://www.ietf.org/rfc/rfc2616.txt, (last access: June 2012)
6. Hammer-Lahav, E.: The OAuth 1.0 Protocol. RFC 5849 (Informational) (April 2010), http://www.ietf.org/rfc/rfc5849.txt (last access: June 2012)
7. Hausenblas, M.: On Entities in the Web of Data. In: Wilde, E., Pautasso, C. (eds.) REST: From Research to Practice, pp. 425–440. Springer, New York (2011)
8. Liskin, O., Singer, L., Schneider, K.: Welcome to the Real World - A Notation for Modeling REST Services. IEEE Internet Computing 16(4), 36–44 (2012)
9. Lucchi, R., Millot, M., Elfers, C.: Resource Oriented Architecture and REST: Assessment of impact and advantages on INSPIRE. Tech. rep., European Comission Joint Research Centre Institute for Environment and Sustainability (2008)
10. Pautasso, C., Wilde, E.: Why is the Web Loosely Coupled? A Multi-Faceted Metric for Service Design. In: Proceedings of the 18th international conference on World Wide Web, WWW 2009, pp. 911–920. ACM, New York (2009)
11. Pautasso, C., Zimmermann, O., Leymann, F.: RESTful Web Services vs. "Big" Web Services: Making the Right Architectural Decision. In: Proceedings of the 17th International Conference on World Wide Web, WWW 2008, pp. 805–814. ACM, New York (2008)
12. Pautasso, C.: RESTful Web service composition with BPEL for REST. Data Knowl. Eng. 68, 851–866 (2009)
13. Rao, J., Su, X.: A Survey of Automated Web Service Composition Methods. In: Cardoso, J., Sheth, A. (eds.) SWSWPC 2004. LNCS, vol. 3387, pp. 43–54. Springer, Heidelberg (2005)
14. Richardson, L., Ruby, S.: RESTful Web Services, 1st edn. O'Reilly (2007)
15. Webber, J., Parastatidis, S., Robinson, I.S.: REST in Practice - Hypermedia and Systems Architecture. O'Reilly (2010)
16. Zuzak, I., Schreier, S.: ArRESTed Development: Guidelines for Designing REST Frameworks. IEEE Internet Computing 16(4), 26–35 (2012)

Event Aware Workload Prediction:
A Study Using Auction Events

Matthew Sladescu[1,2], Alan Fekete[1,2], Kevin Lee[2,3], and Anna Liu[2,3]

[1] The University of Sydney
[2] National ICT Australia
[3] The University of NSW
{firstname.lastname}@nicta.com.au

Abstract. Workload bursts have become notorious for rendering numerous web information systems unavailable. While cloud computing has the potential to alleviate this problem by offering computing resources on an on-demand basis, important challenges remain in finding the right resource control strategies to scale resources cost-effectively and to overcome the initialization lag associated with resource acquisition. An effective strategy involves predicting workload demand in advance so that resources can be provisioned in a timely manner, but not all prediction approaches are made equal. We argue that while most existing approaches show promising results in predicting average workload, they fail to predict workload bursts that are inherently irregular. This paper formulates a new event-aware strategy to more effectively predict workload bursts by exploiting prior knowledge associated with scheduled events. We evaluate our approach by comparing it to state-of-the-art methods in workload prediction using real-world datasets from the online auction domain, and we show that event-aware prediction is superior to other approaches in terms of burst prediction accuracy.

Keywords: cloud computing, workload prediction, online auctions.

1 Introduction

Workload bursts can lead to web-site unavailability with serious consequences for the site owner. Numerous e-commerce sites like deal-of-the-day [4] and ticketing sites [20] have been reported to crash when workload bursts accompanied promotional events for popular products. Many other web sites like result announcement [13] and auction sites [19] can also experience burst-induced crashes and must consider how these bursts can be managed effectively.

The accurate prediction of the time, magnitude and duration of such workload bursts, and workload prediction in general, brings forward a number of benefits including the ability to:

- pre-provision resources in a cloud-computing environment so that they are available on time to maintain quality of service (QoS) for customers during

X.S. Wang et al. (Eds.): WISE 2012, LNCS 7651, pp. 368–381, 2012.

anticipated workload bursts. (Currently resources in the cloud are not available immediately after resource requests, and must be pre-provisioned early to avoid issues with resource initialization lag [9]).

- reduce resource provisioning costs by reserving [14] instances when resource demand predictions indicate that this would be financially beneficial.
- devise smarter event scheduling policies. (eg: For result announcement web sites like [13]: instead of providing results to all recipients in a single release event, one could schedule several smaller release events at different times. A prediction for the load associated with a release event can be used to estimate the number and scale of releases necessary to deliver all results as quickly as possible, while minimizing the computational resources required, and maintaining QoS for recipients.)

These important benefits have motivated mounting research interest in the workload prediction space, as discussed in section 2. This includes the adaptation of a number of machine learning techniques to predict load based on the observed history. While these methods work well in predicting longer term trends and mean load, our results concur with recent literature [8] to indicate that they are ineffective in predicting the type of catastrophic, irregular and non-periodic load bursts described earlier. An increased understanding of the causes behind load spikes can help in predicting when they will occur. In all of the examples above, load bursts can be associated with one or more events. In this paper's context, we define an event as: a scheduled or unscheduled occurrence, (like an online sale); which can be described by a set of features, (like the event starting and ending times); and can induce load for associated resources (like the web pages or database entries associated with a popular online sale). Although spikes are often associated with events, none of the prior art in load prediction surveyed takes advantage of often readily available foreknowledge about scheduled events.

In this paper we engage with the challenge of translating prior knowledge about scheduled events into an accurate forecast for workload bursts. Our contributions are: (1) an overall event aware prediction (EAP) framework, which prescribes the use of key elements for workload prediction including repositories with knowledge about past and upcoming events, a model which can describe how workload fluctuates for each event, and a prediction module that can incorporate these inputs to forecast workload; (2) a case study where we apply EAP and specify an instance for each key EAP element to forecast real workload from the auction domain; and (3) an evaluation of the prediction accuracy in this case study using existing metrics and variants to compare EAP to existing prediction methods. The results of this evaluation show that EAP is superior to some of the most commonly and recently used workload prediction strategies.

2   Related Work

Previous publications on managing workload changes in cloud computing have generally used only the history of workload for predicting future load, and have used these predictions in planning resource allocation decisions. This builds on extensive work in statistics for predicting future values of time series data [12,16], where examples for load prediction include models based on ARIMA [17] and SVR [2]. In [6], artificial neural network (ANN) and linear regression configurations are compared for load prediction accuracy, with an ANN configuration reported as generally exhibiting better accuracy. A number of prediction approaches [2,6], including the SVR, ARMA, and ANN configurations assessed in this paper make use of a sliding window approach [5] to learn the influence of p contiguous observations in the past on an observation in the future, where p is the window size. A more robust approach combining a variety of AI techniques including AR, MA, SVR and ANN has also been used to predict workload [7]. Our EAP proposal differs from all these, by using extra information from external knowledge of when events are scheduled to occur.

Lassnig et al [8] identified the importance of treating bursts explicitly, rather than just regarding them as values within an arbitrary time-series. The prediction method proposed in [8] identifies preconditions that preceded bursts in the past, and makes a prediction for future burst locations by making use of the assumption that similar preconditions will influence when bursts will occur in the future. In contrast, EAP is told when the event (that triggers the burst) will occur, and our focus is on predicting the values of the resulting workload curve.

A number of generic burst models [3,1] have been proposed to represent the workload changes during a burst, and to simulate bursts for the purposes of load testing. In our earlier work [15] we have shown how knowledge about an active or upcoming event can be used with such a burst model to provide improved QoS during workload bursts in a cloud computing environment. Auctions have particular features in their workload: they exhibit a recurring pattern, where most bids occur towards the end of an auction's lifetime, with a smaller majority occurring at the start of an auction soon after it is first advertised [11]. This pattern is observed in the workload model constructed (as described in Section 4) for applying EAP to bidding workload.

When assessing an algorithm for predicting workload, several metrics can be considered. Traditional measures of prediction accuracy in time-series include the Mean Absolute Error (MAE) and Root Mean Squared Error (RMSE). Other metrics that are specific to bursts have been considered by Lassnig et al [8]. A burst can be labeled as any place in the time-series where the value exceeds a threshold [18]; in [8] this threshold is $t \times$ IQR, where the value of t is specified based on the type of magnitude bursts that need to be identified and IQR refers to the inter-quartile range of the time-series. The Mean Absolute Error Distance metric (MAED) introduced in [8] can report the error with which the location of bursts is predicted. The MAED is measured by pairing each predicted workload burst with the next chronological actual workload burst, and then calculating the time-error distance between each actual-predicted workload burst pair. The final result reported by MAED is the average error distance for all pairs, normalized by the length of the time series being examined. Instances where there are predictions for which no actual burst exists, and instances when there are actual bursts that have not been predicted, both incur a penalty that will increase the average error distance that is reported by MAED. We refer to the workload burst location prediction accuracy [8] (1- MAED) as BLPA.

3   Event Aware Prediction

In Section 1 we showed how workload bursts are often associated with important events, and we defined such events. As part of the main contribution for this paper we advocate a novel event aware prediction framework, which, in contrast to the existing load prediction methods discussed in Section 2, can exploit inherent event-burst associations by using foreknowledge about scheduled events, to provide more accurate predictions for workload during bursts. We now present the main inputs of the EAP framework, and then in Section 4 we provide a concrete case study applying this framework to predict the bidding workload for online auctions.

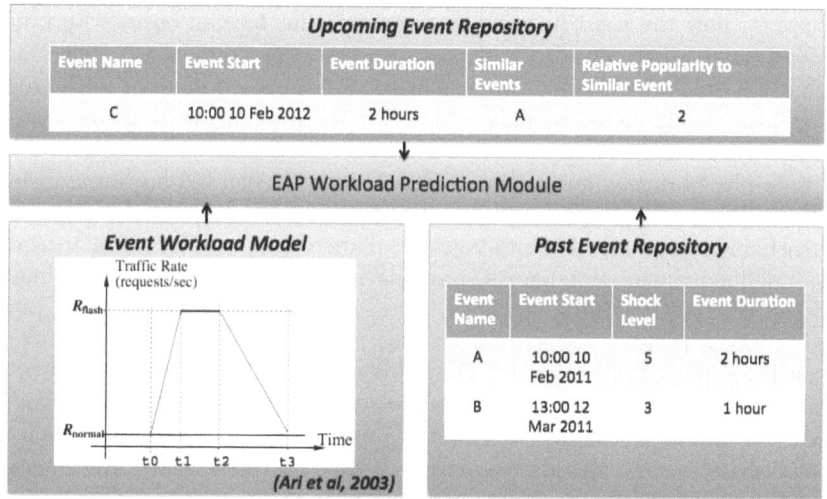

Fig. 1. Main EAP Framework Inputs

As shown in Fig 1, the EAP framework makes use of an event workload model, a repository of past events, and a repository of upcoming events (which can also include details about presently active events) to predict event-associated workload, including workload bursts. Conceptually, EAP makes a prediction for each scheduled future event, by choosing parameters for the workload model, which is a curve that shows how workload is expected to change during an event. Hence, the parameters that must be defined for each event will depend on the workload model being used, such that this model can be fully parameterized. The 3-piece-linear model shown in Fig 1 is taken from [1], and parameters are the event duration (horizontal stretch) and shock-level (which multiplies the non-event-related load to calculate the vertical stretch). In placing information into the repository of scheduled events, the knowledgeable user, such as a marketing manager, might specify some parameters directly (eg: saying when the event will start), or they can indicate some parameters and even curve shape implicitly, by indicating one or more previous events that should be seen as similar. For example, for event C in Fig 1, instead of specifying its shock level directly, the event is linked to similar event A, which has occurred in the past, and C is recorded as expected to be "twice as popular as A", (which can translate to twice the shock level of event A). The EAP framework just discussed describes the key elements necessary for predicting workload using knowledge about scheduled events. In Section 4 we provide instance examples for each of these elements to demonstrate how the EAP framework can be applied for predicting the bidding workload in an auction site.

4   An Instance of EAP for Predicting Auction Workload

This section uses a concrete case study of bidding workload in the auction domain to illustrate how the EAP framework described thus far can be used to predict workload bursts. The repository of past events in this domain is a collection of auctions for which the closing time has elapsed. Each auction event in the repository of past events includes fields to describe the auction starting time, closing time, and associated bid history. The bid history represents a collection of bids that were placed during an auction's lifetime, where each bid has an associated bid time. The workload for auctions is characterized in [11] by identifying how frequently bids occur at different stages of an auction's age. The characterization process defined the auction age as spanning from 0% (auction starting time) to 100% (auction closing time), and then based on the bid time for each bid, placed each bid into a percentile bucket from 1 to 100 (bids occurring in the first age percentile are placed in bucket 1). This is repeated for all auctions in a data set, so that all bids from all auctions in the data-set are placed into a corresponding bucket. The result of this process can be represented on a graph of (%auction age percentile) versus (%bids occurring at this age percentile). We construct a model using this characterization process from the repository of past events, and use it as the event workload model for evaluating EAP performance. The resulting model shows many of the characteristics found in Menasce's characterization [11], including most bids being placed towards the end of an auction's lifetime, with a smaller peak of bids at the start when an auction is first advertised; few bids are placed in the middle period of the auction. The EAP approach works under the assumption that auction events scheduled in the future will exhibit a workload profile similar to the model constructed from the repository of past events.[1] The auction event workload model can be updated incrementally and periodically based on new additions to the repository of past events. The repository of upcoming events includes information about the events for which the closing time has not yet elapsed. This includes the starting and closing time for each event, as well as any bid history that has accumulated for

---

[1] Our use of EAP was not primed with Menasce's or any other model; rather the model was derived from the repository of past events.

auctions where the starting time has passed. Together with the event workload model, the repository of upcoming events can be used to predict auction-event associated workload, including workload bursts.

Now that the main inputs of our EAP framework have been characterized, this section will focus on how the EAP module works in our auction example. We first present a simplified form for the workload prediction algorithm used by the EAP module, known as EAP-core, and later discuss a refinement known as EAP-bb. The EAP-core algorithm is shown in Fig 2 and discusses equations 1 to 8. In abstract terms, EAP-core reports a prediction of bids per minute for a given minute t_i, by summing the predicted bids per minute for all active auctions at minute t_i.

To predict bids per minute at minute t_i:

1. Find all active auction events at t_i. (This corresponds to all auction events where t_i is between their starting time t_s and closing time t_c).
2. For each active auction event:
   (a) Calculate auction event age at t_i as a percentage: $A_i = \dfrac{t_i - ts}{tc - ts} \times 100$
   (b) Calculate expected total % bids, (B_i), that will occur at t_i using A_i from 2 (a) and the auction workload model constructed from the repository of past events, as shown in Equation 7. Here $M[A_i]$ is the workload model that reports the expected B_i at a given integer percentile age A_i; and l represents the length, in minutes, of a percentile for the auction event under consideration. Equation 7 is formulated to estimate the expected B_i for a given minute by interpolating between the expected B_i at adjacent percentile midpoints using the equation of a straight line. B_i for a percentile midpoint is defined as the average percentage of bids per minute expected for an integer age percentile $\left(\frac{M[A_i]}{l}\right)$.
   (c) Calculate the equivalent bid magnitude for 1% of total bids, (P_B), as shown in Equation 8. Here A_C represents the auction age at the present minute; and $T_{\lfloor A_C \rfloor - 1}$ corresponds to the total bids that have occurred up to the last elapsed and completed age percentile. As shown in Equation 8, if past history is available for the event under consideration, the auction event workload model can be used to calculate P_B, otherwise this result can be calculated as the average bid magnitude equivalent to 1% based on information from completed events in the past event repository. (Here T_{avg} corresponds to the average total bid count for all completed auctions).
   (d) Add the predicted bid magnitude for the auction event under consideration, $(C_i = P_B \times B_i)$, to a running bid-count total for t_i.

Fig. 2. EAP-core Algorithm for predicting Bidding Workload

$$m_1(A_i) = \left(\frac{M[\lfloor A_i \rfloor + 1] - M[\lfloor A_i \rfloor]}{l}\right) \tag{1}$$

$$m_2(A_i) = \left(\frac{M[\lfloor A_i \rfloor] - M[\lfloor A_i \rfloor - 1]}{l}\right) \tag{2}$$

$$x_1(A_i) = (A_i - \lfloor A_i \rfloor) \tag{3}$$

$$x_2(A_i) = (A_i - \lfloor A_i \rfloor + 1) \tag{4}$$

$$b_1(A_i) = \frac{M[\lfloor A_i \rfloor]}{l} \tag{5}$$

$$b_2(A_i) = \frac{M[\lfloor A_i \rfloor - 1]}{l} \tag{6}$$

$$B_i(A_i) = \begin{cases} m_1(A_i) \times x_1(A_i) + b_1(A_i) & (\lfloor A_i \rfloor + \frac{1}{2} \le A_i < 99) \vee \lfloor A_i \rfloor = 0 \\ m_2(A_i) \times x_2(A_i) + b_2(A_i) & (1 \le A_i < \lfloor A_i \rfloor + \frac{1}{2} \wedge A_i \ne 100) \\ & \vee \lfloor A_i \rfloor = 99 \\ m_2(99) \times x_2(A_i) + b_2(A_i) & A_i = 100 \end{cases} \tag{7}$$

$$P_B = \begin{cases} \dfrac{T_{\lfloor A_c \rfloor - 1}}{\left(\sum\limits_{j=0}^{\lfloor A_c \rfloor - 1} M[A_j] \right)} & \lfloor A_c \rfloor - 1 \ge 0 \\ \dfrac{T_{avg}}{100} & \lfloor A_c \rfloor = 0 \end{cases} \tag{8}$$

The method for predicting bids per minute for an auction, as described in 2 (d) of Fig 2, can yield a non-integer value. In our experience, this means that potential clusters of integer bids would be spread out in fractions along auction lifetime percentiles, ultimately yielding a smooth prediction, without bursts, similar to those given by ARMA, SVR, and ANN. In order to prevent bursts from being filtered out of our predictions in this way, we introduce a refined prediction approach known as EAP-bb. This approach recognizes that bids are inherently discrete, and seeks to provide better precision for locating bids in time. We observed that this allows for a much clearer prediction for where bids will cluster to form workload bursts. Hence, the EAP-bb approach alters step 2 (d) such that the running bid-count total for minute t_i is only updated when an integer number of bids is predicted to occur. We contribute a new bid-bucket approach, illustrated in Fig 3, to identify when an integer number of bids is expected to occur during an auction event's lifetime.

The bid-bucket approach of EAP-bb divides an auction's lifetime into a number of partitions known as bid-buckets. The first bid bucket begins at the start of an auction's lifetime, and all subsequent bid buckets are contiguous. In the basic case, the length of each bid-bucket corresponds to an event age range that contains 1 integer bid. The percentage age range that corresponds to 1 bid can be calculated using the auction event workload model, as well as the total number of bids for an auction. In cases where some bid history exists for an auction that has not yet completed, the total number of bids in the auction can be estimated using the auction event workload model. Note that as bid-history for an active auction accumulates, the estimated bid count for the auction may change, which means that the bid-bucket lengths and positions can also change. In cases where no bid history exists, the total number of bids can be estimated by looking at the average bid count for completed auction events.

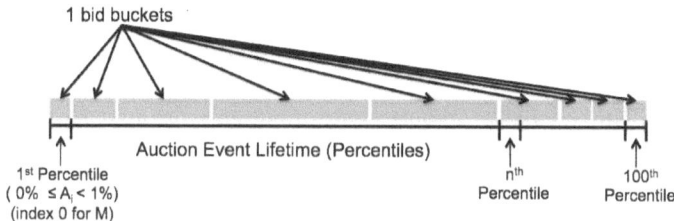

Fig. 3. Bid Bucket Approach: Identifying when bids are expected during an auction events lifetime

The conceptual illustration of Fig 3 shows how the bid-bucket lengths tend to be relatively smaller towards the closing time and starting time of an auction, since these are the times when most bids are frequently placed. So far we have outlined the procedure for determining the position and length of bid buckets. The next step is to identify, for each bid bucket, where an integer number of bids is most likely to occur. For each bid bucket, the event workload model that was constructed from the repository of past events is used to find out at what time (within the boundaries of the bucket range) bids have most frequently occurred. This corresponds to the minute, within each bid bucket, that yields the highest B_i according to Equation 7, and is used as a basis for identifying where integer bids are expected to occur at each bid bucket. The bid-bucket approach does not alter the computational complexity of Fig 2, with respect to the number of auctions considered, and once applied, anticipated workload bursts become much more prominent. We evaluate the EAP-bb approach in the next section.

5   Evaluation

This evaluation compares the load prediction accuracy of EAP-bb to ARMA, SVR, ANN, and the Lassnig prediction techniques that were introduced in Section 2. The workload prediction accuracy of these methods is evaluated using bidding workload and auction event data, described below, which was collected from a popular online auction site.

We assess the burst location prediction accuracy of the compared methods using a variant of BLPA known as BLPAm (defined below), and also assess mean workload prediction accuracy using MAE and RMSE. As described in Section 2, the MAED and related BLPA metrics were proposed in [8] to provide a measure for the average time error distance between matching predicted and actual bursts. For burst prediction methods that are also capable of predicting magnitude, the difference between predicted and actual burst magnitude should also be considered to help prevent incorrectly matched actual-predicted burst pairs from being assessed (eg: a false positive burst prediction paired with an actual burst of much higher magnitude). Hence, for paired actual-predicted bursts, we extend the MAED calculation of [8] such that cases where the predicted burst magnitude does not fall within ±25% of the actual magnitude are penalized in the calculation with the full time-series length. We refer to this extended form of MAED as MAEDm, and we refer to 1-MAEDm as BLPAm. In future work we aim to extend these metrics further to provide a unified measure for burst location and burst magnitude prediction accuracy.

The MAEDm calculation requires two input sets of burst locations for comparison. In order to identify the bursts in the workload from the actual load time series, as well as bursts in the predicted load time series, we employ a variant of the training phase from Lassnig's burst prediction method. The original method identifies bursts by finding all workload values exceeding a given threshold ($t \times$IQR of time series); finding the shape of bursts where each burst has an associated lower time bound, upper time bound, and burst magnitude; as well as merging any overlapping bursts. We follow a similar approach with the exception that the distance between burst bounds and the burst centre location is 0. This allows for a more accurate assessment of individual burst location prediction, without interference from the merging step of the Lassnig training phase. In order to assess the burst prediction accuracy for bursts of differing magnitude, we evaluate BLPAm for values of t varying from 0.25 to 7.75 (after which no more bursts can be identified in the workload to be predicted) in steps of 0.25.

Each assessed prediction method was trained with a training data set containing 87379 auctions, ending within a 2 day period from 11 April to 12 April, 2012, and selected using a restricted set of search keywords. The bidding workload for the training set, constructed from the bid history of all auctions in this set is shown in Fig 4. The only exception to the use of this training set is ARMA, which does not need fixed parameters that must be learned and then used repeatedly; instead ARMA keeps changing the details of its prediction model over time; where each prediction uses recent history obtained from a sliding window of p past values. Hence the ARMA hypothesis is updated with each shift of the sliding window, and so we evaluate this directly on the testing set; if anything, this evaluation provides an advantage to ARMA.

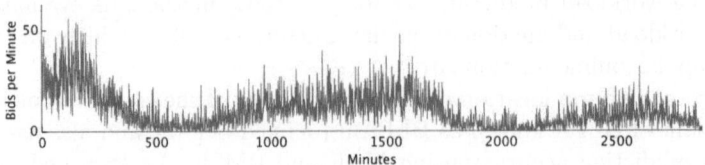

Fig. 4. Bidding Workload from Training Set

The configuration used for the assessed prediction methods is shown in Table 1. The SVR, ARMA, and ANN load prediction methods all employ a sliding window technique, as described in related work. In order to select the size of the sliding window to be used with these methods, we trial sizes 1 to 9, (which were also trialed in [6]), and use the size that yields the lowest MAEDm when predicting workload from the training set. The sliding window size is re-calculated for each value of t that is trialed. Similarly, for calculating the mean prediction accuracy, the window size was adjusted for SVR, ARMA, and ANN to yield the best MAE and RMSE when predicting the workload from the training set. Lassnig's Burst Prediction algorithm [8] outputs a probability series describing the probability of a burst occurring at times within the time-range to be forecasted. A workload burst can be predicted to occur when the probability exceeds a pre-determined threshold of significance; we choose the threshold that yields the lowest MAEDm when predicting workload bursts within the training set.

Table 1. Configuration for Assessed Workload Prediction Methods

Prediction Method	Configuration
SVR	Sequential Minimal Optimization used for training, with the default parameterization used for "SMOReg" in weka 3-6-7 [10].
ANN	A feed forward neural network using the default parameterization for "MultiLayerPerceptron" in weka 3-6-7 [10], apart from using 2 hidden layers as in [8], and a learning rate of 1×10^{-3} (this rate most consistently yielded the lowest MAE from a range of rates trialed varying between the default of 0.3 down through to 1×10^{-5} when running 10 fold cross validation across the training set, for all window sizes of 1 through to 9).
ARMA	An ARMA model is re-created each time a sliding window of size p is formed (or updated), and this model is used for one subsequent prediction.
Lassnig	Based on the value for n used in [8], $n = 3$. The threshold of significance is chosen based on the value that yields the lowest MAEDm when predicting spikes in the training set. The base prediction methods used are the SVR and ANN configurations described above. Configurations corresponding to these base prediction methods shall be referenced as Lassnig-SVR and Lassnig-ANN, respectively.
EAP-bb	Workload prediction using the method described in Section 4.

Once the prediction methods we have discussed are trained, we evaluate them on a testing data set containing 105690 auctions, ending between 30 April and 1 May, 2012, that were selected using the same set of keywords used to select auctions for the training set. This testing set includes a number of significant bursts, where the shaded circles of Fig 5 highlight two of the top bursts. On average there are 4.3 times as many auctions ending per minute within a 20 minute period centered on the most significant burst than the average number of auctions ending per minute for the entire testing set. The presence of a workload spike then becomes justified based on Menasce's finding [11] that most bids are placed towards the end of an auction's lifetime. For each method, we predict workload 15 minutes into the future, where this prediction horizon covers the worst case initialization lag for cloud platforms reported in [9]. In future work we aim to test the hypothesis that long term predictions greater than 15 minutes will favor the accuracy of EAP-bb over the other methods discussed, even further, since EAP-bb is less dependent on recent history.

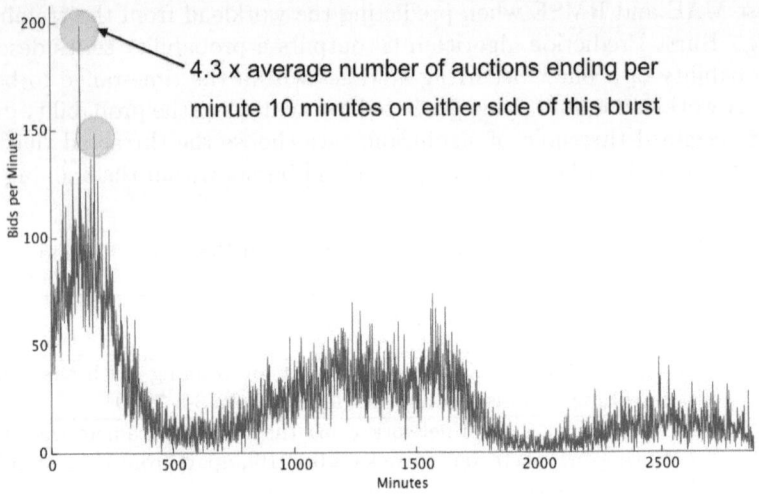

Fig. 5. Bidding Workload from Testing Set

6   Results

This section reports the results for the evaluation described in the previous section. We first report the BLPAm metric for each of the 6 prediction methods assessed. The results shown in Fig 6 were produced by: predicting workload 15 minutes into the future in order to produce a predicted time series; identifying bursts in the workload from the testing set and in the predicted time series as described in the evaluation section; calculating the MAEDm using these two constructed burst collections; and calculating BLPAm from MAEDm.

This process was carried out in a separate trial for each value of t in the range of 0.25 to 7.75 (shown in gray), (after which no more bursts can be identified in the testing workload), in steps of 0.25, in order to assess the burst prediction accuracy for different magnitude bursts. The 95% confidence intervals shown in Fig 6 are calculated based on these multiple trial runs. We also show the burst location prediction accuracy for more significant bursts that are identified by values of t spanning from 4 to 7.75 (in steps of 0.25). If we go to the highest possible t value, where the highest magnitude burst is identified, EAP-bb predicts this burst location with 100% accuracy, Lassnig-ANN with 16.14%, Lassnig-SVR with 12.11%, and SVR, ARMA, and ANN with 0% accuracy.

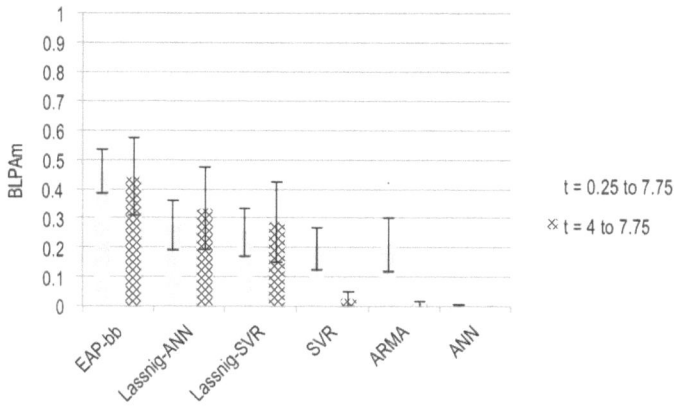

Fig. 6. Predicting Burst Location (BPLAm)

The results thus far have shown that EAP-bb outperforms all other methods discussed for burst prediction accuracy. The results in Table 2 show the traditional MAE and RMSE for all the methods discussed except for Lassnig's method, which does not support workload magnitude prediction. These results indicate that ARMA has the best mean prediction accuracy, whilst EAP-bb provides similar but somewhat worse magnitude prediction across the whole time-series. This aspect is discussed in the next section.

Table 2. Traditional measures of overall workload prediction accuracy

	EAP-bb	SVR	ARMA	ANN
MAE	9.52	7.35	7.25	37.29
RMSE	12.36	11.01	10.43	50.12

7 Discussion

The previous section summarized results for prediction accuracy. When comparing the prediction of burst location, forms of Lassnig's burst prediction method were more accurate than traditional time-series prediction (SVR, ARMA and ANN). This concurs with the results that were published in [8], whose authors use the reason that traditional methods like ARMA and ANN treat workload bursts as outliers in the training phase, to explain their poor performance for burst location prediction accuracy. EAP-bb did even better in predicting burst location than the forms which use Lassnig's approaches.

Catastrophic workload bursts that overwhelm resource provisioning, like those described in the introduction, are expected to be major outliers relative to the average workload. When looking at severe bursts (with higher values of t) we saw that SVR, ARMA, and ANN have much lower burst location prediction accuracy, with EAP-bb again outperforming both forms of Lassnig's prediction method.

As shown in our evaluation results, from all of the methods we assessed, the most accurate burst location prediction was provided by the EAP-bb approach developed in this paper. This superior performance can be attributed to EAP's use of knowledge about events that will occur in the future to predict workload. EAP also has an advantage over Lassnig's proposals, of being able to predict workload burst magnitude, and not just burst location. Although clearly ahead in prediction of bursts, EAP does not do especially well in predicting the mean workload (as measured by MAE or RMSE). This can be attributed to the use of a somewhat generic event workload model that does not fit each individual auction event closely enough. If more individual characteristics of an auction, like auction keywords and seller reputation, were considered in predicting load, one may hypothesize that the load prediction for each auction would be more accurate. In future work, we aim to test this hypothesis using an extensible event workload model, which can consider the influence of an arbitrary number of feature inputs to yield a load prediction. In the auction domain, the extensible model could be trained with additional features, like auction keywords and seller reputation, to yield a more accurate prediction for load. We also aim to investigate how such an extensible model can be used to predict event-associated load for data sets in other domains. Whilst this paper has focused on reporting prediction accuracy results, in future work we also aim to examine the scalability and speed of EAP-bb, and if necessary, investigate potential optimizations.

We started this paper by highlighting the importance of workload burst prediction for maintaining QoS during bursts, as well as for facilitating more cost effective scaling strategies through the use of reserved instances and smarter event scheduling policies. We introduced a new EAP workload prediction approach, which, in contrast to existing approaches, leverages the inherent link often found between important events and workload bursts. The EAP framework introduced can be applied to numerous domains that require event-based burst prediction. A particular model within this framework, EAP-bb, has shown superior burst prediction results when compared to the state of the art, using real workload from the auction domain. Hence these results indicate that, when compared to other prediction methods that use only history, EAP can offer stronger support for realizing the benefits of workload prediction, including effective management of systems during workload bursts.

Acknowledgements. NICTA is funded by the Australian Government as represented by the Department of Broadband, Communications and the Digital Economy and the Australian Research Council through the ICT Centre of Excellence program.

References

1. Ari, I., Hong, B., Miller, E., Brandt, S., Long, D.: Managing flash crowds on the internet. In: International Symposium on Modeling, Analysis and Simulation of Computer Telecommunications Systems (2003)

2. Bermolen, P., Rossi, D.: Support vector regression for link load prediction. Computer Networks 53(2), 191–201 (2009)
3. Bodik, P., Fox, A., Franklin, M.J., Jordan, M.I., Patterson, D.A.: Characterizing, modeling, and generating workload spikes for stateful services. In: Proceedings of the 1st ACM Symposium on Cloud Computing, pp. 241–252. ACM (2010)
4. Chun, J.: Groupon crashes after offering gap deal of the day (August 2010), http://smallbusiness.aol.com/tag/Groupon+site+crash/
5. Dietterich, T.G.: Machine Learning for Sequential Data: A Review. In: Caelli, T.M., Amin, A., Duin, R.P.W., Kamel, M.S., de Ridder, D. (eds.) SSPR & SPR 2002. LNCS, vol. 2396, pp. 15–30. Springer, Heidelberg (2002)
6. Islam, S., Keung, J., Lee, K., Liu, A.: Empirical prediction models for adaptive resource provisioning in the cloud. In: Future Generation Computer Systems (2011)
7. Jiang, Y., Perng, C., Li, T., Chang, R.: Asap: A self-adaptive prediction system for instant cloud resource demand provisioning. In: 2011 IEEE 11th International Conference on Data Mining (ICDM), pp. 1104–1109. IEEE (2011)
8. Lassnig, M., Fahringer, T., Garonne, V., Molfetas, A., Branco, M.: Identification, modelling and prediction of non-periodic bursts in workloads. In: Proceedings of the 2010 10th IEEE/ACM International Conference on Cluster, Cloud and Grid Computing, pp. 485–494. IEEE Computer Society (2010)
9. Li, A., Yang, X., Kandula, S., Zhang, M.: Cloudcmp: comparing public cloud providers. In: Internet Measurment Conference, pp. 1–14. ACM (2010)
10. Hall, M., Frank, E., Holmes, G., Pfahringer, B., Reutemann, P., Witten, I.H.: The WEKA Data Mining Software: An Update. SIGKDD Explorations 11(1) (2009)
11. Menascé, D., Akula, V.: Towards workload characterization of auction sites. In: 2003 IEEE International Workshop on Workload Characterization, WWC-6, pp. 12–20. IEEE (2003)
12. Mills, T.: Time series techniques for economists. Cambridge Univ. Pr. (1991)
13. Nancarrow, D.: Student anger: traffic crashes qtac site (January 2012), http://www.brisbanetimes.com.au/it-pro/government-it/student-anger-traffic-crashes-qtac-site-20120112-1pwbs.html
14. Amazon Web Services: Amazon ec2 reserved instances (June 2012), http://aws.amazon.com/ec2/reserved-instances/
15. Sladescu, M., Fekete, A.: Event aware elasticity control for cloud applications. Tech. Rep. 687, The University of Sydney (2012)
16. Smola, A., Schölkopf, B.: A tutorial on support vector regression. Statistics and Computing 14(3), 199–222 (2004)
17. Tirado, J.M., Higuero, D., Isaila, F., Carretero, J.: Multi-model prediction for enhancing content locality in elastic server infrastructures. In: 2011 18th International Conference on High Performance Computing (HiPC), pp. 1–9. IEEE (2011)
18. Vlachos, M., Meek, C., Vagena, Z., Gunopulos, D.: Identifying similarities, periodicities and bursts for online search queries. In: Proceedings of the 2004 ACM SIGMOD International Conference on Management of Data, pp. 131–142. ACM (2004)
19. Wardrop, M.: ebay facing compensation bill after site crashes (November 2009), http://www.telegraph.co.uk/finance/newsbysector/retailandconsumer/6641314/eBay-facing-compensation-bill-after-site-crashes.html
20. Whyte, S.: One direction fever: Ticketek server crashes in rush to buy tickets, fans say (May 2012), http://www.smh.com.au/entertainment/music/one-direction-fever-ticketek-server-crashes-in-rush-to-buy-tickets-fans-say-20120428-1xrb4.html

Like-Minded Communities:
Bringing the Familiarity and Similarity together

Natwar Modani, Ritesh Gupta, Seema Nagar, Saswata Shannigrahi,
Saurabh Goyal, and Kuntal Dey

IBM India Research Lab

Abstract. Community detection in social networks is a well-studied
problem. A community in social network is commonly defined as a group
of people whose interactions within the group are more than outside the
group. It is believed that people's behaviour can be linked to the behaviour of their social neighbourhood. While shared characteristics of
communities have been used to validate the communities found, to the
best of authors' knowledge, it is not demonstrated in the literature that
communities found using social interaction data are like-minded, i.e.,
they behave similarly in terms of their interest in items (e.g., movie,
products). In this paper, we propose a method for finding communities wherein like-mindedness is an explicit objective. We find small tight
groups with many shared interests using a frequent item set mining approach and use these as building blocks for the core of these like-minded
communities. We show that these communities have higher similarity in
their interests compared to communities found using only the interaction
information. We also compare our method against a baseline where the
weight of edges are defined based on similarity in interests between nodes
and show that our approach achieves far higher level of like-mindedness
amongst the communities compared to this baseline as well.

1 Introduction

Forming communities is one of the fundamental traits of people in social networks. Many methods have been proposed and investigated for finding communities in social networks [1–6]. These communities are based on the communication
patterns in the social networks. A community in social network is defined as a
group of people whose interactions within the group are more than outside the
group. Communities have often been linked to common characteristics of the
community members. In fact, quite often, the presence of a common characteristic among members of a community is taken as an indicator of good quality of
the communities found. For example, the well known Zachary's karate-club and
NCAA college football datasets are used as test of quality of the community finding algorithms (e.g., in [7, 8]), where the membership of the faction/conference
is taken as the ground truth. Similarly, in [5], the fact that the language of majority of the community members is the same is taken as an indicator of good
quality of the found communities.

X.S. Wang et al. (Eds.): WISE 2012, LNCS 7651, pp. 382–395, 2012.

While shared characteristics have been used to evaluate the quality of found communities, they are not taken as input while determining communities in the approaches known to the authors. In fact, the authors are not aware of any formal notion of like-mindedness for communities. In this paper, we take interest in items along with the underlying friendship graph as an explicit input while detecting the communities. In addition, unlike the single attribute characterization of the solution, we take the interest of social network participants in multiple items into account. We also formalize the notion of like-mindedness for the communities. The importance of such communities from a practical point-of-view is significant, since the members of a community are both well connected and like minded.

Recently, rich data has become available containing information about the social network as well as purchase or like/dislike data for the social network participants. Examples include filmtipset [9] which contains information about viewers's rating about movies, as well as the friendship relation amongst them. Another example is telecom domain where call detail records (CDR) as well as purchase/subscription to value added services (VAS) data are available.

It is believed that people's behaviour can be linked to the behaviour of their social neighbourhood. The influence of social neighbourhood was demonstrated in the churn behaviour of telecom subscribers in [10]. This would suggest that information about interest in products should be taken into account (when available) while finding communities. However, to the best of authors' knowledge, the information about the interests of the people are not used in community finding algorithms.

We define a community finding problem suitable for the setting, where the participant of a community also share many interest. We provide an approach to find such communities. Our approach is flexible and is at a meta level. Our approach consists of finding tight communities, performing frequent itemset mining to find core members of the like-minded communities and finding the communities amongst the core community members. We show that the communities we detect based on our method exhibit a stronger tendency to like/dislike items *together* compared to communities detected using only the social interaction data or compared to communities found where the weight of edges of the graph are defined based on similarity in interests between nodes.

In summary, our contributions in this paper are:

- We characterize the desirable properties of like-minded communities. We also formalize the notion of like-mindedness of communities.
- We provide an approach to find like-minded communities. We apply our technique to find such communities on real life dataset.
- We show that the like-minded communities detected using our method exhibit a stronger tendency to like/ dislike items *together* compare to communities found using only social interaction information or compared to communities detected using baseline where the weight of edges of the graph are defined based on similarity in interests between nodes.

The rest of the paper is organized as follows. In Section 2, we survey the related literature. In Section 3, we describe the dataset used in the experiments. In Section 4, we present our approach to finding like-minded communities. In Section 5, we describe our experiments and results. Finally, we conclude in Section 6.

2   Related Work

Identification of communities in social networks has been an active area of research. There have been multiple schools of thoughts and approaches towards defining and identifying communities. Girvan and Newman [4] defined communities as subsets of vertices within which inter-vertex connections are dense but between which connections are less dense. This work was followed up further in [1]. The objective of these approaches are to maximize the modularity of the graph using a greedy strategy of grouping subsets of vertices together. Another body of work investigates structural communities which include clique, quasi-clique, k-core, k-clique, k-club and k-plex [2, 3][11, 12]. Overlapping communities have been studied in [5][13][7]. Communities of similar people are found using interests in items as explicit input, but without using social connections [14].

All the above work have used either connections or interests of users to find communities. Little attention has been paid to integration of connections among the people and their shared interests to find communities while quite common in other purposes like recommendation [15], information retrieval and link prediction [16, 17]. Silva et al. [18] proposed a structural correlation pattern mining problem which aims to find pairs (S, V) in which S is a frequent attribute set and V is a dense subgraph where each node contains all the attributes in S. However, there are two main shortcomings of this work. First, it only finds quasi-cliques as communities. Second, and more important, one needs to know the set of attributes that are to be shared by the community a priori, which is not practical in the community finding task in a social network setting.

The relevance of social ties with behavior (e.g., churn) of customers has been discussed in [10]. In this paper, it was shown that people who know past churners (who relinquish the service of a telecom service provider) are more likely to churn themselves. Moreover, it was also shown that if the past churners are mutually known, their influence on common friends is even more pronounced.

Frequent itemset (and association rule) mining is a popular technique to find group of objects that repeat often [19–23]. Typically, frequent itemset mining is used for market basket analysis. Here, we use the frequent itemset mining as a way to compute the core of multiple overlapping cliques by treating the cliques as transactions and trying to find the nodes which participate in many cliques as frequent itemsets. Some well known algorithms for frequent itemset mining are Apriori [19], Eclat [23] and FP-Growth [20].

We experiment on a dataset that consists of the friendship graph and users' ratings of movies in a movie recommendation website *www.filmtipset.se*. Data from www.filmtipset.se has been analyzed in [9]. Movie recommendation websites have been topic for research recently because of the availability of data [9]. Examples of such websites are MovieLens, GroupLens and Netflix. For creating better movie recommendations, these websites have been previously analyzed in [24–28]. One popular recommendation technique is collaborative filtering which is based on the similarity of user interests/ratings. This is a well-studied area [29, 30]. However, collaborative filtering does not take user to user social interactions into account. [31] and [9] were among the ones to have used social graphs to improve movie recommendations. However, the problem of finding communities is not addressed in this literature.

3   Notation and Datasets

Let $G = (V, E)$ be a simple, undirected, unweighted graph representing the social network, where $v \in V$ represents a person and $e \in E$ represents a relation between a pair of people (u, v). Let the set of items of interest be $T = \{t_1, t_2, \ldots, t_m\}$. Let the bipartite graph $B = (V, T, R)$ represent the level of interest of people for items. We call the graph B as the rating graph of users V on items T. Here, V is the set of people, T is the set of items and R is the set of level of interest of users on the items T (users giving ratings to movies is an example of level of interests). The edges in graph B are given as ordered triplets (v_i, t_j, r_{ij}), where a user v_i expresses a level of interest r_{ij} for an item t_j.

We use the notion of compatibility of interest in an item for the users. The compatibility can be defined in any way suitable for the problem domain. The compatibility relation is a reflexive and symmetric relation.

Consider an item t_j. From the base graph G, we derive a subgraph $G_j = (V_j, E_j)$ such that for an edge $e = (u, v) \in E_j$, the nodes u and v have compatible interest in the item t_j and are connected in the base graph G. V_j is the set of vertices v such that $\exists e = (v, w) \in E_j$, i.e., that each vertex in V_j has at least one compatible interest edge for t_j incident on it.

We use data from a Swedish movie recommendation website *www.filmtipset.se* for the experiments reported in this paper. People using this website rate movies on a scale of 1 to 5, with 1 being the lowest and 5 being the highest rating. In addition, users can make *friends* among themselves. We construct the social network representation from the friendship relation. We identify the people who are connected to at least one other user since users who are socially disconnected do not affect the results of our analysis. There are $24,834$ such socially connected users having $61,775$ undirected edges between them. We call this graph as base graph. We consider $1,212$ movies rated sufficiently often by these socially connected users.

Although we do not insist on the transitivity property for the compatibility relation, the example compatibility relations we show in our experiments are transitive, and hence are equivalence relations. We use the following three compatibility relations between interests with respect to a movie:

1. **Equivalent:** Two persons have equivalent type of compatible interest if they rate the same movie with the same rating.

2. **Near-equivalent:** Two persons have near-equivalent type of compatible interest if they rate the same movie with the similar rating. Ratings 1 and 2 are considered to be similar to each other (low rating), and ratings 4 and 5 are considered to be similar to each other (high rating). All rating are also considered similar to themselves.
3. **Shared Interest:** Two persons have shared interest type of compatible interest if both of them rate the same movie (with any rating).

For each movie, we construct 3 graphs based on the 3 compatibility relations between the ratings of users. We retain only the edges in a graph if the interests of the two persons (on the two ends of the edge) for this movie are compatible. We retain only the persons on whom at least one such edge is incident. We denote the graphs derived using the equivalent, near-equivalent and shared interest compatibility of interest for a movie i as G_i^e, G_i^n and G_i^s, respectively.

4   Method to Find Like-Minded Communities

We define like-minded community as the group of social network participants that are socially well connected and share many interests. By socially well connected, we mean that the social network participants have a lot of connections between themselves but fewer connections with outsiders. By sharing interests, we mean that their interest levels in the same product are compatible (which, as explained earlier, can be defined suitably for the problem domain).

Let C denote a set of communities on a social network graph G. We use set of communities and community structure interchangeably in the paper. The modularity for C as used in [1] is a fair indicator of social *well connectedness*, which has also been generalized/modified. In particular, we will use the generalized modularity as defined in [13] which can account for overlapping communities and is equivalent to the conventional definition for non-overlapping communities.

$$Q(C) = \frac{1}{2m} \sum_{uv} (A_{uv} \delta(u, v) - \frac{d_u . d_v}{2m} . |L_u \cap L_v|)$$

where m is the number of undirected edges in the graph (i.e., $|E|$), A_{uv} is the edge weight of the edge between u and v and d_u is the degree of node u. L_u is the set of communities which u is a member of. $\delta(u, v)$ is 1 if $|L_u \cap L_v| > 0$, and 0 otherwise.

To formalize the notion of like-mindedness, we propose a new measure to capture the level of compatible interests amongst the community members. Let C denote a set of communities on a social network graph G and bipartite rating graph B. Let $c \in C$ be a community. Let u and v be two members of community c. We define the similarity between u and v as the cosine similarity of their ratings, i.e.,:

$$Cos(u, v) = \frac{R_u . R_v}{||R_u|| . ||R_v||}$$

where R_u is the rating vector of user u. The numerator is the dot product of the rating vector (please note that it would be dependent on the interest compatibility relation used) and the denominator is product of norms of these rating vectors.

When the two users do not rate enough items together, the cosine similarity may give a wrong picture. For example, if two people only rate one movie in common and they rate it exactly the same way, assigning them a similarity value of 1 may give misleading picture. Also, it is not very robust, since a change in only one rating can change the similarity value by a large amount. In [32], the cosine similarity is scaled down if there are not enough ratings in common. If the number of movies rated in common by u and v is $w < 50$, then the cosine similarity is scaled by $w/50$. That is, the weighted cosine similarity between u and v is given as:

$$Cos_{wt}(u,v) = \frac{w}{50} \times \frac{R_u.R_v}{||R_u|| \cdot ||R_v||} \quad if \ \ w < 50$$

$$= \frac{R_u.R_v}{||R_u|| \cdot ||R_v||} \quad if \ \ w \geq 50$$

However, one can use a rating similarity definition that one prefers, e.g., L_1 or L_2 distance between the ratings if it is suitable for the problem domain.

We propose the like-mindedness of a set of communities C as:

$$S(C) = \frac{1}{\sum_{u,v} \delta(u,v)} \sum_{u,v} Sim(u,v) \times \delta(u,v)$$

where $\delta(u,v)$ is 1 if u and v belong to the same community and 0 if they belong to different communities. It represents the like mindedness of the people who belong to the same community, summed over. It is reminiscent of the first term in modularity and uses the similarity of the two users in place of edge weight (which may be binary for unweighted graphs). However, unlike modularity, the notion of expected similarity is not required here. Please also note that the proposed definition of like-mindedness works for overlapping communities, too. Of course, by taking the similarity between u and v as $Cos(u,v)$ and $Cos_{wt}(u,v)$, we get two versions of similarity score for the set of communities C, which we denote as $S_{cos}(C)$ and $S_{wc}(C)$, respectively.

Next we describe our approach to find like-minded communities. One highlight of our approach is that we have a lot of flexibility regarding the choices that need to be made. Our approach is at a meta level, wherein suitable definitions, measures and algorithms can be chosen depending on the problem domain and the preferences of the user.

Recall that like-minded communities are socially tightly knit groups which share many interests. Let $G = (V, E)$ be the base graph that describes the social interaction. Let $B = (V, T, R)$ be the bipartite graph describing the interests R of people V in items T. Let r denote the compatibility of interests of people in items. If u and v have compatible interest for item i, then we denote this by

$(u \; r_i \; v)$. We derive $|T|$ graphs from this information for these two graphs in the following manner. For each item t_i, we construct a graph G_i^r as $G_i^r = (V_i^r, E_i^r)$ such that $e = (u, v) \in E_i^r$ iff $e \in E$ and $(u \; r_i \; v)$. Further, V_i^r is the set of vertices v such that $\exists e \in E_i^r$ and e is incident on v. In other words, for each item, we retain the set of edges from the base graph such that the interests of the nodes on the two sides of the edge are compatible for the chosen compatibility relation. We find maximal cliques in these graphs G_i^r ($i = 1, 2, , |T|$). A set of vertices K is called as a clique if all the vertices in K are connected to each other. Further, if there is no $K' \supset K$ such that K' is also a clique, then K is called a maximal clique. These maximal cliques are used as the building block for the like-minded communities.

We consider each maximal clique as a transaction with the people as items. We attach the item for which the maximal clique is found to the transaction as an additional attribute. Now we run frequent itemset mining on this dataset to find all the closed frequent itemsets. An itemset is called as closed frequent itemset if it has the minimum desired support (i.e., this set of items occurs together in sufficient number of transactions) and there is no itemset which contains this itemset and has equal support.

While finding the closed frequent item sets, we also keep track of distinct items t_j in which this item set appears together. After we have found the closed frequent itemsets, we can filter these item sets to keep only the item sets that have number of distinct items t_j of compatible interest (besides the required support in conventional sense).

Next, we take the union of all the closed frequent itemsets, which in turn is a set of people and take the induced subgraph of these people. We find communities on this graph using a community finding method. We do not insist on a particular community finding method.

In our experiments, we show the results for two community finding methods. The communities thus found are the like-minded communities. In the next section, we will show that these communities are indeed more similar in their rating behaviour as compared to the communities found using only the social interaction information (i.e., on the base graph). This implies that our method is effective in finding like-minded communities.

Example: Let us consider a social network consisting of 10 people, $u1$, $u2$, ..., $u10$ as shown in Figure 1. Let there be 4 products, $P1$, $P2$, $P3$ and $P4$. Table 1 shows which person has purchased (or shown interest in) which product.

Table 1. Purchase/interest data for the social network participants. Y indicates purchase/interest and N indicates lack thereof.

Product	$u1$	$u2$	$u3$	$u4$	$u5$	$u6$	$u7$	$u8$	$u9$	$u10$
P1	Y	Y	Y	Y	N	N	Y	Y	Y	Y
P2	Y	Y	Y	N	Y	Y	Y	Y	N	N
P3	N	Y	Y	N	Y	Y	Y	Y	N	Y
P4	N	Y	Y	N	Y	N	N	N	N	N

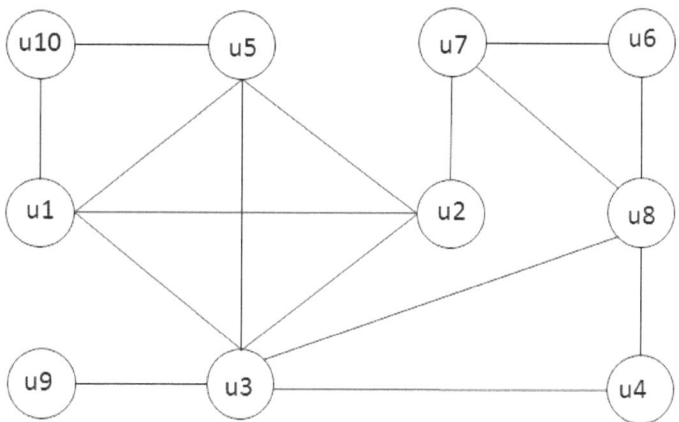

Fig. 1. An example: The social network of 10 participants

The first step of our algorithm will construct 4 induced subgraphs from the base graph corresponding to the 4 products. For example, to construct the subgraph corresponding to $P1$, we consider only the people who have shown interest in $P1$, in this case $u1$, $u2$, $u3$, $u4$, $u7$, $u8$, $u9$ and $u10$ (i.e., everyone except $u5$ and $u6$). Now, we find maximal cliques of size 3 and above in each of these subgraphs. We get $\{u1, u2, u3\}$ and $\{u3, u4, u8\}$ as the maximal cliques corresponding to product $P1$, $\{u1, u2, u3, u5\}$ and $\{u6, u7, u8\}$ corresponding to $P2$, $\{u2, u3, u5\}$ and $\{u6, u7, u8\}$ corresponding to $P3$ and $\{u2, u3, u5\}$ corresponding to $P4$. Now, we treat each of these maximal cliques as a transaction and people in those cliques as items and run frequent item set mining. The number of transactions under consideration would be 7, with 8 distinct items since $u9$ and $u10$ do not appear in any maximal clique in the product induced subgraphs. This is noteworthy since $u10$ does form a maximal clique of size 3 in the base graph along with $u1$ and $u5$, but they do not have any shared interests.

When we take minimum support as 2, the closed frequent item sets found are $\{u1, u2, u3\}$ (with a support of 2), $\{u2, u3, u5\}$ (with a support of 3) and $\{u6, u7, u8\}$ (with a support of 2). The support in terms of distinct products purchased together by people for all these frequent item sets is equal to the support based on number of transaction they appear in together (reported in parenthesis) in this example. Hence, taking a support value of 2 in terms of number of products purchased together results in the same item sets. Now we take the union of all the people who are part of the frequent item sets $\{u1, u2, u3, u5, u6, u7, u8\}$ (everyone except $u4$ from the people who were part of maximal cliques found on the product induced subgraphs). When we run CNM method [1] of community finding on it, we find the two like-minded communities as $\{u1, u2, u3, u5\}$ and $\{u6, u7, u8\}$.

In contrast, communities found from CNM directly on the base graph are $\{u1, u2, u5, u10\}$, $\{u3, u4, u9\}$ and $\{u6, u7, u8\}$. The modularity for the communities we find is 0.28099 and for the community found by CNM method is 0.27539. The like-mindedness based on cosine similarity of communities found by us is 0.77157 and for the communities found by *CNM* method is 0.66553. The like-mindedness based on weighted cosine similarity of communities found by us is 0.04017 and for the communities found by *CNM* method is 0.02398. The like-mindedness using the weighted cosine similarity is much lower than usual cosine similarity based like-mindedness as we have only 4 products here.

5   Experiments

Now we describe the experimental setup and results. We use three types of compatibility in user interests, namely, *Equivalent, Near-Equivalent* and *Shared Interest*. All the further steps are repeated for each of these three types of compatibilities. The compatibility relations give us one graph per movie. We find maximal clique on all these $1,212$ graphs. For finding the maximal cliques, we take the minimum desired clique size as 3. We find all such maximal cliques using the SEL*MaC*2 algorithms proposed in [12].

After the maximal cliques, we find the closed frequent itemsets using the method proposed in [19]. We take the minimum support level as 5 for *equivalent* and *near-equivalent*, and 10 for *shared interest*. We take the minimum size of the frequent itemset as 3. We then take different values for minimum number of items of shared interest and plot against number of core community members (Figure 2).

We pick 5 different support levels of number of distinct movies the group of people (who form the frequent item set) have compatible interest in. For each support level, we retain only the people who are part of at least one frequent item set with this support. We call this group of people as core community members. We find the like-minded communities by performing community finding on these core community members. Again, we use both *CNM* method [1] and *cFinder* method [5] to find communities. For *cFinder* method, we vary the value of k (the size of clique which is used for constructing the communities) from 3 to 5 and report the numbers for all of these. Our results are presented in Table 2 and Table 3. As the measure of goodness of results, we compute the modularity and like-mindedness of the community structures as defined in Section 4. Higher value of modularity indicates more familiarity amongst community members (traditional measure of goodness of community structures with respect to a graph). Higher value of the two like-mindedness scores indicate higher level of similarity of interests amongst the community members.

First, we compare the communities found by our method against the communities found on the base graph using *CNM* and *cFinder*, respectively. We measure goodness of communities on two dimensions: modularity Q and like-mindedness S. For like-mindedness, we present the two variants based on cosine similarity and weighted cosine similarity of the users, i.e., $S_{cos}(C)$ and $S_{wc}(C)$. Note that while the base line community structure remains the same (and thus the number of nodes, communities and modularity remain the same) across the choices of interest compatibility, the like-mindedness score is dependent on the compatibility relation and hence would change.

Table 2. Comparison of goodness of communities found using *CNM* method both for base line and for like-minded communities. Sup is the minimum item level support, #Nodes is the number of nodes in the graph, $|C|$ is the number of communities found, $Q(C)$ is the modularity of the communities, $S_{cos}(C)$ and $S_{wc}(C)$ are the cosine and weighted cosine based like-mindedness of the communities.

Sup	#Nodes	$|C|$	$Q(C)$	$S_{cos}(C)$	$S_{wc}(C)$
Compatibility Relation: Equivalent					
74	2,049	146	0.715728	0.595467	0.595467
50	4,228	194	0.733075	0.535283	0.535262
25	8,371	222	0.739255	0.454128	0.453543
12	12,334	278	0.741095	0.399507	0.394484
6	15,112	309	0.747143	0.358336	0.344198
Base	24,834	726	0.838898	0.307665	0.283171
WtBL	24,820	671	0.768685	0.307905	0.283763
Compatibility Relation: Near-equivalent					
112	2,082	149	0.687382	0.595956	0.595956
80	4,017	183	0.727668	0.542125	0.542113
40	8,501	217	0.732624	0.455592	0.455059
20	12,451	273	0.735585	0.394139	0.388546
10	15,344	310	0.743731	0.354661	0.339600
Base	24,834	726	0.838898	0.307665	0.283171
WtBL	24,825	669	0.766182	0.307941	0.283872
Compatibility Relation: Shared Interest					
220	2,335	151	0.615243	0.600574	0.600573
160	4,043	179	0.670256	0.546479	0.546478
80	8,330	215	0.704675	0.460333	0.459591
40	12,193	262	0.727381	0.403169	0.397680
15	15,464	325	0.736291	0.357331	0.342518
Base	24,834	726	0.838898	0.313281	0.289282
WtBL	24,832	671	0.763807	0.313182	0.289270

Since the baseline above completely ignores the interests of the people, we take another baseline, which tries to incorporate the rating information in the social network. Given the base graph G and interest graph B, we construct a weighted graph $W = (V_w, E_w)$ by retaining only the edges for which the people on the two sides have at least one compatible interest and assign the edge weight as the number of compatible interests between the two people. Now we run the *CNM* method of community finding on this weighted graph. We measure the like-mindedness and modularity and present the results in Table 2 with method name as *WtBL* (for weighted baseline). Please note that the modularity value reported is still based on the unweighted base graph. Also note that for this baseline, the number of nodes and communities would change based on the interest compatibility relation, unlike the other baseline.

One can see that using the method proposed here, we get communities which are socially well-connected as well as like-minded. While the communities found using

Table 3. Comparison of goodness of communities found using *cFinder* method both for base line and for like-minded communities. K is the size of cliques used in *cFinder* method and all other parameters are same as Table 2.

K	Sup	#Nodes	$\|C\|$	$Q(C)$	$S_{cos}(C)$	$S_{wc}(C)$
Compatibility Relation: Equivalent						
3	74	2,036	334	0.56589	0.57388	0.57387
3	50	4,189	614	0.54721	0.51195	0.51190
3	25	8,278	1,133	0.54264	0.43011	0.42920
3	12	12,121	1,651	0.53819	0.37426	0.36766
3	6	14,485	2,056	0.55483	0.34267	0.32928
3	Base	16071	2371	0.57038	0.32118	0.30207
4	74	1,211	207	0.73475	0.60667	0.60667
4	50	2,663	459	0.77846	0.54420	0.54420
4	25	5,501	918	0.81196	0.45052	0.44995
4	12	7,959	1339	0.81539	0.39405	0.38847
4	6	9,283	1,599	0.82529	0.36833	0.35686
4	Base	9,663	1632	0.82970	0.36438	0.35095
5	74	651	94	0.76386	0.62596	0.62596
5	50	1,480	215	0.79879	0.57768	0.57767
5	25	3,121	460	0.82570	0.50359	0.50333
5	12	4,391	650	0.83744	0.45367	0.45046
5	6	5,104	779	0.84440	0.43706	0.43084
5	Base	5,141	774	0.84553	0.43249	0.42455
Compatibility Relation: Near-equivalent						
3	112	2,066	319	0.52554	0.57508	0.57508
3	80	3,977	593	0.53054	0.51964	0.51962
3	40	8,403	1,168	0.53696	0.43179	0.43095
3	20	12,230	1,681	0.53722	0.37433	0.36763
3	10	14,710	2,106	0.56086	0.34061	0.32650
3	Base	16,071	2371	0.57038	0.32118	0.30207
4	112	1,259	213	0.71582	0.60119	0.60119
4	80	2,502	425	0.76606	0.55216	0.55216
4	40	5,515	917	0.81351	0.45439	0.45396
4	20	7,998	1,354	0.81369	0.38998	0.38361
4	10	9,366	1,618	0.82479	0.36724	0.35557
4	Base	9,663	1,632	0.82970	0.36438	0.35095
5	112	681	102	0.74225	0.62462	0.62462
5	80	1,341	193	0.79222	0.58401	0.58399
5	40	3,125	451	0.82429	0.50163	0.50133
5	20	4,364	645	0.83585	0.45299	0.44971
5	10	5,129	789	0.84537	0.43819	0.43205
5	Base	5,141	774	0.84553	0.43249	0.42455

our method are smaller (but one can choose the size) and slightly lower modularity (as expected) compared to the base line methods, the increase in like-mindedness is significant. In fact, surprisingly, the baseline approach that assigns weights according to the similarity between nodes on the familiarity edges does not perform

Table 3. *Continued*

		Compatibility Relation: Shared Interest				
3	220	2,335	420	0.45954	0.59147	0.59147
3	160	4,043	736	0.51153	0.54161	0.54160
3	80	8,330	1,330	0.54736	0.44928	0.44838
3	40	12,193	1,853	0.54730	0.38926	0.38296
3	15	15,464	2,282	0.56409	0.33936	0.32279
3	Base	16,071	2,371	0.57038	0.33018	0.31139
4	220	860	150	0.56659	0.60833	0.60833
4	160	1,727	320	0.66411	0.57503	0.57503
4	80	4,378	792	0.77448	0.48921	0.48890
4	40	6,867	1,197	0.80263	0.42343	0.41918
4	15	9,273	1,576	0.82580	0.38063	0.36898
4	Base	9,663	1,632	0.82479	0.37402	0.36085
5	220	350	50	0.54057	0.62415	0.62415
5	160	688	105	0.66259	0.60615	0.60615
5	80	2,028	306	0.77838	0.52958	0.52944
5	40	3,389	508	0.81681	0.48954	0.48836
5	15	4,885	732	0.84173	0.44090	0.43483
5	Base	5,141	774	0.84553	0.43521	0.42772

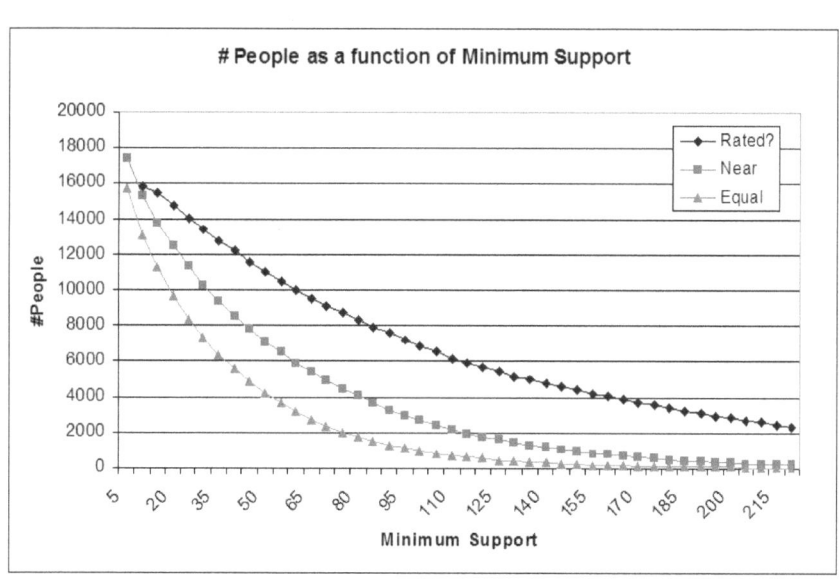

Fig. 2. Number of core community members as a function of minimum item support level

better than the communities found using only the familiarity edges. The increase in similarity is minimal and the reduction in modularity is significant.

6   Conclusions

In this paper, we propose a novel approach for finding like-minded communities. In this approach, we begin with finding social communities around induced subgraphs formed by considering every pair of socially connected people who have shared interest in an item, and do this for all items. We take all such communities found across all items and follow this up by finding closed frequent itemsets in which each community is treated as a distinct transaction and each community member is a distinct item in the transaction. We now get communities which frequently had shared interests across products. We then filter these found frequent communities and retain those who had shared interests across some minimum number of products. We define this as core set of people, which we show have a stronger tendency in terms of shared interests across items, indicating like-mindedness. We find communities amongst the like-minded people. We also demonstrate experimentally that the communities thus found have significantly higher like-mindedness and very good social connectivity.

References

1. Clauset, A., Newman, M.J., Moore, C.: Finding community structure in very large networks. Phys. Rev. E 70(066111) (2004)
2. Dourisboure, Y., Geraci, F., Pellegrini, M.: Extraction and classification of dense communities in the web. In: WWW, pp. 461–470 (2007)
3. Gibson, D., Kleinberg, J., Raghavan, P.: Inferring web communities from link topology. In: HYPERTEXT, pp. 225–234 (1998)
4. Girvan, M., Newman, M.E.J.: Community structure in social and biological networks. Proc. Ntl. Acad. Sci, USA 99(7821) (2002)
5. Palla, G., Derenyi, I., Farkas, I., Vicsek, T.: Uncovering the overlapping community structure of complex networks in nature and society. Nature 435(7043), 814–818 (2005)
6. Fortunato, S.: Community detection in graphs. Physics Reports 486(3-5), 75–174 (2010)
7. Sun, H., Huang, J., Han, J., Deng, H., Sun, Y.: SHRINK: A Structural Clustering Algorithm for Detecting Hierarchical Communities in Networks. In: CIKM 2010, Toronto, Canada (October 2010)
8. Newman, M.E.J.: Fast algorithm for detecting community structure in networks. Phys. Rev. E 69, 66133 (2004)
9. Said, A., Luca, E.W.D., Albayrak, S.: How social relationships affect user similarities. In: SRS (2010)
10. Dasgupta, K., Singh, R., Viswanathan, B., Chakraborty, D., Mukherjea, S., Nanavati, A.A., Joshi, A.: Social ties and their relevance to churn in mobile telecom networks. In: EDBT, pp. 668–677 (2008)
11. Hanneman, R.A., Riddle, M.: Introduction to Social Network Methods. University of California, Riverside (2005)
12. Modani, N., Dey, K.: Large maximal cliques enumeration in sparse graphs. In: CIKM, pp. 1377–1378 (2008)
13. Chen, W., Liu, Z., Sun, X., Wang, Y.: A game-theoretic framework to identify overlapping communities in social networks. Data Min. Knowl. Discov. 21(2), 224–240 (2010)

14. Jaho, E., Karaliopoulos, M., Stavrakakis, I.: Iscode: a framework for interest similarity-based community detection in social networks. In: IEEE INFOCOM Workshops, pp. 912–917 (2011)
15. Guy, I., Werdling, N., Carmel, D., Ronen, I., Uziel, E., Yogev, S., Ofek-Koifman, S.: Personalized recommendation of social software items based on social relations. In: RecSys, pp. 53–60 (2009)
16. Cohn, D., Hofman, T.: The missing link - a probabilistic model of document content and hypertext connectivity. Advances In Neural Information Processing Systems 21, 430–436 (2001)
17. Nallapati, R.M., Ahmed, A., Xing, E.P., Cohen, W.W.: Joint latent topic models for text and citations. In: Proc. of the 14th ACM SIGKDD (KDD 2008), pp. 542–550. ACM, New York (2008)
18. Silva, A., Wagner Meira, J., Zaki, M.J.: Structural correlation pattern mining for large graphs. In: 8th Workshop on Mining and Learning with Graphs (with SIGKDD) (July 2010)
19. Agrawal, R., Srikant, R.: Fast algorithms for mining association rules in large databases. In: VLDB, pp. 487–499 (1994)
20. Han, J., Pei, J., Yin, Y., Mao, R.: Mining frequent patterns without candidate generation. Data Mining and Knowledge Discovery 8, 53–87 (2004)
21. Srikant, R., Agrawal, R.: Mining quantitative association rules in large relational tables. In: SIGMOD, pp. 1–12 (1996)
22. Wang, J., Jiawei, H., Pei, J.: Closet+: searching for the best strategies for mining frequent closed itemsets. In: SIGKDD, pp. 236–245 (2003)
23. Zaki, M.J.: Scalable algorithms for association mining. IEEE Transactions on Knowledge and Data Engineering 12(3), 372–390 (2000)
24. Ahn, S., Shi, C.: Exploring movie recommendation system using cultural metadata. In: Proc. of the 2008 Intl. Conf. on Cyberworlds, pp. 431–438 (2008)
25. Amatriain, X., Pujol, J.M., Tintarev, N., Oliver, N.: Rate it again: increasing recommendation accuracy by user re-rating. In: Proc. of the Third ACM Conf. on Recommender Systems, pp. 173–180 (2009)
26. Golbeck, J., Hendler, J.: Filmtrust: movie recommendations using trust in web-based social networks. In: CCNC, pp. 282–286 (2006)
27. Ono, C., Kurokawa, M., Motomura, Y., Asoh, H.: A Context-aware Movie Preference Model Using a Bayesian Network for Recommendation and Promotion. In: Conati, C., McCoy, K., Paliouras, G. (eds.) UM 2007. LNCS (LNAI), vol. 4511, pp. 247–257. Springer, Heidelberg (2007)
28. Piłászy, I., Tikk, D.: Recommending new movies: even a few ratings are more valuable than metadata. In: Proc. of the Third ACM Conf. on Recommender Systems, pp. 93–100 (2009)
29. Sarwar, B., Karypis, G., Konstan, J., Reidl, J.: Item-based collaborative filtering recommendation algorithms. In: WWW, pp. 285–295 (2001)
30. Breese, J.S., Heckerman, D., Kadie, C.M.: Empirical analysis of predictive algorithms for collaborative filtering. In: Uncertainty in Artificial Intelligence, pp. 43–52 (1998)
31. Bonhard, P., Sasse, M.: Knowing me, knowing you using profiles and social networking to improve recommender systems. BT Technology Journal 24(3), 84–98 (2006)
32. Herlocker, J.L., Konstan, J.A., Borchers, A., Riedl, J.: An algorithmic framework for performing collaborative filtering. In: Proc. of the SIGIR, pp. 230–237. ACM, New York (1999)

Enterprise COllaboration and INteroperability (COIN) Platform: Two Case Studies in the Marine Shipping Domain

Achilleas P. Achilleos[1], Georgia M. Kapitsaki[1], George Sielis[1], Michele Sesana[2], Sergio Gusmeroli[2], and George A. Papadopoulos[1]

[1] Department of Computer Science, University of Cyprus,
1 University Avenue, Nicosia, Cyprus
http://www.cs.ucy.ac.cy
[2] TXT e-solutions S.p.A., TXT Group
27 Via Frigia, Milano, Italy
http://www.txtgroup.com

Abstract. Enterprise Collaboration and Enterprise Interoperability are two key aspects of networked enterprises, which proceed along parallel tracks with rare opportunities to convene, support and influence each other. To resolve this issue we propose the use of the COIN platform, which allows exposure, combination and integration of Web Services to support these aspects in different business sectors. Suitable COIN collaboration and interoperability services were selected based on the requirements analysis performed with Donnelly Tanker Management, our industrial partner in Cyprus. These services adhere to requirements of the shipping sector, such as negotiation of voyage terms, trusted information sharing, document management and user management. Using COIN, a web-based enterprise system was created that supported two case studies. Useful results were obtained by marine experts, which revealed the positive impact of the platform in managing and reducing the time to execute these processes.

Keywords: Web-based Enterprise Systems, Web Services, Enterprise Collaboration, Enterprise Interoperability, Web-based Business Processes.

1 Enterprise COllaboration and Interoperability

Two fundamental aspects of networked enterprises [1] are Enterprise Collaboration (EC) and Enterprise Interoperability (EI). Juncal et al. [2] state in their work that although these key aspects are independent, in most occasions they are simultaneously present. First, EC signifies the business perspective and identifies the enterprises (mainly SMEs) process. This process allows setting up and managing cross-enterprise business relations in response to business opportunities. On the other hand, EI originates from ICT and identifies the capability of enterprise software and applications to be integrated at the level of data, applications, processes and models of each enterprise. These aspects proceed along

X.S. Wang et al. (Eds.): WISE 2012, LNCS 7651, pp. 396–410, 2012.

parallel tracks with rare opportunities to convene and mutually influence each other. Hence, research in EC lacks support of innovative and advanced ICT, while research in EI lacks concreteness and real-life business applications [3].

The COIN research project examined and investigated this problem to design, develop and prototype an open, self-adaptive service platform that offers business collaboration and interoperability services. Foremost, the COIN platform offers EC and EI services developed during the course of the project in the form of Web Services. These services were developed taking into consideration their applicability in various business sectors [3]. Secondly, the COIN service platform enables the exposure, combination, integration and use of these services in different business sectors.

The availability and adoption of innovative COIN ICT services by business consumers is expected to foster and promote interoperability amongst collaborative enterprises to support various business forms such as supply chains and business networks [4]. Furthermore, the open and modular architecture of the COIN platform permits the addition of newly developed services by service providers. The open, self-adaptive nature of the platform allows also the customisation of existing resources (e.g. when applied to a specific business domain), to develop additional highly-specialised services. Thus, the key goal was to promote via the COIN service platform the notion of "Software as a Service" (SaaS) [5] to support the vision: "Interoperability and collaboration services will become a pervasive knowledge and business utility at the disposal of European networked enterprises from any industrial sector and domain" [6].

In this work the key task is to study and extract important business requirements, which necessitate the support of COIN services for advancing collaboration and interoperability in the marine shipping sector. The second task is to utilise the technical capabilities of the COIN platform to implement marine processes in the form of business pilots that can be executed directly using the runtime environment offered by the platform. The key objective is to execute, validate and establish the efficiency of the COIN service platform in terms of reducing the development time and also in terms of managing more efficiently business processes in the marine shipping sector. In particular, the involvement of the industrial project partner Donelly Tanker Management (DTM)[1] aided in the identification of the business requirements and the development of the pilots by considering highly the business perspective. Hence, ICT-based developments were driven by the business requirements. Two business use cases that lack support of ICT capabilities were identified and analysed, so as to select a subset of baseline and innovative services offered by the COIN platform aiding the development and execution of the pilots.

The paper is structured as follows. Section 2 introduces briefly the COIN service platform, while Section 3 presents the Cyprus marine shipping sector, the motivations behind the developments performed and the marine business cases. In the following section we introduce the developed extensions for adapting the selected COIN services strictly to the marine domain. Section 5 presents

---

[1] http://www.donnellytanker.com.cy/

the COIN experience, based on the executions of the pilots. Initial results are presented that outline the business indicators and reveal the benefits obtained based on the feedback received by DTM experts. Finally, Section 6 presents conclusions derived from the feedback received.

2   The COIN Service Platform

The COIN service platform forms the backbone, integrating services for enterprise collaboration and interoperability. It fulfils the COIN objective of providing a pervasive service platform to host Baseline and Innovative COIN services for EI and EC, which can be used by European enterprises for running their business in a secure, reliable and efficient way. The platform is developed on top of the Liferay portal [7], which is an enterprise-based web platform for building technology-oriented, business applications that deliver immediate results and long-term value. Using the COIN platform we have implemented the business use cases using the *ProcessMaker* business process management and workflow software [8] offered by the platform. Via *ProcessMaker* we are able to invoke the necessary Baseline and Innovative COIN services that allow executing the required business tasks of the use cases.

The following COIN services correspond to the essential services as derived from the requirement analysis performed with the project industrial partners:

- Collaboration Visualization Tool (CVT): Formulation and visualization of human collaboration networks, including users and their discovered relations (e.g. joint activities) [9].
- Trusted Information Sharing (TIS): Flexible sharing of business related information (e.g. documents) on the basis of CVT relations [10], [11].
- Interoperability Space Service (ISS): A negotiation tool for exchanging and negotiating business documents in standardized Universal Business Language (UBL) format [12].
- Baseline Communication Services: A suite of services that include Skype call, instant messaging and notification [12].

A detailed description of the COIN platform is out of the scope of this paper and thus interested readers are referred to [6].

3   Introduction to the Marine Business Pilots

3.1   The Cyprus Marine Shipping Sector

In terms of the overall Cyprus economy, the shipping industry is contributing 5% of the Cyprus Gross Domestic Product (GDP). The strength of the marine shipping sector in the country's economy, including both ship management and ship owning, is a result of a number of advantages that Cyprus can uniquely offer: (i) *Excellent geographical position at the crossroads of Europe, Africa and the Middle East, (ii) "Open Registry" (one of two countries in the European Union) allowing non-Cypriot citizens to register ships under the Cypriot flag, (iii) advanced maritime infrastructure (two deep sea multipurpose ports in Limassol and Larnaca) and (iv) favourable taxation regime for ship-owners and crew members.*

The Cyprus Shipping Registry is classified as the 10th largest fleet globally and the 3rd in the European Union. Limassol is considered the largest third party ship-management centre in the EU and one of the largest in the world (in excess of 130 ship-owning, management and other shipping related companies maintain offices there). Nearly 1,000 ships exceeding 21 million gross tonnages are registered in Cyprus and the Cyprus Shipping Sector employs approximately 4000 people ashore and approximately 40,000 sea farers onboard Cypriot flag ships. The European fleet capacity has significantly increased upon Cyprus accession to the EU, with the island contributing approximately 20% of the EU fleet.

3.2   Motivation for Developments

The business scenarios examined refer to the maritime shipping domain and the processes that need to be executed for the establishment and management of the vessel's voyage, so as to transport specific goods. The primary target is to improve collaboration and interoperability amongst the involved parties of a highly distributed team, which includes charterers, shipping captains, port agents, accountants etc. As stated also in [13], this is possible through service workflows that are collaborative in nature and allow for direct interactions between parties involved. In particular, the goal is to simplify these processes and reduce the time for the accomplishment of these processes. This can be achieved by providing the technological capabilities that simplify complex tasks such as continuous communication between parties typically located in different countries, negotiations of voyage terms, trusted sharing of sensitive documents, etc.

In both use cases various enterprise interoperability and collaboration tasks need to be carried out. These refer to the negotiation of details and terms for the voyage setup, exchanging documents and information related to the marine shipping domain via a trusted and secure sharing method, identifying business relationships and tracking information on past interactions with partners of the marine domain and direct communication of a geographically distributed team. Thus, the use of COIN services, such as ISS and TIS described previously, is essential in both business cases.

In this work the necessary developments for adapting the ISS and TIS services were performed, in order to achieve integration and use successfully these services from the implemented workflows. This was required since both services accept as input specific data formats. Foremost, the ISS service accepts input data in the form of standardised UBL documents to enable negotiation tasks in marine processes, while the TIS service accepts data in XML format, so as to enable the sharing of documents in the marine processes in a secure and trusted way. In particular, the development of extensions to COIN services is motivated by: (i) the need to achieve integration with the Da-Desk system used in the marine domain by transforming its legacy format to the UBL format and (ii) to achieve the adaptation of the UBL format, using existing UBL fields, so as to support specific marine information included in standardised documents. Therefore, the introduced extensions to COIN services would contribute to the simplification of the marine processes and the reduction of the processes execution time.

In order to generate the input required from the COIN ISS and TIS interoperability services in the appropriate format and thus support integration with these services, the needed extensions were developed in the form of transformation scripts. These scripts were implemented in the form of PHP based templates (i.e. transformation scripts) and were assigned as triggers to the necessary tasks of the developed workflows. When the trigger associated to a task is executed, the data derived from the completed fields of the form are transformed to the respective UBL and XML data formats. Three different transformation scripts have been defined for the two use cases. Each script is executed upon completion of the specific task to which each script is assigned as a trigger.

3.3   Overview of the Business Use Cases

The marine shipping use cases described next define the processes that are currently executed without the necessary technological support, which will allow reducing their execution time and simplify their management.

Workflow 1 - Negotiations between United Product Tankers (UPT) and Charterers for the pre-fixture queries and the voyage's fixture accomplishment

The establishment of a vessel's voyage is a highly complex procedure. The complexity is due to the involvement of several parties that must communicate with each other before any action is performed. The most important action that has to be completed before the beginning of a vessel voyage is the preparation of the recap. The recap is the outcome of negotiations and logical amendments agreed between UPT and charterers, for the decision of which charterers will get involved in the trip. These negotiations are done following formatted documents, called "standard charter party forms". The aim of these negotiations is the formulation of the final recap document. The recap includes all the negotiation results and it is used by the DTM operators as a "guide" for the successful execution of the vessel voyage. The formulation of the recap document is a process that requires the continuous communication between UPT and charterers during the negotiations phase. In addition, the recap should be forwarded to the vessel's captain via the DTM operators. The operator and the vessel captain follow the voyage instructions included in the recap document and the DTM operator confirms that the vessel captain is aware of these details and any updates during the trip.

Workflow 2 - Creation of the Proforma Disbursement Account

This business scenario refers to the steps undertaken by the DTM operator, port agent and the captain for the formulation of the Proforma Disbursement Account (PDA). After the selection of the right port agent (by the DTM operator), the initial PDA is created as a result of the negotiations between the operator and the agent. The PDA is then shared amongst the involved parties and the DTM operator is responsible to establish direct communication between the captain and the appointed agent. The appointment is setup using the DA-Desk legacy web system (including inputting manually the PDA contents) and thus a notification is sent to the accounting department about the agent appointment for settling the agent's fees. At the same time the port agent and the captain communicate directly in order to achieve smooth completion of the voyage while sending daily updates to DTM operator. Upon completion of the voyage the port agent sends the final PDA to the DTM operator. The operator "closes" the voyage in DA-DESK legacy system.

4   Extensions to COIN Services

The examined processes are currently carried out in an ad-hoc way that involves the exchange of information and documents for the voyage establishment and management using emails, telephone calls and facsimile. Only the second use-case involves the use of the Da-Desk legacy system, which refers to a notification management system. Hence, this ad-hoc procedure creates the following issues: *(i) the actors needed to be highly trained and acquainted with the process to avoid errors and omissions, (ii) difficulty in management and coordination of these highly interactive and mostly dependent tasks, (iii) a huge volume of emails is created for each case (actors are typically involved in different voyages), which makes it difficult to handle the vast amount of information and (iv) slow response time due to email overkill or because a person might not be available for a telephone call.*

The COIN platform provides a portal that allows users to login, develop, execute and monitor processes based on their access privileges. Hence, with the aid of the COIN service platform, two marine processes have been implemented in the form of workflows. These workflows are developed using the platform's capabilities by modelling visually electronic forms, implementing document (e.g. PDF, DOC) generation templates, associating standard input forms to tasks that allow uploading and sharing documents, invoking COIN services using code triggers, etc. These forms, templates and code triggers are associated with different tasks of the processes and each task is assigned to a user or a group of users. Using this approach it is possible to execute the business processes using the runtime environment of the platform.

The only requirement is to develop extensions (i.e. transformation scripts) to COIN services in order to support data transformations. This is required in order to associate these extensions with specific tasks of the processes, so as to transform the data completed using electronic forms to the standardised UBL data formats required as input by the COIN ISS and TIS services. In this way these services can be invoked at the subsequent tasks. Finally, integration with the Da-Desk legacy system can be also achieved via these transformation scripts.

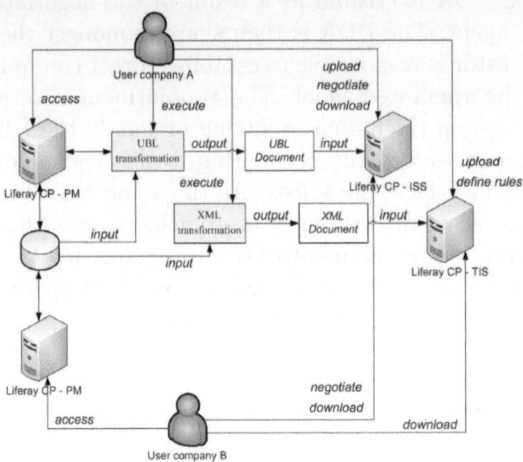

Fig. 1. Extensions to COIN Services

Fig. 1 showcases the developed extensions and the interaction of the user with these extensions, so as to facilitate integration with COIN services. The required extensions are developed in the form of transformation scripts assigned to specific tasks of the processes. Two extensions are developed in the case of the first workflow, since it uses both the ISS and TIS services. For the second workflow, presented in detail in the following section, one extension is developed, since it uses only the ISS service. In this section, we present one of the developed extensions due to space limitations. This extension refers to the script that transforms the PDA form data to the UBL data format required as input by the ISS service. This transformation script is associated with the Recap negotiation task (i.e. Create PDA) of the modelled workflow illustrated in Fig. 2.

The UBL specification was studied in order to identify data fields, which allow representing the PDA document in the form of UBL data. On the basis of the identified data fields (i.e. BillOfLading.xsd) the transformation script is defined in the form of PHP code shown partly in Listing 1. For instance, the first two lines of the script generate the UBL tag "**p2:IssueDate**" shown in Listing 2. Moreover, the following two lines of the script allow retrieving the data from the PDA form (i.e. **@@pda-issue-date**) and appending the data value retrieved to the UBL output shown in Listing 2. Using the same reasoning the rest of the transformation script is defined as shown in the following lines of Listing 1 and the corresponding output is generated as illustrated in Listing 2.

Listing 1. PDA transformation script.

```
$item = $dom->createElement("p2:IssueDate"); // create UBL tag
$root->appendChild($item); // append tag name

$text = $dom->createTextNode(@@pda-issue-date); // retrieve data and create text node
$item->appendChild($text); // append value

$shipment = $dom->createElement("p4:Shipment"); // create UBL tag
$root->appendChild($shipment); // append tag name

$item = $dom->createElement("p2:ID"); // create UBL tag
$shipment->appendChild($item); // append tag name

$name1 = @@vessel_name; // retrieve data from PDA form
$name2 = @@pda_ref; // retrieve data from PDA form
$str = "$name1 - $name2"; // concatenate data

$text = $dom->createTextNode($str); // create text node
$item->appendChild($text); // append value

$item = $dom->createElement("p2:DeclaredCustomsValueAmount"); // create UBL tag
$shipment->appendChild($item); //append tag name

$currencyID = $dom->createAttribute("currencyID"); // create UBL attribute node
$item->appendChild($currencyID); // append attribute node name

$currencyIDValue = $dom->createTextNode("EUR"); // create text node
$currencyID->appendChild($currencyIDValue); // append attribute node value

$text = $dom->createTextNode(@@customs_clearance); // retrieve data and create text node
$item->appendChild($text); // append value
```

Listing 2. Generated UBL format of the PDA document.

```
<p2:IssueDate>
2011-06-25
</p2:IssueDate>
 <p4:Shipment>
   <p2:ID>
   Mount Olympus - tt1234tt
   </p2:ID>
   <p2:DeclaredCustomsValueAmount currencyID="EUR">
   212
   </p2:DeclaredCustomsValueAmount>
```

The resulting UBL document allows using the COIN ISS service in the subsequent task of the process. As illustrated in Fig. 1, user A is able to upload and apply rules for negotiating the details of the PDA document. User B is also able to use the ISS service to view the PDA document (see Fig. 6) and modify the rules in case user B does not agree with the terms of user A. Once agreement is reached, the two users are able to download the finalised charter party document. It is important to note that the COIN platform and the ISS, CVT, TIS and Baseline Communication services are deployed on top of the Liferay portal.

5   The COIN Experience

The application of the described tools of the COIN platform are demonstrated and evaluated in this section through a reference to the *Creation of the Proforma Disbursement Account* case study. The test-bed environment for the scenario was prepared and executed at the premises of Donelly Tanker Management in Limassol, Cyprus. A number of COIN tools are exploited for the PDA creation as detailed next. The workflow modelled in Process Maker highlights five distinct sub-processes with individual tasks as illustrated in Fig. 2. Each task is associated with electronic forms, triggers, input and output documents, as well as with a user or a user group.

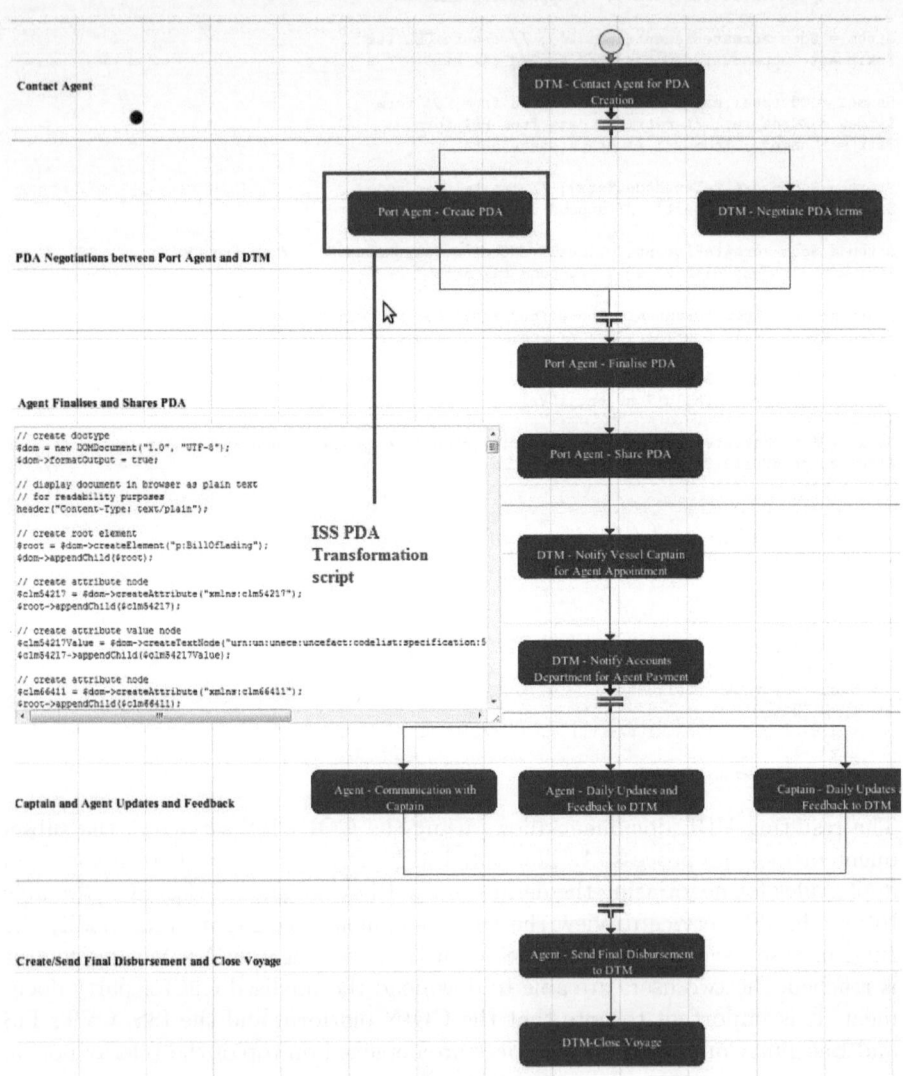

Fig. 2. Extensions to COIN EI services for the second business use case

In the first sub-process, i.e. *Contact Agent*, DTM uses the platform with the aim to identify an appropriate agent for the specific voyage. Using COIN this is achieved through COIN CVT service accessed as a Web application. CVT enables the selection of the optimal agent for the scope of "voyage management" based on past interactions and collaborations recorded between the DTM operator and port agents. The selection is based on: 1) the partner property that reflects the agent's "skill level" and 2) the link property between the two parties, i.e. "personal trust". Both are visualised in relation to the scope of "voyage management". In Fig. 3 the skills of agent "Evans" are higher than those of agent "Smith'. The same applies for personal trust and, therefore, agent "Evans" is appointed as the most suitable partner. The DTM operator is then able to contact the agent and make an assignment for the specific task.

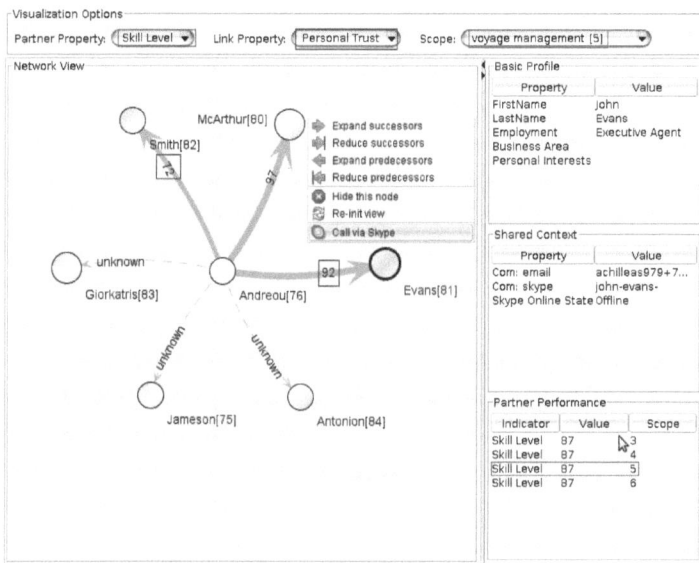

Fig. 3. Port agent selection through CVT

Using COIN the next step of PDA creation is performed by the port agent using an electronic form (Fig. 4). The form is completed by the agent and the transformation script (see Fig. 2) associated to the task is executed upon submission, generating a UBL-based PDA document. This format is uploaded into the COIN ISS. The DTM operator is notified that the negotiation task is pending and accesses the ISS service (Fig. 5). If the operator agrees to the terms, the agent is notified and a settlement is reached. Otherwise, the terms are changed until an agreement is reached. At the same time the involved parties are able to communicate via COIN Baseline communication services (i.e. Skype, email and instant messaging) in order to complement the actions of the ISS service and improve the efficiency of the negotiation process.

The next tasks are to finalise and distribute the PDA document. Hence, as presented partially in Fig. 6, the agreed terms are generated by the ISS service

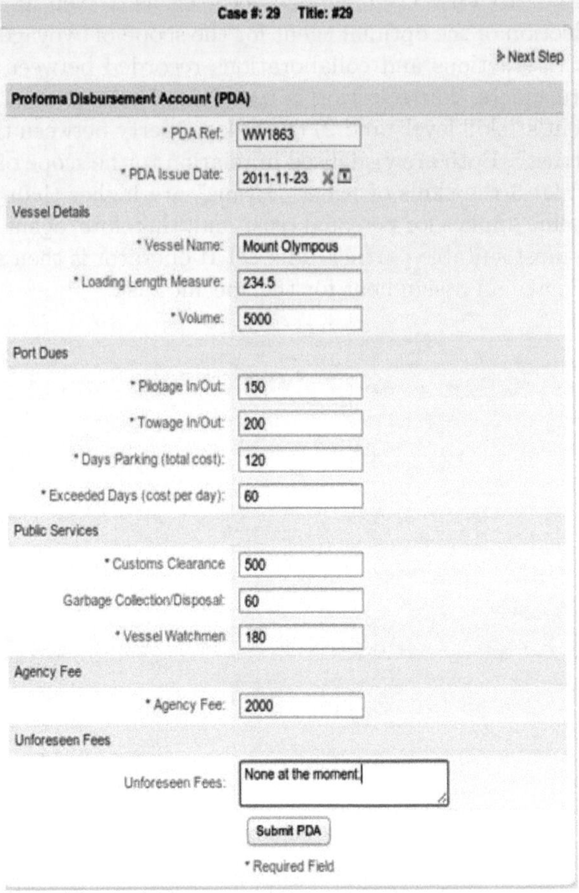

Fig. 4. PDA electronic form

in the form of a document. The PDA is then exported in PDF format and shared using the platform capabilities. Note that in the current scenario the use of the COIN TIS service is not necessary, since the PDA needs to be made available to all participating partners. The PDA distribution to the DTM operator, the vessel's captain and the DTM accountant is performed through ProcessMaker. The subsequent tasks are executed by the DTM operator, which is responsible for notifying the captain and the accounts department about the agent appointment and the agreed PDA terms. Instead of relying on email and fax documents along with DA-Desk, the Baseline Services are used for communicating with the captain, the agent and the accountant. For the accountant selection the CVT service can be exploited similarly as in the port agent selection case.

The following tasks (i.e. *Captain and Agent Updates and Feedback*) are executed in parallel through COIN Baseline Communication services. The next task is the responsibility of the port agent, who needs to modify the PDA in case

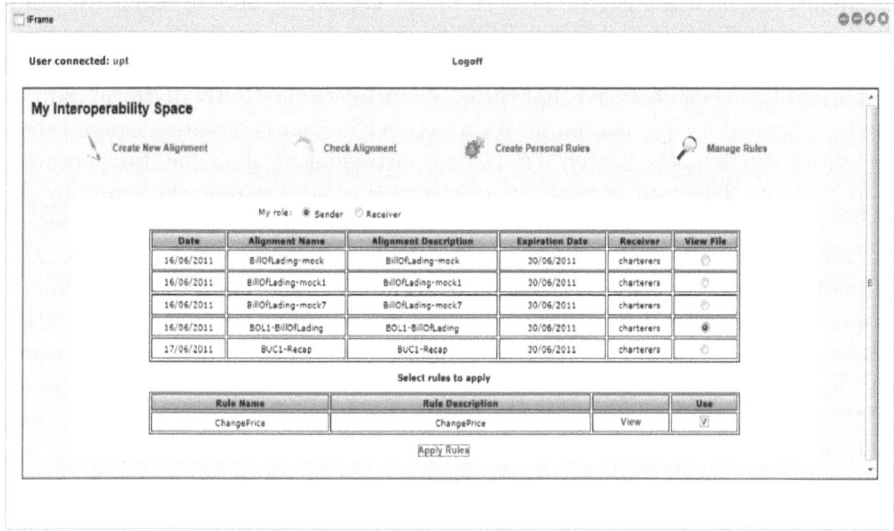

Fig. 5. Using the COIN ISS service

Fig. 6. Output document of the COIN ISS service

unforeseen situations have been encountered during the voyage loading, transport and discharge procedures. Through COIN all actions involve electronic versions of the final disbursement account with all parties being immediately aware of any changes performed. The final step is to close the voyage in DA-DESK. Note that the DA-DESK legacy system needs to be used at specific steps of the process and its usage is enabled via the data transformations.

Table 1 summarizes the differences in the exploited tools between the standard procedure followed for the PDA preparation, where common communication means are used, and the COIN-based procedure. Through the described scenario it has been observed that the exploitation of the COIN platform reduces the time needed for the communication between the parties and assists in avoiding errors, whereas the search for trusted port agent or an accountant becomes more efficient. These advantages were verified through the remarks stated by the DTM employees that participated in the evaluation and pointed out the overall simplification of the distribution and the negotiation process. As attested also in the food industry factories case study [14], the use of information technologies (i.e. enterprise collaboration and interoperability services) replacing routine communication operations reveal essentially a decrease of expenditures including elimination of human work. Consequently, in the COIN world all actions are performed electronically reducing the effort required for preparing and faxing multiple hard copies, while the COIN communication services are highly interactive and provide means for direct communication. Approximately 25% and 28% efficiency improvement was obtained respectively by recording the time needed to execute the business processes (i.e. workflows 1 and 2); 11 and 10 executions were performed respectively.

Table 1. Tools employed in PDA creation

Step	Standard tools	COIN tools
Contact Agent.	phone, email	CVT
PDA Negotiation.	hard copy, fax, phone, email	electr. PDA, ISS, Baseline Communication Services
PDA Finalization and Distribution.	hard copy, fax, phone, email	PDF, ProcessMaker
Captain and Account's Department Notification.	phone, email, fax	Baseline Communication Services, CVT
Updates and Feedback.	phone, email, fax	Baseline Communication Services
Final Disbursement and Close Voyage.	hard copy, fax	Baseline Communication Services

6   Conclusions

The goal of this work is to utilise and evaluate the efficiency of the COIN platform in terms of supporting processes in the marine domain. Hence, two business use cases were identified, analysed and implemented in the form of pilots. Moreover, the necessary extensions to COIN TIS and ISS services were developed in the form of transformation scripts. This was performed in order to facilitate the transformation of the electronic forms data to the required data formats accepted as input by these services. The pilots provided the capability to evaluate the efficiency of executing the marine processes through the use of the COIN platform and its services. These processes were executed directly by the project industrial partners, which aided in quantifying the efficiency improvement in terms of reducing the time to carry out these processes. The feedback of the participating DTM employees has revealed that the COIN platform and its offered services can assist, accelerate and support efficiently organizational workflow processes. Thus, COIN has proven a necessary companion towards the transition of enterprises to fully ICT-based operations.

Acknowledgement. This work is supported by the European Commission as part of the Sub-Project 7 (SP7) extension of the COIN Integrated Project (IP) funded by the Seventh Framework Programme - EU FP7 Project 216256.

References

1. Thompson, K.: The Networked Enterprise: Competing for the Future Through Virtual Enterprise Networks. Meghan-Kiffer Press (2008)
2. Alonso, J., de Soria, I.M., Orue-Echevarria, L., Vergara, M.: Enterprise Collaboration Maturity Model (ECMM): Preliminary Definition and Future Challenges. In: Enterprise Interoperability IV, Part VII, pp. 429–438. Springer, London (2010), doi:10.1007/978-1-84996-257-540
3. Grosso, E.D., Gusmeroli, S., Olmo, A., Garcia, A., Busen, D., Trebec, G.: Are Enterprise Collaboration and Enterprise Interoperability enabling Innovation scenarios in industry? In: The COIN IP perspective in Automotive, e-challenges (2010)
4. Karvonen, I., Conte, M.: Supporting and facilitating the Enterprise Collaboration (EC) and Enterprise Interoperability (EI) solution take-up. In: ICE 2010 16th International Conference on Concurrent Enterprising, Lugano (2010)
5. Buxmann, P., Hess, T., Lehmann, S.: Software as a Service. Gabler Verlag 50(6), 500–503 (2008), doi:10.1007/s11576-008-0095-0
6. Facca, F.M., Komazec, S., Guglielmina, C., Gusmeroli, S.: COIN: Platform and Services for SaaS in Enterprise Interoperability and Enterprise Collaboration. In: IEEE International Conference on Semantic Computing, September 14-16, pp. 543–550 (2009), doi:10.1109/ICSC.2009.72
7. Liliedahl, D.: Is Liferay Right for your Organisation? Whitepaper, http://www.liferay.com/documents/guest/Whitepapers/Business/ (retrieved June 7, 2012)
8. Colossa, Inc., ProcessMaker Workflow Simplified, http://www.processmaker.com (retrieved June 7, 2012)
9. Skopik, F., Schall, D., Dustdar, S.: Supporting Network Formation through Mining under Privacy Constraints. In: 10th IEEE/IPSJ International Symposium on Applications and the Internet, SAINT 2010, pp. 105–108 (2010)
10. Skopik, F., Schall, D., Dustdar, S.: Trusted Interaction Patterns in Large-scale Enterprise Service Networks. In: 18th Euromicro IEEE International Conference on Parallel, Distributed and Network-Based Computing (PDP), Pisa, Italy, February 17-19 (2010)
11. Skopik, F., Truong, H.L., Dustdar, S.: Trusted Information Sharing Using SOA-Based Social Overlay Networks. International Journal of Computer Science and Applications 9(1), 116–151 (2012)

12. Jansson, K., Sesana, M., Skopik, F., Olmo, A.: COIN Innovative Enterprise Collaboration Services. In: The COIN Book: Enterprise Collaboration and Interoperability, Mainz, 200p. (November 2011) ISBN 3-86130-713-8
13. Georgiana, S., Mihai, S.A., Ioan, S., Mihnea, M.: Dynamic Interoperability Model for Web Service Choreographies. In: 6th International Conference on Interoperability for Enterprise Systems and Applications, I-ESA (2012)
14. Kiauleikis, V., Romeika, G., Morkevicius, N., Kiauleikis, M.: Collaboration and Interoperability Services vs. Traditional Communication Means. Journal Science and Processes of Education (2011)

VAM-aaS: Online Cloud Services Security Vulnerability Analysis and Mitigation-as-a-Service

Mohamed Almorsy, John Grundy, and Amani S. Ibrahim

Computer Science and Software Engineering Center
Swinburne University of Technology
Melbourne, Australia
{malmorsy,jgrundy,aibrahim}@swin.edu.au

Abstract. Cloud computing introduces a new paradigm shift in service delivery models. However, the potential benefits reaped from the adoption of this model are threatened by public accessibility of the cloud-hosted services and sharing of resources with other service tenants. This increases the potential for exploitation of newly discovered vulnerabilities that usually take a long time to discover and to mitigate. On the other hand, existing cloud platforms do not provide a means to validate the security of offered cloud services or mitigating security vulnerabilities that arise at runtime. We introduce VAM-aaS, Vulnerability Analysis and Mitigation as-a-service, as a novel, integrated, and online cloud-based security vulnerability analysis and mitigation service. VAM-aaS performs online service analysis to pinpoint new vulnerabilities and weaknesses. It then uses this information to generate security control integration and configuration scripts to block these discovered security holes at runtime. Our approach is based on a new vulnerability signature and mitigation-actions specification approach. We introduce our approach, describe implementation details, and describe an evaluation of our prototype on a set of .NET benchmark applications.

Keywords: SaaS Security, Vulnerability Analysis, Vulnerability Mitigation.

1 Introduction

The cloud computing model [1] introduces a new paradigm shift in computing platforms with an emphasis on increasing business benefits. The cloud model is based on service outsourcing of application hosting on third-party platforms outside of the enterprise network perimeter. It uses a new pay-as-you-go payment model where customers can rent services occasionally and pay only for amount of resources they use. However, the cloud model also introduces new opportunities for attackers to exploit, such as publicly accessible valuable business services. Adopting the multi-tenancy model increases the exploitability of service vulnerabilities because one of the service tenants, who have privileged access to the cloud service, may be a malicious user. This means that they can exploit complicated vulnerabilities that require higher privileges rather than being a public user. Moreover, the number of newly discovered vulnerabilities is increasing rapidly. Web applications, the most prominent application

X.S. Wang et al. (Eds.): WISE 2012, LNCS 7651, pp. 411–425, 2012.

delivery model used in SaaS applications, continue to make up the largest percentage (75%) of the total reported vulnerabilities over the last three years [23].

Commercial vulnerability scanners such as IBM-AppScan, HP-Web inspect, McAfee tools focus mainly on black-box vulnerability analysis to avoid being limited to specific programming languages or platforms. However, none of these scanners cover all known vulnerability types [2]. On the other hand, existing research efforts [3-8] focus on discovering specific vulnerability types including SQLI [9, 10], XSS [9, 11, 12], or input sanitization [13, 14] using static analysis [3, 15], dynamic analysis [9], or hybrid techniques [16, 17].

The key problems with these efforts include: they provide specific techniques for specific vulnerability types; the techniques apply only to specific platforms or languages; and they do not usually support analysis for new vulnerability types. On the other hand, these limitations are key requirements for cloud services vulnerability analysis tools. An *online* vulnerability analysis approach that supports locating well-known as well as new vulnerabilities without waiting for new tool patches is a must to have in the cloud computing environment.

Mitigating application vulnerabilities is usually done manually by modifying application source code and deploying new patches; however, this takes a long time as shown in Fig. 1. This lagging time between vulnerability detection and patch means that the service remains vulnerable to security breaches exploiting such vulnerabilities. The possibility of vulnerability exploitation increases dramatically in the cloud, given the public accessibility of the cloud services and the sharing of services with multiple tenants. Thus, the cloud computing model requires an online vulnerability patching approach that can block such vulnerabilities once reported.

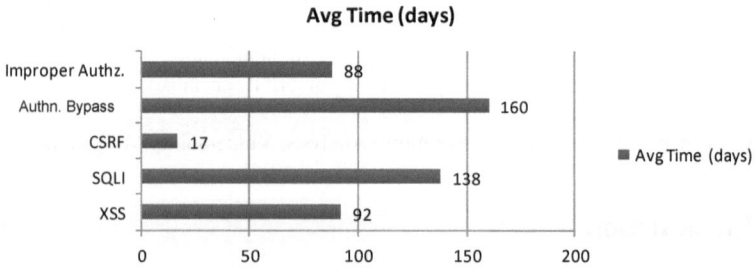

Fig. 1. Average time to fix security vulnerabilities (in days)

In this paper, we introduce VAM-aaS, a new integrated solution to cloud-based services vulnerability analysis and mitigation problems. Our approach is based on a new formal approach to specifying vulnerability signatures as well as enumerated mitigation actions to block such vulnerability. Vulnerability signatures are specified as invariants. When one is matched in a target application it means that the specified vulnerabilities exist in the application or service under analysis. We adopt Object Constraint Language (OCL), as a declarative and formal language based on first order logic and set theory, to capture vulnerability signatures. Such signatures are validated against a comprehensive system description meta-model (represents language semantics) covering most of the object oriented program concepts and entities.

We developed a vulnerability analysis service that locates OCL-based vulnerabilities' signatures in a given SaaS application source code. To support locating new types of vulnerabilities, security experts need to update the vulnerability analysis service repository with OCL-signatures for the new vulnerability. Our vulnerability analysis component can be used to analyze both cloud services and traditional web applications.

The vulnerability mitigation actions specify a set of security solutions that can be used to provide "virtual patching" of the discovered vulnerabilities reported by the analysis component. They also specify configurations and rules that should be applied when activating security controls. We have developed a vulnerability mitigation component that uses these vulnerability mitigation actions to plug-in specified mitigation security controls within the target vulnerable service or application online. Our mitigation component is not limited to or hardcoded to specific security controls. It depends on a simplified security interface that security controls should satisfy in order to be integrated with our mitigation component. Thus, new security controls can be plugged into the system vulnerable entity at runtime to mitigate the vulnerability.

We evaluated our approach and its prototype realization service in capturing the well-known Top10 vulnerabilities reported by OWSAP [24]. We have validated our toolset in locating and mitigating these vulnerabilities on a set of open source web applications.

In Section 2, we describe our approach, vulnerability signatures, and mitigation actions. In Section 3, we describe our OCL-based vulnerability analysis component. In Section 4, we describe our vulnerability mitigation component. In Section 5, we describe our prototype implementation details. In Section 6, we discuss our experimental evaluation and results. In Section 7, we discuss the implications of our work and key directions for further research. In Section 8, we review the key related work.

2   Our Approach

Our security vulnerability analysis and mitigation approach is based on (i) a formalized vulnerability signature and potential mitigation actions specification; (ii) a vulnerability analysis tool that performs OCL-based vulnerability signature-based program analysis; and (iii) a vulnerability mitigation component that blocks service or application security vulnerabilities by generating configuration and integration scripts that integrate security controls at the application or service reported vulnerable points. In Fig. 2, we summarize the possible interactions between the vulnerabilities definition repository, analysis and mitigation components, applications or services, and the hosting service (Web Server, Operating System, etc).

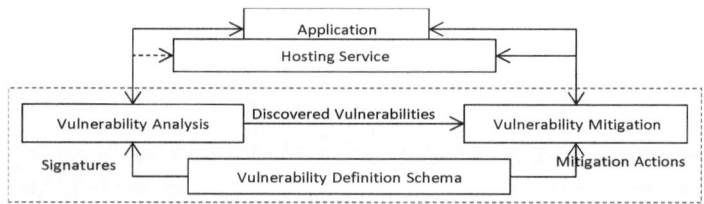

Fig. 2. VAM-aaS Key components, relations and possible interactions

2.1   Vulnerability Signature Specification

Existing software security weakness, or vulnerability definitions, in the Common Weakness Enumeration (CWE) [25] database help in understanding the nature of a given vulnerability. However, these vulnerabilities' definitions are informal. This requires manual analysis (by security experts) to locate such vulnerabilities in the applications under analysis.

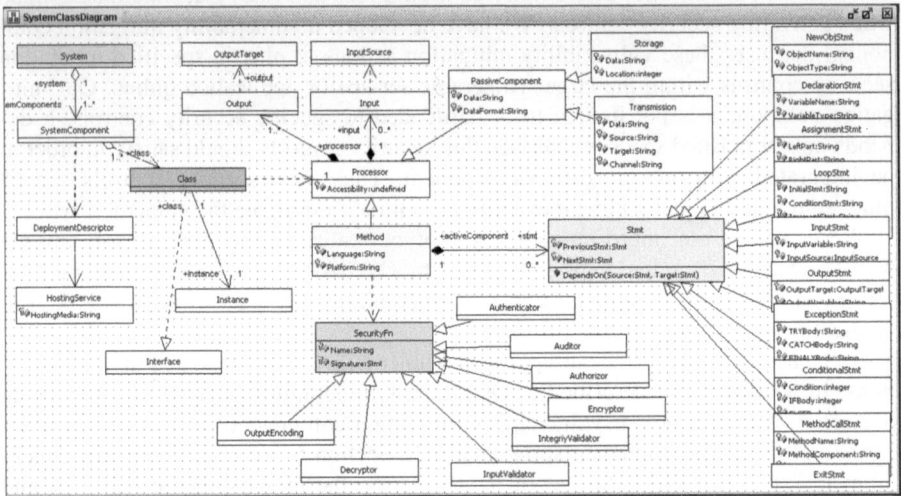

Fig. 3. System description meta-model used to specify OCL-based vulnerability signatures

Formalizing these descriptions – i.e. the vulnerability signatures – allows automation of the vulnerability analysis process. Ideally, a formal vulnerability signature should be specified at an abstract level far from the source code and programming language de-tails, enabling locating of possible vulnerability instances in different programs written in different programming languages. We use the Object Constraint Language (OCL) as a well-known, extensible, and formal language to specify semantic rather than syntac-tical signatures of security weaknesses. To support specifying and validating OCL-based vulnerabilities' signatures, we have developed a system-description meta-model, shown in Fig. 3. This model is inspired from our analysis of the nature of the existing security vulnerabilities. It captures the main entities in any object-oriented program and relationships between them including components, classes, instances, inputs, input sources, output, output targets, methods, method bodies, statements e.g. if-else, loops, new objects, etc. Each entity has a set of attributes such as method name, accessibility, variable name, variable type, method call name. This model helps conducting semantic analysis of the specified vulnerability signatures.

Some vulnerabilities require checking the existence of a security control that au-thenticates, authorizes, audits, etc. at specific locations in the program e.g. before a critical method call, their use should be authenticated and authorized. Fig. 3 shows security functions as part of the system model. They inherit from the Method entity and thus can be checked in method call statement – i.e. check if the invoked method is a security method or not. An analysis tool should have different profiles for different languages and platforms (ASP.Net, PHP, C#, Java, etc.). Thus vulnerabilities with signatures containing input source or output target can be interpreted differently based on the program platform or language. Moreover, security authentication, authorization, sanitization and other functions will be interpreted according to the target system and the underlying platform. In case of custom security functions, system developers have to manually mark their security functions. Table 1 shows examples of vulnerability signatures specified in OCL and using our system description model (Fig. 3).

Table 1. Examples of OCL-specified vulnerability signatures

Vul.	Vulnerability Signature
SQLI	Context Method Inv SQLICheck: self.Statements->exists(S	S.StatementType = 'MethodInvocation' and S.MethodName = 'ExecuteSQL' and S.Parameters.exists(P	self.IsTainted(P.ParameterName) = true)
XSS	Context Method Inv SQLICheck: self.Exists(S	S.StatementType = 'Assignment' and S.RightPart.Contains(InputSource) and S.LeftPart.Contains(OutputTarget))
Authn. Bypass	Context Method Inv SQLICheck: self.IsPublic == true and self->Exists(S	S.StatementType = 'MethodInvocation' and S.IsAuthenitcationFn == true and S.Parent == IFElseStmt and S.Parent.Condition.Contains(InputSource))
Improper Authz.	Context Method Inv SQLICheck: self.IsPublic == true and self.Contains(S| S.Exists(X| X.StatementType = 'InputSource' and X.IsSanitized = false or X.IsAuthorized == False)

SQLI: any *"MethodInvocation"* statement *"S"* where the callee function is *"ExecuteQuery"* and one of the *parameters* passed to it is assigned to *"identifier"* coming from one of the input sources. Taint analysis *"IsTainted"* can be defined as an OCL function that adds every variable assigned to a user input parameter to a suspected list.

XSS: any method statement *"S"* of type assignment statement where left part is of type *"output target"* e.g. text, label, grid, etc. and right part uses input from the input sources or tainted identifier as just discussed.

Authn. Bypass: any public method that has statement *"S"* of type *"MethodInvocation"* where the callee method is marked as Authentication function while this method call can be skipped using user input as part of the bypassing condition.

Improper Authz.: any public method that has statement "S" that uses input data X without being sanitized, authorized.

These exemplar signatures focus on static vulnerabilities signatures and do not consider security solutions applied beyond the system source code either using proxies to filter SQL queries or using security controls deployed on the web server as an http handler. These can be handled by appending a dynamic signature forming a sequence of OCL constraints to be checked – i.e. to check if requests and/or responses contain certain vulnerability pattern. Weak signatures result in more false positives, which may annoy developers, or more false negatives, which may harm customers.

2.2 Mitigation Actions

Discovered application or service security vulnerabilities can be mitigated by different approaches including modifying application source code to block the identified problems (patches). However, this solution is hard to use in the cloud model as it may take a long time to deliver patched versions, as shown in Fig. 1. One solution is to use Web application firewall (WAF) to filter requests and responses that exploit such vulnerabilities. However, WAF has many limitations. These include does not helping with output validation, cryptographic storage, and mitigating improper authorization.

Table 2. Examples of vulnerability mitigation actions

Vul.	Security Control	Entity Level
SQLI	Input sanitization	Method level
XSS	Input encoding	Component level
Authn. Bypass	WAF	Component level
Improper Authz.	Authorization	Method Level

We introduce a new approach that supports integration of different security controls including identity management, authentication controls, authorization controls, input validation, output encoding, WAF, cryptography controls, etc. In our approach, each vulnerability mitigation action specifies a security control type or family to be used in mitigating the related vulnerability, its required configurations, and application or service entity where the security control will be integrated (e.g. hosting service – web-server or operating system, components, classes, and methods). Thus, a reported SQLI vulnerability in a method (M) that belongs to component (C) can be mitigated by adding input sanitization control (Z) on component (C) that removes SQL keyword from every single request to the method (M). In Table 2, we show examples of mitigation actions for some of the known security vulnerabilities. These actions should be specified in XML and included as a part of the formalized vulnerability definition.

3   Our OCL-Based Vulnerability Analyzer

Given that vulnerability signatures are now formally specified using OCL, the static vulnerability analysis component simply traverses the given program looking for code snippets with matches to the given vulnerabilities' signatures. The architecture of our formal and scalable static vulnerability analysis component, as shown in Fig. 4, is based on our formalized vulnerability signature concept.

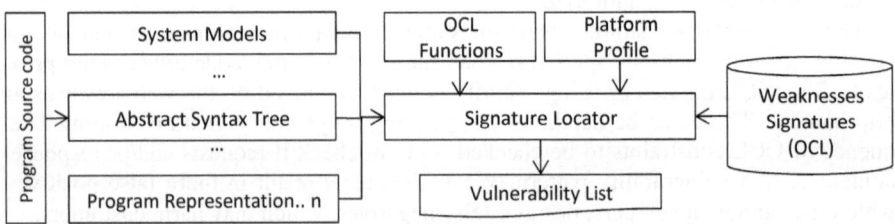

Fig. 4. OCL-based vulnerability analysis component

Program Source Code: Our analysis approach works on source code level. In case of system binaries are only available (dlls, or exes), we use de-compilation techniques to reverse engineer source code from the application to be analyzed.

Abstract Program Representation: Source code is transformed into an abstract syntax tree (AST) representation. This abstracts language-specific source code details away from specific language constructs. Extracting source code AST requires using different language parsers (currently support C++, VB.Net and C#). Then, we perform more abstract transforming from AST to system description model that conforms to the model introduced in Fig. 3.

OCL Functions: represent a library of predefined functions that can be used in specifying vulnerability signatures and in identifying matches to these signatures. This includes control flow, data flow, string patterns, program taint analysis, etc.

Signature Locator: This is the main component in our vulnerability analysis tool. It receives the abstract service or application model and outputs the list of discovered vulnerabilities in the given system along with their locations in code. At analysis time, it loads the platform (C#, VB, PHP) profile based on the details of the program under analysis. Then, it loads the existing weaknesses defined in the weaknesses' signatures database, based on the target implementation program platform or language. The signature locator transforms these signatures into C# methods that check different program entities based on the specified vulnerability signature.

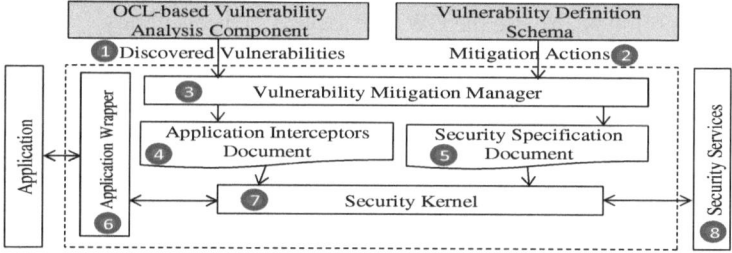

Fig. 5. Vulnerability Mitigation Component

4   Vulnerability Mitigation Component

The analysis component outputs a list of the newly discovered vulnerabilities in each of the cloud hosted SaaS applications (Fig. 5-1). Each entry in this list has a service or application "vulnerable entity" (e.g. a method, class, or component), along with a list of discovered vulnerabilities for this entity. Given this list of vulnerabilities, the security vulnerability mitigation manager queries the vulnerability definition schema database (Fig. 5-2) to retrieve the appropriate actions to be taken in order to mitigate each of such reported vulnerabilities. Examples of such retrieved actions are shown in Table 2. Using these two lists (vulnerable entities and mitigation actions), the vulnerability mitigation manager (Fig. 5-3) decides the patching level (component level, class level, or method level) using e.g. HttpModules, object interceptor using dependency injection, or method level interception using dynamic weaving AOP respectively. These details are maintained in a security point-cut specification document for each application (Fig. 5-4). Moreover, the mitigation manager uses the registered security services' properties to decide which security service realizes what security control type specified in the mitigation action. Finally, the vulnerability mitigation manager updates the security specification document (Fig. 5-5) with the list of actual security services to be triggered whenever the application receives request to every vulnerable resource. The application wrapper (Fig. 5-6) is responsible for intercepting requests to entities specified in the application interceptors' document – i.e. vulnerable entities. These requests will be redirected to the security kernel (Fig. 5-7). The security kernel queries the security specification document to get the security controls or services to be enforced to secure the requested resource. Then, it generates a set of calls to these security services (Fig. 5-8). When these security services return, the security kernel returns the control back to the called resource.

To automate the integration of security services with our mitigation component, which means being implicitly integrated with the managed service or application, we have developed a simple common security interface. Security controls' connectors implement this interface in order to support integration with our mitigation service. For example, an authentication security control should implement *AuthenticateUser* and *IsAuthenticated* functions; an input validation control should implement the *ValidateInput* function; and an output encoder should implement the *EncodeOutput* function. Security controls used in the vulnerability mitigation process can be part of a standard security controls library provided by the service provider or the cloud provider. Moreover, they may be external security controls hosted on other cloud platforms. In our prototype, we use the OWASP security controls library.

5   Implementation

We developed a GUI to assist security experts in capturing vulnerability signatures' in OCL. This provides vulnerability signature editing, validity checking, and testing these signatures' specifications on simple target applications. We use an existing OCL parser to parse and validate signatures against our system description meta-model. Once validated, the vulnerability signature is stored in our weakness signatures database. To parse the given program source code and generate a system abstract model, we use *NReFactory* .NET parser Library [26], which parses source code and generates its corresponding AST (it supports VB.Net and C#. We are currently working on parsers for PhP and Java). Applications without source code - i.e. only binaries are available – are decompiled using *ILSPY*. This is currently supported for C# and VB.NET. We developed a class library to transform the generated AST into a more abstract (summarized) representation that conforms to our system description model. Our signature locator has an OCL translator that translates a given OCL signature into a corresponding C# class with a signature matching method that checks the passed in system entity looking for matches to specified signatures.

The OCL functions library maintains a set of functions that extend the system description meta-model entities capabilities and can be used during the vulnerability analysis phase. This includes control-flow analysis (CFA), data-flow analysis (DFA), and tainted-data analysis. These functions can be extended with further analysis functions based on future vulnerability analysis needs. The OCL to C# transformer performs a transformation for these functions as well as new OCL signatures once defined. Program slicing and taint analysis techniques (core techniques in program and security analysis area) can be easily captured in OCL. Platforms' profiles are specified in XML documents that contain information about specific platforms' details. It is used to set the context of the signature locator according to the system platform.

The vulnerability mitigation component was developed using C#. It uses the Microsoft Unity application block to support configurable runtime dependency injection. This enables injecting interceptors on class and method level. To support legacy services, we use the Yiihaw static AOP [27] to inject aspects into legacy code. Such aspects redirect requests and responses to default security handler class library (we call it Security Enforcement Point - SEP). The SEP is responsible for injecting calls to security controls at runtime based on the specified mitigation actions.

6   Experimental Evaluation

The key objectives of these experiments were to assess the soundness of our VAM-aaS in capturing different vulnerabilities' signatures, detecting these vulnerabilities in given applications, and in mitigating them. We apply the OCL-based vulnerability signatures examples and mitigation actions discussed in Section 2.We selected seven web-based, open source applications developed using ASP.NET and MVC as a benchmark to evaluate our approach. These applications cover a wide business spectrum including: Galactic, an ERP system; SplendidCRM, an open source CRM; KOOBOO, an open source Enterprise CMS; BlogEngine, an open source ASP.NET 4.0 blogging engine; BugTracer, an open-source, web-based bug tracking application; NopCommerce, an open-source eCommerce solution; and Webgoat, developed by OWSAP for security testing purposes. In Table 3, we summarize these benchmark applications' characteristics: number of downloads, LOCs, files, classes, methods, and time (ms) to extract system model from source code (using source code AST).

To assess the effectiveness of our approach we use precision, recall and f-measure to measure our approach's soundness and completeness. These metrics depend on basic measures shown in Table 4. The analysis component results are compared with the actual vulnerabilities discovered by manual analysis using existing open source vulnerability analysis tools. As shown in Table 4-A, True Positive (TP) counts number of vulnerabilities correctly discovered by the analysis component, False Positive (FP) counts number of vulnerabilities incorrectly reported as vulnerability, False Negative (FN) counts number of vulnerabilities missed by the analysis component. As for the mitigation component (Table 4-B), the TP counts number of vulnerabilities that have been correctly blocked, FN counts number of vulnerabilities failed to block, and FP counts entities that have been secured without being a vulnerability.

Table 3. Benchmark applications properties

Benchmark	Downloads	KLOC	Files	Classes	Method	Model time
Galactic	-	16.2	99	101	473	187
SplendidCRM	>400	245	816	6177	6107	765
KOOBOO	>2,000	112	1178	7851	5083	78
BlogEngine	>46,000	25.7	151	258	616	163
BugTracer	>500	10	19	298	223	93
NopCommerce	>10 Rel.	442	3781	5127	9110	484

Table 4. Analysis and mitigation results classification. **A**: analysis, **B**: mitigation

A	Actuals		B		Actuals	
	Vulnerability	N-Vulnerability			Blocked	N-Blocked
Vulnerability	TP	FP		Blocked	TP	FP
N-Vulnerability	FN	TN		N- Blocked	FN	TN

$$Precision = \frac{TP}{TP+FP} \quad (1) \qquad Recall = \frac{TP}{TP+FN} \quad (2) \qquad F-Measure = 2\,\frac{Precision*Recall}{Precision+Recall} \quad (3)$$

Table 5. Experimental results of VAM-aaS using benchmark apps. **TP**: No. of true positives, **FP**: No. of false positives, and **FN**: No. of false negatives.

Component	App	SQLI			XSS			Authn. Bypass			Improper Authz.		
		TP	FP	FN	TP	FP	FN	TP	FP	FN	TP	FP	FN
Analysis	Galactic	2	0	0	3	1	1	4	0	0	2	1	0
	Splendid	13	2	1	7	1	0	3	0	0	3	0	0
	KOOBOO	14	2	0	10	2	0	4	1	0	11	2	1
	BlogEngine	3	0	1	3	0	1	0	0	0	4	0	0
	BugTracer	9	0	1	0	0	1	3	0	1	1	1	0
	NopCommerce	19	2	0	4	0	1	0	0	0	0	0	1
	Webgoat	8	0	1	5	1	0	3	0	1	3	0	0
Total		68	6	4	32	5	4	17	1	2	24	4	2
Mitigation	Galactic	2	1	0	4	1	0	4	0	0	2	1	0
	Splendid	14	0	0	7	1	0	3	2	0	3	0	0
	KOOBOO	14	2	0	10	3	0	4	1	0	12	0	0
	BlogEngine	4	0	0	4	2	0	0	0	0	4	2	0
	BugTracer	10	0	0	1	0	0	4	1	0	1	1	0
	NopCommerce	19	0	0	5	0	0	0	0	0	1	0	0
	Webgoat	9	0	0	5	1	0	4	2	0	3	1	0
Total		72	3	0	36	8	0	19	8	0	30	5	0

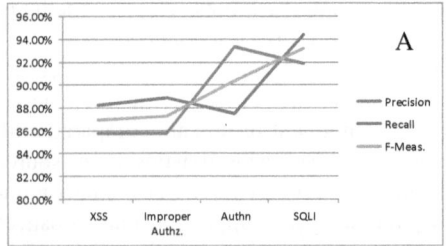
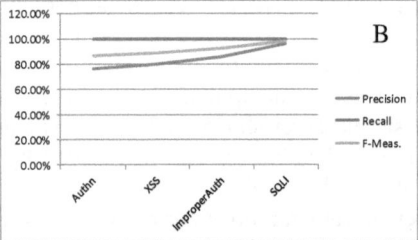

Fig. 6. Precision, Recall, and F-measure metrics of the analysis and mitigation components

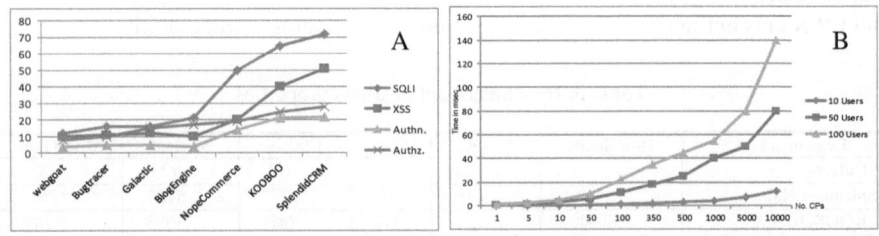

Fig. 7. Performance of vulnerability analyzer (per vulnerability type), and mitigation components

A high precision means that the approach gives more valid results (TP) than invalid results (FP). Thus, the maximum precision is achieved when no false positives given (see Equation 1). The recall metric is used to assess the completeness. A high recall means more valid results (TP) than missed valid results (FN), see Equation 2. The F-measure metric combines both precision and recall. It is used to measure the overall effectiveness of the approach (weighted harmonic mean). It depends on the importance of the recall rate and the precision rate e.g. if we are interested in high precision (more valid results) then we give precision factor high weight, and vice-versa. In our evaluation, we assume both are equally important, see Equation 3. In Table 5, we summarize our experiments' results applied on benchmark suite to identify four of the Top10 web applications vulnerabilities (OWSAP2010 report). For each component, we count number of TP, FP, and FN.

Fig. 6-A shows precision, recall, and F-measure for vulnerability analysis component on different vulnerability types using the benchmark applications. The average precision of the analysis component is around (90%). This means that in every identified (100) vulnerability instances, we have (10) false positives. Its recall is around (92%). This means that in every (100) vulnerabilities, we correctly identify (92) and miss (8). This can be improved by specifying more sound vulnerability signatures. Fig. 6-B, shows the precision, recall and F-measure for the mitigation component, we have incorporated the results returned from the analysis component with the FN missed vulnerabilities. It shows that the recall is (100%) which means that we did not miss any of the specified mitigations; however, we have an average of (85%) precision – i.e. high FP - as we may secure entities that have no security problem. This depends on the accuracy of the specified mitigation actions.

Fig. 7-A shows time (in sec) took in analyzing benchmark applications to pinpoint existing vulnerabilities. The SQLI vulnerability takes much more time to identify than XSS and authorization bypassing. The authentication bypass takes the lowest time. The time required to identify vulnerability instance depends on the complexity of the specified OCL signature. Fig. 7-B shows performance overhead of the mitigation component with different numbers of concurrent users and numbers of critical or vulnerable points - CPs (on a Core2Duo desktop PC with 4GB Memory). Overhead is equal to time spent by the security kernel to query security specification document to get security controls to be employed in securing intercepted point, and time spent in calling these security controls. Time spent by the security controls themselves we do not factor in, as this needs to be spent whether using our approach or traditional hard-coded security. Performance can be further improved using replicas of the mitigation component with different services and platforms.

7   Discussion

We developed a formal vulnerability definition schema including vulnerability signature and mitigation actions, extensible vulnerability analyzer based on the proposed signature specification schema and applied on abstract program representation, and a vulnerability mitigation approach based on dynamic and runtime injection of security controls into vulnerable entities. Use of OCL allowed us to make use of existing validation and query parsing tools. Use of this abstract representation helped us to generalize our analysis away from programming language and platform details. It also helps make our approach scalable for larger applications. Use of a common security interface allows integration of different security controls without a need to develop new system-security control connectors.

From our experience in developing signatures of the Top10 vulnerabilities and our experiments we determined that: **(i)** it is better to use dynamic analysis tools with certain vulnerabilities, such as Cross site reference forgery (CSRF), because these vulnerabilities can be handled by the web server. This means static analysis may result in high FP, if used; **(ii)** some vulnerabilities can be easily identified and located by static analysis such as SQLI and XSS vulnerabilities; **(iii)** some vulnerabilities such as DOM-based SQL and XSS vulnerabilities need a collaborating static and dynamic analysis to locate them. We believe that combining static and dynamic analysis is needed to increase the precision and recall rates.

Our vulnerability analyzer achieves a precision rate of (90%) and recall rate of (92%). These figures are actually impacted by the accuracy of the specified vulnerability signatures. Static analysis approaches usually result in high false positives as they work on source code level – i.e. the vulnerability may be addressed on the component or the application level. This problem can be solved by employing dynamic vulnerability analysis. However, dynamic vulnerability analysis approaches cannot help locating specific code snippets where vulnerabilities exist. Moreover, they do not help testing code coverage by generating all possible test cases.

From our experiments in the mitigation actions and security controls integrations, we found that although the use of web application firewalls is a straight forward solution, it is not always feasible to use WAF to block all discovered vulnerabilities. The selection of the entity level to apply security controls on (application, component, method, etc.) impacts the application performance – i.e. instead of securing only vulnerable methods, we intercept and secure (add more calls) the whole component requests. A key point that worth mentioning is that the administration of security controls should be managed by the service or cloud provider admins. We focus on integrating controls within vulnerable entities.

Our vulnerability mitigation component works online without a need for manual integration with applications and services under management. The overhead added by the mitigation action can be reduced if service developers add a new service patch. In this case, the vulnerability analysis component will not report a vulnerability. Thus, the mitigation component will not inject security controls.

8   Related Work

We are aware of no existing efforts that introduce an integrated solution to the vulnerability analysis and mitigation problem. Most focus on either vulnerability analysis or vulnerability mitigation, although vulnerability mitigation seems of less interest to date. Existing vulnerability analysis efforts can be categorized in static analysis-based, dynamic analysis, and hybrid approaches. Broadly, static analysis techniques work on source code level while dynamic analysis works on application as black-box.

NIST [18] has been running a security analysis tools assessment project (SAMATE). A part of this project is to specify a set of functional requirements that any source code security analysis approach should support. These include a set of weaknesses that an analyzer should be capable of identifying, including SQL injection, XSS, OS command injection, etc. They have also developed a set of test cases that help in assessing the capabilities of a security analysis tool in discovering such vulnerabilities. Halfond et al. [10] introduce a new SQL injection vulnerability identification technique base on positive tainting. They identify "trusted" strings in an application and only these trusted strings to be used to create certain parts of an SQL query, such as keywords or operators. Martin et al [7, 8] introduce a program query language PQL that can be used to capture definition of program queries that are capable to identify security errors or vulnerabilities. A PQL query is a pattern to be matched on execution traces. They focus on Java-based applications and define signatures in terms of code snippets. This limits their capabilities in locating vulnerabilities' instances that matches semantically but not syntactically. Wassermann et al. [12] introduce an approach to finding XSS based on formalizing security policies using W3C recommendation. They conduct a string-taint analysis using CFG to represent sets of possible string values enforced on web pages to assure no untrusted scripts. Ganesh et al [15] introduce string constraint solver to check if a given string have a substring with a given set of constraints. They used it in white box SQLI testing.

Kals et al [3] introduce a vulnerability scanner that uses a black-box approach to scan web sites for the presence of exploitable SQLI and XSS vulnerabilities. They do not depend on a vulnerability signature database, but they require attacks to be implemented as classes that satisfy certain interfaces. Felmetsger et al [4] use an approach for automated logic vulnerabilities detection in web applications. They depend on inferring system specifications of a web application's logic by analysing system execution traces. They then use model checking to identify specification violations; however, they assume that collected traces represent real correct system behaviour.

Existing efforts for vulnerability mitigation can be categorized as code modification guidelines, server-side mitigation, and client side (Browser) approaches. Wurzinger et al [19] introduce SWAP as a server-side solution for detecting and preventing XSS. SWAP works as a reverse proxy intercepting HTML responses and forward them to a java-script detection component. Bisht et al [20] introduce a new XSS prevention solution that works by dynamically learning the set of scripts that a web application intends to create for any HTML request. Ravi et al [21] introduce an analysis of two mitigation approaches for XSS. The first one is a server-side approach intercept requests and perform string analysis looking for XSS signatures. The second approach is browser-based replacing java keywords in the responses with words that have same pronunciation. Vogt et al [22] introduce a new approach to block XSS on the client side by tracking the sensitive information flow inside the web browser.

9   Summary

We described a new integrated, automated online vulnerability analysis and mitigation solution as a service (VAM-aaS). We use a formalized vulnerability definition schema including vulnerability signature and mitigation actions. An OCL-based vulnerability signature specifies a set of invariants that verifies the existence or absence of a given vulnerability in a target program. We developed a static vulnerability analysis tool that uses these signatures to locate possible matches in a target system. Vulnerability mitigation actions specify what to do whenever a vulnerability instance is reported, including security controls to be plugged-in to block discovered vulnerabilities. We validated our approach on a set of seven benchmark applications. Our results show that the OCL-based analysis tool achieves (90%) precision rate and (92%) recall rate while our mitigation component achieves (100%) recall and (85%) precision rate.

References

1. Almorsy, M., Grundy, J., Mueller, I.: An analysis of the cloud computing security problem. In: Asia Pacific Cloud Workshop, APSEC 2010, Sydney, Australia (2010)
2. Bau, J., et al.: State of the Art: Automated Black-Box Web Application Vulnerability Testing. In: 2010 IEEE Symposium on Security and Privacy (2010)
3. Kals, S., et al.: SecuBat: a web vulnerability scanner. In: Proc. of 15th Int. Conf. on World Wide, Web 2006, pp. 247–256. ACM, Edinburgh (2006)
4. Felmetsger, V., et al.: Toward automated detection of logic vulnerabilities in web applications. In: 19th USENIX Conf. on Security, Washington, DC (2010)
5. Jovanovic, N., Kruegel, C., Kirda, E.: Pixy: a static analysis tool for detecting Web application vulnerabilities. In: IEEE Symposium on Security and Privacy (2006)
6. Dasgupta, A., Narasayya, V., Syamala, M.: A Static Analysis Framework for Database Applications. In: 2009 IEEE Int. Conf. on Data Engineering (2009)
7. Martin, M., Livshits, B., Lam, M.: Finding application errors and security flaws using PQL: a program query language. In: 20th Conf. on Object-oriented Programming, Systems, Languages, and Applications, CA, USA (2005)
8. Lam, M.S., et al.: Securing web applications with static and dynamic information flow tracking. In: Symposium on Partial Evaluation and Semantics-based Program Manipulation, California, USA (2008)
9. Kieyzun, A., et al.: Automatic creation of SQL Injection and cross-site scripting attacks. In: 31st Int. Conf. on Software Engineering (2009)
10. Halfond, W.G.J., Orso, A., Manolios, P.: Using positive tainting and syntax-aware evaluation to counter SQL injection attacks. In: 14th Int. Symposium on Foundations of Software Engineering, Oregon, USA (2006)
11. Weinberger, J., Saxena, P., Akhawe, D., Finifter, M., Shin, R., Song, D.: A Systematic Analysis of XSS Sanitization in Web Application Frameworks. In: Atluri, V., Diaz, C. (eds.) ESORICS 2011. LNCS, vol. 6879, pp. 150–171. Springer, Heidelberg (2011)
12. Wassermann, G., Su, Z.: Static detection of cross-site scripting vulnerabilities. In: 30th Int. Conf. on Software Engineering. ACM, Leipzig (2008)
13. Hooimeijer, P., et al.: Fast and precise sanitizer analysis with BEK. In: 20th USENIX Conf. on Security 2011, p. 1. USENIX Association, San Francisco (2011)
14. Balzarotti, D., et al.: Saner: Composing Static and Dynamic Analysis to Validate Sanitization in Web Applications. In: IEEE Security and Privacy (2008)
15. Ganesh, V., Kieżun, A., Artzi, S., Guo, P.J., Hooimeijer, P., Ernst, M.: HAMPI: A String Solver for Testing, Analysis and Vulnerability Detection. In: Gopalakrishnan, G., Qadeer, S. (eds.) CAV 2011. LNCS, vol. 6806, pp. 1–19. Springer, Heidelberg (2011)
16. Monga, M., Paleari, R., Passerini, E.: A hybrid analysis framework for detecting web application vulnerabilities. In: 2009 ICSE Workshop on Software Engineering for Secure Systems, pp. 25–32 (2009)
17. Zhang, R., et al.: Static program analysis assisted dynamic taint tracking for software vulnerability discovery. In: Computers & Mathematics with Application, pp. 469–480 (2012)

18. NIST: Source Code Security Analysis Tool Functional Specification Version 1.1. In: NIST Special Publication 500-268 (May 2007) (accessed 2011)
19. Wurzinger, P., et al.: SWAP: mitigating XSS attacks using a reverse proxy. In: ICSE Workshop on Software Engineering for Secure Systems, Vancouver, pp. 33–39 (2009)
20. Bisht, P., Venkatakrishnan, V.N.: XSS-GUARD: Precise Dynamic Prevention of Cross-Site Scripting Attacks. In: Zamboni, D. (ed.) DIMVA 2008. LNCS, vol. 5137, pp. 23–43. Springer, Heidelberg (2008)
21. Kotha, R., Prasad, K., Naik, D.: Analysis of XSS attack Mitigation techniques based on Platforms and Browsers. In: SEA, CLOUD, DKMP, CS & IT, vol. 5, pp. 395–405 (2012)
22. Vogt, P., et al.: Cross-Site Scripting Prevention with Dynamic Data Tainting and Static Analysis. In: Network and Distributed System Security Symposium, San Diego, CA (2007)
23. CENZIC: Web Applications Security Trends Reports Q1-Q2 2010 (2010), http://cenzic.com/downloads/Cenzic_AppSecTrends_Q1-Q2-2010.pdf
24. OWASP: Open Web Application Security Project, https://www.owasp.org
25. CWE: Common Weaknesses Enumeration, http://cwe.mitre.org
26. SharpDevelop, http://wiki.sharpdevelop.net/
27. Yiihaw:YIIHAW Is an Intelligent and High-performing Aspect Weave, http://yiihaw.tigris.org/

The Role of Twitter
in YouTube Videos Diffusion

George Christodoulou, Chryssis Georgiou, and George Pallis

Department of Computer Science
University of Cyprus
{gchris04,chryssis,gpallis}@cs.ucy.ac.cy

Abstract. Understanding the effects of social cascading on streaming media is of great importance to Web information system engineering. Given the large amount of available videos, it is often difficult for users to discover interesting content. Relying on the suggestions coming from friends seems to be a popular way to choose what to watch. Taking into account the increasing popularity of Online Social Networks and the growing popularity of streaming media, in this paper we present a detailed analysis of social cascading exchange of YouTube videos among Twitter users. Using a real data set we have recently collected, our analysis highlights several important aspects of social cascading, including its impact on YouTube videos popularity, dependence on users with a large number of followers, the effect of multiple sharing follows and the distribution of cascade duration.

Keywords: Social Video Sharing, Social Web, Social Cascading, YouTube, Twitter, Internet Measurements.

1   Introduction

The rapid proliferation of online social networking sites like Facebook and Twitter has made a profound impact on the Internet and tends to reshape its structure, design, and utility [10]. Industry experts [15] believe that Online Social Networks (OSNs) create a potentially transformational change in consumer behavior and will bring a far-reaching influence on traditional industries of content, media, and communications.

Motivation. Two recent trends in Web information system engineering motivate this work: the increasing popularity of OSNs and the growing popularity of streaming media. In contrast to traditional methods of content discovery such as browsing or searching, OSN sites have recently emerged as a popular way of discovering information on the Web through information dissemination along user's social links. Today OSN sites have been noted as being the primary causes behind the recent increases in HTTP traffic observed in measurement studies. According to Hitwise[1], 8.6% of traffic to news sites now comes from Facebook,

---

[1] http://www.experian.com/hitwise/index.html

X.S. Wang et al. (Eds.): WISE 2012, LNCS 7651, pp. 426–439, 2012.

Twitter and smaller social media sites, which is a 57% percent increase since 2009. The percentage coming from search engines, at the same time, is declining; according to Hitwise, it has been observed a drop of 9% since 2009. Social media, in other words, now bring in almost half as much traffic to news sites as search does. A second emerging trend is the growing popularity of streaming media services. The amount of Internet traffic generated every day by online multimedia streaming providers, such as YouTube, is extremely high [18]. Although it is difficult to estimate the proportion of traffic generated by social cascading, it is observed that there are more than 400 tweets per minute with a YouTube link [6].

While the real numbers are debatable, it is clear that the evolution of OSNs and streaming media play a crucial role on Internet traffic, since social cascades (information (i.e., text, image, video) dissemination along links in a OSN [8]) affect the users' navigation behavior. In a recent study [4], authors measured the role that social cascading impacts the diffusion of information. Their experiments showed that social cascades affect significantly the browsing of users. At the same time, YouTube is the most popular and bandwidth intensive service of today's Internet [6]. The mix of the two phenomena has serious implications and presents new challenges for Internet services and content providers towards improving the effectiveness of several services, including caching, content delivery networks, searching and content recommendation [21].

Contributions. In this work we address the following question: *What is the role of social cascading in YouTube video diffusion?* In order to answer this question, we study a large corpus of YouTube videos. For capturing the social cascading effects, we use Twitter, which is one of the most popular OSNs and its core functionality, tweeting, is centered around the idea of spreading information by word-of-mouth [19]. Specifically, Twitter provides mechanisms like retweet (act of forwarding other people's tweets), which enable users to propagate information across multiple hops in the network through cascading. According to a recent announcement[2], Twitter is sharing more than 340 million tweets per day, where 25% of tweets contain links. Overall, this paper makes the following contributions:

- We present the methodology that we have followed in order to collect the Twitter dataset. Our study is based on a newly real dataset from Twitter containing geographic location, follower lists and tweets for 37 million users. Then, we tracked the spreading of more than one million of YouTube videos over this network, analyzing a corpus with more than 2 billions messages and extracting about 1,3 millions single messages with a video link.

- We examine the role of social cascading in YouTube video diffusion. Our analysis highlights several important aspects of social cascading, including its impact on YouTube videos popularity, dependence on users with a large number of followers, the effect of multiple sharing follows and the distribution of cascade duration. Our analysis provides valuable results so as to better

---

[2] http://mashable.com/2012/03/21/twitter-has-140-million-users/

understand how the retweeting mechanism affects the spread of YouTube videos. To this respect, we introduce a new metric, called *video retweet likelihood*, that measures the likelihood of a user retweeting a video. Although our work has focused only on YouTube videos, its wide popularity and its massive user base allow us to gain insights on user navigation behavior on other similar media platforms.

Roadmap. The rest of this paper is organized as follows. Section 2 reviews previous related work. Our data collection methodology is described in Section 3, whereas, our main findings are presented in Section 4. Section 5 concludes the paper and discusses directions for future work.

2   Related Work

Many studies have been carried to analyze the users' behaviors in different media services [1, 18, 22]. Early work in this area is focused on the analysis and characterization of streaming services in the Internet [22]. In [22], the authors explored workload characteristics based on logs from internal media servers at Hewlett-Packard. Recently, YouTube has been a popular research topic in the Internet measurement community [6, 11]. Several studies [1, 7, 12–14, 17] have been conducted to investigate the traffic characteristics of YouTube users. These works are focused on the characteristics of YouTube content, such as file size, bitrate, usage patterns and popularity. After an extensive analysis of the YouTube workload in [13], authors found that there are many similarities to traditional Web and media streaming workloads. From another perspective, the authors in [9] studied YouTube videos and found that the videos have strong correlations with each other since the links to related videos generated by uploaders have small-world characteristics. In [11], it is characterized the growth patterns of video popularity on YouTube and analyzed how the popularity of individual videos evolve since the video's upload time.

OSNs are focused on sharing information and as such, have been studied extensively in the context of information diffusion. For instance, the authors in [19] found how propagation of YouTube videos on Twitter is spread among users who are geographically close together. In the same context, the authors in [21] studied how geographic information extracted from social cascades of Twitter can be exploited to improve caching of multimedia files in a Content Delivery Network (CDN). Similarly, in [23], the authors developed a system that exploits information available from Twitter and regularity of activity patterns so as to distribute long-tailed content while decreasing bandwidth costs. In [6], it is measured the popularity distribution of YouTube videos across different geographic regions and analyze how social sharing affects their spatial popularity. According to this study, it is observed that the impact of social sharing on the geographic properties of YouTube video views is significant. Our findings confirm these results with respect to our investigation of the impact of social cascading regarding geographic popularity.

Various studies in the context of social networks have been conducted to predict properties of the social cascading process. The authors in [3] exploited the social cascades in order to identify influencers in Twitter. The authors in [20] focus on characterizing and modeling the information cascades formed by the individual URL mentions in the Twitter so as to predict which users will predict which URL. In a recent study [4], the authors subject 250 million Facebook users to a controlled experiment in order to measure the role that social cascading impacts the diffusion of information.

The present work builds on these earlier contributions in the following key issues. First, whereas the focus of previous studies [1, 12, 14, 17, 18, 22] has been on the analysis and characterization of streaming services in the Internet, we are interested in the analysis of YouTube videos, taking into account the word-of-mouth diffusion of Twitter. Second, although previous works [4, 19] study the impact of social cascading for YouTube videos, they do not focus on a Twitter data set. As we mentioned above, the core functionality of Twitter is centered around the idea of social cascading.

3   Methodology

In this section we first present the methodology we followed in collecting our Twitter data set and then we extract data characteristics from the obtained set. We note that the data collection was not a straightforward process, mainly due to Twitter's recently modified policy of limiting the number of search requests per hour from a given IP address. Below we explain how we managed to collect our data set while respecting, in our opinion, this policy.

3.1   Data Collection

Data collection took over five months using four Cloud infrastructures (Nephelae[3], Okeanos[4], Amazon EC2 and Rackspace). The data are stored locally in a database. Specifically, a Twitter user keeps a brief profile about each user. The public profile includes the full name, the location, a web page, a short biography, and the number of tweets of the user. The people who follow the user and those that the user follows are also listed. In order to collect user profiles, we search for HTTP URLs that were posted on Twitter.

Totally, the data comprises profiles of 37,343,273 users, 6,820,494,777 directed follower links among these users. Due to computation limitations, we select uniformly at random from the above data set 1,384,758 users (247,399,334 directed follower links among these users), whom we focus on in the remainder of this paper. Specifically, 299,071,571 public tweets were posted by these users. The tweets are from December 2011 until April 2012. The period of study allows us to avoid any seasonal side effects exhibited by users navigation behavior. The data set does not include any tweet information about a user who had set his

---

[3] Nephelae. http://grid.ucy.ac.cy/Nephelae/

[4] Okeanos. https://cms.okeanos.grnet.gr/about/

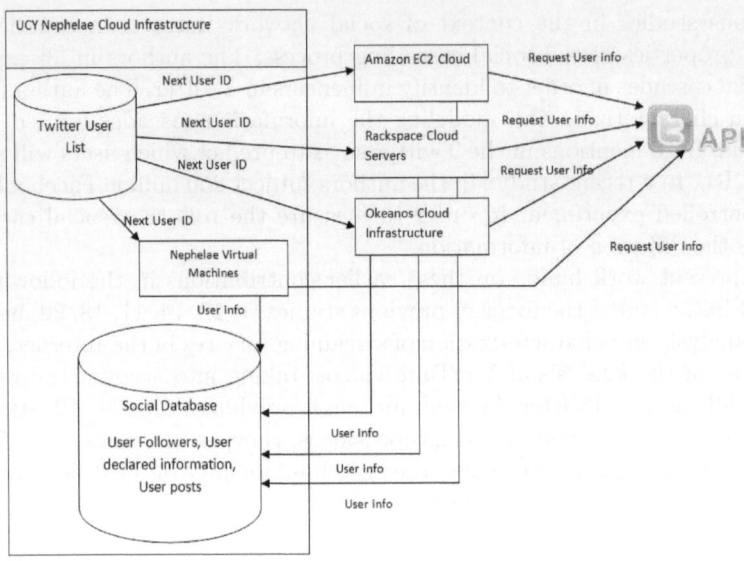

Fig. 1. Data Collection System for Twitter

account private. The large-scale of our dataset captures the geographic location diversity of Twitter users. Figure 1 presents the data collection system that we developed in order to collect the Twitter data. In the following paragraphs, we present the methodology that has been followed.

Using the Twitter API[5], we collected user profiles and tweets for each user and then analyze tweets to get HTTP URLs. Twitter imposes rate limiting in the number of search requests per hour from a given IP address. Specifically, one request can fetch up to 5,000 followers for a user, or 200 tweets for a user, or the information of 200 users. To respect this policy, we used several Twitter accounts. Additionally, in order to increase the number of users we used the Twitter social graph obtained in 2009 [16] and requested for Twitter users that were in the Twitter user pool. During our collection period we managed to collect more than 300 million tweets containing HTTP URLs. For each tweet we have crawled the author, the time when it was sent and the actual content of the message. In order to capture the geographic location of users, we used the Geocoding API of Google Maps. The Google Geocoding API provides a direct way to access a geocoder via an HTTP request. Tables 1 and 2 present the information that we collected for each user profile and each tweet respectively.

The next step was to pre-process the content in tweets. While analyzing the URLs within tweets we have found that the majority of the Web links are from URL shortening services (e.g., bit.ly), which substantially shorten the length of any URL. Thus, we used URL shortening services, such as unshort.me, in order to unshort the URLs. Then, the final step was to gather all the URLs and filter out all the URLs except the YouTube ones.

---

[5] Twitter API. https://dev.twitter.com/

Table 1. Twitter User Profile Information

Verified	user's identity has been verified by email account
followers_count	the number of users that follow the user
Protected	user's account is private and only their approved followers can read their tweets or see extended information about them
listed_count	the number of lists the user is a member of
friends_count	the number of users the user follows
Location	the location of user
geo_enabled	if enabled allows applications to send tweets with a geographic location attached
Lang	the language of user
favourites_count	the number of tweets the user has classified as favorites
created_at	the date that the account has been created
time_zone	the time zone of each tweet

Table 2. Tweet Information

ID	the unique ID of the tweet
Text	the text of tweet (typically up to 140 characters)
created_at	the date that the tweet has been published
Retweeted	if it is new tweet or a retweet
in_reply_to_status_id	the ID of an existing status that the update is in reply to
in_reply_to_user_id	the user ID that the tweet replies
urls	the url of the tweet
retweet_count	the number of times that a tweet has been retweeted

3.2   Data Set Characteristics

Unlike other OSNs, a Twitter user may follow another user to receive his/her tweets, forming a social network of interest. Furthermore, it is not necessarily the case that two users are mutual followers. Thus, Twitter is represented by a directed graph, where nodes represent the users and a direct link is placed from a user to another user, if the first follows the tweets of the latter. Figure 2 depicts an example of a Twitter social graph. Users A and G are mutual followers, while users A and B are not (A follows B but not vice-versa). According to [19] the node in-degree and out-degree distributions measured on this network are heavy-tailed, and the network topology is similar to those of other OSNs like Facebook. Although a very small fraction of users have an extremely large number of friends, the majority of users have only a few friends. The most popular users act as authorities and are usually either public figures or media sources. These observations are confirmed by the data set we have collected. Another interesting observation from our Twitter data set is that the high in-degree

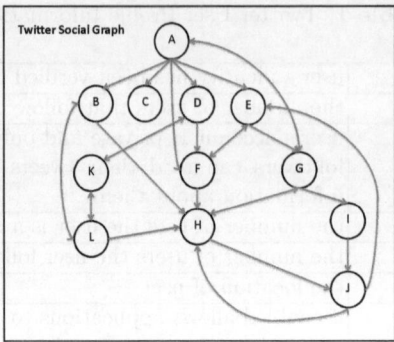

Fig. 2. An example of the representation of Twitter followers as a social graph

nodes are not necessarily high out-degree nodes since a small number of links are bidirectional (that is, followers are mutual).

4   Effects of Social Cascading

In this section we investigate the role of the retweeting mechanism in YouTube video diffusion. To this respect, social cascades include only users that have tweeted a certain video link. Also, without loss of generality, we make the assumption that every video contained in a Twitter message has been viewed by the user who retweet it (in other words, we assume that users do not "blindly" retweet videos).

To measure the impact of social cascading, and more specifically of video retweeting, we define the *video retweet likelihood* metric over our data set, as the likelihood of retweeting a YouTube video. More formally, we define the set *OutgoingVideos* to be the set of videos (over all users) that were *retweeted* by some user to another user (multiplicities are not counted). Similarly, we define the set *IncomingVideos* to include all videos that a user received by some other user, as a result of a *retweet* of the latter. Then, the *video retweet likelihood* is computed by the following expression:

$$\frac{|OutgoingVideos \cap IncomingVideos|}{|IncomingVideos|}.$$

Note that this metric captures only the videos that have been retweeted. The cardinality of the intersection gives the number of videos that users retweeted among the ones that were retweeted to them. Dividing this number with the total number of "received retweeted" videos gives the likelihood of a video being retweeted.

4.1   Impact of Twitter Users

First, we study how retweeting influences the diffusion of YouTube videos. Our aim is to investigate the impact of social cascading on the users' navigation

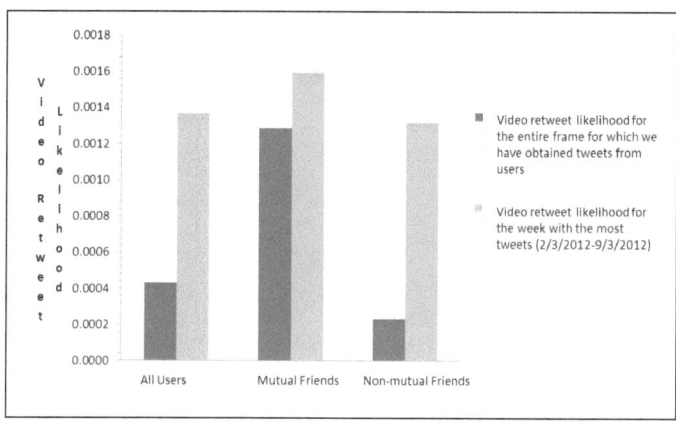

Fig. 3. Influence of *video retweets* with respect to the relationship between Twitter Users

behavior. In general, it is more likely to view and retweet a video that has been tweeted by a follower. Considering that each tweet can be viewed by all the followers of the author, the potential audience that a YouTube video may reach via retweeting is much larger, even if only few users are involved. As displayed in Figure 3, an interesting observation is that the *video retweet likelihood* is 6 times larger for users that are mutual followers. Recall that two users are mutual followers if the relationship of following and being followed is reciprocal (like users A and G in Figure 2). Moreover, we studied the *video retweet likelihood* for the week with the most tweets. Comparing with the results that we took for the whole period, to our surprise, we observe a different behavior for the non-mutual followers. On the other hand, the results for mutual followers are quite similar for both periods.

Furthermore, in Figure 4 we show how the *video retweet likelihood* is affected taking into account the number of users that have shared a tweet. We observe that the *video retweet likelihood* is increased with the number of users' follows who have already shared the same tweet. This increase seems to be exponential when the same tweet is shared by more than 8 follows. This is consistent with recent observational studies in other OSNs, such as Facebook [4].

4.2   Impact of Geographic Popularity

Our next study is to investigate the impact of social cascading regarding the geographic popularity. To capture the geographic popularity, we use the *time-zones* of users. Twitter enables users to declare their time zone. However, instead of using the UTC time zone system (where the globe is divided into 24 time zones), Twitter uses its own time zone system which divides the globe into 142 zones. In our study we consider that the users with the same Twitter time-zone have the same geographic location. Then, according to our data set, the average number of users per zone is 93,076.3, the median 7,167 and the standard deviation is

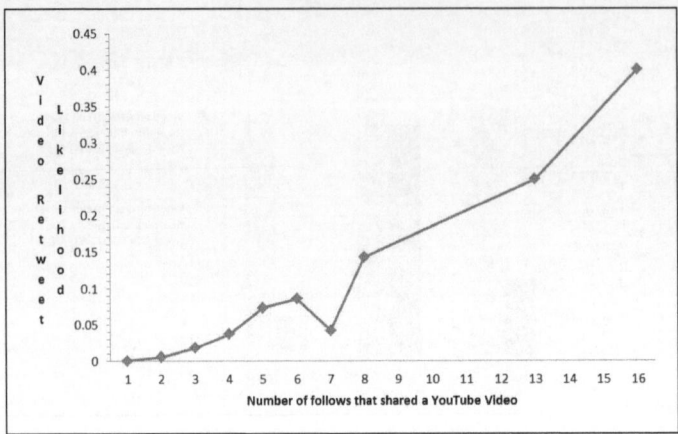

Fig. 4. The *video retweet likelihood* with respect to the number of follows that share a YouTube video

369,823.6 (this large number is due to the different population distribution of different time zones: some zones expand over entire countries, where others over small cities). Figure 5 depicts the *video retweet likelihood* with respect to the population in logarithmic scale. Each point in the plot depicts a group of users that belong in the same time zone, since users are mainly influenced by follows who are in the same geographic location. Our results show that the smaller the population is, the larger the *video retweet likelihood* is. This means that the social cascading effect has high impact on a more focused and less diverse set of geographic regions. These findings confirm recent results [2, 6] which found that geographic distance affects social interaction on OSNs. Specifically, the highest *video retweet likelihood* (0.1) has been observed in the region of Astana. Astana is a place with 700,000 habitants, whereas, the average likelihood is 0,0016.

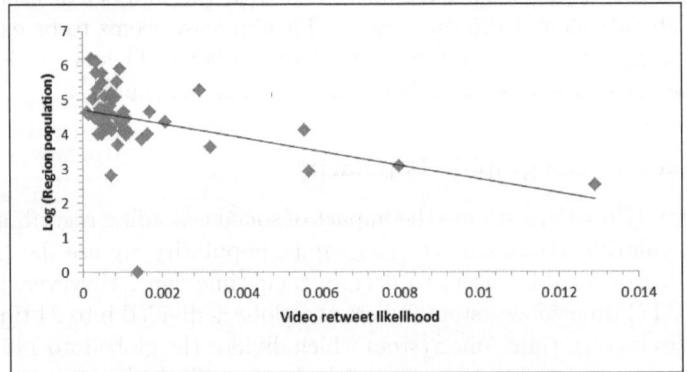

Fig. 5. The log plot geographic location population of YouTube videos with respect to the *video retweet likelihood*

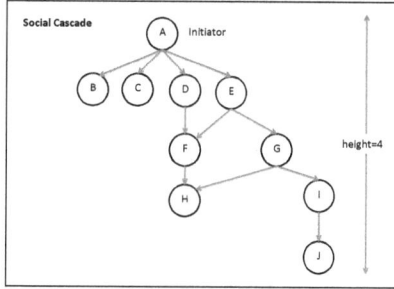

Fig. 6. An example of a cascade represented as a tree rooted at the initiator

Furthermore, we study the relationship between the popularity of videos with respect to the number of retweets. In order to understand the impact of social cascading for videos with higher number of views we have classified all videos regarding their popularity (this statistic is given by the YouTube API) into 16 segments (of views). The results (Table 3) show that the popularity versus the number of retweets is not trivial. The effect of social cascading for these groups of videos is different across these 16 categories. The general trend is that the more popular a video is, the more retweets has. From Table 3, it occurs a surge in the number of retweets for very popular videos. Also, it is interesting that the average number of retweets does not exceed an upper limit, which is 26 cascades in our case.

4.3   Impact of Social Cascade Length

The next experiment is to study the *video retweet likelihood* with respect to the length of the cascade. Cascades can be represented as rooted directed trees where the initiator of the cascade is the root of the tree [3]. Figure 6 depicts an example of a cascade, initiated by user A over the Social Graph of Figure 2. Then, the *length* of the cascade is the *height* of the resulting tree (which is 4 in Figure 6).

Figure 7 depicts the distribution of the cascade length (given in log scale), which is approximately power-law. This measure of popularity demonstrates that it is rare to have large cascades, but when they do take place they can become extremely large. This implies that the vast majority of posted YouTube videos do not spread at all.

4.4   Impact of Time

In Figure 8, we illustrate the distribution of cascade duration (in hours) from the first tweet to the last tweet for each cascade with at least 2 users, not counting the initiator. This result shows how YouTube links can spread on Twitter on a time scale. About 70% of the cascades end within 24 hours. In particular, about 25% of the cascades occur within the first hour, in 3 hours the spread reaches

Table 3. Popularity of YouTube videos

Views	Number of videos	Number of retweets	Avg. Number of retweets
1000	68655	85252	1.24
5000	37899	45640	1.20
20000	43014	49197	1.14
50000	34855	40515	1.16
200000	53509	68099	1.27
400000	23544	34638	1.47
700000	16010	27011	1.68
1 million	8571	16452	1.92
2 millions	13332	27721	2.08
5 millions	11183	30245	2.70
10 millions	4641	18924	4.07
20 millions	2205	12967	5.88
50 millions	1210	10775	8.90
100 millions	393	10519	26.76
200 millions	21	553	26.33
350 millions	7	176	25.14

to 40% and about 85% of the cascades end by the third day (72 hours). This indicates that links to videos can quickly spread over the social network, leading to many views in a short period of time. This information could be exploited, for example, in improving the efficiency of Content Delivery Networks, as discussed in the next section.

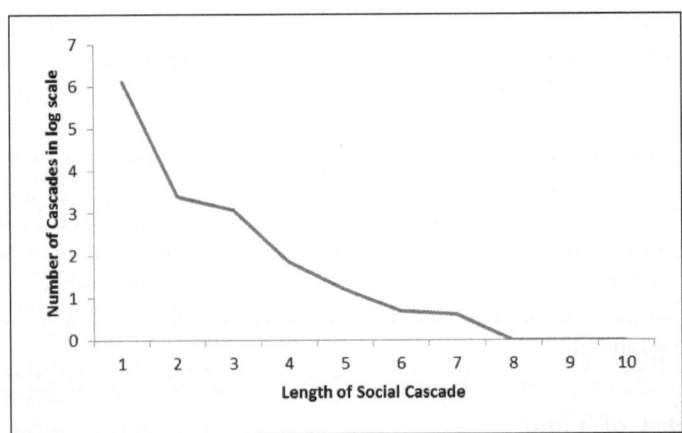

Fig. 7. Number of social cascades in log scale with respect to the length of the social cascade

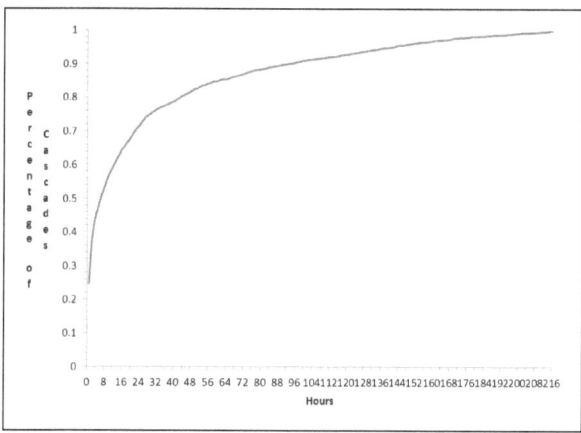

Fig. 8. The cumulative distribution function of the social cascades with respect to the time passed

5   Conclusion and Future Work

The widespread adoption of OSN sites has significantly altered the information diffusion through the Web. In this work, we have presented how the retweeting influences the diffusion of YouTube videos. Using an experimental approach on Twitter, we are able to quantify the effect of social cascading on video spread. This study is useful for Internet service and content providers, who can exploit these findings towards improving the effectiveness of their services.

One of the most sound observations of our study is that the social cascading effect has high impact on a more focused and less diverse set of geographic regions. Also, the social cascading effect ends within 24 hours. These findings are useful for large-scale systems whose traffic is driven by online social services. For instance, Content Delivery Networks (CDNs) can take advantage of the fact that social cascades can spread in a geographically limited area to decide whether a YouTube video is disseminating locally or globally.

Another interesting observation is that social cascading affects the users navigation behavior. In the case of Twitter, users are influenced more from the follows who also follow them as well. This finding may be used to study the role of influencers in Twitter. A related work in this area has been presented in [3]. Also, our analysis showed that most events through Twitter do not spread at all, and even moderately lengthed cascades are extremely rare.

For the future, we plan to further investigate the impact of social cascading in YouTube video diffusion. Specifically, we will study the retweeting influence with respect to the popularity dynamics of YouTube videos over the time [5]. An implication of this study is the improvement of Internet-based content delivery. The rapid proliferation of OSNs opens new perspectives in Internet-based content technologies, raising new issues in the architecture, design and implementation of existing CDNs. In this context, we plan to develop a realistic media workload generator that would reflect the dynamics and evolution of content at media sites and the change of access rate to this content due to the role of social networks in information diffusion. The media workload generator will produce synthetic traces with desired distributions and controllable parameters for performance experiments studying effective streaming content delivery approaches. The ultimate goal of this generator is to be used as a valuable tool in order to study efficient algorithms towards predicting social cascades and improving the performance of CDNs.

Acknowledgement. The authors thank the anonymous reviewers for their constructive comments and suggestions. This work is supported in part by the Startup Grant [George Pallis, 2011], funded by the University of Cyprus.

References

1. Abhari, A., Soraya, M.: Workload generation for youtube. Multimedia Tools Appl. 46(1), 91–118 (2010)
2. Backstrom, L., Sun, E., Marlow, C.: Find me if you can: improving geographical prediction with social and spatial proximity. In: Proceedings of the 19th International Conference on World Wide Web (WWW 2010), pp. 61–70. ACM, New York (2010)
3. Bakshy, E., Hofman, J.M., Mason, W.A., Watts, D.J.: Everyone's an influencer: quantifying influence on twitter. In: Proceedings of the 4th ACM International Conference on Web Search and Data Mining (WSDM 2011), pp. 65–74. ACM, New York (2011)
4. Bakshy, E., Rosenn, I., Marlow, C., Adamic, L.: The role of social networks in information diffusion. In: Proceedings of the 21st International Conference on World Wide Web (WWW 2012), pp. 519–528. ACM, New York (2012)
5. Borghol, Y., Mitra, S., Ardon, S., Carlsson, N., Eager, D., Mahanti, A.: Characterizing and modelling popularity of user-generated videos. Perform. Eval. 68(11), 1037–1055 (2011)
6. Brodersen, A., Scellato, S., Wattenhofer, M.: Youtube around the world: geographic popularity of videos. In: Proceedings of the 21st International Conference on World Wide Web (WWW 2012), pp. 241–250. ACM, New York (2012)
7. Cha, M., Kwak, H., Rodriguez, P., Ahn, Y., Moon, S.: I tube, you tube, everybody tubes: analyzing the world's largest user generated content video system. In: Proceedings of the 7th ACM SIGCOMM Conference on Internet Measurement (IMC 2007), San Diego, California, USA, pp. 1–14. ACM (2007)
8. Cha, M., Mislove, A., Adams, B., Gummadi, K.: Characterizing social cascades in flickr. In: Proceedings of the First Workshop on Online Social Networks (WOSN 2008), Seattle, WA, USA, pp. 13–18 (2008)
9. Cheng, X., Dale, C., Liu, J.: In: Proceedings of the 16th International Workshop on Quality of Service (IWQOS 2008), Enskede, The Netherlands, pp. 229–238. IEEE (2008)
10. Datta, A., Dikaiakos, M.D., Haridi, S., Iftode, L.: Infrastructures for online social networking services. IEEE Internet Computing 16, 10–12 (2012)
11. Figueiredo, F., Benevenuto, F., Almeida, J.M.: The tube over time: characterizing popularity growth of youtube videos. In: Proceedings of the 4th ACM International Conference on Web Search and Data Mining (WSDM 2011), pp. 745–754. ACM, New York (2011)

12. Finamore, A., Mellia, M., Munafò, M.M., Torres, R., Rao, S.G.: Youtube every-where: impact of device and infrastructure synergies on user experience. In: Pro-ceedings of the 2011 ACM SIGCOMM Conference on Internet Measurement Con-ference (IMC 2011), pp. 345–360. ACM, New York (2011)
13. Gill, P., Arlitt, M., Li, Z., Mahanti, A.: Youtube traffic characterization: a view from the edge. In: Proceedings of the 7th ACM SIGCOMM Conference on Internet Measurement (IMC 2007), San Diego, California, USA, pp. 15–28. ACM (2007)
14. Gill, P., Arlitt, M., Li, Z., Mahanti, A.: Characterizing user sessions on youtube. In: Proceedings of the ACM/SPIE Multimedia Computing and Networking Conference (MMCN 2008), San Jose, USA (2008)
15. Goel, S., Hofman, J.M., Sirer, M.I.: Who does what on the web: A large-scale study of browsing behavior. In: Proceedings of the 6th International AAAI Conference on Weblogs and Social Media (June 2012)
16. Kwak, H., Lee, C., Park, H., Moon, S.: What is Twitter, a social network or a news media? In: Proceedings of the 19th International Conference on World Wide Web (WWW 2011), pp. 591–600. ACM, New York (2010)
17. Mitra, S., Agrawal, M., Yadav, A., Carlsson, N., Eager, D., Mahanti, A.: Char-acterizing web-based video sharing workloads. ACM Trans. Web 5(2), 8:1–8:27 (2011)
18. Rao, A., Legout, A., Lim, Y.-S., Towsley, D., Barakat, C., Dabbous, W.: Network characteristics of video streaming traffic. In: Proceedings of the 7th Conference on Emerging Networking Experiments and Technologies (CoNEXT 2011), pp. 25:1–25:12. ACM, New York (2011)
19. Rodrigues, T., Benevenuto, F., Cha, M., Gummadi, K., Almeida, V.: On word-of-mouth based discovery of the web. In: Proceedings of the 2011 ACM SIGCOMM Conference on Internet Measurement Conference (IMC 2011), pp. 381–396. ACM, New York (2011)
20. Romero, D.M., Galuba, W., Asur, S., Huberman, B.A.: Influence and passivity in social media. In: Proceedings of the 20th International Conference on World Wide Web (WWW 2011), pp. 113–114. ACM, New York (2011)
21. Scellato, S., Mascolo, C., Musolesi, M., Crowcroft, J.: Track globally, deliver locally: improving content delivery networks by tracking geographic social cascades. In: Proceedings of the 20th International Conference on World Wide Web (WWW 2011), pp. 457–466. ACM, New York (2011)
22. Tang, W., Fu, Y., Cherkasova, L., Vahdat, A.: Medisyn: a synthetic streaming media service workload generator. In: Proceedings of the 13th International Work-shop on Network and Operating Systems Support for Digital Audio and Video (NOSSDAV 2003), pp. 12–21. ACM, New York (2003)
23. Traverso, S., Huguenin, K., Triestan, I., Erramilli, V., Laoutaris, N., Papagiannaki, K.: Tailgate: handling long-tail content with a little help from friends. In: Proceed-ings of the 21st International Conference on World Wide Web (WWW 2012), pp. 151–160. ACM, New York (2012)

Evaluating Web Archive Search Systems

Miguel Costa[1,2] and Mário J. Silva[3]

[1] Foundation for National Scientific Computing, Lisbon, Portugal
[2] LaSIGE, Faculty of Science, University of Lisbon, Lisbon, Portugal
[3] IST/INESC-ID, Lisbon, Portugal
`miguel.costa@fccn.pt, mjs@inesc-id.pt`

Abstract. The information published on the web, a representation of our collective memory, is rapidly vanishing. At least 77 web archives have been developed to cope with the web's transience problem, but despite their technology having achieved a good maturity level, the retrieval effectiveness of the search services they provide still presents unsatisfactory results. In this work, we propose an evaluation methodology for web archive search systems based on a list of requirements compiled from previous characterizations of web archives and their users. The methodology includes the design of a test collection and the selection of evaluation measures to support realistic and reproducible experiments. The test collection enabled, for the first time, to measure the effectiveness of state-of-the-art IR technology employed in web archives. Results confirm the poor quality of search results retrieved with such technology. However, we show how to combine temporal features, along with the regular topical features, to improve the search effectiveness on web archives. The test collection is available to the research community.

1 Introduction

Every day millions of web documents become inaccessible. Some contain unique information that might become as valuable as ancient manuscripts are today. For instance, the speech of a president after winning an election or the announcement of an imminent invasion of a foreign country. Together, these documents form a comprehensive picture of our cultural, commercial, scientific and social history, expressed by all kinds of people. It is therefore important to preserve and make these data accessible, not only for historical research [1], but also to support current technology, such as assessing the trustworthiness of statements [2], detecting web spam [3] or improving web information retrieval (IR) [4].

Recently, UNESCO endorsed the Universal Declaration on Archives[1], which states that "archives play an essential role in the development of societies by safeguarding and contributing to individual and community memory." At least 77 initiatives[2] undertaken by national libraries, national archives and consortia of organizations are archiving parts of the web to cope with this problem. In total, more than 181 billion web documents (6.6 PB) are already archived and

---

[1] See `http://www.ica.org/6573/reference-documents/universal-declaration-on-archives.html`

[2] See `http://en.wikipedia.org/wiki/List_of_Web_archiving_initiatives`

X.S. Wang et al. (Eds.): WISE 2012, LNCS 7651, pp. 440–454, 2012.

these numbers, as well as their historic interest, are growing over time [5]. A new challenge is how to make historical analysis possible on all the data that has been accumulated over the years.

Full-text search has become the dominant form of finding web information, as notoriously seen in online search engines. It gives users the ability to quickly search through vast amounts of unstructured text, powered by sophisticated ranking tools that order results based on how well they match user queries. However, the poor quality of search results still remains a major hurdle in the way of turning web archives into an usable source of information. Users have to spend too much time and effort exploring retrieved documents in order to satisfy their information needs. As the amount of archived data continues to grow, this problem only tends to aggravate.

The quality of search results greatly depends on the availability of suitable evaluation methodologies and test collections. These resources have been a driver of research and innovation in IR throughout the last decades [6], enabling to: (1) compare multiple systems and approaches, demonstrating their effectiveness and robustness; (2) measure progress and produce sustainable knowledge for future development cycles; (3) predict how well a system will perform when deployed in an operational setting; (4) research under a set of controlled conditions. Unfortunately, existing evaluation methodologies and test collections from evaluation campaigns, such as TREC [6], are not useful for web archives, because they have different task goals and characteristics. For instance, existing collections do not have a temporal dimension, where each document may have several versions throughout time and their relevance depends of the user's period of interest.

In this work, we propose an evaluation methodology to measure the search effectiveness of web archive information retrieval (WAIR) systems. We believe that this methodology, along with the test collection created to support it, are essential pieces of technology to improve WAIR effectiveness. The methodology takes the findings of recent characterizations on web archives and their users in consideration, which is a requirement to providing reliable results tailored for the user information needs. We demonstrate the usefulness of the methodology through an experiment where we measured, for the first time, the search effectiveness of web archives using state-of-the-art methods. We also have been able to significantly improve the observed effectiveness by exploring temporal features intrinsic to web archives.

The remainder of this paper is organized as follows. In Section 2, we cover the related work. In Section 3, we describe the web archive characteristics that guide the design of the evaluation methodology proposed in Section 4. In Section 5, we present a case study applying the methodology and Section 6, the obtained results. Section 7, finalizes with the conclusions.

2   Related Work

2.1   Web Archives Access

Much of the current effort on web archive development focuses on acquiring, storing, managing and preserving data [7]. However, this is just the beginning.

The data must be accessible. Recently, 82% of the European web archives considered the improvement of access tools a high priority [8]. Due to the challenge of indexing all the collected data, the prevalent access method in web archives is based on URL search, which returns a list of chronologically ordered versions of that URL. A recent survey reported that 89% of the world-wide web archives support this type of access [5]. However, this type of search is limited, as it forces the users to remember the URLs, some of which refer to content that ceased to exist many years ago. Another type of access is meta-data search, for instance by category or theme, which was shown to be provided by 79% of web archives. Full-text search has become the dominant form of information access, specially in web search systems, such as Google, which has a strong influence on the way users search in other settings. This explains why full-text search was reported as the most desired web archive functionality [9] and the most used when supported [10]. Even with the high computational resources required for this purpose, 67% of world-wide web archives surveyed support full-text search for at least a part of their collections [5]. In another survey of European web archives this percentage is 70% [8]. As a result, in this work we focus on full-text search.

The large majority of web archives that support full-text search are based on the Lucene search engine[3] or extensions of Lucene to handle web archives, such as NutchWAX[4]. The search services provided by these web archives are visibly poor and frequently deemed unsatisfactory [5]. Cohen et al. showed that the out-of-the-box Lucene produces low quality results, with a MAP (Mean Average Precision) of 0.154, which is less than half the MAP of the best systems participating in the TREC Terabyte track [11].

2.2 IR Evaluations

IR evaluations straddle two opposite, but complementary views: a user-centered and a system-centered [12]. The goal of user-centered evaluations is to measure how people can use a system to retrieve relevant documents. These evaluations provide rich qualitative data about user interactions with the system, for instance, from experiments with users in a laboratory [13] or in their natural environment (in-situ) [14]. The goal of system-centered evaluations is to quantify the extent to which a system retrieves relevant documents, independently of how well users interact with it. The most popular example is the Cranfield paradigm established in the 1960s by Cleverton. This paradigm defines the creation of test collections for evaluating retrieval results composed by three parts: (1) a **corpus** representative of the items (often documents) that will be encountered in a real search environment; (2) a set of **topics** describing user information needs; and (3) **relevance judgments** (a.k.a. *qrels*) indicating the degree of relevance of each document retrieved for each topic. The effectiveness of an IR system is then measured by comparing its results against the known relevant documents for each topic. Our proposed methodology follows the Cranfield paradigm, extending it to address the specificities of web archives.

---

[3] See http://lucene.apache.org/

[4] See http://archive-access.sourceforge.net/projects/nutch/

3   Web Archive Characteristics

3.1   Corpus

A web archive corpus is composed by a stack of content collections harvested from the web over time. These collections are typically very heterogeneous in scope and size. Still, we found some common characteristics across the content collections of web archives:

selective and broad national crawls. 80% of the 42 world-wide web archive initiatives surveyed, exclusively hold content related to their country, region or institution [5]. All initiatives performed selective crawling, for instance, focusing in one sub-domain or topic. These collections are narrower, but deeper, trying to crawl every URL about the topic. 26% of the initiatives also performed broad crawling, including all documents hosted under a country code top-level domain or geographical location. These collections are wider, but shallower. In another survey of European web archives, 71% of them operate selective crawls and 23% broad domain crawls [8].

a variable number of versions per document. Some documents and sites are visited more often by crawlers due to digital preservation policies and, as result, are more frequently collected. The genre of document also influences the number of versions. For instance, newspapers have a higher change rate, while scientific articles tend to be static for long periods.

a diverse set of media types. The characterization of web collections shows that all media types are included in web archive collections, such as text, image, sound and video, but with predominant presence of HTML, PDF, JPEG and GIF formats that comprise over 95% of all web contents [15].

a volume of data between 1TB and 100TB. 81% of web archive collections have a volume of data smaller than 100TB [5]. The predominant volume of data is between 1TB and 10TB (31%) or between 10TB and 100TB (31%).

between 100 million and 1 billion documents. 78% of web archive collections contain less than 1 billion documents (i.e. files) [5]. The predominant number of documents is between 100 million and 1 billion (43%).

a large temporal span of at least 7 years. Four web archives were created in 1996 and their number has been growing since then. Assuming that the oldest web collections are from the creation year of web archives, 58% of the web archives contain collections up to 7 years old [5]. The corpus should have a large time span to not bias future WAIR technology to a specific period when some design patterns and technologies prevailed.

3.2   Search Topics

The evaluation of an information system, such as a web archive, must take into account the characteristics and needs of its user community. Characterizations of web archive users exhibit some characteristics that topics should include:

generic use cases. Despite some professional categories being more prone to use web archives, such as historians, average people also access them occasionally. There are numerous everyday life use cases that web archives can fulfill, as exemplified by Ras and Bussel [9] and log analysis has shown [16].

navigational and informational queries. The information needs of web archive users are mostly navigational, i.e. users intend to see how a web page or site was in the past or how it evolved throughout time [16]. The second most usual information need is informational, i.e. users intend to collect information about a topic written in the past, usually from multiple pages without a specific one in mind. Both represent more than 90% of all information needs.

1/3 of queries restricted by date range. Despite user information needs being focused on the past, the ratio of queries temporally restricted in web archives is only 1/3 [10]. Another aspect is that older years are more likely of being included in such queries.

queries without temporal clues. Only 3% of queries have expressions that could indicate a temporal dependent intent, such as *Euro 2004* [10].

short queries, each with 1 to 3 terms. A typical full-text session is composed by 1 or 2 queries, each having 1 to 3 terms [10]. Queries and terms follow a power law distribution, which means that a small fraction of each is submitted many times, while a large fraction is submitted just a few times.

The last four characteristics have been obtained from studies conducted on the Portuguese Web Archive (PWA). However, we believe that they are general, because it has been shown that users from the PWA and a Portuguese web search engine have a similar search behavior [17]. Thus, the differences between both systems do not affect the way users search in them. Additionally, the results compiled about web search engine users across the U.S. and Europe, including Portugal, were also similar [17,18]. Thus, the users' distinct language, vocabulary and culture have a small impact in the users' search behavior. In conclusion, despite some nuances, it seems that users from both types of systems and different countries, have similar search behaviors.

3.3   Relevance Propagation

A document d collected at n periods has n archived versions $\{v_{t_1}^d, ..., v_{t_n}^d\}$. A web archive enables searching over all these versions and may retrieve one or multiple versions of d. This deeply influences our understanding of relevance in two ways. First, the relevance granularity is the document's version identified by the pair <URL, timestamp>. Second, the relevance is bi-dimensional. Each version has associated a temporal relevance along with a topical relevance.

Topical Relevance. A navigational query intends to find an archived document for some purpose. Thus, if one version of a document d is relevant, we may assume that any version $v_{t_i}^d$ of d has the same topical relevance. Knowing this, we can propagate the topical relevance between versions of the same document. Only one version of each document needs to be assessed for navigational queries. All the other versions receive the same relevance degree.

For informational queries, the topical relevance of a version $v_{t_i}^d$ is measured according to how well it describes the searched topic in detail. Hence, since all versions $v_{t_i}^d$ of a document d may be different, they all may have different topical relevance. We cannot propagate the topical relevance between versions of the same document, except when the content of versions $v_{t_i}^d$ is very similar (e.g. near-duplicates).

Temporal Relevance. The relevance of archived versions depends also on the period of interest of the user query. Users explicitly express a date range that acts as a filter and exclude all versions with timestamps outside this range. This is the users' expected behavior, so we assume that the excluded versions are temporally non-relevant. All the others are considered equally relevant in the temporal dimension, because in web archives: (1) there is not a preference by a period within the date range (e.g. older or newer). This behavior is different from the observed in other search services, such as in news search, where recent and updated information is preferred [19]; (2) highly relevant documents for a topic may exist throughout the entire search period, despite being known that some periods tend to concentrate more relevant documents [20].

Summarizing what was previously discussed, we assume that two versions $v_{t_i}^d$ and $v_{t_j}^d$ of a document d, where $i \neq j$, have identical:

- topical relevance for a given navigational topic.
- topical relevance for a given informational topic if their content is very similar (e.g. near-duplicates).
- temporal relevance for a given topic if the timestamps t_i and t_j are both inside or outside the search interval.

If two versions $v_{t_i}^d$ and $v_{t_j}^d$ have the same topical and temporal relevance for a topic u, we define them as **redundant** for u.

4   Evaluation Methodology

Our proposed methodology, depicted in Figure 1, extends the Cranfield paradigm to support the ad-hoc retrieval task for web archives. The methodology has the following steps:

1. Characterization of web archives along with their collections and users. With the knowledge compiled in the previous section, we are able to build a representative test collection to draw valid conclusions.
2. Selection of a representative corpus of the documents that will be encountered in a real search environment. The corpus must fit the characteristics observed in world-wide web archives, such as their size and temporal span.
3. Selection of topics based on the users' information needs and search behavior. Topics are created from queries sampled from a query log of an operational web archive. These queries represent real and diverse information needs.

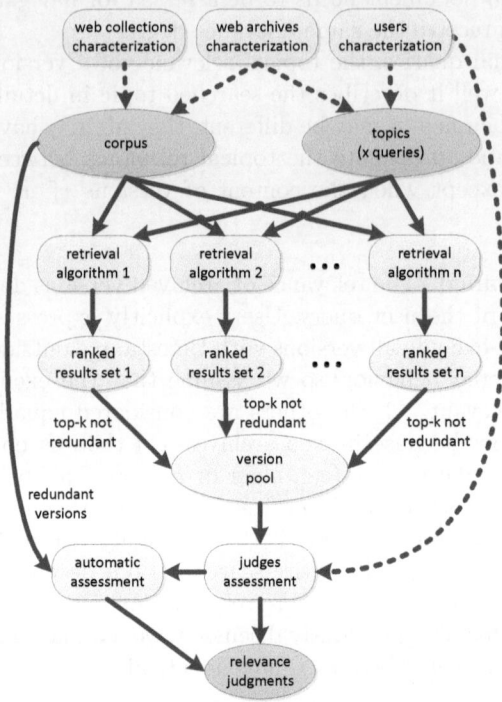

Fig. 1. Methodology for building a WAIR test collection

4. Development of several and diversified retrieval algorithms for matching and ranking document versions for each topic. These algorithms should contemplate topical and temporal features to explore both search dimensions.

5. Aggregation of all top-k versions returned by each retrieval algorithm for each topic into a version pool, ignoring the redundant versions. The aggregated versions have their timestamps within the search interval of interest specified on topics. The versions with timestamps outside the interval are ignored, since they are considered temporally not relevant.

6. Manual assessment of all items in the version pool by a set of judges according to the user information need defined for each topic. The information needs are defined taking into account the characteristics of the user community when using a web archive. All versions in the pool are within the search interval and, thus, are assumed as temporally relevant.

7. Automatic assessment of all versions of a document d with a manually assessed version $v_{t_i}^d$. Each version $v_{t_j}^d$ of d receives the same topical relevance degree given to $v_{t_i}^d$ if their relevance can be inferred (i.e. if they are redundant).

4.1   Evaluation Metrics

The manual and automatic assessments form the ground-truth used to evaluate the effectiveness of all retrieval algorithms and systems. There is now the issue of selecting evaluation measures that reflect the users' search behavior. The measures should consider the relevant versions ranked ahead of the non-relevant and the dependency between redundant versions. It is certainly unpleasant for an user to see multiple versions of the same document in a results page. If this is the case, the typical web archive user interface has associated to each result a link to show all versions of the respective document in a chronological view.

We have two choices to model this dependency. The first, is to design or adopt a measure, such as α-nDCG, that penalizes the relevance of redundant versions [21]. The second, is to use a standard measure after ignoring the redundant versions. We chose the second case, because it is: (1) preferable to use standard measures widely adopted within the community that were already thoroughly researched; (2) easier to optimize an IR system for one objective, than for a bi-objective where relevance is traded-off with diversity. Notice that, search result diversification is an NP-hard optimization problem [22]. As a drawback, the WAIR systems should collapse these redundant versions before presenting the results to the users. However, this corresponds to the common behavior already implemented in the user interfaces of existent WAIR systems.

Concluding, we promote diversity in search results by ignoring easily identifiable redundant versions before applying a standard evaluation measure. Any measure that can make use of these relevance judgments can be used. However, these measures should have a maximum cut-off of k (e.g. nDCG@k), where k is the number of top ranked results assessed. Otherwise, if the number of WAIR systems is small, it is likely that a significant number of relevant versions will not be found beyond rank k and the results biased.

5   Test Collection Construction

In this section we present a case study to empirically validate the proposed evaluation methodology. We detail the design of a test collection for the Portuguese Web Archive (PWA).

5.1   Corpus Selection

Our corpus is composed by six crawls of the Portuguese web, broadly considered the subset of the web of interest to the Portuguese. Since the goal is to create a corpus representative of the documents encountered in a real search environment, we only included collections indexed and searchable through the public access given by the PWA at `http://archive.pt`. The corpus' main characteristics are detailed in Table 1, showing a significant heterogeneity in age, size and type. They result from different crawls, which obtained 256 million documents, corresponding to 6.2 TB of compressed data in ARC format (8.9 TB uncompressed) [23]. This corpus contains some of the first documents published

Table 1. Web crawls that compose the corpus

#	Years	# Documents (K)	Size (GB)	Description
1	1996	75	0.316	selective crawl of most popular sites
2	1996 - 2000	5 047	48	broad crawls periodically made by the Internet Archive
3	2000 - 2008	118 842	1 900	broad crawls periodically made by the Internet Archive
4	2004 - 2006	14 374	165	selective crawls made by the Portuguese National Library
5	2008	48 718	1 600	exhaustive crawl of mostly the .pt domain
6	2009	68 776	2 500	exhaustive crawl of mostly the .pt domain
	Total	255 832	6 213	

in the Portuguese web in 1996 and go until 2009. It includes all common types of textual formats, such as HTML, PDF and Microsoft Office, and other media formats (image, video and audio) to support a faithful rendering of document versions, which are no longer available on the live web. We consider this corpus sufficiently comprehensive and representative, but not too large to discourage its use. The ClueWeb09[5] is the largest corpus made available to support research on IR. It contains over 1 billion web pages, which sums 5 TB compressed (25 TB uncompressed). This size is superior to the size of our corpus and several research groups have demonstrated that their IR systems scale to this order of magnitude, for instance, in the TREC web tracks since 2009.

5.2   Search Topics Selection

We randomly sampled queries from the PWA's query log fitting the general search patterns presented in Section 3. From these queries we created 50 navigational topics, where one third have temporal restrictions. IR evaluation campaigns generally use 50 topics, since this number gives a high confidence in the comparison between evaluated systems, especially for statistically significant differences [24]. We tried to select topics of different difficulties for IR systems, guaranteeing that a substantial part of the query terms are not present in the title or URL of the searched versions, nor all queries try to find site homepages, despite these being common. We also guaranteed that all topics have at least one relevant document archived and are not ambiguous in any sense.

The advantage of selecting queries instead of creating topics from scratch is that these capture the real and diverse user information needs, as opposed to manually creating artificial needs. The disadvantage is that the original intent of queries is not directly available. Topic creators had to examine each query within its user session, together with all the other queries and clicks, to infer the query's underlying need. Topic creators also browsed results from related queries to identify possible interpretations of the selected query.

Each topic is composed by three fields: query, period and description. The query is the set of terms entered by a user when searching in the web archive. The period defines the range of dates of interest to the user. These two fields are the ones submitted to the WAIR system. The description specifies the user information need. This field is important to help assessors judging the relevance

---

[5] See http://lemurproject.org/clueweb09/

of a version and aid future experimenters understanding the topic. An example of a navigational topic with a search period would be:

```
<topic number="1" type="navigational">
  <query>benfica</query>
  <period>
    <start format="dd/mm/yyyy">01/01/2007</start>
    <end format="dd/mm/yyyy">31/12/2007</end>
  </period>
  <description>
    Sport Lisboa e Benfica sports club in 2007.
  </description>
</topic>
```

A set of informational topics could be created in an analogous way.

5.3   Retrieval

WAIR System. The corpus was indexed by the IR system of the PWA, which has been released as an open source project at http://code.google.com/p/pwa-technologies/ The PWA IR system executes three steps in pipeline after receiving a topic's query: (1) versions are topically matched with the query's terms; (2) matched versions are temporally filtered according to the topics search period; (3) the remaining versions are ranked by topical and temporal similarity.

Ranking Models. A ranking model computes a score to each matching version that is an estimate of its relevance to a query. Matching versions are then ranked by score. We implemented 9 models. The first was the Lucene's term-weighting function[6], which is computed over 5 fields (anchor, content, title, hostname, url) with different weights. The second was a small variation of this function used in NutchWAX, with a different normalization by field length. These two models can be considered the state-of-the-art of IR in web archives, since most of the IR technology is based on the Lucene search engine and NutchWAX. As a baseline and third model, we selected the Okapi BM25 with parameters k1=2 and b=0.75 [25].

We also implemented two time-aware models that give a higher score to: (1) documents with more versions; (2) documents with a larger time span between the first and last archived versions. Both are defined by the same function:

$$f(v_{t_i}^d) = \frac{\log_{10}(x)}{\log_{10}(y)} = \log_y(x) \tag{1}$$

where, for the first case, x is the number of versions of document d and, for the second case, x is the number of days between the first and last versions of document d. y is the maximum possible x for normalization. Each of these functions,

---

[6] See http://lucene.apache.org/java/2_9_0/api/all/org/apache/lucene/search/Similarity.html

```
1. Imagine that to find the page of:
       José Saramago, Nobel Prize-Winning Writer in 1998.
2. You submit the query:
       josé saramago
3. And you obtain as result the:
       archived page of 03-24-2007 with the http://www.caleida.pt/saramago/ address.
4. Open the archived page and evaluate its relevance as:
       * Highly relevant: it is exactly the page I was searching for.
       * Relevant: it is a good alternative, but it is not the page I was searching for.
       * Not relevant: it is not the page I was searching for.
       * Don't know / Can not answer.
5. Justify your judgment. Your comments are valuable to us (optional):
```

Fig. 2. Form to assess navigational topics

$f1$, was linearly combined with the NutchWAX's term-weighting function, $f2$, using three different weights (0.1, 0.25, 0.5). That is, functions f_1 and f_2 were combined in three models: (1) $0.1*f_1 + 0.9*f_2$; (2) $0.25*f_1 + 0.75*f_2$; and (3) $0.5*f_1 + 0.5*f_2$. All functions were normalized to a value between 0 and 1. We generally denote these linearly combined models by TVersions and TSpan.

5.4   Relevance Assessment

Manual Assessment. Three judges, including the topics creator, assessed each of the 2 065 <URL, timestamp, topic> triplets aggregated in the version pool. They followed strict guidelines and document versions were presented in a random order, hiding from the judges the algorithm that retrieved the versions and their ranking order. Figure 2 shows the form used for collecting the relevance assessments for the navigational topics. We used a three-level scale of relevance.

The usefulness of the test collection depends heavily on the reliability of relevance judgments. Hence, we analyzed their level of consensus. The inter agreement between judges measured by Fleiss's kappa was 0.46 when considering a ternary relevance scale or 0.55 when considering a binary scale (the highly and partially relevant were considered relevant). This shows a moderate level of agreement, lending confidence to the judgment quality. These inter agreement values are inline with the ones of TREC judges [26].

Automatic Assessment. The relevance assessment is the most time-consuming part of creating a test collection. Hence, we took advantage of the characteristics of the collection to automatically assess 267 822 versions, such as described in Section 3.3. For each version manually assessed, we used the PWA IR system to find all redundant versions of the same document for each topic. Then, we propagated the same topical relevance degree to all these redundant versions.

Extrapolating from the time spent in manual assessments, the automatic assessments enabled us to save more than 4 000 hours per assessor.

6   Experiments and Results

Table 2 presents the results of the ranking models described above. The bold entries indicate the best result for each measure. We can see that BM25 and

Table 2. Results for the tested ranking models

Metric	time-unaware			time-aware	
	BM25	Lucene	NutchWAX	TVersions	TSpan
nDCG@1	0.250	0.220	0.250	0.430 †	**0.450** †
nDCG@5	0.145	0.157	0.215	**0.266** †	0.263 †
nDCG@10	0.119	0.133	0.174	**0.202** †	0.193
P@1	0.300	0.280	0.320	0.500 †	**0.520** †
P@5	0.140	0.164	0.236	**0.264**	0.256
P@10	0.108	0.132	0.168	**0.172**	0.158
S@1	0.300	0.280	0.320	0.500 †	**0.520** †
S@5	0.480	0.500	0.680	**0.780** †	0.760
S@10	0.620	0.600	0.780	**0.840**	0.760

† shows a statistical significance of $p<0.05$ against NutchWAX

Lucene present the worst results and their effectiveness is close. The NutchWAX model has a nDCG@1, nDCG@5 and nDCG@10 superior in 3%, 5.8% and 4.1%, respectively, when compared with the Lucene model. The other measures used, Precision at cut-off k (P@k) and Success at rank k (S@k), show similar results.

We can now determine, for the first time, how effective is the IR technology typically used in web archives. For instance, the Lucene and NutchWAX's results achieved an S@1 value of 0.28 and 0.32, respectively, which is less than half of the best results achieved in the 2004 Web Track, an S@1 of 0.65 [27]. Despite these values not being directly comparable due to the different test collections, there is a considerable gap to the S@1 value of 0.84 obtained by Google [28].

An interesting finding is that the time-aware models are significantly better than the time-unaware. The best configuration of the two models, TVersions and TSpan, presented better nDCG@1, nDCG@5 and nDCG@10 values than the BM25 and Lucene models, for a statistical significance level of 0.01 using a two-tailed paired Student's t-test. When compared with NutchWAX, the TVersions model achieved nDCG@1, nDCG@5 and nDCG@10 values of 18%, 5.1% and 2.8% higher, respectively. These increases have a statistical significance of $p<0.01$, which strongly indicates that the use of temporal information improves the effectiveness of web archives. Notice that, these models could only be evaluated with a multi-version corpus as the one we built.

6.1   Topic Difficulty

Figure 3 plots the nDCG@5 and nDCG@10 averages over the 9 tested ranking models for each of the 50 navigational topics. The topics are sorted by nDCG@5 and it is visible that the topic difficulty varies significantly, between 0 and 0.54. This variance is desirable for a test collection in order to provide topics with different levels of challenge. For instance, there are topics that present very poor results, because the query terms did not match the searched document. The query of topic 21 was *Dona Maria Segunda (second) Theatre*, but the text and link references only contained the terms *Dona Maria II Theatre*.

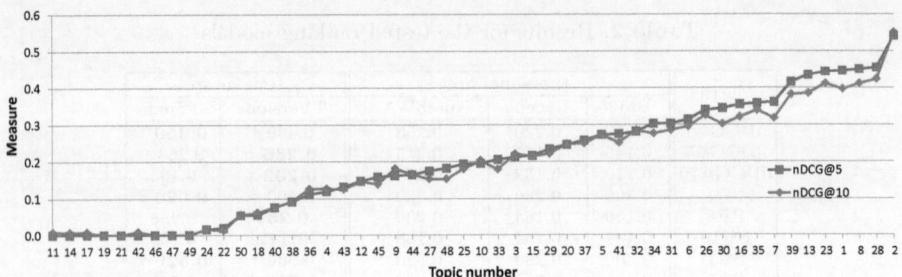

Fig. 3. Navigational topics sorted by the average of the 9 tested ranking models

6.2 Reusability

A test collection is reusable if it provides accurate measurements of the search effectiveness of systems that did not contribute with their results to the document pool. Otherwise, a new system returning relevant documents not previously identified would have its effectiveness underestimated. A test collection using only one IR system, such as this, is very likely to miss many relevant documents and is biased toward that system. Nevertheless, researchers can use this collection to accurately evaluate a new system after assessing their results and adding them to the version pool. The fact that the pool will have versions assessed by different judges over time is not a problem. The ranking between the judged systems will be the same as if judges would have assessed all documents in the same day [29].

Our test collection is available for research at `http://code.google.com/p/pwa-technologies/wiki/TestCollection`. Despite its specificities, such as the language, we believe that this collection could be used as a starting point to tune the WAIR technology handling other national webs.

7   Conclusions and Future Work

Billions of past web documents containing our history are currently archived. However, their retrieval is still in an early stage, preventing users from unfolding the full potential of web archives. Other IR domains have shown that the quality of search results depends greatly on the availability of suitable resources for evaluation. Such resources have been missing for WAIR systems, which could explain why no evaluation has ever been conducted on them. In this work we describe the methodology employed in a test collection based evaluation for WAIR systems. In the end, we were able to measure, for the first time, the effectiveness of state-of-the-art WAIR technology. As anticipated, the results were poor, which motivates the development of a common evaluation framework to foster research in WAIR. We expect that our research may lead to a novel IR task in a major evaluation campaign, such as TREC.

We also experimented two time-aware ranking models for navigational queries. They are based on the idea that the more versions a document has or the longer they existed, the more likely it is of being relevant. We achieved statistically significant improvements in both models over the state-of-the-art IR typically used in web archives. This is just the first step in leveraging temporal information to improve WAIR systems. In the future, we intend to create a dataset for *learning-to-rank* experiments from our test collection, to combine temporal evidences implicitly hidden in the corpus and query matches. The automatic assessment obtained with our methodology provides a fast mean of generating vast amounts of labeled data for machine learning optimization.

Acknowledgments. This work could not be done without the support of FCCN and its Portuguese Web Archive team. We thank FCT for its Multiannual Funding Programme.

References

1. Kitsuregawa, M., Tamura, T., Toyoda, M., Kaji, N.: Socio-Sense: A System for Analysing the Societal Behavior from Long Term Web Archive. In: Zhang, Y., Yu, G., Bertino, E., Xu, G. (eds.) APWeb 2008. LNCS, vol. 4976, pp. 1–8. Springer, Heidelberg (2008)
2. Yamamoto, Y., Tezuka, T., Jatowt, A., Tanaka, K.: Honto? Search: Estimating Trustworthiness of Web Information by Search Results Aggregation and Temporal Analysis. In: Dong, G., Lin, X., Wang, W., Yang, Y., Yu, J.X. (eds.) APWeb/WAIM 2007. LNCS, vol. 4505, pp. 253–264. Springer, Heidelberg (2007)
3. Chung, Y., Toyoda, M., Kitsuregawa, M.: A study of link farm distribution and evolution using a time series of web snapshots. In: Proc. of the 5th International Workshop on Adversarial Information Retrieval on the Web, pp. 9–16 (2009)
4. Elsas, J., Dumais, S.: Leveraging temporal dynamics of document content in relevance ranking. In: Proc. of the 3rd ACM Inter. Conference on Web Search and Data Mining, pp. 1–10 (2010)
5. Gomes, D., Miranda, J., Costa, M.: A Survey on Web Archiving Initiatives. In: Gradmann, S., Borri, F., Meghini, C., Schuldt, H. (eds.) TPDL 2011. LNCS, vol. 6966, pp. 408–420. Springer, Heidelberg (2011)
6. Voorhees, E., Harman, D.: TREC: Experiment and evaluation in information retrieval. MIT Press (2005)
7. Masanès, J.: Web Archiving. Springer-Verlag New York Inc. (2006)
8. Foundation, I.M.: Web archiving in Europe. Technical report, CommerceNet Labs (2010)
9. Ras, M., van Bussel, S.: Web archiving user survey. Technical report, National Library of the Netherlands (Koninklijke Bibliotheek) (2007)
10. Costa, M., Silva, M.J.: Characterizing search behavior in web archives. In: Proc. of the 1st International Temporal Web Analytics Workshop (2011)
11. Cohen, D., Amitay, E., Carmel, D.: Lucene and Juru at Trec 2007: 1-million queries track. In: Proc. of the 16th Text REtrieval Conference (2007)
12. Kelly, D.: Methods for evaluating interactive information retrieval systems with users. Foundations and Trends in Information Retrieval, vol. 3. Now Publishers Inc. (2009)
13. Aula, A., Khan, R.M., Guan, Z.: How does search behavior change as search becomes more difficult? In: Proc. of the 28th International Conference on Human Factors in Computing Systems, pp. 35–44 (2010)

14. Kellar, M., Watters, C., Shepherd, M.: A field study characterizing Web-based information-seeking tasks. American Society for Information Science and Technology 58(7), 999–1018 (2007)

15. Baeza-Yates, R., Castillo, C., Efthimiadis, E.: Characterization of national web domains. ACM Transactions on Internet Technology 7(2) (2007)

16. Costa, M., Silva, M.J.: Understanding the information needs of web archive users. In: Proc. of the 10th International Web Archiving Workshop, pp. 9–16 (2010)

17. Costa, M., Silva, M.J.: A search log analysis of a Portuguese web search engine. In: Proc. of the 2nd INForum - Simpósio de Informática, pp. 525–536 (2010)

18. Jansen, B., Spink, A.: How are we searching the World Wide Web? A comparison of nine search engine transaction logs. Information Processing and Management 42(1), 248–263 (2006)

19. Dong, A., Chang, Y., Zheng, Z., Mishne, G., Bai, J., Zhang, R., Buchner, K., Liao, C., Diaz, F.: Towards recency ranking in web search. In: Proc. of the 3rd ACM International Conference on Web Search and Data Mining, pp. 11–20 (2010)

20. Jones, R., Diaz, F.: Temporal profiles of queries. ACM Transactions on Information Systems (TOIS) 25(3) (2007)

21. Clarke, C., Kolla, M., Cormack, G., Vechtomova, O., Ashkan, A., Büttcher, S., MacKinnon, I.: Novelty and diversity in information retrieval evaluation. In: Proc. of the 31st International ACM SIGIR Conference on Research and Development in Information Retrieval, pp. 659–666 (2008)

22. Agrawal, R., Gollapudi, S., Halverson, A., Ieong, S.: Diversifying search results. In: Proc. of the 2nd ACM International Conference on Web Search and Data Mining, pp. 5–14 (2009)

23. Burner, M., Kahle, B.: The Archive File Form (September 1996), http://www.archive.org/web/researcher/ArcFileFormat.php

24. Voorhees, E.: Topic set size redux. In: Proc. of the 32nd International ACM SIGIR Conference on Research and Development in Information Retrieval, pp. 806–807 (2009)

25. Robertson, S., Zaragoza, H.: The Probabilistic Relevance Framework. Foundations and Trends in Information Retrieval, vol. 3. Now Publishers Inc. (2009)

26. Al-Maskari, A., Sanderson, M., Clough, P.: Relevance judgments between TREC and Non-TREC assessors. In: Proc. of the 31st Annual International ACM SIGIR Conference on Research and Development in Information Retrieval, pp. 683–684 (2008)

27. Craswell, N., Hawking, D.: Overview of the TREC-2004 Web Track. NIST Special Publication, 500–261 (2005)

28. Lewandowski, D.: The retrieval effectiveness of search engines on navigational queries. Aslib Proceedings 63, 354–363 (2011)

29. Blanco, R., Halpin, H., Herzig, D., Mika, P., Pound, J., Thompson, H., Tran Duc, T.: Repeatable and reliable search system evaluation using crowdsourcing. In: Proc. of the 34th International ACM SIGIR Conference on Research and Development in Information, pp. 923–932 (2011)

Improving Recall of Regular Expressions for Information Extraction

Karin Murthy, Deepak P., and Prasad M. Deshpande

IBM Research - India, Bangalore, India
{karin.murthy,deepak.s.p,prasdesh}@in.ibm.com

Abstract. Learning or writing regular expressions to identify instances of a specific concept within text documents with a high precision and recall is challenging. It is relatively easy to improve the precision of an initial regular expression by identifying false positives covered and tweaking the expression to avoid the false positives. However, modifying the expression to improve recall is difficult since false negatives can only be identified by manually analyzing all documents, in the absence of any tools to identify the missing instances. We focus on partially automating the discovery of missing instances by soliciting minimal user feedback. We present a technique to identify good generalizations of a regular expression that have improved recall while retaining high precision. We empirically demonstrate the effectiveness of the proposed technique as compared to existing methods and show results for a variety of tasks such as identification of dates, phone numbers, product names, and course numbers on real world datasets.

1 Introduction

Unstructured data such as emails, blogs, web pages, and chat conversations increasingly form an integral source of information for data analytics. An important step to make unstructured data available for analytics is Information Extraction (IE), which extracts structured information (that is, concepts such as phone numbers, email addresses, person names, and SSNs) from text.

Learning-based IE (for example, [15] and [13]) relies on humans to mark enough examples of the concept that has to be extracted; the labelings are then used to build a mathematical model for IE. Rule-base IE (for example, CPSL [1] and SystemT [4]) in contrast relies on humans to construct appropriate machine-understandable rules to extract specific concepts from a text. Regular expressions (regexes) are the dominant building block for a large class of extraction tasks in rule-based IE. To build effective information extraction engines, regexes that cover almost all instances of a specific concept (high recall) and have few false positives (high precision) are required.

Data-driven fine-tuning of regexes to render them more accurate for information extraction tasks is a highly challenging task. An intuitive method of modifying a regex is to specialize or generalize it. Specializing a regex involves restricting its scope so that the modified regex matches only a subset of earlier matches; an example is to modify $\backslash w\{3\}$, the regex that matches any sequence of three alpha-numeric characters to $\backslash d\{3\}$, a regex that matches any sequence of three numeric characters. An example of generalization is the reverse modification.

X.S. Wang et al. (Eds.): WISE 2012, LNCS 7651, pp. 455–467, 2012.

If all the matches of a regex are labeled as true matches (those that indeed denote the concept of interest) or false matches, the regex itself can be specialized [11] to cover as many of the true matches as possible, while excluding as many of the false matches as possible. However, the analogous task of generalizing a regex to render it more accurate for IE, has not been well addressed in literature. Generalizing a regex is much harder since the scope of labeling is not limited to instances covered by the existing regex. Instead, to identify false negatives additional labeling of the text corpus is necessary.

Consider the regex, 1-$\d\{3\}$-$\d\{3\}$-$\d\{4\}$, intended to extract phone numbers. It is possible that the intent was to also extract phone numbers of the form 1-800-$COMPANY$ (sometimes used for advertisements). However, no labeled instances of this kind were given to the regex learner or the regex author was unaware of such phone numbers. In such a case, it would be beneficial to learn that a slightly generalized expression such as 1-$\d\{3\}$-$\w\{3\}$-$\{0,1\}\w\{4\}$ covers such phone numbers. In the new regex the last two groups of digits ($\d$) were generalized to also allow letters ($\w$) and the second '-' was made optional.

We focus on regex generalization as a way to improve the recall of a regex and exploit the fact that in general humans can easily decide whether newly covered instances are true or false. In particular, we present a technique to incrementally navigate the exponential space of possible generalizations of an initial regex and identify good quality generalizations while soliciting minimal user feedback on correctness of newly discovered matches.

2   Related Work

Learning a regular expression from scratch using a list of true matches has been explored well [5–7]. It has also been found to have applications in learning other patterns like DTDs from XML data [3, 8].

Data-driven modification of a seed pattern to arrive at more accurate patterns has also been subject of some research. Outside the realm of regexes, an early work [12] proposes techniques to generalize a pattern in a data-driven manner; this, however, assumes the availability of a comprehensive set of true and false matches which is harnessed by the search procedure to fine tune the generalization. In particular, the information whether any matching instance is a true match or not, is deemed to be available beforehand. [2] proposes a clustering-based method for a similar generalization task for regexes under such an all-labels-available-beforehand assumption. Our problem addresses the more realistic scenario where only a few labels, those of instances that match the seed regex, are available. Adapting the techniques in [2, 12] to solicit labels on the fly would demand significant manual labeling effort, whereas the manual labeling effort is typically the point of optimization. We show in Section 5.5 that the technique we propose incurs up to orders of magnitude less labeling effort than methods that rely on all-labels-available-beforehand assumptions

If the seed regex is already generic enough, the technique proposed in [11] may be employed to find more accurate specializations of the regex. It works by using a labeling of all instances covered by the seed regex, to arrive at specializations that cover as many of the true matches and as few of the false matches as possible. Our work is complementary in that our goal is to discover generalizations.

[16] presents an active-learning approach to identify instances of concepts that span multiple tokens in the text. This approach makes use of lexical features, and produces a sequence of part-of-speech (POS) tags that are likely to *contain* the concept of interest. Such sequences of POS tags, unlike regexes, cannot be readily used in rule-based IE systems. But the enhanced generalized regex that we identify can be used as input to the work in [16] to discover such POS sequences that could contain the concept.

3   Generalizing a Regex

3.1   The Space of Possible Generalizations

Regexes have two main building blocks: characters (or classes of characters) and quantifiers. For example, the character class [a-zA-Z] matches any alphabet character. Quantifiers specify the number of matches expected for a set of characters. For example, $\backslash d\{5,10\}$ denotes that between 5 to 10 contiguous digits are expected.

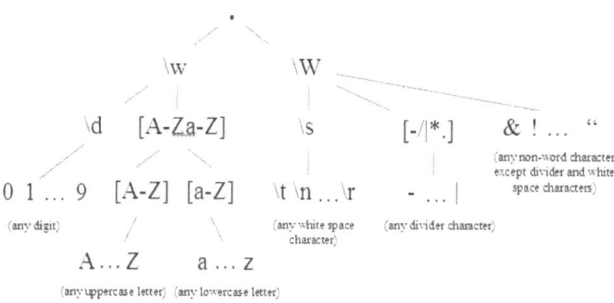

Fig. 1. Character Class Hierarchy

To generalize a regex, we generalize each component of the regex separately and incrementally. Each character class is generalized using the character class hierarchy shown in Figure 1 by incrementally going up the hierarchy until the *root* node (that is, the *dot* that matches any character) is reached. We generalize quantifiers by separately decreasing (and increasing, respectively) the lower and the upper bound of each quantifier. Capturing groups that denote a set of strings (e.g., (India|United States)) are generalized only if an appropriate dictionary or concept hierarchy covering the strings is available.

The example regex $\backslash d\{1,3\}$ has three positions: the character class digit, the lower, and the upper bound of the quantifier. $\backslash w\{0,3\}$ is derived by generalizing the first as well as the second position once each (and thus, is intuitively represented by the vector $[1,1,0]$). Figure 2 shows the space of all possible generalizations for the regex $\backslash d\{1,3\}$, up to a quantifier upper bound of 4.

Definition 1. *Let r_1, r_2, and r_3 be three regexes (with the same number of positions to generalize). r_2 is a <u>direct generalization of r_1</u> if r_1 can be transformed into r_2 by generalizing a single position in r_1 exactly once. We call r_3 a <u>generalization of r_1</u> if there exists a sequence of direct generalizations $[r_1, g_1, g_2, \ldots, g_n, r_3]$.*

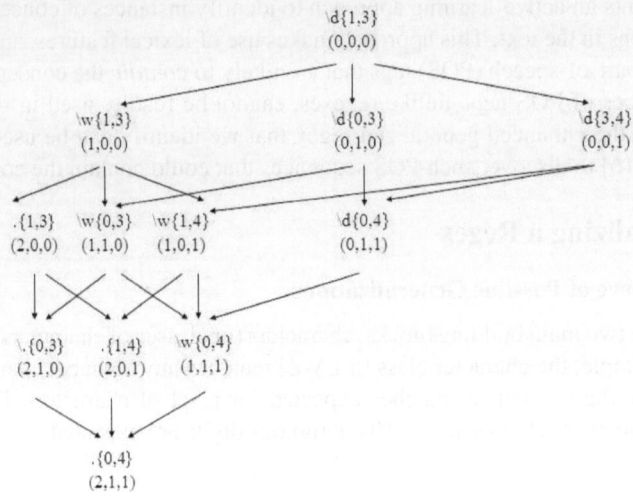

Fig. 2. Generalization Space for \d{1,3}

\w{1,3}, \d{0,3}, and \d{1,4} are direct generalizations of \d{1,3} produced by a single generalization each. \w{0,6} is a generalization of \w{1,4} with the intervening sequence \w{0, 4}, \w{0, 5} of direct generalizations.

Definition 2. *Let r be an initial regex. The generalization space contains all regexes r_i where r_i is a generalization of r. The is-a-direct-generalization-of relation defines a partial order over all the regexes in the generalization space.*

For a regex with k positions, each node in the generalization space can have at most k direct generalizations. When considered up to a depth of n, the number of possible generalizations is up to $\mathcal{O}(k^n)$. It is infeasible to exhaustively search this exponential space. However, we show how to efficiently explore promising regions of the generalization space by employing periodic user feedback.

3.2   Quality of a Generalization

We would like to generalize an initial regex r such that the generalized regex r' has higher recall (that is, it covers more possible variations of the given concept) while still ensuring a user-specified precision for the generalized regex (that is, covering more variations does not significantly increase the amount of false matches). Coverage denotes the number of true matches covered by a regex.

We assess the quality of the new matches (that is, *Matches(r')−Matches(r)*) by soliciting user feedback on whether they are true or false matches. Let the precision (fraction of true matches) for the new matches, M_{new}, be p'; then the coverage and precision of r' is:

$$cov(r') = p * |M| + p' * |M_{new}| \qquad (1)$$

$$prec(r') = cov(r')/(|M| + |M_{new}|) \qquad (2)$$

where p and M denote the precision and the number of matches for the initial regex r. However, as the goal is to minimize manual labeling effort, we do not want to label all matches in M_{new}; instead we would like to *estimate* the quality of r' even before any instances in M_{new} are labeled.

Often, we may be able to roughly estimate an upper bound on the number of possible correct matches of the concept, in a document collection. For example, a corporate email is likely to on average contain two telephone numbers, as a sender may include his/her office phone number(s), and periodically the text may contain information on how to reach a third person. Thus, we may estimate that a collection of n corporate emails contains approximately $2n$ correct matches for the phone number concept. We use such an estimate (called tcm, for total correct matches) to estimate the precision of r'.

The intuition is that even if M_{new} is much larger than tcm, the coverage will not exceed tcm. Based on this we define the upper bound on the precision of r' as:

$$ub\_prec(r') = \frac{min\{p * |M| + |M_{new}|, tcm\}}{|M| + |M_{new}|}$$

We further refine the estimate using the intuition that the generalization of the ith position for the jth time is likely to introduce new matches of similar quality regardless of the number of generalizations performed on other positions. For a given regex, we average all historical precisions of the generalization of the ith position for the jth time (which we denote by $hp(i, j)$) and use it to estimate the precision of a direct generalization as follows:

$$est\_prec(r') = \frac{min\{p * |M| + hp(.,.) * |M_{new}|, tcm\}}{|M| + |M_{new}|}$$

$hp(.,.)$ is the appropriate historical precision of the direct generalization. We now outline the problem of finding good quality generalizations.

Definition 3. *Given an initial regex r and a precision threshold τ, find a set of generalizations $\mathcal{R}$ such that $\forall r' \in \mathcal{R}$, $cov(r') > cov(r)$ and $prec(r') > \tau$.*

Since a generalization r' of r subsumes the matches of r, we retain only r' in $\mathcal{R}$. We now discuss an approach to find such generalizations based on the introduced concepts and definitions.

4   Interactive Regex Generalization Algorithm

Overview: Algorithm 1 describes our best-first-style search [14] that starts by initializing the candidate set with a user-specified regex. At any point, the algorithm picks the best candidate regex according to the estimated precision and calculates the candidate's actual precision by soliciting user feedback on previously unseen matches. If the calculated precision exceeds the user-specified precision threshold τ, it moves the candidate to the result set and adds its direct generalizations whose estimated precision exceeds τ to the candidate set. Upon addition of a regex to the result set, its ancestors in the

Alg. 1. *Interactive Regex Generalization Overview*

Input. Initial regex r, Document set S, Minimal precision τ
Output. Set of generalized regexes $\mathcal{R}$

$Solutions\ \mathcal{R} = \{r\};\ Candidates\ \mathcal{C} = \{c | c \in children(r) \wedge\ est\_prec(c) \geq \tau \}$
$while(\mathcal{C} \neq \phi)$
    $pick\ most\ promising\ (highest\ estimated\ precision)\ candidate\ c\ from\ \mathcal{C}$
    $aks\ user\ to\ label\ new\ matches\ of\ c\ and\ compute\ precision$
    $if(prec(c) > \tau \wedge cov(c) > cov(parent(c)))$
        $\mathcal{R} = \mathcal{R} \cup \{c\}$
        $remove\ obsolete\ solutions\ from\ \mathcal{R}$
        $\mathcal{C} = \mathcal{C} \cup \{c' | c' \in children(c) \wedge est\_prec(c') \geq \tau \wedge no\ other\ condition\ violated \}$

generalization space are purged from the result set. The search progresses as long as the candidate set is non-empty or until termination by the user, upon which the result set is output.

Input Parameters: The user is expected to supply a regex that satisfies her precision requirements. If a precise regex is unavailable, the user may first specialize it (see Section 2) to achieve the desired precision. Apart from the precision threshold τ, the user is also expected to provide a corpus of documents (that is, *training data*) to evaluate regexes during the search.

Navigating the Search Space: Since our generalization space forms a lattice, we are likely to reach the same generalized regex through different paths. To avoid repetetive work in search, we ensure that no regex is examined twice, by efficiently keeping track of which regexes should no longer be explored. The intricate details are omitted due to space constraints.

Picking the Most Promising Candidate: If there are multiple candidates with the same estimated precision, we break ties based on the following criteria:

1. Highest number of new matches (encourages more user feedback early on)
2. Highest number of matches of its parent (through which we reached it)
3. Smallest number of changes (prioritizes breadth first search)

Soliciting User Feedback: To compute the precision of a chosen regex, we solicit user feedback for newly covered matches. We present those instances including a snippet of surrounding text to the user and ask her to label them true or false. In case of many extra matches, we solicit user feedback only on a sample. We empirically show in Section 5 that a sample size of 10 is sufficient to achieve good results.

Maintaining the Candidate List: Once we explore a candidate, we add each direct generalization c' to the candidate set if all of the following holds:

1. c' has an estimated precision of at least τ
2. c' has never been added to the candidate set
3. no parent of c' has been evaluated with a precision below τ
4. no expression in $\mathcal{C}$ is a generalization of c'

Maintaining the Result List: We add each generalized regex whose precision is at least τ and which covers at least one new true instance with respect to its parent, to the result set. While adding a new regex to the result, we purge all regexes in any path from the root to the newly inserted regex and all other regexes that cover only a subset of matches of the new regex.

5 Experiments

5.1 Experimental Setup, Datasets and Evaluation Criteria

All experiments were run on a computer with a 1.07 GHz dual core CPU and 3 GB RAM. If not otherwise mentioned, we ran all experiments with $\tau = 95\%$ and assuming *tcm* to be two times the coverage of the input regex. If the extra matches for a regex exceeded 10, we choose a random sample of 10 instances for user feedback. We stopped each experiment once the user had labeled 100 examples or the run-time (excluding user interaction) exceeded five minutes.

Our datasets comprise random samples from the Enron Email Corpus [9] and data under categories comp.windows.x (Windows) and misc.forsale (ForSale) in the 20 Newsgroups data set [10]. To test our approach for tasks on web pages, we use a dataset of product webpages retrieved from the IBM Intranet and course webpages from the WebKB dataset[1]. We generated separate training and test data sets.

In addition to quantifying the labeling effort incurred during the search, we also evaluated the quality of the generalized regexes. Labeling the test data to include all variations of the concept is very labor-intensive (see Section 5.5). Thus, we use the fractional increase in recall to evaluate the generalized regexes. Let M_t and M_t' be the true matches of r and r' ($|M_t'| \geq |M_t|$ where r' is a generalization of r) and M be the actual true matches (which is unknown), then the fractional increase in recall $\mathcal{R}$ is:

$$\Delta \mathcal{R}(r \to r') = \frac{(M_t'/M) - (M_t/M)}{(M_t/M)} = \frac{M_t' - M_t}{M_t}$$

5.2 Regex Generalization Evaluation

We evaluate our approach on various datasets for four tasks, pertaining to extraction of *dates, phone numbers, product names,* and *course numbers,* respectively.

Evaluation Outline: Each generalization task is first evaluated for the *training* phase where an input regex is generalized by periodically soliciting user feedback on regex matches on the training dataset. In the *testing* phase the quality of the regexes is evaluated on a different *test* dataset that comes from the same distribution as the training dataset. This two-phase approach is used in evaluating most supervised-learning approaches. *The quality of our approach for a particular task is inversely related to the manual labeling effort in the training phase and directly related to the fractional increase in recall measured on the test dataset.*

---

[1] http://www.cs.cmu.edu/~webkb/

Table 1. Results for Initial Regex r_1: $\backslash b \backslash d\{2\}/\backslash d\{2\}/\backslash d\{4\}\backslash b$

Training Data (#docs)		Test Data (#docs)			
	Labeled	r_1	New	Precision of	Fractional Increase
Generalized Expressions	Instances	Matches	Matches	New Matches	in Recall (%)
Enron1 (10k)		EnronTest1 (0.1k)			
$R_1 = \backslash b \backslash d\{1,2\}/\backslash d\{2,2\}/\backslash d\{4,4\}\backslash b$	10	16	21	1.0	131
$\mathbf{R_2 = \backslash b \backslash d\{1,2\}/\backslash d\{1,2\}/\backslash d\{4,4\}\backslash b}$	20		96	1.0	600
Enron2 (1k)		EnronTest1 (0.1k)			
$R_1 = \backslash b \backslash d\{2,2\}/\backslash d\{1,2\}/\backslash d\{4,4\}\backslash b$	10	16	21	1.0	131
$R_2 = \backslash b \backslash d\{1,2\}/\backslash d\{1,2\}/\backslash d\{4,4\}\backslash b$	20		96	1.0	600
$\mathbf{R_3 = \backslash b \backslash d\{1,2\}[-/|.]\backslash d\{1,2\}[-/|.]\backslash d\{4,4\}\backslash b}$	38		96	1.0	600
Enron3 (1k)		EnronTest1 (0.1k)			
$R_1 = \backslash b \backslash d\{1,2\}/\backslash d\{2,2\}/\backslash d\{4,4\}\backslash b$	10	26	21	1.0	81
$R_2 = \backslash b \backslash d\{1,2\}/\backslash d\{1,2\}/\backslash d\{4,4\}\backslash b$	20		21	1.0	81
$R_3 = \backslash b \backslash d\{1,2\}[-/|.]\backslash d\{1,2\}[-/|.]\backslash d\{4,4\}\backslash b$	50		96	1.0	369
$\mathbf{R_4 = \backslash b \backslash d\{1,2\}/\backslash d\{1,2\}/\backslash d\{2,4\}\backslash b}$	60		120	1.0	462

For each task starting with an input regex r_i, we prepare tables summarizing results for generalizing the regex. We report the successively found solutions R_i and the number of examples labeled by the user until a solution was found.

Illustrative Example: We illustrate how to interpret the results taking the first result rows in Table 1 as an example. Each training run starts with a regex r_i. We first focus on the multi-column *Training Data*; the regex generalization approach started with r_1 on dataset *Enron1*, and the first row shows that the approach solicited user feedback on *10* instances before it added the generalization R_1 to the result set. According to the second row, R_2 was added to the result set after feedback on *10* more instances (that is, a total of *20* instances). The fact that R_1 is not shown in bold, denotes that a subsequent more general solution (in this case, R_2) purged R_1 from the result set.

The multi-column *Test Data* evaluates the performance of the three regexes, r_1, R_1 and R_2, on the test set *EnronTest1* containing 100 documents. The number of instances in the test set matching r_1 is 16, as shown in the first column in the *Test Data* multi-column. The second column and all further columns needs to be read in conjunction with the corresponding R_is from the *Training Data* multi-column. R_1 resulted in 21 extra matches on *EnronTest1* beyond the 16 matches of r_1. The set of 21 extra matches has a precision of 1.0 resulting in a 131% increase in recall, over r_1. Similarly, R_2 resulted in 96 additional matches with a precision of 1.0 resulting in a 600% increase in recall.

Date Task: We performed two runs with two separate initial regexes:

$r_1 = \backslash b \backslash d\{2\}/\backslash d\{2\}/\backslash d\{4\}\backslash b$

$r_2 = \backslash b \backslash d\{2,2\}\backslash W\backslash(\text{January}|...|\text{December})\backslash W\backslash d\{4\}\backslash b$

r_1 finds dates of the form $dd/dd/dddd$ where d is any digit. It does not cover single digit days and months, two-digit years, or divider characters other then /. Table 1 shows that r_1 is easily generalized to cover single-digit days in most datasets with a rather small labeling effort of 10 instances, whereas single-digit months are covered with another 10 labelings, on each dataset. The recorded fractional increase in recall across various datasets for the final (bold) regexes averages 500%. r_2 expects alphabetical representations of months and allows for more generic delimiters. Similar to r_1, r_2 was generalized to include single digit days after just 10 labeled instances, with the final regexes showing a similar fractional increase in recall. Details excluded due to space limitations.

Table 2. Results for Initial Regex r_3: $\backslash b\, 1 - \backslash d\{3\} - \backslash d\{3\} - \backslash d\{4\}\backslash b$

Generalized Expressions	Training Data		Test Data			
	Labeled Instances	r_3 Matches	New Matches	Precision of New Matches	Fractional Increase in Recall (%)	
Enron1 (10k)			EnronTest2 (32k)			
$R_1 = \backslash b\,1 - \backslash d\{3,3\} - \backslash d\{3,3\} - \backslash w\{4,4\}\backslash b$	4	374	3	1.0	1	
$R_2 = \backslash b\,1 - \backslash d\{3,3\} - \backslash w\{3,3\} - \backslash w\{4,4\}\backslash b$	6		12	1.0	3	
$R_3 = \backslash b\,1 - \backslash d\{3,3\} - \backslash w\{3,3\} - \{0,1\}\backslash w\{4,4\}\backslash b$	16		57	1.0	15	
$R_4 = \backslash b\,1 - \backslash d\{3,3\} - \backslash w\{3,3\} - \{0,1\}\backslash w\{4,5\}\backslash b$	18		60	1.0	16	
$R_5 = \backslash b\,1 - \backslash d\{3,3\} - .\{3,3\} - \{0,1\}\backslash w\{4,5\}\backslash b$	25		75	1.0	20	
$R_6 = \backslash b\,1 - \backslash d\{3,3\} - \backslash w\{2,3\} - \{0,1\}\backslash w\{4,5\}\backslash b$	25		74	0.99	20	
$R_7 = \backslash b\,1 - \backslash d\{3,3\} - \backslash w\{3,4\} - \{0,1\}\backslash w\{4,4\}\backslash b$	31		59	0.98	15	
$\mathbf{R_8 = \backslash b\,1 - \backslash d\{3,3\} - .\{3,4\} - \{0,1\}\backslash w\{4,4\}\backslash b}$	31		75	0.99	20	
$R_9 = \backslash b\,\backslash d - \backslash d\{3,3\} - .\{3,3\} - \{0,1\}\backslash w\{4,5\}\backslash b$	34		76	0.98	20	
$\mathbf{R_{10} = \backslash b\,\backslash d - \backslash d\{3,3\} - \backslash w\{2,3\} - \{0,1\}\backslash w\{4,5\}\backslash b}$	50		76	0.86	17	
Enron3 (1k)			EnronTest2 (32k)			
$R_1 = \backslash b\,1 - \backslash d\{3\} - \backslash w\{3\} - \backslash d\{4\}\backslash b$	2	374	3	1.0	1	
$R_2 = \backslash b\,\backslash d\{1,2\} - \backslash d\{3\} - \backslash w\{3\} - \backslash d\{4\}\backslash b$	3		22	0.64	5	
$\mathbf{R_3 = \backslash b\,\backslash d\{1,2\}\backslash W\backslash d\{3\} - \backslash w\{3\} - \backslash d\{4\}\backslash b}$	4		25	0.68	5	
ForSale (1k)			ForSaleTest (1k)			
$R_1 = \backslash b\,1\backslash W\backslash d\{3\} - \backslash d\{3\} - \backslash d\{4\}\backslash b$	3	17	3	1.0	18	
$R_2 = \backslash b\,1\backslash W\backslash d\{3\}\backslash W\backslash d\{3\} - \backslash d\{4\}\backslash b$	6		5	1.0	29	
$R_3 = \backslash b\,1\backslash W\backslash d\{3\}\backslash W\backslash d\{3\}[-/|.]\backslash d\{4\}\backslash b$	8		7	1.0	41	
$R_4 = \backslash b\,1\backslash W\backslash d\{3\}\backslash W\backslash d\{3\}\backslash W\backslash d\{4\}\backslash b$	11		10	1.0	59	
$R_5 = \backslash b\,1\backslash W\backslash d\{3\}\backslash W\backslash d\{3\}\backslash W\backslash w\{4\}\backslash b$	12		11	1.0	65	
$\mathbf{R_6 = \backslash b\,\backslash W\{1,2\}\backslash d\{3\}\backslash W\{1,2\}\backslash d\{3\}\backslash W\backslash w\{4\}\backslash b}$	24		13	1.0	76	
Windows (1k)			WindowsTest (1k)			
$R_1 = \backslash b\,1 - \backslash d\{3\} - \backslash w\{3\} - \backslash w\{4\}\backslash b$	1	5	1	1.0	20	
$\mathbf{R_2 = \backslash b\,1 - \backslash d\{3\} - \backslash w\{3\} - \{0,1\}\backslash w\{4\}\backslash b}$	2		1	1.0	20	

Phone Number Task: We used the following initial regex:

$$r_3 = \backslash b\, 1 - \backslash d\{3\} - \backslash d\{3\} - \backslash d\{4\}\backslash b$$

r_3 covers most US phone numbers, while missing out on international phone numbers and descriptive phone numbers such as 1-800-COMPANY. Despite the fact that descriptive phone numbers are rarely used in Enron mails (there are only 55 instances in the 32,000 mails) as well as the 20 Newsgroup data, our approach is able to generalize r_3 to cover such phone numbers. Labeling less than 50 instances (not full documents) is sufficient to discover a generalized regex that covers most descriptive phone numbers while barely compromising the precision of the original expression.

Table 2 shows how the initial regex is incrementally generalized to cover additional phone numbers. For example, for Enron1 the following phone numbers are included step-by-step: 1-800-982-BEVO (R_1), 1-800-The-Card (R_2), 1-800-FLOWERS (R_3), 1-800-NEWPOWER (R_4), and 1-800-97-ENRON (R_6). For Enron3, we additionally generate a regex that covers international phone numbers (e.g. 44-207-783-4040). Note that this requires three changes to the initial expression ($1 \rightarrow \backslash d \rightarrow \backslash d\{1,2\}$) where none of the individual changes produces any new matches. For the ForSale dataset, we generate a new regex that covers phone numbers with a different structural form (e.g. 1-(419)-756-2950). The variety of variants that are covered by our generalizations illustrate the effectiveness of our approach.

Product Name Task: We describe the extraction of IBM product names from a collection of IBM intranet web pages. Our initial regex r_4 is best explained by describing its three constituent parts:

(IBM): This first part enforces that the product name starts with *IBM*.

Table 3. Results on Product Name and Course Number Tasks

Task	Regex		Labeled Instances	Precision of New Matches	Fractional Recall Increase (%)
Product Names	$r_4 = $ \b(IBM)([ ][A-Z][a-z]+)?[ ][0-9]{1,2}(\.[0-9])?\b		0	-	-
	$R_1 = $ \b(IBM)([ ][A-Z][a-z]+){0,2}[ ][0-9]{1,2}(\.[0-9])?\b		10	1.0	297
	$\mathbf{R_2} = $ \b(IBM)([ ][A-Z][a-z]+){0,2}[ ]{1,2}[0-9]{1,2}(\.[0-9])?\b		11	1.0	300
Course Number	$r_5 = $ \bCS\d{3,3}	(e.g., CS215)	0	-	-
	$R_1 = $ \bCS?\d{3,3}	(e.g., C123)	10	1.0	23
	$R_2 = $ \bCS?\d{2,3}	(e.g., CS51)	20	1.0	31
	$\mathbf{R_3} = $ \bCS?\d{2,4}	(e.g., CS1270)	45	1.0	34

([ ][A-Z][a-z]+)?: This second part denotes a capitalized word (e.g., *Java, Lotus*) that is allowed to occur at most once (? in regexes denote $\{0, 1\}$).

[0-9]{1,2}(\.[0-9])?: The last part captures version numbers that can contain an optional sub-version, and can capture strings such as 10 and 1.7.

r_4 is formed by putting these three expressions together with intervening whitespace. It matches product names such as *IBM Java 1.7*. Table 3 shows that the fractional increase in recall improves by 300% by querying the user for as little as 11 samples. In the process, the regex is generalized to include two optional capitalized words instead of at most one (as in r_4), thus enabling matching of product names such as *IBM Directory Server 5.1*.

Course Number Task: We provide an initial regex r_5 for the course number task that matches course numbers of the form *CS*215, accounting for a large fraction of the course numbers in the *WebKB* dataset. Our approach generalized r_5 to include course numbers of the form *CS1270* and *C512*, as shown in Table 3. The fractional increase in recall is low since most course numbers were already covered by the initial regex r_5.

5.3 Scalability

We now analyze the time taken and the memory used when generalizing r_1 and r_3 (see Section 5.2) to find the first through fifth solution. We observed similar trends for other tasks. The time is reported in seconds and includes only the runtime of our algorithm and not the time spent on user interaction (see Figures 3(a) and 3(c)). The space requirement is dominated by the size of the candidate list as well as the number of future candidates kept to ensure that no candidate is explored twice. Thus, the memory usage is reported as the maximal number of candidates and future candidates kept at any point in time to find the respective solution. Figures 3(b) and 3(d) show that the maximum number of items kept in memory is only of the order of thousands.

5.4 Effect of Changing Input Parameters

The precision threshold (τ) simply excludes regexes below a certain precision from the result. In the following we assume $\tau = 0.95$ and study the effect of varying the estimated number of correct matches. We describe our analysis for r_1 on Enron2; similar results were observed for other regexes and data sets.

The parameter tcm influences the amount of exploration of the search space: larger values of tcm induce a larger search. For example, for tcm at 110% of the coverage of the initial regex, only R_1 is found by the search. When increased to 125%, R_2 is still not found but the search discovers $\backslash b\backslash d\{1,2\}/\backslash d\{2,2\}/\backslash d\{4\}\backslash b$; this regex is later subsumed by R_2 for larger tcm values. Any tcm value over 140% produces R_1, R_2, and R_3.

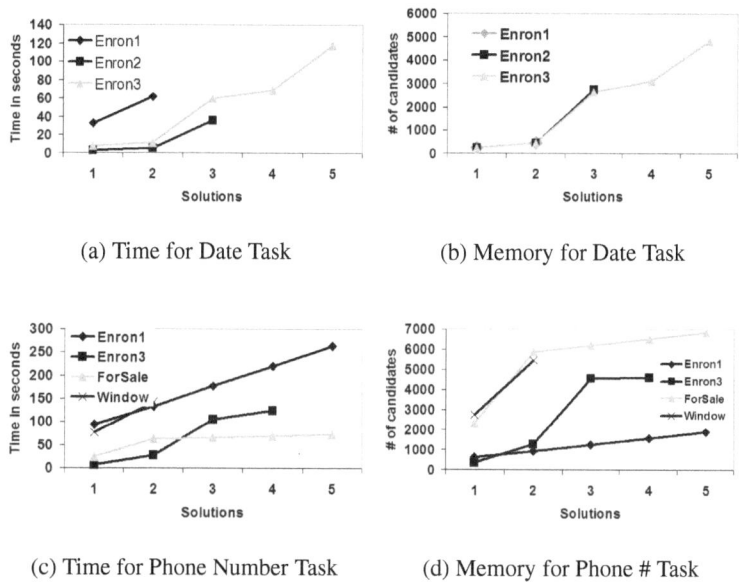

(a) Time for Date Task

(b) Memory for Date Task

(c) Time for Phone Number Task

(d) Memory for Phone # Task

Fig. 3. Time and Memory Usage

It is generally safe to overestimate tcm but a very high tcm may lead to an increase in runtime. Though we found 200% to be empirically effective across many settings, a user may want to first run experiments with lower values of tcm and increase it incrementally if no good results are found within the first few minutes.

Table 4. Comparison of Labeling Efforts

Data Set (#docs)	Regex Generalizations	True instances of new kind	Instances labeled in our approach	Estimated documents to label for learning methods
Enron1 (10k)	$r_1 \rightarrow R_2$	1711	20	6
	$r_3 \rightarrow R_8$	26	31	385
	$r_3 \rightarrow R_{10}$	25	50	400
Enron2 (1k)	$r_1 \rightarrow R_3$	59	38	17
Enron3 (1k)	$r_1 \rightarrow R_3$	55	50	19
	$r_1 \rightarrow R_4$	420	60	2
	$r_3 \rightarrow R_3$	3	4	333
ForSale (1k)	$r_3 \rightarrow R_6$	13	24	77
Windows (1k)	$r_3 \rightarrow R_2$	1	2	1000

5.5   Comparison with Related Work

Approaches for learning regexes from scratch (see Section 2) need at least one example of each kind of instance. The labeling effort to cover less common variants is typically significant. For example, the 10k sample of Enron emails contains only 26 instances of the new kind of phone numbers covered by our generalization from r_3 to R_8. In order to label at least one instance, a human may need to label 385 (that is, $10,000/26$) emails under random sampling. In contrast, labeling 31 instances (not complete emails) was sufficient to derive R_8 using our approach. Table 4 compares the labeling effort in our approach against the effort of labeling at least one instance of the new kind.

We also compare our results against the clustering-based approach (CBA) [2] which works by generalizing an initial regex guided by the set of all true matches. We provide all true examples procured during our interactive generalization process as correct examples to CBA. As expected, we observe that CBA works well if enough labeled instances of each kind exist. CBA generalizes $r1$ (on Enron1) to *three expressions* that together cover exactly the same instances as R_2 (see Table 1) though it does not discover R_2 itself. Thus, even while taking advantage of our sampling, CBA generates result sets that contain many expressions that together are only as powerful as a single regex in our approach. CBA fails to work in cases where only a few true matches exist and is unable to generalize r_3 for any of the datasets.

6   Conclusion and Future Work

We have described a technique to find generalizations of a regular expression that improve its recall without much affect on precision. We have demonstrated empirically that it significantly reduces the labeling effort involved in arriving at regexes that also cover uncommon forms of a concept. We were able to identify good generalizations of an expression within a few minutes and with labeling only tens of instances as opposed to examining hundreds of documents.

As described in Section 2, our work is orthogonal to the work presented in [11] that specializes a regex. It will be interesting to explore how to combine these two approaches into an integrated method that uses both generalizations and specializations. For example, in the search process we may initially over-generalize and then specialize to reduce the false positives.

References

1. Appelt, D.E.: Introduction to information extraction. AI Commun. 12(3), 161–172 (1999)
2. Babbar, R., Singh, N.: Clustering based approach to learning regular expressions over large alphabet for noisy unstructured text. In: AND Workshop, pp. 43–50 (2010)
3. Bex, G.J., Neven, F., Schwentick, T., Tuyls, K.: Inference of concise DTDs from XML data. In: VLDB, pp. 115–126 (2006)
4. Chiticariu, L., Krishnamurthy, R., Li, Y., Raghavan, S., Reiss, F., Vaithyanathan, S.: SystemT: An algebraic approach to declarative information extraction. In: ACL, pp. 128–137 (2010)
5. Ciravegna, F.: Adaptive information extraction from text by rule induction and generalisation. In: IJCAI, pp. 1251–1256 (2001)
6. Denis, F.: Learning regular languages from simple positive examples. Machine Learning 44(1/2), 37–66 (2001)
7. Fernau, H.: Algorithms for learning regular expressions from positive data. Inf. Comput. 207(4), 521–541 (2009)

8. Garofalakis, M.N., Gionis, A., Rastogi, R., Seshadri, S., Shim, K.: XTRACT: A system for extracting document type descriptors from XML documents. In: SIGMOD, pp. 165–176 (2000)
9. Klimt, B., Yang, Y.: Introducing the Enron corpus. In: CEAS (2004)
10. Lang, K.: 20 Newsgroups (1997),
    `http://people.csail.mit.edu/jrennie/20Newsgroups`
11. Li, Y., Krishnamurthy, R., Raghavan, S., Vaithyanathan, S., Jagadish, H.V.: Regular expression learning for information extraction. In: EMNLP, pp. 21–30 (2008)
12. Mitchell, T.M.: Generalization as search. Artif. Intell. 18(2), 203–226 (1982)
13. Nie, Z., Wen, J., Zhang, B.: 2D conditional random fields for web information extraction. In: ICML, pp. 1044–1051 (2005)
14. Pearl, J.: Heuristics: intelligent search strategies for computer problem solving. Addison-Wesley Longman Publishing Co., Inc., Boston (1984)
15. Sarawagi, S., Cohen, W.W.: Semi-markov conditional random fields for information extraction. In: NIPS, pp. 1185–1192 (2004)
16. Wu, T., Pottenger, W.M.: A semi-supervised active learning algorithm for information extraction from textual data: Research articles. JASIST 56(3), 258–271 (2005)

Trade-Off Analysis of Elasticity Approaches for Cloud-Based Business Applications

Basem Suleiman[1,2], Sherif Sakr[1], Srikumar Venugopal[1], and Wasim Sadiq[2]

[1] School of Computer Science & Engineering, Uni. of New South Wales, Australia
[2] Social Business Network Research Practice, SAP Research Australia
{basems,srikumarv,ssakr}@cse.unsw.edu.au, wasim.sadiq@sap.com

Abstract. Infrastructure as a Service (IaaS) providers, such as Amazon Web Services, offer on-demand access to computing resources at pay-as-you-go prices. The key benefit of IaaS is elasticity, i.e., the ability to provision and de-provision resources at will. This feature makes IaaS infrastructure as the best platform for hosting web applications, e.g. e-business, that are subjected to highly-variable request patterns. However, elasticity can be triggered either on the basis of resource utilization or for meeting service level objectives (SLOs). In this paper, we extensively evaluate these two types of elasticity rules using the TPC-W benchmark on Amazon IaaS infrastructure. From this experimental data, we evaluate the performance of these rules against the primary metric of service level satisfaction for web applications, and secondary metrics such as resource utilization and cost. Through our inferences, we present a number of recommendations that would enable practitioners and cloud consumers using Amazon to define appropriate elasticity rules to meet their SLOs and other metrics.

1 Introduction

Infrastructure as a Service (IaaS) Provider, such as Amazon Web Services (AWS), Rackspace, and GoGrid offer internet-based access to a variety of computing resources on-demand and on a pay-as-you-go pricing model. This is also a part of cloud computing. A key benefit of this resource model is *elasticity*, wherein resources can be provisioned and de-provisioned on-demand through a self-service or a programming interface. Elasticity is highly suitable for different application classes [1], particularly web-based transactional business (or e-business) applications such as online shops and Customer Relationship Management (CRM) that are often subjected to intensely variable workloads. Therefore, deploying such applications on an IaaS provider can help consumers to dynamically instantiate computing resources according to their applications workloads at usage-based prices [17].

IaaS providers enable users to control elasticity through different means. Amazon Auto Scaling[1] allows users to set elasticity rules that define actions to be

---

[1] Amazon Auto Scaling: http://aws.amazon.com/autoscaling/

X.S. Wang et al. (Eds.): WISE 2012, LNCS 7651, pp. 468–482, 2012.

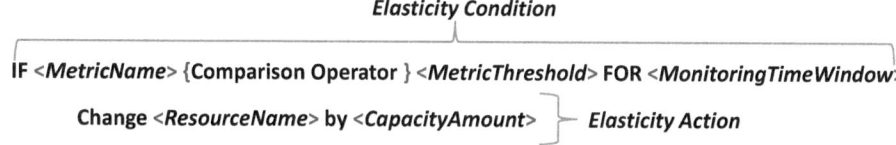

Fig. 1. Common Structure of Amazon Auto Scaling Elasticity Rule

executed in response to user-defined triggers. As shown in Figure 1, an elasticity rule consists of two main parts; condition and action. The condition specifies the metric (<Metric-Name>) to be evaluated against a specific threshold value (<MetricThreshold>). The evaluation is performed at regular intervals specified by <MonitoringTimeWindow>(e.g., 1 minute, 5 minutes). When the condition evaluates to true, the action part is executed. For example, an action could be to increase or decrease the capacity of specific resource (<ResourceName>) by amount specified in <CapityAmount>. Such actions may include starting up a new instance (*scale out*) or shutting down an existing instance (*scale in*). The conditions within the triggers can be defined based on resource-specific metrics, such as CPU utilization, or application-specific metrics, such as response time, where the application is executing on the said resource. Amazon also provides a monitoring interface (CloudWatch[2]) that allows the users to collect data in order to inform these metrics.

The application-specific metrics may be more important to an cloud consumer than resource-specific metrics as the former forms the basis of Service Level Objectives (*SLOs*) that the consumer may be required to meet as part of Service Level Agreements (SLAs) to its customers. Maintaining appropriate SLA levels for e-business applications is crucial as such applications have high business value and any SLA degradation translates into financial loss and/or customer dissatisfaction. However, IaaS providers only provide limited resource-level SLA guarantees such as resource availability [17,4] and do not offer or support any guarantees on consumer-specific SLOs. Also, it has been proven that resources from IaaS providers exhibit variable performance over time [15,7].

In the face of these challenges, it becomes the responsibility of cloud consumers to ensure appropriate SLA satisfaction for their cloud-based applications. Therefore, from the perspective of a cloud consumer, there is an important need for investigating different approaches towards elasticity and how they influence performance of their web applications. In this paper, we evaluate the use and performance of elasticity rules in meeting application-specific SLA metrics, using Amazon Auto Scaling. Particularly, we compare elasticity rules based on resource-specific metrics (resource-based elasticity) against those based on application-specific metrics (SLA-based elasticity). Specifically, we address the following questions:

---

[2] Amazon CloudWatch: `http://aws.amazon.com/cloudwatch/`

1. How well do elasticity rules perform in terms of satisfying application SLA, server costs, response time and percentage of served requests?
2. How well do they perform with servers with different capacity profiles?
3. How consistent or variable is the performance of these elasticity rules?

The main contribution of this work is twofold. First, this analysis equips cloud consumers with quantifiable evidences that can help them evaluate different elasticity rules with important parameters. Second, it provides valuable insights on a number of important factors that could influence achieving acceptable SLA levels. These can guide cloud consumers in designing and building effective resource-based and SLA-based elasticity mechanisms for their cloud-based applications.

The rest of the paper is structured as follows. Section 2 describes the design and approach elasticity or all experiments. The results of all experiments along with statistical analysis are summarized in Section 3. A number of key lessons learned from the experiments are also discussed in Section 4. Section 5 discusses related research. Conclusions and future work are drawn in Section 6.

2   Experimental Approach and Design

In this section we describe the design of our experiments to compare different types of elasticity rules and justify our design decisions and technology choices.

2.1   TPC-W: e-Business Application and User Emulation

We have chosen TPC-W, an industry standard for transactional Web benchmark that has been designed to evaluate performance of e-business systems [2]. Particularly, TPC-W offers a comprehensive specification of functionality of an online bookstore application and the behaviour of its users. Recently, TPC-W has been applied to cloud-related performance studies [10,18]. There are a number of open source implementations of TPC-W specifications. We have chosen the implementation developed by Horvath [6].

The basic scenario of user interaction in the TPC-W benchmark is described as follows. Each emulated user opens a session that consists of a sequence of interactions (or requests) such as Best Sellers, Item Details, Search, Add to Cart and Buy Confirm. Each emulated user waits for certain time (think time) before issuing the next interaction to the web server. The transition from one interaction to another is determined by a state transition matrix which determines the transition probabilities from one interaction to another. Based on user interactions, TPC-W specifications differentiate between three workload profiles; Browsing, Shopping and Ordering profiles as shown in table 1. These profiles vary based on the percentage of each interaction in the *Browse* (read operations) and *Order* (write operations) groups as shown in table 1. In all experiments we use Browsing workload profile as it stresses the application tier (i.e.,95% read operations) and does not stress the Database tier (5% write operations).

Table 1. TPC-W Workload Profiles and Web Interaction Groups

Web Interaction Group/Workload Profile	Browsing	Shopping	Ordering
Browse (Read Operations)	95%	80%	50%
Order (Write Operations)	5%	20%	50%

Table 2. Processing Time SLOs for all TPC-W Requests at the Application Tier

Request Type	Home	Item Detail	New Products	Best Sellers	Search	Execute Search	Shopping Cart	Buy Confirm
SLO	149 ms	81 ms	1362 ms	1276	143 ms	1051 ms	132 ms	105 ms

2.2 Service Level Objectives for TPC-W Requests

We have followed an experimental approach to determine the processing time limits (SLO thresholds) for all TPC-W request types. First, we have run a number of experiments with initial deployment settings, i.e., load balancer, an application server (small server) and a database server (extra-large sever). We gradually increased the workload (number of concurrent users) until the system started dropping requests. This resulted in the maximum number of concurrent users, i.e., 65 concurrent users, which can be served by our system without rejecting any request. We then run four separate experiments, each for one hour, with the same deployment settings and 65 concurrent users workload and collected processing times (in milliseconds) of all request types at the application server. For each request type, we then calculate the 90th percentile of processing times of each request type. We then calculated the average of the resulted four processing time values for each request type as shown in table 2. These SLO thresholds are used for calculating percentage of SLA satisfaction as it will be described in the next section.

2.3 Deployment Architecture

We have adopted a 3-tier deployment architecture for our TPC-W application (i.e., Web, Application and Database), a typical architecture for e-business applications [10,18,16]. With 3-tier architecture, performance bottlenecks can be traced to a certain tier when application workload increases. For example, a workload that includes large volumes of search and browsing key pages will put performance overhead on the application tier. In this study, we focus on scaling the Application tier as the web tier does simple processing (forwarding requests to the application tier and responses to the end user) we also see less necessity to focus on the Web tier. Scaling the Database tier requires specialised mechanisms that take care of data replication and consistency [10] that are outside the scope of this study.

We have deployed TPC-W application on Amazon cloud infrastructure in the following manner. We created a pool of servers with the same capacity and run the same Amazon pre-packaged Linux distribution. On each server, we installed

Table 3. Elasticity Rules Used in all Experiments

Rule	Elasticity Rules (Conditions and Actions)
	Resource-based Elasticity Rules
CPU75	*ScaleOut–* If CPU Utilization >75% for 5 minutes, add 1 app. server
	ScaleIn– If CPU Utilization ≤ 30% for 5 minutes, remove 1 app. server
CPU80	*ScaleOut–* If CPU Utilization >80% for 5 minutes, add 1 app. server
	ScaleIn– If CPU Utilization ≤ 30% for 5 minutes, remove 1 app. server
CPU85	*ScaleOut–* If CPU Utilization >85% for 5 minutes, add 1 app. server
	ScaleIn– If CPU Utilization ≤ 30% for 5 minutes, remove 1 app. server
	SLA-based Elasticity Rules
SLA90	*ScaleOut–* If SLA <90% for 5 minutes, add 1 app. server
	ScaleIn– If SLA ≥ 90% for 5 minutes, remove 1 app. server
SLA95	*ScaleOut–* If SLA <95% for 5 minutes, add 1 app. server
	ScaleIn– If SLA ≥ 95% for 5 minutes, remove 1 app. server

and configured JBoss 2.3.2 as the Servlet and EJB container on which we deployed the TPC-W bookstore application. The bookstore database was hosted on a separate Linux server (m1.xlarge instance[3]). The database was based on MySQL5.1.92 and was populated with 10000 books generated randomly according to TPC-W specifications. We used Amazons Elastic Load Balancer to distribute user requests among the instances of the TPC-W bookstore running on separate servers.The TPC-W client emulation application was deployed on a separate Linux server (m1.xlarge) instance to ensure all clients interactions are generated naturally. All the servers were located in the same Amazon geographic region, US East (Virginia), to ensure reducing network overhead occurring when servers sends and receives data.

2.4   Elasticity Rules

Table 3 summarizes the resource-based and SLA-based elasticity rules used in our experiments. In all elasticity rules (resource and SLA) we focus only on horizontal scaling actions, i.e., scale out and scale in, as it provides more reliable solution than vertical scaling, i.e., scale up/down. This is mainly because horizontal scaling provides transparent fail-over strategy when a server becomes unresponsive or fails [16]. For resource-based rules, we chose CPU utilization as the primary metric as it is a load-dependent metric commonly used in different performance and monitoring studies [5]. We define 3 example rules that vary in CPU utilization threshold value only (i.e., 75%, 80% and 85%). The threshold values have been chosen based on Haines [5] performance testing guidelines. As CPU utilization increases from 80% to 95%, the system begins to thrash, the response time increases and requests are dropped. The objective of the ScaleOut rules is to instantiate new resources when the CPU utilization reaches this range.

---

[3] Capacity:15GB memory, 8 EC2 Compute Units (CUs)(4 virtual cores with 2 EC2 CUs each), High I/O Performance.

The load balancer delegates new requests to the new resources, when they are online, thereby reducing load on the existing resources. We consider a resource with less than 30% CPU utilisation as under-utilised. To reduce costs and improve return on investment, the ScaleIn rule shuts down servers with below 30% utilisation over a 5-minute period.

In the SLA-based elasticity rules we use application SLA satisfaction as the metric to be evaluated the scaling conditions. Similarly, we only change the SLA satisfaction threshold values. The SLA satisfaction metric is based on the processing time of each request spent at the application tier (as we focus on scaling the application tier). TPC-W specifications [2] require that 90% of the requests have to meet its response time limits (SLOs). Therefore, we use 90% as one of the SLA thresholds for triggering a *ScaleOut*. We have also chosen 95% as a more stringent SLA threshold for testing durability and reaction times of the elasticity rules. It also guarantees if one server fails or degrades then not all requests will be influenced especially until requests are redirected to another server. To reduce costs, if the SLA threshold is maintained throughout the monitoring period, then an excess instance is shut down.

2.5   Metrics and Data Collection

As shown in Table 3, percentages of CPU utilization and SLA satisfaction are the primary metrics to enable elasticity rules. Therefore, we configured AWS Cloud-Watch and Auto Scaling APIs to collect CPU utilization of each application server in the application tier. CPU utilizations were aggregated at minute-basis and two alarms (one for scale out and another for scale in) were set to execute scaling actions when the scaling condition is triggered. We configured each application server to log requests details including processing time for each request. We also developed a program to read the logs, evaluate request processing times against SLOs and calculate percentage of SLA satisfaction. This is then stored in AWS CloudWatch as a custom metric for which we define two alarms (scale out and scale in alarms) to execute scaling actions when corresponding scaling conditions are held. In addition to CPU utilization and SLA satisfaction metrics, we have configured the load balancer to collect:

- *End-to-end response time*: request response time seen from the load balancer.
- *Percentage of errors*: % of requests that are not served due to error responses from either application servers (server error 5XX series and client error 4XX series) or at the load balancer (due to unregistered or unhealthy application servers or request rate exceeds the load balancers current capacity).
- *Server usage costs*: computed based on the server-hours of all servers instantiated in each experiment. In all experiments, we used instances from US East N. Virginia region and hence the hourly charges were $0.08 for small and $0.165 for medium Linux servers. SLA penalty costs and AWS CloudWatch metrics and monitoring are not within the scope of this paper. AWS CloudWatch metrics (resource and user-defined metrics) are charged on per-metric-month basis and therefore metric charges are not incurred in our experiments as we use metrics for few hours (not a whole month).

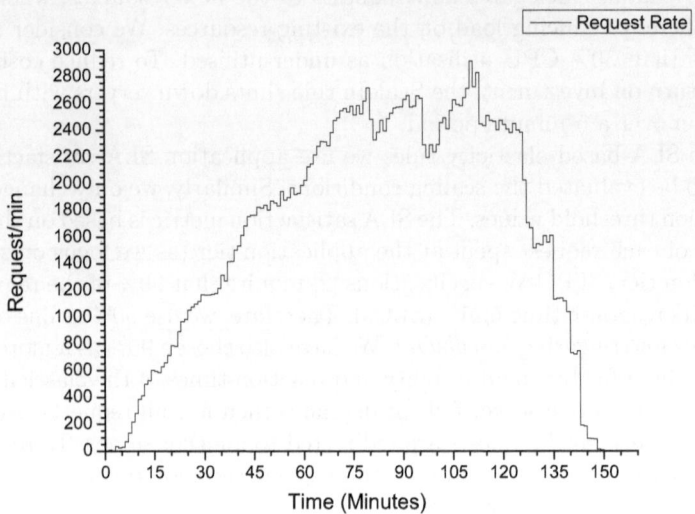

Fig. 2. Generated Workload for TPC-W Application

3 Evaluation Methodology and Results

In this section we describe how we have conducted experiments to evaluate the performance of different elasticity rules.

3.1 Experiment Design

As discussed previously, TPC-W user emulation is based on precise user behavior model and web interaction mixes. The emulation program requires generating workload density (variation of number of concurrent users over time). For all experiments we generated workload density by using Power Law and Poisson distribution functions which have been used in the literature to generate web application workloads [3,12]. The number of concurrent user sessions has been generated using Zipf function (one type of power-law functions) with 65 as number of user sessions and 0.05 as exponent. The inter-arrival time (time to wait before generating new number of concurrent user sessions) has been generated using Poisson function with 7 minutes as a mean value. We fed the number of concurrent user sessions and inter-arrival time to the TPC-W emulation program. Figure 2 shows the resulting workload in terms of request rate.

We ran two experiments with different comparison objectives. The first experiment compares the resource-based and SLA-based rules in Table 2 using metrics such as SLA satisfaction, CPU utilization, requests served and cost. The second experiment then picks up two of the best performing rules and compares their performance on cloud servers with different capacity profiles. For both experiments, the monitoring time period was 5 minutes. However, it is important to notice that choosing optimal combination of values for elasticity rule's metric threshold and monitoring time window is not the focus of this work.

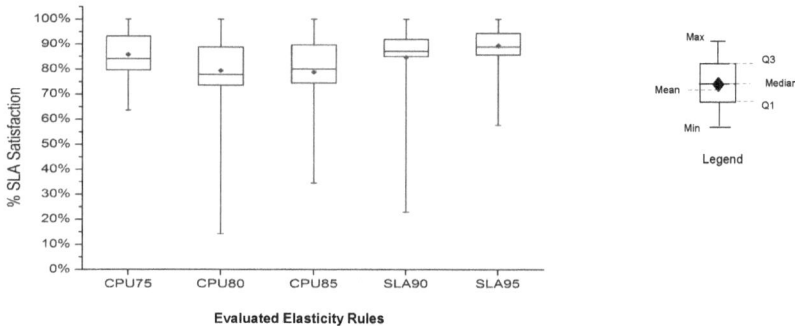

Fig. 3. SLA Satisfaction from Resource-based and SLA-based Elasticity Experiments

Table 4. Summary Statistics of Resource-based and SLA-based Elasticity Experiments

Exp./Metric	CPU Util.(%)	Resp. Time(ms)	%Serv. Reqs.	Costs($)
CPU75	69.36 ± 3.77	351 ± 17	99.996	1.04
CPU80	71.10 ± 2.98	404 ± 29	99.999	0.80
CPU85	78.08 ± 2.89	372 ± 22	99.998	0.80
SLA90	49.74 ± 3.68	299 ± 29	99.998	1.68
SLA95	46.64 ± 3.57	295 ± 10	99.999	2.40

For all experiments, we use box-and-whisker (or box plot) method to sum-marize the data points of SLA satisfaction for each elasticity rule. Box plot provides robust way to show important statistics of data points and its distribu-tion at glance [13]. These statistics are minimum (Q0), lower quartile (Q1=25th percentile), median (Q2=50th percentile), upper quartile (Q3=75th percentile), maximum (Q4) as well as the mean value. In addition, we compute average CPU utilization and average response time along with confidence intervals. We also compute total server costs (at the application tier where elasticity is applied) and percentage of requests served successfully. The results of our experiments are presented in the following Sections.

3.2 Evaluating Resource-Based and SLA-Based Elasticity Rules

Figure 3 shows box plots for the resulted SLA satisfaction of experiments using the elasticity rules specified in Table 3. The higher values for Q0-Q4, the better the SLA satisfaction[4]. Among all resource-based rules, CPU75 has resulted in the highest SLA satisfaction statistics (Q0-Q4). CPU75 also has resulted with SLA satisfaction distributed within range (about 65%-100%) better than SLA satisfaction ranges of CPU80 (about 15%-100) and CPU85 (about 35%-100%). Interestingly, CPU85 has resulted in SLA satisfaction statistics and range either

---

[4] The legend used in here will be the same for all Boxplot diagrams in this paper.

equal to or better than corresponding ones resulted from CPU80 rule. This shows that a decrease in CPU utilization threshold (within 5% magnitude) does not necessarily yield in better SLA satisfaction. In terms of SLA-based elasticity, SLA90 and SLA95 has resulted in almost equivalent 75th percentile and median SLA satisfaction. SLA95, however, resulted in 25th percentile value higher than corresponding one in SLA90. Furthermore, SLA satisfaction data points are distributed within better range (about 55%-100%) when compared to SLA satisfaction range of SLA90 (about 22%-100%). However, comparing both sets of elasticity rules, CPU75 with a mean satisfaction of approx. 87% beats SLA90 with approx. 85% and is beaten by SLA95 by only 2%. The reason for this result is that the changes in utilization are far more sensitive to workload than those in average response time. When the conditions for SLA-based elasticity are triggered, the application is already violating SLOs and this situation continues until a new instance is added after the rule is invoked. However, the CPU75 rule is triggered earlier than the SLA90 rule, when the SLO violations have not yet happened.

Table 4 presents the average CPU utilization over all the resources instantiated during the experiment, average response time, percentage of served requests and server costs for each elasticity rule. As expected, resource-based rules resulted in better average CPU utilization than SLA-based rules. Also, the cost for the SLA-based rule with the best satisfaction is almost twice that of similar resource-based rules. One may think that with a lower trigger threshold, CPU75 should have resulted in high costs. However, this is again a function of the workload sensitivity and the time required to bring new servers online. Since CPU75 triggers early, a new server is able to start up and participate in the load distribution by the time the utilization reaches dangerous levels. With SLA-based rules, a new instance is started while the SLOs are already being violated. Therefore, in this case, multiple new servers are instantiated at a time and hence, the higher costs. Based on the observations above, we can infer that resource-based rules have better performance to cost trade-off than SLA-based rules. Clearly, CPU75 has the best performance in terms of meeting SLOs while keeping costs low. Among the SLA rules, SLA90 is has a lower cost to SLA95 while having a comparable SLA satisfaction.

Due to multi-tenency nature of Amazon EC2 it might be possible that one instance shows CPU utilization below the CPU threshold (specified in a CPU-based rule) while the physical CPU utilization has already become over that threshold and the application SLA starts to be violated. We argue that cloud consumers do not have control beyond the physical cloud host and they only can configure actions to be taken based on the metrics defined in the elasticity rules.

3.3   Elasticity Rules Performance on Different Server Instances

To evaluate the performance of resource-based and SLA-based elasticity rules on cloud servers with different capacity profiles, we first choose the best elasticity rule from each type, i.e., CPU75 and SLA90. For each elasticity rule, we run three experiments with two different application server instance types; i.e.,

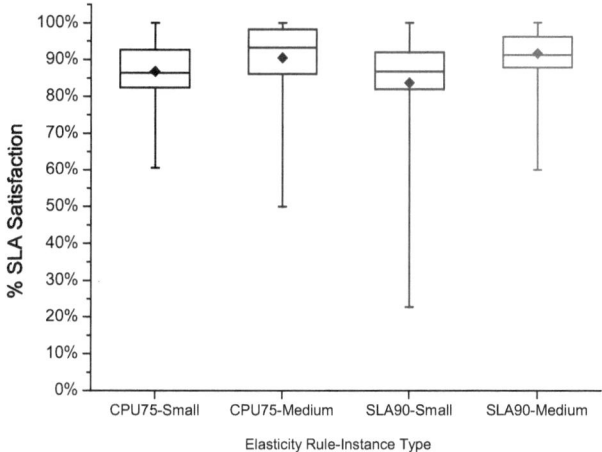

Fig. 4. SLA of CPU75 & SLA90 Elasticity Experiments on Small & Medium Instances

Table 5. Statistics of CPU75 & SLA90 Experiments on Small & Medium Servers

Exp./Metric	CPU Util.(%)	Resp. Time(ms)	%Serv. Reqs.	Costs($)
CPU75-Small	64.01 ± 1.94	321 ± 8	99.999	1.07
CPU75-Medium	46.10 ± 1.86	279 ± 9	100	0.48
SLA90-Small	58.41 ± 2.38	357 ± 16	99.999	1.55
SLA90-Medium	26.23 ± 1.45	245 ± 4	99.999	1.60

Small instance (AWS default) and High-CPU Medium instance[5]. We have chosen medium server instance as it has high CPU capacity that matches TPC-W Browsing workload profile requirements (95% of the Browsing profile web interactions are Browse operations which stress the application tier). Figure 4 shows SLA satisfaction summary statistics resulted from CPU75 and SLA90 experiments on both small and medium server instances. From Figure 4 we see a number of observations worth discussing. Medium server instances have resulted in better SLA satisfaction than small instances. This is because medium instances have CPU capacity two times (2 virtual cores) of small instances. CPU75 experiments with medium instances have median and 75th percentile SLA satisfaction slightly higher than median and 75th percentile SLA satisfaction of SLA90 with medium instances. The improvement of SLA satisfaction on medium instances is also due to the lowered server provisioning time that we collected during all experiments. On average, it took a medium instance about 2 minutes (compared to about 5 minutes for small instances) to become operational and starts serving requests. As a result, the SLA violation time in medium instances experiments is reduced and leaded to better SLA satisfaction.

---

[5] m1.small instance: 1.7GB memory, 1 EC2 Compute Units (CU) (1 virtual core), and moderate I/O Performance. c1.medium instance: similar to small instance but with 5 EC2 CU(2 virtual cores with 2.5 EC2 CU each).

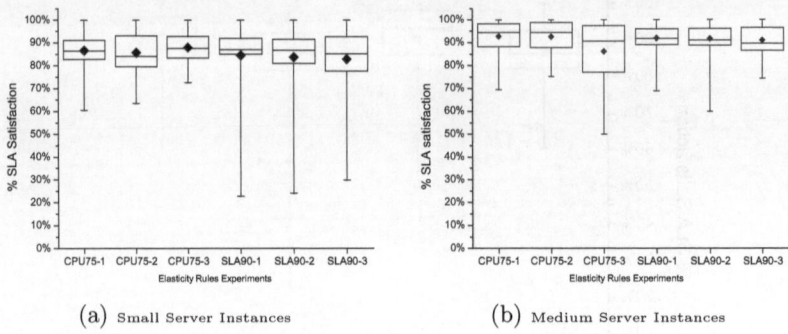

(a) Small Server Instances   (b) Medium Server Instances

Fig. 5. SLA of CPU75 & SLA90 Elasticity Experiments on Small & Medium Instances

While CPU75 and SLA90 experiments with small instances have almost similar SLA satisfaction statistics, SLA90 noticeably has SLA satisfaction points
spread over a larger range than CPU75. Table 5 highlights a number of important observations. CPU75 and SLA90 experiments with small server instances
have better average CPU utilization better than with medium instances. CPU75
experiments with small and medium instances have higher average CPU utilization than SLA90. In both experiments, CPU75 and SLA90, medium instances
have resulted in considerable response time improvements compared to small
instances on account of the higher computation capacity. As a medium instance
takes longer to reach the CPU75 threshold, fewer servers are started and thus
the costs are nearly 50% less as compared to using small instances. The same
relation does not, however, hold for SLA90, as the effect of waiting for an instance to be available while SLOs are being violated, is the same for both small
and medium instances.

3.4   Consistency of SLA Satisfaction with Different Elasticity Rules

In this section, we use the data from the six experiments performed to compare
CPU75 and SLA90 rules in the previous section, to analyse the variability of
SLA satisfaction across the duration of the experiments. For both CPU75 and
SLA90 rules, we grouped the box plots into two groups; experiments with small
instances and experiments with medium instances. The juxtaposition of box
plots then provide a good way to investigate if there are differences between
data sets [13] obtained from individual experiments. Figure 5 summarizes SLA
satisfaction data resulted from the three experiments we run with both CPU75
and SLA90 elasticity rules on (a) small server instances (left figure) and (b)
medium server instances (right figure). From both figures, we can see SLA90
and CPU75 exhibit only slight variations on both small and medium instances.
While CPU75 with medium instances has almost up to %6 difference in average
SLA satisfaction SLA90 with medium has almost up to 2% difference. Both
figures reveal that average SLA satisfaction is consistent with both SLA90 and
CPU75 rules, and on small and medium instances. Therefore, cloud consumers

can rely on both elasticity rule types on both small and medium instances unless their application is not mission-critical, e.g., stock trading.

4   Discussion

From the above experiments and results we draw a number of important inferences that could help cloud consumers to make appropriate decisions for defining robust elasticity rules for their cloud-based business applications.

−Resource-based elasticity rules performed better than SLA-based elasticity rules in terms of server costs, SLA satisfaction and CPU utilization but not end-to-end response time and % of served requests. This is because in resource-based elasticity CPU utilization is influenced by workload changes quicker than application response time and therefor it triggers elasticity actions before SLA violations becomes sever.

−Resource-based and SLA-based rules performed better on medium server instances. CPU75 has resulted in considerable cost reduction (about 55%) on medium instances. CPU75 and SLA90 have also shown relatively average SLA satisfaction improvements (3% and 8% respectively) and good average response time improvements (13% for CPU75 and 30% for SLA90) on medium instances.

−Metric thresholds influence SLA satisfaction, response time and server costs but not % of served requests. Noticeably the highest impact is on server costs where it increased up to 30% and 23% in SLA-based and CPU-based rules respectively. There was also up to 13% increase in average response time with CPU-based rules. Therefore, we recommend cloud consumers to choose lower metric threshold values to reduce such impact. Cloud consumers, for example, can choose CPU75 rule as if they mainly interested in achieving best median SLA satisfaction and average response time at server costs higher than CPU80 and CPU85. However, we do not recommend using SLA95 as it has resulted in very high costs (30% higher than SLA90 costs) although SLA95 has slightly improved median SLA satisfaction and barely average response time.

−Server provisioning time also influences SLA satisfaction. Medium server instances needed average provisioning time 3 minutes less than small instances and therefore it improved overall SLA satisfaction and end-to-end response times.

−Performance of resource and SLA-based elasticity rules is mostly consistent. Our experiment results have also shown that both CPU75 and SLA90 on small and medium server instances exhibit only slight inconsistency in SLA satisfaction across multiple executions with the same workload. However, we see such slight variations in SLA levels cannot be tolerated for mission-critical applications such as stock trading.

5   Related Work

A number of research studies [3,8,11] have recently benchmarked performance of different cloud infrastructure services and its impact on web applications. Lenk et. al. [11] proposed a method to support cloud consumers to measure and compare performance-costs of virtual cloud servers from different providers. Jackson et. al. [8] presented a comprehensive evaluation of high performance computing (HPC) applications on conventional virtualized HPC and Amazon EC2. They also provided analysis of Amazon cloud performance variations and its impact on application performance. Dejun et. al. [3] evaluates (i) performance stability and (ii) performance homogeneity of AWS small instances in different availability zones based on CPU-intensive, database read-intensive and write-intensive web applications. The evaluation was based on server CPU, I/O and applications response time metrics. Our work has unique scope from these works as it focuses on evaluating different resource-based and SLA-based elasticity rules for web transactional applications. In contrast to [3,8], we evaluate e-business application with 3-tier architecture for which defining suitable elasticity rules is crucial. We also provide trade-off analysis between SLA, CPU, response time and cost metrics from a cloud consumer perspective.

Kossamann et. al. [10] evaluated the scalability of web applications but with focus on the database tier. They evaluated scaling out capabilities of different cloud database services (e.g., AWS Relational Database Service, SimpleDB, and MySQL) with different database architectures. Scaling the database tier is beyond the scope of our study as it involves many challenges such as data consistency and replication. Our work can be complementary to Kossamann [10] as it focuses on evaluating scaling out/in the application tier. Some other research work [14,19] presents and evaluates SLA-based approaches to manage cloud resources for transactional business applications deployed in cloud environments. Similar to our evaluation, the authors [14,19] evaluated the performance of their approaches based on SLA satisfaction and cost metrics but with focus on the database tier as well. In contrast to the evaluation method in [10,14,19], we evaluate a number of resource-based and SLA-based rules to dynamically add/remove application servers.

Much research effort has been recently done in performance variability of different cloud infrastructure services including [7,15]. Iosup et. al. [7] shows that some weekly pattern of variability for Amazon EC2 small instances and they stressed the importance of having more steady performance especially for applications use auto scaling. Schad et. al. [15] also observed considerable CPU performance variability for both AWS small and large instances during different weeks. Jayasinghe et. al. [9] evaluated performance and scalability variations of n-tier web applications on different clouds including Amazon EC2. One part of their evaluation is based on scale out scenario with different predefined server configurations. None of these studies have employed resource-based or SLA-based elasticity rules in their evaluation. We evaluated the consistency of application SLA satisfaction with different elasticity rules on Amazon small and medium server instances.

6   Conclusions and Future Work

In this paper we have presented experimental evaluation of different resource-based and SLA-based elasticity rules for e-business applications on Amazon EC2.

Based on this, we have provided an analysis on how well these rules performed in terms of application SLA, response time, percentage of served requests, server costs and CPU utilization. We have shown how metric thresholds (CPU and SLA thresholds) influence almost all these metrics. Cloud consumers need to carefully decide on which elasticity rule to use and its metric threshold values based on the SLA and costs requirements of their application. We have also inferred how both CPU-based and SLA-based elasticity rules on Amazon medium instances (over small ones) have led to quite high improvements in SLA satisfaction, response time and exceptionally server costs. Therefore, it is better for cloud consumers to use medium instances with both CPU-based and SLA-based rules. Furthermore, our experimental analysis has revealed that CPU-based and SLA-based exhibit only slight variations in satisfying application SLA on Amazon small and medium instances and therefore the rules can be reliable.

As future work, we are planning to investigate different workload patterns and its impact on the performance CPU and SLA-based rules. With different workloads we can also study the effectiveness of scale out and scale in and its impact on SLA and server costs. We also plan to investigate the impact of CPU-based and SLA-based elasticity rules on SLA satisfaction at fine-granular level, i.e., request level. This is important as some requests may have more intensity than others and hence it may have much more SLA violations.

Acknowledgements. The cost of using cloud computing (EC2) resources for the experiments in this paper were covered by a grant provided by Amazon Web Services. However, Amazon was neither involved in nor had any influence on the research reported in this paper.

References

1. Amazon Web Services LLC. Case studies (January 2012),
   http://www.aws.amazon.com/solutions/case-studies/
2. T. P. P. Council. Tpc benchmark web commerce specification (tpc-w). Technical Report 202 (February 2002), http://www.tpc.org/tpcw/
3. Dejun, J., Pierre, G., Chi, C.-H.: EC2 Performance Analysis for Resource Provisioning of Service-Oriented Applications. In: Dan, A., Gittler, F., Toumani, F. (eds.) ICSOC/ServiceWave 2009. LNCS, vol. 6275, pp. 197–207. Springer, Heidelberg (2010)
4. Durkee, D.: Why cloud computing will never be free. Commun. ACM 53(5), 62–69 (2010)
5. Haines, S.: Pro Java EE 5 Performance Management and Optimization. Apress, USA (2006)
6. Horvath, T.: Tpc-w java implementation (November 2008),
   http://www.cs.virginia.edu/~th8k/downloads/
7. Iosup, A., Yigitbasi, N., Epema, D.: On the performance variability of production cloud services. In: CCGRID 2011, pp. 104–113 (2011)
8. Jackson, K.R., Ramakrishnan, L., Muriki, K., Canon, S., Cholia, S., Shalf, J., Wasserman, H.J., Wright, N.J.: Performance analysis of high performance computing applications on the amazon web services cloud. In: CLOUDCOM 2010, pp. 159–168 (2010)

9. Jayasinghe, D., Malkowski, S., Wang, Q., Li, J., Xiong, P., Pu, C.: Variations in performance and scalability when migrating n-tier applications to different clouds. In: CLOUD 2011, pp. 73–80 (2011)

10. Kossmann, D., Kraska, T., Loesing, S.: An evaluation of alternative architectures for transaction processing in the cloud. In: SIGMOD 2010, pp. 579–590 (2010)

11. Lenk, A., Menzel, M., Lipsky, J., Tai, S., Offermann, P.: What are you paying for? performance benchmarking for infrastructure-as-a-service offerings. In: CLOUD, pp. 484–491 (2011)

12. Li, H., Venugopal, S.: Using reinforcement learning for controlling an elastic web application hosting platform. In: ICAC 2011, pp. 205–208 (2011)

13. Massart, D.L., Smeyers-Verbeke, J.A., Capron, X., Schlesier, K.: Presentation of data by means of box plots. LC GC Europe 18(4), 215–218 (2005)

14. Moon, H.J., Chi, Y., Hacigümüş, H.: Performance evaluation of scheduling algorithms for database services with soft and hard slas. In: DataCloud-SC 2011, pp. 81–90 (2011)

15. Schad, J., Dittrich, J., Quiané-Ruiz, J.-A.: Runtime measurements in the cloud: observing, analyzing, and reducing variance. VLDB 3(1-2), 460–471 (2010)

16. Shoup, R.: More best practices for large-scale websites: Lessons from ebay (November 2010), http://www.infoq.com/presentations/Best-Practices-eBay

17. Suleiman, B., Sakr, S., Jeffery, R., Liu, A.: On understanding the economics and elasticity challenges of deploying business applications on public cloud infrastructure. JISA 2(3), 1–21 (2011)

18. Tak, B.C., Tang, C., Zhang, C., Govindan, S., Urgaonkar, B., Chang, R.N.: vpath: precise discovery of request processing paths from black-box observations of thread and network activities. In: USENIX 2009, p.19 (2009)

19. Xiong, P., Chi, Y., Zhu, S., Moon, H.J., Pu, C., Hacigumus, H.: Intelligent management of virtualized resources for database systems in cloud environment. In: ICDE 2011, pp. 87–98 (2011)

Extraction and Evaluation of Candidate Named Entities in Search Engine Queries

Areej Alasiry, Mark Levene, and Alexandra Poulovassilis

Birkbeck, University of London,
United Kingdom, London
{areej,mark,ap}@dcs.bbk.ac.uk

Abstract. *Named Entity Recognition* (NER) has recently been applied to search queries, in order to better understand their semantics. We present a novel method for detecting candidate named entities (NEs) using grammar annotation and query segmentation with the aid of top-n snippets from search engine results, and a web n-gram model to accurately identify NE boundaries. We then evaluate this method automatically using DBpedia as a rich data source of NEs, with the aid of a small representative random sample that is manually annotated. Finally, an analysis of the types of named entities that often occur in a query log is conducted, from which a search query driven named entity taxonomy is presented.

Keywords: Named Entity Recognition, Query Log, DBpedia.

1 Introduction

Named Entity Recognition (NER) is the task of extracting from text instances of different classes such as person, location, or company. Having access to lists of NEs (known as gazetteers) can be used in web search tasks such as query suggestion [6] and question answering [9]. Recent work has proposed several approaches to mine resources to extract NEs for various classes. For example, the DBpedia project is dedicated to parsing Wikipeida articles in order to extract structured information and organize it into a variety of categories such as 'American pop singers' or broader classes of entities such as 'artist' [2]. Investigation into the extraction of NEs from web documents has also been carried out, as in the work of Etzioni et al. [5]. Recently, query logs have also been utilized in order to extract NEs, and Guo et al. have reported in [6] that approximately 71% of queries contain NEs. Mining query logs for NEs is especially interesting, as these include NEs as expressed from the users' perspective, which may be associated with other features or NEs. For example, a query containing the movie NE 'The Avengers' would have features such as 'trailer', 'cast', 'show time' associated with it, which Yin and Shah in [12] refer to as *intent phrases*. Moreover, a query containing a location such as 'london' can be associated with other NEs such as 'Birkbeck', and 'Olympics 2012'. Detecting NEs in queries and identifying their corresponding classes can help in predicting the user's search intent, that can in turn improve a user's search experience [12]. A recent related development

X.S. Wang et al. (Eds.): WISE 2012, LNCS 7651, pp. 483–496, 2012.

is Google's *Knowledge Graph*[1] that makes use of about 500 million entities and enables users to search for 'things' rather than 'strings'.

Typically, a web search query consists of only few words and does not provide enough context nor surface clues such as capitalisation to accurately detect NEs. Recent work such as that of Paşca [10], and Jain and Pennacchiotti [8], presented approaches that rely only on the query to identify NEs. In contrast, in this paper, we present an approach that uses snippets of top-n related search engine results to grammatically annotate the words in a search query, and identify candidate NEs that are contained within the boundaries set by a query segmentation process.

The contributions of this paper are threefold:

1. A NER method for search queries that grammatically annotates query tokens, and sets the boundaries of NEs using query segmentation.
2. Evaluation of NER in search queries using DBpedia and a manually annotated random sample of queries.
3. A taxonomy for named entities derived from search engine query logs.

The rest of the paper is organised as follows. In Section 2, we review the main approaches proposed in the literature to identify and classify named entities in query logs. In Section 3, we introduce our NER method for search engine queries. In Section 4, the data on which the method is tested on and the experimental settings are described. The evaluation of our method and the analysis of the results are presented in Section 5. Section 6 details an analysis of the classes pertaining to the extracted candidate NEs using manual classification and the DBpedia ontology, to derive a search engine query log NE taxonomy. We draw the conclusions from our work and present directions of future work in Section 7.

2   Related Work

Recent approaches to automatically identify NEs from query logs involve either mining the query log to construct a knowledge base of NEs [10, 8], or training a model for online prediction of the NEs contained in a query [6, 4].

Both Paşca [10] and Guo et al. [6] applied a relatively simple yet effective method to identify NEs for a set of predefined categories. Using seed instances for each category, the query log is scanned to find the context of each seed instance, where a context is a prefix and/or suffix surrounding the seed instance. The set of contexts are used later to find new instances for each NE category. The problem with these approaches is that they solely rely on the query string to identify NEs and in general, the queries do not provide enough context to accurately detect or classify NEs. For example, the word 'lyrics' can be a context for a song name. At the same time, it appears in queries such as 'song lyrics' where 'song' is not a NE, 'James Brown lyrics', where 'james brown' is not a song name, or 'Music and Lyrics' where the context is a part of a movie name.

---

[1] http://www.google.com/insidesearch/features/search/knowledge.html.

Instead of just using the current user's query, Du et al.[4] used the search session to detect and classify NEs. Two search session features were used, namely the 'class' feature of the class of the NE appearing in the previous query in the session, and the 'overlap' feature of the words between the previous and the current query. Similar to the approach presented by Guo et al., they used only the string of the query to identify the NEs.

Another approach was presented by Jain [8] which is based on the simple assumption that users usually copy-paste existing phrases as their search query. Thus, orthographic features of query tokens are preserved such as capitalised initials. Based on this assumption, a set of candidate NEs are extracted that are the longest consecutive sequence of capitalised words in a query. Although confidence scores were assigned to each extracted NE, based on its presence in a web corpus, relying on capitalisation will often miss many potential NEs in the query log that were not typed in with capital letters.

3 Query Log Candidate Named Entity Recognition

The NER method we propose here for identifying candidate NEs in a search engine query log involves the following four main stages:

1. The query log is pre-processed to remove noise and URLs before any further natural language processing.
2. Each query token is grammatically annotated with its part of speech and orthographic features.
3. The query is segmented to set boundaries for candidate NEs.
4. Candidate NEs are extracted using a set of hand-crafted rules.

We describe each of these stages in the subsections below.

3.1 Pre-processing

First the search engine query log is processed to exclude queries consisting only of non-alphabetic symbols (considered as *noise*) or URLs, and to remove all duplicate queries. In addition, the spelling of each query is verified assuming that queries are posted in English.

3.2 Grammar Annotation

Each query is grammatically annotated using the snippets of the top-n related search engine results. Adopting Bendersky et al.'s [1] bag-of-words approach, each token is annotated with the most common part-of-speech (POS), e.g. proper nouns, and orthographic features (ORTH) e.g. capital initials, according to its corresponding annotations in the snippet set. In contrast to [1], the POS tagging used is more specific, since we differentiate between common and proper nouns. Identifying proper nouns is particularly beneficial, since NEs usually consist of proper nouns. The method proposed operates offline, and therefore the snippets are retrieved and stored in advance. Snippets usually provide enough context to grammatically annotate query tokens, avoiding the cost of parsing complete web documents, as in [8].

3.3 Query Segmentation

Many approaches to query segmentation have been proposed in the literature, for example Hagen et al. [7] used raw n-gram counts to segment queries. Although Hagen et al. achieved high accuracy, in our case we need to segment queries according to their appearance in the top-n related snippets. Previously, Brenes et al. [3] used snippets to segment queries on the fly. In contrast, as mentioned above, in our work snippets are stored in advance, and we use both snippets and a web n-gram model to estimate segment probabilities.

We define query segmentation as follows. For a query Q consisting of tokens $t_1, ..., t_n$, the set of all possible segmentations is $\mathbf{S(Q)} = \{S_1, ..., S_m\}$, where $m \leq 2^{n-1}$. Each segmentation $S_i \in \mathbf{S(Q)}$ consists of one or more segments, and each segment, say s_{ij}, is a sequence of query tokens that obeys the original order of the tokens. We find the best segmentation S^β as the most probable one over all $S_i \in \mathbf{S(Q)}$. The probability of each S_i is calculated as $Pr(S_i) = \prod_{s_{ij} \in S_i} Pr(s_{ij})$, where $Pr(s_{ij})$ is estimated using a *local* n-gram model ($M_{snippet}$) created from the set of retrieved snippets for the query Q. This probability is smoothed by the probability of the segments given a *web* n-gram model (M_{web}) using an empirically set parameter λ between 0 and 1, to obtain

$$Pr(s_j) = \lambda Pr\left(s_j | M_{snippet}\right) + (1 - \lambda) Pr\left(s_j | M_{web}\right) \ . \tag{1}$$

3.4 Recognition of Candidate NEs

A small set of rules for recognising candidate NEs has been defined by examining a random sample of grammatically annotated and segmented queries, as described above. In the sample, three main cases were observed:

1. A sequence of proper nouns contained in a segment, representing approximately 93% of candidate NEs. This case includes nouns that are all capitalised such as 'USA', and mixed letter cases such as 'eBay'. In addition, it includes proper nouns followed by an apostrophe and 's' such as 'Chili's'.
2. The conjunction '&', or the preposition 'of' followed or preceded by proper nouns such as 'University of Wisconsin' (approximately 2% of candidate NEs).
3. A sequence of proper nouns that include a token whose kind is a number, such as 'Microsoft Office Professional *2003*', or a token whose POS is a number and whose kind is a word or alphanumeric with a capital initial, or all capitalised letters such as '*First* Citizens Bank', and 'Fit4less' (approximately 5% of candidate NEs).

Nine rules were created utilising the Java Annotation Patterns Engine (JAPE) to reflect these cases. Applying these rules with respect to query segmentation boundaries will result in, for example, the detection of two NEs from a segmented query such as "[marriot][new jersey]".

4 Data and Experimental Settings

The method was applied to a 2006 MSN query log consisting of approximately 15 million queries. After removing duplicates approximately 5.5 million queries were left. Thereafter, the spelling of each query was checked and corrected using Yahoo API (Spelling Suggestion YQL). In addition, for each query, the top eight snippets were retrieved and stored using Google's Custom Search API. Only eight snippets were used, since it was the maximum number of results that could be retrieved per API request. The snippet set for each query was grammatically annotated using the GATE toolkit (http://gate.ac.uk), where each snippet was processed through a pipeline consisting of a sentence splitter, a tokeniser that assigns the orthographic (ORTH) features for each word, and a POS tagger that is a modified version of the Brill tagger. Each query token was then annotated with the most probable POS and ORTH features according to its corresponding annotations in the snippet set. We used the Bing web n-gram to find the probability of each distinct segment, and smoothed the probability of the same segment given the local n-gram model, using (1), with $\lambda = 0.6$ being empirically set. Making use of the web n-gram model is particularly beneficial as the snippets on their own may not contain enough information to attain a good segmentation of the query. We note that the Bing web n-gram model contains only up to 5-grams, so we estimate larger segments using the chain rule together with the Markov assumption. Finally, candidate NEs were extracted using the set of hand-crafted NER rules.

5 Evaluation and Results Analysis

Table 1 presents a sample of segmented queries and the corresponding extracted candidate named entities, giving an illustration of how query segmentation and grammatical annotation were used to identify candidate named entities in queries. When there is a consecutive sequence of proper nouns in a query as the queries presented in Table 1, segmentation sets the most probable boundaries of the candidate named entities in the query.

Table 1. Sample of extracted candidate NEs

Segmented Query	Extracted NEs
[microsoft] [speech to text] [windows xp]	microsoft , windows xp
[leon] [final fantasy]	leon , final fantasy
[mercedes benz] [used parts]	mercedes benz

In previous work, NER in search queries was evaluated by taking a random sample [8] or the top-n extracted NEs [10, 6] and each checked whether correct or not. However, there are three main issues when only the extracted NEs are considered during evaluation. First, the extracted NEs are evaluated out of context. For example, 'safari' is a NE in the query 'download safari', while it is not in the query 'safari trips'. Second, the NER system is evaluated using only the set of extracted instances without measuring the *False Negatives*, which would reflect the number of instances the system missed. Third, since the extraction is evaluated out of context, there is no reference to the accuracy of the extraction i.e. whether the boundaries of the extracted named entities are set correctly or not. To overcome these issues, our method was evaluated from two perspectives:

Query-Level Evaluation: A random sample from the query log is taken in order to evaluate the extraction of candidate NEs given their representation in the queries from which they were extracted.

Named Entity-Level Evaluation: Each individual extracted NE is checked whether correct or not using DBpedia automatically, and also manually using a small representative random sample. We discuss these evaluations in the following subsections.

5.1   Query-Level Evaluation

Evaluating the method at the query level provides insight into the accuracy of the extracted NE boundaries, the frequency of missed NEs, and the accuracy of NE extractions given their context in the queries from which they were extracted. In addition, since hand crafted rules were used to extract NEs from a query, the accuracy and the coverage of these rules should be measured.

We manually checked a uniform random sample of 1000 queries, where True and False Positives (TP, FP), and True and False Negatives (TN, FN) were counted. These metrics were defined as follows:

TP: The query has one or more NEs and all were correctly detected.
TN: The query has no NE and none were detected.
FP: One or more query tokens were incorrectly tagged by the rules as NEs.
FN: One or more query NEs were incorrectly missed by the rules.

Table 2 presents the accuracy, recall and precision achieved by our method. Two evaluation settings were used: *Evaluation I*, where the detection of a NE is tagged as correct regardless of its boundary, e.g. extracting 'Sony PS3 news'; and a stricter version, *Evaluation II*, where the boundary of a NE must be accurate, e.g. extracting 'Sony PS3'.

Examining the sample, we found that the correctly annotated queries i.e the true positives such as the examples in Table 1, represent 67.7% in Evaluation I, and 54.9% in Evaluation II. On the other hand, the percentage of true negatives, which include queries having no NEs and none being detected, is 12.9% in both evaluations. This results in a total accuracy of 80.6% in Evaluation I, and 67.8% in Evaluation II.

Table 2. Query-level NER evaluation

	Accuracy $\frac{TP+TN}{TP+FP+TN+FN}$	Precision $\frac{TP}{TP+FP}$	Recall $\frac{TP}{TP+FN}$
Evaluation I	0.806	0.794	0.973
Evaluation II	0.678	0.643	0.966

The main boundary errors encountered in the evaluation include cases where only part of the actual NE is detected, e.g. 'britney' was extracted as a candidate NE from the query 'britney spears pictures' or the case where the extraction crosses the boundary of the actual NE, such as extracting 'britney spears pictures' as a candidate NE. These types of error occur due to incorrect POS tagging of query tokens such as tagging 'spears' as a common noun or 'pictures' as a proper noun. In addition, for queries having more than one NE, the type of error we commonly observed was that of detecting a sequence of NEs appearing in a query as a single NE, meaning that the boundaries set by query segmentation were incorrect. All boundary related errors were counted as FPs in Evaluation II, resulting in a decrease in the precision from 79.4% in Evaluation I to 64.3%.

Furthermore, for both evaluations, when a query word that is a common noun appears more often in the snippets set as a proper noun, this will result in extracting this word as a candidate NE e.g. the word 'architect' was POS tagged in the snippets set as a proper noun more often than a common noun, resulting in detecting it as a candidate NE from the segmented query '[daniel nyante] [ny] [architect]' in addition to the NEs 'daniel nyante' and 'ny'.

The percentage of queries having a NE that was missed by the rules was only 1.9%. NEs such as movie names and song names that include words that are not proper nouns, e.g. propositions or verbs, are missed by our rules. For example, in movies such as 'Thank you for Smoking' the NE recognition rules we have defined missed these NEs or inaccurately set their boundaries. However, the rules we have defined achieved high recall between, i.e. 97.3% in Evaluation I and 96.6% in Evaluation II.

5.2 Named Entity-Level Evaluation

In previous approaches the evaluation was conducted at the NE level, using two metrics *precision* as in [10, 6, 8], which is the number of correct NE extractions divided by the total extracted NEs in the sample, and *recall* (also known as *coverage*), which is the fraction of NEs extracted that are also found in a comprehensive source such as Wikipedia [8]. For our 1000 random sample, we first manually examined the candidate NEs extracted (*Phase 1*), and then, using DBpedia, examined them automatically (*Phase 2*), in order to compare the results of both phases.

DBpedia is a project whose aim is to extract structured information from Wikipedia [2]. Currently DBpedia describes 3.64 million 'things' from which 1.83 million instances are classified into a hierarchical ontology including entities such as Persons, Places, and Organisations (www.dbpedia.org). DBpedia provide a lookup service that can be used to locate DBpedia URIs, where each URI identifies a 'thing' in the DBpedia dataset; the URI refers to a Wikipedia article from which that data is extracted. For each extracted NE we can validate it using this look up service, where a set of results related to the extracted NE is retrieved. Each related result consists of a label, which is the string representing the instance found, a description, a class in the DBpedia ontology, a Wikipedia category, and the number of inlinks within Wikipedia pointing to this result. For each candidate NE, out of all related results only the one whose label is most similar to the extracted candidate NE is used. We note that DBpedia lookup may include results that are not necessarily NEs as we have defined them, such as 'gravity'. Although 'gravity' is a common noun i.e. not a NE, it has a matching DBpedia result. In this section we consider only whether the instance being evaluated has a related DBpedia result or not, and in Section 6 we will address the classification of the related DBpedia results within the DBpedia ontology.

Phase 1 (*Manual*): From the uniform random sample of 1000 queries, 1172 candidate NEs were extracted using our method. Each candidate NE was examined manually, and labelled as either (i) an *Accurate NE Extraction* (A-NE), in the case where the candidate NE is accurately extracted by our method, (ii) an *Inaccurate NE Extraction* (I-NE), where the candidate NE is extracted but its boundaries were incorrect as described in the previous section; or (iii) *Not a NE* (N-NE), in the case where the NE extraction is incorrect.

Phase 2 (*DBpedia*): Each NE extraction was automatically checked against DBpedia, and it was labelled as either (i) an Exact Match (EM) when a DBpedia result is found whose label is identical to the extracted NE string, (ii) a Partial Match (PM) when a DBpedia result is found whose label is part of the extracted NE or vice versa e.g. the extracted NE is 'Rhodes & Rhodes' which is a company name and the most similar DBpedia result label is 'Rhodes' which is a location name; or (iii) a No Match (NM) when there are no related DBpedia results.

Table 3 shows the results of the manual evaluation and DBpedia automatic evaluation. From the 1172 extracted instances we found that approximately 34.3% had a DBpedia result whose label exactly matches the string of the NE extracted by our method. From the exactly matching DBpedia results 24% were Accurately extracted NEs while 1.7% were inaccurate NE extractions.

Using the results presented in Table 3, we can evaluate DBpedia as a source for validating NER using the manually checked sample as a gold standard list to which the DBpedia results are compared. For this we have assumed the following

Table 3. Manual evaluation versus DBpedia results (%)

		DBpedia			
		Exact Match	Partial Match	No Match	Total
Manual Evaluation	Accurate NE	24.317	17.235	22.184	**63.736**
	Inaccurate NE	1.792	5.546	5.546	**12.884**
	Not NE	8.191	11.433	3.754	**23.378**
	Total	**34.300**	**34.214**	**31.484**	

definitions of True and False Positives (TP, FP), and True and False Negatives (TN, FN):

TP: A candidate extraction by the system that is a NE according to the manual evaluation and has a corresponding DBpedia result.

TN: A candidate extraction that is not a NE and has no DBpedia result.

FP: A candidate extraction that is not a NE yet has a DBpedia result.

FN: A candidate extraction that is a NE and has no DBpedia result.

In the evaluation, each examined instance is True or False according to the manual evaluation, and the same instance is labelled as Positive when DBpedia has a result related to it, and is labelled Negative otherwise. Two evaluation settings were used: *Evaluation I* is conducted regardless of the boundaries of the candidate NE extracted and how similar it is to the DBpedia result label, while in *Evaluation II* the extracted NE should be accurate and should match the DBpedia result label exactly. Accordingly TP, TN, FP, and FN are counted for each evaluation as illustrated in Table 4.

Table 4. Breakdown of Evaluation I and Evaluation II

Evaluation I

		DBpedia		
		EM	**PM**	**NM**
	A-NE	TP	TP	FN
Manual	**I-NE**	TP	TP	FN
	N-NE	FP	FP	TN

Evaluation II

		DBpedia		
		EM	**PM**	**NM**
	A-NE	TP	FN	FN
Manual	**I-NE**	FP	TN	TN
	N-NE	FP	TN	TN

Table 5, presents the precision, recall, and F-measure of the evaluation of DBpedia. In Evaluation I, we found that 71.4% of the related DBpedia results labeled (EM, PM), were similar to candidate NEs that are accurately or inaccurately detected by our system. In Evaluation II, an instance is considered correct (TP) only when the DBpedia result label exactly matches the accurately extracted instance resulting in a precision of 70.9%. On the other hand, recall, which is the fraction of the extracted NEs that were found in DBpedia, was 63.8% where I-NE extraction and PM DBpedia results are considered correct, otherwise the recall dropped to 38.6% when only those whose label exactly matches the DBpedia result are considered correct.

Table 5. Evaluation of Dbpedia as a source for NER

	Precision $\frac{TP}{TP+FP}$	Recall $\frac{TP}{TP+FN}$	F-Measure $2 \cdot \frac{Precision \cdot Recall}{Precision + Recall}$
Evaluation I	0.714	0.638	0.674
Evaluation II	0.709	0.386	0.496

Given that the sample is representative of the query, the system's performance can be estimated when it is tested on a larger query set using Table 3. For example, the percentage of accurate NE extractions that have exact matching DBpedia results is approximately 71% (24.317 divided by 34.3). Out of all extractions that had an exact match in DBpedia, we can estimate that approximately 24% are Not NEs (8.191 divided by 34.3). As observed from the sample, this is due to the ambiguity of some NEs. For example, the common noun 'meatballs' was incorrectly extracted as a candidate NE from the query 'gif or picture of spaghetti and meatballs'. Although it was labeled as Not NE in the manual evaluation, there was a DBpedia result whose label exactly matches it referring to a Canadian comedy film. Furthermore, we can assume that approximately 70% of candidate NE extractions that had no DBpedia results are accurately extracted NEs (22.184 divided by 31.484). In the sample, the NEs that fall in this category include non-celebrities, facility names such as restaurants and hotels, and small companies and businesses.

Running our candidate NER system on a larger uniform random sample of 100,000 queries resulted the extraction of 110,113 candidate NEs. Checking each automatically against DBpedia, we found that the percentages of the DBpedia evaluation cases (EM, PM, and NM) were 34.967%, 33.800%, and 31.231% respectively. These are close to the results found in the 1000 query sample, shown on the last line of Table 3. This provides further evidence that the above estimates are representative of the query set.

6   Query Log-Driven Named Entity Taxonomy

Early work on named entity recognition and classification used coarse-grained NE classes such as person, location, organisation, which have been referred to as "enamex" since the Message Understanding Conference (MUC-6). For domain specific applications, new named entity classes were studied, such as disease names, project names, proteins, and addresses. Sekin and Nobata [11] designed a hierarchal named entity taxonomy consisting of 200 types of named entities that targeted open-domain applications, consisting of very specific named entity types such as tunnel bridge, spaceship, and color. When NER was first applied to search queries, the named entities classes chosen were only used to evaluate the specific methods proposed. The classes chosen were either fine-grained classes such as athlete, actor, as in [8], cities, countries and drugs as in [10], car models as in [4], and movies, games, and books as in [6], or more general classes such as location and person as in [10].

As far as we know, there has been no general study to date of the types of named entities that users would generally use in search queries, and an estimation of each class frequency. Given our definition of candidate named entities, we can provide some insight of the named entity classes that users usually express in queries. Thus in this section we examine the classes of named entities that were correctly extracted by our method and compare it with the DBpedia ontology classes for the exactly matched instances.

6.1   Manual Annotation of Named Entities

We manually annotated each correctly extracted candidate NE that is either manually checked as A-NE or I-NE with a NE class, by examining the context in which it appears in both the query and the corresponding snippets. Table 6 presents the types of NEs and their percentages detected in the 1000 query sample. The most frequent NE type observed was 'Location' that include names of, for example countries, cities, towns, districts, and province. In the sample, Location NEs were usually associated with NEs of type Facility such as 'riviera maya hotel' or common facility names such as 'resorts in tulum'. All NEs such as hotels, resorts, hospitals, universities, and restaurant were manually annotated as being type of Facility. In addition, NEs of type Person include all athletes, actors, historical figures, TV presenters, and people's names found in Linkedin or Facebook. The NE type TV includes channels, series, and programs, while Music includes album, song, and band names. The list of NE types in Table 6 represents 96.89% of the candidate NEs extracted from the 1000 query sample. Approximately 3% were labelled as Other which include instances such as:

(a) Animal species such as 'yellow finches', and 'bluetick coonhounds';
(b) Events such as 'World War I';
(c) Languages and Nationalities that are proper nouns such as 'Latin', 'English', 'Chinese'; and
(d) Religions as well are proper nouns that are detected as candidate NEs.

Due to their very small overall percentage compared to the more frequent NE types these were labeled as Other.

Table 6. Manual analysis of candidate named entities classes

NE Class	Example	%
Location	alabama, australia, georgia	26.8293
Facility	casa grande restaurant, hard rock casino & resort	18.1818
Company	hummel bros, lexus, nelson industries	15.5211
Person	ben affleck, christina aguilera, donna lopez	10.7539
Product	1997 chevrolet monte carlo, ipod, motorola 6208	7.0953
Website	careerbuilder, gamespot, msn	3.6585
Software	windows xp, powerpoint, corel photo album 6	3.2151
TV	abc, american idol, sienfield	2.439
Music	arctic monkeys, blue, pink floyd	2.3282
Medical	crohns, dawson's, juvenile diabetes	1.7738
Video Game	halo 2, final fantasy, god of war cheats	1.2195
Newspaper/Magazine	arizona, graham newspaper, the detroit	1.1086
Sport	major league baseball all-star game, nascar 1983	1.1086
Movie	curse of the omen, little mermaid, match point	0.8869
Book	marvel mangaverse	0.7761
Other	Animal, Scientific Concept, Event, Language	3.1042

In order to compare the query log-driven NE taxonomy to the DBpedia ontology, the class of each DBpedia result whose string matches a candidate NE is analysed in the following section.

6.2   DBpedia Annotation of Extracted Candidate Named Entities

The DBpedia ontology has a hierarchical structure, for example, the extracted NE 'vivien cardone' is classified as an actor that is an artist, that is a person, and is finally classified to a root class, thing. Table 7 presents the percentages of NE instances that are classified by the DBpedia ontology for the 24% instances whose manual evaluation is an 'Accurate NE' and have DBpedia results whose label exactly matches the extracted NE (see Table 3).

Table 7. Exactly matched DBpedia results classification

Class Name	%	Class Name	%	Class Name	%
administrative region	13.475	athlete	13.475	beverage	13.475
city	7.801	broadcast	7.801	body of water	7.801
company	6.738	educational institution	6.738	disease	6.738
artist	4.255	musical work	4.255	film	4.255
band	3.191	website	3.191	language	3.191
actor	2.837	colour	2.837	mountain range	2.837
television show	2.482	comics character	2.482	organisation	2.482
country	2.128	office holder	2.128	reptile	2.128
place	2.128	software	2.128		
person	1.773	album	1.773		
		Total			58.657

Examining the classified instances, we found that only 58% of all instances that have an exactly matching DBpedia result were classified into the DBpedia ontology. Instances such as 'minnesota department of agriculture' 'fashion expo' 'qwest wireless' and 'garden city hotel' have a DBpedia result whose label exactly matches the extracted candidate NE, yet they are not classified.

Furthermore, other candidate NEs can belong to more than one class in the DBpedia ontology. This is due to the fact that NEs are analysed or evaluated without considering their context which may resolve such ambiguity. Table 8 shows the DBpedia classes for the instances of NE type Location and Person. Under the manual annotation for Location, we found that in addition to administrative region, city, and country, there were NEs such as 'Las Vegas' which could be a Location, as well as a TV series. In addition, for a query such as 'cheap tickets from okc to atl' the candidate NE 'atl' is classified by DBpedia as a film or a band name.

For each class in our taxonomy we checked the corresponding DBpedia ontology classification. Table 9 shows the percentage of instances that were correctly classified, misclassified, and not classified by the DBpedia ontology. Out of the 31.56% instances that were manually labelled as a NE of type Location, 24.8%

Table 8. Mapping the manual classification of NEs to the DBpedia ontology (%)

Manual Classification	Location		Person	
	Correct Classification		**Correct Classification**	
	administrative region	41.573	artist	28.5714
	city	24.719	actor	19.048
DBpedia	country	6.742	person	11.905
Ontology	place	4.494	athlete	9.524
Classification	body of water	1.124	office holder	4.762
	Misclassification		**Misclassification**	
	television show	2.247	administrative region	2.381
	film	1.124	band	2.381
	organisation	1.124		
	Not Classified	16.854	**Not Classified**	21.428

were correctly classified while 1.4% was misclassified. Instances such as NEs of the type Book had a DBpedia result that was not classified within the DBpedia ontology.

Table 9. Results of DBpedia and manual classifications (%)

Manual Classification		DBpedia Ontology Classification		
Class Name	Percentage	Correctly Classified	Misclassified	Not Classified
Location	31.56	24.823	1.418	5.319
Person	14.894	10.993	0.709	3.191
Facility	13.83	2.482	1.064	10.284
Company	12.766	6.738	0.000	6.028
Music	4.61	3.191	0.355	1.064
TV Show	4.255	2.128	0.000	2.128
Product	3.191	0.000	0.709	2.482
Software	3.191	0.709	0.355	2.128
Website	2.837	1.064	0.355	1.418
Medical	2.482	0.355	0.355	1.773
Video Game	1.064	0.000	0.355	0.709
Book	0.709	0.000	0.000	0.709
Newspaper/Magazine	0.355	0.000	0.000	0.355
Sport	0.355	0.000	0.000	0.355
Other	3.901	0.355	0.000	3.546
Total	100	52.838	5.675	41.489

To conclude, we have illustrated two main issues when evaluating a NER system using a source such as DBpedia as a gazetteer of named entities:

(a) The gazetteer may not contain the NE, which was also observed by Jain and Pennacchiotti in [8], where 61% of their NE extractions were correct although Wikipedia had missed them; and
(b) The candidate NE may belong to more than one NE class.

7 Concluding Remarks and Future Work

In this paper we have presented a method for candidate named entity extraction from web search query logs, using grammar annotation and query segmentation. Previously, named entity recognition approaches for query logs were evaluated out of context and without reference to the named entities that were missed by the system. Therefore, we have used two types of evaluations: (i) at query level, to evaluate extracted instances with respect to their context in the query, and (ii) at named entity level, where each instance extracted is evaluated both manually and using DBpedia. In addition, a taxonomy of named entities was devised, driven by a manually annotated representative random sample of queries.

We are currently working on a classification method for the extracted candidate named entities that utilises snippets to obtain context and to resolve the ambiguity of named entities that may belong to more than one NE class. Grammar annotation and query segmentation are both major techniques for analysing search queries that were used here to identify candidate named entities. Here we only measured their performance at the final stage of the named entity extraction. We leave as future work the find-grained evaluation of these methods, and how different approaches to these may affect the final results of the named entity extraction.

References

[1] Bendersky, M., Croft, W.B., Smith, D.A.: Joint annotation of search queries. In: Proc. of ACL, pp. 102–111 (2011)

[2] Bizer, C., Lehmann, J., Kobilarov, G., Auer, S., Becker, C., Cyganiak, R., Hellmann, S.: Dbpedia - a crystallization point for the web of data. Web Semant, 154–165 (2009)

[3] Brenes, D., Gayo-Avello, D., Garcia, R.: On the fly query segmentation using snippets. In: Proc. of CERI, pp. 259–266 (2010)

[4] Du, J., Zhang, Z., Yan, J., Cui, Y., Chen, Z.: Using search session context for named entity recognition in query. In: Proc. of SIGIR, pp. 765–766 (2010)

[5] Etzioni, O., Banko, M., Soderland, S., Weld, D.S.: Open information extraction from the web. Commun. of the ACM, 68–74 (2008)

[6] Guo, J., Xu, G., Cheng, X., Li, H.: Named entity recognition in query. In: Proc. of SIGIR, pp. 267–274 (2009)

[7] Hagen, M., Potthast, M., Stein, B., Bräutigam, C.: Query segmentation revisited. In: Proc. of WWW, pp. 97–106 (2011)

[8] Jain, A., Pennacchiotti, M.: Domain-independent entity extraction from web search query logs. In: Proc. of WWW, pp. 63–64 (2011)

[9] Moll, D., Zaanen, M., Cassidy, S., Ryde, N.: Named entity recognition for question answering. In: Proc. of ALTA, pp. 51–58 (2006)

[10] Paşca, M.: Weakly-supervised discovery of named entities using web search queries. In: Proc. of CIKM, pp. 683–690 (2007)

[11] Sekine, S., Nobata, C.: Definition, dictionaries and tagger for extended named entity hierarchy. In: Proc. of LREC, pp. 1977–1980 (2004)

[12] Yin, X., Shah, S.: Building taxonomy of web search intents for name entity queries. In: Proc. of WWW, pp. 1001–1010 (2010)

Automated Information Extraction from Web APIs Documentation

Papa Alioune Ly[1,2], Carlos Pedrinaci[1], and John Domingue[1]

[1] Knowledge Media Institute, The Open University
[2] School of Computer and Communication Sciences,
École Polytechnique Fédérale de Lausanne (EPFL)
{alioune.ly,carlos.pedrinaci,john.domingue}@open.ac.uk

Abstract. A fundamental characteristic of Web APIs is the fact that, de facto, providers hardly follow any standard practices while implementing, publishing, and documenting their APIs. As a consequence, the discovery and use of these services by third parties is significantly hampered. In order to achieve further automation while exploiting Web APIs we present an approach for automatically extracting relevant technical information from the Web pages documenting them. In particular we have devised two algorithms that automatically extract technical details such as operation names, operation descriptions or URI templates from the documentation of Web APIs adopting either RPC or RESTful interfaces. The algorithms devised, which exploit advanced DOM processing as well as state of the art Information Extraction and Natural Language Processing techniques, have been evaluated against a detailed dataset exhibiting a high precision and recall–around 90% for both REST and RPC APIs–outperforming state of the art information extraction algorithms.

Keywords: Web API, RESTful service, Web Page Segmentation, Information Extraction, Service Discovery.

1 Introduction

On the Web, service technologies are currently marked by the proliferation of Web APIs, also called RESTful services when they conform to REST principles [1]. Major Web sites such as Facebook, Flickr or Amazon provide access to their data and functionality through Web APIs. This trend is impelled by the simplicity of the technology stack, compared to WS-* Web services [2], as well by the simplicity with which such APIs can be offered over preexisting Web sites infrastructure [3]. Thanks to these APIs, we have seen in the last years a proliferation of applications and Web sites that exploit and combine them to provide added-value solutions to users.

When building a new service-oriented application, it is fundamental to be able to swiftly discover existing services or APIs, to figure out what operations and resources they offer as well as to understand the ways in which one can use them. Unfortunately, supporting the aforementioned steps over the Web APIs one can

X.S. Wang et al. (Eds.): WISE 2012, LNCS 7651, pp. 497–511, 2012.
© Springer-Verlag Berlin Heidelberg 2012

find on the Web is most challenging nowadays since most often the only way to carry this out requires interpreting highly heterogeneous HTML pages that are intended for humans and that provide no convenient means for supporting their identification and interpretation by machines. In fact, although REST principles contemplate mechanisms that could help circumvent these issues, e.g., the use of *Hypermedia as the Engine of Application State* (HATEOAS), our previous research revealed that REST principles are seldom strictly followed [4]. Similarly, although there have been a number of languages and formalisms suggested for explicitly describing Web APIs very few API providers actually use them. Thus, while providing support for discovering and invoking Web services and their operations can directly be solved by locating and parsing WSDL documents, discovering Web APIs is far more complex a task that is yet to be adequately addressed.

Driven by the aforementioned observations in our ongoing work on iServe [3], a public platform for service publication and discovery, we are approaching the discovery of Web APIs as a three steps activity, whereby we first carry out a targeted crawling of HTML Web pages providing the technical documentation of Web APIs, we subsequently analyse the pages obtained to detect and extract relevant technical information, e.g., operation names, URLs, etc, and we ultimately provide search functionality over the extracted information. In earlier research we have reported our approach and the results obtained in the first step [5]. In this paper we focus on the second step which is in charge of processing the obtained Web pages to extract as much information as possible that shall serve as a basis for advanced Web API discovery. In particular, we present two novel algorithms for supporting the extraction of key information from Web pages documenting Web APIs providing both Remote Procedure Call (RPC) and RESTful interfaces. The algorithms combine state of the art Information Extraction and Natural Language Processing (NLP) techniques to extract features such as operation names, operation descriptions or URI templates. Our algorithms have been evaluated against a dataset containing 40 highly heterogeneous Web APIs documentation pages covering both RPC and RESTful APIs. The evaluation highlights the superior performance of our approach over state of the art Information Extraction algorithms, reaching a precision of 89% when detecting operations and their corresponding information for RPC Web APIs, and 93.4% when detecting blocks describing methods[1] for RESTful Web APIs.

The remainder of this paper is organised as follows. In Section 2 we cover the related work and provide additional background information relevant for the topic at hand. We next explain the approach we have followed and describe the algorithms we have developed. In Section 4 we present the experiments we have carried out and discuss the evaluation results, and finally in Section 5 we present the main conclusions we have drawn and introduce lines for future research.

---

[1] In Section 3 we explain what we understand by method block in this context and why it is relevant.

2   Background and Related Work

2.1   Service Discovery

Service discovery has been the subject of much research and development. The most renown work is perhaps Universal Description Discovery and Integration (UDDI) [2], while nowadays Seekda[2] provides the largest public index with about 29,000 WSDL Web services. Research on semantic Web services has generated a number of ontologies, semantic discovery engines, and further supporting infrastructure over the years, see [6] for an extensive survey. Despite these advances, however, the majority of these initiatives are predicated upon the use of WSDL Web services, which have turned out not to be prevalent on the Web where Web APIs are increasingly favoured [3].

A fundamental characteristic of Web APIs is the lack of standardisation both concerning their implementation as well as concerning their publication and documentation. An analysis we carried out manually highlighted the heterogeneity existing both in terms of the type of interfaces provided with only 32% apparently RESTful, and the remaining 68% being either purely RPC or hybrid [4]. Additionally, although a few providers do follow REST principles like HATEOAS that enable to some extent the automated discovery and invocation of these services, most do not and provide instead weakly semistructured Web pages documenting the APIs with a highly variable degree of detail (e.g., the HTTP method used is not always indicated, etc) [4]. Furthermore, while the documentation of a single Web API typically follows a certain pattern locally, the structure adopted by different APIs documentation pages is highly heterogeneous which prevents the direct application of general pattern-based solutions. To address these issues, researchers have proposed a number of languages and formalisms for describing APIs, e.g., WADL [7] or for annotating the Web pages documenting them such as SA-REST [8] and hRESTS/MicroWSMO [9]. However, their adoption remains minimal outside academic environments.

As a consequence, there has not been much progress on supporting the automated discovery of Web APIs. The main means used nowadays by developers for locating Web APIs are the use of traditional search engines like Google or searching through dedicated and registries. The most popular directory of Web APIs is ProgrammableWeb[3] which, as of June 2012, lists about 6,200 APIs and provides rather simple search mechanisms based on keywords, tags, or a simple prefixed categorisation. Based on the data provided by ProgrammableWeb, APIHut [10] increases the accuracy of keyword-based search of APIs compared to ProgrammableWeb or plain Google search. Unfortunately, on the one hand, general purpose Web search engines are not optimised for this type of activity and often mix relevant pages documenting Web APIs with general pages, e.g., blogs. On the other hand, current registries provide more focussed information, but still present a number of issues. First and foremost, more often than not, these registries contain out of date information or even provide incorrect links to

---

[2] http://webservices.seekda.com/
[3] http://www.programmableweb.com

APIs documentation pages. Indeed, the manual nature of the data acquisition in APIs registries aggravates these problems as new APIs appear, disappear or change. Secondly, the fact that the data listed is often not that accurate and rather coarse grained hampers significantly the development of advanced search functionality since automated algorithms are mislead and miss relevant information such as the operations provided.

Therefore, despite the increasing relevance of Web APIs, there is hardly any system available nowadays that is able to adequately support their discovery. The main obstacles in this regard concern first of all, the automated location of Web APIs, and subsequently, the gathering, interpretation and extraction of relevant information concerning these APIs which is the main focus of this paper. In this regard, to the best of our knowledge besides our own work we are only aware of another initiative which has carried out some initial steps in a similar direction [11]. However, at the time of this writing, details on the experiments and the results obtained are hardly available and, according to the authors, require further refinement.

2.2   Web Page Analysis and Information Extraction

As previously argued, obtaining the necessary information for discovering or invoking the vast majority of Web APIs nowadays requires interpreting highly heterogeneous HTML pages providing documentation for developers. Although, to the best of our knowledge, no other approaches to automating the extraction of information from Web APIs documentation have been devised so far, considerable effort has been devoted to extracting information from Web pages which is relevant to this work.

Tag-based Segmentation approaches are used to analyse the DOM tree of Web page and automatically divide it into subtrees in order to eventually extract information. The essence of this approach relies on the observation that useful information is usually wrapped into so-called *important blocks*. These techniques usually rely on specific HTML tags, e.g., `<table>` in the case of [12], as block separators. This approach is, however, not performant when dealing with heterogeneous cases where various tags are used as separators. To address this, proposals like [13] contemplate a wider range of tags e.g., `<tr>`, `<hr>` .

Template-based Segmentation techniques rely on the fact that, for repetitive information Web pages often use a recurring structure to capture information [15]. These approaches exploit templates provided either by humans or derived automatically by machine learning techniques in order to extract information from the Web pages. Although, these techniques are performant when dealing with Web pages that share (at least partly) a common structure, in cases where the heterogeneity of the pages is very high the performance is considerably affected. As we shall see, although it is possible to exploit local patterns within a Web API documentation, currently diverse Web APIs use highly diverging structured which prevents us from applying successfully this approach.

Vision-based Segmentation is another popular family of approaches that exploits the technique the "visual" boundaries between blocks of content as a means to segment Web pages into blocks of information. In [16] the authors present the popular Vision-based Page Segmentation Algorithm (VIPS) which views Web pages as (a collection of) images and traverses the DOM tree detecting the "visual" boundaries between blocks. VIPS's granularity of segmentation is based on a *predefined degree of coherence* (PDoC) which acts as a lower bound for the computed *degree of coherence* (DoC) of the identified page blocks. A number of researchers have used VIPS as the underlying page segmentation algorithm, see for instance [17–19]. One weakness of these approaches is due to the fact that the visual boundaries of Web pages are often not clear cut. Additionally, solutions based on VIPS sometimes face granularity problems, which may lead to either too granular segmentations when the PDoC is high, or too coarse grained segmentations when the PDoC is low.

3   Information Extraction from Web API Documentation

Given a Web page documenting a Web API our approach to extracting these features consists on two main steps. On a first step, as is common practice when processing Web pages, we pre-process the Web page to i) circumvent any issues with incorrect HTML pages[4], and ii) to remove scripts and images which will not be taken into account for feature extraction. The second step, takes care of processing and analysing the cleaned Web pages to extract the main technical features of the Web APIs such as the operation names, URI templates, etc.

As we introduced earlier, currently two main styles for implementing Web APIs coexist on the Web, i.e., REST and RPC. Both styles embed a number of architectural and design decisions that condition the way service interfaces are defined and documented, and which, as we shall see, condition the kinds of techniques we can successfully apply to extract information. In the remainder of this section we describe the approach and algorithms we have devised for extracting information from each of these types of interfaces.

3.1   Information Extraction from RPC-Style Web APIs

RPC is a communication style that is centred on the notion of operations or procedures which essentially define the actions that remote components can trigger. Web APIs adopting this style of communication offer programmatic access to the data and functionality hosted on the underlying Web site by means of an arbitrary number of operations. Additionally, the documentation sometimes indicates further details such as the invocation endpoint (a URL) or the HTTP method to be used. Figure 1 shows two examples with the relevant blocks of information extracted from RPC-style Web APIs whereby dashed border rectangles represent the operation identifiers and plain border rectangles represent the information blocks that provide further details, e.g., parameters, and describe what the operations do in natural language. Even though the blocks are

---

[4] We used the Jericho Parser `http://jericho.htmlparser.net/docs/index.html`

both taken from RPC-style Web APIs documentation, the underlying structure and their visual appearance differ significantly, providing an example of the heterogeneity among Web APIs documentation.

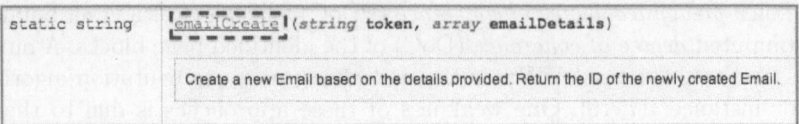

(a) Block extracted from http://www.benchmarkemail.com/API/Library

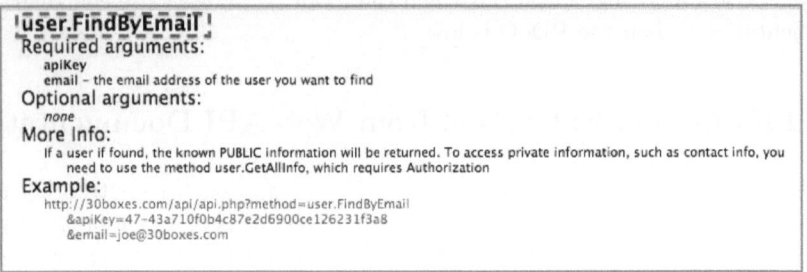

(b) Block extracted from http://30boxes.com/api/

Fig. 1. Examples of block content in RPC-style Web APIs

In extracting relevant information from RPC-style Web APIs the essence of our approach in centred on the detection of blocks providing information about the different operations. This detection relies on two main characteristics we have observed while analysing a plethora of Web APIs documentation. First and foremost, we exploit the fact that operation identifiers typically use words in CamelCase[5] notation whereby the first part is a verb and the rest a noun, e.g., *getArtist*. Secondly, since documentation pages are designed for humans, a single Web API documentation often uses some recurrent visual clue to provide a distinct visual appearance to the different elements, e.g., operation names are all within an <H3> tag, the description within a <p> tag, etc. Thus, although structural patterns across Web APIs documentations cannot be exploited given the existing heterogeneity we previously highlighted, we do utilise the fact that there typically exist local patterns within the same API documentation.

Algorithm 1 details how we extract operation description blocks from RPC-style Web APIs. In a first step we extract all the CamelCase words we can find from the page and track the immediate tag they belong to. In this step, see line 4, we only keep track of the CamelCase words that are composed of a verb and a noun, which, as we shall see in Section 4, is a significant improvement. To this end we use the Log-linear Part-Of-Speech Tagger implemented by the Stanford NLP group[6]. This tagger reads text in english and assigns a part of speech to

---

[5] http://en.wikipedia.org/wiki/CamelCase
[6] http://nlp.stanford.edu/software/tagger.shtml

each word such as verb, noun, and adjective. Doing so allows us to distinguish *getArtist* which denotes potentially an operation identifier as it is made of the verb *get* and the noun *Artist*, from *pageNumber* which is composed of two nouns, *page* and *Number*, and should hence be discarded.

```
input  : Source page of the Web API
output : Operation description blocks
1  foreach tagContent in page source do
2  |    camelCaseList ← tagContent.getCamelCaseWords();
3  |    foreach ccWord in camelCaseList do
4  |    |    if isOperation(ccWord) then
5  |    |    |    tagMap.Add(Pair<tagName, ccWord >);
6  |    |    end
7  |    end
8  end
9  electedTag ← getMostPopularTag(tagMap);
10 operationList ← getOperations(tagMap, electedTag);
11 operationMap ← ∅;
12 foreach tagContent in page source do
13 |    operations ← tagContent.getCamelCaseWords() ∩ operationList;
14 |    foreach ccWord in operations do
15 |    |    if getTag(ccWord) == electedTag then
16 |    |    |    if getTag(other words in operations) ≠ electedTag then
17 |    |    |    |    operationMap.Add(ccWord → tagContent);
18 |    |    |    end
19 |    |    end
20 |    end
21 end
22 return operationMap;
```

Algorithm 1: Block detection for RPC-style Web APIs.

In a second step, we find out the HTML tag that has most commonly been used for the operation identifiers candidates found in the first step. This tag, which we refer to as the *elected tag*, is obtained by function `getMostPopularTag` in line 9. Next, out of all the operation identifiers candidates, we only retain the operation identifiers candidates that were within the elected tags. Finally, the tag scope for each of the operations found is eventually used in order to segment the page into blocks of additional information related to a single operations (e.g., operation description, parameters, etc), see Figure1 where the outer rectangle delimits a single block and the dashed inner rectangle highlights the operation indentifier detected.

3.2   Information Extraction from RESTful Web APIs

Web APIs that conform to REST principles [1] are characterised by a communication model whereby requests and responses are built around the transfer of representations of resources and a prefixed set of operations one can carry over these. The documentation of RESTful Web APIs is thus centred on the notion of resource and the parameters required for identifying these resources as well as on the actual operations allowed over the resources which in the case of Web APIs is indicated by the HTTP method to be used. Figure 2 shows two examples of blocks of information extracted from RESTful Web APIs whereby the inner dashed rectangles represent the resources URI (templates), circles identify the HTTP method to be used and the outer rectangles capture the entire segment of information.

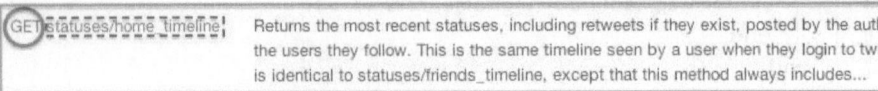

(a) Block extracted from https://dev.twitter.com/docs/api

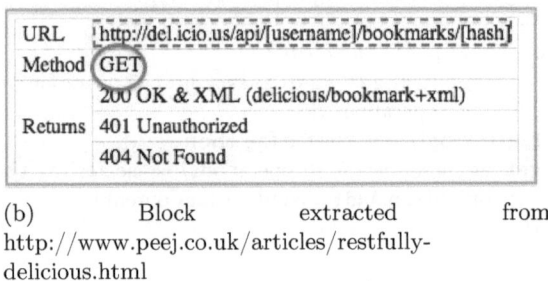

(b) Block extracted from http://www.peej.co.uk/articles/restfully-delicious.html

Fig. 2. Examples of block content in RESTful Web APIs

Given the characteristics of RESTful interfaces, the main goal in this case is to detect and extract the resources, their URIs (templates) along with the HTTP method to be used and the corresponding description (see plain border rectangles in Figure 2). As opposed to the previous case, naming conventions here are hardly usable since there is no usual criterion for defining resource identifiers. Instead, the most characteristic elements in this case are URI templates and HTTP methods although they occur in several places throughout the documentation (e.g., one may GET and POST a resource) which hampers significantly the segmentation of the Web page into coherent sets of resources and operations. In order to deal with this, we exploit in this case the fact that within a single API developers often adopt the same repetitive pattern for documenting each resource and the operations available.

In particular, our algorithm, see Algorithm 2, first analyses the structure of the API documentation in order to segment the Web page into similar structural blocks and keeps those blocks that are believed to provide technical documentation since they contain URIs and, whenever available, HTTP methods, e.g., GET, POST, PUT, DELETE, see line 1. In a second step, see lines 2 onwards, the algorithm computes the similarity of the blocks detected in order to retain those

```
   input  : Source page of the Web API
   output : URI description blocks
 1 blockPattern ← detectBlockPattern();
 2 currentEntropy ← 1; oldEntropy ← 2;
 3 electedDescriptionMap ← ∅; descriptionMap ← ∅;
 4 while currentEntropy < oldEntropy do
 5 │   descriptionMap ← ∅;
 6 │   oldEntropy ← currentEntropy;
 7 │   foreach tagContent in page source do
 8 │   │   tagStructure ← getStructure(tagName);
 9 │   │   if tagStructure starts with blockPattern then
10 │   │   │   while not tagContent contains a URI do
11 │   │   │   │   Tag ← parentTag; Update tagContent;
12 │   │   │   end
13 │   │   │   descriptionMap.Add(tagStructure → tagContent)
14 │   │   end
15 │   end
16 │   structureList ← descriptionMap.keySet();
17 │   currentEntropy ← getEntropy(structureList);
18 │   if currentEntropy < oldEntropy then
19 │   │   electedDescriptionMap ← descriptionMap;
20 │   end
21 end
22 return electedDescriptionMap;
```

Algorithm 2: Block detection for RESTful Web APIs

that appear to exhibit the same structure. Once the blocks have been filtered, we extract the internal features contained, e.g., URIs, HTTP methods, etc, and treat them as a coherent set of information related to a given resource and the operations it offers.

To compute the similarity of the blocks detected the algorithm exploits the notions of **entropy** and **node internal structure**. Entropy calculation is used to quantify strong local patterns exhibited in a page segment, whereby a high entropy indicates important disorder and a low entropy indicating strong similarity.

The main concepts underlying the notion of entropy in our approach are as follows. Given a set of nodes in the DOM tree structure of a Web page, f_i denotes the frequency with which the node name i appears and p_i the probability of node i.

p_i is hence calculated via the formula:

$$p_i = \frac{f_i}{n} \tag{1}$$

where, n is the total number of nodes. Using equation (1), the entropy of node v is defined as:

$$E(v) = -\sum_{i \in V} \frac{p_i log(p_i)}{log(|V|)} \qquad (2)$$

where, V is the set of unique node names that appear in the subtree of the DOM rooted at node v. We can easily extend this concept in calculating the entropy of a list of elements.

We consider the **internal structure** of a node A as being the concatenation of the node name with all of its children traversed top-down left-right in the subtree rooted at A, e.g if we consider part of the source code being `<div><a>title</a><p>body of the div <br></div>`, the internal structure of the `<div>` node is `<div><a><p><br>`.

In order to detect repetitive block patterns within the structure of the HTML page, see function `detectBlockPattern` on line 1, we base our approach on the ideas behind the "Repetition-based Web Page Segmentation" (REPS) algorithm [19] which proposes a flexible approach for recognising repetitive tag patterns in the DOM tree structure of a Web page, and the approach proposed in [20] by Gujjar Vineel who proposed a DOM tree mining approach for page segmentation where he introduced the concept of *page segment nodes* that would characterise nodes that divide the DOM Tree into logical blocks.

In essence, we use the approach in [20] to segment the Web page into logical blocks (e.g., header, footer, left column, right column, content), then in order to find repetitive patterns within the DOM we exploit the approach proposed by REPS and apply it on the each of the logical blocks. This way, we can keep track of logical blocks that turn out to have repetitive patterns with URI within. Eventually, we keep the most repeated pattern and refer to it as *blockPattern*. Therefore, we can exploit the patterns detected to divide the logical block containing the *blockPattern* into sub items that should in principle represent the blocks describing each resource and their operations.

4   Evaluation

In order to evaluate our approach, we manually generated a dataset based on the documentation of 40 Web APIs. The APIs selected were such that 20 had an RPC interface and the other 20 a RESTful interface. APIs) and documentations within each of these were had a significantly different structure (e.g., documentation structured on the basis of tables, HTML headers, flat documentations, etc). Each of the selected Web API documents were manually analysed in order to extract the information captured within them. The resulting dataset contains 355 URIs for resources of RESTful Web APIs and 515 operations for RPC-style Web APIs. For RPC-sytle Web APIs, we stored for each encountered operation: i) the URL of the Web API documentation, ii) the operation identifier, and iii) the description block that also contains the (required) parameters, and other information regarding what the operation does. For RESTful Web APIs we captured for each resource: i) the URL of the Web API documentation, ii) the URI (template), iii) the HTTP method, and iv) the description block. The interested reader can find concrete details about our dataset as well as subsequent evaluations of our algorithms in the website of iServe[7].

On the basis of the dataset generated, we evaluated the algorithms developed. Notably, we applied the algorithms over each Web API documentation to automatically extract the main technical features of the Web APIs. We then computed the overall precision and recall [8] by contrasting the automatically extracted information with the one we manually extracted and recorded in the dataset.

4.1   Evaluation Results for RPC-Style Web APIs

In order to evaluate our algorithm for extracting information from RPC-style Web APIs documentation we tested our own algorithm against the dataset and compared it with other 2 techniques that we used as baseline. Given that no other information extraction techniques have been devised for this purpose, we compared the different techniques with respect to their accuracy for detecting operation identifiers. The subsequent extraction of the operation description block was in all cases based on the same common procedure we devised for Algorithm 1 taking the detected operation identifiers as a starting point. The first technique–T_1–was solely based on the detection of CamelCase words, whereas T_2 was based on the combination of CamelCase word detection and the use of Part-of-Speech tagging to filter out the operation identifiers that were not composed of a verb and a noun. Finally, our algorithm which extends T_2 with the notion of *elected tag*, see Algorithm 1, is referred to in our evaluation as T_3. Table 1 provides a summary of the results for the three techniques. Regarding the results obtained for technique T_3, at the time we were experimenting this approach, page 20[9] could not be tested as it was no longer documenting a Web API. This shows how fast references to Web APIs documentation are changing which motivates further the work presented herein.

Overall, T_1 obtained an average precision of 54.68% and an average recall of 94.55%. These results confirm the fact that most operation identifiers are CamelCase words (see the high recall), but also the fact that there are many CamelCase words that are not necessarily operation identifiers (hence the low precision). A detailed look at the extracted operation names for instance included words like USgov, pageNumber and LinkedIn. The second technique, which exploits Part-Of-Speech tagging, exhibited a significant precision improvement over T_1 of about 20 percentage points, with a minor decrease in recall of about 3.5 percentage points. These results thus highlight the considerable benefits that are obtained by using Part-Of-Speech tagging, but still the overall precision, 75%, remains rather low. Finally, our algorithm exhibited the best precision with 92.71%, i.e., an increase of 38 percentage points over T_1 and 17 over T_2. Similarly, the overall recall exhibited by our algorithm, although minimally worse

---

[7] http://iserve.kmi.open.ac.uk/datasets/apis-information-extraction.html
[8] http://en.wikipedia.org/wiki/Precision_and_recall
[9] http://api.conceptshare.com/API/API_V2.asmx

than the one for the other techniques, remains very good with 90.59%. This result thus highlights the fact that using the notion of *elected tag* which exploits the recurrent use of tags for structuring documents is a good means to fine tune the extraction of information from RPC-style Web APIs documentation. The results illustrate that, by combining CamelCase word detection, Part-of-Speech tagging and the notion of *elected tag*, we succeeded in getting better precision and recall as well as on providing a more stable algorithm.

Table 1. Operation detection from RPC-style Web APIs

Tech	Pages																				Avg
	1	2	3	4	5	6	7	8	9	10	11	12	13	14	15	16	17	18	19	20	
T_1	46.8	31.2	100	71.4	37.5	58.3	80	66.6	62.5	100	65.9	78	27.6	16.6	26	12.	9.5	66.2	90.4	46.1	54.68%
T_2	68.1	62.5	100	92.6	72.7	87.5	100	83.3	83.3	100	90.6	92.9	44.5	31.8	46.2	26.2	50	88.6	95	85.7	75.08%
T_3	68.2	100	96	88.9	93.3	100	83.3	100	100	96.9	97.5	97	63.6	81.8	100	100	94.7	100	100	-	92.71%

(Precision)

Tech	Pages																				Avg
	1	2	3	4	5	6	7	8	9	10	11	12	13	14	15	16	17	18	19	20	
T_1	68.2	83.3	98.9	100	90	93.3	100	100	100	100	86.1	100	94.2	100	100	94.4	100	82.4	100	100	94.55%
T_2	68.2	83.3	76.3	100	80	93.3	100	83.3	100	100	80.5	100	92.9	100	92.3	88.9	100	81.6	100	100	91.04%
T_3	57.7	83.3	100	80	93.3	100	83.3	100	100	77.5	100	92.9	100	69.2	90	100	93.9	100	100	-	90.59%

(Recall)

Based on the operations identifiers detected, we also evaluated the performance of our algorithm for detecting the entire operation description blocks, see lines 11 to 21 in Algorithm 1. The results were very similar with a minor decrease both in precision and recall, notably for T_3 we obtained a good performance with a precision of 89.09%, and a recall of 84.56%. This slight performance decrease when extracting operation description blocks is due to the fact that in a few cases the Web pages do not strictly partition operations' details into different blocks. In these cases, the next level up in the DOM structure with respect to the tag holding the operation identifier contains several operation description blocks rather than just the one sought.

4.2   Evaluation Results for RESTful Web APIs

Like in the previous case we evaluated our algorithm for RESTful APIs and compared it with 2 techniques that we used as baseline. Given that no other information extraction techniques have been devised for this purpose, the comparison was based in this case on the application of diverse heuristics for detecting block patterns since this is the key step prior to extracting the relevant information about resources and how to interact with them, see line 1 of Algorithm 2. The first technique–T_4–was based on the direct application of the approach in [20] for block pattern detection. T_5, on the other hand was based on the use of REPS [19] exclusively. Finally, our algorithm which combines RESP and the approach in [20] to detect block patterns, see Algorithm 2, is referred to in our evaluation as T_6. The subsequent processing exploiting the notion of entropy to select the most promising blocks was in all cases based on our own algorithm.

Table 2 summarises the results obtained in terms of precision and recall when applying T_4, T_5 and T_6 on the RESTful APIs recorded in our dataset. The results obtained show that our approach outperforms state of the art techniques (i.e. T_4 and T_5) in both terms of precision and recall. Notably, we obtain an average precision of 93.4% and an average recall of 83.53% whereas for T_4 and T_5 obtain respectively average precisions of 88.91% and 92.52% and average recalls of 69.18% and 72.46%. The results illustrate that, in most cases, by combining ideas from [20] and [19] we succeeded in getting better precision and recall as well as on providing a more stable algorithm.

Table 2. Block detection from RESTful Web APIs

Precision

Tech	Pages																				Avg
	1	2	3	4	5	6	7	8	9	10	11	12	13	14	15	16	17	18	19	20	
T_4	100	**71.4**	**100**	100	100	50	100	**100**	100	23.7	100	100	100	100	10.8	100	100	**100**	100	100	88.91%
T_5	100	46	**100**	100	100	**100**	100	80	100	**100**	100	100	100	**100**	**100**	100	9.4	100	100	100	92.52%
T_6	100	27	93.8	100	100	**100**	100	85.7	100	48.2	100	100	100	100	**100**	100	100	**100**	100	100	93.4%

Recall

Tech	Pages																				Avg
	1	2	3	4	5	6	7	8	9	10	11	12	13	14	15	16	17	18	19	20	
T_4	100	100	3	80	**100**	100	17.4	84.6	65	**100**	28.2	33.1	31.7	100	100	100	53.6	100	100	25.2	69.18%
T_5	100	100	18.9	80	39.15	100	**17.6**	**92.3**	65	49.9	37	**45.5**	48.6	100	85.7	100	**96.4**	100	100	**92.6**	72.46%
T_6	100	100	**100**	80	**100**	100	10.8	**92.3**	65	**100**	44.3	33.8	**100**	100	85.7	100	**96.4**	100	100	**92.6**	83.53%

5   Conclusion

Despite the availability of a number of best practices, e.g., REST principles, and a plethora of software components and technologies, discovering and exploiting Web APIs requires a significant amount of manual labour. Notably developers need to devote efforts to interacting with general purpose search engines, filtering a considerable number of irrelevant results, browsing some of the results obtained and eventually reading and interpreting the Web pages documenting the technical details of the APIs in order to develop custom tailored clients.

In an attempt to provide further automation while carrying out these activities in this paper we have presented an approach for extracting automatically relevant technical information from Web pages documenting APIs. Notably, we have devised two algorithms that automatically and accurately extract Web APIs features such as operation names, operation descriptions or URI templates from Web pages documenting APIs adopting both RPC-style interfaces or RESTful interfaces. We have manually generated a detailed evaluation dataset which we have used to test and compare our approach. The evaluation results show that we achieved a high precision and recall–around 90% in both cases–outperforming state of the art information extraction algorithms.

In our future work we plan to apply our algorithms over a large scale Web crawl in order to develop a fully-fledged Web APIs search engine based on the features extracted. This work is expected to eventually support a greater search accuracy as well as finer grain discovery support than what is currently available nowadays.

References

1. Fielding, R.T.: Architectural Styles and the Design of Network-based Software Architectures. PhD thesis, University of California, Irvine (2000)
2. Erl, T.: SOA Principles of Service Design. The Prentice Hall Service-Oriented Computing Series. Prentice Hall (July 2007)
3. Pedrinaci, C., Domingue, J.: Toward the Next Wave of Services: Linked Services for the Web of Data. Journal of Universal Computer Science 16(13), 1694–1719 (2010)
4. Maleshkova, M., Pedrinaci, C., Domingue, J.: Investigating Web APIs on the World Wide Web. In: European Conference on Web Services (ECOWS), Ayia Napa, Cyprus (2010)
5. Lin, C., He, Y., Pedrinaci, C., Domingue, J.: Feature lda: a supervised topic model for automatic detection of web api documentations from the web. In: The 11th International Semantic Web Conference (ISWC), Boston, USA (2012)
6. Pedrinaci, C., Domingue, J., Sheth, A.: Semantic Web Services. In: Handbook on Semantic Web Technologies. Semantic Web Applications. Springer (2010)
7. Richardson, L., Ruby, S.: RESTful Web Services. O'Reilly Media, Inc. (May 2007)
8. Sheth, A., Gomadam, K., Lathem, J.: SA-REST: Semantically Interoperable and Easier-to-Use Services and Mashups. IEEE Internet Computing 11(6), 91–94 (2007)
9. Kopecky, J., Vitvar, T., Pedrinaci, C., Maleshkova, M.: RESTful Services with Lightweight Machine-readable Descriptions and Semantic Annotations. In: Wilde, E., Pautasso, C. (eds.) REST: From Research to Practice. Springer (2011)
10. Gomadam, K., Ranabahu, A., Nagarajan, M., Sheth, A.P., Verma, K.: A faceted classification based approach to search and rank web apis. In: ICWS 2008: Proceedings of the 2008 IEEE International Conference on Web Services, pp. 177–184. IEEE Computer Society, Washington, DC (2008)
11. Steinmetz, N., Lausen, H., Brunner, M.: Web Service Search on Large Scale. In: Baresi, L., Chi, C.-H., Suzuki, J. (eds.) ICSOC-ServiceWave 2009. LNCS, vol. 5900, pp. 437–444. Springer, Heidelberg (2009)
12. Lin, S., Ho, J.: Discovering informative content blocks from Web documents. In: Proceedings of the Eighth ACM SIGKDD International Conference on Knowledge Discovery and Data Mining, pp. 588–593 (2002)
13. Debnath, S., Mitra, P., Pal, N.: Automatic Identification of Informative Sections of Web Pages. IEEE Transactions on Knowledge and Data Engineering 17(9) (2005)
14. Chakrabarti, D., Kumar, R., Punera, K.: Page-level template detection via isotonic smoothing. In: Proceedings of the 16th International Conference on World Wide Web, pp. 61–70 (2007)

15. Hammer, J., Garcia-Molina, H., Cho, J., Aranha, R., Crespo, A.: Extracting Semistructured Information from the Web. In: Proceedings of the Workshop on Management of Semistructured Data (May 1997)
16. Cai, D., Yu, S., Wen, J.: Vips: a visionbased page segmentation algorithm. Technical Report MSR-TR-2003-79, Microsoft Research (2003)
17. Liu, Y., Wang, Q., Wang, Q., Liu, Y., Wei, L.: An Adaptive Scoring Method for Block Importance Learning. In: IEEE/WIC/ACM International Conference on Web Intelligence, WI 2006, pp. 761–764 (2006)
18. Wan, X., Yang, J., Xiao, J.: Block-based similarity search on the Web using manifold-ranking. In: Semantic Web: Research and Applications, Proceedings, Peking Univ, Inst Comp Sci & Technol, Beijing 100871, Peoples R China, pp. 60–71 (2006)
19. Kang, J., Yang, J., Choi, J.: Repetition-based Web Page Segmentation by Detecting Tag Patterns for Small-Screen Devices. IEEE Transaction on Consumer Electronics 56(2) (May 2010)
20. Vineel, G.: Web page DOM node characterization and its application to page segmentation. In: 2009 IEEE International Conference on Internet Multimedia Services Architecture and Applications (IMSAA), pp. 1–6 (2009)

Exploring Content Dependencies
to Better Balance Performance and Freshness
in Web Database Applications

Stavros Papastavrou[1], Panos K. Chrysanthis[2,*], George Samaras[1,**]

[1] Dept. of Computer Science, University of Cyprus, Nicosia, Cyprus
stavros@schoolfortheblind.net,
cssamara@ucy.ac.cy
[2] Dept. of Computer Science, University of Pittsburgh, Pittsburgh, PA, USA
panos@cs.pitt.edu

Abstract. In this paper, we present a novel approach for materializing dynamic web pages by exploiting content dependencies and user access patterns. We introduce two new semantic-based data freshness metrics and show that our approach out-performs traditional balancing QoS-QoD approaches in terms of server throughput, increased data freshness and scalability. In our evaluation we use a real-world experimental system that resembles an online bookstore web database application.

Keywords: Caching, Dynamic Web Content, QoD, QoS, Data Freshness.

1   Introduction

Our work focuses on e-commerce web applications such as an online bookstore. Those applications are implemented by dynamic web pages that are generated on-demand by executing resource-hungry template scripts that access local or remote databases to produce html content. Reportedly, billions of dollars are lost every year due to excessive delays in e-commerce web pages that force users to abandon their session [1]. The study in [2] presents a comprehensive and comparative listing of early approaches for enhancing QoS (user-perceived latency) under heavy workload in the blind expense of QoD (freshness of data served). Improving on [2], the approaches in [3, 4, 5, 6, 7, 8] attempt to balance QoS and QoD by re-using from the cache as much as necessary stale content in order to spare computational resources and boost QoS. However, an open challenge has been the quantification of data freshness (QoD) of content and how this can be traded with QoS. In other words, which pages or parts of pages (also known as *content fragments*) are "less important" at a given time for "that particular user" so that they can be re-used from cache.

*   Partially Supported by the USA National Science Foundation Award OIA-1028162.
** Co-funded by the EU Project CONET (INFSO-ICT-224053) & The Project FireWatch (#0609-BIE/09), Sponsored by the Cyprus Research Promotion Foundation.

X.S. Wang et al. (Eds.): WISE 2012, LNCS 7651, pp. 512–525, 2012.
© Springer-Verlag Berlin Heidelberg 2012

In this paper, we pose that current QoS-QoD balancing approaches fail to meet the requirements of modern Web database applications for the following two reasons:

— **Link Dependencies**. There is no consideration for the navigation needs of a user: If a content fragment is reused from cache, then it may be missing a needed valid html link for further user navigation at that given point in time. For example, a link on the upper right part of a web page may be recommending to the user to add the current book in the shopping cart, however, that link may be invalid since its containing fragment was reused from cache.
— **Set-View Dependencies**. There is no consideration for content fragments that must be synchronized (i.e., present consistent information) at the same time. For example, a part of the web page is showing book search results while another part is showing irrelevant suggested book listings from a previous search.

Contributions. We enhance the notion of QoD with the inclusion of the above *content dependencies* (i.e., dependencies of web page content fragments). To encapsulate link dependencies, we introduce the metric of QoLF that considers the freshness of links in the content served and, thus, the ability of the user to navigate to the next page. To encapsulate set-view dependencies, we introduce the metric of QoSF that measures the degree of synchronization between content parts served. We present two content materialization algorithms that balance QoS with data freshness in terms of the proposed QoLF and QoSF metrics. Our experimental findings show that our algorithms outperform traditional QoS-QoD approaches in terms of throughput (i.e., better server-side response time), increased data freshness and scalability by sustaining more user sessions. Our performance evaluation is carried out using a real-world bookstore Web database application, which is the canonical example of the majority of e-commerce web applications and online stores.

Roadmap. Next, we present the underlying assumptions of our work and existing content materialization approaches. In Section 3, we present our approach for QoS-QoD balancing for materialization and in Section 4, our materialization algorithms. In Section 5, we discuss our performance evaluation and conclude in Section 6.

2   System Model and Related Work

2.1   Basic Assumptions

The system model for user-driven, personalized e-commerce web database applications with infrequent database updates is based on the typical client / proxy / web server / application server / application database(s) architecture. All the components may have a cache, however, we focus on the cache of the application server (middle-tier) which is the module responsible for content materialization as well as regulating QoS by varying the quantity of cached content served [9, 10, 11]. Moreover, we do not assume a common shared cache across all user sessions. We distinguish between individual user sessions with the use of cookies in user browsers.

The web server is the public entry point of the application and immediately serves request for static content (style sheets, images). Requests for a dynamic web page are routed to an application server that executes the corresponding template file. Template files include script blocks that relate to web page content fragments that are either materialized from scratch or reused from the cache. The materialization of a fragment includes queries on the application database(s) and formatting/wrapping of their results with HTML. Finally, all the fragments, cached or freshly materialized, are assembled together according to the template file and transmitted to the user through the web server.

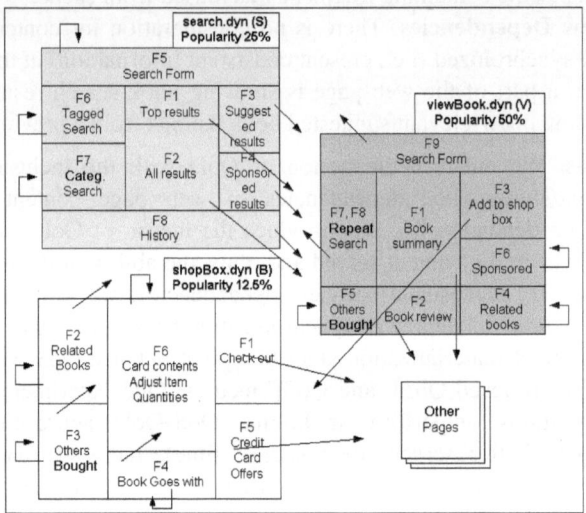

Fig. 1. Breakdown of the bookstore application

Let us take, for example, an online bookstore application with 20 different templates for dynamic web pages of which 4 account for the 95% of user accesses (Fig. 1). The most popular page is a search page (template: search.php) that provides search results, category listings and suggested books in various fragments. The second most popular page is the book viewing page (template: viewBook.php) that presents all information about a selected book in two fragments, related books listing in other 5 fragments plus 1 fragment for adding the book into the shopping card with different options (F_{add}). Typically, the user can navigate from search.php to viewBook.php by picking up a book link. The third book is the shopping card page (template: shopBox.php) that confirms the addition of a book into the shopping card and provides additional suggested listings for direct addition into the shopping card or for viewing. Typically, the user navigates from viewBook.php to shopBox.php by clicking on a link from within the F_{add} fragment).

Fragment materialization in our model is analogous to virtual WebViews [12]. However, WebViews fragments are oblivious to their contents and usage. In our context, the fragments are assumed to contain html form and url links with dynamic parameters that provide the user with the means of navigating between dynamic web pages. Links point statically to a target template and have appended, dynamic parameters, according to the application semantics, i.e., the link "/doBook.php?bookid=2345& action=changeQuantity& value=-1" instructs the target template to perform specific tasks. We assume that a fragment reused from cache always contains outdated links since their parameters would refer to a previous user application-specific state and, therefore, would be invalid. Hence, according to our system model, fragments that are reused from cache do not have any freshness weight (importance). On the other hand, a fragment that is materialized upon a user request is considered fresh within is containing web page.

Definition 1. (Freshness of a Content Fragment) *A dynamic web content fragment is considered fresh if it has been materialized on a user request, according to the user-submitted parameters.*

Definition 2. (Freshness of a Dynamic Web Page) *A dynamic web page is considered fresh if all the fragments in its corresponding template are served fresh.*

2.2   Current Approach (QoIF Approach) and Shortcomings

Current approaches balance QoS and QoD by varying the number of fresh fragments per template request according to the *individual* importance of their containing fragments [5, 6, 7, 8]. The "less important" fragments are the first to be reused from cache when server workload increases. The importance factor or weight of a fragment is template-specific and measures only the fragment's contribution to the overall freshness of its containing template. The sum of the weights of all fragments inside a template sums up to 1, which is the maximum value of freshness when all the fragments of a requested template are materialized. A fragment F contributes to the freshness of a template T, if it is materialized when T is requested.

Definition 3. (weightIF(F,T)) *Let weightIF(F,T) be the freshness importance factor of an individual fragment F in template T. If F_1, F_2, ..., F_n are all the member fragments of a template T, then weightIF(F,T) $\in (0,1)$ and*

$$\sum_{i}^{n} weightIF(F_i, T) = 1$$

Definition 4. countIF(F,T) *The countIF(F,T) of a fragment F in template T is*

$$countIF(F,T) = \begin{cases} 1, & if\ fragment\ F\ is\ materialized\ in\ T \\ 0, & if\ reused\ from\ cache. \end{cases}$$

Given that the current approach focuses on the importance of individual fragments, we refer to their adopted QoD metric as QoIF (Quality of Individual Fragments) and to the current approach as the *QoIF approach*.

Definition 5. (QoIF) *Let F1, F2, ..., Fn be all the member fragments of template T. QoIF(T) is the freshness of template T whose value is*

$$QoIF(T) = \sum_i^n weightIF(F_i, T) \times countIF(F_i, T)$$

The problem with the traditional QoIF approach is that, it considers the templates and their fragments as independent by ignoring content dependencies within and across templates. More specifically the problems are:

1) No Provision for Link Dependencies.

Definition 6. (Link Dependency) *A fragment F_{source} is link-dependent on a template T_{dest}, if there is at least one link inside fragment F_{source} that links to T_{dest}.*

The links between templates are dynamic, in the sense that their parameters are not hardcoded. If a fragment that includes a needed link for navigation is reused from cache, because of its relative low QoIF importance weight, then it does not contain valid links for the user to navigate. This unsatisfied dependency (also called 'broken link') stalls the user session until a fresh version of the fragment is received.

2) No Provision for Set-View Dependencies.

Definition 7. (Set-View Dependency) *A fragment F_i in template T is set-view dependent on fragment F_j in the same template T if both fragments must present consistent (synchronized) information.*

The QoIF approach, which handles fragments independently, fails to synchronize the materialization of set-view dependent fragments since the importance factor employed is fragment-wise and does not force two fragments to be materialized or reused from cache at the same user request for their template.

3   Our Approach for QoS-QoD Balancing

Our approach for balancing QoS with data freshness takes into account link and set-view content dependencies when materializing a dynamic page, thus reducing broken links and unsynchronized content. In a nutshell, our goal is to select the right set of fragments to materialize per page request, given the current server workload constrains. Under light workload, all fragments are materialized and all content dependencies are met. Under heavier workload, the right set of cached fragments is reused so that the most important-to-the-user link and set-view dependencies at met at that particular point in time. Our approach is broken down into the following three sub-goals (or components):

- **Ensure QoS.** Constantly calculate the maximum possible quantity of fragments per template request that must be materialized in order to keep the average response time below a predefined QoS threshold (measured in ms),
- **Speculation.** Employ user access patterns to 'guess' the next template that a user will request,
- **Ensure QoD.** Indicate the appropriate mixture of fragments per template request that satisfy link and set-view dependencies to the highest possible degree in order to reduce broken links and unsynchronized content.

Since our main focus in this paper is content dependencies, we discuss only in brief how we regulate QoS and the methodology of request speculation. The detailed descriptions on QoS regulation and user speculation can be found in [20].

3.1   Ensuring QoS – The QoS Controller

Two essential parameters for ensuring QoS are the maximum tolerable response time (*QoS-threshold*) and the average response time of currently active user sessions (*QoS-average*). The former defines a threshold for the latter which, when violated/crossed, triggers the QoS Controller to take a corrective action. At run time, the QoS-average of active user sessions is computed every tuning period of W seconds. If it is found steadily higher than QoS-threshold, the QoS Controller attempts to lower it by issuing a controlled decrease on the suggested maximum number of fragments that are materialized per template until the next tuning period. This decrease is progressively applied to a percentage of active user sessions per tuning period. In addition, all active user sessions must be affected at least once before the QoS Controller issues any additional decreases as necessary. As soon as the average response time is stabilized below the threshold, the decrease is suspended. In this case, the procedure can be reversed by issuing an increase on the maximum number of fragments per template for materialization. We refer to the action of applying a decrease to a user session as "degrading the user" or "dropping the user". We refer to "upgrade" for the opposite action.

In order to implement this QoS policy, we use two QoS level indexes. The first is called *Global QoS Level* and indicates the suggested number of fragments to be reused from cache per template request. The second is called *User QoS Level* and indicates the actual number of fragments per template to be reused from cache for a particular user. Initially, at light workload, the Global QoS Level is set to 0. The User QoS Level is also set to 0 for all currently active users. Every W seconds, the QoS-average is checked. If it is found to be steadily below the QoS-threshold, then the Global QoS Level is decreased to -1. If workload continues to increase, then a requirement for any extra decrease to the Global QoS Level to -1 is that, all current users have been degraded to -1. In Section 4, we examine how the materialization algorithm regulates the User QoS Level and the directive flags.

3.2   Speculation – The Usage Plans

The second sub-goal of our approach is the speculation on the next template that a user will request. Since user speculation is not the main focus of this paper, we only briefly discuss here a simple speculation scheme based on data mining findings, which we use to implement the speculation module of our materialization algorithms.

According to [13], the popularity of dynamic pages (and of templates) obeys a zipf-like distribution similar to static documents and media files. In other words, fewer templates account for more requests in a structured, almost predictable manner: the most popular template is accessed roughly at a rate of 50%, the second most popular at a rate of 25% and so on. It has also been shown that for web database applications, a small set of templates (approximately four) account for almost 95% of the requests [14], where this set of templates is stable over time [15]. Similarly, [16], [17] refer to "mostly working" user sessions, in which users exhibit a very strong temporal locality in their request patterns on a small set of documents.

In order to encode recurrent access patterns, we introduce the notion of *Usage Plans* (UP) that encapsulate looping user behavior. Figure 2 presents five Usage Plans of the bookstore application of the three most popular templates of the application. Note how two Usage Plans do not share the same template transition. In other words, every transition between two templates is a member of only one UP. This restriction is very important because it allows us to define a session to consist solely of a sequence of non-overlapping UP. For example, a user initially performs a search for a book three times in a row using template S, views a couple of books using V and adds the last viewed book in the shopping basket using B. Then, from within B, the user picks a suggested book to view using V, and then adds it to the shopping basket using B. This sequence of template requests is shown in Figure 3 along with the projected Usage Plans that emerge (S*, (SV)* etc.).

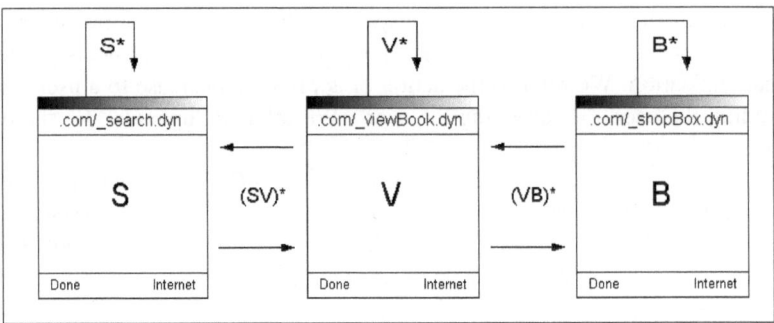

Fig. 2. Five Usage Plans of the Bookstore Application: Three uni-usage plans S*, V*, B*, and two bi-usage plans (SV)* and (VB)*

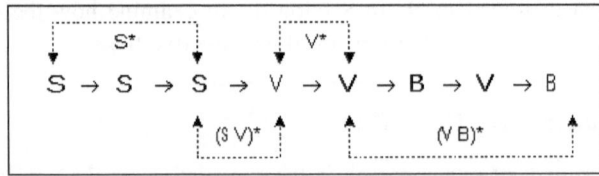

Fig. 3. A Session Illustrated as a Sequence of Usage Plans. Note that each usage plan is immediately followed by another one.

Having established that a user session consists of non-overlapping UPs, we present in brief our simple speculation methodology. We distinguish the UPs to uni-UPs and bi-UPs. The former involve only one-template looping, such as S* and V*. The latter involve two templates such as (SV)* or (VB)*. For each user, we use a FSM Module with three states: (a) "the user is on a uni-UP", (b) "the user is on a bi-UP, and (c) "the user is a moving from a uni-UP to a bi-UP". As the user links between templates, the state on the FSM is changed accordingly. To "speculate" on the next template that the user will request, we use pattern matching that encodes the typical behavior of a user using the FSM and the user's pervious behavior as inputs.

3.3   Ensure QoD – The New QoLF and QoSF Data Freshness Metrics

To better weight the importance of content fragments, we introduce two new semantics-based metrics as follows:

Quality of Link Fragments (QoLF): This metric quantifies the existence of freshly materialized fragments inside a template T_s with link dependencies toward a target template T_d. QoLF applies importance weights on fragments toward link-dependent templates.

Definition 8. (weightLF(F_i,T_s,T_d)) *Let weightLF(F_i,T_s,T_d) be the QoLF importance factor of fragment F_i in template T_s toward template T_d. For all F_i in T_s with a link dependency to template T_d, weightLF(F_i, T_s, T_d) $\in (0, 1)$ and*

$$\sum_{i}^{n} weightLF(F_i, T_s, T_d) = 1$$

In other words, weightLF(F_i,T_s,T_d) measures the navigation/linking importance of fragment F_i in template T_s toward template T_d. In this way, the importance of F_i is dynamic since it depends on a target template T_d. If all fragments inside template T_s with link dependencies to T_d are materialized when T_s is requested by a user, then the QoLF for template T_s toward T_d has the maximum value of 1.

Definition 9. (QoLF(T_s,T_d)) *For all fragments F_i in template T_s with link dependency to T_d, then*

$$QoLF(T_s, T_d) = \sum_{i}^{n} weightLF(F_i, T_s, T_d) \times countIF(F_i, T_s)$$

If a linking fragment from T_s toward T_d is not materialized, then the QoLF value is reduced according to the QoLF importance weight of that fragment toward Td.

Quality of Set-view Fragments (QoSF): The metric of QoSF quantifies the overall set-wise consistency of set-view dependent fragments inside a template. Similarly, we use an importance weight that measures the importance of materializing two set-wise dependent fragments in a template.

Definition 10. *(weightSF(Fi,Fj,T)) Let weightSF(F_i,F_j,T) be the QoSF importance weight between fragments F_i and F_j in template T. For all F_i and F_j which are set-view dependent in T, weightSF(F_i, F_j, T) $\in (0, 1)$ and*

$$\sum_{i,j}^{n} weightSF(F_i, T_j, T) = 1$$

Given that QoSF considers pairs of fragments, only synchronized pairs contribute to and counted toward the freshness of their template.

Definition 11. (countSF(F_i,F_j,T)) The *countSF(F_i,F_j,T) of a pair of fragments F_i and F_j in template T is*

$$countSF(F_i, F_j, T) = \begin{cases} 1, & \text{if fragments } F_i \text{ and } F_j \text{ are synchronized in } T \\ 0, & \text{otherwise.} \end{cases}$$

Fragments F_i and F_j are synchronized in template T if both are materialized or reused from cache. When all set-view dependent fragments of a template T are synchronized then T is fully set-view consistent and its QoSF has the maximum value of 1.

Definition 12. (QoSF(T)) *For all fragment pairs F_i and F_j in template T, then*

$$QoSF(T) = \sum_{i,j}^{n} weightSF(F_i, F_j, T) \times countSF(F_i, F_j, T)$$

4   Materialization Algorithms

In the previous section, we examined how QoS is regulated by the increase or decrease of the Global QoS Level index and introduced the notion of Usage Plans and the new metrics of QoLF and QoSF for measuring data freshness, given the link and set-view dependencies of content fragments. In this section, we explain how we organize QoS Level index, Usage Plans and the new data quality metrics into one convenient structure called MP Selection Table and show how it is used by our materialization algorithms.

4.1   Putting It All Together: The MP Selection Table

The MP Selection Table is a structure that summarizes all combinations of fresh/cached fragments, for a specific template, into groups according to a QoS Level Index. Those combinations are called *Materialization Plans* (MP). For example, at level -1, the table lists 4 possible MP of 3 fresh and 1 cached fragments. Figure 4 shows the MP Selection Table for template search.php (S) (for ease of presentation we show only 4 fragments). The MP '1111' of a template with four fragments implies that all fragments are materialized. The MP '1101' implies that all fragments are materialized except the third one which is retrieved from cache.

For each MP, a QoLF value for each template to which template S links is computed (see Definition 9). In our example, template S links to its self and template viewBook.php (V). In the right-most column (Figure 4), the QoSF for each MP is computed (see Definition 12).

The QLS Algorithm. We first present the QLS algorithm that considers only link dependencies. One instance of the materialization algorithm (shown in Figure 5) is attached to every user session. On each user request for template T_x, the algorithm first secures QoS for the user by increasing or decreasing the User QoS Level, if necessary (lines 4 to 5). Using the user's new QoS level, the algorithm isolates the group of candidate MPs from the MP Selection Table (line 6).

To ensure QoD, the algorithm uses the FSM Module to speculate on the next template that the user will link to from template T_x (line 8-9). Then, the algorithm selects the MP from the group of candidate MPs with the user's QoS Level that maximizes QoLF toward the speculated template. Finally, the fragments that correspond to '1' in the selected MP are materialized from scratch while the fragments with '0' are reused form cache.

MP Selection Table for template search.dyn (S)				
QoS Level	Materialization Plan F1 | F2 | F3 | F4	QoLF Index Values		QoSF Index Values
		UP: S* Target: S	UP: (SV)* Target: V	
0	1 1 1 1	1	1	1
-1	1 1 1 0	0.8	→ 1	0.6
	1 1 0 1	1	0.7	0.4
	1 0 1 1	0.7	→ 0.9	1
	0 1 1 1	0.5	0.4	0
-2	1 1 0 0	0.8	0.7	0
	1 0 1 0	0.5	0.9	0.6
	1 0 0 1	0.7	0.6	0.4
	0 1 0 1	0.5	0.1	0
	0 1 1 0	0.3	0.4	0
	0 0 1 1	0.2	0.3	0
-3	1 0 0 0	0.5	0.6	0
	0 1 0 0	0.3	0.1	0
	0 0 1 0	0	0.3	0
	0 0 0 1	0.2	0	0

Performance Increases (left axis) *Linking Ability Decreases* (right axis)

Fig. 4. The MP Selection Table for template search.php

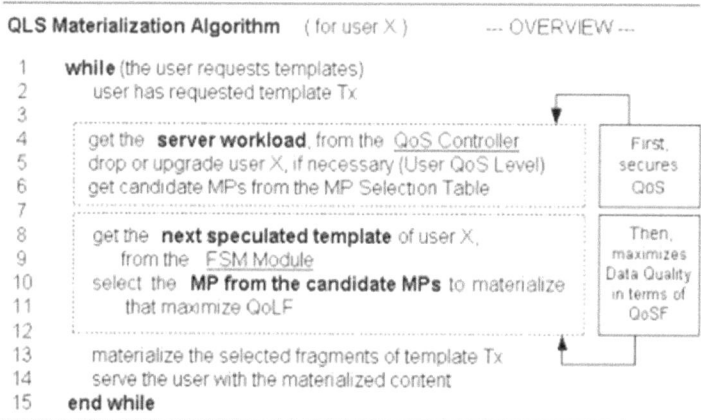

Fig. 5. The QLS Materialization Algorithm

For example, if a user with QoS level equals to -1 has requested template search.php (S), and the speculation FSM Module returns that the user will subsequently request template viewBook.php (V), the algorithm will examine the candidate MPs with QoS Level -1 index and select the MP '1110' which has the highest QoLF value.

The QLSV Algorithm. The QLSV variation of the QLS algorithm considers additionally the QoSV value of the candidate MPs toward increasing the synchronization of set-view dependent fragments. An additional Relax Factor is used to indicate the tolerance on the loss of the QoLF of the selected MP. In our example above, with a Relax Factor of 0%, the algorithm would have select the MP '1110' with a relatively low QoLF value of 0.6 while with a Relax Factor of only just 10%, the MP '1011' with much higher QoSV is selected. In other words, the QLSV variation increases the synchronization of content in a dynamic web page at the expense of linking dependencies.

5   Evaluation

Setup. The evaluation is performed on an experimental platform that emulates a real-world bookstore web database application. Our main server machine (a dual CPU, 2GB RAM, RAID 0) hosted our Java-based web server structured according to the multi-threaded system model. On the same machine, we deployed an application server according to our proposed architecture in Section IV. The application database runs on a separate machine (also a dual CPU, 2GB RAM, RAID 0) on the same local network and it is implemented on SQL Server 2008. The database holds the data for a bookstore with more than a hundred thousand books, in addition to data for book availability, authors, shopping baskets, orders etc. We prepared a mixture of templates, each containing eight to ten fragments. The fragments and their content dependencies are setup according to the bookstore application. Every fragment contains script code that manipulates the results of one read-only query on the application database. In addition, one fragment of the shopBox.php template executes one update on the application database for placing (or removing) a book in a user's shopping box.

For the client workload, on a separate machine, we developed and deployed a multi-threaded User Generator engine capable of emulating a large number of user browsers. We chose to create our own user generator engine in order to have greater control over our experiments in terms of user statistical traces and fragment handling. Specifically, our browser emulators can issue a special HTTP GET request for only receiving a fresh version of a fragment that was served from cache. Our synthetic workload follows basic principles according to the transactional web e-Commerce benchmark (TPC-W) [18], In particular: (a) the popularity of documents follows a zipf-like distribution, (b) a small set of documents (around four) account for at least 95% of total user requests, (c) this set is stable over time, (d) consecutive user requests occur about every ten seconds [19].

Evaluation of the QLS Algorithm. Our first set of experiments compares QLS to the current QoIF approach on the percentage of pages served with broken links. The results of the experiment (Figure 6a, dotted lines) show that this percentage is proportional to the workload. This is because increased workload implies that more users are dropped toward lower QoS levels, and therefore more fragments are served from cache with outdated links. Moreover, the results clearly state that QLS generates approximately 50% less pages with broken links than the QoIF approach, even at high workload. This is because QLS selects the fragments for materialization with link dependencies on the next speculated template of the user. Our analysis has shown that the Speculation Module used by QLS has a hit ratio of 86% in speculating correctly the next template that the user will request. However, Figure 6a (solid lines) plots the performance of QLS by setting the speculation hit ratio manually. The results suggest that our QLS outperforms QoIF even at such a low speculation hit ratio of 40%.

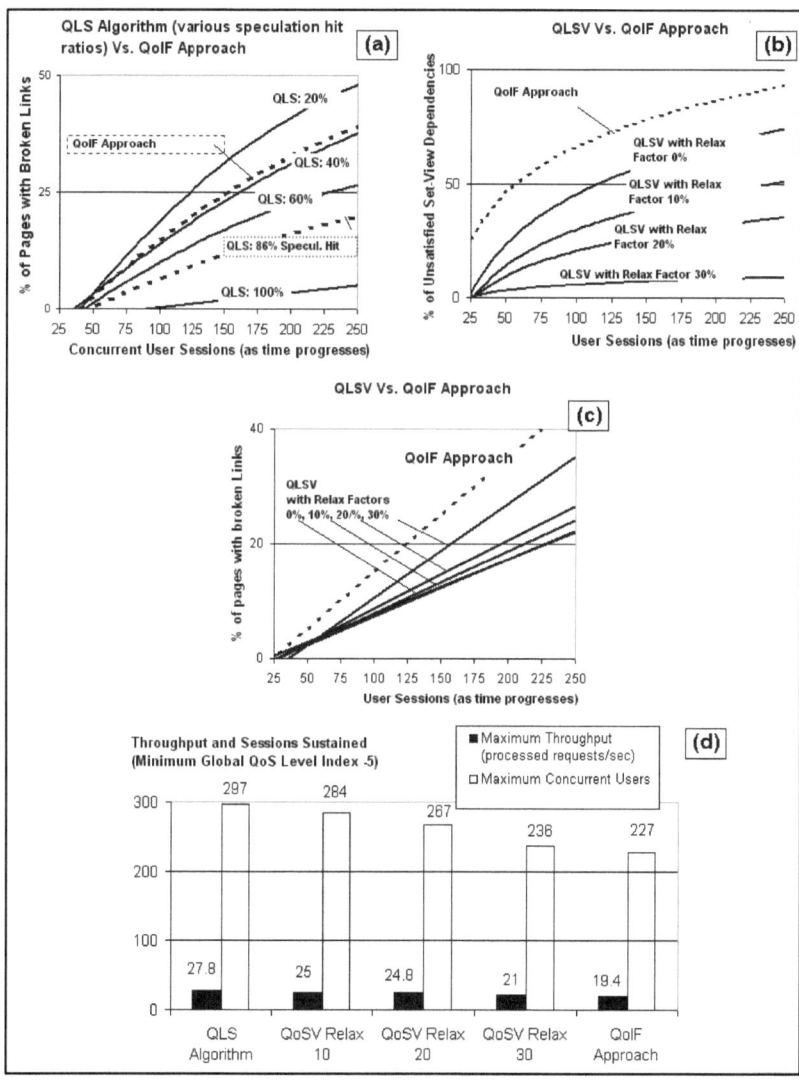

Fig. 6. The Performance Results

Evaluation of the QLSV Variation. In this set of experiments, we compare our QLSV algorithm to the QoIF approach. First, we compare the two approaches on the percentage of unsatisfied set-view dependencies. That is pairs of set-view dependent fragments that are served to the user unsynchronized. Then, we compare them on the percentage of broken links. For these experiments, we run QLSV with relax factors for QoLF equal to 0%, 10%, 20% and 30%. Recall that, the relax factor reduces the maximum possible QoLF of materialization plans in order for the algorithm to select the plan with the maximum possible QoSF value. The results (Figure 6b) show that QLSV serves less unsynchronized set-view dependent fragments than the QoIF approach that has no related provision whatsoever. The gains are greater by using a higher QoLF relax factor of 30%. However, the results come at a cost for the QoLF. Figure 6c plots the percentage of broken links for the four runs of QLSV. The obvious reductions on the previous gains of QLS are attributed to the reduced QoLF imposed by the QoLF relax factor.

Throuput and Maximum Sessions Sustained. Our last experiment measures the maximum throughput and concurrent users that can be sustained by the QoIF approach, QLS algorithm and its QLSV variation using QoLF relax factors of 0%, 10%, 20% and 30%. In other words, this experiment measures the "industrial potential" of our algorithms. This experiment differs from the previous since it provides support for handling broken links in cached fragments. To implement this, we alter the normal request sequence of a user when a template with a cached fragment containing a needed link is received. When this occurs, the user issues an extra special HTTP GET special request to the server in order to receive fresh only the missing fragment that contains valid links. Subsequently, the user resumes its template request sequence. The results of this experiment (Figure 6d) show that both QLS and QLSV outperform the QoIF approach. QLS in particular achieves higher throughput by sustaining about 25% more concurrent users than the QoIF approach. This is attributed to 50% less extra load at the server to handle the special HTTP GET request issued by users for missing fragments. Subsequently, the gains are reduced for QLSV since a higher relax factor generates more broken links than QLS.

6   Conclusion

In this paper, we considered the problem of meeting user QoS expectations in dynamic web database applications under heavy load and we identified the shortcomings of current approaches, which trade QoD for QoS. To mitigate these shortcomings, we proposed two new materialization algorithms, namely QLS and QLSV, for dynamic web pages that can meet user QoS requirements while incurring less impact on the QoD compared to previous QoS-QoD balancing methods. As opposed to QLSV, the QLS algorithm is more suitable in situations characterized by more frequent user clicks - more impatient users - where response time matters the most. Our proposed algorithms achieve their performance by considering content dependencies and user access patterns when selecting which fragments to materialize and which to reuse from the cache when generating a web page. The performance advantages of our two materialization algorithms, including their scalability, were experimentally demonstrated by using a real web_database application of an online bookstore. Although the online bookstore is the canonical example of the majority of e-commerce web applications and online stores, our next step is to evaluate our approach in the context of other web database applications with larger web sites and larger databases such as technical forums and newsgroups.

References

1. Olshefski, D.P., Nieh, J., Nahum, E.: Ksniffer: Determining theremote client perceived response time from live packet streams. In: OSDI 2004, pp. 333–346 (2004)
2. Papastavrou, S., Samaras, G., Evripidou, P., Chrysanthis, P.K.: Adecade of dynamicweb content: A structured survey on past and present practices and future trends. IEEE CS & T 8(2), 52–60 (2006)
3. Schroeder, B., Harchol-Balter, M.: Web servers under overload: Howscheduling can help. ACM Trans. Inter. Tech. 6(1), 20–52 (2006)
4. Guirguis, S., Sharaf, M.A., Chrysanthis, P.K., Labrinidis, A., Pruhs, K.: Adaptive scheduling of web transactions. In: ICDE, pp. 357–368 (2009)
5. Bright, L., Raschid, L.: Using latency-recency profiles for datadelivery on the web. In: VLDB, pp. 550–561 (2002)
6. Labrinidis, A., Roussopoulos, N.: Exploring the tradeoff between performance and data freshness in database-driven web servers. VLDB J. 13(3), 240–255 (2004)
7. Li, W.S., Po, O., Hsiung, W.P., Candan, K.S., Agrawal, D.: Engineeringand hosting adaptive freshness-sensitive web applications on data centers. In: WWW, pp. 587–598 (2003)
8. Qu, H., Labrinidis, A.: Preference-aware query and update scheduling in web databases. In: ICDE, pp. 1–10 (2007)
9. Larson, P.-A., Goldstein, J., Zhou, J.: Mtcache: Transparent mid-tier database caching in sql server. In: ICDE, pp. 177–189 (2004)
10. Luo, Q., Krishnamurthy, S., Mohan, C., Pirahesh, H., Woo, H., Lindsay, B.G., Naughton, J.F.: Middle-tier database caching for ebusiness. In: SIGMOD 2002, pp. 600–611 (2002)
11. Labrinidis, A., Luo, Q., Xu, J., Xue, W.: Caching andMaterialization in Web Databases. Foundations and Trends in Databases 3(2), 169–266 (2009)
12. Labrinidis, A., Roussopoulos, N.: Webview materialization. SIGMOD Rec. 29(2), 367–378 (2000)
13. Wang, Q., Makaroff, D., Edwards, H.K., Thompson, R.: Workloadcharacterization for an e-commerce web site. In: CASCON, pp. 313–327 (2003)
14. Arlitt, M.: Characterizing web user sessions. SIGMETRICS Perform. Eval. Rev. 28(2), 50–63 (2000)
15. Padmanabhan, V.N., Qiu, L.: The content and access dynamics of a busy web site: findings and implications. SIGCOMM Comput. Commun. Rev. 30(4), 111–123 (2000)
16. Cunha, C., Bestavros, A., Crovella, M.: Characteristics of www client-based traces. Boston University, Tech. Rep. TR-95-010 (1995)
17. Oke, A., Bunt, R.B.: Hierarchical workload characterization for abusy web server. In: OOLS, pp. 309–328 (2002)
18. Menasce, D.A.: Testing e-commerce site scalability with tpc-w. In: CMG Conference, pp. 457–466 (2001)
19. Mah, B.A.: An empirical model of http network traffic. In: INFOCOM, p. 592 (1997)
20. Papastavrou, S.: Semantics-based metrics and algorithms for dynamic content in web database applications. PhD dissertation, LC: TK5105.5.P37, CSD, University of Cyprus (2009)

Improving On-Demand Learning
to Rank through Parallelism

Daniel Xavier De Sousa[1], Thierson Couto Rosa[2], Wellington Santos Martins[2],
Rodrigo Silva[3], and Marcos André Gonçalves[3]

[1] Instituto Federal de Goiás, Anápolis, Brazil
dxsousa@gmail.com
[2] Instituto de Informática – UFG, Goiânia, Brazil
{thierson,wellington}@inf.ufg.br
[3] Departamento de Ciência da Computação – UFMG, Belo Horizonte, Brazil
{rmsilva,mgoncalv}@dcc.ufmg.br

Abstract. Traditional Learning to Rank (L2R) is usually conducted
in a batch mode in which a single ranking function is learned to order results for future queries. This approach is not flexible since future
queries may differ considerably from those present in the training set
and, consequently, the learned function may not work properly. Ideally,
a distinct learning function should be learned on demand for each query.
Nevertheless, on-demand L2R may significantly degrade the query processing time, as the ranking function has to be learned on-the-fly before
it can be applied. In this paper we present a parallel implementation
of an on-demand L2R technique that reduces drastically the response
time of previous serial implementation. Our implementation makes use
of thousands of threads of a GPU to learn a ranking function for each
query, and takes advantage of a reduced training set obtained through
active learning. Experiments with the LETOR benchmark show that our
proposed approach achieves a mean speedup of 127x in query processing time when compared to the sequential version, while producing very
competitive ranking effectiveness.

Keywords: Learning to Rank, Distributed Information Retrieval.

1 Introduction

The quality of results returned by a search engine in response to user queries
is mainly determined by the ranking function that it uses. Recently, machine
learning approaches for learning ranking functions have gained increasing interest
from the information retrieval community as an alternative to enhance search
engine results. These approaches are known as *learning to rank* (L2R) and make
use of a training set to learn a ranking function that is then used to order results
for future queries. The training set is composed of triples of the form $\langle q, d, r \rangle$,
where q is a query, d is a document identification and r is a relevance value.

Many L2R approaches have been proposed that use distinct machine learning
paradigms and strategies as neural networks [3], genetic programming [6] and

X.S. Wang et al. (Eds.): WISE 2012, LNCS 7651, pp. 526–537, 2012.

support vector machines [16], among others. Most of these studies are concerned mainly with deriving a better ranking function, in terms of quality of the results ordering, neglecting flexibility and efficiency issues in most cases.

However, flexibility and efficiency are increasingly important issues in L2R [4]. Flexibility is related to how the search engine adapts to the necessity of obtaining new training sets. Due to the fact that the Web changes very frequently, ranking functions need to be re-learned repeatedly, thus a flexible machine learning approach that avoids re-training at all is much desirable. Efficiency, on the other hand, has to do with the time spent on the training, to obtain a learning function, and the time needed by the learned function to process new queries. Thus, these two operations must be efficient in order to not deteriorate the total processing time for each query.

In this paper we present a combination of techniques that include association rules, active learning, and the use of parallel computation to obtain a flexible, scalable, very efficient and effective L2R method. We propose a parallel implementation of a recent L2R method that computes a ranking function on demand for each query. The method is completely flexible, given that no model is constructed a priori, thus there is no need to re-train when new samples are incorporated in the training set. Our implementation is very efficient because the computation of the demand-driven rules composing the ranking function is massively parallelizable and can benefit from current many core processors, such as the modern Graphics processing units (GPUs).

GPUs are specialized architectures originally designed as special-purpose co-processors for dedicated graphics rendering. Due to the high computation power and improved programmability, they have recently become a powerful accelerator for general purpose computing (GPGPU). GPUs can be regarded as massively parallel processors with approximately ten times the computation power and memory bandwidth of CPUs. Moreover, the computational performance of GPUs is improving at a rate higher than that of CPUs and at an exceptionally high performance-to-cost ratio.

Parallel processing has been used in many tasks related to information retrieval, including query processing [7], partition and intersection of inverted lists [5,2], index construction and compression [15], as well as in learning to rank [14,13,11]. However, the use of parallelism in learning to rank (L2R) methods has concentrated on reducing the training time since this correspond to the most demanding part of the majority of the methods, that is, those based on a batch strategy. In this work we propose to use parallelism to improve the performance of both the training time and the query time. This is because our method is on demand, that is, it learns the ranking function for each query, and consequently, it is more demanding for efficiency because training and testing occur together. We take advantage of the power of current many core GPUs to learn to rank by mining association rules on the fly. In this respect, another somewhat related work is the parallel implementation of itemset computation [10,17], although, in this case, the focus is on computation of itemsets used for association rules in general and not specifically to the L2R problem.

In summary, the main contributions of this paper is are series of solutions to support a parallel implementation of a very flexible and effective on-demand Learning to Rank method and a series of experiments that show that our proposed approach achieves a mean speedup of 127x in query processing when compared to the sequential version of the method.

The rest of the paper is organized as follows. In Section 2, we present the on-demand driven approach for generating rule-based learning functions, and the active learning strategy to reduce the size of the training set. In Section 3, we discuss the parallel implementation of the on-demand driven approach in the GPU architecture. In Section 4, we present results of experiments using our methods in distinct benchmark collections. Finally, in Section 5 we present some conclusion and give some suggestions for future work.

2 L2R with Association Rules

The task of L2R is defined as follows. A *training set* $\mathcal{D}$ is given, which is composed of records of the form $\langle q, d, r_d \rangle$, where q is a query, d is a document and $r_d \in \{r_0, \ldots r_k\}$ is the relevance value of d to q. A document d is represented as a list of m attribute-values or features $\{f_1, f_2, \ldots, f_m\}$. The training set is used to learn a *ranking function* f that maps features of a document to a relevance value $r \in \{r_0, \ldots r_k\}$. A *test set* $\mathcal{T}$ is also given and consists of records of the form $\langle q, d, ? \rangle$, for which only the query q and the document d are known. The learned function f is then applied do to each record $\langle q, d, ? \rangle \in \mathcal{T}$ to estimate relevance value of document d to the query q.

Veloso et al [12] used association rules to derive ranking functions. A *rule-set* $\mathcal{R}$ is composed of association rules of the form $f_j \wedge \ldots \wedge f_l \xrightarrow{\Theta} r_i$, where the combination of features at left of the arrow is named *antecedent* and the right part of the rule is known as *consequent*. Θ is known as the *confidence* of the rule [1] and corresponds to the conditional probability of the consequent given the antecedent.

In [12], a demand-driven rule extraction method named *Learning to Rank with Association Rules* - LRAR [9] is proposed where computation of rules is delayed until a set of documents is retrieved for a given query in $\mathcal{T}$. For each individual document d in $\mathcal{T}$ a projected training data $\mathcal{D}_d$ is obtained by LRAR, including in $\mathcal{D}_d$ only examples in $\mathcal{D}$ that shares at least one attribute-value with d. Then, a specific rule-set $\mathcal{R}_d$, extracted from $\mathcal{D}_d$, is produced for each document d in $\mathcal{T}$.

Once the set $\mathcal{R}_d$ is obtained for a document d, a score $s(d, r_i)$ is computed for each level of relevance r_i, regarding document d, according to Equation (1)[1]:

$$s(d, r_i) = \frac{\sum_{(\mathcal{X} \rightarrow r_i) \in \mathcal{R}_d} \Theta(\mathcal{X} \rightarrow r_i)}{\log |\mathcal{R}_d|} \tag{1}$$

---

[1] As proposed in [12], the denominator in Equation 1 is the number of rules in $\mathcal{R}_d$. Here we use the log of that value because the number of rules may become huge when we use rules with many features as their antecedent.

In (1), $\Theta(\mathcal{X} \rightarrow r_i)$ corresponds to the confidence of rule $\mathcal{X} \rightarrow r_i$. The likelihood of d having a relevance level r_i is obtained by normalizing the scores, as expressed by $\hat{p}(r_i|d)$, as shown in (2):

$$\hat{p}(r_i|d) = \frac{s(d, r_i)}{\sum_{j=0}^{k} s(d, r_j)} \tag{2}$$

The relevance of document d is finally estimated by a linear combination of the likelihoods associated with each of the k relevance levels, as expressed by the ranking function $rank(d)$, which is shown in (3):

$$rank(d) = \sum_{j=0}^{k} (r_j \times \hat{p}(r_j|d)) \tag{3}$$

The final ranking is formed by the list of documents in decreasing order of $rank(d_i)$ values, for $1 \leq i \leq |\mathcal{T}|$.

Active Learning. Now we briefly describe SSAR (Selective Sampling using Association Rules) which is an active learning method proposed by Silva et al to be applied to the L2R problem, that usually generates very small training sets. More details about the method can be found in [9]. Considering a large set of unlabeled elements $\mathcal{U} = \{u_1, u_2, \ldots, u_n\}$, where each element i is a triple $\langle q_i, d_i, ? \rangle$, SSAR works according to the following steps:

1. Choose the document d among documents in triples of $\mathcal{U}$ that maximizes the size of the projected data in $\mathcal{U}$, i.e., d is the document for which $\mathcal{U}_d$ is the largest. Thus, d is the document that shares more feature-values with all the other documents of the collection $\mathcal{U}$ and can be considered the best representative of it. The triple $\langle q, d, ? \rangle$ is the first triple to be inserted in the initially empty training set $\mathcal{D}$. The chosen triple is not removed from $\mathcal{U}$.
2. Choose document d_i from $\mathcal{U}$ for which the set of rules $|\mathcal{R}_{d_i}|$ over the current projected training set is the smallest one, i.e., choose $d_i \in \mathcal{U} : \forall j, j \neq i, d_j \in \mathcal{U}, |\mathcal{R}_{d_i}| \leq |\mathcal{R}_{d_j}|$. Then, copy the triple $\langle q, d_i, ? \rangle$ in $\mathcal{D}$, maintaining the original triple in $\mathcal{U}$. The rationale here is that document d_i is the one that differs most from documents in current $\mathcal{D}$ and thus is a document that should be included to expand $\mathcal{D}$.
3. Repeat step 2 until some document d_i is chosen again.

3 On-Demand L2R Using Active Learning and Parallelism

In this section we present a methodology that allows for a viable on-demand approach to L2R. Our method is composed by three stages described as follows.

Obtaining a Reduced Training Set. The first stage consists in obtaining a reduced, but representative training set by using SSAR active learning strategy.

This is important because of issues related to the latency of the GPU global memory, as we should discuss next. SSAR may also be applied to an existing training set $\mathcal{D}$ aiming to reduce its size while keeping the most representative triples of $\mathcal{D}$. Experiments with SSAR in the LETOR set generated reduced training sets whose sizes varied between 1.1% and 1.4% of the original training sets. Consequently, the size of the set of rules $\mathcal{R}_d$ for each test document d is also reduced in the same proportion. Besides generating a reduced training set which has implications in the performance of our methods, as we shall see, this reduced set has also a good side effect of removing noise for the generation of the rules, as discussed in [9].

Copying Training Data to the GPU Memory. The second stage starts by obtaining a set of inverted lists of features derived from the training set. An inverted list for a feature contains of a list of identifications of the training set documents in which the feature occurs. These inverted lists are obtained, in a straightforward way, as the training set file is read, and thus is processed by the CPU. However, these lists need to be copied to the GPU global memory so that they can be used by GPU threads to compute the projected training set $\mathcal{D}_d$ for a given test document d. Due to the great latency of the GPU global memory and the need to optimize access to this memory, the size of these inverted lists must be kept small. To this end, we make intensive use of bitmaps to represent the inverted lists. An inverted list is represented by a bitmap structure which is an array of unsigned integers that is used to express the list of document identifiers associated with a feature. Each feature has its own bitmap array and each bit in the bitmap represents whether the corresponding document has that feature. The set of inverted lists for the training set as well as document related data, i. e., document relevance value, features count and the features of the document, are all copied to the GPU global memory at the end of this stage.

Parallel LRAR – PLRAR. In the third stage, we use PLRAR, our parallel implementation of LRAR, to compute the set $\mathcal{R}_d$ on-the-fly for each document d of a triple $\langle q, d, ? \rangle$ of the test set. We explore the fact that the set of rules $\mathcal{R}_d$ can be computed in parallel for each document and assign each document to a thread. In this way, a kernel can be launched with as many blocks of threads as necessary to account for all documents present in the test set of a given query. Since there are usually a considerable number of documents to be processed per query, there will be thousands of threads in flight on the GPU, resulting in better latency hiding. As shown in Fig. 1, blocks of threads have access to the GPU global memory which stores the inverted list for each feature and the documents' data. Each square inside a block represents a thread that is responsible for processing a document.

Each thread in PLRAR is responsible for processing a document. Since each thread has a unique index in the GPU, we associate this index to a document index. Each thread computes the rules containing the features of its assigned document. We say that a rule has *size k* if it its antecedent is composed by $k-1$ features. PLRAR starts by computing rules of size two and proceeds until a rule

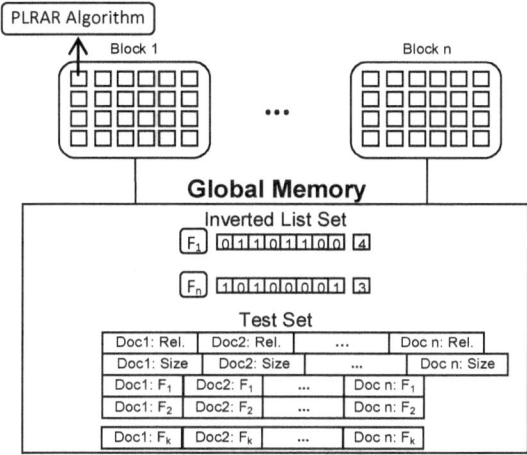

Fig. 1. GPU Architecture and Data Structures

size determined by the user via a parameter. A thread has to calculate rules whose support and confidence are above pre-defined thresholds. The PLRAR algorithm is shown in Algorithm 1. PLRAR uses as input the set of inverted lists and the data associated with a test document, and produces the ranking of that document. The algorithm consists of four nested loops. For each rule size, all features of the document are combined, taking into account each relevance value, and the support and confidence are calculated. Finally, the ranking of the document is produced.

Algorithm 1. PLRAR – Parallel L2R with Association Rules.

Require: Inverted List Set, Test Document Data, MAX_RULE_SIZE, Support and Confidence Threshold
Ensure: The ranking number
   {The next lines of code process the $document_i$ by the $thread_i$}
1: for s=1 to MAX_RULE_SIZE do
2:    for all Feature f in document d_i do
3:       for all combination c wiht size s that includes f do
4:          for all relevance $rel \in r_0, r_1, ..., r_k$ do
5:             compute support for rule: $c \rightarrow rel$
6:             compute confidence for rule: $c \rightarrow rel$
7:          end for
8:       end for
9:    end for
10: end for
11: compute the ranking for the d_i from all rules generated

To calculate the support and confidence of a rule, each thread has to scan the bit map for features composing the rule. This represents the processing part of a thread that is more computationally intensive, and in addition, concentrates most of the accesses to the (training) data. To get the frequency of two features we make a bitwise AND operation, and count the number of resulting 1's. This is done in as many steps as it is required to process the array elements (32-bits blocks) used in the bit map representation. When more than two features are considered, that is, rules with 3-itemsets or more, we use an auxiliary array to contain the intermediate pairwise results of successive bitwise operations. Fig. 2 illustrates the case where three features (F_1, F_3 and F_4) are processed using an intermediate array of bitmap (F_x). For the sake of simplicity, 16-bits blocks are shown in the figure. The Count column indicates the number of documents that have that feature.

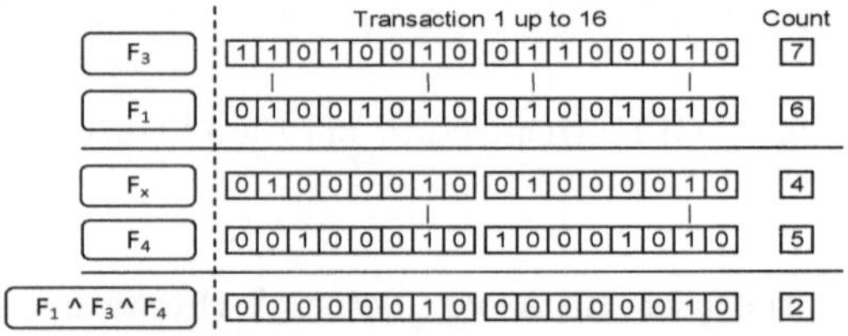

Fig. 2. Bitmap Structure

Besides reducing the number of memory accesses, it is important to coalesce the reads and writes so that they access linear memory in a well-defined stride. This implies in less memory transactions and consequently less time to fetch data. To deal with that, we transformed Array-of-Structures (AoS) into Structure-of-Arrays (SoA) in the implementation. For example, a query's document contains the following information: features, relevance value and number of features (Size). So, instead of creating an array of structures contain this information, we defined a structure containing arrays, with each array holding information of different documents. In addition, we got the document's features sorted. This way, threads accessing a given information (e.g. relevance value) will generate less memory transactions since the information will be contiguous in memory.

4   Experimental Results

The Computational Platform. The experiments we present in this section were conducted using two machines. The first one is a dual-core Intel 2.00GHz processor, 3GB of main memory, 4MB of L2 cache memory with Linux (Ubuntu 11.04) operating system. This machine was used to process all serial tests. The second one is equipped with an NVIDIA Tesla C1060 GPU and Linux operating system. The Tesla consists of an array of 30 Streaming Multiprocessors (SMs), each with eight scalar processor cores (total of 240 cores) with 1.3GHz of clock speed, 4 GB of global memory and 16 KB of shared memory. Our algorithms are programmed in C/C++ and CUDA C/C++ 4.0.

Collections Used. We experimented with different collections of version 3.0 of LETOR benchmark[8] which are traditionally used on L2R research. LETOR is composed by collections containing triples of the form $\langle q, d, r \rangle$, where q is a query, d is a document and r is the relevance value which varies from one to two. Each document d is represented by a set of features covering a wide range of properties, such as term frequency, BM25, PageRank, HITS etc. Each collection is arranged in five folds, including training, validation and test data, which were used in our 5-fold cross-validation experiments. Table 1 summarizes the set of collections we used, including the number of queries in each one of them, the sizes of the test and training sets, as well as the percentage of the training sets chosen by SSAR, on average, in the first stage of our method for each collection.

Table 1. Characteristics of the LETOR Collections used

	Collections					
	TD2003	TD2004	NP2003	NP2004	HP2003	HP2004
Queries	50	75	150	75	150	75
Test Set	49,058	74,146	148,657	74,834	14,760	74,409
Tr. Set	29,435	44,488	89,195	44,301	88,564	44,646
Red. Tr. Set	2.3%	1.4%	1.1%	1.5%	1.2%	1.9%

4.1   Time Results

In this section we present results of experiments using PLRAR in the many collections we used. We start by presenting results in terms of execution time and compare PLRAR performance using both, the full and the reduced training sets. We also show comparisons between PLRAR and its serial version, LRAR, in terms of execution time. Next we analyze the results of PLRAR in terms of effectiveness, measured using the traditional MAP measure, which is much used in L2R literature.

Impact of a Reduced Training Set. Table 2 shows the mean query processing time (in seconds) achieved by PLRAR when using both the original and the reduced training set generated by SSAR in the many collections of the LETOR benchmark. As can be seen, PLRAR achieves high speedup in many of the collections. The mean average speedup is about 127x in execution time when using the reduced training set. These execution times are proportional to the reduction on the size of the training set obtained due to the use of SSAR. Although Table 2 shows only results for rules of size three, similar differences in speedup are obtained for rules of sizes two, four and five. Since results using the reduced training set are better for all the collections used, in what follows we only present results of experiments using these reduced sets.

Table 2. Mean execution time to process queries on demand, using 3-sized rules

	Collections					
	TD2003	TD2004	HP2003	HP2004	NP2003	NP2004
Red. Tr. Set	0.74	0.43	0.51	0.52	0.48	0.43
Norm. Tr. Set	7.70	11.12	145.14	12.52	176.52	11.53
Speedup	16.2	25.90	287.10	24.10	366.00	26.80

Serial vs. Parallel implementation. In Table 3 we present the mean query processing time (in seconds) of PLRAR (on GPU) and LRAR (on CPU only), both using the reduced training sets. We show average execution time to process queries, considering, in turn, rules of sizes varying from two to four. The computation of rules with sizes greater than four are not viable in LRAR. Table 3 also presents the speedups achieved by PLRAR over LRAR in each collection and for each rule size considered. Note that except for rules of size two, PLRAR achieves gains up to two orders of magnitude in execution time, when compared to LRAR. Speedup is smaller with rules of size two, because in this case bitwise operations do not occur since there is only one feature in the antecedent part of these rules. Moreover, by increasing the rule size we achieve greater gains with GPU processing. This occurs because with the increasing of the rule size, the processing time for each document also increases, thus the serial processing time of a query is much affected since it corresponds to the accumulated time to process all the documents. This does not happen with the parallel execution where the documents are processed concurrently. As a consequence, we can afford to compute rules with greater sizes using PLRAR, what is difficult to obtain using LRAR.

Table 3. Execution time of PLRAR and LRAR with rules of varying sizes

	Time Execution using Rule Size 2					
	TD2003	TD2004	NP2003	NP2004	HP2003	HP2004
PLRAR	0.23	0.22	0.17	0.22	0.17	0.22
LRAR	2.10	2.03	2.17	2.15	2.15	2.12
SpeedUp	9.03	9.41	12.72	9.73	12.56	9.80
	Time Execution using Rule Size 3					
	TD2003	TD2004	NP2003	NP2004	HP2003	HP2004
PLRAR	0.47	0.43	0.51	0.52	0.48	0.43
LRAR	38.14	42.84	34.88	40.89	34.59	32.87
SpeedUp	80.38	99.72	68.98	78.72	71.74	76.54
	Time Execution using Rule Size 4					
	TD2003	TD2004	NP2003	NP2004	HP2003	HP2004
PLRAR	7.23	7.64	10.95	10.65	10.22	7.61
LRAR	1147.24	1331.17	1010.37	1207.24	998.86	952.21
SpeedUp	158.62	174.32	92.31	113.37	97.77	125.14

4.2 Effectiveness Evaluation

In this section we report an experiment that evaluates the variation of effectiveness of PLRAR as rule sizes vary. In the experiments we use the MAP metric computed by the evaluation script provided with the LETOR package. Table 4 shows the MAP values for each collection, for different rule sizes. As we can see from Table 4, rules of size three present effective gains over rules of size two. However, with rules with sizes greater than three the gains are not as consistent. Specially, rules of size five presented results that are inferior to those of rules of size three for the great majority of the collections. We believe that this happens because rules with many features tend to have high confidence, but a low support.

Rules of greater sizes did not enhance effectively the results, however results in Table 4 indicate that some large rules are in fact effective in ranking documents. Thus, the results suggest future work on learning how to choose the best rules, among rules of different sizes, to rank documents. Nevertheless, the parallel implementation proposed in this work has given us the scalability to process more rules of large sizes.

Table 4. Effectivity comparison of PLRAR with different rule sizes

	TD2003	TD2004	NP2003	NP2004	HP2003	HP2004
Rule Size 2	0.25	0.23	0.74	0.60	0.68	0.57
Rule Size 3	0.27	0.22	0.75	0.67	0.70	0.59
Rule Size 4	0.27	0.22	0.76	0.67	0.70	0.61
Rule Size 5	0.24	0.21	0.76	0.65	0.68	0.57

Table 5. Effectivity comparison of PLRAR with several strategies

	TD2003	TD2004	NP2003	NP2004	HP2003	HP2004
PLRAR	0.27	0.22	0.75	0.67	0.70	0.59
SVMRank	0.26	0.22	0.70	0.66	0.74	0.67
Regression	0.24	0.21	0.50	0.53	0.56	0.51
SVMMAP	0.25	0.21	0.74	0.72	0.69	0.66
AdaRankMAP	0.23	0.22	0.72	0.67	0.68	0.62
ListNet	0.28	0.22	0.77	0.69	0.69	0.67
Average	0.25	0.22	0.70	0.65	0.68	0.62

We now compare the effectiveness of PLRAR with those of many other methods, using the LETOR benchmark. In Table 5 we show the results of PLRAR using rules of size three, because it was with this kind of rules that PLRAR achieved its best results. Note that PLRAR performs better than Regression and SVMMAP methods in the majority of the collections, and performed better than SVMRank in three collections and loses in two collection. Also, PLRAR performed better than the average of all methods in each collection. Although ListNet presented better results among all methods, it is not an on-demand method and, consequently, it is not adaptable to each test element as is the case of PLRAR. This is an important aspect to L2R, specially in web search engine, where new queries appear constantly and the ranking function need to be flexible in order to generate appropriate rank of documents to each query.

5   Conclusion

In this paper, we have proposed a parallel implementation in GPUs of an on-demand L2R method that uses association rules and have proposed the combination of this method with an active sampling technique proposed in recent work. We have shown through experiments conducted with the LETOR benchmark that this combination results in a L2R method that provided considerable gains in query processing time that largely outperforms the serial implementation.

We are planning to extend this work in several directions. First we intend to investigate enhancements to our parallel implementation. One possible approach is to exploit another level of parallelism by assigning a document to a block of threads. This way, different rule sizes can be generated in parallel, or/and the support and confidence can be calculated in parallel. Additionally we want to investigate feasible techniques that allow for choosing among rules of different sizes those which are the best for ranking documents in response to a test query.

References

1. Agrawal, R., Imieliński, T., Swami, A.: Mining association rules between sets of items in large databases. In: Proceedings of the 1993 ACM SIGMOD International Conference on Management of Data, SIGMOD 1993, pp. 207–216. ACM, New York (1993)
2. Barroso, L.A., Dean, J., Hölzle, U.: Web search for a planet: The google cluster architecture. IEEE Micro. 23(2), 22–28 (2003)
3. Burges, C., Shaked, T., Renshaw, E., Lazier, A., Deeds, M., Hamilton, N., Hullender, G.: Learning to rank using gradient descent. In: Proceedings of the 22nd International Conference on Machine Learning, ICML 2005, pp. 89–96. ACM, New York (2005)
4. Chapelle, O., Chang, Y., Liu, T.-Y.: Future directions in learning to rank. Journal of Machine Learning Research - Proceedings Track 14, 91–100 (2011)
5. Clarke, C.L.A., Terra, E.L.: Approximating the top-m passages in a parallel question answering system. In: Proceedings of the Thirteenth ACM International Conference on Information and Knowledge Management, CIKM 2004, pp. 454–462. ACM, New York (2004)
6. de Almeida, H.M., Gonçalves, M.A., Cristo, M., Calado, P.: A combined component approach for finding collection-adapted ranking functions based on genetic programming. In: Proceedings of the 30th Annual International ACM SIGIR Conference on Research and Development in Information Retrieval, SIGIR 2007, pp. 399–406. ACM, New York (2007)

7. Ding, S., He, J., Yan, H., Suel, T.: Using graphics processors for high-performance ir query processing. In: Proceedings of the 17th International Conference on World Wide Web, WWW 2008, pp. 1213–1214. ACM, New York (2008)

8. Qin, T., Liu, T.-Y., Xu, J., Li, H.: Letor: A benchmark collection for research on learning to rank for information retrieval. Inf. Retr. 13, 346–374 (2010)

9. Silva, R., Gonçalves, M.A., Veloso, A.: Rule-Based Active Sampling for Learning to Rank. In: Gunopulos, D., Hofmann, T., Malerba, D., Vazirgiannis, M. (eds.) ECML PKDD 2011, Part III. LNCS, vol. 6913, pp. 240–255. Springer, Heidelberg (2011)

10. Teodoro, G., Mariano, N., Meira Jr., W., Ferreira, R.: Tree projection-based frequent itemset mining on multicore cpus and gpus. In: Proceedings of the 2010 22nd International Symposium on Computer Architecture and High Performance Computing, SBAC-PAD 2010, pp. 47–54. IEEE Computer Society, Washington, DC (2010)

11. Tyree, S., Weinberger, K.Q., Agrawal, K., Paykin, J.: Parallel boosted regression trees for web search ranking. In: Proceedings of the 20th International Conference on World Wide Web, WWW 2011, pp. 387–396. ACM, New York (2011)

12. Veloso, A.A., Almeida, H.M., Gonçalves, M.A., Meira Jr., W.: Learning to rank at query-time using association rules. In: Proceedings of the 31st Annual International ACM SIGIR Conference on Research and Development in Information Retrieval, SIGIR 2008, pp. 267–274. ACM, New York (2008)

13. Wang, B., Wu, T., Yan, F., Li, R., Xu, N., Wang, Y.: Rankboost acceleration on both nvidia cuda and ati stream platforms. In: Proceedings of the 2009 15th International Conference on Parallel and Distributed Systems, ICPADS 2009, pp. 284–291. IEEE Computer Society, Washington, DC (2009)

14. Wang, S., Gao, B.J., Wang, K., Lauw, H.W.: Parallel learning to rank for information retrieval. In: Proceedings of the 34th International ACM SIGIR Conference on Research and Development in Information Retrieval, SIGIR 2011, pp. 1083–1084. ACM, New York (2011)

15. Wei, Z., JaJa, J.: A fast algorithm for constructing inverted files on heterogeneous platforms. In: Proceedings of the 2011 IEEE International Parallel & Distributed Processing Symposium, IPDPS 2011, pp. 1124–1134. IEEE Computer Society, Washington, DC (2011)

16. Yue, Y., Finley, T., Radlinski, F., Joachims, T.: A support vector method for optimizing average precision. In: Proceedings of the 30th Annual International ACM SIGIR Conference on Research and Development in Information Retrieval, SIGIR 2007, pp. 271–278. ACM, New York (2007)

17. Zhang, F., Zhang, Y., Bakos, J.: Gpapriori: Gpu-accelerated frequent itemset mining. In: Proceedings of the 2011 IEEE International Conference on Cluster Computing, CLUSTER 2011, pp. 590–594. IEEE Computer Society, Washington, DC (2011)

Topic-Sensitive Hidden-Web Crawling*

Panagiotis Liakos[1] and Alexandros Ntoulas[1,2]

[1] National and Kapodistrian University of Athens, Greece
[2] Zynga, San Francisco, USA
grad0990@di.uoa.gr, ntoulas@gmail.com

Abstract. A constantly growing amount of high-quality information is stored in pages coming from the Hidden Web. Such pages are accessible only through a query interface that a Hidden-Web site provides and may span a variety of topics.

In order to provide centralized access to the Hidden Web, previous works have focused on query generation techniques that aim at downloading all content of a given Hidden Web site with the minimum cost. In certain settings however, we are interested in downloading only a specific part of such a site. For example, in a news database, a user may be interested in retrieving only sports articles but no politics. In this case, we need to make the best use of our resources in downloading only the portion of the Hidden Web site that we are interested in.

In this paper, we study how we can build a topically-focused Hidden Web crawler that can autonomously extract topic-specific pages from the Hidden Web by searching only the subset that is related to the corresponding category. To this end, we present query generation techniques that take into account the topic that we are interested in. We propose a number of different crawling policies and we experimentally evaluate them with data from two popular sites.

Keywords: Crawling, Hidden Web, Focused Crawling.

1 Introduction

An ever-increasing amount of high-quality information on the Web today is accessible through Web pages pulling information from data sources such as databases or content management systems. Such pages are collectively called the Hidden Web because they are hidden from the search engine crawlers but are typically only accessible after issuing one or more queries to a search interface.

In most cases, the information existent in the Hidden Web is of high quality as it has been carefully reviewed, edited or annotated before being stored in a database or a content management system and may span a multitude of topics ranging from sports and politics to different medical treatments of a particular disease.

In order to facilitate the discovery of information on the Web, search engines and content-aggregation systems could greatly benefit from approaches that would allow them to collect and download the Hidden Web pages. Having information from the Hidden Web all in one place can be of great benefit to both users (as they can have a

* Partially supported by PIRG06-GA-2009-256603.

X.S. Wang et al. (Eds.): WISE 2012, LNCS 7651, pp. 538–551, 2012.

one-stop shop for their information needs) and for the search engines (as they can serve their users better).

Since a Hidden Web site can be accessed by search engine crawlers only through a search interface, previous work has investigated ways of collecting the information from the Hidden Web sites. In most cases [17,18,5,6,11,21,22] previous work has focused on generating queries that are able to download *all* of (or as much as possible) a given Hidden Web site, with the minimum amount of resources spent (e.g. queries issued). For example, the technique in [17] iteratively issues queries to Hidden Web sites and can download about 90% of some sites using about 100 queries.

Although such approaches can work well for the cases where we are interested in doing a comprehensive crawl of a Hidden Web site, there are cases where we may be interested only in a specific portion of a Hidden Web site. For example, a search engine that specializes in traveling may be interested in picking only news articles that pertain to traveling from a general-purpose news database. A portal talking about politics may want to identify the politics-related articles from a blogs database and leave out the sports-related ones. Or, a mobile application focusing on night-life in San Francisco may want to pull only the related articles from all the postings on events in the wider Northern California.

In this paper, we study the problem of building a topic-sensitive Hidden Web crawler, that can automatically retrieve pages relevant to a particular topic from a given Hidden Web site. One way to achieve this goal would be to employ previously-developed techniques (e.g. [17,18,5]) to retrieve the majority of the Hidden Web site and then only keep the content that we are interested in. Since this approach may lead to downloading a number of pages that we are ultimately not interested in, it may also lead to a great waste in resources, measured in time, money, bandwidth or even battery life in the case of a mobile setting.

To this end, the goal of our crawler is to retrieve from a Hidden Web site as many pages related to a given topic as possible with the minimum amount of resources. Our main idea is to proceed in an iterative fashion issuing queries to the Hidden Web site that are very relevant to the topic that we are interested in.

In summary, this paper makes the following contributions:

- We formalize the problem of topically-sensitive Hidden Web crawling, i.e. downloading the pages from a Hidden Web site that pertain only to a given topic.
- We present an algorithm for performing topically-sensitive Hidden Web crawling. Our idea is to identify candidate keywords from the crawled documents that are relevant to the topic of interest.
- We propose a number of different policies that can be used to decide which of the candidate queries we can issue next. As we will show later in our experimental section some polices are much better in producing good keywords than others.
- We evaluate our algorithm using the different policies on two real *Hidden Web* sites and we showcase the merits of each of the policies.

2   Topic-Sensitive Hidden Web Crawling

At a high level, the goal of a general topic-sensitive (or focused) Web Crawler [9] is to retrieve pages from the Web that pertain to a given topic. The crawler has to evaluate the contents of each downloaded page to decide whether it is relevant to a particular topic. Hence, the crawler aims at examining as small of a subset of the total search space as possible in order to avoid wasting resources by downloading pages that are irrelevant to the topic.

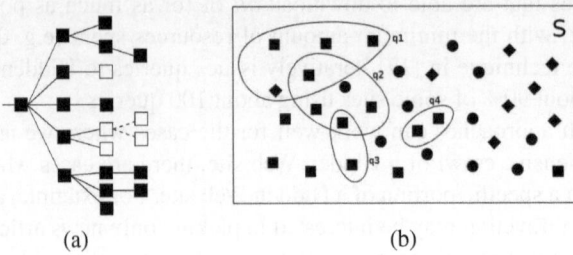

(a)   (b)

Fig. 1. (a) Topic-Sensitive Crawling in the publicly indexable Web, (b) Representing a Hidden Web site as a set

Figure 1a illustrates this process in the publicly indexable Web. Pages that are relevant to our topic of interest are represented with black boxes while the irrelevant ones are represented with white ones. The goal of the crawler is to follow only links with relevant pages, in order to use its resources efficiently. In the case of a Hidden-Web crawler however the pages are not directly accessible by following links as in [9]. Since the only access point that the crawler has is a search interface, it needs to identify good keywords that will return pages pertaining to a given topic.

The definition of what constitutes a "topic" may be different depending on the application at hand. For example, if we are interested in loosely collecting Web pages that are about football, then any page that contains the words "football", "quarterback", "touchdown", etc., may be relevant to the topic. Alternatively, in the cases when we are interested in discerning more accurately between topics (e.g. external affairs vs. internal politics) a more elaborate mechanism (e.g. an SVM-based classifier) may be used. Our work presented here does not depend on how exactly we detect whether a particular page belongs to a given topic. Instead, we make two generic assumptions regarding the topic definition that allows for a variety of different topic detection techniques to be employed. First, we generally assume that a topic can be described by a distribution of keywords. This is essentially the same assumption as most text classifiers are based on. Second, we assume that not all of the keywords in the distribution are known beforehand, as in this case the problem becomes trivial (i.e. the crawler simply needs to iterate over all the relevant keywords). We should note that not knowing all of the keywords is the typical practical scenario. In our football example above, we may know that "quarterback" or "touchdown" are good initial keywords but we may not necessarily know that "huddle" or "hailmary" are also good keywords for the crawler to use.[1] In our work, we typically provide the crawler with a small set of keywords that are relevant

---

[1] This scenario is also typical for the cases where one may want to perform a topic-sensitive crawling-by-example, i.e. provide an example document as input and have the crawler crawl documents that are similar to the input document.

to begin with and the crawler discovers more keywords to use as it progresses through the Hidden-Web site.

At a high level, in order to access the pages from a Hidden Web site we typically need to apply the following steps:

1. First, submit a query through the interface provided by the Web Site, that characterizes what we want to locate.
2. Then we receive a result index page, that contains links and possibly additional information of the pages that matched the submitted query.
3. Finally, having identified a promising page in the results (we can skip pages already downloaded based on their URL), we follow the link and eventually visit the actual Web page.

2.1   A Topic-Sensitive Hidden Web Crawling Approach

As we discussed in the previous section, we gain access to the contents of a Hidden Web site only by issuing queries through a search interface and obtaining a result index page. Therefore, a Hidden Web crawler should automatically issue queries to the site and then retrieve the pages included in the results.

One of the biggest challenges in implementing such a crawler is that we need to pick the most appropriate terms that will retrieve the pages pertaining to the desired topic in the most effective way. If the crawler can pick terms that are very well-suited to the topic we are interested in, the crawler will be able to also retrieve pages that are of the given topic. On the other hand, if the crawler issues terms that are completely irrelevant, it will simply waste its resources by downloading pages that are of no interest to the users (and potentially also degrading the quality of the search engine that employs the crawler).

To illustrate, we assume that the crawler downloads pages from a Hidden Web site that contains a set of pages S (the rectangle in Figure 1b). These pages might cover a vast area of topics, most of which may potentially be irrelevant to our search. The results of each potential query q_i can be considered as a subset of S that contains the pages that the site would return as a result.

In practice, we are interested only in pages relevant to a specific topic (i.e. the squares in Figure 1b). To this end, in order to minimize the total cost of crawling, we need to discover queries that focus on the topical area of our interest, and in that way avoid the extra cost of issuing queries that return irrelevant results.

In order to select the best keywords, at each step of the algorithm, we examine the candidate keywords that we have retrieved from the collection so far and we pick one to be issued as our next query. We describe the keyword extraction and selection processes in section 2.2. In order to bootstrap our algorithm, we consider an initial description of the topic as a set of terms. This initial description of the topic can either be one or more terms that are relevant to the topic, or one or more documents relevant to the topic. After the first step, and after we have downloaded each batch of results based on the term that we queried, we augment our list of candidate terms to include terms that were newly found. This however, implies that we need to be able to determine whether the downloaded page pertains to the desired topic or not. For this step, we also employ a classifier as we will discuss in more detail in Section 2.3.

Algorithm 2.1

```
(1)  WordCollection = di ;
(2)  while (available resources) do
     // extract the terms and build a WordCollection
(3)    if(cnt++ mod N == 0)
       T(WordCollection) = ExtractTerms(WordCollection);
     // select a term to send to the site
(4)    qi = SelectTerm(T(WordCollection));
     // send query and acquire result index page
(5)    R(qi) = QueryWebSite(qi);
     // download and evaluate the pages of interest
(6)    DownloadAndEvaluate(R(qi), pi, WordCollection);
     done
```

Fig. 2. Topic-Sensitive Crawling of a Hidden Web Site

Figure 2 shows the algorithm for a Topic-Sensitive Hidden Web Crawler. First, the Word Collection is initialized with an exemplary document, d_i. Step (3) of the algorithm is only issued once every N times –to reduce the number of required computations– and it extracts the best terms of the Word Collection. This is done by calculating the *TF/IDF* weight for every term of the collection. The process is explained in detail in Section 2.2. Step (4) picks the best of the terms that Step (3) extracted from the document, that has not been used so far, while step (5) uses this term to issue a query and retrieves the result index page. Finally, Step (6) downloads the Hidden Web pages that were included in the results and had not been previously downloaded and evaluates the contents of the results using one of the policies (p_i) presented in this work. The evaluation process of this step is responsible for the maintenance of the collection of words used for keyword extraction. This process depends heavily on the policy that is to be followed. The different policies are explained in Section 2.3. This process is repeated (Step (2)) until we use our available resources (e.g. number of queries allowed, storage, bandwidth, etc.)

2.2   Word Collection

In this section we provide the details of the Word Collection that serves as a pool of possible queries for our algorithm.

This pool of words is the document d_i used in Step (1) of the Algorithm 2.1. It consists of text that is related to the topic in search and is initialized with an exemplary document that describes that topic. So, if for instance we wanted to crawl for sports articles from a news site, we could provide the algorithm with an input document (or snippet, or a small set of keywords) that would consist of a few sport-related articles. However, the Word Collection cannot remain static during the execution of the algorithm for a variety of reasons.

First, the input document given to the algorithm during the initialization of the *Word Collection*, may not be enough for the extraction of all (or enough) terms that are needed for the retrieval of a sufficient amount of Web Pages. No matter how good that initial document may be in capturing the essence of the topic in search, it may manage to provide only a limited number of terms. Second, the initial *Word Collection* may be too specific, in a way that the terms extracted would not be general enough to capture the whole topical area of the document. For instance, if those sport-related articles mentioned earlier, were taken from a WNBA fan-site, the terms extracted from the *Word Collection* would retrieve results concerning women's sports and basketball. We are interested in matching the input document with a broad topic, in order to retrieve all the related Web Pages. Therefore, it is necessary to broaden our *Word Collection* during the execution of our algorithm. Finally, to successfully retrieve the maximum amount of Web Pages from a Hidden Web site, it is essential that we adapt to its terminology. For instance, we cannot retrieve but a subset of Web Site that indexes multilingual content if all the words in our *Word Collection* are English.

Therefore, it is clear that for effective Topic-Sensitive Hidden Web Crawling, the pool of words that is used for term extraction must be enriched continuously as well as adapted to the site in search. To face this issue, we exploited the contents of the results as potential parts of the *Word Collection*. Each result page is evaluated using one of the policies described in Section 2.3 and the contents of the ones relevant to the topic in search are added to the *Word Collection*.

In that way, the *Word Collection* is able to provide a sufficient amount of terms to be issued as queries. Furthermore, since the *Word Collection* is updated with content directly from the site in search, it can provide the algorithm with terms that not only are relevant to a specific topic, but have a higher significance for that site.

In order to effectively retrieve results, the queries picked from the Word Collection must be focusing on the topic that we are interested in. To successfully address this matter, in Step (4) we use the well-known *TF/IDF* term weighting system that measures accurately the general importance of a term in a collection.

2.3   Result Evaluation Policies

In this section we provide the details of the various evaluation policies that we employ in our work. These policies are used in order to decide if each page contained in the results is relative to the topic in search, and therefore will be helpful later. The contents of the pages that are considered in-topic are added to the *Word Collection*, and consequently take part in the keyword selection process.

The policies that we consider in our work are the following:

- *perfect:* We use the categorization information directly from the site in search. Each document of *dmoz* and *Stack Exchange* is classified into topics and this policy utilizes this knowledge. Of course such information is not available in most cases. In our work, we will use this policy as a benchmark to determine how well the rest of the policies can do relative to this one that has perfect information regarding the topic that every result document belongs to.
- *do-nothing:* We accept all of the returned pages as in-topic. This policy updates the *Word Collection* with every page the crawler manages to discover. Since the first few terms are extracted from the input document it is expected that the first queries that will be submitted will be meaningful and so the corresponding result pages will have a high chance of being in-topic. However, since the results are never filtered it is expected that a lot of out-of-topic pages will find their way to the Word Collection and affect the term selection process significantly. Thus, this policy can also be used as a comparison point for the other policies.

- *NaiveBayes:* We use a Naive Bayes classifier that decides if a result is relevant or not to the topic in search. This policy examines the effects of using a topic classifier during the crawling process. The classifier needs to go through a training phase before it can be used. We feed the classifier with samples of documents belonging to a certain topic and then are able to test if other documents should be classified under this topic or not. Therefore, it is clear that this method can only be used for topics that the classifier is already trained for. Obviously, the better the classifier is trained the less erroneous pages will be added to the *Word Collection*.
- *CosineSimilarity:* We examine the cosine similarity of every result page with the initial example document and accept only a small percentage of the closest pages. In this way we ensure that the pool of words will be enriched only with terms closely related to the topic defined by our input. This policy is more flexible compared to the NaiveBayes policy in terms of adaptability as it requires no training. However, since this method depends heavily on the input document, we will examine how the quality of the input document affects performance in our experimental section.

3   Experimental Evaluation

3.1   Experimental Setup

We start our experimental evaluation by presenting the datasets used in our experiments. Then, we demonstrate the performance of the proposed policies as described in Section 2.3 for a variety of topics. For the experiments we present here, we set variable N of Algorithm 2.1 to 7. That is, we update the *Word Collection* of Algorithm 2.1 every 7 queries. We experimentally tested different values for N and we found that 7 is essentially the breaking point after which we would observe a significant degradation in terms of the performance of our algorithms. Additionally, during the operation of the *CosineSimilarity* policy, we kept the top 1% of the returned documents.

For our experiments we used two different datasets, namely the Open Directory Project [1] and all the public non-beta *Stack Exchange* sites [2] contents. *Dmoz* indexes approximately 5 million links that cover a broad area of topics. Each link is accompanied with the site's title, a brief summary and its categorization. The links are searchable through a keyword-search interface and *dmoz* enforces an upper limit on the number of returned results (10,000 results). The Stack Exchange sites contain a total of 391,522 questions over twenty different topics. In order to further examine the performance of our algorithms, we enforced an upper limit of 1,000 results per query for this dataset as well. We considered the titles and summaries of each indexed link of the *dmoz* website and the questions of the *Stack Exchange* sites as documents, and allowed our policies to utilize them during the keyword extraction process.

For every topic we studied here, we needed an initial exemplary document to serve as a representative of that topic. In the following, we used random documents from the sites from the respective topics to serve as such exemplary documents. We do not report on how selecting a different initial document affects performance as we experimentally confirmed the findings of [17] that the selection of the first document does not significantly affect the returned results. However, we do study how the *number* of initial exemplary documents affects performance in Section 3.5.

3.2 Evaluation of the Different Policies over the Same Topic

We start our evaluation by studying the effectiveness of the policies proposed in Section 2.3 in retrieving pages of a given topic from *dmoz* and *Stack Exchange*. We show the fraction of documents returned for *dmoz* and *Stack Exchange* as a function of the queries issued in Figures 3a and 3b respectively. For *dmoz* we report results for the topic *Sports* while for *Stack Exchange* we report results for the topic *Wordpress*. The results over topics other than Sports and Wordpress, are very similar to figures 3a and 3b. We study the performance of our policies on additional topics in Section 3.4.

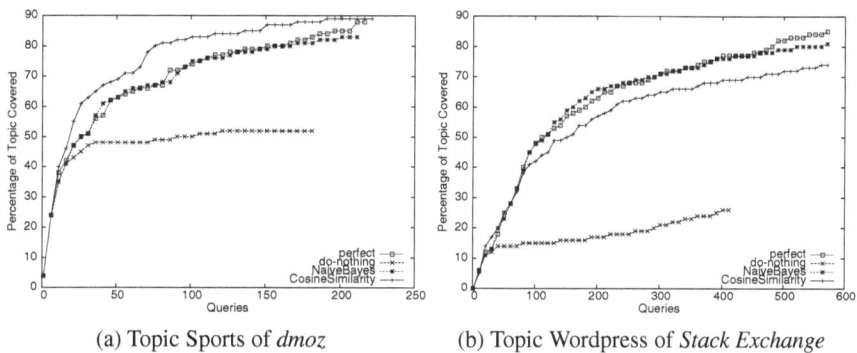

(a) Topic Sports of *dmoz*   (b) Topic Wordpress of *Stack Exchange*

Fig. 3. Results of the four policies over two datasets

For the topic *Sports dmoz*, contains a total of 90,693 pages. As Figure 3a shows, all four policies behaved identically for the first seven queries, where only the initial document is being used for keyword extraction. *Do-nothing*'s performance was enormously affected after those first few queries, when the *Word Collection* was updated with all the returned results, since as the name of the policy implies, no action was taken to filter them. As a consequence, *Do-nothing* performed badly, as expected. Policies *Perfect* and *NaiveBayes* behaved similarly with the first one managing to retrieve a slightly bigger percentage of *Sports*-related documents. This can be explained from the fact that the two policies had a high percentage of common queries issued (28%). After 210 queries, the *Perfect* policy retrieved 86.43% of the total relevant documents, while the *NaiveBayes* policy managed to collect 83.42% of them.

The *CosineSimilarity* policy managed to outperform all three of them by retrieving 89.66% of the documents categorized under the Topic *Sports* after issuing about 180 queries. 19% of the terms used for query submission were common with the *Perfect* policy and 20% of them were common with the *NaiveBayes* policy. This implies that *CosineSimilarity* does a better exploration of the keyword space compared to the other policies and it can find keywords leading to more relevant documents earlier compared to the Perfect policy which takes about 220 queries to achieve the same performance.

We also retrieved all the documents of the *Stack Exchange* sites relevant to the *Wordpress* publishing platform. There was a total of 17,793 questions categorized under this topic. The results are illustrated in figure 3b. After 572 queries, the *Perfect* policy retrieved 85% of the total relevant documents, the *NaiveBayes* policy retrieved 81% and the *CosineSimilarity* policy managed to collect 71% of them. The significant amount of more queries needed for the retrieval of documents from this dataset is explained by the much smaller upper limit of returned documents per query.

3.3   Queries Issued and Topic Accuracy

In order to investigate more closely the performance of our policies we further examined the actual queries issued by each policy. We present a sample of queries issued by each policy together with the precision achieved (i.e. fraction of in-topic documents) in Figure 4.

One thing that we should note is that there is a lot of overlap in the results of each term. Although every policy manages to discover meaningful terms that return a good number of results, a large portion of them has been already discovered by previous queries. Additionally, the *Do-nothing* policy is the most successful in finding "popular" terms. Most of the terms illustrated in Figure 4 returned the maximum of 10,000 results, while the average after 210 queries was 8,650. This is due to the fact that the Word Collection of this policy was eventually enriched with terms from every possible topic of the Open Directory Project. The *NaiveBayes* policy was second with 7,927, the *Perfect* policy third with 7,530 and the *CosineSimilarity* policy last with 7,426 results per query.

Using the *dmoz* categorization information for every downloaded document, we also measured the precision of the different policies as shown in Figure 4. Overall, the *Perfect* policy was the most successful one since it is allowed to use the class information. Of course, in practice such information is not available and thus this policy serves as a benchmark in terms of classification precision. The *Perfect* policy achieved a 68.2% precision on average.

The *Do-nothing* policy chooses to accept every document it downloads as relevant to the topic in search, so its errors are equal to the total number of links retrieved minus the links that were actually in-topic. It's accuracy was 7.8% on average. The classifier used in the *NaiveBayes* policy on the other hand, proved to be very effective and achieved a 64.4% precision on average. However, the impact of all errors is not the same for our algorithm. An in-topic document that is classified as irrelevant is not added to the Word Collection and does not affect the term extraction process. On the other hand, an irrelevant document that "sneaks" its way into the Word Collection, may cause the selection of inappropriate terms for query submission. For the *NaiveBayes* policy 45% of the documents added to the Word Collection, were actually categorized under another topic in the *dmoz* classification taxonomy.

Finally, the *CosineSimilarity* achieved a 58.6% precision on average. However, this did not affect the policy in a negative way. The retrieved documents, despite the fact that they belong to a different topic, had very high cosine similarities with the input document, so naturally, the Word Collection was not altered in an undesirable way. As a result, the *CosineSimilarity* policy outperformed the other policies in terms of recall as we showed in the previous section.

No	Term	Precision	Term	Precision	Term	Precision	Term	Precision
1	results	42.83%	results	42.83%	results	42.83%	results	42.83%
2	statistics	43.61%	statistics	43.61%	statistics	43.61%	statistics	43.61%
3	roster	71.36%	roster	71.36%	roster	71.36%	roster	71.36%
10	men	27.26%	schedules	10.31%	schedules	10.31%	tables	5.61%
15	scores	27.89%	church	0.00%	standings	67.96%	player	24.36%
20	players	30.62%	coaching	17.89%	baseball	38.29%	players	33.70%
25	hockey	41.85%	methodist	0.00%	records	8.38%	hockey	44.08%
30	tennis	7.43%	beliefs	10.11%	membership	1.52%	baseball	32.20%
40	rugby	14.43%	stellt	0.00%	county	0.39%	race	14.56%
60	sport	3.82%	bietet	0.00%	fc	5.45%	conference	1.73%
100	competition	12.28%	nach	0.00%	standing	26.66%	competitive	11.61%

(a) *Perfect*   (b) *Do-nothing*   (c) *NaiveBayes*   (d) *CosineSimilarity*

Fig. 4. Queries issued by the different policies

3.4 Comparison of Policies under Different Topics of dmoz

In this section, we present the performance of the policies *Perfect*, *NaiveBayes* and *CosineSimilarity* over five different topical areas belonging to the classification taxonomy *dmoz* uses. To this end we used the following topics: Computers (103,336 documents), Recreation (91,931 documents), Shopping (87,507 documents), Society (218,857 documents) and Sports (90,639 documents).

Figure 5a illustrates the behavior of our approach for these five different categories of *dmoz* using the *Perfect* policy. Topics *computers* and *sports* proved to be the easiest to retrieve while *society* needed a significantly bigger amount of queries to surpass the 80% barrier. This is due to the fact that *society* is a much larger category compared to the rest. More specifically, topic *Computers* returned 92.17% of the total documents after 210 queries. Topics *Sports* and *Recreation* discovered 86.43% and 80.76%, respectively, with the same amount of queries. Finally for the topics *Shopping* and *Society* the policy collected 77.82% and 62.06%, respectively.

The results for the said topics using the *NaiveBayes* policy are presented in Figure 5b. This policy is performing slightly lower than *Perfect* for each of the five topics. The ordering of the topics is, however, a little different, since the *NaiveBayes* policy behaved very poorly for the topic *Recreation*, which was ranked fourth below the topic *Shopping*. More explicitly, after 210 queries, topics *Computers* collected 89.81% of the documents, topic *Sports* 83.42%, topic *Shopping* 73.34%, topic *Recreation* 70.86% and topic *Society* 54.86%. This is due to the fact that *Recreation* is a much broader topic than the rest and thus *Perfect* can benefit more by knowing the topic of the documents beforehand.

In Figure 5c we observe similar results for the *CosineSimilarity* policy. Topics *Computers* and *Sports* were again first with 91.84% and 89.66%, respectively, after issuing 210 queries. Topic *Recreation* collected 83.55% of the related documents, while topics *Shopping* and *Society* returned 79.02% and 62.97%, respectively, after the same amount of queries.

At a high level, the *CosineSimilarity* policy performed slightly better than the *Perfect* policy in 4 of the 5 topics examined, *Computers* being the only exception. Additionally,

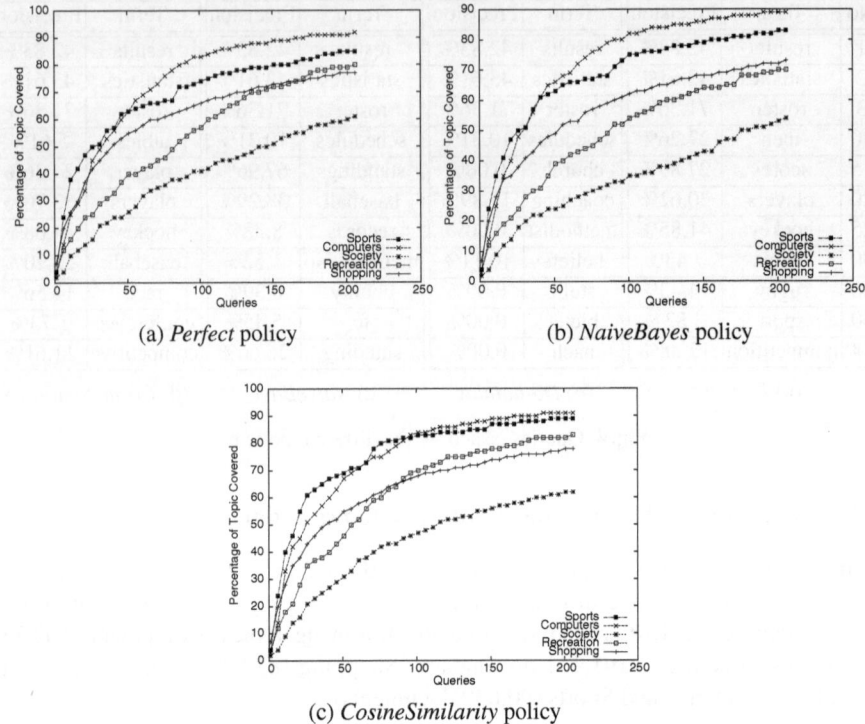

(a) *Perfect* policy

(b) *NaiveBayes* policy

(c) *CosineSimilarity* policy

Fig. 5. Comparison of polices on different topics

it out-scored the *NaiveBayes* policy for every one of the five different topics. Topic *Recreation* behaved better than *Shopping* after the 87th query, as it did with the *Perfect* policy after the 162nd query.

3.5 Impact of Input Document Size

The *CosineSimilarity* policy depends heavily on the input document, since it does not use it to only extract the first few queries to be issued, but to evaluate the result documents retrieved as well. To this end, we examine the impact of the size of the initial document on the behavior of this policy. The other three policies use the initial document only for the first step of the process, so they are not affected as much by the size of the initial document.

Figure 6 illustrates the results for *CosineSimilarity* under three different input documents of variable size, while retrieving documents from *dmoz* under the *Computers* category. As we can see, as we limit the input size, the performance deteriorates. However, even for a very small initial document we can still get reasonably good results. More specifically, using an input document that consists of 1,000 titles and summaries of links indexed by *dmoz*, *the CosineSimilarity* policy retrieved 91.14% of the relevant documents after 190 queries. In comparison, with only 100 titles and summaries, the policy discovered 87.77% of the documents with the same number of queries. With an input document of 50 titles and summaries, it retrieved a sizeable 86.67% of them.

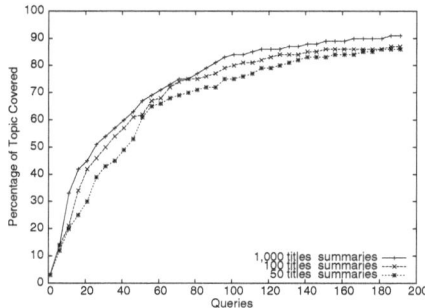

Fig. 6. Impact of the input document size using *CosineSimilarity* on topic *Computers*

3.6   A High-Level Comparison to Generic Hidden Web Crawling

In [17], an experiment on crawling all of the *dmoz* site's documents is presented. It takes about 700 queries to retrieve a little over 70% of its contents and after that we would have to analyze the pages to identify which ones belong to the topic at hand. The *CosineSimilarity* policy proposed in this paper was able to download 70% of topics *Sports* and *Computers* after 52 and 60 submitted queries respectively. Therefore, we can significantly reduce the cost of crawling when we are interested in a specific topic by using our techniques.

4   Related Work

Research on the Hidden Web has emerged during the last years. In [18], Raghavan et al. first introduced the problem of Hidden Web crawling by presenting an architectural model for a Hidden Web crawler. Their efforts in this work, mainly focus on learning Hidden-Web query interfaces. The potential queries are either provided manually by users or collected from the query interfaces. Similar work is presented in [6] where some new strategies for effective discovery of Hidden Web forms are presented, while [14] discusses the approach used from Google in filling such Web forms.

   In [5,17,21] the problem of automatically producing meaningful queries that can return large fractions of a document collection is examined. [17] provides a theoretical framework for analyzing the process of generating queries for a document collection as well as examining the obtained result. In addition, the framework is applied to the problem of *Hidden Web crawling* and the efficiency of the approach is quantified. In [5], a number of methods for building multi-keyword queries is presented and experimentally evaluated to show that a large fraction of a document collection can be returned. Wu et al. [21] focus on the core issue of enabling efficient Web Database crawling through query selection and propose a theoretical framework that transforms the database crawling procedure into a graph-traversal problem; in the latter, the proposal is based on following up "relational" links. [20] developed a new set covering algorithm to address the problem of web database crawling and in [13], the incremental web database crawling problem is examined. Our work differs from the above efforts as we retrieve only portions of a *Hidden Web* site that are relevant to a specific topic.

Chakrabarti et al. studied ways to selectively search for pages relevant to a pre-defined set of topics [9]. A complete process for discovering topic-specific resources from the Web is presented. A classifier and a distiller are used for the evaluation of crawled pages based on their hypertext. However, this methodology is only applicable on the publicly indexable web, since hyperlinks are not used at all on the Hidden Web. Since then, there has been many similar work made [16,10,15,4], again for the publicly indexable web. In [7] the authors present a method for determining the topic of a given page based on keywords found in its URL. In our case, we opted to also use content from the page since we download such content anyway in order to get access to keywords that will be used to update our Word Collection.

In [23], Yang et al. propose a method for retrieving documents relevant to a query document. Candidate phrases are extracted from an input document and a score is assigned to each one of them, using two different policies. The first is a linear combination of the total *TF/IDF* score of all the terms of the phrase and their degree of coherence. The second is based on the probability of pairs of terms occurring in the same document (mutual information based). Our goal is different in that we aim at retrieving as many documents as possible around the topic of the input document. In addition, we proceed iteratively and learn from previously downloaded documents in order to determine our queries.

In [3], a focused crawler for the Hidden Web is presented. Its main focus is on extracting information from labels of form elements, to associate Hidden Web sites to topic domains and to analyze the forms to find ways to execute queries automatically. However, the queries to be issued are not produced automatically, but are picked from predefined resources. Bergholz et al. have also addressed this matter in [8]. However, as the case was before, they use predefined sets of keywords named "querying resources" in order to retrieve results from the collections it discovers. In [12], Ipeirotis et al. attempted to categorize Hidden Web Databases by probing them with queries and evaluating the amount of the results. They used pre-classified documents to train a rule-based document classifier and transformed the rules it produced, to queries for use. Similarly, [19] presents a technique called informed probing where potentially available metadata (e.g. title, or meta tags) from the entry page to a Hidden Web site can be used as probe keywords in order to categorize the Hidden Web sources.

5   Conclusion and Future Work

We examined how we can build a Topic-Sensitive Hidden Web Crawler that, given an exemplary document, can effectively retrieve those documents that are relevant to a certain topic. We presented an algorithm for a Topic-Sensitive Hidden Web Crawler and proposed and experimented with a variety of policies deployed to help us measure the relevance of the returned documents with the searched topic. Our experimental evaluation indicates that our suggested algorithm has great potential for harnessing topic-specific documents.

In the future, we plan to exploit diverse query formulations in order to further reduce the overheads involved in the process, to experiment with additional Hidden Web sites that may freely provide their data, and finally, to explore techniques that may incrementally retrieve recently updated content.

References

1. The Open Directory Project, http://www.dmoz.org
2. Stack Exchange, http://stackexchange.com/
3. Álvarez, M., Raposo, J., Pan, A., Cacheda, F., Bellas, F., Carneiro, V.: Deepbot: a focused crawler for accessing hidden web content. In: Proc. of Int. Workshop on Data Enginering Issues in E-commerce and Services, DEECS 2007, NY, USA (2007)
4. Angkawattanawit, N., Rungsawang, A.: Learnable crawling: An efficient approach to topic-specific web resource discovery (2002)
5. Barbosa, L., Freire, J.: Siphoning hidden-web data through keyword-based interfaces. In: Proceedings of SBBD, Brazil, (2004)
6. Barbosa, L., Freire, J.: An adaptive crawler for locating hidden-web entry points. In: Proceedings of the WWW Conference, NY, USA (2007)
7. Baykan, E., Henzinger, M., Marian, L., Weber, I.: Purely url-based topic classification. In: Proceedings of the WWW Conference, Madrid, Spain (2009)
8. Bergholz, A., Chidlovskii, B.: Crawling for domain-specific hidden web resources. In: Proceedings of WISE, DC, USA (2003)
9. Chakrabarti, S., van den Berg, M., Dom, B.: Focused crawling: a new approach to topic-specific web resource discovery. In: Proceedings of the WWW Conference, NY, USA (1999)
10. Diligenti, M., Coetzee, F., Lawrence, S., Giles, C.L., Gori, M.: Focused crawling using context graphs. In: Proceedings of VLDB, CA, USA (2000)
11. Ipeirotis, P.G., Gravano, L.: Distributed search over the hidden web: hierarchical database sampling and selection. In: Proceedings of VLDB, Hong Kong (2002)
12. Ipeirotis, P.G., Gravano, L., Sahami, M.: Probe, count, and classify: categorizing hidden web databases. SIGMOD Rec. 30, 67–78 (2001)
13. Liu, W., Xiao, J., Yang, J.: A sample-guided approach to incremental structured web database crawling. In: Proceedings of ICIA, Harbin, China (2010)
14. Madhavan, J., Ko, D., Kot, L., Ganapathy, V., Rasmussen, A., Halevy, A.: Google's deep web crawl. In: Proceedings of VLDB, Auckland, New Zealand (2008)
15. Menczer, F., Pant, G., Srinivasan, P.: Topic-driven crawlers: Machine learning issues. ACM TOIT (submitted, 2002)
16. Noh, S., Choi, Y., Seo, H., Choi, K., Jung, G.: An Intelligent Topic-Specific Crawler Using Degree of Relevance. In: Yang, Z.R., Yin, H., Everson, R.M. (eds.) IDEAL 2004. LNCS, vol. 3177, pp. 491–498. Springer, Heidelberg (2004)
17. Ntoulas, A., Zerfos, P., Cho, J.: Downloading textual hidden web content through keyword queries. In: Proceedings of JCDL, NY, USA (2005)
18. Raghavan, S., Garcia-Molina, H.: Crawling the hidden web. In: Proceedings of VLDB, San Francisco, CA, USA (2001)
19. Seshadri, S., Cooper, B.F.: Routing queries through a peer-to-peer infobeacons network using information retrieval techniques. IEEE TPDS 18, 1754–1765 (2007)
20. Wang, Y., Lu, J., Chen, J.: Crawling deep web using a new set covering algorithm. In: Proceedings of the ADMA Conference, Berlin, Heidelberg (2009)
21. Wu, P., Wen, J.-R., Liu, H., Ma, W.-Y.: Query selection techniques for efficient crawling of structured web sources. In: Proceedings of the ICDE, Washington, DC, USA (2006)
22. Wu, W., Yu, C., Doan, A., Meng, W.: An interactive clustering-based approach to integrating source query interfaces on the deep web. In: Proceedings of SIGMOD, NY, USA (2004)
23. Yang, Y., Bansal, N., Dakka, W., Ipeirotis, P., Koudas, N., Papadias, D.: Query by document. In: Proceedings of WSDM, NY, USA (2009)

Can Social Features Help Learning
to Rank YouTube Videos?

Sergiu Viorel Chelaru, Claudia Orellana-Rodriguez,
and Ismail Sengor Altingovde

L3S Research Center, Hannover, Germany
{chelaru,orellana,altingovde}@L3S.de

Abstract. We investigate the impact of social features (such as likes, dislikes, comments, etc.) on the effectiveness of video retrieval in YouTube video sharing system using state-of-the-art learning to rank approaches and a greedy feature selection algorithm. Our experiments based on a dataset of 3,500 annotated query-video pairs reveal that social features are promising to improve the retrieval performance.

1   Introduction

What happens when a user clicks on a like or dislike button or posts a comment for a digital object, say a blog post, interesting photo or funny video displayed in her favorite Web 2.0 platform? Other than being shown to the future visitors of the same object (and serving as a medium of user participation/interaction), can these social signals help the underlying search systems for guiding its users to reach to a better quality or more relevant content? Despite the rapid and growing interest for Web 2.0 applications from both the industry and researchers from various disciplines, these questions are still not clearly answered.

While we witness some recent moves from big players towards a more social search (such as the Google+ application and Bing's expansion of results with those "liked by" the users' Facebook friends[1]), the ways search engines and/or Web 2.0 applications exploit social signals (if they ever do) are usually not disclosed. In terms of the academic research, there exists a large body of work analyzing the rich content posted on Web 2.0 platforms [4,1,18] that also fueled research in recommendation systems, opinion mining, trend analysis, etc. However, to the best of our knowledge, there is no study that systematically investigates the impact of various forms of social signals on the retrieval performance in a realistic and state-of-the-art framework.

In this work, we focus on the keyword-based video search for YouTube video sharing site and investigate the impact of *social* features on the retrieval performance. Social features, as we call here, refer to the information that is created by some explicit or implicit user interaction with the system (such as likes, dislikes, favorites, comments, etc). On the other hand, we call the features that would be typically involved in a keyword search scenario, such as the textual similarity

---

[1] http://www.pcmag.com/article2/0,2817,2380874,00.asp

X.S. Wang et al. (Eds.): WISE 2012, LNCS 7651, pp. 552–566, 2012.

of the user queries to the video titles and tags, as the *basic* features. Our work essentially explores whether the social features can help retrieving more relevant videos for the queries when they are combined with the basic features. Note that, while our choice of YouTube is based on the availability of a rich set of social features in this platform, we believe that our findings are applicable to the text, image and/or video search in other platforms that support similar kinds of social features. In particular, we seek to answer the following research questions:

- *What are the characteristics of the YouTube query results with respect to the social features?* The first contribution of our work is providing an in-depth analysis of the social features associated with the top-ranked videos retrieved by YouTube for 1,450 real user queries. The queries are obtained from a major search engine's auto-completions specialized for YouTube and their top-300 results are retrieved from YouTube API, making our dataset a unique and valuable collection to work on. In Section 2 we describe the details of our dataset and in Section 3 we report several interesting statistics regarding the queries, their resulting videos and the characteristics of these videos in terms of the associated social features.

- *How effective is each individual feature for ranking videos for a given query? How similar are the rankings created by the different features?* Our second contribution is investigating the individual performance of each basic and social feature for ranking videos. To this end, we conducted a detailed user study and obtained relevance annotations for 3,500 query-video pairs. Furthermore, by using a variant of Kendall's Tau metric [7], we measure the overlap in top-10 query results between the pairs of features. Our findings reported in Section 4 show that while the basic features are the most effective ones on the average, certain social features are also quite promising and yield the best ranking for 48% of the queries. We also find that some of the feature pairs retrieve quite similar results and there is room for a feature selection strategy to eliminate such redundant features.

- *Can social features help improving the video retrieval performance in a learning to rank framework?* This is the central question that we seek to answer in this work. We investigate the potential gains of using social features along with the basic features for a number of different learning to rank algorithms [2]. Our experimental results reported in Section 5 show that the subsets of basic and social features identified using a greedy feature selection algorithm [10] have the potential to improve the retrieval effectiveness, although the gains vary among different algorithms.

We conclude the paper in Section 6, where we summarize our major findings and point to various future research directions.

Related Work. Cheng et al. provides a large-scale analysis of the content in YouTube and provide statistics related to the videos, such as the distribution of categories, duration, size, bit rate and popularity [4]. An analysis of the video characteristics, such as the popularity distribution and evolution over time are addressed in [1]. Vavliakis et al. compare YouTube and two other data sharing platforms in terms of several factors and identify the correlations between

Table 1. Metadata fields stored for each video v in V

Metadata	Notation	Metadata	Notation
No. of views	$W(v)$	Title	$TitleText(v)$
No. of likes	$L(v)$	Tags	$TagText(v)$
No. of dislikes	$D(v)$	Comments	$CommentText(v)$
No. of favorites	$F(v)$	Uploader	$U(v)$
No. of comments	$C(v)$	Age	$G(v)$

these factors via regression analysis. Various properties of the comments posted for YouTube videos are analyzed in [16,17]. In [15], comments are leveraged for the aesthetic-aware re-ranking of image search results. The closest work to ours is [19], which utilizes YouTube comments for the video retrieval. However, their work is only limited to the comment feature for the known-item retrieval scenario. To the best of our knowledge, no previous work investigates the potential of social features for improving the video search effectiveness, as we do here.

2   Data Collection, Methods and Characteristics

Query Set (Q). We first obtained around 7,000 queries using the auto-completion based suggestion service specialized for the YouTube domain from a major search engine. In particular, we submitted all possible combinations of two-letter prefixes in English (like aa, ab, ..., zz, etc.) and collected top-10 query suggestions for each such prefix (e.g., "aaliyah", "aaron carter", "abba dancing queen", etc.) in a similar fashion to [3]. From this initial set, we sampled a subset of 1,450 queries, denoted as Q, which constitutes the query set for this study. Note that, different from all earlier studies that seeded their crawlers with some generic queries (e.g., queries from Google's Zeitgeist archive [16] or terms from the blogs and RSS fields [17]), we employ a set of real YouTube queries, as the collected query suggestions are based on real and potentially popular queries previously submitted to YouTube.

Query Results (R). For each q in Q, we obtained the top-300 result videos (denoted as R_q) from YouTube API along with the available metadata fields (see Table 1) in late 2011. This process resulted in a superset of 380K videos, i.e., around 262 videos per query are retrieved. Among these videos, 365K of them are unique (i.e., only 4% of all videos overlap among different query results). The set of unique videos is denoted as V in the paper.

In addition to the metadata fields directly available via API, we crawled up to 10,000 most recent comments that are posted for each video from actual HTML responses of YouTube (API can provide only up to 1,000 comments). Due to the difficulties of crawling HTML, we could obtain around 33 million comments posted for 86K unique videos in our dataset. This is a fairly large set of comments as the recent works also employ similar (e.g., up to 1,000 comments for 40K videos in [17]) or smaller number of comments (e.g., a total of 6.1 million comments in [16]).

Table 2. Metadata statistics per each video v in V

	No. of views	No. of likes	No. of dislikes	No. of favorites	No. of comments	Age (days)
Avg.	274,019	910	64	786	505	503
Max.	675,909,639	1,123,471	2,093,280	1,088,770	7,000,143	2,332
Stddev.	3,274,834	8,945	4,002	8,453	13375.31	478.52

Finally, we also constructed the profiles of users who uploaded the videos in V. To this end, for each user u, we again crawled HTML pages to obtain the number of uploaded videos, number of subscribers (i.e., the number of users that is following the user u), and total number of views for the content uploaded by the user u. We ended with the profiles for 208K unique users, denoted as U.

Note that, the metadata fields in Table 1 (other than $TitleText$ and $TagText$ that are related to the basic features) constitute the raw social features, i.e., the basis of our features used for ranking. We defer the formal definitions of social features derived from the raw ones for ranking purposes to Section 4.

In Table 2, we provide the basic statistics on the appropriate metadata fields computed over the set of 365K unique videos, V. As V includes the videos retrieved for mostly popular search suggestions, some of the popularity-related metadata statistics seem to be much higher than those obtained for the YouTube crawls that are closer to random distributions (i.e., those based on generic queries). For instance, the average of the view counts is 274K for our dataset, which is an order of magnitude larger value than those reported in some recent works (e.g., see Table 1 in [18] and Table 2 in [4]).

Table 2 also shows that the average number of comments in our collection is 0.18% of the average number of views, i.e., very close to 0.16% reported by Cha et al. in 2009 [1], but interestingly, less than 0.5% reported by another recent study [17]. As a final remark, we observe that the standard deviation values are rather high for all metadata fields presented in Table 1, a result again in line with the previous findings [18].

3 Query Result Characterization Using Social Features

Characteristics of YouTube Queries. In contrast to web search, for which publicly available query logs allow analyzing issues like the user search intentions, result distributions, etc., there exist no public query logs that can be exploited for characterizing the users interests for video search in platforms like YouTube. So, we begin with a short analysis of the queries in Q, as the way our query set is constructed allows us to shed light on the real interests of YouTube users.

We classified the 1,450 queries in Q based on the YouTube provided category of their resulting videos. In particular, a query's category is designated as the most popular category among those of the videos retrieved for this query. Not surprisingly, the majority of the queries fall into the "music" category. The other popular query categories are "entertainment", "gaming" and "sports", as shown in Figure 1a. In addition to the automatic categorization, we also conducted a

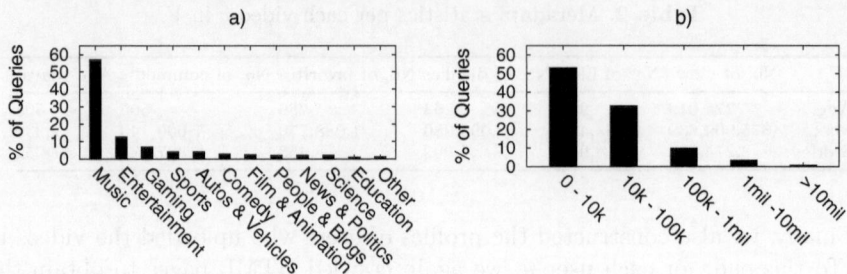

Fig. 1. Query characteristics: (a) Category distribution, and (b) No. of results

manual analysis to detect named entities appearing in the queries and found that 46% of the queries include a person entity (e.g., a singer, movie star, music band, YouTube user, etc.) and another 9% of them include a product entity.

In Figure 1b, we provide the distribution of the number of results for our queries. The plot shows that for almost half of the queries YouTube reports more than 10K resulting videos, which is rather expected, as our queries are composed of popular query suggestions.

Characteristics of the Top-Ranked Results. In this section, we present the basic characteristics of top-50 query results with respect to the raw social features such as the number of views, likes, dislikes and comments. While earlier studies about YouTube (e.g., [4,1,18]) also report statistics for some of these features, their analyses are usually over a *set* of videos (e.g., such as those we provide in Table 2). To the best of our knowledge, ours is the first study that provides an analysis for social features taking into account the rank of the videos in the query results.

Figure 2a shows the average number of views for the videos that are ranked at the i-th position in the query results, where $1 \leq i \leq 50$. In general, the number of views is quite high (around 300,000 views even at rank 50), which is not surprising due to the way we choose our queries, as discussed before. The number of views for the top-ranked video is considerably higher than that for the others and indeed the videos that are in top-7 results are viewed more than 1 million times on the average.

A common behavior of YouTube users is rating the viewed video, usually expressed by clicking like/dislike buttons. We find that the videos in top-10 have higher number of likes and dislikes in comparison to the rest of the videos (see Figures 2b and 2c, respectively). Moreover, there is an order of magnitude difference between the number of likes and dislikes: the former starts around 16,000 and goes down to 2,000 whereas number of dislikes starts from 1,800 and goes down to 200 for top-50 videos. Note that, for such popular videos with hundred of thousands of views, it should not be surprising that there are some dislikes, as well. However, on the average, we observe that more than 93% of the ratings for top-50 (and even for top-300) videos is positive. In addition to likes/dislikes, some videos are marked as favorite by some users. We observed a similar trend to that in Figure 2b for the favorite counts (plot not shown here).

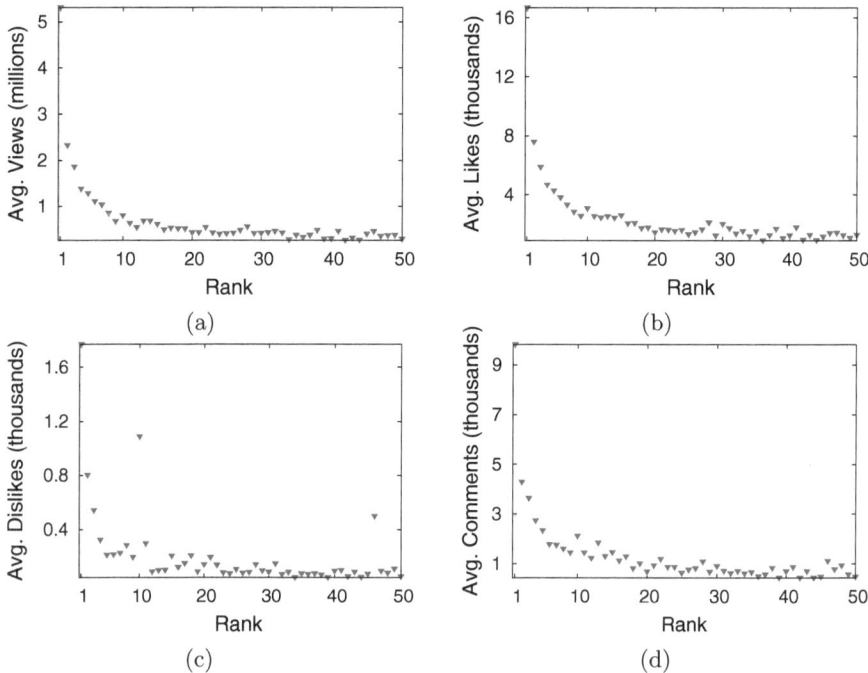

Fig. 2. Avg. no. of (a) views, (b) likes, (c) dislikes and (d) comments vs. video rank

An even stronger form of the user interaction and participation in YouTube is posting comments for videos [16,17]. Figure 2d depicts the average number of comments for videos at each result position. As in the previous cases, the top results have attracted considerably more attention than others, and the average number of comments drops under 3,000 for the videos ranked below top-5.

To summarize, we find out that the videos in top-10 YouTube results for our queries are usually those that attracted very high user interest as expressed by the number of views, likes, favorites and comments. For the rest of the results, the differences among the videos in terms of the values for these social features seem to be rather negligible. The high values of social features for the videos in top-10 results can also be attributed to well-known Yule process (or rich-get-richer principle) [1], as the videos that appear in the first page are more likely to be viewed and interacted. On the other hand, when all other features (such as the textual relevance to a query, etc.) are equal, it seems intuitive to put a video that is highly viewed/liked/commented etc. at a higher rank than one that attracted no interest. While it is not possible to draw a conclusion whichever of these explanations (or, maybe both) holds for YouTube (and indeed, it is not the goal of this paper), it seems as a worthwhile direction to further investigate the retrieval potential of the social features in more depth.

4   Effectiveness and Similarity of the Individual Features

In this section, we seek the answer for the following two questions: 1) How effective is each individual feature for ranking videos? and, 2) How similar are the rankings generated by the different pairs of features? To answer these questions, for each query q, we need to re-rank the retrieved videos $v \in R_q$ (where $|R_q| \leq 300$) with respect to each feature f in our feature set, F. So, we begin with formally defining the basic and social features that are used for ranking videos.

Basic Features. We define two basic features that represent the similarity of a query to video titles and tags, as typically employed in a video retrieval system. The features "title similarity" (f_{Title}) and "tag similarity" (f_{Tags}) represent the vector-based similarity score of the query text, q, to a video's title, $TitleText(v)$, and tags, $TagText(v)$, respectively. In our setup, we create two indexes, namely for video titles and tags, using Lucene 3.5 library (www.lucene.apache.org) and employ its default retrieval function (based on the TF-IDF weighting model) to obtain the title and tag similarity scores for each (q, v) where $v \in R_q$. For each query, Lucene similarity scores are normalized to $[0, 1]$ range.

Social Features. In this paper, we call the features that are formed due to some interaction with the users *after* the video becomes available as social features. In this sense, as also discussed in the previous section, we first exploit the raw features provided by the system. For the metadata fields shown in Table 1, namely, number of views ($W(v)$), likes ($L(v)$), favorites ($F(v)$), and comments ($C(v)$), we create the corresponding features f_W, f_L, f_F, f_C. However, to be able to use them for ranking videos, we normalize each feature value with the age of the video, $G(v)$. For the sake of completeness, we also consider the age of a video as a possible ranking feature and denote as f_G, while it is not a truly social feature (i.e., it is not based on the user interaction). Furthermore, we derive the following social features from the raw features and available data for our videos in V (all of these feature values are further normalized into $[0, 1]$ range based on the maximum score observed for a given query):

- *Normalized no. of ratings (f_R)*: This feature represents the total number of ratings per video. The ranking criteria is: $(L(v) + D(v))/G(v)$.
- *Normalized ratio of likes (f_{RL})*: This feature captures the fraction of likes over all ratings for a video. The ranking criteria is: $(L(v)/(L(v)+D(v)))/G(v)$.
- *Normalized no. of comment authors (f_{CA})*: We extract the username fields from the crawled comments to capture the number of different users who commented on a video. The ranking criteria is: $A(v)/G(v)$ where $A(v)$ is the number of unique users who posted a comment for v.
- *Comment similarity (f_{Com})*: We first aggregate the top-25 most popular comments (i.e., those with the highest number of likes) of each video into a single document and index these documents using Lucene. Then, the Lucene score between q and the comment document is computed for each $v \in R_q$.
- *Average Comment Positivity (f_{Pos})*: We analyze the sentiment expressed in the comments by using a public vocabulary based tool, SentiWordNet [6],

as in [16]. Simply, this tool assigns a triplet representing the objectivity, negativity and positivity scores for each word in a comment, which are then averaged to obtain the overall scores for the comment. For ranking purposes, we only consider the average positivity score over all comments of a video, for which the tool can generate a score. The ranking criteria is: $\sum Pos(c_i)/C(v)$ where $Pos(c_i)$ is the positivity score for the comment c_i of a video v.

- *Uploader popularity (f_{Up}):* The ranking criteria for a video v with an uploader u is $\sum W(v_j)/|Videos(u)|$, where $v_j \in Videos(u)$ and $Videos(u)$ includes the videos uploaded by u.

To sum up, our feature set consists of two basic and eleven social features in total, i.e., $F = \{f_{Title}, f_{Tags}, f_W, f_L, f_F, f_C, f_G, f_R, f_{RL}, f_{CA}, f_{Com}, f_{Pos}, f_{Up}\}$.

User Study. To compute the effectiveness of each individual feature for ranking videos, we need the relevance annotations for (q, v) pairs, where $v \in R_q$. As this task requires serious human effort, we sampled 50 queries from our query set, Q. Furthermore, since most users do not go beyond the first page of the results, we focused on the effectiveness of the top-10 results using each feature. Therefore, for each query q, we ranked $v \in R_q$ using each feature $f \in F$ and obtained top-10 videos, denoted as $R_{q,f}$. Thus, the videos to be annotated for a query q in our annotation set is $\cup R_{q,f}$, $f \in F$. Due to the overlaps among the rankings, we ended up with around 70 videos per query and 3,500 query-video pairs in total.

To obtain the relevance judgments, we conducted a user study that involves 19 participants who evaluated 3 queries on the average. Four of the participants were female and the rest were males, and the age range was 20-35. All participants were from computer science related disciplines and 3 of them are undergraduates, 12 of them are graduate students, and the rest are post-docs. The participants were physically located in Germany, Turkey and USA.

We asked each participant to choose a few queries that are interesting for them from our set of queries. Next, we assigned each participant 2 or 3 queries, and asked to annotate the videos displayed for a given query using a 5-point rating scale, i.e., in the order of highly irrelevant, irrelevant, undecided, relevant, and highly relevant. The annotation process was carried out using our Web site.[2] Since videos are not downloaded but streamed directly from YouTube, it turned out that a small percentage of them have disappeared in time, i.e., deleted by the uploader or not displayed in certain countries due to copyright violation issues. The participants were asked to annotate such videos with rating 0. Finally, to avoid any bias, no social features were displayed along with the videos, but titles and tags are kept to facilitate the judgment task.

Among these 3,500 relevance annotations, 37% and 18% of the videos are judged as highly relevant and relevant, respectively; whereas only 21% of them are found irrelevant or highly irrelevant. We noted that only 10% of the videos are graded as 0, i.e., not accessible. As most of these are due to copyright issues, i.e, potentially official videos for songs, movies, etc., we take an optimistic approach

---

[2] http://tomcat.l3s.uni-hannover.de:8080/Evaluation2/welcome.xhtml

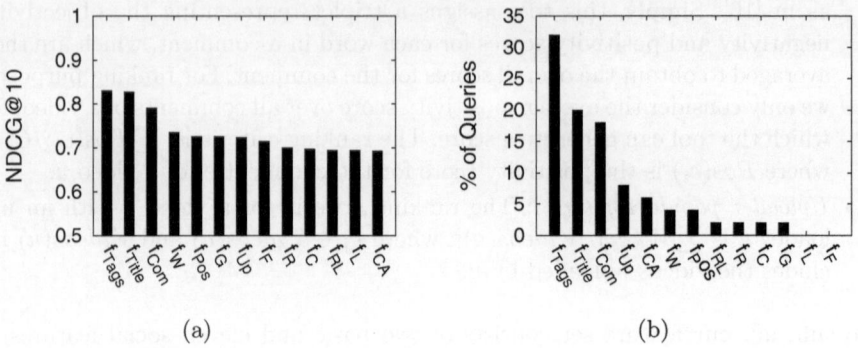

Fig. 3. (a) Avg. NDCG@10 for top-10 videos per feature, (b)Fraction of queries for which a given feature yields the ranking with the highest NDCG@10

and assume their labels as relevant (otherwise, for some rankings, we could end up with less than 10 videos, which is not preferable for the comparison purposes).

Effectiveness of the Individual Features. To evaluate the performance of each individual feature, we compute the Normalized Discounted Cumulative Gain (NDCG) metric using the well-known trec_eval software package[3]. For each query q and top-10 ranking $R_{q,f}$, where $f \in F$, we compute the $NDCG$@10 score using the annotations from our user study as the golden data.

Figure 3a reports the average $NDCG$@10 scores over all 50 queries for each feature. We see that top-3 most effective features are f_{Tags}, f_{Title}, and f_{Com}. In addition to the average values, we explore how successful each feature in a micro level and compute the percentage of queries for which a particular feature yields the highest $NDCG$@10 score. As shown in Fig. 3b, the above three features provide the best rankings for 32%, 20% and 18% of the queries, respectively. Given that f_{Tags} and f_{Title} are basic features, summing up to 52%, a non-trivial 48% of the queries can benefit from social features. To obtain the upper bounds for using basic and all features, we assume an oracle ensemble method choosing for each query the best ranking from basic features and all features (i.e., basic + social), respectively. In this idealistic case, the average NDCG scores would be 0.8644 for using only basic features, and 0.9036 for using all available features. This means that there is a non-trivial potential of improving retrieval quality using social features along with the basic features, as we explore in Section 5.

Similarity of the Individual Features. We compute the pair-wise overlap between the features by averaging the similarity of their top-10 rankings $R_{q,f}$, $(f \in F)$ over all 50 queries. A typical method for measuring the similarity of two ranked lists is using Kendall's Tau metric [7]. Since we focus on the top-10 rankings, we employ a special variant of Kendall's Tau [7], which can handle the cases when a video appears in one of the top-10 rankings but not in the other.

---

[3] http://trec.nist.gov/trec_eval/

Table 3. Kendall's Tau similarity between each pair of features

	f_W	f_L	f_{RL}	f_R	f_F	f_C	f_{CA}	f_{Title}	f_{Tags}	f_{Com}	f_G	f_{Up}	f_{Pos}
f_W	1	.73	.24	.74	.78	.69	.34	.05	.04	.13	.12	.42	.24
f_L		1	.39	.98	.80	.79	.37	.05	.04	.12	.17	.37	.26
f_{RL}			1	.39	.26	.34	.15	.04	.04	.05	.39	.08	.03
f_R				1	.80	.80	.37	.05	.04	.11	.17	.37	.26
f_F					1	.72	.35	.05	.04	.12	.14	.40	.26
f_C						1	.41	.05	.04	.11	.20	.35	.24
f_{CA}							1	.06	.06	.17	.41	.18	.33
f_{Title}								1	.28	.08	.08	.06	.04
f_{Tags}									1	.08	.08	.05	.03
f_{Com}										1	.02	.11	.14
f_G											1	.04	.01
f_{Up}												1	.19
f_{Pos}													1

In Table 3, we provide the Kendall's Tau scores that are normalized to [0-1] range where 0 means completely different rankings and 1 means equal rankings. The scores imply some redundancy: for instance top-10 rankings provided by the feature pairs (f_L, f_F), (f_L, f_R), (f_R, f_F), (f_C, f_R) all yield the similarity scores higher than 0.8. In Section 5, we will use a feature selection approach to filter the redundant features.

5   Learning to Rank Videos Using Social Features

In the light of the above findings, it is promising to combine basic and social features to optimize the retrieval performance. Moreover, as there is a high overlap between the rankings generated by certain pairs of the features, it also seems reasonable to apply a feature selection algorithm. In what follows, we present our video retrieval framework involving a number of state-of-the-art learning to rank (LETOR) strategies and a greedy feature selection strategy adapted from [10]. In this framework, we explore the impact of social features on the video retrieval.

Video Retrieval Framework. In recent years, traditional ranking approaches based on manually designed ranking functions (such as BM25, TF-IDF, etc.) are replaced or complemented by the rankers built by machine learning strategies [2]. In a typical LETOR framework, a machine learning algorithm is trained using a set of triples of $(q, \boldsymbol{F}, r)$, where q is the query id, $\boldsymbol{F}$ is the m-dimensional feature vector for a result object retrieved for q, and r is the relevance score. The learnt model is used to predict the relevance score for each pair $(q, \boldsymbol{F})$ in the test set, which is then sorted with respect to these predicted scores. The success of the ranking model is evaluated using the measures like NDCG [2].

The LETOR algorithms proposed in the literature fall into three categories, namely, point-wise, pair-wise and list-wise [2]. In this paper, we employ six LETOR approaches that cover all of these categories. We provide a concise description of each approach and refer the readers to the literature for details.

- *RankSVM*: This approach extends traditional SVM by utilizing instance pairs and their labels during training. In this work, we use the implementation[4] by Joachims [12].
- *RankBoost*: First introduced by [8], this algorithm also employs a pair-wise boosting technique for ranking.
- *CoordinateAscent*: This is a list-wise linear model which uses coordinate ascent technique that optimizes multivariate objective functions by sequentially doing optimization in one dimension at a time [13]. For RankBoost and Coordinate Ascent approaches, we use the RankLib package[5] (also see [5]).
- *Gradient Boosted Regression Trees (GBRT)*: This is a simple yet very effective point-wise method for learning non-linear functions [9] and indeed, said to be the current state-of-the-art learning paradigm [14].
- *Random Forests (RF)*: Random Forests is a point-wise ranking approach based on the bagging technique, i.e., applying the learning algorithm multiple times on different subsets of the training data and averaging the results [14].
- *Initialized Gradient Boosted Regression Trees (iGBRT)*: This approach uses the predictions from the RF algorithm as a starting point for the GBRT algorithm [14]. We use the RT-Rank library[6] for the GBRT, RF and iGBRT.

For the above algorithms, we tried a range of parameters and report the results for the best-performing setup. The final parameters are usually close to their default values; still, we provide the exact settings at our web site[7] along with all other data and result files used in this study.

Feature Selection for LETOR Approaches. Feature selection is a well-known approach in machine learning for enhancing accuracy (e.g., by preventing over-fitting) of the learnt model and efficiency of the learning process [10,5]. Geng et al. address the vitality of the feature selection issue for machine learning based approaches to the ranking problem and propose a greedy feature selection strategy that fits perfectly to our framework [10]. Formally, given a set of features $\{f_1, ..., f_m\}$ and the target number of features, k, the goal is selecting k features that would yield the maximum performance for a LETOR algorithm. Each feature is associated with an importance score, $Imp(f)$, which is an indicator of the retrieval effectiveness of f. Furthermore, for each feature pair (f_i, f_j), similarity of their top-N rankings, is computed. The optimization problem is defined as choosing a set of k features that maximizes the sum of the feature importance scores and minimizes the sum of the similarity scores between any two features.

The greedy search algorithm in [10] starts with choosing the feature, say f_i, with the highest importance score into the k-feature set. Next, for each of the remaining features f_j, their importance score is updated with the statement $Imp(f_j) \leftarrow Imp(f_j) - Sim(f_i, f_j) \cdot 2c$, where c is a constant to balance the importance and similarity optimization objectives. The algorithm proceeds with

---

[4] http://www.cs.cornell.edu/people/tj/svm_light/svm_rank.html
[5] http://people.cs.umass.edu/~vdang/ranklib.html
[6] http://research.engineering.wustl.edu/~amohan/
[7] http://139.179.21.106/~ismaila/social.html

choosing the next feature with the highest importance score and updating the remaining scores, until k features are determined.

In our case, following the practice in [10], the feature importance score, $Imp(f)$, is set to $NDCG@10$ score of f obtained over the queries in the training set. The similarity score $Sim(f_i, f_j)$ between any two features f_i and f_j is computed by the variant of Kendall's Tau metric (as described in Section 4) between their top-10 rankings, again, over the queries in the training set.

Experimental Results for Feature Selection. In our LETOR framework, all experiments are conducted using five-fold cross validation over the set of 50 queries described in the previous section. For each fold, we first used the training set of 40 queries (i.e., around 2,800 annotations) to determine the k-feature sets (where $1 \leq k \leq 13$) from the set of all basic and social features, i.e, F, using the greedy selection algorithm. Next, for each value of k, the LETOR algorithms in our repository are trained with the same set of instances and these k features; and tested on the remaining 10 queries (700 annotated instances). The average $NDCG@5$ and $NDCG@10$ scores are computed using `trec_eval` software for the test queries (similar results for metrics $P@5$ and $P@10$ are discarded due to lack of space). The final scores are obtained by averaging over the folds.

In Figure 4, we provide the performance of each LETOR algorithm with respect to number of selected features. As in [10], the performance fluctuates as new features are added to the set. Nevertheless, for each algorithm, there exists a set of features, so-called the *best-k* set, that yields a higher performance than using all of the features, which justify our use of a feature selection algorithm.

Experimental Results for the Impact of Social Features. To expose the potential of social features for video ranking, we compare the retrieval performance of using best-k feature sets (from Figure 4) to the performance of using only basic features. For the latter case, we only use features f_{Tags} and f_{Title} for training and testing all the LETOR algorithms. Our findings are reported in Table 4. Note that, all best-k sets include some social features as k is always found to be greater than the number of basic features, i.e.,2.

Before discussing our results, please note that we avoid ranking the LETOR algorithms among each other in our framework. Because, the one-way ANOVA test for comparing $NDCG@10$ scores of 50 queries for these six algorithms reveals that the performance differences among them are usually not significant, regardless of the feature set they employ (i.e., basic or best-k features). In other words, it is not accurate, from statistical point of view, which of these algorithms performs best in our video retrieval scenario, and thus it is important to improve the performance of any of these algorithms using the social features.

As Table 4 reveals, for all of the algorithms, the best-k features can improve the $NDCG@10$ scores that are obtained by the basic features alone. For some cases, the improvement is numerically small (though, so are most of the results reported in the LETOR literature, e.g., see [2,14]) while in some other cases, using particular social features in combination with the basic features can add up to an absolute 8% to the effectiveness. The gains in $NDCG@10$ obtained by using

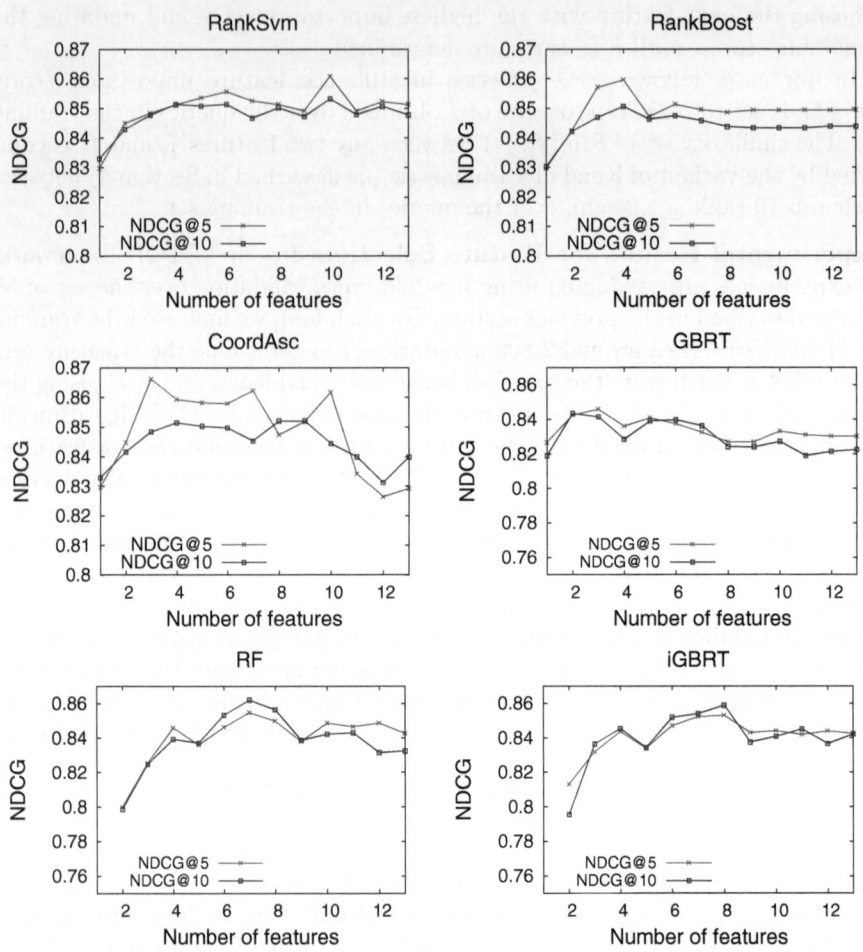

Fig. 4. *NDCG* scores for the LETOR algorithms w.r.t. the number of features

Table 4. Avg. *NDCG*@10 scores for LETOR algorithms using basic and best-k features (for bold cases, differences from the baseline are statistically significant)

	RankSvm	RankBoost	CoordAsc	GBRT	RF	iGBRT
Basic features	0.8511	0.8423	0.8510	0.8447	0.7775	0.8019
Best-k features	0.8532	0.8506	0.8540	0.8458	**0.8536**	**0.8512**
No. of features (k)	10	4	9	3	7	8

the best-k sets with social features are found to be statistically significant on a 95% confidence level for the RF and iGBRT approaches. The smaller numeric improvements in *NDCG*@10 scores, observed for the other algorithms, are not statistically significant.

To sum up, we conclude that social features (especially those related to the comments) have the potential to improve the video retrieval performance when combined with the basic features, which indeed constitute a very strong baseline.

6   Conclusion and Future Work

We show that while basic features using the query similarity to video titles and tags are most effective for the video retrieval, social features are also valuable and can yield the best rankings for 48% of the queries, which indicates a potential to improve the retrieval effectiveness. Our evaluations using a greedy feature selection algorithm and six state-of-the-art LETOR algorithms support our hypothesis: the rankers based on the subsets of features including both basic and social features outperform those built by using only basic features.

For future research, we plan to address some limitations of the current study. First, while we have a fairly large set of annotated data (i.e., 3,500 (query, video) pairs), it is still mandatory to have larger annotated datasets to draw more general conclusions on the value of social features for ranking. This issue can be tackled by the emerging crowdsourcing solutions. Secondly, while we consider our query set (based on the query suggestions for YouTube) as an important asset for research in this field; on the downside, this set essentially includes the popular queries. As a remedy, we plan to extend our set with the tail queries, which can be obtained via user studies or crowdsourcing. Finally, we plan to enrich our feature set with the other comment-related features proposed in [11].

Acknowledgments. This work is partially supported by EU FP7 Projects CUBRIK (contract no. 287704) and GLOCAL (contract no. 248984).

References

1. Cha, M., Kwak, H., Rodriguez, P., Ahn, Y.Y., Moon, S.: Analyzing the video popularity characteristics of large-scale user generated content systems. IEEE/ACM Trans. Netw. 17(5), 1357–1370 (2009)
2. Chapelle, O., Chang, Y.: Yahoo! learning to rank challenge overview. Journal of Machine Learning Research - Proceedings Track 14, 1–24 (2011)
3. Chelaru, S., Altingovde, I.S., Siersdorfer, S.: Analyzing the Polarity of Opinionated Queries. In: Baeza-Yates, R., de Vries, A.P., Zaragoza, H., Cambazoglu, B.B., Murdock, V., Lempel, R., Silvestri, F. (eds.) ECIR 2012. LNCS, vol. 7224, pp. 463–467. Springer, Heidelberg (2012)
4. Cheng, X., Dale, C., Liu, J.: Statistics and social network of youtube videos. In: Proc. of IEEE IWQoS 2008 (2008)
5. Dang, V., Croft, W.B.: Feature selection for document ranking using best first search and coordinate ascent. In: Proc. of SIGIR 2010 Workshop on Feature Generation and Selection for Information Retrieval (2010)
6. Esuli, A., Sebastiani, F.: Sentiwordnet: A publicly available lexical resource for opinion mining. In: Proc. of LREC 2006, pp. 417–422 (2006)
7. Fagin, R., Kumar, R., Sivakumar, D.: Comparing top k lists. SIAM J. Discrete Math. 17(1), 134–160 (2003)

8. Freund, Y., Iyer, R., Schapire, R.E., Singer, Y.: An efficient boosting algorithm for combining preferences. J. Mach. Learn. Res. 4, 933–969 (2003)
9. Friedman, J.H.: Stochastic gradient boosting. Comput. Stat. Data Anal. 38(4), 367–378 (2002)
10. Geng, X., Liu, T.Y., Qin, T., Li, H.: Feature selection for ranking. In: Proc. of SIGIR 2007, pp. 407–414 (2007)
11. Hsu, C.F., Khabiri, E., Caverlee, J.: Ranking comments on the social web. In: Proc. of CSE 2009, pp. 90–97 (2009)
12. Joachims, T.: Training linear svms in linear time. In: KDD 2006, pp. 217–226 (2006)
13. Metzler, D., Bruce Croft, W.: Linear feature-based models for information retrieval. Inf. Retr. 10(3), 257–274 (2007)
14. Mohan, A., Chen, Z., Weinberger, K.: Web-search ranking with initialized gradient boosted regression trees. Journal of Machine Learning Research 14, 77–89 (2011)
15. San Pedro, J., Yeh, T., Oliver, N.: Leveraging user comments for aesthetic aware image search reranking. In: WWW 2012, pp. 439–448 (2012)
16. Siersdorfer, S., Chelaru, S., Nejdl, W., San Pedro, J.: How useful are your comments?: analyzing and predicting youtube comments and comment ratings. In: WWW 2010, pp. 891–900 (2010)
17. Thelwall, M., Sud, P., Vis, F.: Commenting on youtube videos: From guatemalan rock to el big bang. JASIST 63(3), 616–629 (2012)
18. Vavliakis, K.N., Gemenetzi, K., Mitkas, P.A.: A correlation analysis of web social media. In: WIMS 2011, pp. 1–5 (2011)
19. Yee, W.G., Yates, A., Liu, S., Frieder, O.: Are web user comments useful for search? In: Proc. of SIGIR 2009 Workshop on LSDS-IR (2009)

An Empirical Study for Determining Relevant Features for Sentiment Summarization of Online Conversational Documents

Gino Mangnoesing[1], Arthur van Bunningen[2], Alexander Hogenboom[1],
Frederik Hogenboom[1], and Flavius Frasincar[1]

[1] Erasmus University Rotterdam
PO Box 1738, NL-3000 DR, Rotterdam, The Netherlands
gvh.sing@gmail.com, {hogenboom,fhogenboom,frasincar}@ese.eur.nl
[2] Teezir BV
Wilhelminapark 46, NL-3581 NL, Utrecht, The Netherlands
Arthur.van.Bunningen@teezir.com

Abstract. The phenomenon of big data makes managing, processing, and extracting valuable information from the Web an increasingly challenging task. As such, the abundance of user-generated content with opinions about products or brands requires appropriate tools in order to be able to capture consumer sentiment. Such tools can be used to aggregate content by means of sentiment summarization techniques, extracting text segments that reflect the overall sentiment of a text in a compressed form. We explore what features distinguish relevant from irrelevant text segments in terms of the extent to which they reflect the overall sentiment of conversational documents. In our empirical study on a collection of Dutch conversational documents, we find that text segments with opinions, segments with arguments supporting these opinions, segments discussing aspects of the subject of a text, and relatively long sentences are key indicators for text segments that summarize the sentiment conveyed by a text as a whole.

1 Introduction

In the last decade, the World Wide Web has exponentially grown to a network with more than 555 million websites and over 2 billion users worldwide [11]. The Web has become an influential source of information with an increasing share of user-generated content (UGC) from many contributors. This content has taken many different forms, such as forums, wikis, (micro)blogs, review sites, podcasts, or parts of a website, e.g., reviews on Amazon.com or Booking.com. Amongst all this content are many opinions which can carry a specific sentiment, for example about products, brands, or politics. People complain or recommend what products or services to buy or not to buy, which movies to see, or what places to go to. Consequently, the Web as a medium has become a strong influencer of purchasing decisions and a platform that reflects consumer preferences, which is interesting for both consumers and producers.

X.S. Wang et al. (Eds.): WISE 2012, LNCS 7651, pp. 567–579, 2012.

However, it has become difficult to use the Web as a helping hand for making decisions, as it has become much harder to keep track of all available data online. Today's data is often unstructured, scattered all over the Web, and expanding extremely fast. This phenomenon is also referred to as big data [7]. There is simply too much information to process, as well as a lack of filters to extract the parts that are relevant and informative with respect to one's requirements.

This situation has led to a great need for aggregation for a better information overview and making big data insightful and eventually profitable. Fortunately, there are ways to accomplish aggregation of opinions by means of sentiment summarization. Whereas sentiment analysis computes a score to indicate the attitude people have towards a certain topic, sentiment summarization takes it one step further by extracting the most important text fragments that sufficiently represent the sentiment of the text as a whole. In this light, sentiment summarization could help consumers to quickly discover the pros and cons of products and services and it could support companies with brand monitoring and customer relationship management.

Existing work in the field of sentiment summarization identifies relevant text fragments by using one or just a few characteristics (features), such as product aspects [3, 6, 12–14], intensity [6], or a fragment's position within a document [2]. In our current work, we take a much broader approach, attempting to learn about the relative importance of a larger set of features. Additionally, rather than focusing on automatic detection of features, we let users annotate features in order to find out which features apply to a specific summary. Another distinctive factor in our study is our focus on conversational documents about a brand, taken from forums, whereas many existing studies use opinion-focused documents like movie reviews or restaurant reviews. Conversational documents are texts that have the characteristics of a dialogue, for example texts in which people ask or answer questions, give comments, or express complaints. Conversational documents are typically found on forums and social media platforms.

The remainder of this paper is structured as follows. Section 2 discusses related work on sentiment summarization. Section 3 discusses the features we consider as proxies for the relevance of sentences in terms of the extent to which they reflect the overall sentiment of a text. Section 4 discusses our method for feature evaluation of sentiment summaries. The evaluation of the proposed method is discussed in Section 5. Finally, we present our conclusions and directions for future work in Section 6.

2   Sentiment Summarization

Sentiment analysis typically aims to examine "what other people think" about a specific entity or topic [9, 10] and allows for determining a score for the polarity of text. These sentiment scores are important for sentiment summarization, as the aim here is to extract text fragments that reflect the overall sentiment of a text, while using less words than the original text. The key to generating such a summary is distinguishing relevant sentences from irrelevant sentences.

Existing work is typically focused on sentences that discuss aspects of a topic and often summarize sentiment found in reviews. For instance, Blair-Goldensohn et al. [3] present a sentiment summarizer for user reviews of local services like restaurants or hotels. Their method extracts the snippets from the text that include sentiment-carrying text. The snippets are then examined on the presence of service aspects, such as the food, service, or price. Subsequently, the sentiment per aspect is aggregated, based on the snippets that include these aspects. Blair-Goldensohn et al. [3] observe that most services share similar basic aspects and that acceptable summaries can be created when guiding the selection process of text fragments by the presence of topic aspects in these fragments. However, other work has shown that other features have their merit too [2, 6].

Lerman et al. [6] propose several sentiment summarizers that include various summary features. One of the used features besides aspects is the intensity of a text fragment, capturing the magnitude of its conveyed sentiment. Another considered feature is the mismatch of a text fragment, which measures the difference between the sentiment of the fragment and the known overall sentiment of the topic. The results of this study indicate that none of the proposed sentiment summarizers is strongly preferred over any other. However, users prefer summarizers that account for aspects and sentiment over those that do not.

Beneike et al. [2] aim to extract a single fragment from a movie review that reflects the sentiment of the author towards the movie. This fragment is referred to as a quotation. Beneike et al. [2] show that three features appear to be predictive of whether a text fragment is chosen as a quotation. The first feature is the location of the fragment within the paragraph. Quotations occur most often at the ends of paragraphs. The second feature is the location of the fragment within a document, often early in the document or the final 5% of the text. The third feature is the word choice. In quotations, the most used words often express emotion directly and/or are interchangeable with the topic, e.g., a reviewer with a positive opinion on a movie may refer to the movie as "a piece of art".

As such, existing work only considers few features as proxies for the relevance of a text fragment for a sentiment summary. Existing work shows only few common denominators in considered features, with one of the most widely used features being the discussion of certain aspects of a topic. We aim to investigate which of the existing as well as new features can identify relevant fragments for sentiment summaries. In addition, the focus of our work is on conversational documents captured by forum messages, which differ from the reviews typically used in existing work in that they are less explicitly focused on expressing opinions and in that they are structured as conversations rather than as single messages.

3   Selecting Features for Sentiment Summarization

In this research, we perform a user-based evaluation of summary sentences and various summarization features. Users evaluate these sentences in terms of the extent to which they would fit in a summary reflecting the sentiment conveyed by the text as a whole. The features we explore in this research are listed below.

1. The sentence contains an opinion about the topic.
2. The sentence is rather positive or rather negative (high intensity).
3. The sentence includes one or more (sub)aspects of the topic.
4. The sentence is part of the introduction of the document.
5. The sentence is part of the conclusion of the document.
6. The sentence contains an adjective.
7. The sentence contains an adverb.
8. The sentence addresses an event or experience described in the document.
9. The sentence contains an advice or recommendation.
10. The sentence contains an argument supporting an opinion, vision, or statement in the document.
11. The sentence contains or is part of a comparison in the document.
12. The sentence contains words that are also present in the document title (with the exception of definite and indefinite articles).
13. The sentence contains a list or sequence.
14. The sentence is relatively long.
15. The sentence is relatively short.

Some of these features are inspired by existing work, discussed in Section 2, in order to be able to validate existing results on conversational documents. Yet, most of our considered features are contributions of our current endeavors. We include feature 8, as we hypothesize that discussions of events are relevant on forums, because people tend to use forums to complain about something, which is typically a recent experience or event. Feature 9 is considered, because we hypothesize that a recommendation can be important for a summary. For example, an advice may reflect or be supported by someone's sentiment towards the brand. Similarly, arguments can further motivate an opinion or statement, thus rendering feature 10 an interesting feature. In this light, we also consider feature 11, as people tend to motivate why they favor one brand over another in a comparison. Additionally, feature 12 is included, because the subject of forum conversations is typically mentioned in the title of a document. Titles may hence form a concise representation of the most important message of a document and may thus be useful in identifying relevant fragments in the body of a document. Feature 13 is included as people tend to list their complaints and possible advantages or disadvantages. Last, features 14 and 15 are included in order to determine whether relevant summary sentences are typically relatively long or short sentences.

4   Feature Evaluation for Sentiment Summarization

In order to support our current research goals, we propose a method for Feature Evaluation for Sentiment Summarization (FESS). The goal of this method is to collect user evaluations of sentences with respect to the relevance of the sentence for inclusion in a sentiment summary. By doing so, we aim to find out which features are good indicators for the relevance of a sentence. In the following sections, we first present an overview of our method design. Then, we elaborate on the implementation of our method.

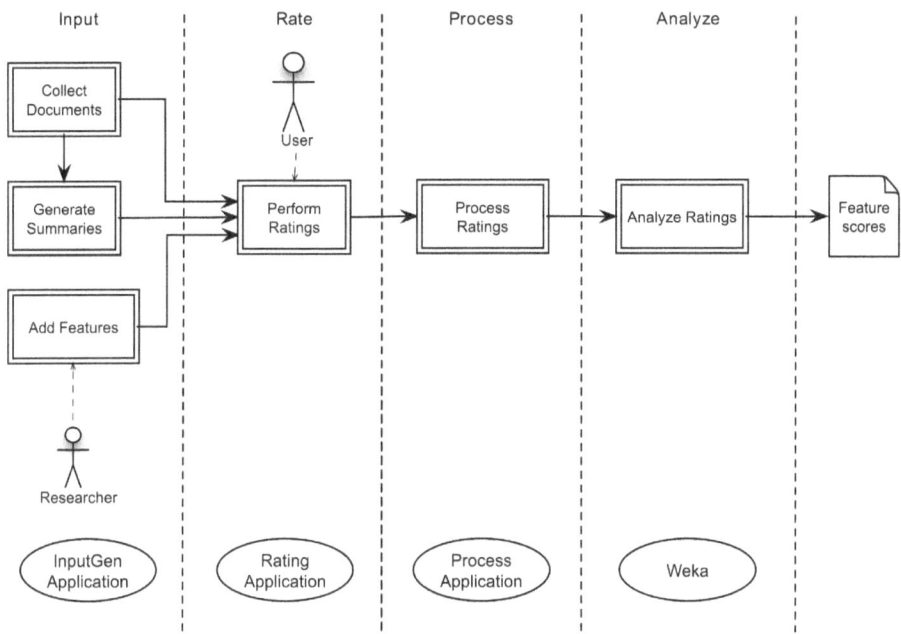

Fig. 1. A schematic overview of our proposed method

4.1   Method Design

Figure 1 demonstrates the design of our proposed method. FESS consists of four main steps. First, we prepare the inputs for the second step, in which we collect evaluations for these inputs from human raters. We then process these ratings and analyze the results in order to be able to provide recommendations on which types of features to use in sentiment summarizers for conversational documents.

Input. The first step of our method is intended to provide input to the second step in which we retrieve user evaluations. The first input we collect is a set of documents, crawled from the Web and filtered on several specific characteristics, e.g., the topic and the source of the document. Our method supports several types of conversational texts, e.g., tweets, comments, or forum posts.

The second input consists of candidate summary sentences. A collection of such sentences is generated for every single document in our collection. In order to do so, we first preprocess the text by detecting sentences and words. After preprocessing, we select summary candidates by flagging (marking) sentences that are likely to contain one or more of our considered summary features. Sentences may carry multiple flags. For every document, we then select a predefined number of summary sentences that are randomly selected from a collection of candidate sentences. Flags may be taken into account in this process in order to ensure that no features are under-represented in the final selection.

Last, a list of considered features of sentences is required as an input for the rating phase. We propose to use the features detailed in Section 3. When rating the selected sentences with respect to the extent to which they are relevant for inclusion in a summary, users can annotate these sentences for the presence or absence of our considered features.

Rate. In the rating process, all documents, summary sentences, and features are presented to human raters. We propose to divide users into groups in order to spread the time-consuming evaluation work among users. Each group of raters is presented a unique set of documents, with several summary sentence candidates per document. The evaluation scores can be saved for further processing.

The interaction model of FESS for collecting feature evaluations is composed of four steps. First, users are presented a random document related to a predefined topic, e.g., a document relevant for a search query for a brand. After having read the document, users are presented a potential summary sentence extracted from the same document. Subsequently, we ask users to classify the sentiment conveyed by the document as well as the sentiment conveyed by the sentence. Here, one can consider classes like negative (-1), neutral (0), and positive (1). Additionally, we ask users to evaluate the relevance of this sentence in terms of the extent to which it summarizes the sentiment conveyed by the document as a whole. This can be done by means of binary classification or by means of classification on an ordinal or continuous scale that can be mapped to binary classes. Subsequently, the users are asked to select the features applying to this specific sentence from a list.

Process. In the processing phase of our method, the evaluation data is transformed into a collection of data points, where each data point represents a single summary sentence with the evaluation scores for the document sentiment, the summary sentiment, the summary relevance score, and a binary representation of features, indicating whether the users selected these features (1) or not (0). In order to generate these data points, the user ratings for each sentence are aggregated. Sentiment evaluations are averaged over all raters, whereas majority voting is applied to both the (binary) relevance scores and the features.

Analyze. The collection of data points thus obtained can subsequently be analyzed in order to identify important proxies for the relevance of a sentence in a sentiment summary. First, we propose to use the collected data in order to determine the information gain [8] of each considered feature with respect to the binary classification of the relevance of our evaluated sentences. As an alternative method, we propose to use a feature selection method in order to identify the most informative subset of features, by considering the predictive power of each individual feature along with the degree of redundancy between the features [5]. Subsets of features that are highly correlated with the class while having low inter-correlation are preferred. In order to validate the results, we propose to perform stratified 10-fold cross-validation on these analyses.

4.2   Implementation

We have implemented our method using programming languages ASP.NET and C# in combination with an SQL database. As depicted in Fig. 1, each step of our framework is performed by a separate application.

Input. In order to provide inputs for the rating process, we have developed the InputGen Application, which collects documents, selects candidate summary sentences, and provides a list of considered features. In the document collection step, our implementation focuses on crawling popular Dutch forums for messages about Ziggo, a national media and communications services provider in the Netherlands. We filter the data for documents with a maximum length of a predefined number of characters. All documents are saved into a database.

Sentences that are potential candidates for inclusion in a sentiment summary are selected by the InputGen Application by first tokenizing the text into separate sentences and words. Then, we create a sub-collection of candidate sentences for each document and automatically flag the sentences, when they include a feature from a subset of features we scan for in advance. The collection of candidate sentences is retrieved by automatically filtering sentences with a predefined minimum amount of words. We make use of this minimum in order to filter out the majority of presumably meaningless short sentences.

Additionally, we automatically flag sentences that match the criteria of some of our features, such that features are as well-represented as possible in our selection of candidate sentences. The first feature we explicitly look for in sentences is the presence of aspects of our topic. In our application, sentences that contain words matching a pre-compiled list of lexical representations of aspects of our topic are flagged for this feature.

Another feature we explicitly look for is whether sentences are part of a conclusion. In order to accomplish this, we flag sentences in the last 25% of a document as being part of a conclusion. By doing so, we assume that conclusions typically occur at the end of the relatively short documents in our collection.

A third feature our application automatically scans for is sentences with high intensity, as high intensity often signals a strong presence of opinion. To this end, our application scores the sentences in our documents for their conveyed sentiment and assumes high absolute sentiment scores to signal high intensity. The sentences in our data set are analyzed for the sentiment conveyed by their text by means of an existing framework for lexicon-based sentiment analysis [1], which is a pipeline in which each component fulfills a specific task in analyzing the sentiment of a document. It first prepares documents by cleaning the text and performing initial linguistic analysis by identifying each word's part-of-speech as well as by distinguishing opinionated words and their modifiers from neutral words. Sentences are subsequently scored by sum-aggregating the sentiment scores of its opinionated words, while accounting for their modifiers, if any.

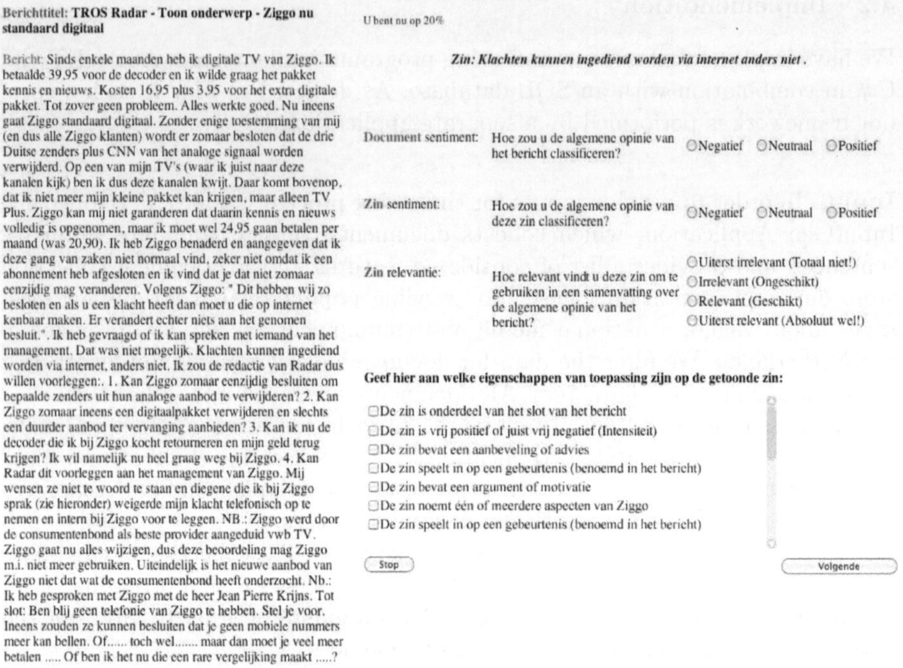

Berichttitel: **TROS Radar - Toon onderwerp - Ziggo nu standaard digitaal**

U bent nu op 20%

Bericht: Sinds enkele maanden heb ik digitale TV van Ziggo. Ik betaalde 39,95 voor de decoder en ik wilde graag het pakket kennis en nieuws. Kosten 16,95 plus 3,95 voor het extra digitale pakket. Tot zover geen probleem. Alles werkte goed. Nu ineens gaat Ziggo standaard digitaal. Zonder enige toestemming van mij (en dus alle Ziggo klanten) wordt er zomaar besloten dat de drie Duitse zenders plus CNN van het analoge signaal worden verwijderd. Op een van mijn TV's (waar ik juist naar deze kanalen kijk) ben ik dus deze kanalen kwijt. Daar komt bovenop, dat ik niet meer mijn kleine pakket kan krijgen, maar alleen TV Plus. Ziggo kan mij niet garanderen dat daarin kennis en nieuws volledig is opgenomen, maar ik moet wel 24,95 gaan betalen per maand (was 20,90). Ik heb Ziggo benaderd en aangegeven dat ik deze gang van zaken niet normaal vind, zeker niet omdat ik een abonnement heb afgesloten en ik vind dat je dat niet zomaar eenzijdig mag veranderen. Volgens Ziggo: " Dit hebben wij zo besloten en als u een klacht heeft dan moet u die op internet kenbaar maken. Er verandert echter niets aan het genomen besluit.". Ik heb gevraagd of ik kan spreken met iemand van het management. Dat was niet mogelijk. Klachten kunnen ingediend worden via internet, anders niet. Ik zou de redactie van Radar dus willen voorleggen:. 1. Kan Ziggo zomaar eenzijdig besluiten om bepaalde zenders uit hun analoge aanbod te verwijderen? 2. Kan Ziggo zomaar ineens een digitaalpakket verwijderen en slechts een duurder aanbod ter vervanging aanbieden? 3. Kan ik nu de decoder die ik bij Ziggo kocht retourneren en mijn geld terug krijgen? Ik wil namelijk nu heel graag weg bij Ziggo. 4. Kan Radar dit voorleggen aan het management van Ziggo. Mij wensen ze niet te woord te staan en diegene die ik bij Ziggo sprak (zie hieronder) weigerde mijn klacht telefonisch op te nemen en intern bij Ziggo voor te leggen. NB.: Ziggo werd door de consumentenbond als beste provider aangeduid vwb TV. Ziggo gaat nu alles wijzigen, dus deze beoordeling mag Ziggo m.i. niet meer gebruiken. Uiteindelijk is het nieuwe aanbod van Ziggo niet dat wat de consumentenbond heeft onderzocht. Nb.: Ik heb gesproken met Ziggo met de heer Jean Pierre Krijns. Tot slot: Ben blij geen telefonie van Ziggo te hebben. Stel je voor. Ineens zouden ze kunnen besluiten dat je geen mobiele nummers meer kan bellen. Of...... toch wel....... maar dan wel je veel meer betalen ..... Of ben ik het nu die een rare vergelijking maakt .....?

Zin: Klachten kunnen ingediend worden via internet anders niet .

Document sentiment:	Hoe zou u de algemene opinie van het bericht classificeren?	○ Negatief ○ Neutraal ○ Positief
Zin sentiment:	Hoe zou u de algemene opinie van deze zin classificeren?	○ Negatief ○ Neutraal ○ Positief
Zin relevantie:	Hoe relevant vindt u deze zin om te gebruiken in een samenvatting over de algemene opinie van het bericht?	○ Uiterst irrelevant (Totaal niet!) ○ Irrelevant (Ongeschikt) ○ Relevant (Geschikt) ○ Uiterst relevant (Absoluut wel!)

Geef hier aan welke eigenschappen van toepassing zijn op de getoonde zin:

☐ De zin is onderdeel van het slot van het bericht
☐ De zin is vrij positief of juist vrij negatief (Intensiteit)
☐ De zin bevat een aanbeveling of advies
☐ De zin speelt in op een gebeurtenis (benoemd in het bericht)
☐ De zin bevat een argument of motivatie
☐ De zin noemt één of meerdere aspecten van Ziggo
☐ De zin speelt in op een gebeurtenis (benoemd in het bericht)

(Stop)   (Volgende)

Fig. 2. Rating Application user interface

Except for collecting documents and automatically selecting candidate summary sentences, our InputGen Application also enables one to specify a list of considered features. These features can be manually added through an interface.

Rate. After providing all the necessary input, we let users evaluate the selected summary sentences. First, users are assigned a group number, which they can fill in at the start screen. The set of documents and summary sentences presented to a user depends on the group number. Users provide their ratings through the Rating Application, depicted in Fig. 2.

On the left hand side, we display a document. On the right hand side, we ask the user to rate the given document and a selected sentence (displayed in the upper right corner) for sentiment and relevance with respect to sentiment summarizations, as well as to select the features that apply to the selected sentence. In our current endeavors, we only focus on the relevance score and the features – sentiment scores, which can be either negative, neutral, or positive, are collected for future research purposes. The relevance of candidate summary sentences can be scored as either very irrelevant, irrelevant, relevant, or very relevant, yet these scores are mapped to relevant and irrelevant in the processing phase.

Process. In order to transform the evaluation data from the rating phase into usable data, our Process Application first retrieves all evaluations from our database. Then, we generate data points by applying the majority rule, as detailed in Section 4.1. The data thus generated is saved in a format that allows for easy analysis in our application.

Analyze. The analysis of our data is performed by means of the Weka software package [4]. We compute the information gain using the *InfoGainAttributeEval* method and we select subsets of the most relevant features using the *CfsSubsetEval* method combined with the *ExhaustiveSearch* method. Both analyses are performed with stratified 10-fold cross-validation.

5   Evaluation

Following the method described in Section 4, we have performed an empirical study in order to determine relevant features for sentiment summarization of online conversational documents. The experimental setup and results of this study are detailed below.

5.1   Experimental Setup

In our current endeavors, we focus on a set of 60 Dutch forum posts about the Dutch company Ziggo. We limit the size of the documents in our set to 2,500 characters. The minimum length of a candidate summary sentence is assumed to be five words. For each of our 60 documents, we select candidate summary sentences to be presented to users.

Ideally, we would present all sentences to a user. However, this would render the human evaluation phase a very time-consuming process. Therefore, we present each user only seven sentences per document, i.e., two sentences flagged for the aspects feature, one sentence flagged for intensity, one sentence flagged for being part of a conclusion, and three random sentences. Through this preselection process, we aim to reduce the risk of features being under-represented in the final selection.

For our evaluations, we divide our collection of 60 Dutch conversational documents and their associated candidate summary sentences (seven per document) into three equally-sized groups. Each group of documents is rated by a group of three human annotators. As such, we have nine human raters that evaluate 20 documents each, with seven summary sentences per document. This yields a total of 1,260 ratings, which are represented by 420 data points. These data points represent the evaluations of our human annotators for each of our considered sentences, as determined by means of majority voting.

5.2   Experimental Results

By using our application, we have collected user evaluations of seven candidate summary sentences for each of our 60 documents. The distribution of the features

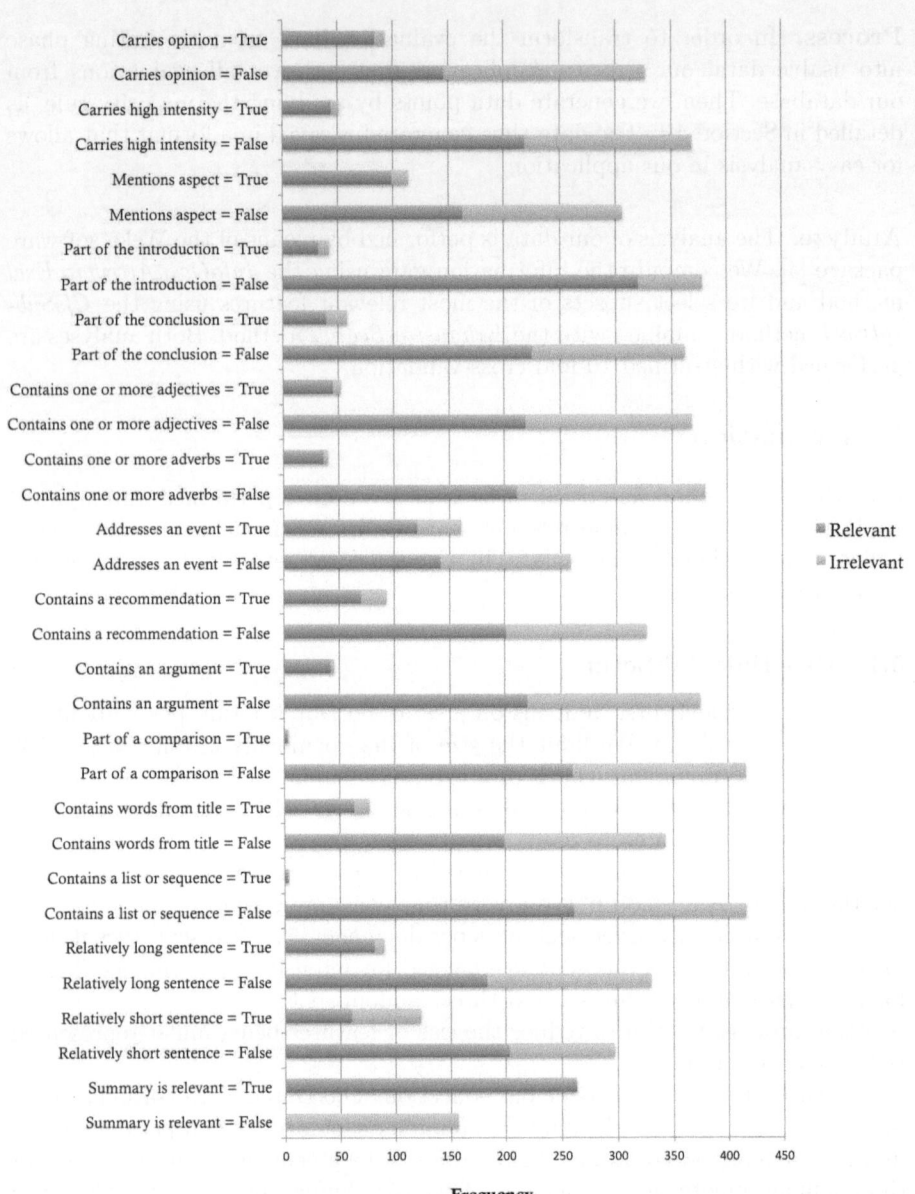

Fig. 3. Distribution of features over our data set

over our data set, according to our human annotators, is visualized in Fig. 3. Most features appear to be sufficiently represented in our data. The comparison and sequence features do however appear to be very rare in our data.

When we analyze the features in our data set in terms of their associated information gain, we can clearly distinguish useful from less useful features. Figure 4 suggests that in our data, three features contain relatively much information that

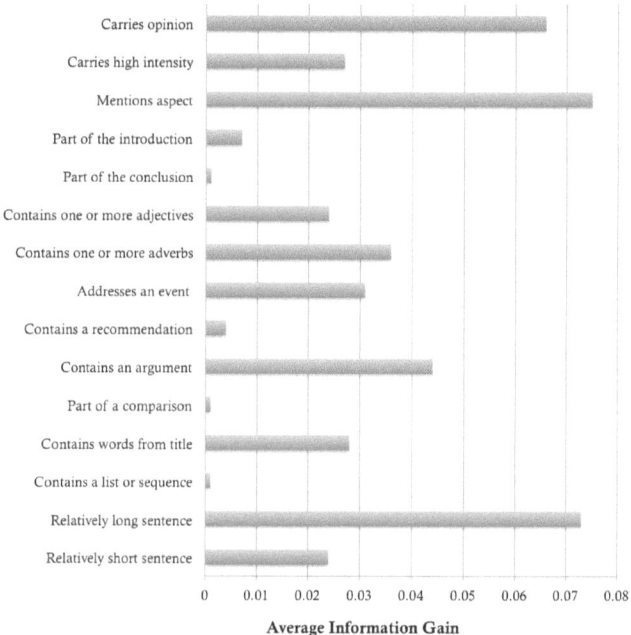

Fig. 4. Information gain of our considered features, averaged over 10 folds

can be used to distinguish relevant from irrelevant text fragments for sentiment summaries. Aspects provide the highest information gain, closely followed by sentences that are relatively long and sentences that carry opinion. Conversely, three features seem to have a rather marginal relevance, i.e., conclusion sentences, sentences with lists, and sentences that are part of a comparison. This may be related to their relatively low frequency in our data set.

Another analysis, in which we apply a feature selection method in order to identify the most informative subset of features in each of our 10 folds by considering the predictive power of each individual feature along with the degree of redundancy between the features [5], we obtain similar results. Figure 5 shows that four features are always selected, i.e., sentences with opinions, sentences with aspects, sentences with arguments, and long sentences. Sentences containing adverbs and sentences containing words that also occur in the title of a document are selected relatively often, yet not in all cases.

As such, our results indicate that, according to our human annotators, four features are relatively important proxies for a text fragment's relevance in sentiment summaries of conversational documents, i.e., fragments discussing aspects of a text's subject, fragments that are relatively long, fragments with opinions, and fragments containing arguments supporting these opinions. Especially features referring to arguments and long text fragments are remarkable, as they are, to the best of our knowledge, not used in existing sentiment summarizers.

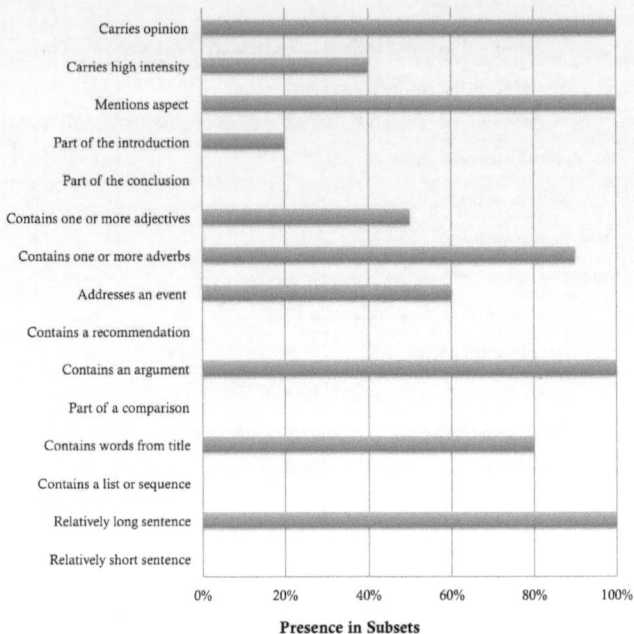

Fig. 5. Relative frequency of selection of our considered features in 10 folds

6   Conclusions and Future Work

When it comes to summarizing the sentiment conveyed by a piece of conversational text, we have shown that relatively long, opinionated fragments are good candidates for inclusion in a summary. Additionally, our results indicate that in our corpus of Dutch conversational documents, it is not so much the absolute position of text fragments – e.g., fragments' occurrence in an introduction or conclusion – that distinguishes relevant from irrelevant fragments. Conversely, it is rather the role sentiment-carrying fragments play – e.g., arguments supporting the overall message, or fragments discussing different aspects of the topic – that renders them useful in summaries reflecting the sentiment of a document.

In this light, we plan to validate our findings on other corpora and languages, as well as to further investigate how we can account for structural features (e.g., argumentation structures) and semantic features (e.g., distinct aspects of a topic) of content when summarizing its conveyed sentiment. In future work, we additionally aim to investigate the link between sentiment of relevant fragments and sentiment of a text as a whole. Furthermore, we aim to find combinations of sentences constituting a good summary. Last, we plan to implement our findings and to assess different (weighted) combinations of features in order to improve the state-of-the-art in sentiment summarization for conversational text.

Acknowledgments. We would like to thank Teezir (http://www.teezir.com) for their technical support, fruitful discussions, and for supplying us with valuable data. The authors of this paper are partially supported by the Dutch national program COMMIT.

References

1. Bal, D., Bal, M., van Bunningen, A., Hogenboom, A., Hogenboom, F., Frasincar, F.: Sentiment Analysis with a Multilingual Pipeline. In: Bouguettaya, A., Hauswirth, M., Liu, L. (eds.) WISE 2011. LNCS, vol. 6997, pp. 129–142. Springer, Heidelberg (2011)
2. Beineke, P., Hastie, T., Manning, C., Vaithyanathan, S.: Exploring Sentiment Summarization. In: AAAI Spring Symposium on Exploring Attitude and Affect in Text: Theories and Applications, pp. 12–15. AAAI Press (2004)
3. Blair-Goldensohn, S., Hannan, K., McDonald, R., Neylon, T., Reis, G., Reynar, J.: Building a Sentiment Summarizer for Local Service Reviews. In: WWW 2008 Workshop on NLP Challenges in the Information Explosion Era (NLPIX 2008) (2008), http://www.cl.cs.titech.ac.jp/~fujii/NLPIX2008/paper3.pdf
4. Hall, M., Frank, E., Holmes, G., Pfahringer, B., Reutemann, P., Witten, I.: The WEKA Data Mining Software: An Update. SIGKDD Explorations 11(1), 10–18 (2009)
5. Hall, M., Smith, L.: Practical Feature Subset Selection for Machine Learning. In: 21st Australasian Computer Science Conference (ACSC 1998), pp. 181–191. Springer, Heidelberg (1998)
6. Lerman, K., Blair-Goldensohn, S., McDonald, R.: Sentiment Summarization: Evaluating and Learning User Preferences. In: 12th Conference of the European Chapter of the Association for Computational Linguistics (EACL 2009), pp. 514–522. Association for Computational Linguistics (2009)
7. Madden, S.: From Databases to Big Data. IEEE Internet Computing 16(3), 4–6 (2012)
8. Mitchell, T.: Machine Learning. McGraw-Hill Series in Computer Science. McGraw-Hill (1997)
9. Pang, B., Lee, L.: Opinion Mining and Sentiment Analysis. Foundations and Trends in Information Retrieval 2(1), 1–135 (2008)
10. Pang, B., Lee, L., Vaithyanathan, S.: Thumbs up? Sentiment Classification using Machine Learning Techniques. In: Empirical Methods in Natural Language Processing (EMNLP 2002), pp. 79–86. Association for Computational Linguistics (2002)
11. Pingdom: Internet 2011 in Numbers (2012), http://royal.pingdom.com/2012/01/17/internet-2011-in-numbers/
12. Popescu, A., Etzioni, O.: Extracting Product Features and Opinions from Reviews. In: Conference on Human Language Technology and Empirical Methods in Natural Language Processing (HLT 2005), pp. 339–346. Association for Computational Linguistics (2005)
13. Titov, I., McDonald, R.: A Joint Model of Text and Aspect Ratings for Sentiment Summarization, pp. 308–316 (2008)
14. Zhu, J., Zhu, M., Wang, H., Tsou, B.: Aspect-Based Sentence Segmentation for Sentiment Summarization. In: 1st International CIKM Workshop on Topic-Sentiment Analysis for Mass Opinion (TSA 2009), pp. 65–72. Association for Computing Machinery (2009)

A Meta-plugin
for Bespoke Data Management in WordPress

Stefania Leone, Alexandre de Spindler, and Moira C. Norrie

Institute for Information Systems, ETH Zurich
CH-8092 Zurich, Switzerland
{leone,despindler,norrie}@inf.ethz.ch

Abstract. WordPress is a powerful and extensible platform for web-based information publishing and management. While the WordPress core is targeted to the publication of chronologically ordered textual articles typical of blogs, users have developed plugins as well as themes to support the data management requirements of specific domains such as e-commerce or e-learning. However, the creation of such plugins requires development skills and effort. We present a meta-plugin that automatically generates bespoke plugins for data management based on user-defined ER models. We illustrate the approach using an example of creating a WordPress site for managing information about courses.

Keywords: Wordpress, Meta-Plugin, Data Management Platform

1   Introduction

WordPress is a powerful information management and publishing platform that allows end-users to set up their web sites by selecting and adapting shared themes. Through the administrator interface, users can customise their theme of choice as well as authoring content, uploading media and integrating a wide variety of plugins. The success of the approach is reflected by the large number of over 50 million online web sites[1] based on WordPress.

Although WordPress is commonly associated with blogging sites, nowadays it is widely used as a general information management and publishing system. Examples include museums[2], corporate websites[3] and more domain-specific applications, such as online stores[4] and e-learning systems[5]. However, while WordPress provides a powerful infrastructure for web publishing, the core data model is very limited since it is based on a simple pages and posts paradigm for the publication of semi-structured text and embedded media.

The WordPress core model and functionality can be extended through plugins. Thousands of plugins have been developed by the community with examples

---

[1] http://en.wordpress.com/stats
[2] http://wordpress.org/showcase/the-toledo-museum-of-art
[3] http://wordpress.org/showcase/atlantic-southeast-airlines
[4] http://kartellstorela.com
[5] http://testdatei.schatzverlag.ch

X.S. Wang et al. (Eds.): WISE 2012, LNCS 7651, pp. 580–593, 2012.

including an e-commerece plugin[6] and the Buddypress[7] plugin for the design of social networking sites. Plugins can easily be shared among the user community, but users with very specific data management requirements may have to develop their own plugins which requires knowledge of PHP as well as a detailed understanding of the WordPress platform and its inner workings.

To simplify the task for end-users, we show how the concept of a meta-plugin can be used to generate bespoke plugins for data management. The meta-plugin allows end-users to specify an ER model in the WordPress administrator interface and automatically generates a plugin based on this model. This enables users to profit from the infrastructure provided by WordPress, while being able to tailor the underlying model to their needs rather than having to work around the limitations of a core model originally designed for blogging sites.

Section 2 discusses support for end-user development of data-intensive web sites. The core model of WordPress is presented in Sect. 3 and an extension to support ER models in Sect. 4. In Sect. 5, we detail the concept of meta-plugins and demonstrate their use in Sect. 6. Concluding remarks are given in Sect. 7.

2 Background

Nowadays, many professional as well as private web sites are developed by single users, either as end-users or developers with a small amount of technical knowledge combined with some design skills. In line with research on end-user development, it is therefore important to consider how to make web information systems not only easy to use, but also easy to develop [1].

Within the web engineering research community, model-driven approaches to web site development such as [2–4] have been advocated strongly. These methodologies offer systematic approaches based on models defining the structural, navigational and presentation aspects of a web information system. In the case of WebML [2], a developer designs an ER data model, followed by navigation and presentation models. The associated tool Web Ratio [8] can then generate and deploy the application. While model-driven approaches are powerful and can generate complex web sites with little or no programming by the developer, they are not targeted at end-user development since they still require detailed knowledge of the models and how the web functions. Rather, they were aimed at supporting teams of developers where there should be a clear separation of concerns between database developers, web architects, programmers and designers.

End-users typically use a platform that offers document-based content publishing and allows users to design their web sites by configuring the content and structure of the site in terms of general publishing units and presentation styles. Popular platforms include WordPress and Drupal[9].

---

[6] http://wordpress.org/extend/plugins/wp-e-commerce
[7] http://buddypress.org
[8] http://www.webratio.com
[9] http://drupal.org

In the case of WordPress, the core model was targeted at blogging sites and the content is organised in terms of the two basic textual publishing units posts and pages that can be enriched with media. Further, WordPress users can employ a design-by-example approach [5] by selecting one of many web site themes developed and shared by the user community. This combination of a simple model and design-by-example enables users to set up a blogging site and start producing content within minutes. A major benefit of such a platform is that both the configuration and content of a web site can be updated dynamically. While Drupal is less blogging-specific, so called distributions provide pre-configured installations with similar support for setting up web sites.

WordPress features a plugin mechanism with which the original blogging model can be extended in terms of additional entity types, management operations and user interface widgets. A similar mechanism is present in Drupal, allowing developers to bundle reusable application components into modules. As a result, the development of web applications not only consists of extending and configuring a basic initial site with existing plugins or modules, but also optionally includes the development of such extensions.

Web information systems usually have a data management component and some researchers have addressed the problem of end-user development of data-intensive web sites. WYSIWYG application editors [6, 7] have been proposed to allow end-users to specify custom data. For example, the editor presented in [6] supports a top-down approach where a user specifies the presentation layer by creating forms representing domain entities. Based on these forms, an ER graph is extracted and the corresponding database schema is created automatically along with the presentation views. Visual mashup editors such as MashMaker [8] and Mash-o-matic [9] have been designed to create web information systems by integrating existing data sources. However, they do not provide the basic infrastructure to facilitate the design of new web information systems.

Drupal offers a module that supports the definition of custom data types, together with the generation of user interfaces to manage data. Similar support is provided in WordPress through plugins[10]. However, in both cases, custom data types are not explicitly represented in the database back-end. WordPress plugins typically make use of a single key-value table per data type instance to attach attributes to individual entities in a semi-structured manner. In Drupal, attribute declarations and values are represented on a meta level using two tables, one containing all declarations and another containing all values with references to the entities containing these values. Note that the WordPress plugin referenced does not support the association of entities. Drupal entities may reference each other by means of dedicated attributes containing entity identifiers. This effectively realises a single generic relationship construct rather than enabling custom relationships with constraints over source and target entities. Furthermore, since generic relationships are not explicitly represented in the database, referential integrity must be maintained as part of the application code.

---

[10] e.g. Ultimate Post Type Manager:
   http://wordpress.org/extend/plugins/ultimate-post-type-manager

These approaches make it much more difficult to write application-specific queries than in the usual representation of ER models and may lead to poor performance. Moreover, it is difficult to reuse and extend custom types as the application evolves and to integrate custom data with external systems. While plugins and modules offer general extension mechanisms and developers could customise how they represent application data, such extensions require programming skills and a deep understanding of the platform-specific programming model. Consequently, their development is not a suitable option for end-users.

We decided to investigate ways in which platforms such as WordPress and Drupal could be extended to support the development of web information systems with application-specific data management requirements. In web information system development, data requirements are typically modelled using a structured data model such as ER. The general idea is similar to the work in [10], where they introduce a domain-specific language to support the generation of corporate web sites on top of popular wiki software. The approach that we adopted was to build on the powerful concept of plugins in WordPress and produce a meta-plugin that can generate data management plugins based on user-defined ER models.

3 WordPress Data Model

In this section, we will have a detailed look at the concepts supported by WordPress and its core data model. Note that, although the main concepts and terminology are introduced in a document describing the WordPress Semantics[11], the details of the core data model can only be established by examining the underlying database as well as extracting bits and pieces from various articles in the WordPress documentation [11]. Figure 1 gives a conceptual overview of the WordPress core concepts and how they relate to each other.

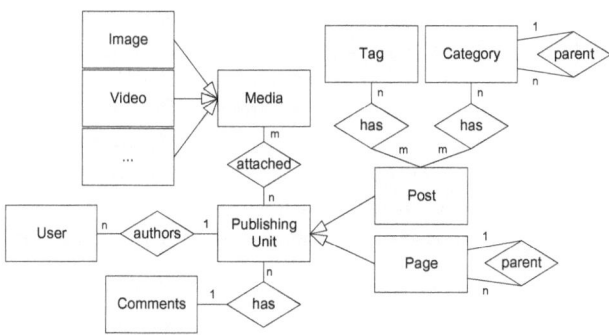

Fig. 1. ER model of WordPress concepts

WordPress distinguishes between static and dynamic content using two types of publishing unit—pages for static content and posts for dynamic content. When

---
[11] http://codex.wordpress.org/WordPress_Semantics

designing a web site, pages are typically used for content that has a fixed location within a web site and changes rarely. In the case of a blogging site, this could be information about the author. A page has a title, content and date. Content may be pure text, which is often enriched with HTML for structuring, or may also embed media, such as images, videos and audio files. Pages are usually accessed over a navigation menu and can exhibit nested structures.

In contrast, posts are used to publish new content. Although the content of an individual post is usually static, the collection of posts is dynamic and often presented in reverse chronological order with only the latest posts being visible in prominent positions such as the home page. This means that the location of a post within a web site changes over time and it becomes less and less visible to the users. Similar to pages, a post has a title, content and date. Both pages and posts may have comments, which can be configured by the developer.

To structure and organise posts, users can tag them or assign them to specific categories. Both tags and categories are user-defined. While tags represent a flat, user-defined taxonomy, categories typically have a hierarchical structure. The categories may be used to provide further navigational structures within a web site by creating corresponding menu items and showing only posts belonging to that category on the associated web page. Details of the structure and layout is controlled by the selected theme.

While Fig. 1 shows the view of the WordPress model presented to end-users, the internal metamodel that would be used by the developers of plugins is somewhat different as indicated in Fig. 2. We have based this developer model on the OMG Meta Object Facility (MOF) [12], where a model is represented by means of metamodel concepts, model concepts and data.

The first thing to note is that while the end-user model distinguishes the concepts of posts and pages, internally these are both instances of a general `PostType` in the M-2 metamodel. For the sake of clarity, we have therefore labelled the corresponding classes for these publishing units as `BlogPost` and `Page` in the developer model. A `PostType` defines a name and a list of attributes where `Attribute` is defined by a name and a type.

The data model on level M-1 represents the WordPress core data model. Here, we focus on the various post types since this is the extension point for data management plugins. We have therefore shaded out the parts of the core model dealing with other concepts such as taxonomies and users on the left of the figure and will not deal with them in detail.

WordPress actually offers five default post types for publishing content, which include attachments (any media file), revisions and navigation menus as well as pages and blog posts [11]. Furthermore, developers are free to create customised post types for publishing content that is structured differently. The design of customised post types is done programmatically and WordPress offers a number of methods in their API for the creation and registration of such custom types. Technically, the design of custom types is encapsulated and realised as plugins, which define the data types as well as the associated behaviour and presentation. Plugins will be discussed in more detail in Sect. 5.

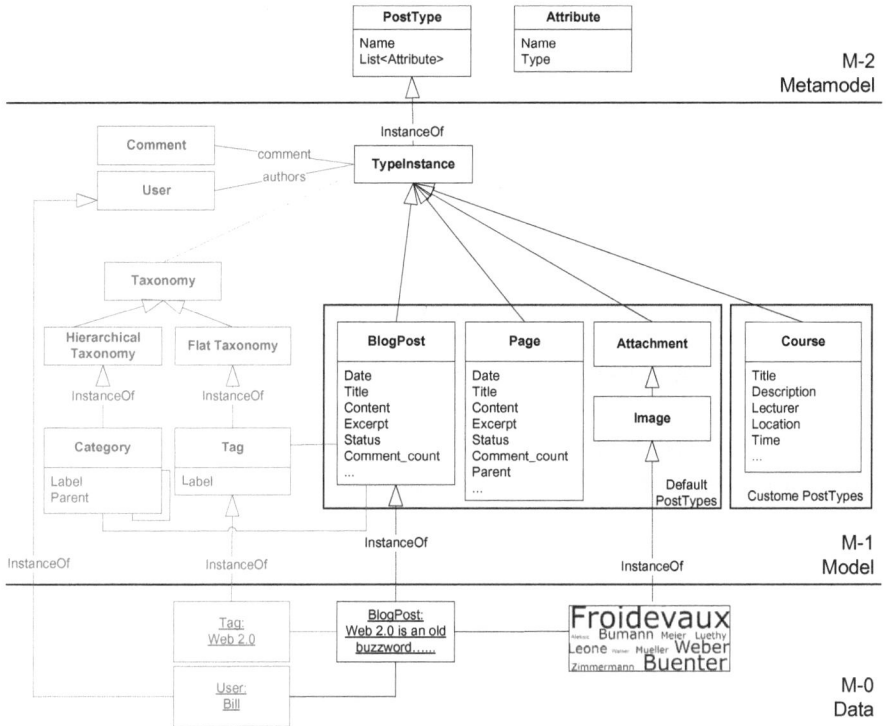

Fig. 2. WordPress developer model

In the centre of Fig 2, we show three of the five default post types, namely
`BlogPost`, `Page` and `Attachment`. There are some interesting differences between
pages and blog posts that can also be seen in the model. Note that only blog
posts can be associated with tags and categories while only pages can be nested.

On the M-0 level, we indicate actual data instances such as a blog post about
Web 2.0 with an image of a tag cloud as an attachment. The article has been
written by a user named Bill and tagged with the term 'Web 2.0'.

Now consider the case where a developer might want to publish some structured data on a web site. For example, they might have the task of creating a
WordPress site to publish and manage information about courses offered by a
university. Rather than trying to manage the data about courses as text contained within pages or posts, a developer could create a plugin defining a custom
post type `Course` as shown in Fig 2. This post type is specifically targeted at
publishing information about university courses, specifies attributes such as title, description, time and location and is involved in a relationship associating
courses with their lecturers.

Our goal was to enable end-users to create such plugins in the administrator interface without requiring detailed knowledge of the internal WordPress
model or programming effort. We did this by extending the metamodel and then

creating a meta-plugin that could generate plugins automatically based on an
ER model defined by the end-user through a form-style interface. We will first
detail the extension to the metamodel before describing the meta-plugin.

4   WordPress Core Extension for Supporting ER Models

Having introduced the WordPress core data model, we will now show how the
model can be extended based on user-defined ER models. We will present the
extensions using the example of the course management system. Assume the ER
model for the system is as shown in Figure 3.

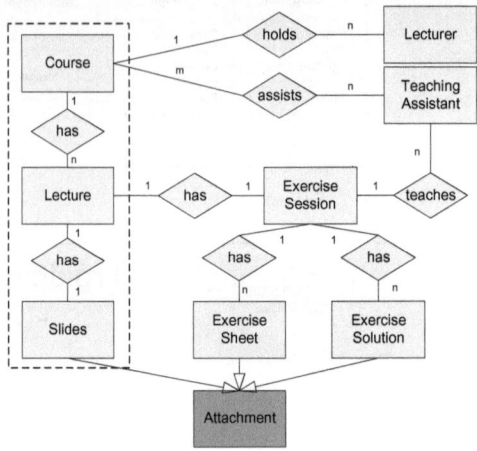

Fig. 3. ER model of university course management

Each course has one or more lecturers and one or more teaching assistants. A
course has a number of lectures with associated slides. Each lecture may have
an exercise session and, for each exercise session, there are exercise sheets and
solutions. Also, an exercise session is taught by one or more teaching assistants.

Figure 3 also indicates how certain concepts of the ER model relate to concepts
in the WordPress core model. Slides, exercise sheets and solutions are sub-entities
of `Attachment` which is shaded to indicate that this is a concept of the core
WordPress model shown in Fig. 2.

Figure 4 shows the extended WordPress model using MOF. On the metamodel
level M-2, we have introduced a number of new concepts which complement the
core concepts post type and attribute with other concepts of the ER metamodel.
The `PostType` corresponds to an entity type and defines a number of attributes.
An `EntitySet` is used to manage entities of a specific `PostType`. Entity sets
can be related to other entity sets via n-ary relationships. A `Relationship`
defines a name and list of attributes as well as cardinalities. While we named
the cardinalities `source` and `target`, it is simply a naming convention since
relationships are not directed. These concepts are used to instantiate an ER
model on the M-1 level.

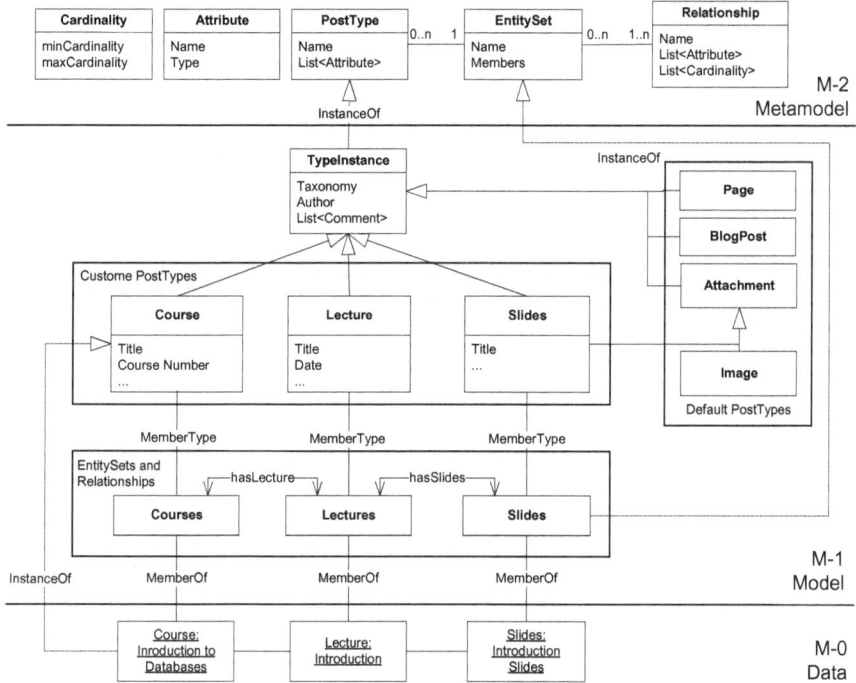

Fig. 4. WordPress model with course extension

The WordPress data model for our course management system is shown on the M-1 level. For the sake of space, we only show the additional concepts corresponding to the part of the ER model contained with the dashed line in Fig. 3, i.e. courses, lectures and slides. We also omit other parts of the core model such as the users and taxonomies.

Three new post types `Course`, `Lecture` and `Slides` have been integrated into the WordPress model, with the type `Slides` as a subtype of `Attachment`. Instances of these types are managed in the corresponding entity sets, namely `Courses`, `Lectures` and `Slides`, as indicated by the `MemberType` association between post types and entity sets. These entity sets are all instances of the `EntitySet` on level M-0. The relationships defined in the ER model can be represented using regular relationships between the entity sets. Note, however, that ER relationships that define attributes may be represented as entities in their own right that associate two entity sets.

On the M-0 level, data objects are shown. On the left, there is a course instance with the title 'Introduction to Databases'. The course has an associated lecture object which represents the introductory lecture of the course and is associated to the introduction slides on the right. These objects are instances of the newly defined post types and members of the corresponding entity sets, as indicated by the `InstanceOf` and `MemberOf` associations.

5   Meta-plugin

We now introduce in detail the notion of a WordPress meta-plugin, which is a plugin that generates new plugins. The idea is similar to that of template-based programming, for example in C++ [13] or XSLT [14].

The meta-plugin introduces a new data design process into the WordPress core. Instead of only being able to create pages and posts, a user can use the meta-plugin to define an ER model specifying application data entities and associations. From the specified ER model, the meta-plugin generates a bespoke plugin that the user simply has to install to create a data-backend for their web site. The generated plugin will allow application data to be created and manipulated based on the defined structure. Of course, the plugin can be used in combination with the powerful plug-n-play infrastructure provided by Word-Press. This means that the structure provided by the generated plugin can be extended with additional pages and posts to refine the design, either using standard WordPress functionality or by installing additional plugins. Since all data entities are realised as post type instances, they can also be classified by means of categories and tags without additional development effort. For the look-and-feel, we rely on WordPress themes, which can be used and adapted by the user.

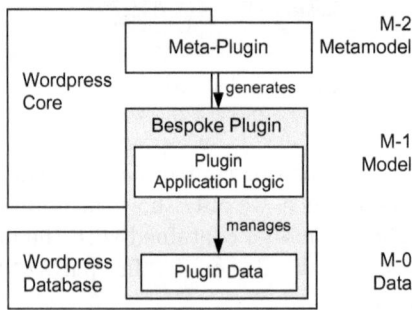

Fig. 5. Meta-plugin, plugin and data

Figure 5 gives an overview of our extension. The WordPress metamodel extension on level M-2 of Fig. 4 has been realised as a meta-plugin that extends the WordPress core with ER modelling capabilities and allows users to define ER models through a graphical user interface. Based on these user-defined ER models, the meta-plugin automatically generates bespoke plugins that realise the user-defined ER models on the M-1 level. Again, the generated plugin extends the WordPress core model, in this case, however, with a bespoke data model. In our example, that would be support for course management. The bespoke plugin consists of application logic and also an extension to the WordPress database to manage the data of the generated plugin. In our example, this would be data about courses, exercises, lecturers, assistants etc. The generated application logic offers functionality to create, manipulate and also view this data.

The meta-plugin as well as the generated plugins are realised as regular Word-Press plugins that can be installed through the WordPress administrator inter-face, i.e. the dashboard. Once installed, the meta-plugin extends the WordPress dashboard with functionality for ER modelling. More concretely, it creates a menu item 'ER Modelling', with sub-menus to view and create entity types, en-tity sets and relationships between them, as shown in the menu bars on the left of Figs. 6 (a) and (b). Using these menus, the user can create their ER model: Fig. 6 (a) shows the interface for creating a new entity type and Fig. 6 (b) for creating a relationship. In the current example, a number of entity sets have already been created and the user can select the source and target entity sets of the relationship from the drop-down menus. Once the user has defined an ER model, they can trigger the generation process of the bespoke plugin us-ing the 'Generate Plugin' menu. The user has to provide a plugin name and a description. The generation of a bespoke plugin is then triggered.

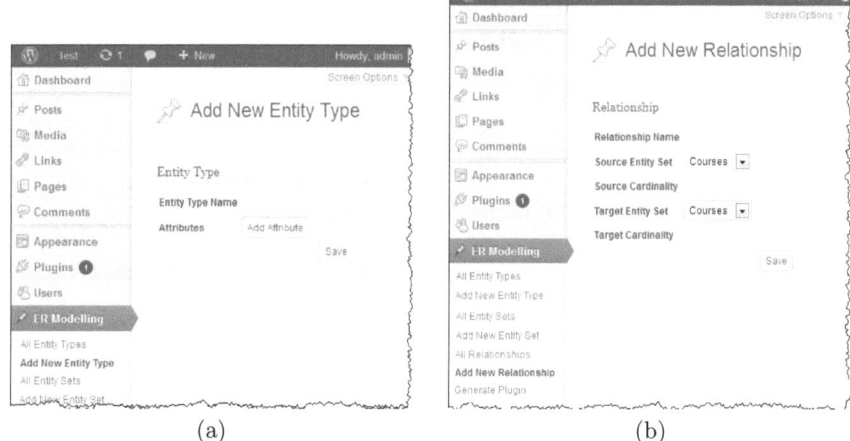

(a)   (b)

Fig. 6. Meta-plugin screenshots

6 Bespoke Plugin Generation

As part of its plugin mechanism, WordPress offers a number of hooks, which al-low users to inject additional functionality, data structures and presentation into execution environment of the WordPress core. Hooks are plugin lifecycle events such as their installation or uninstallation, as well as administrative or end-user activities including the creation, manipulation, retrieval, selection, display and deletion of posts, pages or plugin-specific data. Typically, plugin code includes functions for creating and deleting database tables, for inserting and selecting table data and the assignment of these functions to particular hooks.

To install a plugin, the files containing the plugin code need to be uploaded into the target WordPress platform through the dashboard. The availability of the plugin is then displayed to the user and can be activated. As a result of the activation, the additional functionality, data structures and presentation facilities become part of WordPress and are available for immediate use.

The meta-plugin presented previously is a plugin capable of generating files which constitute bespoke plugins. The code contained in these files is created using parametrised code templates instantiated with information from the ER models. The generated plugin encapsulates functions for the creation of custom types and database relations corresponding to the ER model, the extension of the dashboard with functionality to create data according to the model and functionality to present this data. We will now describe how these functions are bound to the WordPress hooks.

On plugin installation, custom types and database relations need to be created, and the dashboard extended with data creation and management functionality. First, a function is generated, which uses the custom type registration facility in order to register each entity type. As a result of such a registration process, the dashboard is automatically extended with the functionality to manage entity type instances. Then, for each entity type, entity set and relationship, functions containing CREATE TABLE and DROP TABLE statements are generated.

```
...
$plugin.=
"<?php
$postTypeRegistration;
register_activation_hook(_FILE_,
  '".$n."_activate');
...
function ".$n."_activate(){
  global \$wpdb;
  \$wpdb->query(
  'CREATE TABLE ".$tablename."(
    ID INT(6) PRIMARY KEY
      NOT NULL AUTO_INCREMENT,
    post_id BIGINT(20),
    ".$attributes.");'
  );
  ...
}?>"
...
```

```
<?php
register_post_type('Course',$args);
...
register_activation_hook(_FILE_,
  'coursemgt_activate');
...
function coursemgt_activate(){
  global $wpdb;
  $wpdb->query(
  'CREATE TABLE course(
    ID INT(6) PRIMARY KEY
      NOT NULL AUTO_INCREMENT
    post_id BIGINT(20),
    title VARCHAR(45) not null,
    description VARCHAR(256),
    location VARCHAR(45);');
}
...
?>
```

Fig. 7. Parameterised PHP Template   **Fig. 8.** Generated PHP File

Figure 7 shows an excerpt of the PHP template used to generate these functions. Upon template instantiation, variable $postTypeRegistration is replaced by a code snippet that registers all the entity types defined in the ER model as custom types. Then, a parametrised function invocation of the activation function registration is executed, where the activation function name is passed as a string, followed by a parametrised CREATE TABLE statement.

In Fig. 8, we show an excerpt of the resulting code for our course management plugin example. First, the course custom type is registered, followed by the registration of the activate function with the activation hook. On plugin activation, the function `coursemgt_activate()` is invoked and, as a first step, a database table for the course entity is created.

Below, there is an excerpt of the relational model in the underlying WordPress database after the activation. The first relation `WP_posttype` is the WordPress relation used to store all post type instances, be it pages, posts or custom post type instances. The following relations `Course`, `Session` and `HasSession` have been created upon plugin activation. Whenever a new entity, e.g. a course, is created, WordPress automatically generates a new tuple for the `WP_posttype` relation. In addition, our plugin creates a tuple in the `Course` relation containing all the course data values, with a foreign key to the `WP_posttype` tuple.

WP_posttype (<u>postID</u>, title, content, date, postType_id, ...)

...

Course (<u>courseID</u>, postID, description,...)
Lecture (<u>sessionID</u>, postID, title, date, ...)
HasLecture (<u>id</u>, courseID, sessionID)

...

The relation `Lecture` is used to manage lecture tuples and, analogously to the `Course` relation, it defines the `postID` as a foreign key. The `HasLecture` relation represents the relationship between courses and lectures. Note that, in the current implementation, cardinality constraints are handled in the application logic and, therefore, every ER relationship is realised as an M:N relationship. Moreover, entity sets are currently represented by all the tuples of a relation and, hence, a course entity is automatically a member of the `Courses` entity set.

The `coursemgt_activate` function also contains code that extends the dashboard with a new menu item and sub-menus that offer support for the creation and manipulation of entities, as well as the functionality to associate entities from one entity set to entities from another entity set as defined in the ER model. In the screenshot in Fig. 9, the administrator interface of the generated course management plugin is shown.

The screenshot shows the interface to create a new course, by specifying course title, course description and location. In addition, the interface offers the possibility to relate the current entity to other entities according to the data model. In the current example, the generated interface offers the possibility to associate courses to lecturers and assistants directly. Each time a new entity is created, functions that are registered to the `save_post` hook are invoked. In our case, a function `save_entity_info` is registered, which checks for the instance's post type, and creates an entry in the corresponding database table.

For each post type, code is generated that specifies how the post type is displayed. In order to display data, the plugin registers a function to the `the_post` hook, which is invoked when data is loaded. The registered function checks for the post type of the instance that is to be displayed. Based on that, the appropriate data is retrieved from the corresponding database table and displayed

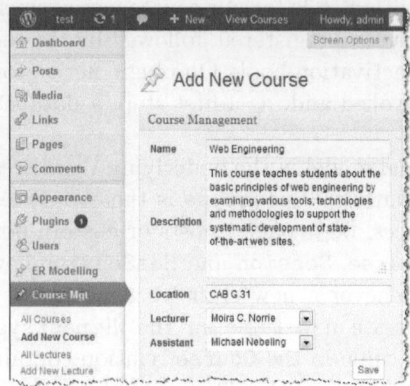   

Fig. 9. Course management admin   **Fig. 10.** Course management example

according to the presentation defined for that specific custom post type. Entity sets are currently represented as pages containing a list of entities. Users can click on any entity to see a detailed view of the entity with all the attribute values. Also, for relationships defined between entities, we provide links in the entity detail view to navigate from one entity to an associated one. For the layout of the application, the user can use the regular WordPress themes to define the look and feel of their application. Given the use of the `the_post` hook, templates are kept completely independent of the post types.

The screenshot in Fig. 10 shows the generated course management web site, in this case for the courses of our research group. Links have been created from the selected course to the lecturer and assistant.

Table 1 provides a complete overview of all WordPress hooks and meta- as well as bespoke plugin functions assigned to them.

Table 1. WordPress hooks and high-level plugin functionality descriptions

Hook	Functionality
Meta-Plugin	
Display administration interface (`admin_menu`)	Add support for the definition of entity types and relationships
Bespoke Plugin	
Installation (`register_activation_hook`)	Create database relations realising the ER model
Display administration interface (`admin_menu`)	Add support for the management of entities and relationships
Create post (`save_post`)	Insert attribute values and relationship tuples into corresponding database relations
Display post (`the_post`)	Retrieve and display data from corresponding database relations
Uninstallation (`register_deactivation_hook`)	Drop relations created during installation

7   Conclusion

We have presented an approach that complements the rich web information publishing infrastructure provided by WordPress with support for application-specific data management requirements. We plan to further investigate how the design process can be facilitated and enriched, for example by introducing the notion of component themes customised to a specific post type that can be generated and composed to an overall theme. Having tightened the notion of relationships within WordPress, we are also beginning to distinguish kinds of relationships such as generalisation and inclusion relationships. Furthermore, we plan to exploit the fact that entity types, entity sets and relationships could be tagged and categorised in WordPress. In this way, it would not only be possible to enhance WordPress with ER modelling capabilities, but also to enhance the ER model with WordPress concepts.

References

1. Lieberman, H., Paterno, F., Wulf, V. (eds.): End User Development (Human-Computer Interaction Series). Springer (2006)
2. Ceri, S., Fraternali, P., Bongio, A.: Web Modeling Language (WebML): A Modeling Language For Designing Web Sites. Computer Networks 33(1-6) (2000)
3. Hennicker, R., Koch, N.: A UML-Based Methodology for Hypermedia Design. In: Evans, A., Caskurlu, B., Selic, B. (eds.) UML 2000. LNCS, vol. 1939, pp. 410–424. Springer, Heidelberg (2000)
4. Vdovják, R., Frăsincar, F., Houben, G.J., Barna, P.: Engineering Semantic Web Information Systems in Hera. Journal of Web Engineering 1(1-2) (2003)
5. Lee, B., Srivastava, S., Kumar, R., Brafman, R., Klemmer, S.R.: Designing with Interactive Example Galleries. In: Proc. ACM Intl. Conf. on Human-Computer Interaction, CHI 2010 (2010)
6. Yang, F., Gupta, N., Botev, C., Churchill, E.F., Levchenko, G., Shanmugasundaram, J.: WYSIWYG Development of Data Driven Web Applications. Proc. VLDB Endow. 1(1) (2008)
7. Karger, D.R., Ostler, S., Lee, R.: The Web Page as a WYSIWYG End-User Customizable Database-backed Information Management Application. In: Proc. ACM Symposium on User Interface Software and Technology, UIST 2009 (2009)
8. Ennals, R., Brewer, E., Garofalakis, M., Shadle, M., Gandhi, P.: Intel Mash Maker: join the web. SIGMOD 36(4) (2007)
9. Murthy, S., Maier, D., Delcambre, L.: Mash-o-Matic. In: Proc. DocEng. (2006)
10. Díaz, O., Puente, G.: A DSL for Corporate Wiki Initialization. In: Mouratidis, H., Rolland, C. (eds.) CAiSE 2011. LNCS, vol. 6741, pp. 237–251. Springer, Heidelberg (2011)
11. Wordpress.org: Wordpress Documentation (2012), http://codex.wordpress.org
12. OMG: Meta Object Facility (MOF) Core Specification Version 2.0 (2006), http://www.omg.org/cgi-bin/doc?formal/2006-01-01
13. Prata, S.: C++ Primer Plus, 5th edn. SAMS (2005)
14. Grossniklaus, M., Norrie, M.C., Büchler, P.: Metatemplate Driven Multi-Channel Presentation. In: Proc. Workshop on Multi-channel and Mobile Information Systems, WISEW 2003 (2003)

mNIR: Diversifying Search Results
Based on a Mixture of Novelty, Intention and Relevance

Reza Taghizadeh Hemayati, Laleh Jafarian Dehkordi, and Weiyi Meng

Department of Computer Science, Binghamton University
Binghamton, NY 13902, USA
{hemayati,ljafari1,meng}@cs.binghamton.edu

Abstract. Current search engines do not explicitly take different meanings and usages of user queries into consideration when they rank the search results. As a result, they tend to retrieve results that cover the most popular meanings or usages of the query. Consequently, users who want results that cover a rare meaning or usage of query or results that cover all different meanings/usages may have to go through a large number of results in order to find the desired ones. Another problem with current search engines is that they do not adequately take users' intention into consideration. In this paper, we introduce a novel result ranking algorithm (mNIR) that explicitly takes *result novelty, user intention-based distribution* and *result relevancy* into consideration and mixes them to achieve better result ranking. We analyze how giving different emphasis to the above three aspects would impact the overall ranking of the results. Our approach builds on our previous method for identifying and ranking possible categories of any user query based on the meanings and usages of the terms and phrases within the query. These categories are also used to generate category queries for retrieving results matching different meanings/usages of the original user query. Our experimental results show that the proposed algorithm can outperform state-of-the-art diversification approaches.

Keywords: Search engine, result ranking, diversification, user intention.

1   Introduction

Search engines are widely used by users to find desired information. Unfortunately, modern search engines still do not satisfy many users' needs. Several reasons contribute to this problem. First, users tend to submit very short queries (most have only 1-3 terms), which can be ambiguous with different interpretations. Second, different users may have different search goals even when they submit the same query. Some search algorithms (e.g., PageRank [13]) tend to retrieve results that cover the most popular meanings/usages of query terms. For example, when "apple" was submitted to Google on May 24th, 2012, all search results in the first result page are related to the company Apple. This means that if a user wants to find results about some relatively rare meanings/usages of a query, the user probably has to go through a long list of results to find what he/she wants. For example, according to Wikipedia, the term "jaguar"

X.S. Wang et al. (Eds.): WISE 2012, LNCS 7651, pp. 594–608, 2012.

may refer to a large cat, a car, a supercomputer, a type of military aircraft (SEPECAT Jaguar), etc. When this term is submitted as a query to Google on May 24th, 2012, the first result relevant to SEPECAT Jaguar was ranked at 333 (in the 33rd page with 10 results on each page).

One way to remedy the above problem is to intentionally select search results covering different possible meanings/usages of a query and show them among the top-ranked results. This is called *diversification*. In order to perform diversification well, we should address the following two issues. First, we need to identify different meanings/usages of a given query so we can find results for each meaning/usage and include them among the top results. Second, we need to know to what extent each of the meanings/usages match the *expected intention* of the user who submitted the query so we can diversify accordingly (e.g., include more results from the more likely intentions in the result lists).

In this paper, we propose a novel algorithm to select and rank search results (called mNIR) which tries to rank the search results by a ***mixture*** of three result ranking preferences: *result **n**ovelty*, *user **i**ntention-based distribution* and *result **r**elevance*. *Result novelty* means that the ranking algorithm tries to include search results covering different meanings/usages among the top-ranked ones. This is also known as *result diversification*, which is beneficial for queries that have multiple meanings/usages. *User intention-based distribution* means that the ranking algorithm takes different possible intentions of the user into consideration when ranking the search results. When a query has multiple possible interpretations, it is often the case the likelihood for different interpretations (intentions) are different. *Result relevance* means that the ranking algorithm attempts to rank search results in descending order of their likelihood to be relevant to the query. In traditional information retrieval, relevance-based ranking is converted to similarity-based ranking in practice. The result ranking methods used in current search engines are mostly relevance-based ranking techniques.

In this paper, we assume that, for each query, its different possible meanings/usages and their likelihoods/probabilities have been estimated using an existing method [9].

This paper has the following contributions:

1. We introduce a novel search result diversification algorithm (called mNIR) by mixing three ranking preferences, namely, result novelty, user intention-based distribution and result relevance. By assigning different percentage values to these ranking preferences, our algorithm tries to produce the results that satisfy those ranking preference (based on the emphases given to them) as much as possible.
2. We perform extensive experiments to evaluate the effect of different mixing percentages on the performance of our proposed algorithm using a diverse set of measures. We also compare our algorithm with two existing state-of-the-art diversification algorithms. Our experimental results show that the new algorithm can achieve better overall performance.

The rest of the paper is organized as follows. Section 2 reviews related work. Section 3 provides a brief review of the background needed for our approach. Section 4 presents the main steps of our approach. Section 5 reports our experimental results. Section 6 concludes the paper.

2   Related Work

Result diversification has received a lot of attention recently. In [3], the authors described the Maximal Marginal Relevance (MMR) problem, and they showed how a trade-of between novelty and relevance of search results can be made explicit through the use of two functions, one measuring the similarity among documents and the other the similarity between documents and the query. Zhai et al. [22] stated that in general it is not sufficient to return a set of relevant results as the correlation among the returned results is also very important. They studied both novelty and relevancy in the language modeling framework. They developed an evaluation framework for subtopic retrieval based on the metrics of subtopic recall and subtopic precision. Agrawal et al. [1] proposed a diversification objective that tries to maximize the likelihood of finding a relevant document among the top-k results given the categorical information of the queries and documents. Furthermore, the authors generalized some classical IR metrics, including NDCG, MRR, and MAP, to evaluate their work for the value of diversification. They showed that their algorithm performs better in these generalized metrics compared to the results produced by commercial search engines.

Gollapudi et al. [8] defined a set of natural, intuitive axioms that a diversification system is expected to satisfy. The authors proved that all axioms cannot satisfy all these axioms simultaneously. Finally, they proposed an evaluation methodology to characterize the objectives and the underlying axioms. Rafiei et al. [12] modeled the diversity problem as expectation maximization and presented algorithms to estimate the optimization parameters. In [4], documents are selected sequentially according to relevance. The relevance is conditioned on documents having been already selected. Words in previous documents are associated with a negative weight to improve novelty. They considered different query intents, so that the result set covers multiple query intents. Clough et al. [7] examined user queries with respect to diversity. The authors found that a broad range of query types may benefit from diversification. They also showed that non-ambiguous queries can also need diverse search results. In [2], the authors studied the problem of diversifying search results by exploiting the knowledge derived from query logs. They presented a general framework for query result diversification using query recommendation algorithms, to detect ambiguous queries that would benefit from diversification, and to devise all the possible common specializations to be included in the diversified list of results along with their probability distribution. The different meanings and facets of queries are disclosed by analyzing user behaviors recorded in query logs. During the analysis, the popularity of the different specializations is also derived. The popularity distribution is then used to maximize the "usefulness" of the final set of documents returned.

Santos et al. [16] introduced the xQuAD probabilistic framework for search result diversification, which explicitly represented different query aspects as 'sub-queries'. They defined a diversification objective based on the estimated relevance of documents to multiple sub-queries, as well as on the relative importance of each sub-query to the original query. Later they [17, 18] proposed a selective diversification approach. In particular, given an unseen query, their approach learns a trade-off between relevance and diversity, based on optimal trade-offs observed for similar training queries. As a result, their approach effectively determines when and how to diversify the results for an unseen query. Sakai et al. [14, 15] proposed an alternative way to evaluate diversified search results, given intent probabilities and per-intent graded relevance assessments.

Diversification has also been studied for purposes different from search engine result diversification. Radlinski et al. [11] directly learned a diverse ranking of results based on users' clicking behavior through online exploration. They maximize the probability that a relevant document is found among the top-k ranked results. Since users tend not to click similar documents, online learning produces a diverse set of documents naturally. This approach cannot readily be used for tail queries since it requires user feedback. Clarke et al. [6] studied diversification in question answering while Ziegler et al. [23] studied the problem from a "recommendation" point of view.

Our approach differs from the above methods in the following aspects. First, we leverage the previous works [9, 19] to find all possible intentions (meanings/usages) of a query based on Wikipedia and WordNet. This method also estimates the importance of each meaning/usage. Second, we introduce different ranking preferences (*result relevancy*, *result novelty*, and *user intention-based distribution*) that a diversification system should satisfy. We introduce an algorithm to satisfy these preferences in different ways. One of the objectives of this paper is to analyze the performance of our method by giving different emphases to each of the ranking preferences. Our algorithm follows a two-step process for selecting and ranking search result records (SRRs; each SRR consists of a URL, a title and a snippet) while taking the preferences into consideration. Finally, we propose a scheme to retrieve SRRs that can guarantee the retrieval of SRRs covering different meanings/usages of each query.

3   Background Review

In this section, we provide a brief review of the background needed by our approach. This review outlines how different meanings/usages of a query are obtained, how their probabilities are estimated and how the initial SRRs for each meaning/usage are retrieved.

1. *Alternative query generation* [9]. For each user query Q, this step generates a set of *alternative queries* (AQs). All AQs contain the same set of query terms that appear in Q but contain different phrases. For convenience of discussion, we will refer both query terms and phrases as *concepts*.

2. *Definition category generation* [9]. A *definition category* (DC) [9] is a combination of *meanings* or *usages* derived from the concepts (query terms/phrases) of an AQ. This step generates all possible DCs for each AQ by combining one meanings/usage from each concept in the AQ.

3. *Definition category probability estimation* [9]. This step estimates the probability of each DC (it represents a meaning/usage of query Q) generated in Step 2. The approach in [9] includes a method to compute the ranking score of any given DC (the score is denoted by $RS(DC)$). Let $DC_1, \ldots, DC_m$ be all the DCs generated in Step 2 and *total_score* = $RS(DC_1)$ + ... + $RS(DC_m)$. Then we define the probability of DC_k with respect to query Q to be $Pr(DC_k | Q) = RS(DC_k) / total\_score$.

4. Definition category label generation [19]. Given a DC as input, we use the method introduced in [19] to obtain a label for the DC. A label for a DC is a set of terms/phrases that summarizes the DC. This method uses Wikipedia and WordNet to generate candidate labels. Wikipedia provides some useful information for each concept (like definitions, categories, disambiguation page, etc.), which can be used to generate candidate labels. WordNet provides information like synonyms, hypernyms, etc. that can also be used as candidate labels. These candidate labels are ranked for each DC and then the top-ranked candidate label is taken as the label for the DC.

5. Input SRR generation. For each DC, we use its label as a query, called *label query* or *category query*. We submit the query to a search engine (Google is used in this paper) to retrieve the top n SRRs (n is 10 in this paper). We denote this list of SRRs as SRR(DC). Note that the order of the SRRs in the list is important.

For each user query Q, the set of DCs, i.e. DC_1, ..., DC_m, and the lists of SRRs retrieved by Q and the label queries of these DCs, i.e. SRR(Q), SRR(DC_1), ..., SRR(DC_m), form the input SRRs to our result ranking algorithms. In our experiments, for comparison purpose, we also use the original user query to retrieve a list of top k SRRs using the same search engine.

To summarize, for the rest of this paper, we assume that for each user query, we have obtained its meanings and usages (i.e., the DCs), and for each such meaning/usage, we have obtained its probability (i.e., the likelihood this meaning/usage captures the intention of the user who submitted the query) and the list of SRRs (retrieved by the label query of the DC). Based on the above information, we develop our mNIR algorithm.

4   Selecting and Ranking SRRs

In this section, we present the algorithm mNIR that tries to mix the relative weights on the three ranking preferences when selecting and ranking search results. In Section 4.1, we define the three preferences more precisely. In Section 4.2, we describe the two-step process of the algorithm mNIR. In Section 4.3, we present our result selection method for the first step. In Section 4.4, we present our result ranking method for the second step.

4.1   Ranking Preferences

We define the ranking requirement for each of the three ranking preferences below. Let Q be the user query under consideration.

Result Novelty Based Ranking. To satisfy this ranking preference, the displayed results should be as diverse as possible.

For example, suppose query Q has 5 meanings/usages. If 5 results are desired, then there should be one result for each of the 5 meanings/usages.

User Intention Based Distribution Ranking. Given a probability distribution of query Q on different meanings/usages of Q, $\{Pr(DC_k \mid Q) \mid k = 1, \ldots, m\}$, to satisfy this ranking preference, the percentage of displayed results for DC_k among all displayed results should be equal to $Pr(DC_k \mid Q)$.

Novelty-based ranking and intention-based distribution ranking are closely related. The latter can be approximately considered as a weighted version of the former, i.e., in the latter approach, categories (DCs) that have higher probabilities will have more related results selected for display while the former selects at least one result from each category (as much as possible). Another difference is that categories with very small probabilities are often ignored by the intention-based distribution ranking but less likely by the novelty-based ranking. Specifically, if $Pr(DC_i \mid Q) * n$ is significantly less than 1, where n is the number of desired results to be displayed, then DC_i will effectively be ignored by the intention-based ranking.

Result Relevance Based Ranking. To satisfy this ranking preference, the results retrieved from a given corpus (e.g., the set of documents indexed by a search engine) need to be ranked in descending order of their likelihood to be relevant to Q. In this paper, the likelihood of a result's relevance to a query will be modeled by the similarity of the result with the query.

Popular similarity functions such as the *Okapi function* [13] and the *Cosine function* can be used to compute the similarity between a retrieved result and a query.

4.2   Two-Step Framework

In this paper, we assume that there is a way for users to specify their search preferences through the query interface of the search engine the user uses. In this paper, users can mix the three ranking preferences described in Section 4.1 by assigning different weights to them. The total weight of all ranking preferences should be equal to one all the time. To better support different preference mixes, we adopt a two-step framework for our algorithm. These two steps are described below.

- Step 1: *Result selection.* The goal of this step is to select a set of SRRs from the initial lists of SRRs, i.e., SRR(DC$_1$), …, SRR(DC$_m$), SRR(Q) (see Section 3).
- Step 2: *Result ranking.* Suppose n SRRs are to be displayed to the user. The goal of this step is to rank the n SRRs from the set of SRRs selected in Step 1, again, based on the ranking preference mix given by the user.

These two steps are introduced in the next two subsections.

4.3   Selection Method (SM)

We will select $m*n + \tau*n$ SRRs by sending the label queries generated from the categories (DC$_1$, …, DC$_m$) and the original query (Q) to a search engine. We retrieve n SRRs for each DC and $\tau*n$ SRRs for the original query (in this paper τ is 5). These SRRs will be used in the next step. The perfect selection will be the list that has the most relevant SRRs for each DC and Q. These SRRs will be ranked in the next step (Section 4.4). If only the original query Q is used to retrieve results, then to ensure results for some rare meanings/usages are obtained, we often need to retrieve hundreds or even thousands of SRRs. To overcome this problem we use the label query generated for each category [19] to retrieve relevant SRRs. But if only the label queries are used, we may miss some important relevant SRRs with respect to the original query. Thus, we use both label queries and the original query to collect initial results.

The ranking method of mNIR (see Section 4.4) will be based on a mixture of three ranking preferences. For each result that is retrieved by a category label query, we know its category (meaning/usage). This allows us to evaluate its novelty and whether it matches any given intention during the ranking process. In order to implement the relevance-based preference, we need to know the relevance of each result. In practice, the likelihood of relevance of a result is estimated based on its similarity with the query (we use the original query Q). If a result SRR* retrieved by a category label query is not retrieved by Q, its similarity is unknown. By retrieving $\tau*n$ SRRs for the original query with $\tau > 1$, we try to increase the chance that SRR* is retrieved by the original query Q. If SRR* is not included in the $\tau*n$ SRRs retrieved by Q, we assign a low relevancy score to it (the score is lower than the lowest score of the SRRs retrieved by Q).

4.4   Ranking Method (RM)

In this method, we assign a percentage value to each ranking preference. Let λ_1, λ_2, and λ_3 denote the percentage values assigned to the preferences of *result novelty*, *result relevancy*, and *user intention-based distribution*, respectively, and satisfy the following conditions:

$$0 \leq \lambda_1, \lambda_2, \lambda_3 \leq 1 \quad \text{and} \quad \lambda_1 + \lambda_2 + \lambda_3 = 1.$$

In this section we introduce a Ranking Method (RM) – a novel probabilistic ranking algorithm that explicitly mixes different preferences into the ranking process. This method is shown in Figure 1.

Given an ambiguous query Q and a set of SRRs R selected in the first step for this query, we build a new SRR ranking list S by iteratively selecting the n highest scored SRRs from R according to the following probability mixture model:

$$\lambda_1 \text{Nov}(DC,Q) + \lambda_2 \text{Rel}(d,Q) + \lambda_3 \text{Int}(DC|Q) \tag{1}$$

In particular, these probabilities can be regarded as modeling *result novelty, result relevancy*, and *user intention-based distribution*, respectively, with a mixing of parameters λ_1, λ_2 and λ_3 controlling the tradeoff among the three. In Expression (1), $\text{Int}(DC|Q)$ is the likelihood of DC being the real intention behind user's query Q (See Steps 2&3 in Section 3), $\text{Rel}(d, Q)$ is the likelihood of a document d's relevance to Q, and $\text{Nov}(DC, Q)$ is a binary value, it is one if there hasn't been any relevant SRRs selected for this DC, and it is zero once an SRR relevant to this DC is selected. Initially $\text{Nov}(DC, Q)$ is one for all DCs. For the above model, when $\lambda_1=1$ and there is a tie among at least two SRRs, we use a secondary preference to break the tie. In this case we select an SRR from the higher weighted DC.

$RM\ (n,\ Pr(DC|Q),\ \lambda_1,\ \lambda_2,\ R)\ //\ \lambda_3 = 1 - (\lambda_1 + \lambda_2)$

1    $S \leftarrow \{\ \}$

2    $Nov(DC,\ Q) \leftarrow 1$

3    $Int(DC|\ Q) \leftarrow Pr(DC|\ Q)$

4   While $|S| < n$ do

5        $d^* \leftarrow arg\ max_{d \in R\backslash S}(\lambda_1 Nov(DC,\ Q) + \lambda_2 Rel(d,\ Q) + \lambda_3 Int(DC|\ Q))$

6        $R \leftarrow R - \{d^*\}$

7   add d^* to the end of S

8        $Nov(DC,\ Q) \leftarrow 0$

9        $\forall DC, if\ (Nov\ (DC, Q)) == 0)$

10            $\lambda_2 \leftarrow ((\lambda_1/2) + \lambda_2);\ \lambda_3 \leftarrow ((\lambda_1/2) + \lambda_3\ )$

11            $\lambda_1 \leftarrow 0$

12        $Int(DC|\ Q) \leftarrow (n^*Int(DC|\ Q)-1)/n$

13   End while

14   Return S

Fig. 1. Ranking Method of mNIR

In each iteration once an SRR relevant to a DC from R is selected, we reduce Int(DC|Q) from that DC using Int(DC|Q)=(n*Int(DC|Q)-1)/n. Novelty for a DC changes from one to zero once an SRR has been selected from that DC. When the novelty score Nov(DC, Q) becomes zero, its percentage value (i.e., λ_1) is evenly distributed to λ_2 and λ_3.

5   Evaluation

In Section 5.1, we describe the dataset used in our evaluation study. We also introduce various performance measures. In Section 5.2, we review the algorithms to be evaluated and compared. In Section 5.3, we present the experimental results.

5.1   Dataset and Performance Measures

The query dataset we use consists of 50 queries. These queries are from TREC 2010 Web track. Among the queries, the numbers of queries having 1, 2, 3 and 4 terms are 26, 11, 10 and 3, respectively. There are 2.59 categories per query on the average. We consider up to top 5 meanings/usages for each query. Some queries may have fewer meanings/usages.

Some traditional IR metrics like MAP and MRR are widely used to measure search quality. However, they do not take the quality of diversification into consideration. Several metrics, including the k-call metric [20] and a-NDCG [21], have been proposed for evaluating the diversity of search results. However, these metrics do not consider the relative importance of different categories (DCs) and how well a result matches a category.

To address this problem in this paper, we use Intent-Aware (IA) measures to evaluate the ranking algorithms with different ranking preferences. We use existing measures MAP-IA@n [1, 3], MRR-IA@n [1, 3] and NDCG-IA@n [3, 18] to evaluate different aspects of our ranking algorithms. We review these measures below.

MAP-IA@n (Intent Aware (IA) MAP) [1, 3]
In order to define MAP-IA@n, we first define AP(DC)@n (Average Precision with respect to category DC) [1]. This measure is an adaptation of the AP measure in IR research [10]. This adaptation takes the results retrieved for different categories (DCs) into consideration, making the measure capable of also measuring the effectiveness of diversification. The average precision of a ranked result set for query Q and category DC_k for n search results is defined as:

$$AP(DC_k)@n = \frac{\sum_{j=1}^{n} P(j) * relevance\ (j)}{N(DC_k)} \tag{2}$$

where j is the ranking position of a retrieved result, *relevance(j)* denotes the relevance of the j-th ranked result (it is 1 if the j-th ranked result is relevant with respect to DC_k, and 0 otherwise), $P(j) = \sum_{i=1}^{j} relevance(i)/j$, and $N(DC_k)$ denotes the number of relevant results for category DC_k among n search results.

Traditional IR metrics focus solely on the relevance of documents. Consider a query that belongs to two categories DC_1 and DC_2. Suppose $Pr(DC_2 | Q) \gg Pr(DC_1 | Q)$, and we have two results, r_1 is rated excellent for DC_1 (but unrelated to DC_2), and r_2 is rated good for DC_2 (but unrelated to DC_1). As a result, the order (r_1, r_2), i.e., ranking r_1 ahead of r_2, will yield a higher AP score. Yet an "average" user will find the order (r_2, r_1) more useful. This is because AP treats all intentions (categories) equally. However, some categories are more likely to match users' expected intentions. In order to take into account the likelihood of user intention, we define intent aware MAP over all intentions of a query Q to be:

$$MAP\text{-}IA(Q)@n = \sum_{k=1}^{m} Pr(DC_k | Q) * AP(DC_k)@n \tag{3}$$

The MAP-IA@n over all queries in the query set (SetQ) is defined as follows:

$$MAP\text{-}IA@n = \left(\sum_{Q \in SetQ} MAP\text{-}IA(Q)@n \right) \Big/ |SetQ| \tag{4}$$

where |SetQ| is the number of queries in SetQ.

NDCG-IA@n (Intent Aware (IA) NDCG) [1, 3]
This measure was used in different works (e.g., [3, 18]) and it is an adaptation of the NDCG (normalized discounted cumulative gain [10]) for measuring the diversification of the search results produced by a search system. This adaptation takes the results retrieved for different categories (DCs) into consideration, making the measure capable of also measuring the effectiveness of diversification. NDCG(DC$_k$)@n is defined as follows:

$$NDCG(DC_k)@n = \frac{\sum_{r=1}^{n} GG(r)/\log(r+1)}{\sum_{r=1}^{n} GG^*(r)/\log(r+1)} \tag{5}$$

where $GG(r)$ is the (cumulative) gain at rank r produced by an algorithm, $GG*(r)$ is the (cumulative) gain at rank r in an ideal ranked list (the ideal list is the list built based on the intention distribution based on the probability of each category). In this paper, we consider two levels for gain value: 0 if the result is not relevant and 1 if it is relevant. In order to take into account the likelihood of user intention, we define intent aware NDCG of a given query Q using the following formula:

$$NDCG\text{-}IA(Q)@n = \sum_{k=1}^{m} P(DC_k \mid Q) * NDCG(DC_k)@n \tag{6}$$

The NDCG-IA@n over all queries in the query set SetQ is defined as follows:

$$NDCG - IA@n = \left(\sum_{Q \in SetQ} NDCG - IA(Q)@n \right) \Big/ |SetQ| \tag{7}$$

MRR-IA@n [1, 3]

The reciprocal rank (RR) is the inverse of the position of the first relevant document in the ordered result list. This value is zero if there is no relevant document among retrieved documents. The mean reciprocal rank (MRR) of a query set is the average reciprocal rank of all queries in the query set. Using the same idea of averaging user intention, we define intent aware MRR for a given query Q to be:

$$MRR\text{-}IA(Q)@n = \sum_{k=1}^{m} \Pr(DC_k \mid Q) * RR(DC_k)@n \tag{8}$$

where $RR(DC_k)@n$ is reciprocal rank of the first result from DC_k. The MRR-IA@n over all queries in the query set SetQ is defined as follows:

$$MRR - IA@n = \left(\sum_{Q \in SetQ} MRR - IA(Q)@n \right) \Big/ |SetQ| \tag{9}$$

5.2   Algorithms to Be Evaluated

In this section, we list the algorithms we will evaluate and compare.

- mNIR: In Section 4 we introduced a two-step framework which diversifies SRRs based on three preferences, with a mixing of parameters λ_1, λ_2 and λ_3 controlling the tradeoff among the three.

We compare the above algorithm with the following two state-of-the-art algorithms:

- xQuAD [16]: Santos et al. [16] introduced the xQuAD probabilistic framework for search result diversification, which explicitly represented different query aspects as 'sub-queries'. They defined a diversification objective based on the estimated relevance of documents to multiple sub-queries, as well as on the relative importance of each sub-query to the original query. By introducing a tuning parameter, their approach has a capability to give different preferences to either relevance or novelty. To enable fair comparison (i.e., all algorithms are given the same set of initial results), in our implementation of xQuAD, we use the label queries generated from categories (DCs) as sub-queries (these label queries are also used in our algorithms to obtain the search results corresponding to different meanings/usages). The confidence of the relevance likelihood of a document to a sub-query is determined by the estimated relevance of the document to the category (DC). The probability likelihood of a sub-query to be the real intention of user's query is determined by the probability likelihood of a category (DC) to be the real intention of user's query.

Here is the brief overview of the adapted algorithm. Given an ambiguous query Q and an initial ranking R produced for this query, a new ranking S will be built by iteratively selecting the n highest scored documents from R, according to the following probability mixture model:

$$(1-\lambda)P(d\mid Q)+\lambda \sum_{DC_k \in Q}\left[P(DC_k\mid Q)P(d\mid DC_k)\prod_{d_j \in S}(1-P(d_j\mid DC_k))\right] \qquad (10)$$

where $P(d|Q)$ is the relevance likelihood of a document to Q, and $P(DC_k|Q)$ is the likelihood of DC_k being the real intention of Q, $P(d|DC_k)$ is the relevance likelihood of document d to DC_k. In particular, these two probabilities can be regarded as modeling relevance and diversity, respectively, with a parameter λ controlling the tradeoff between the two.

- IA-Select [1]: In [1], the results are retrieved using the original user query and are then divided into different categories using a classifier. Again, in order for the comparison to be fair, in our implementation of IA-Select, the results for different categories are directly obtained using the label queries generated from categories (DCs). In [1], one result is selected to be ranked next at a time and, each time, the result with the highest product of its "relevance score" and a weight score of its category is selected. After a result from a category is selected, the weight of the category is reduced to increase the chance for a result from a different category to be selected next. The confidence of the classification of a document to a given class is determined by the estimated relevance of the document to the category (DC) that represents the class.

Algorithm mNIR differs from IA-Select and xQuad mainly on preferences consi-dered. Approximately, IA-Select and xQuad consider only (expected) user intention (they call it novelty in their papers) and relevance while mNIR additionally considers real novelty. These two methods are more likely to miss less likely (rare) mean-ings/usages of queries than mNIR.

5.3   Experimental Results

In this section, we report the experimental results for the algorithms described in Sec-tion 5.2 using the performance measures introduced in Section 5.1. In particular, we aim to answer the following three questions:

1. Can we improve diversification performance by using our proposed algorithm?
2. Can adding intention-based distribution preference to novelty and relevancy pre-ferences improve diversification performance?
3. How do different mixings of the three ranking preferences affect the performance of the algorithm mNIR?

We start our experiments by studying MRR-IA@n results to see which algorithm can diversify more results. Then later report MAP-IA@n and NDCG-IA@n results to see the diversification performance considering both relevancy and novelty for all algorithms. We start executing our algorithm with a full emphasis being given to the novelty preference and then eventually start to reduce the emphasis to the novelty and increase the emphasis to the other two preferences evenly. We repeat the above execution for relevancy preference and intention preference.

MRR-IA@n: The MRR-IA measure checks the position of the first result for each category. This means that this measure emphasizes the quality of diversification. The MRR-IA results for mNIR (different values of λ_1, λ_2 and λ_3 are tested), xQuAD (different values of λ are tested) and IA-Select with different n's are shown in columns 5-7 of Table 1. mNIR performed the best among all algorithms when more emphasis is given to the novelty preference. This is due to the fact that it tries to select and rank SRRs as diversified as possible. When the same maximum emphasis is given to both novelty and intention preferences evenly (i.e., $\lambda_1=\lambda_3=0.5$ and $\lambda_2=0$), the algorithm performs as well as when $\lambda_1=1$. The reason is that we try to satisfy the distribution of different intentions of queries at each rank as much as possible. This produces more diversified results at higher ranks. xQuAD performs relatively better when more preference is given to novelty (larger values of λ place more emphasis on novelty). IA-Select performed relatively poorly compared with other algorithms.

MAP-IA@n: The MAP-IA results for mNIR, IA-Select and xQuAD with different n's are shown in columns 8-10 of Table 1. It can be seen that mNIR performed the best among all algorithms when more emphasis is given to both novelty and intention preferences evenly ($\lambda_1=\lambda_3=0.5$). When more emphasis is given to the novelty preference ($\lambda_1=1$) mNIR also performs very well for MAP-IA@n when n=3 and n=5. mNIR performs the best when $\lambda_3=1$ for MAP-IA@10. We have observed that algorithms generally perform relatively well if they can cover most categories first by showing one result from each category. Once they cover most categories, they will perform better if they rank SRRs from more likely intentions first. SRRs from more likely intentions have a better chance to be relevant to the query. This can explain the behavior we observe in this experiment. This is due to the fact that by giving more preference to intention-based distribution, we are able to show more SRRs from more likely intentions and rank these SRRs higher. This will give us a higher score for NDCG-IA and MAP-IA. The algorithms that consider only the *novelty* and *relevancy* preferences but not have a separate intention-based distribution preference (such as xQuad and IA-Select) do not have this benefit.

NDCG-IA@n: The results for NDCG-IA@n for all algorithms are shown in columns 11-13 of Table 1. In this evaluation, the ideal ranking list is the list built based on the expected intentions of each query. It can be seen that mNIR has the overall best performance for this measure when more emphasis is given to the intention-based distribution preference. xQuad ($\lambda = 0.8$ and $\lambda = 1$) also performed very well overall.

Table 1. MRR-IA@n, MAP-IA@n, nDCG@n

*	n	r	i	MRR	MRR	MRR	MAP	MAP	MAP	NDCG	NDCG	NDCG
	$\lambda 1$	$\lambda 2$	$\lambda 3$	@3	@5	@10	@3	@5	@10	@3	@5	@10
m	1	0	0	**0.654**	**0.688**	**0.688**	**0.512**	**0.486**	0.194	0.554	0.54	0.614
	0.8	0.1	0.1	0.618	0.654	0.654	0.464	0.444	0.178	0.538	0.536	0.602
	0.6	0.2	0.2	0.618	0.654	0.654	0.464	0.444	0.178	0.538	0.536	0.602
	0.4	0.3	0.3	0.492	0.514	0.514	0.464	0.444	0.178	0.464	0.466	0.49
	0.2	0.4	0.4	0.602	0.622	0.622	0.43	0.366	0.188	0.556	0.564	0.598
	0	0.5	0.5	0.582	0.588	0.602	0.394	0.338	0.178	0.526	0.532	0.596
	0	0.1	0.9	0.628	0.636	0.604	0.436	0.356	0.204	**0.588**	0.606	0.652
	0	1	0	0.462	0.472	0.5	0.33	0.31	0.206	0.452	0.474	0.538
	0.1	0.8	0.1	0.486	0.526	0.544	0.34	0.314	0.162	0.46	0.496	0.564
	0.2	0.6	0.2	0.548	0.564	0.584	0.37	0.334	0.172	0.496	0.514	0.584
	0.3	0.4	0.3	0.548	0.58	0.588	0.394	0.35	0.182	0.528	0.53	0.592
	0.4	0.2	0.4	0.62	0.656	0.656	0.466	0.402	0.176	0.54	0.542	0.602
	0.5	0	0.5	**0.654**	**0.688**	**0.688**	**0.512**	**0.446**	**0.196**	0.554	0.546	0.616
	0	0	1	0.622	0.628	0.598	0.438	0.36	**0.218**	**0.586**	**0.608**	**0.658**
	0.1	0.1	0.8	0.644	0.668	0.638	0.47	0.378	0.204	0.57	0.58	0.636
	0.2	0.2	0.6	0.65	0.67	0.664	0.51	0.368	0.202	0.548	0.578	0.622
	0.3	0.3	0.4	0.62	0.642	0.642	0.466	0.41	0.178	0.54	0.554	0.606
	0.4	0.4	0.2	0.584	0.616	0.608	0.464	0.392	0.174	0.54	0.548	0.6
	0.5	0.5	0	0.584	0.62	0.62	0.44	0.434	0.172	0.508	0.53	0.598
X	1	0	-	0.514	0.554	0.576	0.312	0.268	0.202	0.52	0.582	**0.65**
	0.8	0.2	-	0.53	0.57	0.588	0.354	0.326	0.198	**0.564**	**0.594**	0.634
	0.6	0.4	-	**0.59**	**0.596**	**0.616**	0.462	0.324	0.19	0.554	0.56	0.598
	0.5	0.5	-	0.554	0.56	0.58	0.382	0.284	0.17	0.52	0.514	0.576
	0.4	0.6	-	0.52	0.556	0.572	**0.416**	**0.324**	0.168	0.482	0.5	0.572
	0.2	0.8	-	0.462	0.472	0.5	0.33	0.31	0.162	0.452	0.474	0.546
	0	1	-	0.462	0.472	0.5	0.33	0.31	**0.206**	0.452	0.474	0.538
I	-	-	-	**0.514**	**0.554**	**0.482**	**0.312**	**0.268**	**0.202**	**0.52**	**0.582**	**0.408**

* m=mNIR, X=XQUAD, I=IA-Select, n=Novelty, r=relevancy, i=intention

In this section, we investigated the performance of our proposed algorithm along with two state-of-the-art approaches (i.e., xQuAD and IA-Select). We used three major measures (MRR-IA@n, NDCG-IA@n and MAP-IA@n) to compare the performance of these algorithms. Our experiments show that our proposed algorithm can improve diversification performance without sacrificing the relevancy. By defining three parameters to assign weight to our ranking preferences, our proposed algorithm has the capability and flexibility to perform under different objectives. When the goal is to diversify the results as much as possible our algorithm performs the best when more emphasis is given to the novelty preference among all algorithms studied in this paper. When the goal is to have a balance between novelty and relevancy our algorithm still performs the best when more emphasis is given to intention-based distribution preference. We have shown that by adding user intention based preference to our algorithm, we are able to improve the diversification performance.

6   Conclusion

In this paper, we investigated the problem of diversifying search results based on three different ranking preferences (result novelty, user intention and result relevance) and proposed a novel algorithm (called mNIR) to select and rank search results which tries to rank the search results by a mixture of three result ranking preferences. By assigning different percentage values to these ranking preferences, our algorithm has the flexibility to produce the results that satisfy those ranking preference (based on the emphases given to them) as much as possible. We evaluated the proposed algorithm by assigning different percentage values to each of the ranking preferences together with two state-of-the-art algorithms (xQuAD and IA-Select) using a variety of performance measures. Our experimental results show that the proposed algorithm in this paper performs better than both existing algorithms. In the future, we plan to study the performance of our algorithm for different type of queries. In particular, given a diversification approach and an unseen query, we would like to develop a method to predict an effective mixture of the ranking preferences based on previously seen similar queries.

References

1. Agrawal, R., Gollapudi, S., Halverson, A., Ieong, S.: Diversifying search results. In: ACM Intl. Conf. on Web Search and Data Mining (2009)
2. Capannini, G., Nardini, F.M., Perego, R., Silvestri, F.: Efficient diversification of web search results. PVLDB 4(7) (April 2011)
3. Carbonell, J., Goldstein, J.: The use of MMR, diversity-based reranking for reordering documents and producing summaries. In: ACM SIGIR, pp. 335–336 (1998)
4. Chapelle, O., Ji, S., Liao, C., Velipasaoglu, E., Lai, L., Wu, S.: Intent-based diversification of web search results: metrics and algorithms. Information Retrieval Journal (2011)
5. Chen, H., Karger, D.R.: Less is more: probabilistic models for retrieving fewer relevant documents. In: ACM SIGIR, pp. 429–436 (2006)
6. Clarke, C.L., Kolla, M., Cormack, G.V., Vechtomova, O., Ashkan, A., Buttcher, S., MacKinnon, I.: Novelty and diversity in information retrieval evaluation. In: ACM SIGIR, pp. 659–666 (2008)
7. Clough, P., Sanderson, M., Abouammoh, M., Navarro, S., Paramita, M.: Multiple approaches to analysing query diversity. In: ACM SIGIR, pp. 734–735 (2009)
8. Gollapudi, S., Sharma, A.: An axiomatic approach for result diversification. In: WWW Conference, pp. 381–390 (2009)
9. Hemayati, R.T., Meng, W., Yu, C.: Identifying and Ranking Possible Semantic and Common Usage Categories of Search Engine Queries. In: Chen, L., Triantafillou, P., Suel, T. (eds.) WISE 2010. LNCS, vol. 6488, pp. 254–261. Springer, Heidelberg (2010)
10. Järvelin, K., Kekäläinen, J.: Discounted Cumulated Gain. In: Encyclopedia of Database Systems, pp. 849–853 (2009)
11. Radlinski, F., Kleinberg, R., Joachims, T.: Learning diverse rankings with multi-armed bandits. In: ICML, pp. 784–791 (2008)
12. Rafiei, D., Bharat, K., Shukla, A.: Diversifying web search results. In: WWW Conference, pp. 781–790 (2010)

13. Robertson, S., Walker, S., Beaulieu, M.: Okapi at Trec-7: Automatic Ad Hoc, Filtering, Vlc, and Interactive Track. In: 7th Text REtrieval Conference, pp. 253–264 (1999)
14. Sakai, T., Craswell, N., Song, R., Robertson, S., Dou, Z., Lin, C.-Y.: Simple evaluation metrics for diversified search results. In: EVIA 2010, pp. 42–50 (2010)
15. Sakai, T., Song, R.: Evaluating Diversified Search Results Using Per-intent Graded Relevance. In: ACM SIGIR (2011)
16. Santos, R.L.T., Macdonald, C., Ounis, I.: Exploiting query reformulations for Web search result diversification. In: WWW Conference, pp. 881–890 (2010)
17. Santos, R.L.T., Macdonald, C., Ounis, I.: Selectively diversifying Web search results. In: ACM CIKM, pp. 1179–1188 (2010)
18. Santos, R.L.T., Macdonald, C., Ounis, I.: Intent-aware search result diversification. In: ACM SIGIR (2011)
19. Hemayati, R.T., Meng, W., Yu, C.: Categorizing Search Results Using WordNet and Wikipedia. In: Gao, H., Lim, L., Wang, W., Li, C., Chen, L. (eds.) WAIM 2012. LNCS, vol. 7418, pp. 185–197. Springer, Heidelberg (2012)
20. Xu, Y., Yin, H.: Novelty and topicality in interactive information retrieval. J. Am. Soc. Inf. Sci. Technol. 59(2), 201–215 (2008)
21. Zhai, C.: Risk Minimization and Language Modeling in Information Retrieval. PhD thesis, Carnegie Mellon University (2002)
22. Zhai, C.X., Cohen, W.W., Lafferty, J.: Beyond independent relevance: methods and evaluation metrics for subtopic retrieval. In: ACM SIGIR, pp. 699–708 (2003)
23. Ziegler, C.-N., McNee, S.M., Konstan, J.A., Lausen, G.: Improving recommendation lists through topic diversification. In: WWW Conference, pp. 22–32 (2005)

Computing Scientometrics in Large-Scale Academic Search Engines with MapReduce

Leonidas Akritidis and Panayiotis Bozanis

Dpt. of Computer & Communication Engineering, Univ. of Thessaly, Volos, Greece

Abstract. Apart from the well-established facility of searching for research articles, the modern academic search engines also provide information regarding the scientists themselves. Until recently, this information was limited to include the articles each scientist has authored, accompanied by their corresponding citations. Presently, the most popular scientific databases have enriched this information by including *scientometrics*, that is, metrics which evaluate the research activity of a scientist. Although the computation of scientometrics is relatively easy when dealing with small data sets, in larger scales the problem becomes more challenging since the involved data is huge and cannot be handled efficiently by a single workstation. In this paper we attempt to address this interesting problem by employing MapReduce, a distributed, fault-tolerant framework used to solve problems in large scales without considering complex network programming details. We demonstrate that by setting the problem in a manner that is compatible to MapReduce, we can achieve an effective and scalable solution. We propose four algorithms which exploit the features of the framework and we compare their efficiency by conducting experiments on a large dataset comprised of roughly 1.8 million scientific documents.

1 Introduction

Following the evolution of the Web search engines, the scientific databases and academic search engines have significantly enriched the content of their result pages. Therefore, the results of a query for a research paper are now accompanied by information regarding the articles' authors. Some of the most popular scientific search engines such as Google Scholar[1], Microsoft Academic[2], and Scopus[3] extended this information by constructing author profiles where they compute and present their associated *scientometrics*.

The scientometrics are single, scalar values introduced with the aim of evaluating the research work of a scientist. The first and most widespread among them is *h-index*, devised by J.E. Hirsch [9]. This metric assigns a value to each scientist by taking into account not only the number of publications he/she has authored, but also, by considering the number of citations each article received.

---

[1] http://scholar.google.com
[2] http://academic.research.microsoft.com
[3] http://scopus.com

X.S. Wang et al. (Eds.): WISE 2012, LNCS 7651, pp. 609–623, 2012.

After the introduction of h-index, an entirely new line of research was drawn and multiple variants were proposed.

The computation of scientometrics is relatively easy when it is performed on small, well-controlled collections of research papers. For instance, in the case of h-index, the evaluation mechanism just needs to determine the articles each researcher has authored and then enumerate all their incoming citations. However, when the size of the collection increases the evaluation becomes more complex since a single workstation cannot accommodate all the involved data (i.e. documents, authors and citations). Therefore, we either have to use a secondary (and slower) type of storage, or solve the problem in parallel by distributing the data to a number of interconnected machines.

MapReduce is a distribution framework designed for solving problems in large scales. It is mainly oriented towards fault-tolerance, distributed storage, and simple implementation without requiring network programming details. This model has been used extensively by the Web search engines to develop a wide range of parallel algorithms. Examples include data mining tasks, information extraction from graphs, data structures construction, text processing, and others.

In this paper we propose four methods based on MapReduce to compute in parallel the scientometrics in large scientific databases. To the best of our knowledge, this is the first work in the current literature attempting to address this problem in large scales. All previous bibliography does not study in depth the issue in question, since until recently the data collections were small and the problem was not very important. However, the introduction of the large scientific databases and their constantly expanding repositories in combination with the users' increased interest, has rendered the issue much more challenging.

The rest of the paper is organized as follows: In section 2 we describe the most important related work about MapReduce and we refer to some popular scientometrics. Section 3 contains the key elements deriving from the related theory. In section 4 we set the problem in a basis that renders it manageable by MapReduce and we design the execution plan of our proposed algorithms. The experimental evaluation of our methods is given in section 5 and finally, in section 6 we conclude our work.

2   Related Work

In this section we present some fundamental articles about MapReduce and its architecture and we discuss some remarkable works which introduce strategies for solving common problems in parallel. Finally, we refer to a number of scientometrics that have been proposed in the related bibliography.

MapReduce was initially introduced by two Google engineers in [7]. In [8] the authors described GFS, the distributed file system on which the framework operates. A more extended presentation of the components of MapReduce is provided in [4]. The most popular open source implementations of MapReduce and GFS are *Hadoop* and the *HDFS* respectively. A technical overview of their architectural logic and design is provided in [3].

Numerous works have proposed expansions and modifications which allowed the framework to be used in a wider variety of applications. For instance, [2] introduced *HadoopDB*, an architectural hybrid between Hadoop and database management systems. In [16] the authors appended a "Merge" phase to the execution plan of the system with the aim of joining the relational outputs of two separate MapReduce tasks.

Several other research articles have described important problems which were efficiently solved by using MapReduce. For instance, Web search engines have used the framework extensively in data intensive tasks such as inverted index construction [13], and PageRank computation [12]. Text intensive applications include duplicate and near-duplicate document detection [6], language processing algorithms [11] and numerous others. Finally, [17] introduced Pregel, a computational model for processing large graphs. Pregel programs are expressed as a sequence of iterations, in each of which a vertex can receive messages sent in the previous iteration. However, unlike PageRank computation, the evaluation of scientometrics can be performed in a single MapReduce job without requiring multiple iterations.

Now let us refer to some papers which are related to scientometrics. The pioneering article in the area is [9], where J.E. Hirsch introduced *h-index*, a metric which evaluates each scientist by rewarding both productivity and influence. Motivated by the success of h-index, several other metrics were proposed. Examples include the *SCEAS* system [15], *g-index* [5] and *f-index* [10]. Additionally, in [14] the authors describe two time-aware variations of h-index, namely the *contemporary h-index* and the *trend h-index*. The former takes into consideration the elapsed time since an article was published, whereas the latter considers the date an article received each of its citations.

3   Preliminaries

Now we briefly present the principles which characterize the framework and we describe its basic components. Furthermore, we revise some of the most popular scientometrics and we examine their attributes.

3.1   MapReduce Basics

MapReduce builds on the key idea of simplicity; that is, its users should not deal with complex network programming issues [7,4]. Instead, the system provides an abstraction that requires from the algorithm developers to express their solutions by using only two functions: *map* and *reduce*.

The co-ordination of the parallel execution is performed by a single machine, the *Master*. The Master splits the input data into multiple fragments and assigns the processing of each fragment to a number of m *Workers*. The Workers (called *Mappers* in this phase) apply the map function to every key/value pair of their input and generate an arbitrary number of intermediate key/value pairs. When the input is exhausted, the system employs a number of r Workers (now called

Reducers) that apply the reduce function to all values associated with the same intermediate key. Their final output is the solution of the assigned task, also formatted in key/value pairs and partitioned in r shards.

There are two more optional components which can be involved in a MapReduce task: The *Partitioner* and the *Combiner*. The former is used to determine how the intermediate files produced by the Mappers should be transferred to the local file systems of the Reducers. The latter, is used to improve the efficiency of the execution by limiting the size of the data to be transferred from the Mappers to the Reducers by merging the values associated with the same key into associative arrays. The Combiner is deployed by either explicitly declaring a *combine* function, or by properly implementing it within the Mapper itself (*in-Mapper Combiner*). According to [12], the second option is usually preferable.

The MapReduce jobs are executed on top of a distributed file system [7] which transparently addresses all the problems that may occur (e.g., fault tolerance). For instance, in case a worker dies due to a hardware malfunction, Master assigns the job it was processing to another worker without any data loss.

3.2   Scientometrics

Here we provide brief descriptions of some important metrics that have been proposed for evaluating the research work of a scientist. The first and most popular metric among them is *h-index*, defined as follows:

A researcher a has h-index M_h^a, if M_h^a of his/her $|P^a|$ articles have received at least M_h^a citations each and the rest $|P^a| - M_h^a$ articles have received no more than M_h^a citations.

This metric calculates how broad the research work of a scientist is, since it accounts for both productivity and impact. Two interesting generalizations of h-index are the *contemporary* and the *trend* h-indices, both introduced in [14].

The *contemporary* h-index is an attempt to introduce temporal aspects in the evaluation of a scientist's work by taking into account the age of an article. According to its definition, each paper p_i of an author a is assigned a score $S_c^{p_i}$ determined by the following formula:

$$S_c^{p_i} = \gamma \frac{|P_c^{p_i}|}{(\Delta Y_i)^\delta} \tag{1}$$

where γ and δ are two constant coefficients; typical values for them are 4 and -1 respectively. ΔY_i symbolizes the number of the years elapsed since the publication of p_i, whereas $|P_c^{p_i}|$ is the number of the articles citing p_i. This way an old article gradually loses its value, even if it still gets citations. Based on the score $S_c^{p_i}$ the contemporary h-index is defined as follows:

A researcher a has contemporary h-index M_c^a, if M_c^a of his/her $|P^a|$ articles get a score $S_c^{p_i} \geq M_c^a$ and the rest $|P^a| - M_c^a$ articles get a score $S_c^{p_i} < M_c^a$.

Another mechanism for ranking scientists is the *trend* h-index. Here the idea is to assign scores to each paper by considering the year an article acquired a particular citation. This idea is expressed by the following equation:

$$S_t^{p_i} = \gamma \sum_{n=1}^{|P_c^{p_i}|} \frac{1}{(\Delta Y_n)^\delta} \tag{2}$$

where γ and δ are defined as previously. The scores $S_t^{p_i}$ are now used to define the trend h-index:

A researcher a has trend h-index M_t^a, if M_t^a of his/her $|P^a|$ articles get a score of $S_t^{p_i} \geq M_t^a$ and the rest $|P^a| - M_t^a$ articles get a score of $S_t^{p_i} < M_t^a$.

In this paper we examine how the three aforementioned metrics can be evaluated in large scales by employing the MapReduce framework. Nevertheless, the algorithms we present here can also be applied to compute other types of scientometrics (i.e. g-index, f-index) without any additional effort.

4   Computing Scientometrics with MapReduce

In this section we identify the key components of the problem and we design the Map and Reduce functions. In the sequel, we optimize our approaches by introducing in-Mapper Combiners.

4.1   Problem Formulation

Let us begin by introducing P which is the set containing all papers, and A which includes all authors. Each paper $p_i \in P$ contains a reference section encountered towards the end of the manuscript. From this section we extract $P^{p_i} \subset P$ which

Table 1. List of the most frequent symbols

Symbol	Meaning
P	The set containing all papers
A	The set containing all authors
p_i	An arbitrary paper $p_i \in P$
C^{p_i}	The textual content of p_i
a_j	An arbitrary author $a_j \in A$
A^{p_i}	The authors who created p_i
P^{a_j}	The papers authored by a_j
P^{p_i}	The papers referenced by p_i
$S_x^{p_i}$	The score of a paper p_i with respect to the metric x
$M_x^{a_j}$	A metric evaluating the work of a_j $M_h^{a_j}$: h-index of a_j $M_c^{a_j}$: contemporary h-index of a_j $M_t^{a_j}$: trend h-index of a_j

contains all papers referenced by p_i; for each reference $p_j^{p_i} \in P^{p_i}$ we retrieve all
the contributing authors $A^{p_j^{p_i}}$. In Table 1 we summarize all the above notations.

The input of the problem can be considered as a set of (p_i, C^{p_i}) pairs, where
p_i represents the integer identifier of an article and C^{p_i} symbolizes its content.
Our objective is to construct a list of (a, M_x^a) pairs, where x identifies the metric
M employed to evaluate each scientist (see last row of Table 1). According to
the definitions of all three metrics, it is required that we decompose the required
(a, M_x^a) pairs and construct for each author, one pair of the following form:

$$\left(a, SortedList\left[(p_1, S_x^{p_1}), ..., (p_N, S_x^{p_N})\right]\right) \tag{3}$$

where $S_x^{p_i}$ is the score of p_i with respect to the metric x. In case we are interested in computing h-index, this score merely represents the total number of the
incoming citations that p_i received. A detailed analysis on how these scores are
computed is provided in subsection 4.2.

According to 3, to calculate the metric values for an author, we first need to
identify all the publications he/she has authored, and then compute their corresponding scores. Notice that the elements of this $(paper, score)$ list must be
sorted in descending score order to enable fast metric evaluation with a single iteration. In the following section we present four methods to solve this interesting
problem by using MapReduce.

4.2 Basic Algorithm Design

We start by feeding the system with the given set of the publications P. According to our previous discussion, we express the input of the Map function in
a $(key, value)$ manner by defining (p_i, C^{p_i}) pairs. Within the Mapper, we parse
the textual content of each paper $p_i \in P$ and we retrieve all its outgoing references P^{p_i}. For each reference $p_k^{p_i} \in P^{p_i}$ we compute a score $S_x^{p_k}$, according to
the metric x we need to evaluate.

For the plain h-index metric, we set the score equal to 1 for all references,
thus denoting that the paper $p_k^{p_i}$ has one incoming citation (which of course, is
p_i). For the other two metrics, we need to consult equations 1 and 2. Our goal
is to properly set the partial scores in the map phase in order to compute the
final scores in the reduce phase. For this reason, during the map phase, we set
the partial scores recorded in Table 2.

In the sequel, each reference is again parsed and its authors $A^{p_k^{p_i}}$ are identified.
For each extracted author, we create one tuple that will be sent to the Reducer and
there are two options to format this tuple. The first one (called $method\ 1$) dictates

Table 2. Setting the partial paper scores in the map phase for various scientometrics

Metric	Partial Score
h-index	$S_h^p = 1$
contemporary h-index	$S_c^p = \gamma/(\Delta Y_p)^\delta$
trend h-index	$S_t^p = \gamma/(\Delta Y_{p,c})^\delta$

Algorithm 1. Mapper class(es): In case method 1 is used, the framework executes the step 8a. If we use method 2, we need to execute the step 8b.

1: **class** Mapper
2:   **method** map (integer p_i; string C^{p_i})
3:     $P \leftarrow$ ExtractReferences(C^{p_i})
4:     **for all** references $p \in P$
5:       $S^p \leftarrow$ ComputeScore(p)
6:       $A^p \leftarrow$ ExtractAuthors(p)
7:       **for all** authors $a \in A^p$
8a: **emit** $(a, pair[p, S^p])$
8b: **emit** $(pair[a, p], S^p)$
---

that we set the author as the key, and create a pair (*paper, score*) for the value field. Our second option (called *method 2*) is to generate a composite key of the form (*author, paper*), and place the paper score within the value field. Algorithm 1 illustrates a pseudocode for the map function implementing these two methods.

Although the map phase is almost identical for methods 1 and 2, the reduce phases must implement a different strategy. Notice that the MapReduce framework guarantees that the data sent from the Mappers arrives at the Reducers in a sorted key order. This gives method 2 an advantage; method 2 implements a *secondary sort*, that is, the data not only is brought to the Reducers in an ascending author order (as holds for method 1), but also, in ascending paper order. This allows a more robust approach of the reduce phase, since we save the cost of *searching* for the incoming papers. To make our state clearer, we provide Algorithms 2 and 3 for the reduce phase of methods 1 and 2 respectively.

Let us discuss Algorithm 2 first. Since the (*paper, score*) pairs are brought to the Reducers in arbitrary order, we need to store these pairs into a data structure H which will allow us to accumulate the partial paper scores. More specifically, for each value field of the Reducer input, we search in H for the input paper. In case this search fails, we insert the paper along with its corresponding score. In the opposite case, we just accumulate the incoming score to the one which is already stored in H. After all the tuples have been processed, we sort H in a descending score order and we compute the desired metric by iterating through its entries (steps 11–16). The sorting of H can be performed within the main memory of the Reducer since it stores at most a few hundreds entries; the vast majority of the authors has published fewer than 1000 articles.

On the contrary, in Algorithm 3 there is no need for searching; instead, it is only required to allocate an array H to store the paper scores. Since the tuples arrive in sorted order, we just need to compare the paper we are currently processing to the previous one (step 9). In case their identifiers are equal we accumulate their partial scores and update the last record of H. In the opposite case, we store the new paper score in a new position at the end of H. When all the papers of an author have been processed, we repeat the steps 9–17 of Algorithm 2 to compute the desired metric and we proceed with the next author. The final (a, h^a) tuple must be written out in the close method.

Algorithm 2. Method 1, Reducer class

1: **class** Reducer
2:    **method** reduce (**string** a; pairs[**integer** p, **float** S^p])
3:        $H \leftarrow$ new AssociativeArray
4:        **for all** pair $v \in$ pairs[**integer** p, **float** S^p]
5:            **if** $v.p \in H$
6:                $H^p.S \leftarrow H^p.S + v.S^p$
7:            **else**
8:                H.add(v)
9:        **sort** H in descending S order
10:       **integer** papers $\leftarrow 0$, metric $\leftarrow 0$
11:       **for all** pairs $\in H$
12:           papers $\leftarrow$ papers $+ 1$
13:           **if** $H^p.S \geq$ papers
14:               metric $\leftarrow$ metric $+ 1$
15:           **else**
16:               **stop iteration**
17:       **emit** (a, metric)

Finally, notice that the pair values of Algorithm 2 and the pair keys of Algorithm 3 are not included in the basic data types of MapReduce. Consequently, it is required that we implement additional classes which explicitly define how these data types must be read and written by the framework. Nevertheless, the complexity for Algorithm 3 is increased since the custom data type is used in the key; hence, it is required to determine how the system will compare the keys to each other to achieve sorted Mapper output (compareTo method). However, the increased complexity of Algorithm 3 is rewarded with improved execution performance.

4.3 Optimizing the Performance

Despite their difference in tuples formatting, the Mappers of both methods 1 and 2 still emit data to the Reducers each time an author of a paper reference is extracted. Since we do not check whether the key we are currently processing has been previously sent, it is inevitable that we transmit the same key multiple times. This leads to a performance bottleneck due to the increased network traffic caused among the nodes of the system. Here we attempt to address this problem with the support of the Combiners.

The Algorithm 4 shows how we can extend method 1 with the aim of supporting an in-Mapper Combiner. We call this new approach as method 1-C, where the letter "C" signifies the presence of a Combiner. The cornerstone of method 1-C is to avoid multiple emissions of identical author names and thus, save valuable network bandwidth. To achieve this, we first initialize a container data structure H which shall allow us to emit ($author, list[paper, score]$) tuples instead of the simple ($author, (paper, score)$) tuples of method 1. During the references parsing process, each time an author is encountered we perform a

Algorithm 3. Method 2, Reducer class

```
1: class Reducer
2:     method initialize
3:         string a_prev ← ""
4:         integer p_prev ← 0
5:         integer n ← 0
6:         H ← new Array
7:     method reduce (pair[string a, integer p]; float S^p)
8:         if a = a_prev
9:             if p = p_prev
10:                H(n) ← H(n) + S^p
11:            else
12:                H.add(S^p)
13:                n ← n + 1
14:        else
15:            Perform steps 9–17 of Algorithm 2
16:            H.reset()
17:            a_prev ← a
18:            p_prev ← p
19:            n ← 0
20:    method close
21:        emit (a, metric)
```

---

look-up in the container (step 10); in case the author is not present in H we insert the record along with the corresponding (*paper*, *score*) pair (steps 11–13). In the opposite case, we need to check whether the current reference belongs to the (*paper*, *score*) list of the found author. If the search is unsuccessful we store the paper and its score in the list (step 16); otherwise, we update the corresponding list record by summing up the new paper score to the stored one (step 18). After all the input data has been processed, the Mapper emits the tuples stored within H to the Reducer via the Close method.

It is immediately obvious that the method 1 generates an immense number of key-value pairs compared to method 1-C. Method 1-C is much more compact since with method 1, the author is repeated for each reference we send to the Reducer. Nevertheless, we need to mention here that there are two side effects deriving from the usage of a Combiner. The first one is the increased memory footprint of the map function due to the allocation of the container data structure. The second is a possible delay in the execution of the map phase due to the double search we perform (one for the author and one for paper). However, in this specific application that we examine, our experiments reveal that this delay is infinitesimal due to the small length of the (*paper*, *score*) lists, and that the usage of a Combiner definitely leads to significant acceleration of the entire task.

Finally, we introduce method 2-C where we inject the in-Mapper Combiner approach in method 2. The Algorithm 5 illustrates the basic steps which are similar to those of Algorithm 4. In this case however, the container data structure does not store a list of (*paper*, *score*) pairs for each author, but a single

Algorithm 4. Method 1-C: Improved version of method 1 with Combiners

---

1: **class** Mapper
2: **method** initialize
3:      $H \leftarrow$ new AssociativeArray
4: **method** map (integer p_i; string C^{p_i})
5:      $P \leftarrow$ ExtractReferences(C^{p_i})
6:      **for all** references $p \in P$
7:          $S^p \leftarrow$ ComputeScore(p)
8:          $A^p \leftarrow$ ExtractAuthors(p)
9:          **for all** authors $a \in A^p$
10:             **if** $a \notin H$
11:                 $L^a \leftarrow$ new Array
12:                 L^a.add(p, S^p)
13:                 H.add(a, L^a)
14:             **else**
15:                 **if** $p \notin H.L^a$
16:                     $H.L^a$.add(p, S^p)
17:                 **else**
18:                     $H.L^a$.update($p, +S^p$)
19:     **method** close
20:         **for all** authors $a \in H$
21:             emit (a, list(p, S^p))
22: **class** Reducer
23:     **method** reduce (**string** a; list[**integer** p, **float** S^p])
24:         $H \leftarrow$ new AssociativeArray
25:         **for all** pair $v \in$ list[**integer** p, **float** S^p]
26:             **if** $v.p \in H$
27:                 $H^p.S \leftarrow H^p.S + v.S^p$
28:             **else**
29:                 H.add(v)
30:         Perform steps 9–17 of Algorithm 2

---

cumulative score value per each distinct (*author, paper*) pair. This minimizes the benefits of using a Combiner because the (*author, paper*) keys are more numerous than the simple author keys of method 1-C. In addition, notice that the reduce phase in this case is identical to that of method 2.

5   Experiments

For the experimental evaluation of our theoretic analysis we employed Hadoop 0.20.2, an open-source implementation of the Google's MapReduce framework. We begin this section with a brief description of our test dataset and we proceed with data size measurements and efficiency assessments.

Algorithm 5. Method 2-C: Improved version of method 2 with the introduction of in-Mapper Combiners. The Reducer is identical to the one of Algorithm 3.

```
 1: class Mapper
 2:    method initialize
 3:        H ← new AssociativeArray
 4:    method map (integer p_i; string C^{p_i})
 5:        P ← ExtractReferences(C^{p_i})
 6:        for all references p ∈ P
 7:            S^p ← ComputeScore(p)
 8:            A^p ← ExtractAuthors(p)
 9:            for all authors a ∈ A^p
10:                if pair(a, p) ∉ H
11:                    H.add(pair(a, p), S^p)
12:                else
13:                    H.update(pair(a, p), +S^p)
14:    method close
15:        for all pairs (a, p) ∈ H
16:            emit (pair(a, p), S^p)
```

5.1 Dataset Characteristics

Collecting bibliometric data is a challenging task, due to the strict data protection policies applied by the digital libraries. Since crawling is forbidden, we are limited in using only open access document collections. The largest among these collections is the CiteSeerX [1] dataset, an open repository comprised of approximately 1.8 million scientific articles. The dataset is available in three forms: The first one contains the raw text of the publications scattered in 1.8 million plain text files. The other two contain certain meta-data of the documents expressed in SQL and XML formats respectively. The raw text format of the articles requires much and intensive effort towards two directions: a) disambiguation of the authors names and b) references extraction. Although these problems are both interesting and challenging, they are out of the scope of this paper. For this reason, we choose to work with the XML formatted dataset.

For each article of the dataset there are one or more small-sized XML files, each of which represents a different version of the same article. The dataset includes in total 3.9 million XML files, however, in our experiments we use only the latest version; consequently, 1.8 million XML files are used. This large number of small-sized files renders the dataset inappropriate for MapReduce, because

Table 3. Problem input-output statistics

Statistic	Value
Input Records	1,844,272
Input Size	27.6 GB
Output Records	2,865,282
Output Size	39.9 MB

the underlying distributed file system is designed for optimal performance when dealing with considerably larger files. For this reason, we performed a conversion of the dataset by packing thousands of these XML files into larger binary files. After this process, our "new" dataset was comprised of 432 files of 64 MB each.

5.2   Data Sizes

In this subsection we perform measurements of the data sizes exchanged between the Mappers and the Reducers of our proposed methods. Initially we provide method independent numbers indicating the data sizes involved in the examined problem. The first two rows of Table 3 concern the Mapper input, whereas the last two are connected to the Reducer output. As mentioned, the input consists of approximately 1.8 million articles which occupy in total roughly 27.6 GB. After the processing of the dataset with MapReduce, the system outputs a set of about 2.8 million (*author, metric*) pairs the size of which touches 40MB.

Table 4 illustrates various statistics; in the first double column we measure the size of the Mapper output of all four examined methods, expressed in number of records and data size in MB. The latter measurement is essential since it reflects the overall size of the data exchanged among the Mapper and Reducer nodes of the cluster. In the next column we record the counts of the Reducers input groups which represent the number of the unique keys which *arrive* at the Reducers. To acquire these measurements, we executed all four methods by employing only one Worker node; Table 4 derives from the report generated by the framework at the end of the task.

Initially we examine the performance of our methods in terms of sizes of the Mapper outputs. The map phase of methods 1 and 2 transmitted in total 36.7 million records occupying roughly 688 MB. On the contrary, the usage of a Combiner in method 1-C decreased these values by 42% (21.7 million records) and 13% (601 MB) respectively. As we anticipated, method 2-C was not equally efficient despite the usage of a Combiner. Compared to methods 1 and 2 we only achieve a reduction in the size of the outputted data by a margin of 6.5%. This is due to the fact that the Combiner of method 1-C lists (*paper, score*) pairs per each unique author, whereas method 2-C stores one partial score value for each distinct (*author, paper*) key; the latter key type is much rarer than the former.

The counts of the Reducer input groups reveal that the number of the unique keys which arrive at the Reducers of methods 1 and 1-C is equal to the number of records that depart from it (see third row of Table 3). This is due to the fact

Table 4. Record counts and data sizes for the four examined methods

| Method | Mapper Output | | Reducer |
	Records	Size (MB)	Input Groups
method 1	36,687,999	688.4	2,865,282
method 2	36,687,999	688.4	12,260,311
method 1-C	21,736,395	600.8	2,865,282
method 2-C	34,251,437	643.2	12,260,311

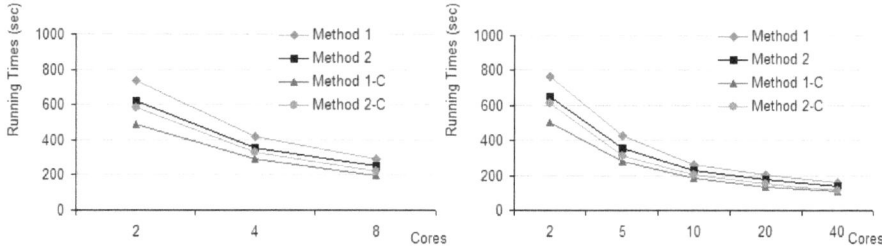

Fig. 1. Running times of the four methods in a small local cluster (Left), and a Web cluster infrastructure (Right)

that the output of the entire task (i.e. (a, h^a) pairs) has the same key as the Mapper output of these two methods. On the other hand, the tuples produced by the Mappers of methods 2 and 2-C are keyed by using $(author, paper)$ pairs, consequently, the unique keys which arrive at the Reducers increase by a factor of approximately 4.2.

5.3 Efficiency Measurements

In this subsection we evaluate the performance of the four methods. To exhaustively attest the scalability of our algorithms, we measured their running times by using two platforms. The first one includes a small-sized lab network, whereas the second one is a larger Web cluster infrastructure. Each experiment was repeatedly performed by employing different numbers of processing cores each time. The results are depicted in Figure 1.

Our first observation is that in both platforms, all of our methods scale well for fewer than 20 cores; the doubling of the cluster size almost leads to halved running times. For more cores the gains are slightly limited, due to the increased network latencies. Notice that the running times between the two clusters are not comparable, since these clusters are equipped with different hardware and they adopt different architectures. In all occasions, method 1-C outperformed the other approaches by a margin ranging between 32% and 35%. Apparently, the existence of the Combiner results in decreased exchange of data among the nodes of the clusters. Although method 2-C also employs a Combiner, it did not perform equally well; compared to method 1-C it was outperformed by about 18%. We have previously explained that the $(author, paper)$ keys of method 2-C are more numerous than the simple author keys of method 1-C, consequently, the benefits of using a Combiner are limited.

Regarding the plain methods 1 and 2, we notice that the latter completed the assigned task slightly faster. Although the amount of data exchanged among the nodes of the system is equal in both methods, method 2 achieves better performance due to the more robust implementation of its Reducer. More specifically, the $(author, paper)$ keys emitted by the Mapper of method 2 are brought to the Reducer in sorted author and paper order (secondary sort), thus saving us the cost of searching for the input papers.

6   Conclusions

In this paper we studied the issue of computing several scientometrics in large-scale academic search engines with MapReduce. The scientometrics are scalar values used to evaluate the research work of a scientist. The large volumes of data employed by the modern academic search engines in combination with their popularity, has rendered the examined problem both interesting and challenging. We introduced four methods to compute three of these metrics, h-index, and two variations, the contemporary and trend h-indexes. However, these methods can be applied to compute a wider variety scientometrics with no additional effort. We proposed optimizations with the aim of decreasing the size of the data exchanged among the nodes of the system, and we conducted experiments with the CiteSeerX dataset, a large repository comprised of about 1.8 million research articles. Our experiments demonstrated the usefulness of method 1-C, a strategy which achieves both effective and efficient execution.

References

1. CiteSeerX Data, http://csxstatic.ist.psu.edu/about/data
2. Abouzeid, A., Bajda-Pawlikowski, K., Abadi, D., Silberschatz, A., Rasin, A.: HadoopDB: An Architectural Hybrid of MapReduce and DBMS Technologies for Analytical Workloads. Proceedings of the VLDB Endowment 2(1), 922–933 (2009)
3. Borthakur, D.: The Hadoop distributed file system: Architecture and design (2007)
4. Dean, J., Ghemawat, S.: MapReduce: Simplified Data Processing on Large Clusters. Communications of the ACM 51(1), 107–113 (2008)
5. Egghe, L.: Theory and Practise of the g-index. Scientometrics 69(1), 131–152 (2006)
6. Elsayed, T., Lin, J., Oard, D.: Pairwise document similarity in large collections with MapReduce. In: Proceedings of 46th Annual Meeting of the Association for Computational Linguistics on Human Language Technologies, pp. 265–268 (2008)
7. Ghemawat, S., Dean, J.: MapReduce: Simplified Data Processing on Large Clusters. In: Symposium on Operating System Design and Implementation (OSDI 2004), San Francisco, California, USA, pp. 137–150 (2004)
8. Ghemawat, S., Gobioff, H., Leung, S.: The Google file system. ACM SIGOPS Operating Systems Review 37, 29–43 (2003)
9. Hirsch, J.: An Index to Quantify an Individual's Scientific Research Output. Proceedings of the National Academy of Sciences 102(46), 16569 (2005)
10. Katsaros, D., Akritidis, L., Bozanis, P.: The f index: Quantifying the Impact of Coterminal Citations on Scientists' Ranking. Journal of the American Society for Information Science and Technology 60(5), 1051–1056 (2009)
11. Lin, J.: Scalable language processing algorithms for the masses: A case study in computing word co-occurrence matrices with MapReduce. In: Proceedings of the Conference on Empirical Methods in Language Processing, pp. 419–428 (2008)
12. Lin, J., Dyer, C.: Data-intensive Text Processing with MapReduce. Synthesis Lectures on Human Language Technologies 3(1), 1–177 (2010)
13. McCreadie, R., Macdonald, C., Ounis, I.: On single-pass indexing with MapReduce. In: Proceedings of the 32nd International ACM SIGIR Conference on Research and Development in Information Retrieval, pp. 742–743 (2009)

14. Sidiropoulos, A., Katsaros, D., Manolopoulos, Y.: Generalized Hirsch h-index for Disclosing Latent Facts in Citation Networks. Scientometrics 72(2), 253–280
15. Sidiropoulos, A., Manolopoulos, Y.: A Citation-Based System to Assist Prize Awarding. ACM SIGMOD Record 34(4), 60 (2005)
16. Yang, H., Dasdan, A., Hsiao, R., Parker, D.: Map-reduce-merge: simplified relational data processing on large clusters. In: Proceedings of the 2007 ACM SIGMOD International Conference on Management of Data, pp. 1029–1040 (2007)
17. Malewicz, G., Austern, M.H., Bik, A.J.C., Dehnert, J.C., Horn, I., Leiser, N., Czajkowski, G.: Pregel: A System for Large-Scale Graph Processing. In: Proceedings of the 2010 ACM SIGMOD International Conference on Management of Data, pp. 135–146 (2010)

Analyzing the Effectiveness of Graph Metrics for Anomaly Detection in Online Social Networks

Reza Hassanzadeh, Richi Nayak, and Douglas Stebila

School of Electrical Engineering and Computer Science, Science and Engineering Faculty,
Queensland University of Technology, Brisbane, Australia
{r.hassanzadeh,r.nayak,Stebila}@qut.edu.au

Abstract. Online social networks can be modelled as graphs; in this paper, we analyze the use of graph metrics for identifying users with anomalous relationships to other users. A framework is proposed for analyzing the effectiveness of various graph theoretic properties such as the number of neighbouring nodes and edges, betweenness centrality, and community cohesiveness in detecting anomalous users. Experimental results on real-world data collected from online social networks show that the majority of users typically have friends who are friends themselves, whereas anomalous users' graphs typically do not follow this common rule. Empirical analysis also shows that the relationship between average betweenness centrality and edges identifies anomalies more accurately than other approaches.

Keywords: Anomaly detection, Graph mining, Data mining, Online Social Networks.

1 Introduction

Online social networks are being used in various domains such as business, education, telemarketing and many others. With increasing use of social networks comes increasing prevalence of illegal activities using social networks [1]. It is critical that methods of anomaly detection in social networks are developed to coincide with developments in usage of social networks.

An online social network can be modelled as graph [2] in which the nodes represent people and the edges represent the links between nodes using a range of relationships such as friendship, affiliation, family and many others. In this paper we propose the use of various graph properties for differentiating people's online behaviour by their usage patterns. If the usage pattern of a user follows common patterns, we describe the usage as *normal*, otherwise the usage is an *outlier* or *anomalous*. Looking at the relationships of users can reveal meaningful patterns: users can hide their identity by supplying false information but they cannot hide certain types of metadata, such as the links that they have established with other users.

We use *local graph properties* to extract common rules. Local metrics refer to a single node (e*go*), its 1-level neighbourhood (an *egonet*) and 2-level neighbourhood (a *super-egonet*). These undeniable relationships can help in spotting behaviours that are abnormal.

X.S. Wang et al. (Eds.): WISE 2012, LNCS 7651, pp. 624–630, 2012.

In particular, we propose the use of *betweenness centrality* and *average betweenness centrality* of a user's egonet, and the *community cohesiveness* of the user's super-egonet as potential measures for identifying anomalies based on the structure of users' links. Additionally, we give a framework for evaluating the effectiveness of various combinations of properties for identifying anomalous nodes in unlabelled datasets.

We evaluate the proposed methods with existing data collected from three online social networks (Facebook, Orkut, and Flickr). Results show that the majority of users follow the "friends of friends are often friends" pattern and a very few users follow either the "cliques or near-cliques" pattern (all the neighbours connected) or the "stars or near-star" pattern (mostly disconnected). Previous works [1, 3-5] have established that these two types of patterns can be connected to abnormalities in the graph, particularly in online social networks. Several graph theoretic metrics, in particular average betweenness centrality give better accuracy in detecting anomalies than existing approaches.

1.1   Related Work

Limited work has been done on applying anomaly detection techniques to online social networks until recently [4, 6]. Recent work can be divided into two categories: *behaviour-based techniques* that consider the dynamic usage behaviour of users; and *structure-based techniques* [1, 3-5, 7] that consider the static structure of the graph. Behaviour-based techniques concentrate on mining users' usage patterns. Although they can help to spot anomalies, they are very technology-dependent. Akoglu et al [4] designed a structure-based approach entitled the OddBall algorithm for analyzing social network graphs. OddBall is based on the power law relation between number of nodes and number of edges and a density-based outlier detection technique to calculate a final anomaly score. However, using only a power law relation is prone to miss some outliers especially for egonets with a high number of nodes and edges. In this paper, we propose an algorithm and a framework for detecting anomalies in an unlabelled social network's dataset based on betweenness centrality.

In traditional data mining, a common method of detecting outlier is identification of *clusters*. Similarly, within the modelled graph, a *community* can be defined as a group of nodes which share common properties. Detecting communities can give us useful information to find if there are any similarities or common interests between the friends of suspected users. Graph-based community detection techniques have been investigated in the literature [8, 9]. Existing algorithms try to find parts of the network that are better connected internally. We propose an alternative method based on the number of external links between two users' egonets.

2   The Proposed Framework

We propose a framework that introduces semi-supervised graph-based anomaly detection with the use of a scoring method to report anomalies. It aims to find the common behaviour that is followed by the majority of nodes. It computes graph metrics of a user's egonet and then examines relationships between these properties. The common patterns are then used in distinguishing users that may be anomalous. Our proposed analysis method consists of the following steps, which will be explained in detail in the rest of this section:

Step 1: Compute graph metrics
Metrics computed include: N: number of nodes in a user's egonet; E: number of edges in a user's egonet; ABC: the average betweenness centrality of all nodes in a user's egonet; and Com: the community cohesiveness of the user's super-egonet.

Step 2: Compute fitting curve
For the relationships between N vs. E, ABC vs. E, and N vs. Com, the fitting curve will be computed. The fit may be linear or power law [10].

Step 3: Compute outlier score
For each relationship, an anomaly score function, which is based on distance from the fitting line, is determined.

Step 4: Label for evaluation
A labelled subset of nodes is obtained.

Step 5: Find threshold
Using the scoring function from step 3, a threshold that minimizes the number of false negatives and false positives rate is determined for the labelled subset of data.

2.1   Step 1: Compute Graph Metrics

A graph $\mathcal{G} = (\mathcal{V}, \mathcal{E})$ consists of a set $\mathcal{V}$ of vertices (nodes or users) and a set $\mathcal{E}$ of edges (links between two users). Given the graph $\mathcal{G}$, an ego i is a user (or node) and $egonet_i = \{i, i_1, i_2, i_3, ..., i_n\}$ consists of the user's neighbours $i_1, ..., i_n$. A user's *super-egonet* includes the user's egonet and the egonets of all its neighbours: the super-egonet of ego i is $super-egonet_i = \{egonet_i, egonet_{i_1}, ..., egonet_{i_n}\}$.

2.1.1   Average Betweenness Centrality

The *betweenness centrality* ($\mathcal{BC}$) of a node in a graph is the number of shortest paths between all pairs of nodes within that graph that go through that node.

Definition 1 (Betweenness centrality). The *betweenness centrality* of a vertex $i \in \mathcal{V}(\mathcal{G})$ is

$$BC_i = \sum_{s \neq i \neq d} \psi_i^{sd} / n_{sd} \qquad i, s, d \in \mathcal{V} \tag{1}$$

where ψ_i^{sd} is the number of shortest paths between s and d passing through node i and n_{sd} to be the total number of shortest paths from s to d. Brandes' algorithm for computing betweenness centrality runs in time $\mathcal{O}(nm)$ and space $\mathcal{O}(n + m)$, where n is the number of nodes and m is the number of edges [11]. Each new edge defining a new shortest path will reduce $\mathcal{BC}$ of the central node

We propose the use of the *average betweenness centrality* ($\mathcal{ABC}$) of a node within the node's egonet. Recall that $\mathcal{BC}$ for each node is computed as the number of shortest paths between all pairs of nodes within the egonet that go through that node. Adding edges between nodes in the egonet reduces the betweenness centrality of the ego. Intuitively, an egonet has higher average betweenness centrality when more nodes are involved in shortest paths.

Definition 2 (Average Betweenness Centrality). The *average betweenness central-ity* of $egonet_i$ is:

$$\sigma_i^{abc} = \frac{f(i) + \sum_{j=1}^{n} f(i_j)}{n}, \quad where \quad n = |\mathcal{V}^{egonet_i}|, i \in \mathcal{V}(\mathcal{G}) \tag{2}$$

We define $f(i_j): \mathcal{V}^{egonet_i} \to \mathbb{R}^{\geq 0}$ as the function that maps each node i_j within $egonet_i$ to its betweenness centrality within its own egonet.

2.1.2   Community Detection

People naturally tend to form communities based on their similarity and common interests. This behaviour stands true in online social networks [9]. The information which can be extracted from communities' structure is useful to analyze the behaviour of a user and can lead towards identifying anomalous behaviour. For community de-tection we examine users' super-egonets, which can give us sufficient information to find if there are any similarity and common interests between their friends by examin-ing their connections. The pattern of communities between friends of friends also can set rules that help us to spot anomalous users.

Definition 3 (External Degree). The external degree of $egonet_{i_m}$ to $egonet_{i_n}$ is defined as:

$$d_i(i_m, i_n) = |\mathcal{V}^{egonet_{i_m}} \cap \mathcal{V}^{egonet_{i_n}}| +$$
$$|ij \in \mathcal{E} : i \in \mathcal{V}^{egonet_{i_m}}, j \in \mathcal{V}^{egonet_{i_n}}| \quad , i_m, i_n \in \mathcal{G} \tag{3}$$

where $\mathcal{V}^{egonet_{i_m}}$ is the set of nodes of $egonet_{i_m}$ and $\mathcal{V}^{egonet_{i_n}}$ is the set of nodes of $egonet_{i_n}$. The *normalized external degree* is defined as follows:

$$d_i(i_m, i_n)_{norm} = \frac{d_i(i_m, i_n)}{\min(|i_m|, |i_n|)} \tag{4}$$

Definition 4 (Community). The egonets of users i_m and i_n form a *community* if at least the half of the nodes of the smaller egonet connect to the other egonet.

$$C_i(i_m, i_n) = \begin{cases} 1, & if\ d_i(i_m, i_n)_{norm} \geq \min(|i_m|, |i_n|) / 2 \\ 0, & otherwise \end{cases} \tag{5}$$

2.2   Step 2: Compute Fitting Curve

Local graph metrics related to a single node, its egonet and its super-egonet are used to identify common patterns. We model the relationships between the local metrics using distribution models such as linear and power law. Coefficient of determination (R^2) of each model is computed as a goodness of fit measure for the fitting curves.

$R^2 = 1 - SS_{residual} / SS_{total}$, $SS_{residual} = \sum_{i=1}^{k}(\mathcal{Y}_i - \mathcal{Y}_i^p)^2$, where $\mathcal{Y}_i^p$ is predicted value of $\mathcal{Y}_i$ and $SS_{total} = \sum_{i=1}^{k}(\mathcal{Y}_i - E(\mathcal{Y}_i))^2$, where $E(\cdot)$ gives expected value.

Table 1 includes fitting line equations and R^2 for each relationship and dataset; plots are omitted due to page limitations.

N vs. E (power law) [4]
Compute a fitting line $E_i \propto N_i^{\,a}$, where $1 \le a \le 2, E_i$ is the number of edges, N_i is the number of nodes, and a is the power law exponent for user i's egonet.

E vs. ABC (power law)
Compute a fitting line $\mathcal{Y} = C\mathcal{X}^\theta$, where $\mathcal{Y}$ is E, and $\mathcal{X}$ is ABC, and θ is the power law exponent for user i's egonet.

N vs. Com (power law)
Compute a fitting line $\mathcal{Y} = C\mathcal{X}^\theta$, where $\mathcal{Y}$ is Com, $\mathcal{X}$ is N, and θ is the power law exponent for user i's super-egonet.

N vs. E (linear)
Compute a fitting line $E_i \propto \beta N_i$, where E_i is number of edges, N_i is number of nodes, and β is the gradient of the fitting line for user i's egonet.

E vs. ABC (linear)
Compute a fitting line $E_i \propto \lambda \sigma_i^{abc}$, where E_i is the number of edges, σ_i^{abc} is ABC and λ is the gradient of the fitting line for user i's egonet. Our experiments show there is a relationship between anomaly and the proportion of E_i to σ_i^{abc} .

2.3   Step 3: Compute Outlier Score

For each power law fitting line from step 2, we used the following anomaly score to determine the distance from the fitting line for ego_i ; the calculating follows the OddBall method [4]:

$$aScore(i) = \frac{max\left(\mathcal{Y}_i, C\mathcal{X}_i^\theta\right)}{min\left(\mathcal{Y}_i, C\mathcal{X}_i^\theta\right)} * \log\left(\left|\mathcal{Y}_i - C\mathcal{X}_i^\theta\right| + 1\right) \tag{6}$$

where $\mathcal{Y}_i$ is the y-value, $\mathcal{X}_i$ is x-value of $egonet\ i$, and θ is a power law exponent. For the power law equation $\mathcal{Y} = C\mathcal{X}^\theta$ this measures "distance to fitting line" by penalizing the number of times that $\mathcal{Y}_i$ deviates from the line.

For each linear fitting line from step 2, we computed $aScore(i)$ in a similar way, but with $\mathcal{Y} = C\mathcal{X} + \theta$ in place of $\mathcal{Y} = C\mathcal{X}^\theta$.

2.4   Step 4: Label for Evaluation

Since the existing datasets were not labelled, we used visual inspection to label anomalies. In particular, we visually examined the egonets of each node and decided whether the node was anomalous our not based on evidence from previous works [1, 3-5]: the majority of users follow the "friends of friends are often friends" pattern and very few users follow either the "cliques or near-cliques" pattern (all the neighbours connected) or the "stars or near-star" pattern (mostly disconnected).

2.5   Step 5: Find Threshold

In this step, we compute determine for each metric a threshold value on the outlier score *aScore* that minimizes the *F-Score*, which is the number of false positives and false negatives in the labelled dataset from step 4. The *F-Score* is calculated as *F-Score =2*Precision*Recall / (Precision +Recall)*; its highest value (1) indicates perfect classification of labelled data, whereas its lowest value (0) indicates completely wrong classification of labelled data.

3   Experimental Results

Our proposed method is evaluated with three real-life datasets Orkut, Flickr, and Facebook. These datasets were collected by crawling techniques in 2008 [12]. The Orkut dataset has 3M nodes and 23M edges; the Flickr dataset has 1.8M nodes and 22M edges; and the Facebook dataset has 64K nodes and 1.5M edges.

We applied the proposed framework to 20,000 randomly sampled egonets from each dataset. After computing the graph metrics (step 1), fitting curves were computed using regression to determine relationships between metrics (step 2). Outlier scores were then computed for each node (step 3). A labelled subset of 100 nodes (step 4) was then used in threshold finding (step 5) to identify a threshold outlier score that minimizes false negatives and false positives. The resulting F-score was calculated to allow comparison of metrics.

Table 1 compares our observed results for the various graph properties for each of our datasets. We compare five metrics: N vs. E (Linear), E vs. ABC (Linear), E vs. ABC (Power law), N vs. Com (Power law), and N vs. E (Power law); the last being the "OddBall" method of Akoglu et al. [4].

Table 1. Comparison of effectiveness graph theoretic properties for anomaly detection in real-life datasets

Dataset	Method	Fitting curve	R^2	Recall %	Precision %	F-score %
Facebook	E vs. N (Linear)	$V = 0.0638 * E + 29.223$	0.80	50.51	100.00	67.11
	E vs. ABC (Linear)	$E = 0.0281 * ABC + 10.553$	0.73	92.45	98.00	95.15
	E vs. ABC (Power law)	$E = 0.3839 * ABC^{0.7019}$	0.86	100.00	100.00	100.00
	N vs. Com (Power law)	$Com = 0.0369 * N^{2.3508}$	0.77	52.08	100.00	68.49
	N vs. E (Power law) [4]	$E = 0.5454 * N^{1.571}$	0.95	49.49	98.00	65.77
Flickr	E vs. N (Linear)	$V = 0.009 * E + 187.39$	0.65	70.00	98.00	81.67
	E vs. ABC (Linear)	$E = 144.77 * ABC - 8272.1$	0.62	70.42	100.00	82.64
	E vs. ABC (Power law)	$E = 0.6151 * ABC^{0.6401}$	0.91	77.78	98.00	86.73
	N vs. Com (Power law)	$Com = 0.1248 * N^{2.0304}$	0.88	57.78	52.00	54.74
	N vs. E (Power law) [4]	$E = 0.3098 * N^{1.6644}$	0.96	48.39	90.00	62.94
Orkut	E vs. N (Linear)	$V = 0.0513 * E + 54.272$	0.64	77.78	75.68	76.71
	E vs. ABC (Linear)	$E = 25.025 * ABC + 177.83$	0.61	94.87	100.00	97.37
	E vs. ABC (Power law)	$E = 0.544 * ABC^{0.6483}$	0.82	96.97	86.49	91.43
	N vs. Com (Power law)	$Com = 0.1045 * N^{2.0689}$	0.89	75.68	75.68	75.68
	N vs. E (Power law) [4]	$E = 0.5362 * N^{1.5676}$	0.90	100.00	83.78	91.18

As we can see from Table 1, our results find that the E vs. ABC (Power law) and E vs. ABC (Linear) methods have the best overall performance across the three datasets. These methods both have higher *F-score* that N vs. E (Power law), which is the Odd-Ball method of Akoglu et al. [4].

4   Conclusion

The direct connectivity of online social networks can facilitate illegal activity. In this paper, we have expanded previous research of static analysis of user relationships for detecting anomalous behaviour in online social networks. We have introduced metrics based on a variety of graph properties and presented a framework for detecting nodes with anomalous relationships with other nodes. We applied our approach to datasets from existing online social networks with a manually labelled set of nodes based on existing observations. We identified several metrics—involving the relationship between number of edges and average betweenness centrality of a user's immediate neighbourhood—that perform better than previous. Interesting future work in this area includes the consideration of other datasets and labelling, specifically datasets with pre-established labelling of anomalies, such as email spam or criminal records.

References

1. Shrivastava, N., Majumder, A., Rastogi, R.: Mining (social) network graphs to detect random link attacks. IEEE (2008)
2. Newman, M.E.J., Watts, D.J., Strogatz, S.H.: Random graph models of social networks. Proceedings of the National Academy of Sciences of the United States of America 99(suppl. 1), 2566 (2002)
3. Tong, H., Lin, C.Y.: Non-negative residual matrix factorization with application to graph anomaly detection. In: SDM (2011)
4. Akoglu, L., McGlohon, M., Faloutsos, C.: oddball: Spotting Anomalies in Weighted Graphs. In: Zaki, M.J., Yu, J.X., Ravindran, B., Pudi, V. (eds.) PAKDD 2010. LNCS, vol. 6119, pp. 410–421. Springer, Heidelberg (2010)
5. Sun, J., et al.: Neighborhood formation and anomaly detection in bipartite graphs. In: Fifth IEEE International Conference on Data Mining (2005)
6. Limsaiprom, P., Tantatsanawong, P.: Social network anomaly and attack patterns analysis. In: 6th International Conference on Networked Computing, INC (2010)
7. Heard, N., et al.: Bayesian anomaly detection methods for social networks. The Annals of Applied Statistics 4(2), 645–662 (2010)
8. Ball, B., Karrer, B., Newman, M.: An efficient and principled method for detecting communities in networks. Arxiv preprint arXiv:1104.3590 (2011)
9. Yang, Y., Guo, Y.C., Ma, Y.N.: Characterization of Communities in Online Social Network. In: Proceedings of 2010 Cross-Strait Conference on InformationScience and Technology, pp. 600–605799 (2010)
10. Clauset, A., Shalizi, C.R., Newman, M.E.J.: Power-law distributions in empirical data. Arxiv preprint arxiv:0706.1062 (2007)
11. Brandes, U.: A faster algorithm for betweenness centrality. Journal of Mathematical Sociology 25(2), 163–177 (2001)
12. Mislove, A., et al.: Measurement and analysis of online social networks. In: Proceedings of the 7th ACM SIGCOMM Conference on Internet Measurement. ACM, San Diego (2007)

An Unsupervised Technique to Extract Information from Semi-structured Web Pages⋆

Hassan A. Sleiman and Rafael Corchuelo

University of Sevilla, Spain
{hassansleiman,corchu}@us.es

Abstract. We propose a technique that takes two or more web pages generated by the same server-side template and tries to learn a regular expression that represents it and helps extract relevant information from similar pages. Our experimental results on real-world web sites demonstrate that our technique outperforms others in terms of both effectiveness and efficiency and is not affected by HTML errors.

Keywords: Web information extraction, unsupervised learning.

1 Introduction

The Web is a huge information repository. Semi-structured web pages are generated by server-side scripts that retrieve information from databases and present it in HTML templates that introduce irrelevant information and attractive styles and layouts. Information extractors help extract the relevant information in a web page by using machine learning techniques.

Many information extractors rely on extraction rules. Although they can be handcrafted, the costs involved motivated many researchers to work on proposals to learn them automatically [4]. These proposals are either supervised, i.e., they require the user to provide a number of samples to be extracted, or unsupervised, i.e., they extract as much prospective information as they can and the user then gathers the relevant information from the results. Some authors have worked on unsupervised proposals that do not rely on extraction rules, but are based on a number of hypothesis and heuristics that have proven effective [1, 7]. Since typical web pages are growing in complexity, some authors are also paying attention to the problem of identifying information regions [11].

In this paper, we introduce a technique that allows to learn a regular expression that describes the structure of the template used to generate some input web pages; this expression can later be used to extract information from similar pages. The idea is to compare the input web pages in order to discover shared patterns that are common to all of them and, thus, are not likely to contain any

⋆ Supported by the European Commission (FEDER), the Spanish and the Andalusian R&D&I programmes (TIN2007-64119, P07-TIC-2602, P08-TIC-4100, TIN2008-04718-E, TIN2010-21744, TIN2010-09809-E, TIN2010-10811-E, TIN2010-09988-E, and TIN2011-15497-E).

X.S. Wang et al. (Eds.): WISE 2012, LNCS 7651, pp. 631–637, 2012.

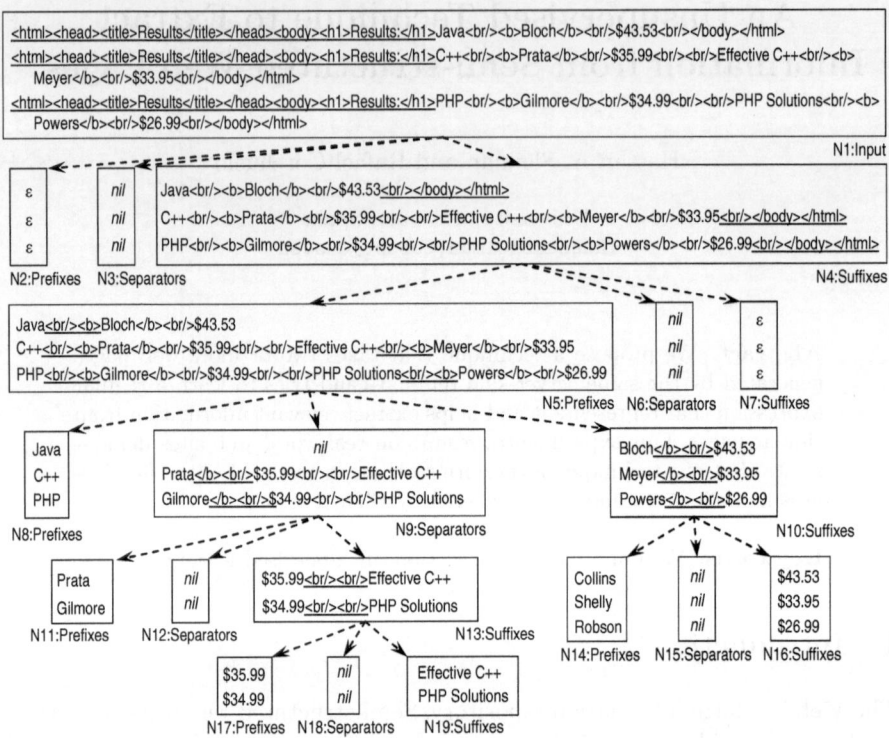

Fig. 1. A sample trinary tree. (Underlined strings represent shared patterns.)

relevant information. The idea of identifying shared patterns lies at the heart of the other related proposals, but the techniques they use differ significantly. EXALG [2] uses two statistical techniques to find the tokens that belong to the template; RoadRunner [5] uses a multi-string alignment algorithm that is exponential in the size of the input documents [6]; FiVaTech [8] relies on tree clustering and an ad-hoc matrix alignment technique. Our proposal relies on a multi-string alignment algorithm that has proven to be very effective and efficient in practice. Contrarily to the previous proposals, ours does not require the input web documents to be translated into DOM trees and thus does not require the input documents to be corrected so that they are well-formed HTML.

The rest of the paper is organised as follows: Section 2 describes the main algorithms of our proposal; Section 3 presents the results of our experimental evaluation; Section 4 draws a few conclusions; the paper finishes with some selected references to the literature.

2   Algorithm

Our algorithm works in two steps: first, it creates a trinary tree that represents the input pages and makes it explicit what fragments are shared patterns and the

```
createTrinaryTree(n: Node; min, max: nat)     expand(n: Node; s: int): boolean
    expanded = false                              result = false
    size = max                                    if size of n ≥ 2 then
    while size ≥ min and not expanded do              map, pattern = findPattern(n, s)
        expanded = expand(n, size)                    if map ≠ {} then
        size = size - 1                                   result = true
    if expanded then                                      set pattern of node to pattern
        leaves = getLeaves(n)                             foreach text in node do
        foreach leaf in leaves do                             using map and text
            createTrinaryTree(leaf, min, size)                    create the prefixes of n
                                                                  create the separators of n
                                                                  create the suffixes of n
                                                  return result
```

Fig. 2. Algorithms to create a trinary tree and to expand a node

prefixes, separators, and suffixes that they induce; later, this tree is transformed into a regular expression with capturing groups that represents the template used to generate the input pages. This expression allows to extract information from other pages that were generated by the same template thanks to the capturing groups.

Creating a trinary tree: Figure 1 presents a sample trinary tree that we use throughout the paper to illustrate our proposal. The tree is composed of Nodes of the form (T, a, p, e, s), where T is a collection of Texts, a is a Text that stores a shared pattern, and p, e, and s are three Nodes called prefixes, separators, and suffixes, respectively. A Text is a sequence of tokens, which represent HTML tags, script blocks, style blocks, and #PCDATA. We use ϵ to denote an empty Text and *nil* to refer to the inexistence of a Text or a Node.

Figure 2 presents our algorithm to create a trinary tree. It works on a Node with Texts that represent the input web pages and two naturals called *min* and *max* that limit the search for shared patterns to those of size *max* down to *min*. The algorithm first attempts to expand the input node, which is a process whose goal is to find a shared pattern in the collections of Texts in this node and use it to split them into new collections of prefixes, separators, and suffixes; if the input node is expanded, then the algorithm is applied recursively to the newly created leaves. To find shared patterns, we rely on Algorithm findPattern, which returns a map from Texts onto lists of naturals and a pattern; the former maps each Text in the input node onto the list of positions at which the latter was found. If the map is not empty, that implies that a shared pattern has been found, in which case, the algorithm sets the shared pattern of the input node to that pattern and then iterates over the Texts it contains and creates the corresponding prefix, separator, and suffix nodes.

To illustrate this algorithm, assume that it is invoked on Node $N1$ in Figure 1. It first searches for shared patterns in the Texts of this node and finds the following: <html><head><title>Results</title></head><body><h1>Results:

```
learnTemplate(n: Node; result: Regex): Regex
    if isOptional(n) then result += "("
    if isLeaf(n) then
        if not every texts in n is empty then
            result += "{" + freshLabel() + "}"
    else
        result += learnTemplate(prefix of n, result)
        result += pattern of n
        if isRepeatable(n, separators of n) then
            result += "(" + learnTemplate(separators of n, result) + getPattern(n)
            if nil is in separators of n then
                result += ")*"
            else
                result += ")+"
        result += learnTemplate(suffix of n, result)
    if isOptional(n) then result += ")?"
    return result
```

Fig. 3. Algorithm to learn a template from a trinary tree

</h1>. It then expands Node $N1$ into three additional nodes, namely: $N2$, $N3$, and $N4$. Since the shared pattern is found at the beginning of the Texts in $N1$ and it is not repeated in any of them, then Node $N2$, which contains the prefixes of the shared pattern, only contains three empty Texts; node $N3$, which contains the separators between the occurrences of the shared pattern in each Text, only contains three *nil* values since there are not any separators; contrarily, there are three suffixes that are stored in Node $N4$. Then, the algorithm is applied recursively to $N2$, $N3$, and $N4$ to search for new shared patterns. $N2$ and $N3$ are not processed again since they only contain empty or *nil* Texts; only Node $N4$, whose Texts share the 3-token pattern
</body></html>, is expanded again to create Nodes $N5$, $N6$, and $N7$. (Note that there are other shared patterns, but the algorithm searches for the longest one and breaks ties arbitrarily.) The same procedure is applied as many times as necessary until no more shared patterns are discovered.

Learning a regular expression: Figure 3 presents our algorithm to learn a regular expression from a trinary tree. It relies on two ancillary concepts: optionality and repeatability. A Node is optional if one or more of its Texts, but not all, are empty; a Node is repeatable if one or more of its non-empty Texts have more than one occurrence of the shared pattern, which implies that the separators Node contains one or more Texts that are not equal to *nil*.

The algorithm proceeds as follows: it works on a node and a regex that is expected to be an empty string initially; the algorithm constructs its result by adding text to this parameter on each recursive invocation. If the node being processed is a leaf and not all of its Texts are empty, the algorithm then adds a capturing group that represents a piece of text that has to be extracted; if it is not a leaf, then the algorithm builds recursively the regular expressions that correspond to the prefixes, the separators, and the suffixes. If the separators node is repeatable, then the decision on whether to use a star or a plus closure is made as follows: if *nil* is included in the Texts in the separators node, then it means that there is at least an input web page in which the separator does not appear, in which case, a star closure must be used; otherwise, a plus closure must be used. The first and the last lines of the algorithm take into account the case in which the node being processes is optional; in such cases, a parenthesis and an optional operator are added to the resulting regular expression.

The template learnt for our running example is the following:

```
(<html><head><title>Results</title></head><body><h1>Results:</h1>) {_A_} (<br/><b>)
( ( {_B_} (</b><br/>) {_C_} (<br/><br/>) {_D_} )? (<br/><b>) )*
{_E_} (</b><br/>) {_F_} (<br/></body></html>)
```

As is the case in other unsupervised techniques, it is the user who must assign a meaning to the capturing groups. In our running example, _A_ and _D_ stand for titles, _B_ and _E_ stand for authors, and _C_ and _F_ stand for prices. The problem of mapping the information extracted by the capturing groups onto structured records was dealt with elsewhere [3].

3   Experimentation

We have developed a prototype of our proposal using the CEDAR framework [10]. We performed a series of experiments on a cloud computer that was equipped with a four-threaded Intel Core i7 processor that ran at 2.93 GHz, had 4 GB of RAM, Windows 7 Pro 64-bit, and Oracle's Java Development Kit 1.7.0_02. The default heap size of the Java Virtual Machine was not modified. The experiments were carried out on a collection of 29 datasets that provide 688 web pages. The datasets were gathered from real-world web sites on books, doctors, movies, and from the EXALG repository [2]. We experimentally found that setting $min = 1$ and $max = \lfloor 0.05m \rfloor$ was the maximum allowable bias to our search procedure, where m denotes the size of the longest input document; this resulted in a significant increase in efficiency without an impact on effectiveness.

We ran our proposal, RoadRunner [5], FiVaTech [8], and WIEN [9] on the datasets in order to learn extraction rules. (Unfortunately, we could not find a public implementation of EXALG, so we added WIEN to our comparison since it ranks amongst the most cited proposals in information extraction.) We then computed the usual effectiveness and efficiency measures. In the case of WIEN, it was easy to compute the effectiveness measures since the technique is supervised, i.e., it requires the user to provide annotations with the information to be extracted so that an extraction rule can be learnt and evaluated. Contrarily, our proposal, RoadRunner, and FiVaTech are unsupervised, i.e., they learn an extraction rule that extracts as much information as possible, give computer-generated labels to the capturing groups, and it is the responsibility of the user to assign a meaning to these labels. We then used the following approach: we compared the information extracted by each capturing group to every annotation

Table 1. Results of our experimentation

Summary	N	E	Trinity				RoadRunner				FivaTech				WIEN			
			P	R	LT	ET	P	R	LT	ET	P	R	LT	ET	P	R	LT	ET
Mean	28.67	37.63	0.98	0.95	0.13	0.01	0.49	0.49	44.40	0.01	0.81	0.92	116.18	0.24	0.68	0.55	7.70	9.83
Standard deviation	8.60	37.98	0.03	0.11	0.18	0.02	0.47	0.46	193.77	0.01	0.19	0.11	192.06	0.44	0.24	0.30	4.93	6.79
Site	N	E	P	R	LT	ET	P	R	LT	ET	P	R	LT	ET	P	R	LT	ET
Books www.abebooks.com	30	2.94	1.00	1.00	0.03	0.00	-	-	-	-	0.92	0.99	15.46	0.12	0.52	0.16	9.66	10.26
www.awesomebooks.com	30	2.16	1.00	0.87	0.03	0.00	1.00	1.00	0.92	0.00	0.85	1.00	8.14	0.14	0.77	0.26	5.01	6.27
www.betterworldbooks.com	30	2.30	0.99	1.00	0.17	0.00	0.00	0.00	0.98	0.00	0.99	0.96	85.32	0.39	0.43	0.35	15.57	17.04
www.manybooks.net	30	6.50	0.99	0.99	0.11	0.02	-	-	-	-	0.77	0.97	65.49	0.12	0.25	0.23	6.74	8.14
www.waterstones.com	30	6.46	0.96	1.00	0.07	0.00	1.00	0.89	1.14	0.02	1.00	0.94	51.53	1.98	0.71	0.67	8.02	8.72
Doctors doctor.webmd.com	30	24.10	1.00	1.00	0.02	0.00	0.00	0.00	0.73	0.00	0.77	1.00	11.81	0.03	0.60	0.60	9.45	14.54
extapps.ama-assn.org	30	36.00	0.98	1.00	0.02	0.02	-	-	-	-	-	-	-	-	0.60	0.60	6.99	7.38
www.dentists.com	30	103.27	0.92	1.00	0.01	0.00	1.00	1.00	867.63	0.00	0.56	0.99	9.94	0.08	1.00	1.00	1.97	1.84
www.drscore.com	30	33.07	1.00	1.00	0.03	0.00	1.00	1.00	3.28	0.02	0.78	1.00	25.35	0.06	0.78	0.80	3.88	3.62
www.steadyhealth.com	30	24.00	1.00	1.00	0.67	0.02	0.00	0.00	0.67	0.02	0.83	0.83	9.59	0.11	0.75	0.75	9.56	9.77
Movies www.allmovie.com	30	59.40	0.97	0.96	0.21	0.00	0.27	0.30	1.64	0.03	0.79	0.74	14.84	0.11	0.13	0.07	7.78	5.91
www.citwf.com	30	20.90	1.00	1.00	0.02	0.00	1.00	1.00	0.69	0.02	1.00	1.00	29.70	0.05	0.39	0.30	2.79	3.79
www.disneymovieslist.com	30	32.33	1.00	1.00	0.07	0.00	0.00	0.00	0.59	0.00	0.71	0.67	259.23	0.08	0.72	0.72	3.95	3.82
www.imdb.com	30	138.80	0.93	0.86	0.45	0.03	0.00	0.00	0.97	0.02	-	-	-	-	0.38	0.38	15.87	16.46
www.soulfilms.com	30	66.13	0.99	0.92	0.03	0.00	0.00	0.00	0.47	0.03	0.59	1.00	17.24	0.05	0.91	0.81	9.73	9.36
EVALG cars.amazon.com	21	20.00	0.93	0.73	0.01	0.00	0.27	0.33	0.55	0.02	0.60	0.67	2.29	0.05	0.97	1.00	19.47	8.10
players.uefa.com	20	10.40	1.00	0.90	0.02	0.02	0.92	0.92	0.51	0.02	0.91	0.94	10.81	0.05	0.92	0.51	3.53	3.70
popartist.amazon.com	19	35.00	1.00	0.98	0.02	0.00	0.99	0.99	0.98	0.03	1.00	1.00	246.97	0.11	0.92	0.58	15.66	10.22
teams.uefa.com	20	32.80	0.99	0.99	0.03	0.00	0.90	0.92	0.28	0.02	0.97	0.99	0.89	0.02	0.64	0.75	1.79	2.28
www.ausopen.com	29	66.73	1.00	1.00	0.26	0.02	0.37	0.39	1.15	0.02	0.24	0.82	132.24	0.27	0.67	0.32	9.67	16.96
www.ebay.com	50	18.12	0.97	1.00	0.51	0.08	0.00	0.00	2.51	0.02	0.83	1.00	577.97	0.48	0.70	0.12	5.44	19.55
www.majorleaguebaseball.com	9	26.00	0.98	0.55	0.04	0.00	0.00	0.00	0.95	0.02	0.99	1.00	158.33	0.06	0.65	0.33	3.60	6.13
www.netflix.com	50	125.86	0.99	0.99	0.18	0.02	-	-	-	-	0.82	0.80	706.64	0.76	0.99	0.99	7.47	31.54
www.rpmfind.net	20	9.90	0.95	0.97	0.03	0.05	0.98	0.99	1.31	0.03	-	-	-	-	0.99	0.99	1.29	10.42

Table 2. Correlation from number of errors to effectiveness

	Trinity		RoadRunner		FiVaTech		WIEN	
	P	R	P	R	P	R	P	R
Coefficient	-0.14	-0.08	-0.11	-0.06	-0.33	-0.08	0.08	0.20
P-value	0.38	0.61	0.04	0.74	0.04	0.64	0.57	0.16

and computed the precision and recall. Then, we considered that the precision and recall of the extracted information corresponds to the extracted piece of text with the highest harmonic mean of precision and recall, i.e., the F_1 measure.

Table 1 shows our results. The first few rows provide a summary in terms of mean and standard deviations of the number of web pages in each dataset (N), the average number of errors in each dataset (E), precision (P), recall (R), rule learning time in CPU seconds (LR), and extraction time in CPU seconds (ET). The remaining rows provide the results we computed for each web site. (A dash, which indicates that the corresponding technique was not able to learn an extraction rule in 15 CPU minutes.) In average, our proposal seems to outperform the other techniques in both precision and recall, with a mean learning and extraction times that are clearly smaller than the other techniques'; the only exception is RoadRunner, whose extraction time seems similar to ours.

We were also interested in determining if there was a correlation from the number of errors in a collection of input pages to the effectiveness of our proposal. Unfortunately, it is not easy to draw an intuitive conclusion from the data in the table. We then conducted a statistical analysis of correlation using the well-known non-parametric Kendall's τ procedure. Table 2 shows the results of the study. Note that there are only two p-values that are smaller than the standard significance level $\alpha = 0.05$: the one that corresponds to the precision of RoadRunner and the one that corresponds to the precision of FiVaTech; this implies that the number of errors in the input web pages seems to have an impact on the precision of these techniques, whereas the impact on our proposal is not significant from a statistical point of view.

4   Conclusions

We have proposed a new and effective unsupervised information extractor that is based on the hypothesis that web pages generated by the same server-side template share patterns that provide irrelevant information. The rule learning algorithm searches for these patterns, builds a trinary tree, which is then used to learn a regular expression that represents the template that was used to generate the input web pages. Our experiments on real-world web pages proved that our technique is highly effective and efficient and that it is not significantly influenced by the presence of errors in the input HTML pages.

References

[1] Álvarez, M., Pan, A., Raposo, J., Bellas, F., Cacheda, F.: Extracting lists of data records from semi-structured web pages. Data Knowl. Eng. 64(2), 491–509 (2008)

[2] Arasu, A., Garcia-Molina, H.: Extracting structured data from web pages. In: SIGMOD Conference, pp. 337–348 (2003)

[3] Arjona, J.L., Corchuelo, R., Ruiz, D., Toro, M.: From wrapping to knowledge. IEEE Trans. Knowl. Data Eng. 19(2), 310–323 (2007)

[4] Chang, C.H., Kayed, M., Girgis, M.R., Shaalan, K.F.: A survey of web information extraction systems. IEEE Trans. Knowl. Data Eng. 18(10), 1411–1428 (2006)

[5] Crescenzi, V., Mecca, G.: Automatic information extraction from large websites. J. ACM 51(5), 731–779 (2004)

[6] Crescenzi, V., Mecca, G., Merialdo, P.: RoadRunner: Towards automatic data extraction from large web sites. In: VLDB, pp. 109–118 (2001)

[7] Elmeleegy, H., Madhavan, J., Halevy, A.Y.: Harvesting relational tables from lists on the Web. PVLDB 2(1), 1078–1089 (2009)

[8] Kayed, M., Chang, C.H.: FiVaTech: Page-level web data extraction from template pages. IEEE Trans. Knowl. Data Eng. 22(2), 249–263 (2010)

[9] Kushmerick, N., Weld, D.S., Doorenbos, R.B.: Wrapper induction for information extraction. In: IJCAI (1). pp. 729–737 (1997)

[10] Sleiman, H.A., Corchuelo, R.: A Reference Architecture to Devise Web Information Extractors. In: Bajec, M., Eder, J. (eds.) CAiSE Workshops 2012. Lecture Notes in Business Information Processing, vol. 112, pp. 235–248. Springer, Heidelberg (2012)

[11] Sleiman, H.A., Corchuelo, R.: A survey on region extractors from web documents. IEEE Trans. Knowl. Data Eng. 99(pre-prints) (2012)

A Space-Efficient Indexing Algorithm
for Boolean Query Processing

Jianbin Qin[1], Chuan Xiao[2], Wei Wang[1], and Xuemin Lin[1]

[1] The University of New South Wales, Australia
[2] Nagoya University, Japan

Abstract. Inverted indexes are the fundamental index for information retrieval systems. Due to the correlation between terms, inverted lists in the index may have substantial overlap and hence redundancy. In this paper, we propose a new approach that *reduces* the size of inverted lists while retaining time-efficiency. Our solution is based on merging inverted lists that bear high overlap to each other and manage their content in the resulting *condensed index*. An efficient algorithm is designed to discover heavily-overlapped inverted lists and construct the condensed index for a given dataset. We demonstrate that our algorithm delivers considerable space saving while incurring little query performance overhead.

1   Introduction

Inverted index is a fundamental indexing data structure for information retrieval and has found its way into database systems. It associates tokens with their corresponding inverted lists; each list contains a sorted array of document identifiers in which the token appears. The primary advantage of the inverted index is that it supports *boolean queries* efficiently. For example, to retrieve documents containing both keywords x and y, we can intersect the inverted lists of x and y.

One issue with the traditional inverted index is its size. Currently, various compression techniques are used to reduce the size of each individual lists. However, little effort is paid to account for the redundancy among the inverted lists. Due to the existence of frequently co-occurring tokens (e.g., phrases), there will be high redundancy due to large overlaps.

In this paper, we propose a novel way to arrange the inverted index physically to achieve reducing the size of the inverted index by exploiting overlaps among inverted lists of groups of tokens. We name the resulting inverted index the *condensed inverted index*. The idea is to form groups of tokens and then explicitly represent the intersections of their corresponding inverted lists such that every document identifier only occurs at most once within the group. This not only reduces the overall size of the index but also accelerates certain queries. We present the query processing algorithm for boolean queries on the condensed index (Section 2).

One technical challenge is how to construct an optimal condensed index. We show that finding the minimum-sized condensed index is a very hard problem, and even a greedy algorithm is typically too expensive to be practical. We propose non-trivial optimizations to the greedy algorithm (Section 3).

X.S. Wang et al. (Eds.): WISE 2012, LNCS 7651, pp. 638–644, 2012.

We conducted experiments with several real-world datasets. It demonstrates the space and time trade-offs of the condensed index and the efficiency of the optimized index construction algorithm (Section 4).

Preliminaries. Let a record r be a *set* of tokens taken from a finite universe $\mathcal{U} = \{ w_1, w_2, \ldots, w_{|\mathcal{U}|} \}$, and R be a collection of records. A boolean query q is a sequence of tokens concatenated by boolean operators, AND, OR, and NOT. The task is to find all records r in R such that r satisfies the query q. The number of tokens in r is denoted as its *size*, or $|r|$.

An efficient way to answer boolean queries is to use *inverted indexes* [1]. An inverted list, l_w, is a data structure that maps the token w to a sorted list of record ids such that w is contained by the corresponding records. $l_w[i]$ denotes the i-th entry in the inverted list of token w.

After the inverted lists for all tokens in the record set are built, we can scan each token in the query q, probe the indexes using every token in x, and obtain a set of posting lists. Merging the posting lists using the boolean operators q will give us the final answer to the query.

2 A New Index Structure for Boolean Queries

We design a new condensed index to exploit the correlation of multiple inverted lists.

We illustrate the idea in Figure 1(b). Consider merging two lists l_A and l_B. It will produce a new list $l_{\widehat{AB}}$. The two tokens A and B will share the new list $l_{\widehat{AB}}$, and it will be traversed when either A or B appears in the query. This reduces the index size and the number of entries to be accessed when the query contains both A and B. However, it will probe more entries and introduce false positives when the query contains only A (or B). To address these issues, we divided the merged lists into *blocks*. Each block indexes a combination of the tokens in this list. For example, the first block maps to the records that contain only A, i.e., p and r. The second block maps to the records that contain only B, and the third block maps to the records that contain both A and B.

Figure 1 shows the structure of the merged inverted lists. We call the lists formed by merging *groups*, and assign a group id to each of them. We keep the *token-group table* that maps token id to group id, so as to locate the group that stores the token's inverted list. At the stage of index probing, the tokens in the query are first collected according to their groups. For each of these groups, we probe the blocks that contain the (combination of) tokens. To handle the boolean operators within a group, we need to probe the following blocks: (1) AND The blocks that index all the tokens (of this group) in the query. (2) OR The blocks the index any of the tokens (of this group) in the query. Then we take the union of the results. Note that we do not need to remove duplicates when processing union within a group since the blocks are disjoint in indexed entries.

In the interest of space, we refer readers to [2] for the detailed query processing algorithm.

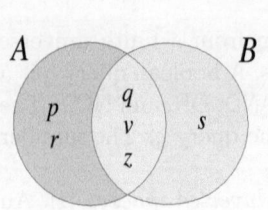

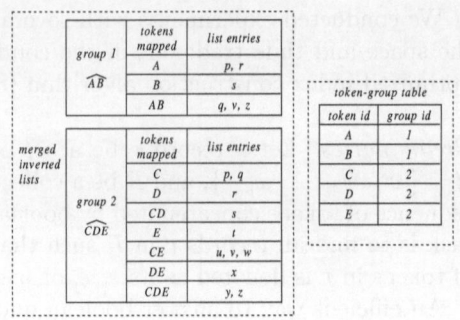

(a) Combining Two Inverted Lists

(b) Data Structure of Condensed Indexes

Fig. 1. Condensed Indexes

3   Choosing Inverted Lists to Merge

The condensed index structure can be implemented using a small amount of application-level code. Both space and time efficiency of the index structure depends on which of lists are chosen to be merged, yet this is not an easy task. In this section, we provide an efficient greedy algorithm that chooses lists to merge considering both space and time factors.

3.1   Greedy Algorithm

We start with a basic greedy algorithm that repeatedly merges the two lists that yield the most space saving.

Algorithm 1 describes the algorithm. Suppose the input lists L have been sorted by increasing token id. We initialize the groups by treating each inverted list as a single group (Line 1), and assign a group id according to their token id. Then we search for the pair of lists with the largest overlap in each iteration (Line 2 and 6), merge them into one group (Line 4 and 5), and assign a new group id, which is required to be greater than all of the current ones. To strike a balance between space saving and time efficiency, we use a parameter M to limit the maximum size of a group. The resulting group serves as a new inverted list to replace the two merged ones. The algorithm repeats until no pair of lists can be found to improve the overall space saving.

Algorithm 2 captures the process searching for the pair of lists with the largest overlap. We scan every inverted list, denoted l_i, searching for the list that has the most overlap with l_i (Line 2). We call this list the *partner* of l_i if we can safely merge the two lists without exceeding the size limit M.[1] We compute the overlap between each l_i and its partner, and arrange them in a max-heap E. The pair of lists at the top of E is the pair that yields the largest overlap.

In order to find the partner of each inverted list l_i, we use an array of counters to calculate the overlap between l_i and the other lists in L. The records indexed

---

[1] Note that the definition of partner is not symmetric.

Algorithm 1: MergeLists (R, L)

---

1   $E \leftarrow \emptyset;$ $g_i \leftarrow 1(1 \leq i \leq |L|)$;   `/* E is a max-heap */`
2   $(l_x, l_y, score) \leftarrow$ SearchListPair (R, L, E);
3   **while** $l_x \neq \emptyset$ **do**
4       $g_{new} \leftarrow g_x + g_y$;   `/* increase group size */`
5       $l_{new} \leftarrow l_x \cup l_y, L \leftarrow L \setminus \{l_x, l_y\} \cup \{l_{new}\}$;
6       $(l_x, l_y, score) \leftarrow$ SearchListPair (R, L, E);
7   **return** L

Algorithm 2: SearchListPair (R, L, E)

---

1   **for** $i = 1$ **to** $|L|$ **do**
2      $(l_i, l_j, score) \leftarrow$ SearchPartner (l_i);
3      $E.push(l_i, l_j, score)$;
4   $(l_x, l_y, score) \leftarrow E.pop()$;
5   **return** $(l_x, l_y, score)$

---

by l_i is sequentially scanned. For each token w in each record, we increase the counter corresponding to l_w by one. The inverted list with the greatest value among the counters is reported as l_i's partner. The pseudo-code is given in Algorithm 3.

Algorithm 3. SearchPartner (l_x)

---

1   $O_{max} \leftarrow 0, l_{max} \leftarrow \emptyset$;
2   $O \leftarrow$ empty map from group id to int;
3   $A \leftarrow$ empty map from group id to record id;
4   **for each** $r \in l_x$ **do**
5   **for each** $w \in r$ **do**
6          $y \leftarrow w$'s group id;
7   **if** $g_x + g_y \leq M$ **and** $A[y] \neq r$ **then**
8              $O[y] \leftarrow O[y] + 1, A[y] = r$;
9   **if** $O[y] > O_{max}$ **then**
10                 $O_{max} \leftarrow O[y], l_{max} \leftarrow l_y$;
11 **return** (l_x, l_{max}, O_{max})

3.2   Further Optimizations

The above greedy algorithm returns the condensed inverted lists. An important issue is that it has to recompute the partner of each list once two lists l_x and l_y are merged. These repeated computations incur significant overhead, and render the algorithm unable to output results for large-scale datasets in reasonable time. Nevertheless, we can avoid such computation by enforcing a constraint that a list's group id should always be greater than its partner's group id. We formally state the principle in the lemma below.

Algorithm 4. OptimizedSearchListPair (R, L, E)

```
1  if this function is called for the first time then
2      for i = 1 to |L| do
3          (l_i, l_j, score) ← SearchPartner (l_i);
4          E.push(l_i, l_j, score);
5  else
6      (l_new, l_j, score) ← SearchPartner (l_new);
7      E.push(l_new, l_j, score);
8  (l_x, l_y, score) ← E.pop();
9  while either l_x or l_y has been merged do
10     if l_x has not been merged then
11         (l_x, l_z, score) ← SearchPartner (l_x) ;    /* search x's new partner */
12         E.push(l_x, l_z, score);
13     (l_x, l_y, score) ← E.pop();
14 return (l_x, l_y, score)
```

Lemma 1. *Let the partner of a list l_i be the list whose (1) group id is smaller than the group id of l_i; (2) group size will not exceed M if it is merged with l_i; (3) overlap with l_i is the largest among all the lists that satisfy the first two conditions. If a list changes its partner after merging l_x and l_y, then the partner of this list must be either l_x or l_y before the merging.*

Algorithm 5. OptimizedSearchPartner (l_x)

```
1  O_max ← 0; l_max ← ∅;
2  O ← empty map from group id to int;
3  A ← empty map from group id to record id;
4  if l_x is the new list formed in the previous iteration then
5      (O_max, l_max) ← GetO_max ForNewList (l_x);
6  for i = 1 to |l_x| − O_max do
7      r ← l_x[i];
8      for each w ∈ r do
9          y ← w' group id;
10         if y < x and g_x + g_y ≤ M and A[y] ≠ r then
11             O[y] ← O[y] + 1, A[y] = r;
12             if O[y] > O_max then
13                 O_max ← O[y], l_max ← l_y;
14 for each y such that O[y] > 0 do
15     O[y] ← |l_x ∩ l_y| ;                              /* evaluate exact overlap */
16     if O[y] > O_max then
17         O_max ← O[y], l_max ← l_y;
18 return (l_x, l_max, O_max)
```

This principle enables us to avoid committing the costly scanning over the set of lists L. Instead, only the lists whose partners are l_x or l_y need to be assigned with new partners. Additionally, we perform a *lazy* update to postpone the searching for such lists' partners. Only if these lists are popped from the max-heap E, we search for new partners for them. The merging algorithm will benefit since these lists may have been merged and discarded from further consideration before we are forced to seek new partners. We give the pseudo-code for the above method in Algorithm 4, and replace Algorithm 2 with it.

Algorithm 6. GetO$_{\max}$ForNewList (l_x)

---

1 $(l_u, l_v) \leftarrow$ the two lists that were merged to form l_x;
2 **if** $u < v$ **then** $w \leftarrow u$; **else** $w \leftarrow v$;
3 $z \leftarrow w$'s partner;
4 **if** z has not been merged and $g_z + g_x \leq M$ **then**
5 $\quad | \quad O_{\max} \leftarrow |l_w \cap l_z|,\ l_{\max} \leftarrow z$;
6 **else**
7 $\quad | \quad O_{\max} \leftarrow 0,\ l_{\max} \leftarrow \emptyset$;
8 **return** $(O_{\max}, l_{\max})$

Another important optimization is to speed up the count algorithm we use in Algorithm 3. Since we are looking for the partner that has most overlap with l_x, a filtering condition can be developed using the current maximum overlap $O_{\max}$. Considering the following prefix filtering principle.

Lemma 2 (Prefix Filtering Principle). *Consider an ordering $\mathcal{O}$ of the token universe $\mathcal{U}$ and a set of records, each sorted by $\mathcal{O}$. Let the p-prefix of a record x be the first p tokens of x. If $|x \cap y| \geq \alpha$, then the $(|x| - \alpha + 1)$-prefix of x and the $(|y| - \alpha + 1)$-prefix of y must share at least one token.*

If there exist l_y such that $|l_x \cap l_y| > O_{\max}$, then l_y must share at least one token with the $(|l_x| - O_{\max})$-prefix of l_x. Therefore, only the first $(|l_x| - O_{\max})$ entries in l_x need to be probed in order to generate the candidate lists that have potential to become l_x's partner. This filtering condition is tightened as $O_{\max}$ increases. Finally, the candidate lists are verified for the exact overlap. The improved algorithm is captured in Algorithm 5, and is used to replace the original partner searching algorithm in Algorithm 3. In addition, we can infer an initial lower bound of $O_{\max}$ before partner search, given that l_x is the list formed by merging two lists during previous iteration, supposing they are l_u and l_v, and $u < v$. Since we have obtained the overlap between l_u and its partner l_i, it is guaranteed the overlap between l_x and l_i is no less than this value. This is because $|l_i \cap l_x| = |l_i \cap (l_u \cup l_v)| \geq |l_i \cap l_u|$. The pseudo-code is given in Algorithm 6, and invoked in Line 5 of Algorithm 5.

4   Experiments

In the interest of space, we briefly present our experimental results in this section. Please refer to [2] for the full version of experimental evaluation.

Three publicly available datasets were used in our experiments: DBLP bibliography records, TREC-9 Filtering Track Collections, and Enron email collection. We generated queries for each dataset by sampling records from the dataset and randomly selecting a number of consecutive tokens containing no stop words.

We evaluated the optimization methods proposed in Section 3. On DBLP, the lazy update technique exhibits a speed-up up to 9.3x over the basic greedy list merging algorithm with the constraint exploiting Lemma 1. Further applying prefix filtering principle achieves an additional speed-up of 2.6x, and runs in 10 to 20 seconds with varying maximum group size.

We compared the condensed index sizes with the original index sizes. The total index sizes decrease as the maximum group size M grows, bottoms at 6 or 7, and then rebounds. The overall space savings against the original inverted index are 16.4% on DBLP, 26.8% on TREC, and 39.2% on ENRON.

We evaluated the query processing time with varying numbers of tokens in a query. The best choice of M increases when more tokens are introduced. $M = 2$ yields the best runtime performance when a query contains two or three tokens.

5   Conclusion

We propose a novel inverted index structure to support boolean queries efficiently. By exploiting the overlaps among inverted lists of groups of tokens, the condensed structure is able to represent the intersections of their corresponding inverted lists, so that the redundancy among the inverted lists of frequently co-occurring tokens can be avoided. We design an efficient greedy algorithm to find a good condensed index. Experimental results show that our condensed index structure occupies less space yet achieves accepatable runtime performance.

References

1. Baeza-Yates, R., Ribeiro-Neto, B.: Modern Information Retrieval, 1st edn. Addison-Wesley (May 1999)
2. Qin, J., Xiao, C., Wang, W., Lin, X.: Condensed inverted index: A space-efficient index for boolean queries. Technical report, University of New South Wales (2012)

Detecting Wikipedia Vandalism
with a Contributing Efficiency-Based Approach

Xiaoyue Tang[1,2], Guofu Zhou[1,*], Yuchen Fu[3], Lin Gan[2],
Wei Yu[2], and Shijun Li[1,2]

[1] State Key Laboratory of Software Engineering, Wuhan University, Wuhan, China
[2] School of Computer, Wuhan University, Wuhan, China
[3] School of Computer Science & Technology, Soochow University, Suzhou, China
{sharontang,gfzhou,yuwei,eriney_gl,shjli}@whu.edu.cn, yuchen@suda.edu.cn

Abstract. The collaborative nature of wiki has distinguished Wikipedia as an online encyclopedia but also makes the open contents vulnerable against vandalism. The current vandalism detection methods relying on basic statistic language features work well for explicitly offensive edits that perform massive changes. However, these techniques are evadable for the elusive vandal edits which make only a few unproductive or dishonest modifications. In this paper we proposed a contributing efficiency-based approach to detect the vandalism in Wikipedia and implement it with machine-learning based classifiers that incorporate the contributing efficiency along with other languages features. The results of extensional experiment show that the contributing efficiency can improve the recall of machine learning-based vandalism detection algorithms significantly.

Keywords: Classification, Vandalism detection, Wikipedia.

1 Introduction

As a prominent online cooperative project, Wikipedia distinguishes itself by encouraging users to edit the open articles collaboratively. However, the openness of this wiki-formed content creating community has brought about the inaccuracies resulting from vandalism, which can be informally defined as any addition, removal, or change of content made in a deliberate attempt to compromise the integrity of Wikipedia [1]. Several studies have shown that vandalism is quite prevalent in Wikipedia, consisting of around 5% of all the edits [6].

Some vandalism detection methods relying upon text patterns or basic statistic features can achieve a high performance in recognizing regular vandal edits (with common swear words or massive changes, etc.). Unfortunately, they become insufficient as the elusive vandal edits are increasingly prevalent nowadays, which sophisticatedly make only a few but dishonest changes to the content. This type of vandalism is hard to identify, and a study [2] has revealed that the number of them has been increasing since 2004 and has remained at about 10% of all vandal edits overall.

---

* Corresponding author.

X.S. Wang et al. (Eds.): WISE 2012, LNCS 7651, pp. 645–651, 2012.

In this paper, we introduce contributing efficiency, a feature that quantifies how efficient a revision (a combination of addition, deletion, and update) is in adjusting the article towards the ground-truth within certain amount of editing cost, to help detecting elusive vandalism. This strategy is motivated by the key observation: while most regular revisions contribute to make the article more close to the ground-truth, the vandal ones are intentionally doing the opposite. Therefore, those revisions of low contributing efficiency are more likely to be vandalism. We incorporate this new feature to a set of basic statistic language features of revisions, and apply the enhanced feature set to various classifiers. Through extensive experimentation, we demonstrate the effectiveness of our approach over our 196,491 Wikipedia revisions extracted from the history pages of 60 popular articles.

2 Related Work

Most early vandalism detecting methods come from the research that assesses the quality and reliability of Wikipedia articles. On the one hand, many researchers focus on utilizing the Wikipedias revision history to evaluate the content of articles [6]. On the other hand, another idea is to build a reputation system for users based on the persistence of their edits [5]. However, the problem of vandalism detection is essentially different from that of building a reputation system because a low-quality edit is not necessarily a vandal edit.

The recent articles directly addressed the detection of vandalism. Typical anti-vandalism methods are either rule-based, or machine learning based utilizing the common features that characterize vandalism patterns including size of changes, statistics in new edits, obscenity of words, and anonymity of users. Potthast et al. [7] leverage logistic regression to classify vandalism instances according to the edit content and the editing category (insertion, replacement, and deletion). Smets et al. [8] implement the classification of revisions occurring in one hour with the Prediction by Partial Match compression model. Some work also considers features from statistical language models [3] or the the meta-information of revisions [4].

However, these approaches have not considered small changes to the content are also likely to generate vandal edits. Therefore, they become ineffective at detecting elusive vandalism and the recall ratios are relatively low.

3 The Contributing Efficiency-Based Approach Overview

Contributing efficiency, defined as how much benefits that a revision has contributed to an article within certain amount of costs to match the article content with the ground truth better, is related to several key factors including the ground truths of articles, the benefits and costs of revisions. Below, we discuss details of the definition by exploring these factors.

3.1   The Ground Truth of an Article

For an article of Wikipedia, the continues process of content varying can be regarded as evolution, in which the editors contribute positive or negative changes to the content and generate new states. All the content states, kept as documents, can be represented in the model of BoW (Bag of Words). Whereas BoW does not preserve the order of words, we consider bi-gram, the sequence of two adjoining words, as the alternative semantic unit.

According to the good faith assumption of Wikipedia, most editors would do the right modifications to achieve the true explanations of items. Therefore, unlikely to be accurately generated though, the ground truth of an article can be approximated by merging the historical documents together appropriately. Here We adopted 2 methods to approximate the ground truth of an article: (1) The mean approximation G_m is an aggregation of all historical documents where the frequency of a bi-gram is the average of all its frequencies; (2) The polling approximation G_p is an aggregation in which the frequency of a bi-gram is replaced by its most frequent frequency among all documents.

Assuming size of the bi-gram dictionary for all documents is N, if we represent the frequency of bi-gram w_j, $1 \leq j \leq N$ in document D_i as $\|w\|_i$, and there are a total of K documents kept in the editing history of an article, then the mean approximation of ground truth, G_m, can be stated as:

$$G_m = (\sum_{j=1}^{K} \|w_1\|/K, \sum_{j=1}^{K} \|w_2\|/K, \ldots, \sum_{j=1}^{K} \|w_N\|/K). \tag{1}$$

To profile the polling article of ground truth, first we define a series of binomial function $A_i(x)$, $1 \leq i \leq K$, of a value x, $x \in X$, to represent whether a bi-gram w_j, appears x times in document D_i or not. Here, $X \subset N$ is the assemblage of all the possible frequency values.

$$A_i(x) = \begin{cases} 1 & \text{if } \|w\|_i = x \\ 0 & \text{else} \end{cases}. \tag{2}$$

If we define a function $F_{1 \leqslant i \leqslant K}(\|w_j\|_i)$ of of bi-gram frequency $\|w_j\|_i$ to capture the most widely appeared value among all the frequency values of bi-gram w_j in documents D_1 to D_K as:

$$F_{1 \leqslant i \leqslant K}(\|w_j\|_i) = \arg \max_{\|w_j\|_i} \sum_{j=1}^{K} A_i(\|w_j\|_i). \tag{3}$$

Then we state the polling approximation of ground truth as following:

$$G_p = (F_{1 \leqslant i \leqslant K}(\|w_1\|_i), F_{1 \leqslant i \leqslant K}(\|w_2\|_i), \ldots, F_{1 \leqslant i \leqslant K}(\|w_N\|_i)). \tag{4}$$

3.2   The Benefit and the Cost of a Revision

First we introduced a few critical terms. Once a word/bi-gram is shown in a revision as well as in the ground truth, we say that is a hit of this document to the ground truth, and the word that makes a hit is called a hitting word/bi-gram. An assemblage of hitting words/bi-grams in a document D_i is stated as:

$$H_G(R_i) = \{w \mid \forall w \in G \text{ and } w \in D_i\}. \tag{5}$$

Where G stands for either type of the ground truth articles. Since both hitting bi-grams number and length can affect the how much a document D_i matches the ground truth, the benefit of revision R_i, $B_{hit\_rate_G}(R_i)$, should be reasonably defined as the increment of the matching-ground-truth degree that R_i has made from two consecutive documents, D_{i-1} and D_i, and be presented as following:

$$B_{hit\_rate_G}(R_i) = \frac{|H_G(R_i)|}{|D_i|} - \frac{|H_G(R_{i-1})|}{|D_{i-1}|}. \tag{6}$$

In which $|\ |$ is an operator for an assemblage to obtain its size. To represent the difference between two documents, D_i and D_j, we define the unique part in D_i that does not appear in D_j as:

$$Q_{i,j} = D_i - D_j = \{w \mid \forall w \in D_i \text{ and } w \notin D_j\}. \tag{7}$$

With the operator $|\ |$, we define the cost of a revision R_i, $Cost(R_i)$, as the amount of differences that the revision R_i has made between D_{i-1} and D_i as following:

$$Cost(R_i) = |\boldsymbol{Q_{i,i-1}} + \boldsymbol{Q_{i-1,i}}|. \tag{8}$$

3.3   Contributing Efficiency and Other Language Features

With the benefit and cost both defined, the contributing efficiency can thus be presented as following:

$$Eff_{hit\_rate_G}(R_i) = \frac{2 \cdot \arctan\left(B_{hit\_rate_G}(R_i)/Cost(R_i)\right)}{\pi}. \tag{9}$$

Here arctan is adopted here for normalization.

Some other language features are also helpful in classifying vandalism and the innocent edits apart. We employ them as the basic features to estimate the baseline performance. Their definitions are provided in Table 1, where rare word/bi-gram refers to words/bi-grams occurring less than 5 times in the history of an article, and diff means the comparison page in which all content differences between two documents are presented.

4   Experiments

For the vandalism detecting task, the biggest obstacle is the skew and ever changing class distribution, while machine learning provides a solution to the problem. In this paper, we employ two machine learning methods, *logistic regression* and *SVM*, to build and train classifiers, and categorize the revisions in Wikipedia.

Table 1. Definition of the basic statistic language features

Features	Definitions
ΔHit	Change in number of hit bi-grams
$\Delta Perplexity$ (word)	Change in number of different words
$\Delta Perplexity$ (bi-gram)	Change in number of different bi-grams
ΔNum (rare word)	Change in number of rare words
$\Delta Percent$ (rare word)	Change in percentage of rare words
ΔNum (rare bi-gram)	Change in number of rare bi-grams
$\Delta Percent$ (rare bi-gram)	Change in percentage of rare bi-grams

4.1 Dataset and Experimental Setup

Our corpus included complete histories (by Aug. 20th, 2011) for 60 Wikipedia articles, the topics of which were the top 60 movies ranked by IMDB. These popular articles had been revised for plenty of times by massive users. We extracted 196,491 historical documents of the 60 articles as well as the difference pages between all pairs of two succeeding documents from Wikipedia. Each document was processed into sentence-per-line form to enable statistic processing. With the reverted revisions tagged as vandalism, 60% of all the revisions were randomly picked out for training and the rest were left for test.

For comparison, We also calculated another feature which is usually the first order measurement in problems concerning the quality of text content, $Info\_gain$ of a revision R_i between documents D_{i-1} and D_i:

$$Info\_gain(R_i) = Entropy(D_i) - Entropy(D_{i-1}). \qquad (10)$$

Where $Entropy(D_i)$ represents the entropy value for document D_i.

We used the CMU-toolkit to build bi-gram statistical language models for each revision, and calculated the basic language features in Table 1 as well as the $Info\_gain$ and contributing efficiency. We leveraged libsvm-3.11 to implement SVM and logistic regression classification. Each classifier was run over the dataset extracted from Wikipedia and evaluated with 10-fold cross-validation. The performance of classifiers with only basic language features was set as baselines. Then with $Info\_gain$ and contributing efficiency added into the feature set, we trained the classifiers again and accomplished the test.

4.2 Results and Analysis

The results of the extensive experiments are provided in Table 2. The $F1$-measure comparison results of classifiers are given $F1$-measure in Fig. 1 and Fig. 2.

Fig. 1 and Fig. 2 show that the $F1$-measures of both classifiers are improved significantly with either $Eff_{hit\_rate_{G_m}}$ or $Eff_{hit\_rate_{G_p}}$. For both classifiers, $Eff_{hit\_rate_{G_p}}$ is the most effective in detecting the vandalism. By contrast, $Info\_gain$ does not work well in discriminating the vandals from the innocent

Table 2. Performances of Classifiers

<table>
<tr><td colspan="4">(a) SVM classifiers</td><td colspan="4">(b) Logistic Regression Classifiers</td></tr>
<tr><td>Features</td><td>ACC</td><td>PRE</td><td>REC</td><td>Features</td><td>ACC</td><td>PRE</td><td>REC</td></tr>
<tr><td>Baseline</td><td>0.925</td><td>0.801</td><td>0.605</td><td>Baseline</td><td>0.927</td><td>0.828</td><td>0.616</td></tr>
<tr><td>$Info\_gain$</td><td>0.918</td><td>0.757</td><td>0.582</td><td>$Info\_gain$</td><td>0.925</td><td>0.767</td><td>0.589</td></tr>
<tr><td>$Eff_{hit\_rate_{G_m}}$</td><td>0.927</td><td>0.840</td><td>0.683</td><td>$Eff_{hit\_rate_{G_m}}$</td><td>0.931</td><td>0.851</td><td>0.692</td></tr>
<tr><td>$Eff_{hit\_rate_{G_p}}$</td><td>0.928</td><td>0.821</td><td>0.713</td><td>$Eff_{hit\_rate_{G_p}}$</td><td>0.936</td><td>0.825</td><td>0.735</td></tr>
</table>

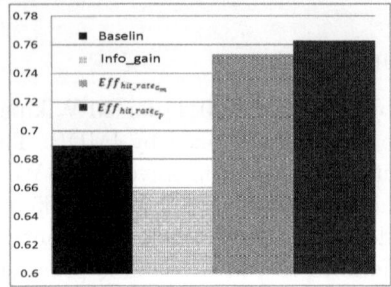

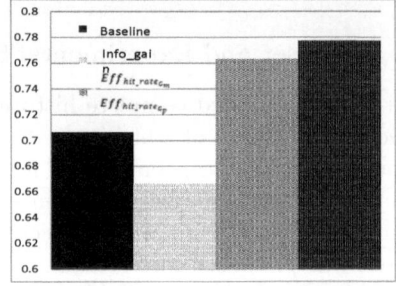

Fig. 1. F1-measure of SVM classifiers

Fig. 2. F1-measure of logistic regression models

due to the lack of considering relevance weight. As for the performances, logistic regression classifiers marginally outperform SVM classifiers. As detailed in Table 2, classifiers trained with $Eff_{hit\_rate_{G_m}}$ have a higher rise in precision, while those trained with $Eff_{hit\_rate_{G_p}}$ are better improved in recall ratio. That is because of the difference between the two ground truths involved in $Eff_{hit\_rate_{G_m}}$ and $Eff_{hit\_rate_{G_p}}$. G_m merges the historical documents together more smoothly by averaging bi-gram frequencies of all, while G_p only takes the most widely approved frequencies of bi-grams, which makes the content of G_p distinguished out of more documents, therefore, when compared with G_p, there are more revisions identified as vandalism. And as a trade-off for the increase of recall ratio, more false positives make the classification less precise.

There are two reasons why the F1-measure cannot be improved above 80%. One is that the features are limited within the statistic attributes of revisions, including the contributing efficiency. The lack of specific information of text content makes the classifiers incapable of differentiating the bad intentional texts from the normally unproductive content. Once the hit rate does not change between two document, the revision would have a contributing efficiency of 0 no matter what the user intention is. And this type of revisions possesses 6.33% among all according to our statistic result. The other reason is that the class of an revision is not always obvious, even for human reviewers. This is due to the either-or nature of binary categorization. With hierarchical ranking results from human reviewers employed into the training process, in this case the ambiguity in classification could be mitigated.

5   Conclusion

In this paper we propose a new approach to detect the vandalism that are missed when using basic statistic language features. We introduce a new feature of contributing efficiency, which characterizes how capable a revision is under the criterion that normal edits adjust the article towards the ground truth while the vandal ones prefer the opposite. We utilize some basic statistic language features as well as the contribution efficiency to drive machine learning-based classifiers. Also, we adopted the feature of information gain for classification for comparison. We evaluate the performance of our approach with 196,491 tagged Wikipedia revisions on the topics of 60 popular movies. The experimental results show that the contributing efficiency-based approach improves the performance of selected machine learning algorithms significantly. Finally we analyze the factors that rust the performance improvement, and point out the necessity of introducing manual hierarchical ranking into the detection process.

For future work, we plan to introduce the manual hierarchical ranking results into the training process, and assess the performance of this machine learning classifier. Also, we are interested in building up the approximated ground truth in a dynamic learning process and compare the qualities of the statically built ground truth approximation and that dynamically built one.

Acknowledgments. This work is supported by the National Natural Science Foundation of China (NSFC-60970018). Thanks to Qiaozhu Mei, Tao Sun, and countless Wikipedia contributors.

References

1. Adler, B.T., Chatterjee, K., de Alfaro, L., Faella, M., Pye, I., Raman, V.: Assigning trust to Wikipedia content. In: WikiSym 2008 (2008)
2. Wu, Q., Irani, D., Pu, C., Ramaswamy, L.: Elusive Vandalism Detection in Wikipedia: A Text Stability-based Approach. In: CIKM 2010 (2010)
3. Chin, S., Street, W.N., Srinivasan, P., Eichmann, D.: Detecting wikipedia vandalism with active learning and statistical language models. In: WICOW 2010 (2010)
4. West, A.G., Kannan, S., Lee, I.: Detecting wikipedia vandalism via spatio-temporal analysis of revision metadata. In: EUROSEC 2010 (2010)
5. Adler, B.T., de Alfaro, L., Mola-Velasco, S.M., Rosso, P., West, A.G.: Wikipedia Vandalism Detection: Combining Natural Language, Metadata, and Reputation Features. In: Gelbukh, A. (ed.) CICLing 2011, Part II. LNCS, vol. 6609, pp. 277–288. Springer, Heidelberg (2011)
6. Priedhorsky, R., Chen, J., Lam, S.K., Panciera, K., Terveen, L., Riedl, J.: Creating, destroying, and restoring value in Wikipedia. In: GROUP 2007: Proceeding of the 2007 ACM Conference on Supporting Group Work, pp. 259–268 (2007)
7. Potthast, M., Stein, B., Gerling, R.: Automatic Vandalism Detection in Wikipedia. In: Macdonald, C., Ounis, I., Plachouras, V., Ruthven, I., White, R.W. (eds.) ECIR 2008. LNCS, vol. 4956, pp. 663–668. Springer, Heidelberg (2008)
8. Smets, K., Goethals, B., Verdonk, B.: Automatic vandalism detection in Wikipedia: Towards a machine learning approach. In: WikiAI 2008: Proceedings of the AAAI Workshop on Wikipedia and Artificial Intelligence (2008)

Definition and Enactment
of Instance-Spanning Process Constraints

Maria Leitner, Juergen Mangler, and Stefanie Rinderle-Ma

University of Vienna
Faculty of Computer Science
Research Group Workflow Systems and Technology
{maria.leitner,juergen.mangler,stefanie.rinderle-ma}@univie.ac.at

Abstract. Currently, many approaches address the enforcement and monitoring of constraints over business processes. However, main focus has been put on constraint verification for intra-instance process constraints so far, i.e., constraints that affect single instances. Existing approaches addressing instance-spanning constraints only consider certain scenarios. In other words, a holistic approach considering intra-instance, inter-instance, and inter-process constraints is still missing. This paper aims at closing this gap. First of all, we show how the Identification and Unification of Process Constraints (IUPC) compliance framework enables the definition of instance-spanning process constraints in a flexible and generic way. Their enactment and enforcement is demonstrated within a prototypical implementation based on a service-oriented architecture.

Keywords: Instance-spanning Process Constraints, Process-based Compliance Management, Process Engine, Web-based Business Processes.

1 Introduction

Process constraints have become an important instrument to define, enact, and enforce regulations, standards, or other requirements that are imposed on business processes and workflows. Powerful approaches have arisen that enable modeling, monitoring, and verifying process constraints throughout the entire process life cycle e.g., [8,4].

Current approaches typically focus on a particular topic, like (1) authorization, (2) security in general, (3) checking of structural requirements at design-time or (4) result-checking at run-time. Additionally, they typically either deal with **rule enactment** (cmp. [9]), covering process specific topics such as providing and monitoring rules in conjunction with process models, or with **rule enforcement** which deals with certain (process) tasks e.g., separation/binding of duties of tasks (e.g. [2]) or synchronization [5,12]. All the above mentioned approaches provide *intra-instance constraints*. They are typically defined in a process schema and enforced in single instances. As stated in [2], these constraints can be enforced statically in the process schema and dynamically during process execution.

X.S. Wang et al. (Eds.): WISE 2012, LNCS 7651, pp. 652–658, 2012.

However, this specialization on particular topics (and related components in Process-Aware Information Systems (PAIS)) for single process instances, poses a problem when considering inter-instance, inter-process or inter-organizational constraints (further denoted as *instance-spanning*). *Inter-instance constraints* apply to multiple instances of a single process schema. Typically, instances of a process are enacted within an organization. *Inter-process constraints* are defined over single or multiple instances of multiple process schemas. *Inter-organizational constraints* are a special case of inter-process constraints; the enforcement of these constraints is managed over multiple organizations.

As most of the above mentioned approaches cover different formalizations, implementations (for specific components), and/or topics; the instance-spanning aspect has to be separately handled for every approach. Examples for existing parallel evolution of instance-spanning approaches for different topics include e.g., inter-instance authorization constraints [13,14] or inter-process task synchronization constraints e.g., [3]. An additional problem is that solutions for inter-process constraints often cannot directly be transferred to instance-spanning scenarios, as there is no standardized way for describing:

1. *Process Scope:* Which set of processes or instances does a constraint refer to?
2. *Constraint Scope:* Which set of tasks does a constraint refer to? I.e., there is no standardized way to describe that a constraint covers e.g., tasks from all instances of a certain process, or tasks from particular instances from different processes.
3. *Enactment & Enforcement Aspects:* How to define certain basics of integrating constraints with processes. This includes (1) referring to processes structure which is defined by e.g., Linear Temporal Logic (LTL) [11] and Compliance Rule Graphs (CRG) [4], or (2) dealing with process data, time and resource identification.

The contribution of this paper is a comprehensive conceptual framework based on [7], for the specification of instance-spanning constraints. Instead of modifying specific approaches to make them instance-spanning aware, we want to introduce (1) a formalism, and (2) an architecture, how to enact and enforce instance-spanning constraints. We think that this can lead to unified enactment and enforcement of process constraints for PAIS, without the need of different infrastructures and/or components for different constraint topics. The semantic understanding of the enforcement has to be concentrated at enforcing components, while enactment can remain generic: checking conditions and forwarding the semantic part of constraints to enforcing components. Moreover, we illustrate our findings with examples for instance-spanning process constraints. In addition, we evaluate our findings with a proof-of-concept prototype for this architecture. In the following, Section 2 shows how instance-spanning constraints are specified in the IUPC framework. Furthermore, Section 3 displays the enactment of these constraints. Section 4 gives an overview on related work and Section 5 concludes the paper.

2 Design of Instance-Spanning Process Constraints

As stated in the introduction, we will formalize instance-spanning constraints based on the Identification and Unification of Process Constraints (IUPC) framework [7]. The purpose of the IUPC framework is to provide a means to integrate existing approaches that deal with various constraints topics (as explained in the introduction). Due to space limitations, we will only provide the extension of the main IUPC concepts (see [7] for a comprehensive definition), enriched with some examples how instance-spanning approaches fit in. We found instance-spanning constraints to typically have three dimensions - **Localization, Span** and **Dependency**:

Property 1 (Localization). If a constraint is connected to a task, this task can basically occur in a process, multiple processes, or processes in multiple organizations. Four different restriction scenarios are possible:

(1) A constraint should only be enacted for a certain instance (typically denoted as **intra-instance localization**). I.e.: $\forall a : a.\text{instance} = \text{CONST}$.

(2) A constraint is enacted for tasks in all instances of a process (typically denoted as **inter-instance localization**). I.e. $\forall a : a.\text{instance.process} = \text{CONST}$.

(3) A constraint is enacted for tasks in instances of many process (typically denoted as **inter-process localization**). I.e. $\forall a : a.\text{organization} = \text{CONST}$.

(4) A constraint is enacted for tasks in instances of processes that are present in more than one organization (typically denoted as **inter-organizational localization**). I.e. $\forall a : a.\text{organization} \neq \text{CONST}$.

Property 2 (Span). A constraint e.g., separation of duty, often affects multiple tasks at once. Five different scenarios are possible:

(5) All affected tasks are in the same instance (typically denoted as **intra-instance constraint**) e.g., $\forall a, b : a.\text{instance} = b.\text{instance} \wedge a.\text{organization} = b.\text{organization}$. As instances are unique, this definition should be sufficient.

(6) Affected tasks are spread over multiple instances of a process (typically denoted as **inter-instance constraint**). This is the typical case of inter-instance synchronization, as described in the related work e.g., $\forall a, b : a.\text{instance.process} = b.\text{instance.process} \wedge a.\text{organization} = b.\text{organization}$.

(7) Affected tasks are spread over multiple instances of multiple processes (typically denoted as **inter-process constraint**). E.g., $\forall a, b : a.\text{organization} = a.\text{organization}$.

(8) Affected tasks are spread over multiple instances of multiple processes of multiple organization (typically denoted as **inter-organization constraint**) e.g., $\forall a, b : a.\text{instance.process} \neq b.\text{instance.process} \wedge a.\text{organization} \neq b.\text{organization}$.

(9) Affected tasks are spread over multiple instances of single processes that exist in multiple organizations (typically denoted as **trans-organizational constraint**) e.g., $\forall a, b : a.\text{instance.process} = b.\text{instance.process} \wedge a.\text{organization} \neq b.\text{organization}$.

Property 3 (Dependency). This characteristic deals with the temporal aspect of constraint enactment. We define **Dependent** constraints to utilize behavior tuples to realize e.g., Case Handling, Retain Familiar patterns (`workflowpatterns.`

com) in conjunction with a worklist. For preceding enactments, a constraint may save a value e.g., $\forall a : a.\texttt{behavior\_data}[count\_invocations]+ = 1$; now it is possible to enact e.g., only every second time. We define **Independent** constraints as being independent of subsequent or preceding enactments.

To **integrate** these characteristics **into the IUPC framework** (cf. [7]), we utilize the following rules: Often, simple intra-instance constraints are defined in the Linkage, Condition and/or Behavior. For example, separation of duty constraints are defined in the Condition as comparison if two tasks are executed in the same instance such as $b.instance = c.instance$. In case of inter-instance constraints, the Context $(\mathcal{P} \times \mathcal{I}_P)$ defines which process $(\mathcal{P} \in \mathcal{P}_n)$ and which instances (i.e. $\mathcal{I}_P \subseteq \mathcal{I}$) are affected by the constraint. On the other hand, the Context$((\mathcal{P} \times \mathcal{I}_P))$ specification of inter-process constraints defines a set of processes and instances i.e. $\mathcal{P}_n \subseteq \mathcal{P}$ and $\mathcal{I}_P \subseteq \mathcal{I}$. As a motivational example shown in Fig. 1, we specify intra-instance $(C1)$, inter-instance $(C2)$ and inter-process $(C3)$ constraints based on the IUPC framework in [7].

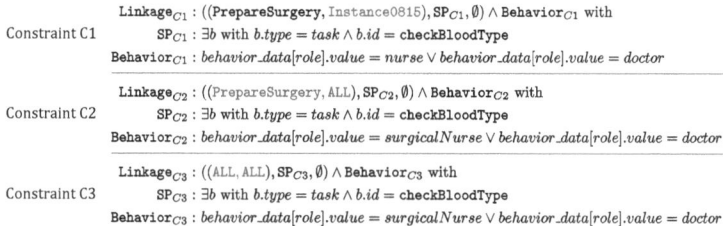

Fig. 1. Constraint Examples

3   Enactment of Instance-Spanning Constraints

Architecture. The architecture of our framework consists of a workflow execution engine (EE), worklist handler (WH), constraint engine (CE), users, external services, and data sources and is shown in Fig. 2. Depending on the type of task, manual or automated, the workflow execution engine delegates tasks either to a WH (manual task) or an external service (automated task). In case of manual tasks, the WH offers tasks to users which further can accept and execute them. On the other hand, automated tasks are delegated to an external service (e.g., a scheduler in scientific workflows) which further distributes tasks to data sources (e.g., nodes). Please note that the steps between a scheduler and a node are often not visible due to encapsulation. But to provide a comprehensive approach, these steps have to be considered for e.g., compliance checking. Throughout these two cases, a CE supports the enactment and monitoring of constraints during process execution.

Enforcement and Monitoring of Process Constraints. For a better understanding, we will show how previously defined constraints (cf. constraint C1-3) are enforced and monitored in the system. The function sequence to enact and monitor resource assignments (constraint **C1**) is shown in Fig. 3. Similar to constraint

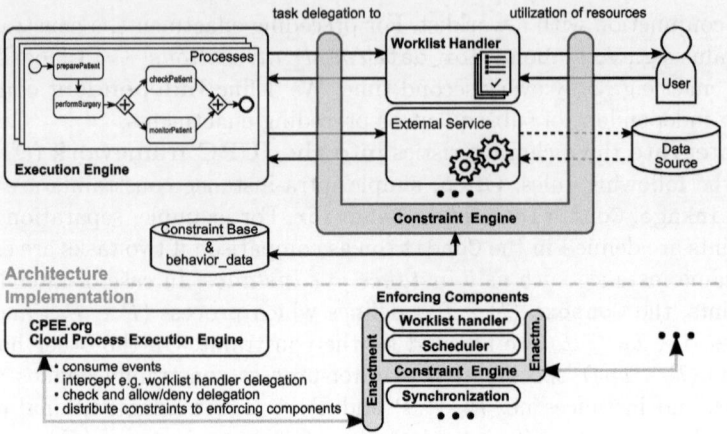

Fig. 2. Architecture of the Compliance Framework

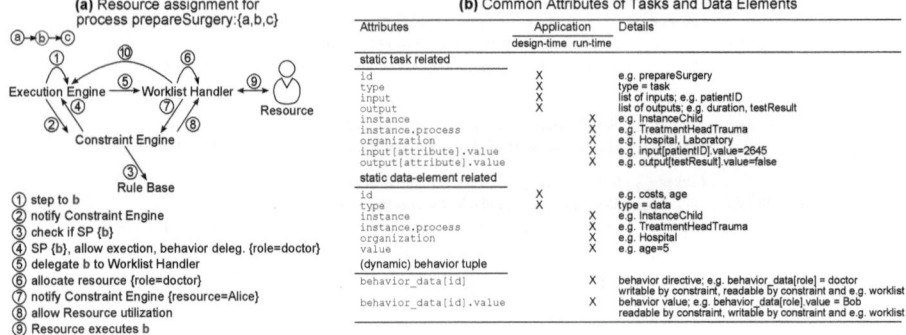

Fig. 3. Resource Assignment and Common Attributes

C1, this function sequence can be adapted to *all* instance-spanning constraints such as constraints C2 and C3.

Prototype. In order to elaborate the prototype, we first define a set of common terms to be used. Processes contain tasks with a well defined input/ouput and data elements (i.e., variables modified at runtime) and are specified in our prototype. In Fig. 3, we give a set of common attributes that, as a precondition, have to be accessible in order to monitor or enact constraints. The static task and data element related attributes, are dealing with typical properties of tasks (*type=task*) and data elements (*type=data*) like id. We also assume that for each task it is possible to find out the instance it belongs to. Please note that process is a property of the instance, as every instance belongs to exactly one process. The attribute organization is independent of instance and process as one process is able to run in multiple organizations. This coordination of processes in different organizations is a desirable side-effect of our approach. Furthermore, the dynamic behavior tuple realizes a shared space to coordinate the enactment of constraints. Constraints can (1) store values that are available for later enactments of constraints on the same task, and (2) share information with enforcing components such as worklists.

In Fig. 2 below the architecture, a short summary depicts the implementation of the prototype. For implementation, we rely on the service-oriented process testbed. As shown in the Fig., each component is a service e.g., the EE carries out tasks and the WH manages the assignment of tasks to users (i.e., manual tasks) or other services (i.e., automated tasks). As an EE we use CPEE (Cloud Process Execution Engine, http://cpee.org) as described in [10,6]. This event based engine allows for a loosely coupled CE that only consumes events during the execution of instances. For the CE, we rely solely on the event type running/syncing_before which allows to delay the process execution.

By requiring only minimal event based interaction with enforcing components, we ensure that our approach can be easily integrated with other existing solutions. Typically, the EE and WH are tightly coupled in PAIS when specifying process constraints (e.g., authorization constraints). In our approach, EE and WH are independent from each other. Hence, a separated process model-related enactment and task-related enforcement of constraints is supported. execution engines in case of inter-organizational business processes.

4   Related Work

Mainly, research centers on intra-instance settings for constraints. First, most approaches for intra-instance constraints enable the definition and enforcement of constraints on structural patterns (e.g., [8]). Furthermore, how the scope can be extended towards data-aware process constraints is presented in e.g., [1]. Lastly, resource assignments such as roles (e.g., [2,13]) are extensively investigated in literature. In case of inter-instance constraints, the workflow role-based access control model [13] defines *inter-case constraints* (e.g., the number of times activities are executed by single users) and *reciprocal separation of duties*. Moreover, the logic-based approach in [14] gives an overview on resource, data, and time inter-instance constraints. Since no generic solution is provided, inter-instance constraints are not comprehensively supported in [14] either.

In summary, the enforcement of constraints has been merely addressed by single approaches for a certain process scope (mostly intra-instance settings). In this paper, we specify an extensive set of instance-spanning constraints which is the first comprehensive approach managing multiple process scopes.

5   Conclusion

In this paper, we provide a comprehensive approach for instance-spanning constraints based on the IUPC framework. Furthermore, we provide an novel SOA-based architecture where a constraint engine is tightly coupled with an execution engine. Moreover, we demonstrate our findings with a proof-of-concept prototype. In future work, we aim at investigating instance-spanning constraints further within inter-organizational settings. Moreover, the IUPC approach and its implementation within the CPEE process engine will be evaluated by means of case studies in different domains such as care and virtual factories within the EU FP7 project ADVENTURE (http://www.fp7-adventure.eu/).

Acknowledgements. This work was partially supported by the Commission of the European Union within the ADVENTURE FP7-ICT project (Grant agreement no. 285220).

References

1. Awad, A., Weidlich, M., Weske, M.: Specification, Verification and Explanation of Violation for Data Aware Compliance Rules. In: Baresi, L., Chi, C.-H., Suzuki, J. (eds.) ICSOC-ServiceWave 2009. LNCS, vol. 5900, pp. 500–515. Springer, Heidelberg (2009)
2. Bertino, E., Ferrari, E., Atluri, V.: The specification and enforcement of authorization constraints in workflow management systems. ACM Trans. Inf. Syst. Secur. 2(1), 65–104 (1999)
3. Heinlein, C.: Synchronization of Concurrent Workflows Using Interaction Expressions and Coordination Protocols. In: Meersman, R., Tari, Z. (eds.) CoopIS/DOA/ODBASE 2002. LNCS, vol. 2519, pp. 54–71. Springer, Heidelberg (2002)
4. Ly, L.T., Rinderle-Ma, S., Knuplesch, D., Dadam, P.: Monitoring Business Process Compliance Using Compliance Rule Graphs. In: Meersman, R., Dillon, T., Herrero, P., Kumar, A., Reichert, M., Qing, L., Ooi, B.-C., Damiani, E., Schmidt, D.C., White, J., Hauswirth, M., Hitzler, P., Mohania, M. (eds.) OTM 2011, Part I. LNCS, vol. 7044, pp. 82–99. Springer, Heidelberg (2011)
5. Mangler, J., Rinderle-Ma, S.: Rule-Based synchronization of process activities. In: 13th Conf. on Commerce and Enterprise Computing, pp. 121–128. IEEE (2011)
6. Mangler, J., Stuermer, G., Schikuta, E.: Cloud process execution Engine-Evaluation of the core concepts. Arxiv preprint arXiv:1003.3330 (2010)
7. Rinderle-Ma, S., Mangler, J.: Integration of process constraints from heterogeneous sources in Process-Aware information systems. In: Int'l Workshop Enterprise Modelling and Information Systems Architectures, EMISA 2011. LNI, GI (2011)
8. Sadiq, S., Orlowska, M., Sadiq, W.: Specification and validation of process constraints for flexible workflows. Inf. Syst. 30(5), 349–378 (2005)
9. Specification, W.M.C.: Workflow management coalition terminology & glossary (Document no. WFMC-TC-1011. document Status-Issue 3.0). Tech. rep., Workflow Management Coalition Specification (February 1999)
10. Stuermer, G., Mangler, J., Schikuta, E.: Building a modular service oriented workflow engine. In: IEEE Int'l Conf on Service-Oriented Computing and Applications, pp. 1–4. IEEE (January 2009)
11. van der Aalst, W.M.P., de Beer, H.T., van Dongen, B.F.: Process Mining and Verification of Properties: An Approach Based on Temporal Logic. In: Meersman, R. (ed.) OTM 2005. LNCS, vol. 3760, pp. 130–147. Springer, Heidelberg (2005)
12. van der Aalst, W.M.P., ter Hofstede, A.H.M., Kiepuszewski, B., Barros, A.P.: Workflow patterns. Distributed and Parallel Databases 14(1), 5–51 (2003)
13. Wainer, J., Barthelmess, P., Kumar, A.: W-RBAC - a workflow security model incorporating controlled overriding of constraints. International Journal of Collaborative Information Systems 12(4), 455–485 (2003)
14. Warner, J., Atluri, V.: Inter-instance authorization constraints for secure workflow management. In: Proc. of the 11th ACM Symposium on Access Control Models and Technologies, pp. 190–199. ACM (2006)

Predicting Application Performance for Multi-vendor Clouds Using Dwarf Benchmarks

Vegard Engen, Juri Papay, Stephen C. Phillips, and Michael Boniface

IT Innovation Centre, University of Southampton, SO16 7NS, U.K.

Abstract. Future Internet applications are becoming increasingly dynamic and can be composed of a wide range of services controlled and hosted by different stakeholders. This paper addresses the challenge of resource provisioning for applications that have specific Quality of Service (QoS) requirements and where consumers of Cloud resources want to avoid lock-in to any specific Infrastructure-as-a-Service (IaaS) provider. Application modelling can be used to predict performance of applications given certain resources, workload and configuration. However, application modelling is a significant challenge for Cloud consumers due to the limited and varying information IaaS providers disclose about infrastructure resources. We demonstrate in this paper how Dwarf benchmarks can be used as a uniform and informative way of characterising compute resources, which is successful for application modelling, achieving high prediction accuracy on a range of applications.

Keywords: Cloud Computing, Future Internet, Resource Estimation, Quality of Service, Application Modelling, Application Benchmarking, Dwarfs.

1 Introduction

Cloud computing offers the potential to dramatically reduce the cost of software services through the commoditisation of information technology assets and on-demand usage patterns. However, the complexity of determining Quality of Service (QoS) requirements for applications in such environments introduces significant market inefficiencies and has driven the emergence of service engineering tools for modelling, analysing and planning the QoS of service based applications deployed within the Cloud [1,2].

In this paper we address the problems of resource provisioning for Software-as-a-Service (SaaS) providers with guarantees on QoS whilst avoiding lock-in to any particular Infrastructure-as-a-Service (IaaS) provider. This is a significant challenge for applications deployed across federated Clouds as the resource offerings by different IaaS providers vary significantly. In practice, the approach to determining resources required for a particular application is often *ad hoc*, most likely requiring SaaS providers to run their application on different resources (on different IaaS providers) and observe the performance. This can be a very time consuming and costly exercise, which typically leads to relying on a single IaaS

X.S. Wang et al. (Eds.): WISE 2012, LNCS 7651, pp. 659–665, 2012.

provider. Dwarf benchmarks have been proposed as a way to describe compute resources in a uniform manner across different IaaS providers whilst also being intended to be sufficiently information-rich to be used directly in application modelling [3].

This paper demonstrates how the Dwarf benchmarks can be used in application modelling to successfully predict the performance of several common multimedia and scientific applications. This use of the Dwarf benchmarks, therefore, enables transferability of service engineering tools to different IaaS providers, opening up the Cloud market and helps SaaS and PaaS providers exploit the potential for multi-vendor Clouds.

2 Background

2.1 Benchmarking Compute Resources

Measuring the performance of computers by benchmarking is a well-established activity and a large collection of benchmarks exists, such as SPEC, EEMBC, LINPACK and LAPACK. The issue with such benchmarks, which are not application-focused, is that the results can be uninformative and misleading [4,5].

Colella [6] proposed a Dwarf taxonomy for benchmarking aiming to capture known computational patterns. The Dwarf taxonomy was furthered developed at UC Berkeley [7,8], now comprising 13 Dwarfs: Finite State Machines, Combinatorial, Graph Traversal, Unstructured & Structured Grids, Dense & Sparse Matrices, Spectral, Dynamic Programming, Particles, MapReduce (Monte Carlo), Backtrack and Branch & Bound, and Graphical Models.

Initial results in [3] indicate that using this taxonomy of Dwarfs is a useful way to describe Cloud compute resources as they expose non-obvious differences in resources deemed to be the same by the IaaS provider.

2.2 Resource Estimation and Application Modelling

One of the motivations of the work discussed in this paper is helping Cloud consumers determine which IaaS provider(s) and specific resources are required to run their applications in the Cloud with particular QoS constraints. This work fits particularly well within the service engineering tools that a Platform-as-a-Service (PaaS) provider can offer as part of the wider role of helping the application provider develop, deploy and manage their application.

Application modelling can be used to predict the performance of an application given some specific resources. A generic application model takes as input a description of the expected static application *workload*, a description of the *resources* (physical or virtual) used to execute the application (including the resource *reliability*) and a description of any expected *user interactions* which contribute to the workload or otherwise affect the process [3]. Using a mathematical process, the model makes a prediction of the application performance.

We focus here on computing the core processing time of components in such a model. In related research, the work described in [9] achieved this by performing extensive benchmarking of the application on the same hardware the application would be run on. These benchmarks, therefore, could not support new hardware, nor be transferable to another IaaS provider.

3 Method

We investigate the use of Dwarf benchmark scores to characterise computational resources, which are used as input to an application model to predict the application performance. Using the Dwarf benchmark scores, we achieve a uniform description of compute resources, which we hypothesise will allow prediction of application performance on unseen resources.

3.1 Benchmark Suite

The benchmark suite we have adopted is described in [3], and therefore not all details are repeated here. The suite currently comprises eight out of the thirteen Dwarfs suggested by Asanovic et al. [8]: Structured & Unstructured Grid, MapReduce, Dense & Sparse Matrix, Graph Traversal, Particle and Spectral.

To calculate the Dwarf scores, we have used the as Phillips et al. [3]. Each Dwarf in the benchmark suite is executed multiple times to obtain a mean performance metric that is used to calculate the Dwarf score [3]. Thus, giving a numerical performance characterisation of a compute resource in the form of eight Dwarf scores.

3.2 Applications

Similarly to Phillips et al. [3] we make use of the following three applications for this investigation: Gromacs v. 4.0.7 (molecular dynamics), FFmpeg v. 0.6.2 (video transcoding) and Blender v. 2.49.2 (3D rendering).

For Gromacs, two different workloads have been chosen. One configuration uses a spherical cut-off for the electrostatic calculations and the other one uses the Particle Mesh Ewald (PME) method. We observe in [3] that these algorithms do correlate differently with the different Dwarfs, although computing an approximation of the same physical property.

The chosen FFmpeg computation is the transcoding of the "Big Buck Bunny" video [10] from M4V (h264 encoded) to OGV (libtheora encoded), and changing the frame size from 640x360 to 480x270. The sound is also changed from AAC to FLAC. As in [3], we have used Blender to render a bespoke animation small enough to process on resources constrained to 1GB RAM.

3.3 Computational Resources

We have conducted this investigation as part of an experiment in the BonFIRE project [11], which offers a multi-site testbed of heterogeneous Cloud resources across Europe for Internet of Services research. At the time, BonFIRE offered four infrastructure testbeds: EPCC, HLRS, IBBT and Inria This investigation also includes five public Cloud providers, all of which have different resource offerings and ways of describing them; Amazon EC2, Rackspace, CloudSigma (Zürich site), ElasticHosts and GoGrid.

The BonFIRE testbeds use a common labelling of *small, medium, large,* etc., which have defined number of cores and RAM size. However, the 100% of the CPU is given, which means the performance of resources with the same label can vary significantly between the testbeds due to heterogeneous hardware [3]. Amazon EC2 also operates with similar labels, but defines CPU performance in ECUs as well as the number of virtual cores and RAM size. Therefore, the performance of resources with the same ECUs should be the same even if they are heterogeneous as Amazon EC2 scale the CPU speed accordingly. This is not the case, however, as demonstrated in [3], as characterising the performance of a compute resource based on one parameter is not sufficient.

Other Cloud providers offer more fine-grained specifications of the VM instances, such as ElasticHosts, allowing you to determine exactly the virtual CPU speed, number of virtual cores, RAM size and storage space. CloudSigma also offers a similar scaling of CPU speed, and for both providers, this is a guaranteed minimum. Rackspace and GoGrid do not allow control of the CPU properties, but offer different server options that vary in RAM and disk space. Since the Dwarf benchmarks are invariant to the number of cores and RAM size [3], we effectively only make use of one resource from each of these providers.

To increase the number of data points for statistical analysis and to build better mathematical prediction models, we have also included five different physical hosts we had access to in-house. For these machines, we have executed the benchmarks and applications on a Ubuntu Maverick VM running on VMWare 4.0 with 1GB RAM. We have used an Ubuntu Maverick image on all the public Cloud providers, and in BonFIRE a Debian Squeeze image. In total, obtaining 23 unique computational resources on which benchmarks and applications have been executed.

3.4   Modelling Techniques and Validation Method

As discussed in Sect. 2.2, we focus on the challenge of calculating the core computation time and ignore the problems of varying application workload and user interactions. We have investigated several mathematical models/functions for predicting the performance of the different applications, some based on a single Dwarf, a combination of two Dwarfs and a combination of all Dwarfs.

Based on a single Dwarf, we have investigated 1st to 5th order functions to determine if there are any performance gains in increasing the complexity of the function. For the sake of brevity, we only report results with a 1st order (linear) function and 5th order polynomial.

Most applications will perform different types of computations and are, therefore, unlikely to be accurately modelled by just a single Dwarf. Therefore, we have investigated a linear combination of two Dwarfs to determine any gain in accuracy. The selection of Dwarfs for a given application could be done in different ways; for example, based on knowledge about the application, code profiling or according to correlation analysis. We present results for the latter here.

The final model we have investigated is the Moore-Penrose inverse matrix calculation [12], which can take as input all Dwarfs. Each mathematical prediction model is built on training data, to create a function that takes Dwarf score(s) as input and outputs application performance. To make best use of the data available, we conduct leave-one-out validation and report the mean percentage error. For each validation step (equal to the number of data points), the percentage error ε is calculated as:

$$\varepsilon = \frac{|m - p|}{m} \times 100$$

Where m is the real measured application performance and p is the predicted performance. The performance for all applications in this investigation is taken as the execution time.

4   Empirical Results

All the mathematical functions examined here are able to successfully predict the performance of all the applications using the Dwarf scores as characterisations of the compute resources. The accuracy varies with the complexity of the mathematical function, as expected. However, even with a simple linear regression based on a single Dwarf, the mean prediction error is as low as 16.72%, as seen in Table 1 (best results highlighted in bold). The improvements achieved with the 5th order function are significant on all applications; as much as 13.01 percentage points on Blender.

The best results obtained with the 1st order function are achieved with the Dwarfs that are correlated very highly with the application, which is to be expected. However, the lowest error achieved with the 5th order function is with a different Dwarf compared with the 1st order function for all but one application.

A linear combination of the two highest correlated Dwarfs can improve the prediction accuracy compared with using only one Dwarf, as seen in Table 2. However, not in all cases. The 5th order function based on one Dwarf does give better results on FFmpeg and Blender (over 10 percentage points lower).

More complex combinations of two Dwarfs may yield better results still, as for the results with the Moore-Penrose inverse matrix. The mean error rates with this function is in the range 4.70% - 6.47%. These results are encouraging, especially considering the statistically low number of data points for this investigation (23 unique compute resources) and that the benchmark suite only considers computational benchmarks, which does not include six Dwarfs that represent patterns that could further improve these predictions.

Table 1. Mean percentage error of the 1st and 5th order functions on single Dwarfs

| Dwarf | Gromacs cut-off | | Gromacs PME | | FFmpeg | | Blender | |
	Linear	5th order	Linear	5th order	Linear	5th order	Linear	5th order
Structured Grid	19.60	15.54	19.10	14.84	22.08	8.63	25.07	15.44
Unstructured Grid	20.80	15.50	18.86	**13.72**	20.70	7.91	**22.24**	**9.23**
MapReduce	**17.98**	15.37	**16.72**	13.50	18.04	**7.61**	23.82	14.41
Dense Matrix	19.15	17.24	18.06	15.36	**17.93**	9.25	22.59	15.17
Sparse Matrix	18.63	**14.11**	18.57	13.85	19.47	7.48	25.19	14.73
Graph Traversal	25.85	20.65	25.37	20.73	19.80	9.71	29.23	22.17
Particle	18.36	16.00	18.76	15.74	20.16	8.07	27.68	17.43
Spectral	25.16	17.00	25.20	15.79	24.89	9.81	29.05	10.83

Table 2. Overview of prediction results (mean percentage error)

Application	1 Dwarf 1st order	1 Dwarf 5th order	2 Dwarfs	All Dwarfs
Gromacs cut-off	17.98	14.11	11.80	5.79
Gromacs PME	16.72	13.72	15.02	4.70
FFmpeg	17.93	7.61	22.29	5.24
Blender	22.24	9.23	23.20	6.47

5   Conclusions and Further Work

Based on an investigation in BonFIRE and five public Clouds, we have demonstrated that the characterisation of compute resources in the form of Dwarf benchmark scores is indeed successful for application modelling to predict the performance of two multimedia applications (Transcoding and rendering) and a scientific application (molecular dynamics).

Ultimately we could imagine each IaaS provider describing the performance of their resources in terms of a standard set of benchmark scores, such as the Dwarfs, or even agreeing SLAs in such terms. Alternatively, a PaaS provider may measure the performance of many IaaS providers, adding to one of the possible services that could be offered. This would avoid consumers of Cloud resources being locked in to a particular IaaS provider, which opens up the Cloud market and helps SaaS and PaaS providers exploit the potential for multi-vendor Clouds.

Further work on this would benefit from extending the benchmark suite by implementing the remaining five Dwarfs and addressing the challenge of using the Dwarfs in modelling and predicting application performance with varying workloads and taking into account the resource reliability in the Cloud from both a computational and networking perspective. Disk and memory performance are also important factors to be included in the future.

Acknowledgements. This work has been carried out in BonFIRE, an EC supported 7th Framework Programme ICT project (FP7- 257386).

References

1. Cucinotta, T., Checconi, F., Kousiouris, G., Kyriazis, D., Varvarigou, T., Mazzetti, A., Zlatev, Z., Papay, J., Boniface, M., Berger, S., Lamp, D., Voith, T., Stein, M.: Virtualised e-Learning with Real-Time Guarantees on the IRMOS Platform. In: IEEE International Conference on Service Oriented Computing and Applications, SOCA (2010)
2. Marquezan, C., Metzger, A., Pohl, K., Engen, V., Boniface, M., Phillips, S., Zlatev, Z.: Adaptive Future Internet Applications: Opportunities and Challenges for Adaptive Web Services Technology. In: Adaptive Web Services for Modular and Reusable Software Development. IGI Global (2012)
3. Philips, S., Engen, V., Papay, J.: Snow White Clouds and the Seven Dwarfs. In: IEEE International Conference and Workshops on Cloud Computing Technology and Science, CloudCom (2011)
4. Seltzer, M., Krinsky, D., Smith, K., Zhang, X.: The case for application-specific benchmarking. In: 7th Workshop on Hot Topics in Operating Systems (1999)
5. Zhang, X.: Application-Specific Benchmarking. PhD thesis, Engineering and Applied Sciences: Harvard University, Cambridge, Massachusetts (2001)
6. Colella, P.: Defining Software Requirements for Scientific Computing. DARPA HPCS Presentation (2004)
7. Asanovic, K., Bodik, R., Catanzaro, B., Gebis, J., Husbands, P., Keutzer, K., Patterson, D., Plishker, W., Shalf, J., Williams, S., Yelick, K.: The Landscape of Parallel Computing Research: A View from Berkeley. Technical Report UCB/EECS-2006-183, Electrical Engineering and Computer Sciences, University of California at Berkeley (2006)
8. Asanovic, K., Bodik, R., Demmel, J., Keaveny, T., Keutzer, K., Kubiatowicz, J., Morgan, N., Patterson, D., Sen, K., Wawrzynek, J., Wessel, D., Yelick, K.: A view of the parallel computing landscape. Communications of the ACM 52(10), 56–67 (2009)
9. Metzger, A., Boniface, M., Engen, V., Phillips, S., Zlatev, Z.: Towards Critical Event Monitoring, Detection and Prediction for Self-adaptive Future Internet Applications. In: 1st International Workshop on Adaptive Services for the Future Internet (2011)
10. Blender: Big Buck Bunny, http://www.bigbuckbunny.org
11. Hume, A., Al-Hazami, Y., Belter, B., Campowsky, K., Carril, L., Carrozzo, G., Engen, V., García-Pérez, D., Ponsatí, J., Kűbert, R., Liang, Y., Rohr, C., Van Seghbroeck, G.: BonFIRE: A Multi-cloud Test Facility for Internet of Services Experimentation. In: 8th International ICST Conference on Testbeds and Research Infrastructures for the Development of Networks and Communities (2012)
12. Penrose, R.: A Generalized Inverse for Matrices, pp. 406–413. Cambridge Philosophical Society (1955)

Experiences in Building an Event-Driven and Deployable Platform as a Service

Dana Petcu, Silviu Panica, Călin Şandru,
Ciprian Dorin Crăciun, and Marian Neagul

Institute e-Austria Timişoara and West University of Timişoara, Romania

Abstract. Conceived to expose remote infrastructures and services to application developers, most of the current Platform-as-a-services are based on proprietary technologies. While several initiatives for building open-source platforms were started only recently, the developers' positive feedback is already reflected in substantial contributions. A such new platform, deployable in private and public Clouds, is presented shortly and details are provided for its event-driven approach and support for web applications.

Keywords: Platform-as-a-Service, Event driven system, Open-source.

1 Introduction

The Cloud computing has promise to address a new programming paradigm: the programmable use of e-infrastructure resources. This promise is still far to be fulfilled. Using IaaS deployment model the application developers still need to manually install and configure the proper software stack on the acquired resources. In the SaaS model, the control on the e-infrastructure resources by the application user is lost. The middle approach, PaaS has a higher potential to bring an important contribution to the concept of programmable e-infrastructure resources. However, due to the complexity of the software stack that is required for the implementation, the providers offer different proprietary solutions (an overview of their APIs is done in [6]). The consequence is the vendor lock-in: the codes developed for one specific service cannot be easily ported to another service. To overcome this problem, several solutions have emerged, from adopting open APIs (like OCCI) or protocols (like jClouds or libcloud), proposing standards (like CDMI or OVF) or reference architectures, as well as using semantic repositories (like UCI) or domain specific languages. Most successful are the efforts for providing uniform interfaces and standards for managing the virtual machines, but none of them is fully accepted at this moment and the migration of application codes is still subject of considerable efforts.

Looking to the Cloud computing landscape, we can easily identify two categories of technologies: hosted services or deployable software. Hosted services are usually encountered in PaaS case and their proprietary interfaces are creating the vendor lock-in problem. A deployable PaaS has the potential to ensure the support for smooth migration of applications from one e-infrastructure service

X.S. Wang et al. (Eds.): WISE 2012, LNCS 7651, pp. 666–672, 2012.

to another and can be used to build Private Clouds. Moreover, if the deployable PaaS is open-source, the developers community can adapt the software to their needs or to new e-infrastructure services. Several open-source and deployable middlewares for PaaS (referred as Cloudwares in what follows) have been developed recently. Only few of them are providing a full stack of software that is deployable on top of well-known IaaS. This paper is dedicated to such Cloudware, namely mOSAIC. Next section is describing its main targets and building blocks, and positioning it in the PaaS landscape. Various components and features of mOSAIC's API and PaaS were detailed in previous papers during their design and development. In this paper we provide a general overview of the software stack of the current version of the Cloudware, and focus on the event-driven architecture, as well as on the support for Web applications.

2   Overview of mOSAIC's Cloudware

Design Principles. mOSAIC Cloudware, result of an European collaborative project (details at www.mosaic-cloud.eu), is designed to support the activities of the developers of software or service products relaying on top of Cloud resources (infrastructure or software), named here Cloud applications. These applications are offered to end-users and consume Cloud resources. Two concerns of the developers are addressed: portability of the applications between Clouds and selection of the Cloud resources. Scenarios in which services from multiple Cloud are involved were defined recently in [5]; mOSAIC is covering the followings:

1 – *Changing the Cloud.* The developers or their clients should be able to change providers to make an optimal choice regarding utilization, expenses or earnings. But the PaaS software stack that allows the Cloud service consumption is usually defined by the resource providers and standards and protocols allowing code portability are not respected or not yet available. The applications designed using mOSAICs APIs are portable from one IaaS provider to another, and the software stack is using open source technologies.
2 – *Service brokerage.* Finding the best fitted Cloud services for a certain application, preferable at the deployment phase, in the conditions of a fast growing market is a challenging problem. The mOSAIC's Semantic Engine, based on a Cloud ontology, supports the matching of requests with offers (functionalities and resources), while a Cloud Agency plays the broker role.
3 – *Development of Cloud applications.* It usually requires an expertise in preparing the execution environment (IaaS) or the programming style (PaaS). The mOSAICs deployable PaaS can be used on a desktop as well as on a local cluster or in a Private Cloud. After intensive testing, the full stack of software can be deployed without changes in a Public Cloud.

Long-running and scalable applications, build from components, and accessible via Internet protocols, are the main targets of the proposed Cloudware. The components are expected to have clear dependencies in terms of communication and data between them. Communication patterns based on message-queues are

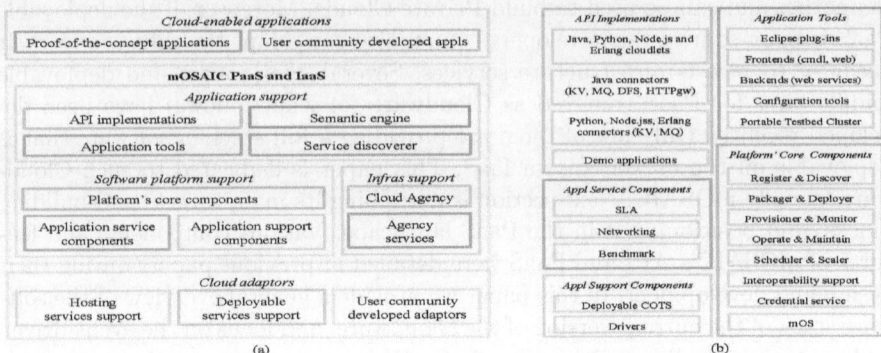

Fig. 1. (a) General architecture of the Cloudware; (b) Details of five sub-systems

supported. The Cloudware is designed to allow elasticity at the level of components: the components are expected to be able to be started, stopped, replaced, or multiplied during the life cycle of applications. However, in order to ensure such type of elasticity of the components, some restrictions are imposed, including: (a) the component is stateless i.e. it must do not store any state in global variables, and is life-cycle controllable (paused, terminated, upgradeable, etc.); (b) the component implementation adopts event-driven programming style.

To ensure the portability, the Cloudware decouples the development from the deployment. Operations on the Cloud resources are described at an abstract level independent from the API of available Cloud services. While the development is done independent from the Cloud, it is aware of its technologies. The deployment consists in the selection of the Cloud resources and start of the components.

Architecture. The Cloudware has a complex architecture. Its main sub-systems are depicted in Fig. 1 (a). The *Application Support sub-system* includes the API Implementations and Application Tools, as well as the Semantic Engine and Service Discoverer (first two detailed in the next paragraph). The Semantic Engine is the sub-system supporting the user in selecting APIs components and functionalities needed for building the application, and the list of needed resources to be acquired from the Cloud providers (details in [3]). The *Infrastructure Support sub-system* refers to the Cloudware services provisioning the concrete Cloud resources. The Cloud Agency is the sub-system for resource provisioning services and for intermediating service level agreements at the deployment time. It is connected to the Cloud providers in order to perform provisioning through Vendor agents (details in [11]). The *Cloud adaptors* are reflecting the connections to various Cloud services. Adaptors for Amazon EC2, Flexiscale (hosting services), Eucalyptus and OpenNebula (deployable software) were already developed and tested. In the time frame of the project adaptors for other seven hosting services and three deployable software are foreseen. The *proof-of-the-concept applications* developed during the life time of the project are referring to Earth observation applications, intelligent maintenance system, information extraction, model exploration service, and civil engineering structure analysis (details in [4]).

Implementation Details. The platform design was presented earlier in [7]. Its implementation required a re-structuring. Fig. 1 (b) offers a detailed view of some current re-structured or improved sub-systems (on which we focus on what follows). The stable versions of the API Implementations, the Software Platform and part of the Application Tools and Support Components are available in an *open source code repository* at `http://bitbucket.org/mosaic`. Documentation and simple examples are available at `http://developers.mosaic-cloud.eu`.

An *API implementation* (currently in Java, Python, Node.js and Erlang) consists in a collection of classes and interfaces used to build Cloudlets, the main abstractions for the user application functionality built in order to access Cloud resources. A Cloudlet includes the user defined functionality of the application and is designed to be monitored by the Software Platform and to be subject of elasticity. A Component provides the applications with the functionality of an already existing deployable software. A Connector abstracts the operations on a Cloud resource; to perform its task is directly using the resource's Driver.

The *API tools* are programs, portals or other software which facilitate the development, deployment or the monitoring of applications. Currently the most used tool is the Portable Testbed Cluster (PTC) which allows the development, testing and debugging of the applications on the developer desktop and the seamless deployment of the application on the Public Clouds, as well as the Web front-end of the platform that allows the control of the components.

The *Software Platform* manages the execution environment of the applications. It receives an application descriptor and an deployment descriptor. Based on the application descriptor, it is able to identify the application components and the needed Cloud resources. The components are retrieved from the appropriate Component servers (specified in the deployment descriptor). Internal components (like Drivers) are also instantiated in order for the Connectors to be able to access the Cloud resources. The deployed Components are monitored in terms of the resources usage, information needed by core Components like the Scaler. The platform performs also the start/stop of the components when such a request is received. Moreover, it schedules application components to run on specific execution virtual machines. Scheduling happens either at the deployment time when all the components should be placed on execution virtual machines or at the Scaler decision (it monitors different parameters related to resources consumption and may decide to scale the components up or down).

The *Core Components* are responsible for the control, scheduling, scaling, monitoring (as mentioned above) as well as for the deployment of the application. The packaged component is retrieved by the Deployer from the received location and it is installed into an appropriate execution environment inside the virtual machine. mOS is a custom operating system, thought to run on the top of virtual machines and to be used by the Software Platform to host Components.

The *Components-of-the-shelf* (COTS) are specific resources to be added to an application and are based on open source technologies. Intensive used COTS are Riak and RabbitMQ Servers. The *Drivers* facilitate the applications' access to Cloud resources at the low level through Connectors. They are specific to providers, i.e. there is a driver for Amazon S3 key-value store, another one for Riak key-value store and so on. Drivers are deployed as regular Components and subject of the same operations as other Components. Different Driver types are included in the Platform. The Riak Driver and the S3 Driver allow the access to two types of key-value store. The JMS Driver and the RabbitMQ Driver support two message queues mechanisms. Other Drivers are useful in order to access other mechanisms: distributed file systems, columnar database, MapReduce, etc.

Event-Driven Approach. mOSAIC's API was designed to be event-driven. The reasons for such an approach are multiple: (a) avoids the expensive pooling on Cloud resources; (b) long running applications have often periods in which changes in state do not occur a considerable period of time; (c) deal with an unlimited number of messages; (d) requirements from scenarios involving mobile phones are pointing to an event-driven model; (e) adaptability is event-driven.

There are few implementations of event-driven approaches in Cloud computing, but the most known are the proprietary Amazon's SNS and Microsoft's Azure, or the open-source Node.js. Amazon SNS is delivering notifications to clients using a push mechanism and is used to build event-driven workflows and messaging applications. Node.js is promoted as Javascript event-driven framework for low-latency real time applications. Unfortunately an event-driven approach has also drawbacks: (a) the need of designing the callbacks by the application developer; (b) data cannot be provided with callbacks due to the access rights; (c) states of a resource should be well defined to trigger a call to the API by the resource provider. A dependence on the provider is created for the callback protocol, as well as a need to trust in the provider callback protocol.

To overcome the above described problems, mOSAIC has proposed an abstraction layer that allows the application developer to follow the concepts of event-driven architecture, while the low level components of the platform are treating the cases of demand-driven approaches in the connection with specific Cloud services. An interoperability component of the Software Platform acts as a proxy between the instances following the two different models of interactions.

Support for Web Applications. The typical architecture of the Web application consuming Cloud resources is including a browser, several load balancers and front-ends, message queues, as well as several workers and their data. The number of load balancers and workers is expected to vary, as well as the message queuing system to cope with a fluctuating number of messages. The mOSAIC's Cloudware includes several components allowing the fast development of Web applications. For example, the HTTP Gateway (HTTP-G) component receives messages from the HTTP channel and sends out messages on queues (it is used to build easily HTTP interfaces to the developed applications); RabbitMQ is embedded to support message queuing and AMQP protocol is supported.

A *demonstration application* (provided in open source repository) is a *Twitter watcher* comprised of several components: gateway, servlet, fetcher, indexer, scavenger, message queue. It uses Riak as key value store, RabbittMQ as embedded

Table 1. Differences between three PaaS: mOSAIC, CloudFoundry and OpenShift

Product	CloudFoundry	mOSAIC	OpenShift
Owner	VMWare	mOSAIC Consortia	RedHat
Site	cloudfoundry.com	mosaic-cloud.eu	openshift.com
Repository	github.com/ cloudfoundry	bitbucket.org/ mosaic	github.com/ openshift
State	Beta	Alfa	Production
Languages	Java, Ruby, Node.js, Groovy	Java, Python, Node.js, Erlang	Java, Python, Perl, PHP, Ruby
Data support	MongoDB, SQLFire, Redis PotsgreSQL	Riak, CouchDB, Redis, MySQL, MemcacheDB, S3	MongoDB, MySQL, Amazon RDS
Virtualization	VMWare	VirtualBox, KVM, Xen	Red Hat
Messaging	RabbitMQ	RabbitMQ	Own design
Clouds tested	VMWare	Amazon, Eucalyptus Flexiscale, OpenNebula	Amazon, Rackspace, SmartCloud,RightScale
Web aps or general	Web apps	General	Web apps
Desktop Cloud	Yes	Yes	No
API access	No	Yes	No
Thread access	Yes	No	Yes
Choose stack	Yes	Yes	No
QoS agreements	No	Yes	No
Private Cloud	Yes	Yes	No
Web control & logs	No	Yes	Yes

component, and Jetty as servlet container. Note that the Jetty HTTP connector was replaced with a custom load-balancer using RabbitMQ. The demo videos on YouTube about mOSAIC (keywords for search: mOSAIC Cloud computing) are based on this watcher application. Tests were performed to study the application response to the variation in the user number: hundreds of requests per seconds were able to be supported using only two virtual machines [8]. Another prototype of a Web application is reported in [9], its subject being a *checkout process*; the tests showed that the throughput and latency are in favor of selecting the unconventional event-driven message-oriented and asynchronous approach, instead the classical threading and synchronous approach.

Related Work. Open-source Cloud services were compared recently by different authors. We mention here for example [2] and [12], the latest taking in consideration also mOSAIC's Cloudware. The most prominent representatives of the open-source and deployable PaaS are currently: VMware's CloudFoundry and WaveMaker, RedHat's OpenShift, as well the results of other research projects, like ConPaaS [10] for PHP applications, or AppScale [1] compatible with Google App Engine. Most complex offers are currently provided by CloudFoundry and OpenShift. The first is in development stage, as mOSAIC, while the second is in production. We have considered these two in our comparison from Table 1. It shows that mOSAIC's Cloudware is comparable with the two proposals from the well-known companies, and moreover is bringing an improvement.

3 Conclusions

Open-source deployable platform as a service able to run on multiple e-infrastructure Cloud services is a feasible approach to solve the vendor lock-in solution encountered in Cloud computing market. The nature of the mainly targeted applications in Cloud computing, namely Web applications, is pointing towards the need of an event-driven architecture for such a platform. The proof-of-concept prototype of the mOSAIC's Cloudware intends to answer to these basic requirements for Cloud application portability. Initial comparisons with the on-going efforts, commercial as well as academics, are revealing several benefits of the proposal, like independence from the e-infrastructure service, support for Private Clouds, or Web control of the running processes of the applications.

Acknowledgments. This research is supported by the grants EC-FP7-ICT-2009-5-256910 (mOSAIC) and RO-PN-II-ID-PCE-2011-3-0260 (AMICAS).

References

1. Bunch, C., Chohan, N., Krintz, C.: Appscale: open-source platform-as-a-service, UCSB Technical Report 2011-01 (2011)
2. Cordeiro, T., Damalio, D., Pereira, N., Endo, P., Palhares, A., Gonçalves, G., Sadok, D., Kelner, J., Melander, B., Souza, V., Møngs, J.E.: Open source cloud computing platforms. In: Procs. GCC 2010, pp. 366–371 (2010)
3. Di Martino, B., Cretella, G.: Towards a semantic engine for cloud applications development support. In: Procs. CSIS 2012 (in print, 2012)
4. Di Martino, B., Petcu, D., Cossu, R., Goncalves, P., Máhr, T., Loichate, M.: Building a Mosaic of Clouds. In: Guarracino, M.R., Vivien, F., Träff, J.L., Cannataro, M., Danelutto, M., Hast, A., Perla, F., Knüpfer, A., Di Martino, B., Alexander, M. (eds.) Euro-Par-Workshop 2010. LNCS, vol. 6586, pp. 571–578. Springer, Heidelberg (2011)
5. Hogan, M., Liu, F., Sokol, A., Tong, J.: Nist Cloud computing standards roadmap-version 1.0, Special Publication 500-291 (2011)
6. Larus, J.: Programming Clouds. In: Gupta, R. (ed.) CC 2010. LNCS, vol. 6011, pp. 1–9. Springer, Heidelberg (2010)
7. Petcu, D., Crăciun, C., Neagul, M., Panica, S., Di Martino, B., Venticinque, S., Rak, M., Aversa, R.: Architecturing a Sky Computing Platform. In: Cezon, M., Wolfsthal, Y. (eds.) ServiceWave 2010 Workshops. LNCS, vol. 6569, pp. 1–13. Springer, Heidelberg (2011)
8. Petcu, D., Frincu, M., Craciun, C., Panica, S., Neagul, M., Macariu, G.: Towards open-source Cloudware. In: Procs. UCC 2011, pp. 330–331 (2011)
9. Petcu, D., Macariu, G., Panica, S., Crăciun, C.: Portable Cloud applications – From theory to practice. Future Generation Computer Systems (2012), doi:10.1016/j.future.2012.01.009
10. Pierre, G., El Helw, I., Stratan, C., Oprescu, A., Kielmann, T., Schütt, T., Artač, M., Černivec, A.: ConPaaS: an integrated runtime environment for elastic cloud applications. In: Procs. Middleware 2011, Art. 5. ACM (2011)
11. Venticinque, S., Tasquier, L., Di Martino, B.: Agents based cloud computing interface for resource provisioning and management. In: Procs. CISIS 2012, pp. 249–256 (2012)
12. Voras, I., Mihaljevic, B., Orlic, M., Pletikosa, M., Zagar, M., Pavic, T., Zimmer, K., Cavrak, I., Paunovic, V., Bosnic, I., Tomic, S.: Evaluating open-source cloud computing solutions. In: Procs. MIPRO 2011, pp. 209–214 (2011)

Integrating Feature Analysis and Background Knowledge to Recommend Similarity Functions

Seung Hwan Ryu and Boualem Benatallah

School of Computer Science & Engineering,
University of New South Wales, Sydney, NSW, 2051, Australia
{seungr,boualem@cse.unsw.edu.au}

Abstract. Existing approaches in similarity analysis is little concerned with the right choice of similarity functions. We present an approach for suggesting which similarity functions (e.g., edit distance) are most appropriate for a given similarity search task. We identify data features (e.g., misspellings) that are considerable when choosing similarity functions. We also introduce the concept of similarity function background knowledge that associates data features with similarity functions, and apply the knowledge to recommend suitable similarity functions.

Keywords: Similarity Function, Recommendation, Feature Analysis.

1   Introduction

Similarity search (or similar entity search) is the task of finding entities (objects, instances) that most closely resemble a given entity [10]. Each entity is described by attributes (e.g., person has name, interest and phone number) and relationships (e.g., *authorBy*). In similarity search, a main issue is that entities may contain different types of attributes. In such a case, no *single* similarity function can perform well for different attributes [5,10]. A common approach for measuring the similarity between two entities is to exploit appropriate similarity functions (e.g., edit distance or jaccard similarity) [8] for different attributes.

Choosing appropriate functions for attributes could depend on the features of attribute values. For example, in Figure 1, assume we find entities similar to a given query entity q by comparing *names*. When e_1 is given as q, if `edit distance` function, which works well for capturing spelling errors, is applied to the names, we get two true similar entities: $(q, e_5, 0.9)$ and $(q, e_2, 0.8)$. On the other hand, if `jaccard` function is applied, we miss the two entities whose scores are $(q, e_5, 0.3)$ and $(q, e_2, 0)$ as the function is not effective in comparing strings that have misspellings.

Most previous approaches for similarity analysis rely on attribute similarity to determine the overall similarity between entities [3,5,10]. Other approaches employ machine learning-based techniques [2,3,9,13]. For example, some of them use training examples to learn the best combination of similarity functions [2,13]. In addition, some works (e.g., [6]) propose mechanisms for using the relationships between entities. Although existing approaches have made significant progress,

X.S. Wang et al. (Eds.): WISE 2012, LNCS 7651, pp. 673–680, 2012.

Person Entity				Result by comparing names			
ID	Name	Interest	Biography	q	Features	Function	Result
e_1	John Smyth	KD	His research interests include data mining and knowledge ...	e_1	Misspell	ED	e_5(0.9), e_2 (0.8)
e_2	Joh Smith	Data mining	He received his PhD from School of CSE, the University of ...	e_2	Misspell	ED	e_5(0.9), e_1 (0.8)
e_3	Susan Kim	Artificial Intelligence	Recent Computer Science graduate with Knowledge Acquisition...	e_3	¬ Misspell, differentOrder	Jaccard	e_4(1.0)
e_4	Kim Susan	AI	She received her PhD from the University ...	e_4	¬ Misspell, differentOrder	Jaccard	e_3(1.0)
e_5	John Smith	Knowledge Discovery	He is a member of ACM, AAAI, Association ...	e_5	¬ Misspell, ¬ differentOrder	Jaro	e_2(0.96), e_1(0.93)

Threshold= 0.5

Fig. 1. Person entities and, for a given q, search result by comparing person *names*

they have not addressed a fundamental issue: which similarity functions should be used in comparing entities. In some cases, even a function that shows good performance for some datasets can perform poorly on new and different ones [3,7].

To address the issue, we provide an approach for suggesting suitable similarity functions by integrating data feature analysis and background knowledge. The approach works in two steps, as shown in Figure 2. Given a dataset (in our case, a set of entities of a same class), for each attribute, the *entity attribute profiling* step (Section 2) takes as input the attribute values and splits them into non-overlapping attribute groups according to their common features, such as length or word frequencies. The *recommendation rule generation* step (Section 3) assigns most appropriate similarity functions to the attribute groups generated from the previous step, based on the similarity function background knowledge that associates data features with similarity functions. Our system *automatically* generates recommendation rules from the mappings between functions and attribute groups. The rules are stored in a rule base and reused when entities are compared. In Figure 2, a rule base is used for managing the rules. Rules in the rule base can be also incrementally refined by domain experts to make more fine-tune function recommendation, using our previous work [11].

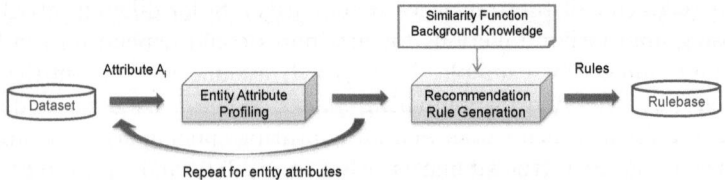

Fig. 2. Overall process

2   Entity Attribute Profiling

This section first describes concepts and terms we use. Then, it explains how to choose attributes playing a vital role for similarity search and present a set of feature extraction operators used for analyzing features from chosen attributes.

Table 1. Sample features

Feature	Description
Misspell	If a string value has spelling errors, YES; else No.
Short/Long	If the length of a string is less than α, Short; else Long.
Abbr/Syn	If a string has an abbreviation or synonym relation, YES; else No.
hiWeight/loWeight	If a string value contains rare terms, hiWeight; else loWeight.
differentOrder	If tokens of a string appear differently in another one, YES; else No.

2.1 Concepts and Terminologies

Entity class (e.g., *Person*) has a set of entities (e.g., *John* and *Susan*) where each entity consists of multiple attributes $A = \{A_1, ..., A_m\}$. We represent the domain of attribute A_i as G_{A_i}, e.g., the domain of person name attribute as G_{name}. Given an attribute A_i, *features* give some indication about common characteristics or properties of A_i attribute values. Table 1 shows some example features. We consider F as a finite set of features $f_1, f_2, ..., f_n$. Once features are identified, we split the input values into several groups where each group only contains the values sharing one or more common features. We call such a group an *attribute group* and denote it as $AG_{A_i}^{F'}$: a group of values of attribute A_i, which have common features $F' \subseteq F$. For example, for **name** attribute, $AG_{name}^{Misspell}$ is a group of names where each name has misspellings.

2.2 Attribute Interestingness Criteria

Comparing all available attributes is not always more effective than comparing a few attributes [9]. Thus, we need criteria for deciding any key attributes that play an important role for similarity search. We take the approach of specifying interestingness of attributes on the basis of what is not interesting [12]. An attribute might be significant when it contains values that are not unique, i.e., they can be found in values of another entities. For example, in Figure 1, the attribute ID can be disregarded as it has all unique values. In contrast, attributes with all the same values or small domains (e.g., **Gender** attribute) might not be interesting as each value will be repeated on a large number of entities. To characterize these properties, we have identified the following two measures:

- $unique\_ratio(A_i) = unique(A_i)/notNull(A_i)$, where $unique(A_i)$ is the number of unique values and $notNull(A_i)$ is the number of non-null values.
- $repeated\_ratio(A_i) = repeated(A_i) / notNull(A_i)$, where $repeated(A_i)$ is the number of repeated values and $notNull(A_i)$ is the number of non-null values.

To determine the importance (interestingness) of an attribute A_i, we combine the two measures described above as follows:

- $Int(A_i) = 1 - \frac{|unique\_ratio(A_i) - repeated\_ratio(A_i)|}{max\{unique\_ratio(A_i), repeated\_ratio(A_i)\}}$.

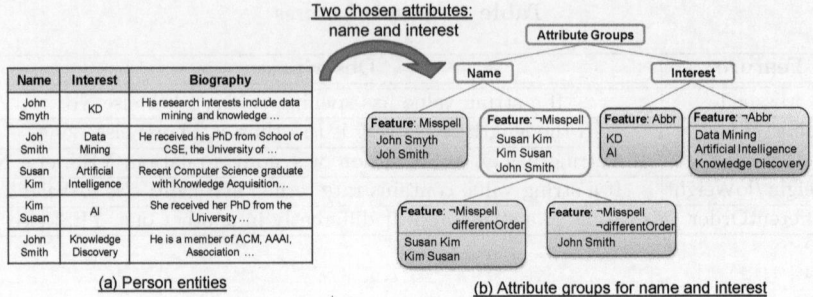

Fig. 3. For `name` and `interest` attributes chosen as important attributes, generate attribute groups of them. The shaded rectangles represent the final attribute groups. In this case, there are: 3 groups for `name` attribute and 2 groups for `interest` attribute.

2.3   Feature Extraction Operators

When an entity attribute is chosen as an important attribute, domain experts use feature extraction operators (extraction operators, for short) to identify features from the attribute values. Generally, an extraction operator can be applied to only certain data types of attributes.

Misspell Extraction Operator: it splits the input strings into two groups: one contains the values having spelling errors and the other contains the values having no errors.

Example 1. In Figure 3(b), person names can be clustered into two groups: one of names having a feature *Misspell* and the other of names having no such feature.

Value Distribution Extraction Operator: it groups the values according to the characteristics of value distributions, such as the length of attribute values or the order of words in a string. For example, for a group of strings in which their word order is different from that of any other strings, we may hypothesize that the token-based functions are more likely to be suitable than the character-based ones which are sensitive to the position of words [14].

Lexical Relation Extraction Operator: it groups the input values by identifying lexical relations (e.g., synonym) or abbreviation relations among them.

Example 2. In Figure 3(b), using this operator, experts can split the interests into two groups: one of values having abbreviation relations and the other of values having no such relations.

Word Weight Extraction Operator: it clusters the input values according to how many important words there are in the values. The importance of a word is based on the concepts of TF-IDF weighting scheme that is widely used in the information retrieval community. The scheme assigns higher weights to rare words.

Pattern Extraction Operator: it identifies the values that satisfy a specific pattern and clusters them into two groups: one with the pattern and the other

Table 2. Example background knowledge

Tuple	Feature Set (S_i^*)	Similarity Function (sf_i^{**})	Ref.
1	$S_1 = \{\text{Misspell}\}$	$sf_1 = $ ED	[3,5,14]
2	$S_2 = \{\text{Short, Abbr}\}$	$sf_2 = $ Transformation+ED	[3,7]
3	$S_3 = \{\text{multipleWord, differentOrder}\}$	$sf_3 = $ Q-grams	[1,7]
4	$S_4 = \{\text{Long, }\neg\text{Misspell}\}$	$sf_4 = $ Jaccard	[3,7,14]

*Each feature set S_i belongs to $\mathcal{P}(\mathcal{F})$, which is a power set of F.
**Similarity function sf_i belongs to SF, which is a set of available similarity functions.

with non pattern. For example, assume we want to obtain the group of strings that start with "Bill", such as "Bill Gate" and "Bill Gate III". The attribute group may be helpful to identify the strings that have synonym relationships.

3 Automatic Recommendation Rule Generation

In this section, we begin by describing the similarity function background knowledge (henceforth background knowledge), then explain how to exploit the knowledge in automatically generating recommendation rules.

3.1 Background Knowledge

We leverage the background knowledge (BK) that represents domain experts' knowledge and experience on a relation between similarity functions and data features. For example, if strings have multiple words and words of certain strings are ordered differently in other strings, Q-grams function is more effective for comparing those strings than edit distance or Jaro function [1]. Table 2 shows sample background knowledge. The BK consists of tuples (S_i, sf_i), where S_i represents features, which should be considered when choosing a function to a certain attribute group, and sf_i is a similarity function chosen for the attribute group having features S_i. For example, in Table 2, tuple 1 states that edit distance is suggested to an attribute group having a Misspell feature. The BK could be reused across different similarity search tasks.

Obtaining the BK: We argue that an initial background knowledge should be shipped with the initial configuration of our system. The BK can be either provided by domain experts, based on their experience or the literature (e.g., [1,3,5,7,9,14]). The BK can be also added or modified as domain experts get to better know domains and their experiences are accumulated over time.

3.2 Generating Recommendation Rules Using the BK

In what follows, we describe how to automatically generate recommendation rules using the BK.

Table 3. Example rule base

ID	Recommendation Rule
R_1	IF C^*= Person and $AG_{name}^{Misspell}$ and $AG_{interest}^{Abbr}$ THEN $AG_{name}^{Misspell} \rightarrow$ ED, $AG_{interest}^{Abbr} \rightarrow$ Transformation + ED
R_2	IF C= Person and $AG_{name}^{\{\neg Misspell, differentOrder\}}$ and $AG_{interest}^{Abbr}$ THEN $AG_{name}^{\{\neg Misspell, differentOrder\}} \rightarrow$ Jaccard, $AG_{interest}^{Abbr} \rightarrow$ Transformation + ED

*C stands for Class.

Rule Generation. We exploit the BK to generate the recommendation rules that contain information about mappings between similarity functions and attribute groups. A recommendation rule consists of two parts: *condition* and *conclusion* [11]. The condition part refers to a function usage context, such as entity class, chosen attributes, and attribute groups. The conclusion part has a list of pairs (attribute group, similarity function) that indicates which functions are most appropriate to which attribute groups. Table 3 shows some example rules generated from the attribute groups in Figure 3.

Example 3. The condition of R_1 checks whether a query entity q is an instance of Person class and the name and interest attribute values of q belong to $AG_{name}^{Misspell}$ and $AG_{interest}^{Abbr}$ respectively. If the condition is satisfied by q, the conclusion says that edit distance and transformation + edit distance functions should be applied to compare the name and interest values of q with the names and interests of the other entities, respectively.

Rule Matching. After an expert selects a certain entity as q, the recommender system matches q against the conditions of recommendation rules to identify the potential similarity functions being applicable to q. For example, given a query entity q: (class: Person, name: John Smyth, interest: KD, biography: NULL), the condition of R_1 is matched and the functions of the matched rule are returned. Experts can use the suggested functions to compare q with other entities.

4   Evaluation

This section presents the evaluation results obtained from applying our approach to entity matching (one of application domains of similarity search).

Datasets: Our evaluations were performed on the following two real-world datasets. Firstly, we consider a restaurant dataset from RIDDLE repository [4]. It is a database of 865 records that contains 112 duplicates. The second dataset is the Census data manipulated synthetically due to privacy concerns [3] .

Evaluation Methodology: We compare our approach with two other methods. The first method is a support vector machine (SVM) that is a currently known best machine learning-based technique for entity matching [4]. The second method is to compute the overall similarity between entities as the average similarities between attributes. We will refer to this method as the AVG method, which applies similarity functions to attributes without considering data features.

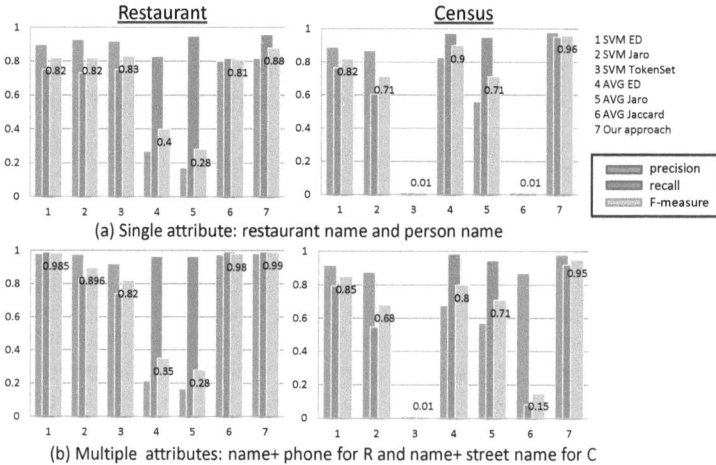

Fig. 4. Results of evaluations: R and C stand for restaurant and census datasets

Results: Figure 4 shows the overall performance (precision, recall, F-measure) results from different methods carried out across the two datasets. The top half shows the results obtained by applying the methods on a *single* attribute. We selected some attributes that look most meaningful from the datasets and then applied three different similarity functions to the chosen attributes (e.g., name for the restaurant dataset). The considered three functions are two character-based functions (edit distance and Jaro) and one token-based function (jaccard), which are commonly used in entity matching [3,9]. The bottom half shows the results achieved by applying the three methods on *two* attributes, e.g., name and phone number for the restaurant data.

From the results on a single attribute, we see that our approach is the top or among the top performing methods for the matching tasks. In Figure 4(a), we can see the AVG method with the jaccard function worked well for the restaurant dataset as long strings like restaurant names tend to favor the function. However, it performed worst for the census dataset since the person names have many misspellings. From the results on two attributes, we could achieve the better quality than using a single attribute for the restaurant dataset. However, for the census dataset, using two attributes is not more effective than using one attribute. Our approach could effectively solve the matching tasks by analyzing features from attributes, such as misspelling, abbreviation/synonym, partOf, differentOrder, etc.

5   Conclusion

In this paper, we proposed an approach for recommending similarity functions that are most suitable for comparing two entities. Particularly, we presented a set of feature extraction operators used for analyzing features from attribute data. We also described how to automatically generate recommendation rules that contain information about mappings between similarity functions and attribute groups, based on the identified features and background knowledge.

References

1. Tailor: A record linkage tool box. In: Proceedings of the 18th International Conference on Data Engineering, ICDE 2002. IEEE Computer Society (2002)
2. Bilenko, M., Mooney, R.J.: Adaptive duplicate detection using learnable string similarity measures. In: KDD, pp. 39–48. ACM (2003)
3. Bilenko, M., Mooney, R.J., Cohen, W.W., Ravikumar, P.D., Fienberg, S.E.: Adaptive name matching in information integration. IEEE Int. Sys. (2003)
4. Chaudhuri, S., Chen, B.-C., Ganti, V., Kaushik, R.: Example-driven design of efficient record matching queries. In: VLDB, pp. 327–338 (2007)
5. Christen, P.: A comparison of personal name matching: Techniques and practical issues. In: ICDM Workshops, pp. 290–294 (2006)
6. Dong, X., Halevy, A.Y., Madhavan, J.: Reference reconciliation in complex information spaces. In: SIGMOD Conference, pp. 85–96 (2005)
7. Elmagarmid, A.K., Ipeirotis, P.G., Verykios, V.S.: Duplicate record detection: A survey. IEEE Trans. Knowl. Data Eng. 19(1), 1–16 (2007)
8. Hall, P.A.V., Dowling, G.R.: Approximate string matching. ACM Comput. Surv. 12, 381–402 (1980)
9. Köpcke, H., Thor, A., Rahm, E.: Learning-based approaches for matching web data entities. IEEE Internet Computing 14(4), 23–31 (2010)
10. Lange, D., Naumann, F.: Efficient similarity search: arbitrary similarity measures, arbitrary composition. In: CIKM, pp. 1679–1688 (2011)
11. Ryu, S.H., Benatallah, B., Paik, H.-Y., Kim, Y.S., Compton, P.: Similarity Function Recommender Service Using Incremental User Knowledge Acquisition. In: Kappel, G., Maamar, Z., Motahari-Nezhad, H.R. (eds.) ICSOC 2011. LNCS, vol. 7084, pp. 219–234. Springer, Heidelberg (2011)
12. Sahar, S.: Interestingness via what is not interesting. In: KDD, pp. 332–336 (1999)
13. Tejada, S., Knoblock, C.A., Minton, S.: Learning domain-independent string transformation weights for high accuracy object identification. In: KDD (2002)
14. Wang, J., Li, G., Feng, J.: Fast-join: An efficient method for fuzzy token matching based string similarity join. In: ICDE, pp. 458–469 (2011)

Securing Data Warehouses from Web-Based Intrusions

Ricardo Jorge Santos[1], Jorge Bernardino[2],
Marco Vieira[1], and Deolinda M. L. Rasteiro[3]

[1] CISUC – DEI – FCTUC – University of Coimbra – Coimbra, Portugal
[2] CISUC – DEIS – ISEC – Polytechnic Institute of Coimbra – Coimbra, Portugal
[3] DFM – ISEC – Polytechnic Institute of Coimbra – Coimbra, Portugal
lionsoftware.ricardo@gmail.com, {jorge,dml}@isec.pt,
mveira@dei.uc.pt

Abstract. Decision support for 24/7 enterprises requires 24/7 available Data Warehouses (DWs). In this context, web-based connections to DWs are used by business management applications demanding continuous availability. Given that DWs store highly sensitive business data, a web-based connection provides a door for outside attackers and thus, creates a main security issue. Database Intrusion Detection Systems (DIDS) deal with intrusions in databases. However, given the distinct features of DW environments most DIDS either generate too many false alarms or too low intrusion detection rates. This paper proposes a real-time DIDS explicitly tailored for web-access DWs, functioning at the SQL command level as an extension of the DataBase Management System, using an SQL-like rule set and predefined checkups on well-defined DW features, which enable wide security coverage. We also propose a risk exposure method for ranking alerts which is much more effective than alert correlation techniques.

Keywords: Database security, Web security, Intrusion detection, Data warehouses.

1 Introduction

Many business models using web-based infrastructures require continuous access to decision support means such as Data Warehouses (DWs). To ensure this kind of access, DWs need to be available at any time from any location through communication infrastructures such as the Internet. This creates a main security issue, since it provides a mean for accessing their databases from outside the enterprise.

Intrusion is as a set of actions that attempt to violate the integrity, confidentiality or availability of a system [8]. Automatic detection of intrusion actions in databases is the main goal of Database Intrusion Detection Systems (DIDS). Since DWs are the core of enterprise sensitive data, quickly detecting and responding to intrusions is critical. However, most DIDS applied to DWs typically spawn too low true intrusion detection rates (*i.e.* false negatives) or too high false alarm rates (*i.e.* false positives) [7, 8, 9]. In the first case, many intrusions pass undetected; in the second case, the number of generated alerts is frequently so large that it leads to wasting vast amounts of time and limited resources, or they are simply just too much to be checked [7, 8]. This jeopardizes the credibility and feasibility of the DIDS [5, 7, 8]. Given the

X.S. Wang et al. (Eds.): WISE 2012, LNCS 7651, pp. 681–688, 2012.

well-defined features intrinsic to DW environments, we argue they require specifically tailored DIDS. To the best of our knowledge, no such DIDS has been proposed.

Although alert correlation techniques [7, 10] have been proposed to decrease false positive rates, we also argue they are not the best choice for alert management in DW environments. These techniques filter alerts to determine those which present a higher probability of referring a true intrusion, given a predefined threshold. Using a threshold implies some alerts are discarded and thus, there is always the risk that a true intrusion may pass undetected. Given the value of DW data, this is not advisable. In our approach, we decide not to correlate/filter alerts, but measure their risk exposure (probability *vs* impact) to the enterprise. Instead of filtering alerts, our approach ranks all alerts by their potential cost to the enterprise, dealing with the most critical intrusions first instead of wasting time checking alerts with low impact for the enterprise.

The main achievements and contributions of our work are: Our DIDS is the first tailored for web-acessible DWs, analyzing each user command both *a priori* and *a posteriori* of its execution; It is also the first to use risk exposure to increase alert management efficiency; We use a very easy to understand and use declarative SQL-like form for defining rules at a fine-grain level for intrusion detection (ID) and response. Their flexibility covers a very large spectrum of possibilities that enables detecting and responding to a wide range of intrusions; The DIDS works as an extension of any DBMS, adding real-time ID and response management to the native database server; It can be easily implemented and used in any web-accessed DW, acting transparently at the application layer between DW user applications and the database.

The remainder of this paper is structured as follows. In section 3, we present our proposal, describing its architecture and each of its components and explaining how intrusion detection and response is managed. In section 4 we describe how each form of attack is dealt with by our solution. Section 5 presents related work on DIDS. Finally, in section 6 we present our conclusions and future work.

2   Data Warehouse Database Intrusion Detection System

Figure 1 shows the typical user action flow in a web-accessible DW, while Figure 2 shows the DIDS architecture, working as an extension of the DBMS.

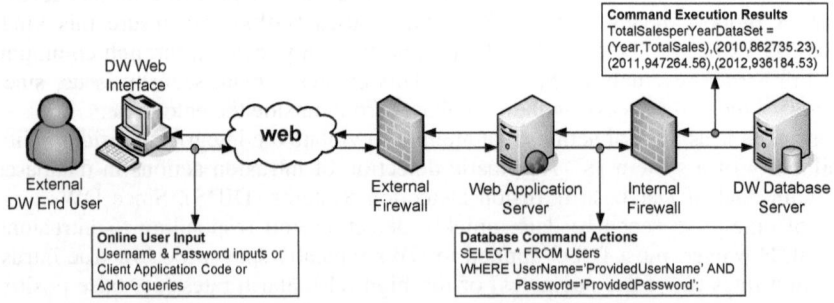

Fig. 1. Typical user action flow in a web-accessible Data Warehouse

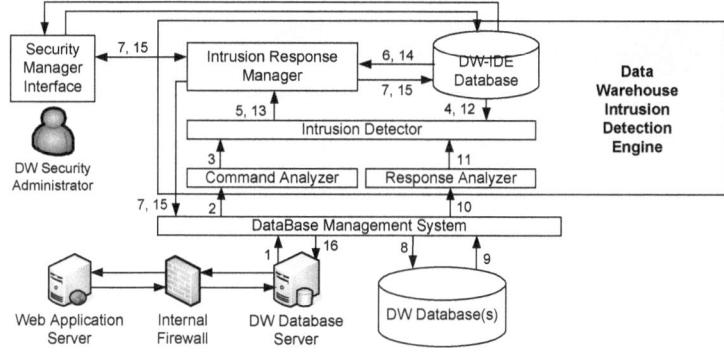

Fig. 2. The conceptual architecture of the DIDS for DWs

The sequence of intrusion detection steps is labeled in the figure and described as: A user requests an action through a *Web Application Server*, arriving at the DBMS for execution (step 1). Before executing it, the *Command Analyzer* retrieves the command text, date/time, and user/IP identification (step 2), parses the command, splits it into the ID features and passes all information to the *Intrusion Detector* (step 3). This component then gets the statistical model values for all features from the *DW-IDE Database* (step 4) and applies the ID algorithms (explained in subsection 2.2) to decide if the command is a potential intrusion. The detector then passes all information (features and respective intrusion detection results) to the *Intrusion Response Manager (IRM)* (step 5). Given each feature's result considering the user's action as an intrusion, the *IRM* retrieves the probability and impact rules, evaluate its risk exposure and generates the resulting alert (step 6), stores the data concerning the alert and the feature(s) that generated it in the *DW-IDE Database* for future reference, takes the appropriate actions to deal with the potential intrusion through the DBMS and notifies the *DW Security Administrator* (step 7). The *IRM* takes action by commiting the command's execution in the DBMS (in case it has been considered a non-intrusion) or by suspending or killing its execution, or killing the user session, either automatically or on request of the *DW Security Administrator* (step 7). If the user action is not considered an intrusion the *IRM* will simply update the feature's statistics in the *DW-IDE Database* (step 7) without notifying the *DW Security Administrator*.

If the *IRM* concludes the user's action is not an intrusion, it notifies the DBMS to normally execute it against the *DW Database(s)* (step 8). After the user command has been computed (step 9), its response is analyzed by the *Response Analyzer* before returning it to the interface which requested it (step 10), extracting the response features and passing them to the *Intrusion Detector* (step 11), which will repeat step 4 to detect possible intrusion action for each response feature. The *Intrusion Detector* will then pass the information to the *IRM*, which will repeat steps 5, 6, 7 as steps 13, 14 and 15. Finally, if the *IRM* concludes that the response is accepted or considered an intrusion, the computed results are respectively either sent back to the user interface which requested them or eliminated (step 16).

2.1   Risk Exposure Assessment

Given a user action, *risk exposure* is a function of both the *probability* it has of being an intrusion and the *impact* it may have, *i.e.*, the potential magnitude of the cost for the enterprise related to the damage or disclosure of the sensitive data which the action affects. Risk analysis consists on ranking the alerts given their computed risk exposure, according to a matrix similar to Table 1.

Table 1. The risk exposure matrix

<table>
<tr><td colspan="2" rowspan="2"></td><td colspan="4">Probability</td></tr>
<tr><td>Very Low</td><td>Low</td><td>High</td><td>Very High</td></tr>
<tr><td rowspan="4">Impact</td><td>Very High</td><td>High</td><td>High</td><td>Very High</td><td>Critical</td></tr>
<tr><td>High</td><td>Low</td><td>High</td><td>High</td><td>Very High</td></tr>
<tr><td>Low</td><td>Very Low</td><td>Low</td><td>High</td><td>High</td></tr>
<tr><td>Very Low</td><td>Very Low</td><td>Very Low</td><td>Low</td><td>High</td></tr>
</table>

To define which responses should be taken given the risk exposure matrix, the *DW Security Administrator* may define rules as the following:

```
GIVEN RISK EXPOSURE AS Low|Medium|High|Critical
ON FEATURE {FeatureName1, FeatureName2, ...}, AllFeatures
TAKE ACTION {DoNothing,Alert,PauseCommand,TerminateCommand,KillSession}
```

The definition of probability and impact rules that make up the assessment of risk exposure measures depend on the chosen intrusion detection features and sensitive data assessment by the DW Security Administrator, and will be explained in the next subsections. All risk exposure, probability and impact rules are stored in the *DW-IDE Database* and used by the *Intrusion Response Manager (IRM)*, as explained formerly.

2.2   Intrusion Detection and Response Management

Intrusion Detection. Given the distinctive assumptions for typical web-accessible DWs [4], in Table 2 we define the relevant ID features from a usability perspective. As shown, several features group values per user/IPAddress, other features are referred to values per command given each user/IPAddress, and further features refer those that are grouped by each session of each user/IPAddress. This allows testing features in different grouping levels (per user / per user session / per SQL command) and thus, widens the detection scope. Our approach adjusts a probabilistic distribution for each feature $\{F_1, ..., F_{29}\}$ for each user, from observations (feature values) during an initial training stage. To obtain those observations, we suppose the existence of an "intrusion-free" database command log. Executing that log's user commands we extract the values, *i.e.*, observations for building each feature's statistical distribution. Statistical adjustment tests are performed to obtain each population's distribution.

For each active user session, we gather each new value generated for each feature and build sample sets. To detect an intrusion, statistical tests are performed: given each feature's original population, a new sample set is built joining that population with the user session sample set for that feature. New statistical tests are performed to adjust a new probability distribution to the former data collection. By testing if the new feature's distribution matches its original one (H_o), using Chi-square, Kolmogorov-Smirnov or Shapiro-Wilk tests, all performed at a level of 5% significance, for each test decision for a certain feature that results in rejecting the distribution's equality (H_o), we consider the user action as a probable intrusion.

Table 2. Intrusion detection features

F#	FeatureName	Description
Features per User/IPAddress		
F_1	#ConsFailedLoginAttempts	The number of consecutive failed database login attempts by a UserID or from an IPAddress (accumulated or in a given timespan)
F_2	#SimultSQLSessions	The number of active simultaneous database connections
F_3	#UnauthorAccessAttempts	The number of consecutive user requests to execute an unauthorized actions (*e.g.* request to modify data when the database is read-only, or requesting to query data to which does not have access privileges)
Features per User/IPAddress per Command		
F_4	CPUTime	CPU time spent by the DBMS to process the command
F_5	ResponseSize	Size (in bytes) of the result of the command's execution
F_6, F_7	#ResponseLines, #ResponseColumns	Nr. of lines and columns in the result of the command's execution
F_8, F_9	#ProcessedRows, #ProcessedColumns	Nr. of accessed rows and columns for processing the command
F_{10}	CommandLength	Number of characters
F_{11}	#GroupBy	Number of GROUP BY columns
F_{12}	#Union	Number of UNION clauses
$F_{13}...F_{17}$	#Sum, #Max, #Min, #Avg, #Count	Nr. of SUM, MAX, MIN, AVG and COUNT functions
F_{18}, F_{19}	#And, #Or	Nr. of AND and OR operators in the command's WHERE clause(s)
F_{20}	#LiteralValues	Nr. of literal values in the command's WHERE clause(s)
Features per User/IPAddress per Session		
F_{21}	#GroupBy	Number of GROUPBY columns in all SELECT statements, p/ session
F_{22}	#Union	Number of UNION clauses in all SELECT statements, per session
$F_{23}...F_{27}$	#Sum, #Max, #Min, #Avg, #Count	Nr. of appearances of SUM, MAX, MIN, AVG and COUNT functions in all commands, per session
F_{28}	TimeBetwCommands	Time period (in seconds) between exec. of commands, per session
F_{29}	#SimultaneousCommands	Number of commands simultaneously executing, per session

Defining Risk Probability and Impact. To determine each feature's individual importance in the overall intrusion detection process (which will be directly related to its risk probability), we attribute a weight to it. To compute its weight, we assume *a priori* each feature has the same relevance and will be incrementally self-calibrated using their respective True Positives *TP* (*i.e.* alerts generated by the feature that were confirmed as true intrusions) and False Positives *FP* (*i.e.* confirmed false alarms). For each feature F_i, its weight W_i is given by:

$$W_i = 0.5 + ((TP_i - FP_i) / (TP_i + FP_i)) / 2 \qquad (1)$$

where TP_i and FP_i are the total number of *TP* and *FP*, respectively, of all alerts generated by feature F_i. Every time an intrusion alert is generated by a given feature F_i, after it is checked the feature will have its *TP* or *FP* rate updated if it respectively refers to a true intrusion or a false alarm and, consequently, its weight W_i is also accordingly updated (increased or decreased). Thus, the self-calibrating formula works smoothly, giving a higher importance to the features that are more accurate.

To define the probability of each intrusion alert given the feature that generated it, our approach allows defining rules with the following syntax (list values with | are to be chosen from, while clauses in brackets are optional):

```
DEFINE PROBABILITY AS None|VeryLow|Low|High|VeryHigh
ON FEATURE {FeatureName1, FeatureName2, ...}, AllFeatures
[WHERE {List of filtering conditions}]
[WHEN {List of time-based conditions}]
```

Using this rule syntax, the intrusion probability of each feature F_i given its W_i as:

```
DEFINE PROBABILITY AS VeryLow ON FEATURE Fi WHERE Weight(Fi)<=0.25
DEFINE PROBABILITY AS Low
    ON FEATURE Fi WHERE Weight(Fi)>0.25 AND Weight(Fi)<=0.50
DEFINE PROBABILITY AS High
    ON FEATURE Fi WHERE Weight(Fi)>0.50 AND Weight(Fi)<=0.75
DEFINE PROBABILITY AS VeryHigh ON FEATURE Fi WHERE Weight(Fi)>0.75
```

The assessment of the impact caused by a user action is based on *which*, *how much*, and *when* sensitive data can be exposed or damaged by the user command, as well as *who* is the user. It is managed by using the following rules:

```
DEFINE IMPACT AS VeryLow|Low|High|VeryHigh
ON FEATURE {FeatureName1, FeatureName2, ...}, AllFeatures,
[WITH COLUMNS {Column1,Column2,...},AllColumns]
[WHERE {List of filtering conditions}]
[WHEN {List of time-based conditions}]
[JOINED WITH {Column1,Column2,...},AllColumns
```

The clauses are used in a similar manner to those in the probability rules, plus the clause distinguishing which is the user command (ON COMMAND) and the clause defining the impact of two or more columns being processed or shown together (WITH COLUMNS). The WHERE clauses in the DIDS rules (as in standard SQL WHERE clauses) allow a wide range of definitions and due to lack of space are not included. We just wish to make clear that the *IRM* algorithms can be easily adapted to cope with a wide range of rule possibilities, providing a very wide ID scope.

3   Experimental Evaluation

We used the TPC-H benchmark [18] to build a 1GB DW using Oracle 11g DBMS on a Pentium 2.8GHz machine with 2GB SDRAM (with 512MB dedicated to the database server), in a scenario with ten open web connections to the DW in which there are 2 "intruders" and 8 "true" DW users (non-intruders). For each "true" DW user's workload, a set of randomly chosen TPC-H benchmark queries were selected, *i.e.*, each user has different queries to execute, as well as a distinct number of queries. In each workload's queries, several were randomly picked for randomly modifying their parameters, to obtain a larger scope of diverse user actions. Each workload also included a random number of *random* queries (randomly picking a set of tables, columns, functions to execute, grouping and sorting, and literal restrictions for columns included in the WHERE clauses). The TPC-H queries represent typical reporting behavior, while *ad hoc* queries were simulated by random queries, in smaller number.

To build the statistical models for each feature of each "true" user, we executed each user's workload 50 times. To build each "intruder" workload, we generated 200 random intrusion queries of several types: SQL injection tautologies; Login/password guessing; inserting, changing or deleting a random number of rows; Selecting a random amount of columns and a random amount of functions (MAX, SUM, etc.) from a random number of tables, with and without a random number of grouping columns, with and without range value restrictions; SQL union queries with a random amount of columns and a random amount of tables; Query flooding; Unauthorized actions (create, drop, etc). These intrusion queries represent a wide variety of attacks.

The TPC-H benchmark has approximately seven years of business data. We consider the data from the most recent year to have high impact due to intrusion actions, the data from the two previous years as high impact, the data from the two years before that as low impact and the remaining as having very low impact.

Table 3 shows the ID results. Figure 3 shows the TP rate is considerably high (89%) while the FP rate is relatively low (5%), with an absolute number of 48 false alarms for a total of 225 generated alerts. Observing Table 4, the absolute number of false negatives is relatively low (23 in a total of 995 non-intrusions). The approach's precision is considerable (79%) and its accuracy is high (94%). Observing Table 4, the most relevant alerts (very high and critical) represent approximately one third of all alerts; these should be the ones first deserving attention on behalf of the security staff, instead of wasting time checking the remaining alerts (two thirds of all alerts), given that they present smaller impact. Finally, we measured an average overhead of 20% on user workload response time due to running the DIDS detection algorithms.

Table 3. Experimental results for the generated alerts (absolute values)

# True user actions	# intruder actions	#TP	#FP	#TN	#FN
1020	200	177	48	972	23

Table 4. Number of generated alerts per risk exposure measure

Very Low	Low	High	Very High	Critical	Total Number of Alerts
36	50	59	49	31	225

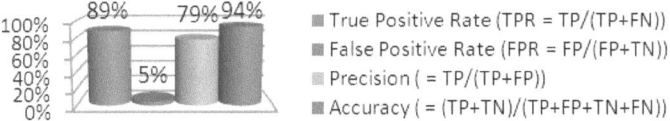

Fig. 3. True positive and false positive rates, precision and accuracy

4   Related Work

In [1], database transactions are defined by directed graphs describing the SQL command types used for malicious data access detection. This approach cannot handle *ad hoc* queries and works at the coarse-grained transaction level as opposed to a fine-grained query level. A Role Based Access Control mechanism for DIDS was proposed by [3]. Data mining techniques are used, namely classification and clustering, against SQL instructions stored in database audit files to deduce role profiles of normal user behavior. A limitation of this approach is that it cannot extract correlation among queries in transactions. Moreover, since this solution is role-based, it works at a higher coarse-grained level than the user-based profiling in our approach. Detecting attacks by comparison, summarizing SQL statements into compact access patterns named as fingerprints, is the focus of [5]. Profiling the data accessed by users to try to determine their intent is an approach used in [6], using statistical learning algorithms. They argue that analyzing what the user is looking for (*i.e.*, what data) instead of analyzing how s/he is looking for it (*i.e.*, which SQL expressions), is more efficient for anomaly detection. In our paper, we integrate both these views. Data correlation using data mining or machine learning techniques is used in [2, 7, 8, 10].

5   Conclusions and Future Work

We have pointed out issues involving ID in web-access DWs and proposed a specific DIDS for these environments. Our DIDS works transparently between user applications and the database server as an extension of the DBMS itself. The SQL-like rule-base allows extending DBMS data access policies and covers an extremely wide range of intrusion attacks. Risk exposure assessment is used for ranking and prioritizing the generated intrusion alerts, presenting clear advantages when compared with correlation techniques. Experimental results show our approach achieves high efficiency and accuracy for the tested setup. As future work, we intend to test our approach in real-world DWs, namely in cloud environments.

References

1. Fonseca, J., Vieira, M., Madeira, H.: Online Detection of Malicious Data Access Using DBMS Auditing. In: ACM Symposium on Applied Computing, SAC (2008)
2. Hu, Y., Panda, B.: A Data Mining Approach for Database Intrusion Detection. In: ACM Symposium on Applied Computing, SAC (2004)
3. Kamra, A., Terzi, E., Bertino, E.: Detecting Anomalous Access Pat-terns in Relational Databases. Springer VLDB Journal 17 (2008)
4. Kimball, R., Ross, M.: The Data Warehouse Toolkit, 2nd edn. Wiley & Sons, Inc. (2002)
5. Lee, S.-Y., Low, W.L., Wong, P.Y.: Learning Fingerprints for a Database Intrusion Detection System. In: Gollmann, D., Karjoth, G., Waidner, M. (eds.) ESORICS 2002. LNCS, vol. 2502, pp. 264–279. Springer, Heidelberg (2002)
6. Mathew, S., Petropoulos, M., Ngo, H.Q., Upadhyaya, S.: A Data-Centric Approach to Insider Attack Detection in Database Systems. In: Jha, S., Sommer, R., Kreibich, C. (eds.) RAID 2010. LNCS, vol. 6307, pp. 382–401. Springer, Heidelberg (2010)
7. Pietraszek, T.: Using Adaptive Alert Classification to Reduce False Positives in Intrusion Detection. In: Jonsson, E., Valdes, A., Almgren, M. (eds.) RAID 2004. LNCS, vol. 3224, pp. 102–124. Springer, Heidelberg (2004)
8. Srivastava, A., Sural, S., Majumdar, A.K.: Database Intrusion De-tection using Weighted Sequence Mining. Journal of Computers I(4) (2006)
9. Treinen, J.J., Thurimella, R.: A Framework for the Application of Association Rule Mining in Large Intrusion Detection Infrastructures. In: Zamboni, D., Kruegel, C. (eds.) RAID 2006. LNCS, vol. 4219, pp. 1–18. Springer, Heidelberg (2006)
10. Valdes, A., Skinner, K.: Probabilistic Alert Correlation. In: Lee, W., Mé, L., Wespi, A. (eds.) RAID 2001. LNCS, vol. 2212, pp. 54–68. Springer, Heidelberg (2001)
11. Transaction Processing Council, TPC Decision Support Benchmark H, http://www.tpc.org/tpch

Majority-Rule-Based Web Service Selection

Karim Benouaret[1], Dimitris Sacharidis[2],
Djamal Benslimane[1], and Allel Hadjali[3]

[1] Claude Bernard Lyon1 University, LIRIS, 69622 Villeurbanne, France
{karim.benouaret,djamal.benslimane}@liris.cnrs.fr
[2] IMIS, Athena Research Center, Marousi 15125, Greece
dsachar@imis.athena-innovation.gr
[3] Enssat, University of Rennes 1, IRISA, 22305 Lannion, France
allel.hadjali@enssat.fr

Abstract. In many Web service selection scenarios, the responsibility to decide which is the appropriate service is shared among multiple parties, e.g., among the department heads of a university. The standard approach is to discard services which are unanimously inappropriate, and return the rest. However, as the involved parties may have conflicting interests, it is possible that only few services are eliminated, and thus almost all discovered services need to be considered. This work addresses this shortcoming, by enforcing the *majority rule*: a service is discarded if the majority of the parties find it inappropriate. We formulate the majority-rule-based service selection problem based on the notions of dominance and skyline. Furthermore, we propose an algorithm that returns a more manageable set of services, eliminating many inappropriate ones, and is more efficient that standard skyline techniques.

1 Introduction

Several techniques for discovering Web services have been recently proposed. As the number of services and service providers proliferate, there is a large number of candidate, most likely competing, services for fulfilling a desired task. Thus, *service selection* is becoming important for helping users to identify desirable services. User preferences play a key role during the selection process [3,14,4]. However, in many practical situations, the responsibility to decide which is the appropriate service is shared among multiple parties, e.g., among the department heads of a university.

The service selection process follows two phases. In the first, given the user's preferences on service description attributes, the degrees of match between a requested and an available service (see e.g., [12,11,8]) are computed. In this work, we assume the Jaccard coefficient for matching service descriptions. If I_1, I_2 are two intervals, their Jaccard coefficient is $J(I_1, I_2) = \frac{|I_1 \cap I_2|}{|I_1 \cup I_2|}$, where $|I|$ measures the length of the interval [9].

The second phase of service selection is to identify the most interesting services w.r.t. users preferences. Most of service selection approaches focus on computing a score for each service as an aggregate of its individual matching degrees.

X.S. Wang et al. (Eds.): WISE 2012, LNCS 7651, pp. 689–695, 2012.

Various approaches for aggregating the matching degrees exist. A common direction is to assign weights over different preference attributes; e.g., [10]. However, when multiple users are involved, it would be difficult to make tradeoffs between different weights. The natural option is to use the skyline operator [5,7,13] to determine an objectively good set of services [15,1,16,2,18,17]. We refer to this set as the *unanimous service skyline*, and it contains all services which are not unanimously dominated. A service *unanimously dominates* another, if the former has higher matching degrees than the latter in all users' preferences.

Computing the unanimous service skyline frees users from assigning relative importance over different preference attributes. However, a major drawback is that, when multiple parties are involved, the number of services in the skyline becomes very large and no longer offers any interesting insights. The reason is that as the number of users and preferences increase, for any services s_i, s_j, it is more likely that s_i and s_j are incomparable, i.e., better than each other in different matching degree. It is thus crucial to further reduce the size of the service skyline.

The core of the above drawback is in the definition of dominance, which requires a unanimous verdict. To mitigate this, we choose to follow the majority rule. Informally, a service *majority-dominates* another, if the former has higher matching degrees than the latter in the *majority* of users' preferences. Then, we naturally define the *majority service skyline* as the services which are not majority-dominated.

To compute the majority service skyline, we make the observation that conventional skyline computation algorithms, with the exception of [6], cannot be adapted, due to the intransitivity of the majority-dominance relationship. Therefore, an extension of the algorithms in [6] can be used to compute the majority service skyline. However, we propose a novel algorithm for the service selection problem and show that it most cases it outperforms the extended algorithms.

The rest of the paper is structured as follows. Section 2 introduces the problem of majority service skyline and describes the majority service skyline computation algorithm. Section 3 presents our experimental study and Section 4 concludes the paper.

2 Computing the Majority Service Skyline

Section 2.1 introduces the problem, while Section 2.2 describes our algorithm.

2.1 Problem Definition

We assume a set of users $\mathcal{U} = \{u_1, u_2, \ldots, u_m\}$, and a set of discovered services $\mathcal{S} = \{s_1, s_2, \ldots, s_n\}$. We use $s_i.u_k$ to denote the matching degrees of service s_i w.r.t. user u_k. Given a user u_k, we say that service s_i *weakly dominates* s_j w.r.t. u_k, denoted as $s_i.u_k \succeq s_j.u_k$, iff s_i has better matching degrees than s_j on all specified preference attributes. A service s_i *dominates* s_j w.r.t. u_k, denoted as $s_i.u_k \succ s_j.u_k$, iff s_i has better matching degrees than s_j on all specified preference attributes, and strictly better matching degree on at least one.

Given a set of users $\mathcal{U}$, we say that service s_i *unanimous-dominates* s_j, denoted as $s_i \succ_U s_j$, iff s_i weakly dominates s_j w.r.t. all users, i.e., $\forall u_k \in \mathcal{U}\ s_i.u_k \succeq s_j.u_k$, and there exists one user, say u'_k, for which s_i dominates s_j, i.e., $\exists u'_k \in \mathcal{U}\ s_i.u'_k \succ s_j.u'_k$. Given a set of discovered services $\mathcal{S}$ and a set of users $\mathcal{U}$, the *unanimous service skyline* $USS(\mathcal{S},\mathcal{U})$ comprises the set of services that are not dominated by any other.

Given a set of users $\mathcal{U}$, we say that service s_i *majority-dominates* s_j, denoted as $s_i \succ_M s_j$, iff (1) there exists a subset $\mathcal{U}' \subseteq \mathcal{U}$ containing more than half of the users such that s_i weakly dominates s_j w.r.t. all users in this subset, i.e., $|\mathcal{U}'| > \lfloor |\mathcal{U}|/2 \rfloor$ and $\forall u_k \in \mathcal{U}'\ s_i.u_k \succeq s_j.u_k$, and (2) there exists one user, say u'_k, for which s_i dominates s_j, i.e., $\exists u'_k \in \mathcal{U}\ s_i.u'_k \succ s_j.u'_k$. Given a set of discovered services $\mathcal{S}$ and a set of users $\mathcal{U}$, the *majority service skyline* $MSS(\mathcal{S},\mathcal{U})$ comprises the set of services that are not majority- dominated by any other.

Problem Statement: Given a set of users $\mathcal{U}$ and a set of discovered services $\mathcal{S}$, compute the *majority service skyline*.

2.2   Majority Service Skyline Algorithm

In this section, we introduce the *Majority Service Skyline Algorithm* (MSA), which is based on the following properties. Note that the proofs of all lemmas and theorems appear in the full version of this paper[1].

Theorem 1. *It is possible to have a set of users $\mathcal{U}$ and a set of discovered services $\mathcal{S} = \{s_1, s_2, \ldots, s_n\}$ such that s_1 majority-dominates s_2, s_2 majority-dominates s_3, ..., s_{n-1} majority-dominates s_n and s_n majority-dominates s_1, i.e., forming a cyclic majority dominance relationship.*

The previous theorem shows that the majority dominance relationship shares the cyclic property of the k-dominance relationship introduced in [6]. Therefore, a service cannot be discarded even if it is majority-dominated because it might be needed for excluding other services. This justifies why the existing algorithms for computing the skyline are not applicable for computing the majority service skyline. However, the one scan algorithm (OSA) and two scan algorithm (TSA) of [6], can be adapted to compute the majority service skyline, by exchanging k-dominance checks for majority dominance checks. In the following, we denote as OSA and TSA the adaptations of the algorithms in [6] to computing the majority service skyline.

The MSA algorithm also takes advantage of the following observations.

Lemma 1. *If s_i unanimous-dominates s_j, then s_i majority-dominates s_j. i.e., $s_i \succ_U s_j \Rightarrow s_i \succ_M s_j$.*

Lemma 2. *If s_i unanimous-dominates s_j and s_j majority-dominates s_k, then s_i majority-dominates s_k. i.e., $s_i \succ_U s_j \wedge s_j \succ_M s_k \Rightarrow s_i \succ_M s_k$.*

---

[1] http://liris.cnrs.fr/Documents/Liris-5691.pdf

Lemma 3. *Let $f : \mathcal{S} \to \mathbb{R}^+$ be a monotone function aggregating the matching degrees of s_i for all users. If s_i unanimous-dominates s_j, then $f(s_i) > f(s_j)$. i.e., $s_i \succ s_j \Rightarrow f(s_i) > f(s_j)$.*

From Lemma 1 and Lemma 2, we can see that it is sufficient to compare each service against the unanimous skyline services to detect if it is part (or not) of the majority service skyline. This essentially reduces the number of comparisons. Specifically, *if a service s_i is unanimous-dominated, then discard it as (1) it is not part of the majority service skyline (Lemma 1), and (2) it is unnecessary for eliminating other services (Lemma 2).*

Lemma 3 also helps reduce unnecessary comparisons. In fact, to exploit this property, we sort the services in non-ascending order of the sum of their matching degrees. Then, given a service s_i, searching for services by which s_i is unanimous-dominated can be limited to the part of the service before s_i. This is the idea behind the SFS algorithm [7], which in this context we apply it for cyclic dominance relationships.

The MSA algorithm leverages the observations made above to compute efficiently the majority service skyline. Based on Lemma 1 and Lemma 2, MSA maintains two sets $\mathcal{R}$ and $\mathcal{T}$, containing respectively the set of intermediate majority skyline services and the set of intermediate unanimous skyline services that are not in $\mathcal{R}$. Thus, $\mathcal{R} \cup \mathcal{T}$ constitutes the intermediate unanimous skyline.

The MSA algorithm operates as follows. First, services in $\mathcal{S}$ are sorted in a non-ascending order of the sum of their matching degrees, and both sets $\mathcal{R}$ and $\mathcal{T}$ are initialized to empty sets. Then, the top service (i.e., the service with the maximum sum of matching degrees), say s_i, is extracted from $\mathcal{S}$. Service s_i is compared against services in $\mathcal{R} \cup \mathcal{T}$, i.e., the set of services that may unanimous-dominate s_i (as the other services cannot dominate s_i from Lemma 3). If s_i is unanimous-dominated, then it is removed from $\mathcal{S}$ as it is not part of the majority service skyline (Lemma 1) and it is unnecessary for eliminating other services (Lemma 2). Otherwise, i.e., when s_i is not unanimous- dominated by any service in $\mathcal{R} \cup \mathcal{T}$, if s_i majority-dominates any service s_j in $\mathcal{R}$ (i.e., s_j is not a service in MSS), then s_j is removed from $\mathcal{R}$ to $\mathcal{T}$, as it is a unanimous skyline service, thus useful for eliminating other services. For the same reason, if s_i is *majority-dominated* by any service in $\mathcal{R} \cup \mathcal{T}$, it is inserted into $\mathcal{T}$ as it is not part of the majority service skyline. Else, s_i is an intermediate MSS service and is thus inserted into $\mathcal{R}$. Once all services in $\mathcal{S}$ have been examined, i.e., $\mathcal{S}$ is empty, services in $\mathcal{R}$ form the majority service skyline, and $\mathcal{R}$ is returned.

3 Experimental Evaluation

In this section, we present an experimental evaluation of our approach. Our objective is to prove the *effectiveness* of the majority service skyline and the *efficiency* of the proposed algorithm. Specifically, we focus on two issues. (1) The size of the majority service skyline (denoted as MSS). To demonstrate that the majority service skyline further reduces the size of the (traditional) skyline, we also compute the size of the unanimous service skyline (denoted as USS) (2) The performance of our algorithm in terms of elapsed time for computing the majority service skyline. For comparison purposes, we also implemented the adaptations of OSA and TSA [6] for computing the majority service skyline.

Table 1. Parameters and Examined Values

Parameter	Symbol	Range	Default
Number of discovered services	n	$[2, 10]$K	5K
Number of users	m	$[3, 7]$	5
Number of preferences per user	d	$[3, 7]$	5

Due to the limited availability of real-world service data, we implemented a service generator that takes as input a (real-world) model service and its associated constraints, representing the requested service and the multiple users preferences, and produce a set of synthetic services, as well as their associated constraints, representing the set of discovered services. The Jaccard coefficient is used for computing the matching degrees between discovered service' constraints and users preferences. The generation of the sets of synthetic services is controlled by the parameters in Table 1, which displays the parameters under investigation, their corresponding ranges and their default values. In each experimental setup, we investigate the effect of one parameter, while setting the remaining ones to its default value.

The service generator and the algorithms, i.e., MSA, OSA and TSA were implemented in Java, and all experiments were conducted on a 2.3 GHz Intel Core i5 processor.

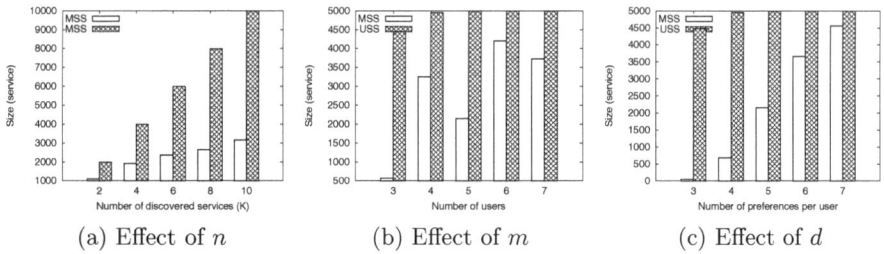

(a) Effect of n (b) Effect of m (c) Effect of d

Fig. 1. Result cardinality

Figure 1 shows the cardinality of MSS and USS w.r.t. n, m and d. Constantly, the size of MSS is less than USS, which is almost equal to the number of discovered services, as the unanimous service skyline cannot discard all inappropriate services, while the majority service skyline includes only the most interesting ones. As shown in Figure 1a, the size of the majority service skyline increases slightly with n. This is because as n varies, it is becoming more difficult to find services which are majority-dominated. Figure 1b shows a fluctuation in the size of the majority service skyline. The fluctuation is related to the definition of the majority dominance relationship. Indeed, we can distinguish two trends. One for the even values of m, and the second for the odd values of m; each trend increases as m increases. This is because, if we have an odd value of m, say m_o, and an even value of m, say m_e, such that $m_o = m_e + 1$, then the percentage of most of users for m_e is greater than that of m_o. For example, for $m = 4$, the percentage is $\frac{3}{4} = 0.75\%$, and for $m = 5$ the percentage is $\frac{3}{5} = 0.60\%$. When this percentage is large, a small number of services is discarded, and vice versa. As depicted in Figure 1c, the size of the majority service skyline increases significantly with d. As d increases, a service has greater probability to not be dominated in all preference attributes w.r.t. a given user.

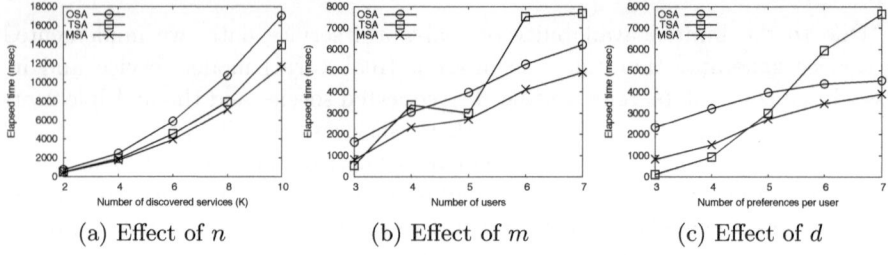

(a) Effect of n (b) Effect of m (c) Effect of d

Fig. 2. Elapsed time

Figure 2 investigates the runtime of OSA, TSA and MSA w.r.t. n, m and d. Overall, MSA outperforms OSA and TSA. Figure 2a shows that the execution time of the algorithms increases with n. However, MSA consistently outperforms OSA and TSA. As shown in Figure 2b, when m increases, the performance of TSA deteriorates due to the second scan performed. However, the execution time of OSA and MSA increases slightly with m. Still, MSA is better. As shown in Figure 2c, TSA is better than OSA and MSA for $d \leq 4$ since the size of the majority service skyline is small, thus a large number of majority-dominated services can be eliminated in the first scan. However, TSA does not scale with d as the size of the majority service skyline becomes large, thus the second scan is very time consuming. The execution time of OSA and MSA, on the other hand, increases slightly with d. Finally, observe that MSA consistently performs better than OSA.

4   Conclusion

We introduce a novel concept for the preference-based Web service selection under multiple users preferences problem based on the majority rule. This allows users to make a "democratic" decision on which services are the most appropriate. We develop a suitable algorithm for the majority-rule-based Web selection problem. Our experimental evaluation demonstrates the effectiveness of the concept and the efficiency of the algorithm.

References

1. Alrifai, M., Skoutas, D., Risse, T.: Selecting skyline services for qos-based web service composition. In: WWW, pp. 11–20 (2010)
2. Benouaret, K., Benslimane, D., Hadjali, A.: On the use of fuzzy dominance for computing service skyline based on qos. In: ICWS, pp. 540–547 (2011)
3. Benouaret, K., Benslimane, D., Hadjali, A., Barhamgi, M.: Fudocs: A web service composition system based on fuzzy dominance for preference query answering. PVLDB 4(12), 1430–1433 (2011)
4. Benouaret, K., Benslimane, D., Hadjali, A., Barhamgi, M.: Top-k web service compositions using fuzzy dominance relationship. In: IEEE SCC, pp. 144–151 (2011)
5. Börzsönyi, S., Kossmann, D., Stocker, K.: The skyline operator. In: ICDE, pp. 421–430 (2001)
6. Chan, C.Y., Jagadish, H.V., Tan, K.L., Tung, A.K.H., Zhang, Z.: Finding k-dominant skylines in high dimensional space. In: SIGMOD Conference, pp. 503–514 (2006)
7. Chomicki, J., Godfrey, P., Gryz, J., Liang, D.: Skyline with presorting. In: ICDE, pp. 717–719 (2003)
8. Dong, X., Halevy, A.Y., Madhavan, J., Nemes, E., Zhang, J.: Simlarity search for web services. In: VLDB, pp. 372–383 (2004)
9. Duda, R.O., Hard, P.E.: Pattern Classifcation and Scene Analysis. A Wiley-Interscience Publication, New York (1973)
10. Lamparter, S., Ankolekar, A., Studer, R., Grimm, S.: Preference-based selection of highly configurable web services. In: WWW, pp. 1013–1022 (2007)
11. Li, L., Horrocks, I.: A software framework for matchmaking based on semantic web technology. In: WWW, pp. 331–339 (2003)
12. Paolucci, M., Kawamura, T., Payne, T.R., Sycara, K.: Semantic Matching of Web Services Capabilities. In: Horrocks, I., Hendler, J. (eds.) ISWC 2002. LNCS, vol. 2342, pp. 333–347. Springer, Heidelberg (2002)
13. Papadias, D., Tao, Y., Fu, G., Seeger, B.: An optimal and progressive algorithm for skyline queries. In: SIGMOD Conference, pp. 467–478 (2003)
14. Wang, H., Xu, J., Li, P.: Incomplete preference-driven web service selection. In: IEEE SCC (1), pp. 75–82 (2008)
15. Yu, Q., Bouguettaya, A.: Computing service skyline from uncertain qows. IEEE T. Services Computing 3(1), 16–29 (2010)
16. Yu, Q., Bouguettaya, A.: Computing service skylines over sets of services. In: ICWS, pp. 481–488 (2010)
17. Yu, Q., Bouguettaya, A.: Efficient service skyline computation for composite service selection. IEEE Transactions on Knowledge and Data Engineering 99(preprints) (2011)
18. Yu, Q., Bouguettaya, A.: Multi-attribute optimization in service selection. World Wide Web 15(1), 1–31 (2012)

Memory-Efficient Index
for Cache Invalidation Mechanism with OpenJPA

Miki Enoki, Yosuke Ozawa, Hiroshi Horii, and Tamiya Onodera

IBM Research, Tokyo, Japan
{enomiki,ozawaysk,hhorii,tonodera}@jp.ibm.com

Abstract. OpenJPA is an implementation of the Java Persistence API (JPA) for Apache, with a caching layer for database queries. However the caching performance is poor when an application includes write transactions, because the OpenJPA cache-invalidation mechanism is coarse-grained and this results in a low cache hit rate. In this research, we implemented an index mechanism for cache invalidation optimized for the data access patterns. The sizes of the index can be adjusted for the available cache memory. The results of our benchmark indicated that the optimized index drastically improved the performance of OpenJPA even with a small index size in various data access pattern scenarios.

Keywords: EJB3.0, JPA, OpenJPA, Index, Cache invalidation.

1   Introduction

The Java Persistence API (JPA) that was defined as a part of the Enterprise JavaBeans (EJB) 3.0 specification standardizes the Object-Relational (O/R) mapping to enhance the capabilities of EJB. There are many JPA implementations with best known being Apache OpenJPA[1], Hibernate[2] and TopLink Essentials[3]. Each of them has a caching layer for database queries in the application server and this is a critical component for high performance, since the caching layer can reduce the database accesses [2,3,4].

When an application includes write transactions, we have to maintain consistency between the cached data and the database. In Apache OpenJPA v2.0, there is no maintenance procedure to identify which data is modified by a write transaction with Java Persistence Query Language (JPQL), so all of the cache entries that are related to the modified table are removed. Such coarse-grained invalidation causes a low cache hit rate and sacrifices much of the benefit of a cache [6].

Database index is usually created using column of a database table for the read transactions, but this mechanism can also be used to maintain the consistency of the cache entries affected by the write transactions. However, having many indexes may reduce the memory available for the data cache itself, since memory is limited and

---

[1] http://openjpa.apache.org/
[2] http://www.hibernate.org/
[3] http://glassfish.java.net/javaee5/persistence/

X.S. Wang et al. (Eds.): WISE 2012, LNCS 7651, pp. 696–703, 2012.
© Springer-Verlag Berlin Heidelberg 2012

much smaller than disk storage [7,11]. If we simply reduce the index size, then more cache entries are left out of the index. In this paper, we describe a size-adjustable index mechanism. Here are the contributions of our work:

- Implementation of a size-adjustable hash index optimized for the data access pattern for fine-grained cache invalidation in OpenJPA.
- Formulation of constructing an optimal size-adjustable index to multi-way number partitioning problem. The optimal index is constructed to use larger space for more important region depending on each data access pattern.
- Empirical results showing that the optimized hash index drastically improved the performance of OpenJPA even with a small index size in various data access pattern scenarios.

2   Apache OpenJPA and Cache

2.1   DataCache

OpenJPA is an implementation of JPA by the Apache Software Foundation. A table in a database corresponds to a Java class called an entity in OpenJPA. The client program processes select, update, insert, and delete operations through entity objects. A row of data in a table can also correspond to an instance of an entity. OpenJPA provides the Java Persistence Query Language (JPQL) which is an SQL-like language to retrieve and manipulate data.The DataCache consists of a key-value store. It stores entities loaded from the database. A key for the DataCache is an entity ID (primary key of the table) and the value is the entity instance.

2.2   Cache Invalidation Mechanism by JPQL

If an entity is updated by specifying a value in a column, then OpenJPA has to get all of the cache entries related to that table and iterates to find the corresponding cache entries, because OpenJPA stores each row of data as an entity object in the value of the DataCache and can obtain entries only with the primary key. However, iterating through the cache is expensive, so the current OpenJPA simply removes all of the cache entries in the DataCache related to the updated table [6].

3   Index for Cache Invalidation

3.1   Index for Cache Invalidation

To improve the OpenJPA cache invalidation mechanism, we devised a cache invalidation index for the OpenJPA caching layer. An index consists of a key/value store, where the key is the parameter of the WHERE clause of an update query and the value is a list of primary keys (IDs). Figure 1 (a) shows how to maintain the cache using an invalidation index.

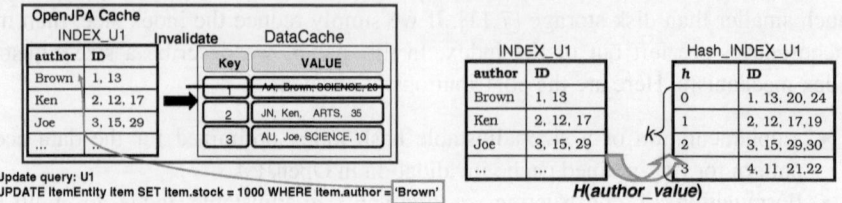

Fig. 1. (a)DataCache maintenance using index, (b) Transformation to hash index

An invalidation index is created for each update query pattern. However, the invalidation indexes may consume the space needed for the DataCache when the available memory is limited, and shrinking the index would leave some cache entries out of the index and eventually lead back to the original coarse-grained approach.

3.2   Size-Adjusted Cache Invalidation Index

3.2.1   Hash Index for Cache Invalidation

The hash index controls the size of the invalidation index. Unlike the index used for read transactions, the invalidation index for write transactions does not need to precisely specify the primary keys to be invalidated. The invalidation index may use a larger number of keys as long as the list includes the primary keys that must be invalidated to maintain database consistency. This allows us to create a small index with a predefined key size k. The space of the small index divided into k-regions. Each region stores the corresponding primary keys for the DataCache.

First, we get the parameters of the WHERE clause and combine and convert them to an integer value (z in this example) for the update query. Each input becomes an integer z in the range 0 to $n-1$, and the output must be an integer h in the range 0 to $k-1$, where n is larger than k. The hash function H could be $H = z \bmod k$. Figure 1(b) is an example of the transformation to a hash index from an invalidation index. For the Hash_INDEX_U1, if OpenJPA returns a "0" after applying 'Brown' to the hash function H, this refers to the ID list whose key is "0" in Hash_INDEX_U1 and the system then invalidate the cache entries with IDs 1, 13, 20, and 24 from the DataCache and also removes those IDs.

As the criteria for data partitioning in an RDBMS, not only a hash, but also list and range partitioning are well known [15]. These methods also can be applied for the invalidation index. In this research, we use hash partitioning to lookup data quickly with the hash function.

3.2.2   Optimized Hash Index for Data Access Pattern

The IDs in the hash index are evenly distributed because the hash function maps the IDs to the hash values as evenly as possible. However realistic Web applications are unlikely to access all of the data equally often. For instance, for Internet shopping, popular items are frequently viewed and purchased. The number of accesses for such items should be relatively high.

Therefore, we optimize the hash index by monitoring the data access pattern. The optimal condition of the optimal hash index is to assign more space to the frequently read region, with a smaller space for the frequently written region.

Step 1. Calculation of weighting value

To identify data access patterns for each region, we introduce a weighting value representing the utilization of the index. The weighting value of each region i is calculated from the frequencies of the read and write transactions:

$$Weight_i = F(r_i, w_i),\tag{1}$$

where r_i is the number of read transactions and w_i is the number of write transactions in region i. For a simple example, the metric $F(r_i, w_i)$ is calculated by r_i / w_i. The weighting value increases as the number of read transactions increases and decreases as the number of write transactions increases. However our approach is not specific to the use of a particular metric for the weighting value, a more sophisticated or complicated metrics can be used as long as the metric is monotonic.

Step 2. Construction of an optimal size adjusted index

We reconstruct the hash index to be optimized with the weighting values and the subdivided hash index. If the weighting values are evenly distributed, it indicates that the index space is effectively.

We expand the size of the hash index to $2k$ with the weighting values. Figure 2 shows this example. We create an expanded hash index for $H = z$ mod $2k$ instead of k and copy each *weight* value to *weight'*.

Based on the sequence of weighting values $\{weight'_0, weight'_1, \ldots, weight'_{2k-1}\}$, we distribute $\{h'_0, h'_1, h'_2, .., h'_{2k-1}\}$ into k disjoint regions $\{h_0, h_1, h_2, .., h_k\}$ to minimize this condition:

$$\max_i\{\sum_{h'_j \in h_i} weight'_j\} - \min_i\{\sum_{h'_j \in h_i} weight'_j\}\tag{2}$$

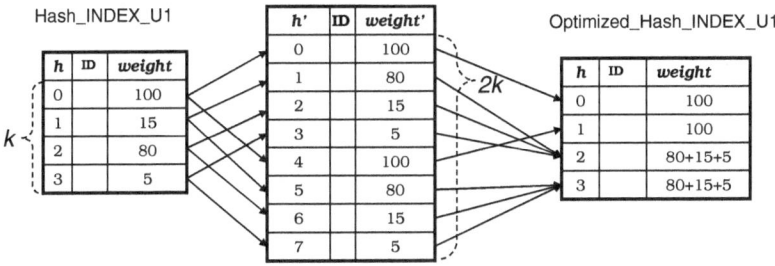

Fig. 2. Transformation to optimized hash index from hash index

We regard this problem as the 2.p 1.itioning n EJto use.tion.multi-way partitioning problem [12] which is known to be NP-complete. The task is to partition a set of numbers into k mutually exclusive and collectively exhaustive subsets, so that the difference between the largest and smallest subset sums is minimized. A generalization of the greedy heuristic to k-way partitioning is described in [13,14]. We sort the *weight'* values in decreasing order, and always place the next unassigned *weight'* value in the subset with the smallest sum so far, until all of the *weight'* values have been assigned. The right-side index of Figure 2 is the result of this greedy heuristic algorithm.

4   Performance Evaluation

4.1   Evaluation Methodology

We measured the throughput and cache hit rate for the DataCache with queries to the ITEM table defined in the TPC-W benchmark [9]. We used 100,000 items. There were 20% writes and 80% read transactions, which is the same as the shopping mix scenario of TPC-W. Each write transaction has an update query modified by an author's id (a_id) column while each read transaction is a select query using the primary key (i_id). An invalidation index is used for the write transactions.

Here are our benchmark scenarios. In addition to the random data access scenario that typical benchmarks usually employs, we define uneven data access scenarios to evaluate the advantage of the optimized hash index. The patterns are based on Pareto principle (also known as the 80-20 rule) [10] to align with realistic Web application scenarios.

Scenario 1: Uniformly access

Both write and read transactions access the items uniformly.

Scenario 2: Read transaction concentration

A total of 80% of the read transaction accesses are concentrated on 20% of the items.

Scenario 3: Read and Write transaction concentration at the same region

Here, 80% of the read and write transactions are concentrated on 20 % of items.

Scenario 4: Read and Write transaction concentration at overlapped region

In this scenario, 20% of the items are frequently read and written, and another 10% of the items are frequently read. This represents best selling or popular items being read frequently and purchased while there are also some new items or special sale items that are frequently looked at, but not purchased as often.

Our test environment has two servers, a 64-bit dual-core AMD Opteron with 4 GB of RAM for the client server and a 64-bit dual-core AMD Opteron with 8 GB of RAM for the database server. OS is Linux 4.2 We installed DB2 UDB v9.1 as the database server and used OpenJPA v2.0. OpenJPA.

4.2   Performance Results

First, we compared the performance with and without the index for cache invalidation. Figure 3(a) shows the throughput and cache hit rate for the DataCache in

Scenario 1. "No Index" means without any index for cache invalidation, which is the original OpenJPA. "Hash Index (25000)" means using a hash index with effectively no size limitation, with 25,000 index keys. This configuration behaves exactly the same as the normal invalidation index and the performance indicates the best one among index configurations. "Hash Index (12500)" is half the size of " Hash Index (25000)".

In Figure 3 (a), both the throughput and cache hit rate with the index were superior to the original OpenJPA. The cache hit rate for "No Index" was 0.1% because each write transaction invalidated all of the cache entries for the ITEM table in the DataCache. In contrast, the cache hit rate for hash index (25000) was 78%. Even the half-size hash index had a cache hit rate of 65% and about 88% of the throughput of the full-size index. Since the data access pattern of Scenario 1 was uniformly data access, the hash index was not recreated to optimize the hash index.

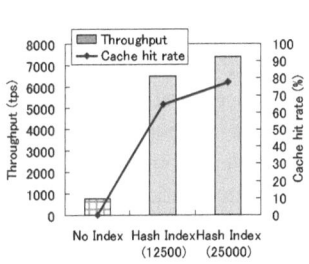 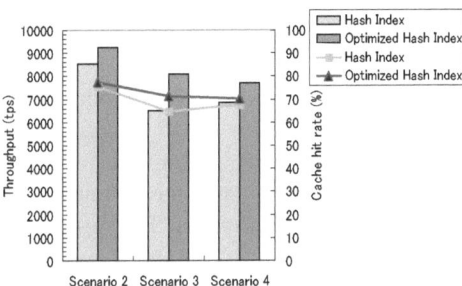

Fig. 3. (a) Performance of Scenario 1, (b) Performances of Scenarios 2, 3, and 4

The throughputs and cache hit rates for Scenarios 2, 3 and 4 are shown in Figure 3 (b). The index size was set to 12,500 for these tests. The performance of the optimized hash index was better than the naïve hash index in every scenario. This indicates that the optimized hash index is efficient when the application data is accessed unevenly.

In the next experiment, we evaluate the performance with various index sizes. Figure 4 shows the results for Scenario 4. The overall throughputs of the optimized hash index were better than the naïve hash index. The throughput of the unlimited hash index(25000) was 8,545 tps, which means the 2500-key optimized hash index retained about 60% of the throughput.

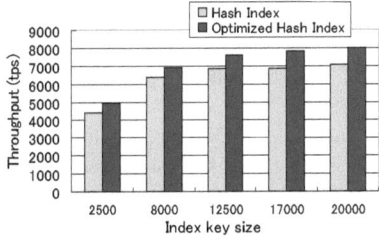

Fig. 4. Index size and throughput for Scenario 4

5   Related Work

There has been extensive study of database query caching, leading to such caches as DBCache [3], DBProxy [2],MTCache [4]. These cache techniques differ from the OpenJPA approach, which stores all of the query results in a key-value store. Examples of related research for cache maintenance include Gupta et al. [5] and Kenneth et al. [8], who used materialized views by incrementally updating the differential data. Our work is different from their techniques, since we used an index for the specific cache entries that are invalidated. Stonebraker [11] suggested a partial index using a predicate-defined subset of a table. Seshadri et al. [7] proposed a heuristic-based method to build partial indexes. Their indexes improve the performance of select queries for various subsets of the results with a complete index.

6   Conclusions and Future Work

In this paper, we implemented an index for cache invalidation in the caching layer of OpenJPA. For effective cache memory usage, we introduced a size-adjustable index mechanism. It is also optimized by learning the data access patterns of each application. We formulated the optimal size-adjustable index problem as a multi-way number partitioning problem. The results of our micro-benchmarks indicated that the OpenJPA with the index for cache invalidation drastically improved the performance and the optimized index is even better than the naïve index when the data accesses are aligned with realistic Web application scenarios.

Various problems remain as future work. An obvious problem is the need for dynamic and automatic recreation of optimized indexes for cases when data access patterns change over time.

References

1. EJB 3.0 Expert Group, JSR 220: Enterprise JavaBeans Version 3.0 Java Persistence API, Sun Microsystems, Santa Clara, CA (2006)
2. Amiri, K., Park, S., Tewari, R., Padmanabhan, S.: DBProxy: A dynamic data cache for Web applications. In: ICDE, pp. 821–831 (2003)
3. Luo, Q., Krishnamurthy, S., Mohan, C., Pirahesh, H., Woo, H., Lindsay, B.G., Naughton, J.F.: Middle-tier database caching for e-business. In: SIGMOD, pp. 600–611 (2002)
4. Larson, P.A., Goldstein, J., Zhou, J.: MTCache: Transparent mid-tier database caching in SQL server. In: ICDE, pp. 177–188 (2004)
5. Gupta, A., Mumick, I.S., Subrahmanian, V.S.: Maintaining views incrementally. In: SIGMOD, pp. 157–166 (1993)
6. Enoki, M., Ozawa, Y., Onodera, T.: Performance Improvement of OpenJPA by Query Dependency Analysis. In: Kitagawa, H., Ishikawa, Y., Li, Q., Watanabe, C. (eds.) DASFAA 2010. LNCS, vol. 5982, pp. 370–379. Springer, Heidelberg (2010)
7. Seshadri, P., Swami, A.: Generalized Partial Indexes. In: ICDE, pp. 420–427 (1995)
8. Ross, K.A., Srivastava, D., Sudarshan, S.: Materialized View Maintenance and Integrity Constraint Checking: Trading Space for Time. In: SIGMOD, pp. 447–458 (1996)

9. Transaction Processing Council. TPC-W specification, `http://www.tpc.org.tpcw`

10. Narula, A.: Ubs Publishers Distributors Ltd, "80/20 Rule of Communicating Your Ideas Effectively" (2005)

11. Stonebraker, M.: The case for partial indexes. In: Proceedings of the 34th Very Large Databases, VLDB 1989 (1998)

12. Garey, M.R., Johnson, D.S.: Computers and Intractability: A Guide to the Theory of NP-Completeness. W.H. Freeman, New York (1979)

13. Karmarkar, N., Karp, R.: The differencing method of set partitioning. Technical Report UCB/CSD 82/113, University of California, Berkeley, CA (1982)

14. Korf, R.: A complete anytime algorithm for number partitioning. Artificial Intelligence 106(2), 181–203 (1998)

15. Partition (database), `http://en.wikipedia.org/wiki/Partition_%28database%29`

Towards Proactive
Cross-Layer Service Adaptation*

Chrysostomos Zeginis, Konstantina Konsolaki,
Kyriakos Kritikos, and Dimitris Plexousakis

Information Systems Laboratory, ICS-FORTH, Heraklion, Greece
{zegchris,konsolak,kritikos,dp}@ics.forth.gr

Abstract. Service-Based Applications (SBAs) enable the automation of business processes. Therefore it is crucial to monitor their non-functional properties and take adaptation actions when QoS violations occur, across all functional layers. In this paper we propose a framework for the proactive cross-layer adaptation of SBAs. We exploit a cross-layer monitoring mechanism to detect a wide range of events, based on which we can both reactively and proactively adapt the system. In particular, the detection of event patterns help us to prevent future faults and failures from happening, by firing specific, dynamically derived rules, that map event patterns to suitable adaptation strategies. Our framework is validated using a traffic management scenario.

Keywords: monitoring, proactive adaptation, cross-layer, pattern, rules.

1 Introduction

Web Services evolution provides testimony of a move towards combining existing and new applications in order to provide more complex SBAs. An SBA is constituted by three main layers [4]: The Business Process and Management (BPM) layer provides the business process along with the activities, the involved roles and the Key Performance Indicators (KPIs), that define and measure progress toward organizational goals. The Service Composition and Coordination (SCC) layer refines this process by combining suitable Web services accompanied by the corresponding SLAs and the Service Infrastructure (SI) layer provides the underlying infrastructure. While Web services evolve in a very dynamic and vulnerable environment, they need to be monitored to detect violations of their functional and non-functional properties. Moreover, whenever a violation is detected, it is crucial to adapt the SBA accordingly. Web service monitoring and adaptation should take place across all the SBA layer, since these layers are closely related with many dependencies among them.

In this paper, we analyze our approach towards proactive cross-layer service adaptation. Firstly, Section 2 summarizes the related work. Section 3 presents

---

* This work constitutes preparatory foreground work on the FP7 IP PaaSage (Model-based Cloud Platform Upperware).

X.S. Wang et al. (Eds.): WISE 2012, LNCS 7651, pp. 704–711, 2012.

an updated version of ECMAF [10] which enables the efficient derivation and management of rules that map event patterns to suitable adaptation strategies, i.e. workflows of adaptation actions. Then, we exemplify the framework's functionality. The proposed monitoring engine captures monitored events from all the SBA layers, while the adaptation mechanism exploits the monitored events to derive new event patterns and the corresponding rules, in the way described in Section 4. Section 5 validates the framework using a specific case study. Finally, some concluding remarks and future work directions are presented in Section 6.

2 Related Work

During the past years some approaches towards cross-layer monitoring and adaptation have been proposed. CLAM [11] is a cross-layer adaptation manager, which identifies the application capabilities affected by the adaptation actions and an adaptation strategy that solves the adaptation problem. In [9] the authors propose a methodology for the dynamic and flexible adaptation of multi-layer applications using adaptation templates and taxonomies of adaptation mismatches. Guinea et al. [3] present an integrated approach for monitoring and adapting multi-layered SBAs, based on a variant of MAPE control loops [5]. Gjørven et al. [2] propose an approach towards cross-layer self-adaptation, which exploits mechanisms across two layers, Service Interface and Application, related to SCC and BPM layers respectively. In [8] the authors present an AOP-based approach towards runtime adaptation of service compositions for preventing SLA violations. Finally, in [6], the authors present the design and implementation of an experimental facility for cross-layer adaptation. The aforementioned approaches are compared in Table 1 according to a set of criteria, that are explained in [10].

Table 1. Comparison of Cross-layer Monitoring and Adaptation approaches

Approaches	Cross-layer	Dynamicity		Intrusiveness		Timeliness		Type of properties	
		Mon.	Adapt.	Mon.	Adapt.	Mon.	Adapt.	Kind	Scope
Zengin et al. [11]	✓	–	✓	–	~	~	R	NF-F	I-C
Popescu et al. [9]	✓	–	✓	–	✓	Im	R	F	I
Guinea et al. [3]	✓	✓	✓	✓	~	Im	R	NF-F	I-C
Gjørven et al. [2]	✓	~	✓	~	–	~	R	F	I
Leitner et al. [8]	✓	✓	✓	✓	✓	Im	R-P	NF	I
Jiang et al. [6]	✓	–	✓	–	~	Im	R	NF	I-C
ECMAF	✓	✓	✓	–	–	Im	R-P	NF-F	I-C

✓: Satisfaction, –: Unsatisfaction, ~: Uncertainty, Im: Immediate, R: Reactive, P: Proactive, F: Functional, NF: Non-Functional, I: Instance, C: Class

From the comparison of the related cross-layer approaches we conclude that none of them satisfies all the criteria. As far as dynamicity is concerned, most of these approaches perform dynamic adaptation unlike dynamic monitoring. In addition, most approaches are non-intrusive regarding monitoring and adaptation, while all of them are reactive rather than proactive. Finally, as far as the type of properties is concerned, they mostly focus on non-functional properties, while the scope of most of them is instance-based.

The main strengths of our proposed framework are the ability to handle both functional and non-functional properties, its proactive adaptation capabilities, by exploiting detection of warning event patterns and the efficient rule management, and its extensibility, as it can integrate new monitoring and adaptation techniques with the existing ones, while preserving its functionality and integrity.

3   The ECMAF Framework

The proposed framework (ECMAF) (Fig. 1), presented in [10], has been slightly changed and partially implemented with the purpose of providing an event-based approach towards cross-layer service monitoring and adaptation. Its main architectural differences with the previous one are the detailed view of the Monitor Manager, the addition of the Rule Manager and the integration of the reasoner in the Event Pattern Detector. Moreover, we have extended the framework's functionality, so as to efficiently address both reactive and proactive adaptation.

We adopt three types of events in our approach. **Successful events** carry information about successful invocations and normal state of the system components. **Warning events** indicate that a component (device, service, software, etc) of our system does not perform normally and a monitoring property has exceeded the warning threshold, which has been defined by the service requester as part of the SLA (e.g. service execution time surpasses warning threshold of 100ms). These events may lead to failing events, helping us to proactively adapt our system. **Failing events** indicate that a failure appeared during the SBA

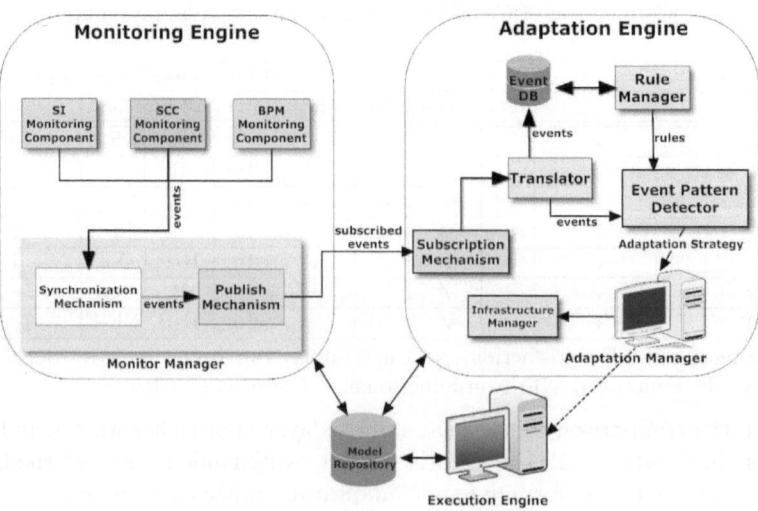

Fig. 1. Architecture of ECMAF

execution. These events are detected by the monitoring engine when a monitoring property is violated. Moreover, failing events, as well as warning events, can capture functional properties, as for example the service input type.

Concerning monitoring of the BPM and the SCC layers, we are using the Astro monitoring tool [1]. As Astro exploits the ActiveBPEL engine, it provides the ability to monitor service compositions implemented in BPEL. It supports both instance and class monitors, as well as the monitoring of functional and non-functional properties. The framework detects violations at the SCC and the BPM layers by comparing the monitoring property values with the predefined thresholds, and records them at a log file. The SI monitoring is performed by the Nagios (`http://www.nagios.org/`) monitoring tool, which offers complete monitoring and alerting for servers, switches, disks and other types of infrastructural components. Both monitoring tools also provide success and warning events. For the definition of the monitored properties we exploit the OWL-Q non-functional service description language [7], which has been extended accordingly so as to allow the definition of functional properties.

As far as adaptation is concerned, our approach relies on detecting patterns of monitored events, in order to prevent future failing events. Initially, the subscribed monitored events are collected by the Siena subscription mechanism (`http://www.inf.usi.ch/carzaniga/siena/`), which, in turn, passes them to the *Translator* and are finally stored in a MySQL *Event Database*. The *Rule Manager* regularly consults the database to derive new event patterns and the corresponding mapping to suitable adaptation strategies, in the form of new rules, which are then sent to the Event Pattern Detector (EPD). This procedure is explained in Section 4. The *Translator*, a Java-based program, transforms the monitored events to the respective format of the *Event Pattern Detector*. For event processing and comprehensive pattern detection, we use Esper (`http://esper.codehaus.org/`), which can detect an event pattern while new monitored events are delivered to the EPD. Then, EPD detects the event pattern that corresponds to the body of a specific rule produced by the Rule Manager and in this way selects the appropriate adaptation strategy that corresponds to the rule head and forward it to the Adaptation Manager. Each adaptation strategy is mapped to a BPEL file which determines which adaptation actions is performed by which component as well as the ordering of the adaptation actions.

Finally, the *Adaptation Manager* is responsible to provide the appropriate BPEL file, which realizes the adaptation strategy, by defining the responsible component for each adaptation action and the ordering of these actions. Complex adaptation strategies are produced from simpler ones in a bottom-up manner by merging the corresponding BPEL files accordingly. The *Infrastructure Manager* is able to treat malfunctions regarding the SI layer, such as memory reallocation, server switching and other. This tool is planned to be a separate one with infrastructure adaptation capabilities. The *Execution Engine*, which is planned to be an extension of the existing Astro BPEL engine with adaptation capabilities, is called to apply the adaptation strategies.

4   Rule Derivation

In order to define the patterns of monitored events for the executed SBA, the Rule Manager is regularly querying the Event Database to discover the event patterns that lead to failing events. A number of SBA instance invocations have to be considered, so as to find these event sequences that are repeated for many times and have always the same SBA failing result. Then, the Rule Manager marks this sequence as an event pattern for this SBA. Unique identifiers, such as business process IDs, service IDs and infrastructure IDs, are exploited for event correlation during pattern derivation.

After defining an application-specific event pattern, that may contain events from all three layers, the Rule Manager maps this pattern to an adaptation strategy in the following way. Some predefined simple rules, mapping events to specific adaptation strategies, are produced manually by the application manager and are passed to the Rule Manager. Each monitored event can be mapped to more than one adaptation strategy, but we keep a priority of these strategies, defined by the application manager, to express their suitability. The Rule Manager extracts only the rule with the highest priority for each monitored event. Simple events are mapped to one or more adaptation strategies, while the strategies of more complicated event patterns (containing more that one event) are produced from the individual strategies of the events that participate in the pattern. We assume that each unique adaptation action has a unique name, in order to facilitate the strategy matching. Finally, the Rule Manager extracts a rule that is passed to the Event Pattern Detector. The following techniques, exploiting the strategy sets that have been mapped to each individual event of the pattern, are used to examine which is the best strategy combination.($S(e)$ symbolizes the set of strategies assigned to the monitored event e):

Case 1: The optimum solution derives from the intersection of the adaptation strategies. If the intersection is non-empty, then the event pattern is associated to all the strategies of the intersection but only one of them is selected to be the most suitable one, according to strategies priority. In particular, the rationale is to multiply the priorities and take the intersection with the highest combined priority. Big products mean high priority. **Case 2:** The worst case is the situation, where there is no intersection among the strategy sets. In this case we choose the sets union, i.e. the adaptation strategy with the highest priority for each one of the monitored events. Moreover, the other strategy combinations are stored and assigned a priority. **Case 3:** In any other case, the overlapping between the sets is considered to decide the adaptation strategies. For instance, if S(e1) is overlapped with S(e2) and S(e2) is overlapped with S(e3), then the adaptation strategy will comprise two adaptation strategies with the highest priorities from the following set $(S(e1) \cap S(e2)) \cup (S(e2) \cap S(e3))$.

5   Case Study

Our work aims at locating the warning event and taking adaptation actions in order to prevent future failing events, as well as to reactively adapt the SBA when proactive adaptation fails to detect the failure. The application domain considered in this paper concerns a traffic management system, which is analyzed in detail in [10] (Fig. 2). At the SCC layer, the oval shapes represent the third-party composite services, the rectangles the internal composite services, while the dashed ones the manual activities. We suppose that the domain expert specifies in a separate document his goals for the quality of the system using a set of KPIs using OWL-Q. In addition, she defines adaptation strategies that can be taken in order to adapt the business process as well as the initial rules. The corresponding SLAs of the web services are stored in the model repository. We assume that the following manual rules have been imported in the Rule Manager:

- *available memory <100MB ⇒memory realloc. (R. 5.1, Pr. 1)*
- *elapsed time of assessment service >500ms ⇒memory realloc. (R. 5.2, Pr. 1)*
- *execution time of assessment service >1sec ⇒DCS migration (R. 5.3, Pr. 1)*
- *process duration >10min ⇒recomposition (R. 5.4, Pr. 1)*
- *available memory < 100MB + elapsed time of assessment service >500ms ⇒ memory realloc.* (derived from Rules 5.1, 5.2) *(R. 5.5, Pr. 1)*

At run time, we continually collect monitored events from the Astro and the Nagios monitoring tools. Successful events indicate that the system is running normally and the adaptation manager does not have to take any actions. After a certain period of time, the monitoring engine detects that the assessment service has currently low available memory (below warning threshold of 100MB). At the same time we notice that this service is not running optimally and the Astro monitoring tool detects a warning event indicating that its elapsed time since its invocation has exceeded the warning threshold of 500ms. These events are passed directly to the Monitor Manager and delivers them to the Adaptation Engine. This event sequence appeared many times in the past SBA execution history leading to the occurrence of the failing event, so the Rule Manager has derived a corresponding event pattern and mapped this pattern to the most suitable adaptation strategy according to the predefined manual rules. Specifically, the exported rule 5.5 stems from the intersection of the adaptation strategies of the

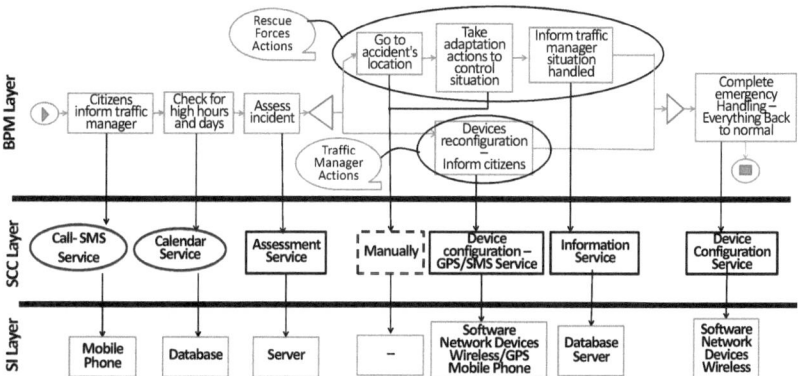

Fig. 2. Critical traffic conditions - Traffic management scenario

two warning events. The strategy's suitability lies on the fact that by executing this service with better memory allocation, the probability that the SLA guarantee corresponding to the service execution time is not violated becomes very high. However, the service execution time is finally violated. Consequently, the respective event is detected and Rule 5.3 is fired, aiming at addressing a future KPI violation, i.e. that the process duration surpasses the 10 minutes threshold, by migrating the Device Configuration Service (DCS) to another more powerful machine to compensate for the additional time spent by the assessment service. This strategy is performed in parallel with the manual activities, which consume most of the process time, so it does not introduce any additional delay.

6   Conclusions and Future Work

To sum up, in this paper we have presented an extended framework, that can efficiently deal with both reactive and proactive cross-layer of adaptation of SBAs. Its main contribution is the pattern-based handling of monitored events in order to perform adaptation. On the one hand, it is both the detection of warning event patterns and the efficient rule management that enables the proactive SBA adaptation, and on the other hand, the efficient rules management reinforces its reactive adaptation capabilities. Furthermore, initial validation of our approach has been presented, based on a traffic management scenario. As future work, we will finalize the framework by implementing adaptation enactment mechanisms, that will experimentally evaluated. In addition, the framework will be enriched with new capabilities to capture the state of the different applications.

References

1. Barbon, F., Traverso, P., Pistore, M., Trainotti, M.: Run-time Monitoring of Instances and Classes of Web Service Compositions. In: ICWS, pp. 63–71. IEEE (2006)
2. Gjørven, E., Rouvoy, R., Eliassen, F.: Cross-layer self-adaptation of service-oriented architectures. In: MW4SOC, pp. 37–42. ACM (2008)
3. Guinea, S., Kecskemeti, G., Marconi, A., Wetzstein, B.: Multi-layered Monitoring and Adaptation. In: Kappel, G., Maamar, Z., Motahari-Nezhad, H.R. (eds.) ICSOC 2011. LNCS, vol. 7084, pp. 359–373. Springer, Heidelberg (2011)
4. Hielscher, J., Metzger, A., Kazhamiakin, R.: Taxonomy of adaptation principles and mechanisms. S-Cube Project Deliverable (2009)
5. Horn, P.: Autonomic Computing: IBM's Perspective on the State of Information Technology. Tech. rep. (2001)
6. Jiang, S., Hallsteinsen, S., Lie, A.: An experimental facility for cross-layer adaptation of service oriented distributed systems. In: NIK, pp. 97–108 (2011)
7. Kritikos, K., Plexousakis, D.: Semantic QoS Metric Matching. In: ECOWS. IEEE, Zurich (2006)

8. Leitner, P., Wetzstein, B., Karastoyanova, D., Hummer, W., Dustdar, S., Leymann, F.: Preventing SLA Violations in Service Compositions Using Aspect-Based Fragment Substitution. In: Maglio, P.P., Weske, M., Yang, J., Fantinato, M. (eds.) ICSOC 2010. LNCS, vol. 6470, pp. 365–380. Springer, Heidelberg (2010)
9. Popescu, R., Staikopoulos, A., Liu, P., Brogi, A., Clarke, S.: Taxonomy-driven Adaptation of Multi-Layer Applications using Templates. In: SASO (October 2010)
10. Zeginis, C., Konsolaki, K., Kritikos, K., Plexousakis, D.: ECMAF: An Event-Based Cross-Layer Service Monitoring and Adaptation Framework. In: NFPSLA-SOC. Springer (2011)
11. Zengin, A., Marconi, A., Pistore, M.: CLAM: Cross-layer Adaptation Manager for Service-Based Applications. In: QASBA 2011, pp. 21–27. ACM (2011)

Search Intent Discovery
by Structurization of Community QA Contents

Soungwoong Yoon[1], Adam Jatowt[1,2], and Katsumi Tanaka[1]

[1] School of Informatics, Kyoto University,
Yoshida honmachi, Sakyo, Kyoto 606-8501 Japan
[2] Japan Science and Technology Agency,
4-1-8, Honcho, Kawaguchi-shi, Saitama 332-0012 Tokyo, Japan
{yoon,adam,tanaka}@dl.kuis.kyoto-u.ac.jp

Abstract. Web search users often suffer from formulating keyword queries although their search intent may be clear. Moreover, it is difficult for search engines to guess search intent from queries only. We propose a new method for discovering search intents and for generating suggested queries of a given input Web search query to address these problems. Precisely, we introduce the process which analyzes and structurizes corresponding Community Question-Answer corpus data: Finding question-answer pairs (QAs) related to a user's query, extracting keywords from QAs related to the user's intent, transforming QAs into a graph, and generating suggested queries using QA graphs.

Keywords: Query intent, Community Question-Answer corpus, Query suggestion.

1 Introduction

The Web is the biggest repository of information, but it is far from being a well arranged 'treasury'. Typical queries submitted to commercial Web search engines (SEs) contain very short keyword phrases [1], thus they are often insufficient to fully describe user's information need, called *intent*. Automatically discovering intents behind Web queries should obviously support users in their search tasks. Previous studies focused on the intent-finding task as classifying queries into some predefined categories through training classifiers [2,3,4]. However, discriminative feature representation and sufficient number of training samples are needed to make this approach succeed [5]. These conditions are hard to be met due to the sparseness of query features combined with the diversity of labeled training data [3], especially when treating rare queries.

If the probable intents associated with an input query can be suggested to users, the searchers could choose the intent suitable for them. Our approach can be intuitively understood as detecting intents underlying input query through virtually *asking* users back. We treat 'querying' the Web as a parallel to 'questioning' the people. The results of direct questioning people are actually recorded in Community Question-Answer corpus (CQA) which usually contain millions of

X.S. Wang et al. (Eds.): WISE 2012, LNCS 7651, pp. 712–718, 2012.

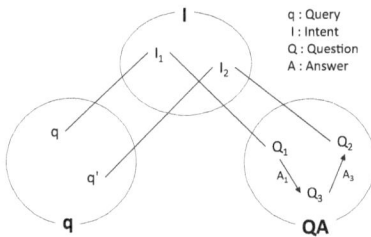

Fig. 1. Concept of intent finding by structurization of QAs

questions with their answers reflecting many kinds of intents which users have. We thus propose to detect possible intents of any input query by mining CQA for query-related content.

In particular, we assume that question-answer pairs (QAs) matched with the keywords of input query contain expressions of intent. Thus words selected from QAs can be used to represent user's search intent. To find such words and suggest them to users, we propose a new approach composed of the following steps: 1) Searching QAs related to a user's keyword query, 2) Extracting words from QAs related to the user's intent through syntactic & statistical analysis, 3) Transforming QAs into the graphs through determining the semantic connections between the answers and the questions, and 4) Generating alternative suggested queries. Especially step 3 relies on forming question-answer-question connections, which can be regarded as modeling user's search intent by detecting the 'next' intent phase through calculating different kinds of similarities within CQA data.

Using our method, the QA graphs related to the input query are generated without any background knowledge. We can then discover underlying intents behind the query and suggest queries following intents. Figure 1 depicts the concept behind our approach. For a given input query q, our system finds a set of probable intents (denoted as $\mathbf{I}$) through the process of QA structurization and suggests alternative queries coordinated with $\mathbf{I}$ (denoted as $\mathbf{q}$).

2   QA Structurization for Suggesting Alternative Queries

Various useful types of Web data could be used for solving our problem but we believe that CQA contain the most direct reflection of typical search intents. Hence we propose to use them. CQA contents matched with q are denoted as $\mathbf{QA}_q$ ($\mathbf{QA}$ for simplicity). We consider that each $QA_{ij} \in \mathbf{QA}$ contains two functional parts from the viewpoint of intent, the subject (a brief statement of the question) and content (additional detail of the question), if any, called *Question part* (Q_i), and the set of answers of this question called *Answer part* ($\{A_{ij}\}, j \geq 1$). We then want to select words from $\mathbf{QA}$ which represent probable user search intents $\mathbf{I}$ ($|\mathbf{I}| \geq 1$) that are behind q. The problem is how to extract useful words from QAs and how to combine them together to discover $\mathbf{I}$.

2.1   Extracting Representative Words from QAs

Let **WQA** be intent defining words in **QA**. Then words extracted from given QA_{ij}, $WQA_{ij} \in$ **WQA** are described as (WQ_i, WA_{ij}) where $WQ_i = \{wq_i^1, ..., wq_i^l\}$ $(l > 0)$ is the set of words in Q_i and $WA_{ij} = \{wa_{ij}^1, ..., wa_{ij}^{l'}\}$ $(l' \geq 0)$ is one in A_{ij}. Below we discuss how to find **WQA** from **QA**.

For finding words in question part, WQs, we use syntactic structure of a question phrases and topical importance given by CQA. First, we assume that the semantic core of a question phrase contains the best representation of questioning intent. Headword extraction [6,7] heuristic is used for extracting the first noun phrase of the question sentence. Second, we calculate TFIDF scores after removing stopwords[1] in each category group given by CQA for measuring the importance of words within question topics. Top k frequent words of each question category group are chosen as the candidate words, and then WQs are generated by selecting candidate words near the headword by applying the passage kernel with given threshold (3 words in our experiment).

However, this method is not useful to find words in answer part, WAs, because answers usually contain more than one sentence and answer contents may be quite diverse, as well as noisy terms may be included. Instead, we use fully unsupervised key phrase extraction method called TextRank [8] with modified conditions according to the following steps: 1) A document (Q_i, A_{ij}) is tokenized excluding stopwords and a single word is considered as a vertex. 2) The vertices form a graph in which links have weights calculated by sentence-based co-occurrence. After the undirected, unweighted graph is constructed, the score associated with each vertex is set to 1 and TextRank is calculated on the graph for several iterations until it converges (we do 20 iterations using empirically set threshold of 0.01.) 3) The top one third (maximum 10) vertices are used as WA_{ij} for each WQ_i.

2.2   Structurization of QA Data

The answers of a question may contain various types of explanations. This implies that WAs may be quite diverse and distant in terms of their meanings, even though their associated WQ may indicate only one particular intent. We then need to group both the answers and questions in order to extract more complete data.

Connecting Answer to Question: To address this issue, we propose to structurize QAs by creating a connected graph. The intuition behind our approach is that in the real world answers to a given question often cause the asker to ask another question(s), which allows to more thoroughly satisfy her information needs (e.g. by clarification, extension, etc). The subsequently asked questions may however trigger answers that again need more information which triggers another question, and so on. This leads to formation of interconnected questions

---

[1] http://www.ranks.nl/resources/stopwords.html

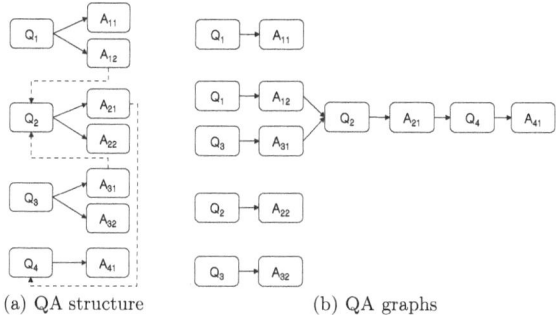

(a) QA structure   (b) QA graphs

Fig. 2. Example of QA connection structure and QA graphs

where the connections are constructed on the basis of received answers. We try
to model this process in CQA contents as follows.

Obviously each question in **QA** is semantically connected with its own an-
swer(s). The problem is rather finding other question(s) concerned with a given
answer. First, we measure the overlaps of WA_{ij} and $WQ_{i'}$s in **QA**. If there
are such overlaps, we calculate the cosine *similarity* between two contents and
then sum the two scores to find the connection strength. Next, if the connection
strength of $Q_{i'}$ is higher than the predefined threshold (empirically set to 0.3),
we calculate the *dissimilarity* between $Q_{i'}$ and Q_i and then choose the most
dissimilar question as the connected question. This procedure is done for every
answer in **QA**. Note that there may be answers to a question which have no
other connected questions. Such answers would correspond to clear answers.

Generating QA Graphs: In result, the bipartite connections of answers and
questions in **QA** are formed (See Figure 2(a)). By using the question and its
answers as the basic semantic connections, we can reflect the sequential connections of intents. Each question is initiated as a starting point of the QA graph
structure. Then another question(s) can be connected using the answer(s) of
a given question based on the answer-question connections. The cycles are removed from the graph. Then we merge subgraphs to generate the QA graphs
as the representation of intents of the original query q. Figure 2(b) shows all
possible QA graphs for the example of Figure 2(a).

2.3   Forming Alternative Suggested Queries

Finally we form suggested queries using $WQAs$ in QA graphs through the following steps: 1) We concatenate the words of the initial query with $WQAs$ in a
node which has no edges from other nodes (called starting node) because these
nodes are the initial points of intent within a QA graph. 2) These $WQAs$ are
then extended with the $WQAs$ extracted from the consecutive nodes until a
certain window. Such words are called *prefix query*.

3   Experimental Evaluation

3.1   Test Queries and Baselines

To evaluate our method, we use the intent-driven evaluation results of NTCIR-9 INTENT task[2] (NTCIR set) as the 'correct answers' of intent. NTCIR set contains 100 queries, descriptions of their intents, intent probabilities and query examples for each particular intent. This data was first provided from a group of human annotators and then corrected by information science specialists.

We choose 51 queries[3] from the NTCIR set as the test queries. For each test query, we collect up to 1,000 QAs using Yahoo! Answers API[4] from May to June 2012. Then we extract WQAs and structurize QA graphs to find alternative suggested queries as described in the previous sections. The prefix window size is chosen to be either 1 and 2, named as prefix1 and prefix2 respectively. For evaluation, we choose the top 10 frequent alternative suggested queries output by our method for each test query and then we manually assign their corresponding intents within the NTCIR set.

As baselines we use combinations of a simple term extraction method from CQA contents based on TFIDF scoring and the following three constraints: words extracted from questions (Q), ones extracted from answers (A) and ones from both (QA).

3.2   Results

Our evaluation is performed by counting the overlaps between intents specified in NTCIR set ($\mathbf{I}_{NTCIR}$) and the ones in target words ($\mathbf{I}_T$). There are four types of target words: suggested queries prefixed with the original query q by using question words ($\mathbf{I}_Q$), ones obtained by using answer words ($\mathbf{I}_A$), ones obtained by QAs ($\mathbf{I}_{QA}$) and ones obtained by WQAs from QA graphs ($\mathbf{I}_{WQA}$). To check the potential of **QA**, we vary $|\mathbf{QA}|$ from 50 to 1000 and investigate the performance of our method.

First, we measure how many different intents are covered by the NTCIR set in the top 10 alternative suggested queries in relation to q. This measure is called *intent diversity*, and is as follows.

$$Diversity_T = \frac{|I_{T,i} \in \mathbf{I}_T \ s.t. \ I_{T,i} \in \mathbf{I}_{NTCIR}|}{|\mathbf{I}_{NTCIR}|} \quad (1)$$

Next, we measure how many alternative suggested queries have matched intents as provided in NTCIR set, called *intent precision*, as follows.

$$Precision_T = \frac{|I_{T,i} \in \mathbf{I}_T \ s.t. \ I_{T,i} \in \mathbf{I}_{NTCIR}|}{|\mathbf{I}_T|} \quad (2)$$

---

[2] http://research.nii.ac.jp/ntcir/workshop/OnlineProceedings9/
NTCIR/Evaluations/INTENT/ntc9-INTENT-eval.htm

[3] NTCIR set was constructed by Japanese and Chinese, so we translate the contents and manually remove queries specific for Japanese or Chinese searchers only.

[4] http://developer.yahoo.com/answers

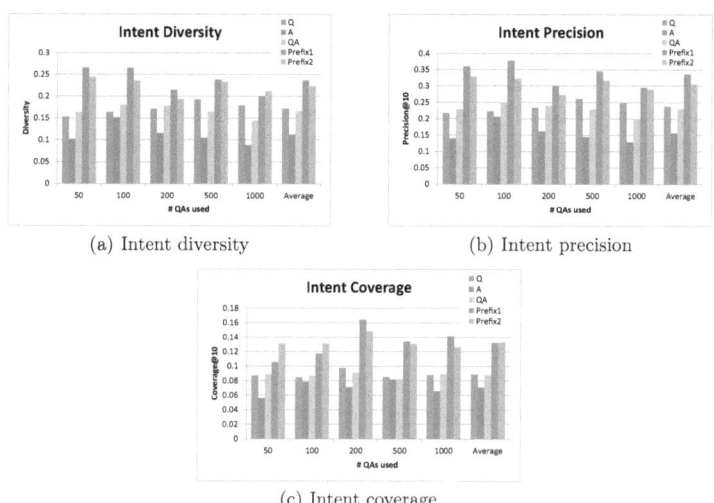

(a) Intent diversity   (b) Intent precision

(c) Intent coverage

Fig. 3. Experimental results of top 10 words

Finally, we measure the accumulated effects of our alternative suggested queries using the probabilities of intents in NTCIR set, called *intent coverage*, as follows.

$$Coverage_T = \frac{\sum_i \left[Pr(I_{T,i}) \, s.t. \, I_{T,i} \in (\mathbf{I}_{NTCIR} \cap \mathbf{I}_T) \right]}{|\{I_{T,i} \, s.t. \, I_{T,i} \in \mathbf{I}_T\}|} \tag{3}$$

Figure 3(a) shows intent diversity through test queries. Generally, the diversity of the proposed method (prefix1 and prefix2) is better than that of baselines, however, on average, it has rather low score. The highest performance, 0.265, is achieved for the case of prefix1 using 50 QAs. Figure 3(b) shows the averaged intent precision through test queries. The proposed method also shows better performance than the TFIDF-based approaches. The best results (0.378) are achieved for the prefix1 using 100 QAs.

Figure 3(c) shows the intent coverage. The precision of prefix2 case shows better performance than the one of prefix1 on average and when less than 100 QAs are used. Note that intent diversity and precision of prefix2 are similar with the ones of prefix1 when 100 QAs are used, which implies that we can choose 100 QAs to achieve the best performance.

4 Conclusion and Future Work

In this paper we propose method for suggesting queries reflecting probable search intents using CQA contents which match user-issued original Web queries. Through statistical and structural analysis using semantic connections of CQA contents we structurize QA graphs and then extract useful words together with the query keywords to suggest intent behind input query. Experimental results show that alternative suggested queries constructed by the proposed method are characterized by both wider diversity and more precision toward probable intents than baselines when compared to human-annotated intents. This indicates that CQA contents constitute useful knowledge for revealing search intent.

In future we need to investigate the diversity problem of QA words for suggesting appropriate queries to users, through synonym / paraphrase extraction. Also we need to reveal the features of QA structurization which affect the generation of intents such as the termination condition of prefix queries. In addition, intent can be changed even during a search session, so analysing and suggesting temporal intent can also be investigated.

Acknowledgment. This work was supported in part by the following projects: Grants-in-Aid for Scientific Research (No. 24240013) from MEXT of Japan, and a Kyoto University GCOE Program entitled "Informatics Education and Research for Knowledge-Circulating Society."

References

1. Jansen, B.J., Spink, A., Saracevic, T.: Real life, real users, and real needs: a study and analysis of user queries on the web. Inf. Process. Manage. 36, 207–227 (2000)
2. Jansen, B.J., Booth, D.L., Spink, A.: Determining the informational, navigational, and transactional intent of web queries. Inf. Process. Manage. 44, 1251–1266 (2008)
3. Beitzel, S.M., Jensen, E.C., Frieder, O., Lewis, D.D., Chowdhury, A., Kolcz, A.: Improving automatic query classification via semi-supervised learning. In: Proceedings of the Fifth IEEE International Conference on Data Mining, ICDM 2005, pp. 42–49. IEEE Computer Society, Washington, DC (2005)
4. Broder, A.Z., Fontoura, M., Gabrilovich, E., Joshi, A., Josifovski, V., Zhang, T.: Robust classification of rare queries using web knowledge. In: Proceedings of the 30th Annual International ACM SIGIR Conference on Research and Development in Information Retrieval, SIGIR 2007, pp. 231–238. ACM, New York (2007)
5. Hu, J., Wang, G., Lochovsky, F., Sun, J.T., Chen, Z.: Understanding user's query intent with wikipedia. In: Proceedings of the 18th International Conference on World Wide Web, WWW 2009, pp. 471–480. ACM, New York (2009)
6. Yoon, S., Jatowt, A., Tanaka, K.: Intent-Based Categorization of Search Results Using Questions from Web Q&A Corpus. In: Vossen, G., Long, D.D.E., Yu, J.X. (eds.) WISE 2009. LNCS, vol. 5802, pp. 145–158. Springer, Heidelberg (2009)
7. Metzler, D., Croft, W.B.: Analysis of statistical question classification for fact-based questions. Inf. Retr. 8, 481–504 (2005)
8. Mihalcea, R., Tarau, P.: Textrank: Bringing order into text. In: EMNLP, pp. 404–411. ACL (2004)

An Investigation on Repost Activity Prediction for Social Media Events*

Juarez Paulino Silva Júnior, Lucas Almeida, Felipe Modesto,
Thiago Neves, and Li Weigang

University of Brasilia, Brasilia, Brazil
{juarez.paulino,lucas.augusto.almeida}@gmail.com,
{felipe,tfn.thiago,weigang}@cic.unb.br

Abstract Repost activity provides a way to measure the rate of information propagation about an event on a microblog service and is a key concept to understand its success. In this paper we deal with the repost prediction challenge proposed on the WISE 2012 conference, which required us to predict repost activities for 33 posts of 6 events within a period of 30 days. To achieve this objective, we propose the construction of a representative model based on semantic relationship between the events within the dataset. Next, we use two state of the art data-mining approaches when estimating the reposting of messages: (*i*) the Logistic Regression and (*ii*) the Conditional Random Fields. We also present a novel simulation framework which best uses the characteristics of our semantic data model in predicting results.

1 Introduction

Social media is an interactive experience, which enables a highly connected user network to interchange information in large scale and real time. Many reasons encompass the current success of social medias, among which we can highlight: (*i*) integration of many users on a single shared network, (*ii*) information propagation through a wide variety of content types, such as text, image and video, (*iii*) enabling communication through many kinds of platforms like smartphones and notebooks. Because of these capabilities, social medias are capable of distributing major event related information faster and more effectively than traditional medias, such as TV or radio. In this frontline, microblog web services like *Weibo* and *Twitter* are getting even more popular.

Related to this scenario, the 13th International Conference on Web Information System Engineering (WISE 2012) proposed a challenge based on *Weibo* Social Network with two tracks: (*i*) the performance track (T1) and (*ii*) the mining track (T2). In the second track, participants are required to predict the reposting activities of thirty-three posts of six events. The reposting prediction problem is defined as the process of estimating the reposting activity over a predefined period of time, given an original *followship network* and an initial seed of posts and repost basic messages elements. In particular, it is

---

* This research has been partially supported by the FAPDF grant.

X.S. Wang et al. (Eds.): WISE 2012, LNCS 7651, pp. 719–725, 2012.

required to predict two measurements: (i) the number of times that the original post is reposted (M1) and (ii) the number of times of possible-view of the original post (M2). In this paper, we face the mining track (T2).

The rest of this paper is organized as follows: Section 2 describes the methodology applied on building the prediction model; Section 3 describes a semantic-driven simulation framework which give us the required resulting metrics M1 and M2; Section 4 contains the simulation results and Section 5 concludes this paper.

2   The Repost Prediction Methodology

In the traditional repost prediction problem, we need to consider a set of features extracted from the posted target-messages and apply a previously trained model to predict the outcome of each considered message. According to our Monte-Carlo simulation framework, the trained models are required to infer the outcome probabilities when predicting the repost of messages. So, in this work, we basically use two state of the art data-mining approaches: (i) Logistic Regression (LR) and (ii) Conditional Random Fields (CRF), which efficiently allow us to calculate the desired probabilities.

Previously, CRF and LR were broadly used in the natural language processing [1]. In this scenario it provided good results because they fit the sequential and hierarchical nature of texts. Since the challenge datasets are basically texts, we chose software that also use linear-chains. For Logistic Regression we used the LIBLINEAR [2] library, a highly used tool for implementing LR learning in linear chains. Amongst the software that implement CRF, we chose the CRF++ [3] because, similarly to Peng et al. [4], it uses an algorithm for fast training and large scale numerical optimization. In this work, we cross-validated our data-mining models and obtained an overall accuracy of 70% when predicting and comparing events within the provided dataset.

Several authors [4,5] classify features used for repost prediction in three categories: (i) Content Based, (ii) Network Based and (iii) Time Based. Similarly to [6], we propose a semantic analysis for events in the posts. From that, we propose to classify all the 45 events presented in the dataset into classes based on their similarities. In this way, when performing repost predictions, we only consider events that belong to the same class. The objective here is to reduce the training datasets and to improve the prediction due the initial intuition that users react similarly in similar events. We empirically confirmed that this initial intuition is correct. So we classify the 45 events in **4 classes**, related mainly with the following topics: tragic events, national news in China, disasters and global news.

For each post we calculate a total of 9 features for repost prediction: (1) the number of common friends between the post sender and the post receiver; (2) the number of followers of the original post author; (3) the time difference between the post receiver and the post sender; (4) the time difference between the post receiver and the original post sender; (5) if or not the post has a URL; (6) if or not the original post has a URL; (7) if or not the post receiver is mentioned by the post sender; (8) if or not the post occurs during the weekend; and (9) if or not the post is related to one of the 4 event classes. Note that each post can be a repost or an original post. We consider both in our feature extraction.

Model Acquisition Methodology

Figure 1 describes all the steps of our methodology for repost prediction. In Step 1 we first obtain the **Raw DB**, which contains the 2 main datasets provided by WISE. The **Pre-Processed DB** is obtained after processing and standardizing all the data in the Raw DB. In this step the huge user and message id numbers are replaced by smaller ones, maintaining the consistency and considerably reducing the file sizes. The objective of Step 1 is to obtain 3 files: the standardized-message, standardized-relationship files and a standardized-message-with-events file which contains only messages with the associated events. In Step 2 we obtain the **Tree DB**. We first split the standardized-message-with-events file into 4 files, each file has messages of a single class of events described in the Section 2. For each of these 4 files we build a forest of post-trees with at most 5 levels above each post. Each post-tree described how a post propagated through time. The algorithm 1 summarizes the forest generation process. It works by iteratively looking on all messages w of a chosen class $\mathbb{C}$ and building a localized 5-level limited tree, taking w as leaf and recovering the root message r through FINDROOT.

Next, we insert an additional data due to a mutation factor of up to 15% on the number of acquired trees. We select these trees from messages that are not related to the current class of events but we choose to require them to be associated with the users ID's gathered so far. We empirically verified that the insertion of these mutational trees implied an important reduction on over-trainning effects, and we chose 15% mutation factor as the ratio which best scored on cross-validation results of our trained models. Finally, the EXPANDTREEWITHNEGATIVELABELS routine works by going downward from the root to leafs and inserts negative label data (non-reposted messages) on each tree on ratio 2 : 1 of the number of positive data. Inserting negative data is obligatory on reposting prediction problem since we also need to train the model to predict messages that have lower chances to be reposted.

In Step 3 we use the 4 post-tree files to generate 4 feature files, or the **Feature DB**. We use the features previously described in this section. In Step 4 we use

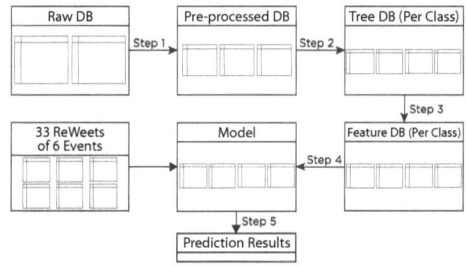

Fig. 1. Methodology for data prediction – all steps

Algorithm 1. Forest Generation Process.

Require: $\mathbb{W} := \{w | w \in \mathbb{C}\}$: Set of messages related to events on class $\mathbb{C}$.
Ensure: $\mathbb{F}$: Forest of post-trees related to class $\mathbb{C}$.
    $U \leftarrow \emptyset;$ {Set of user id's gathered so far on the retrieving process}
  while $(\exists w \in \mathbb{W})$ **do**
    $\mathbb{W} \leftarrow \mathbb{W} \setminus w,\ r \leftarrow$ FindRoot$(w, U);$
    if $(r \notin \mathbb{F})$ **then** $\mathbb{F} \leftarrow \mathbb{F} \cup r;$
  end while
  for $i = 1$ to $(0.15 * \text{size}(\mathbb{F}))$ **do**
    $u \leftarrow$ SelectUserNonDeterministically$(U);$
    $v \leftarrow$ SelecMessageNonDeterministically$(u);$
    if $(v \notin \mathbb{W})$ **then** $r \leftarrow$ FindRoot$(v, U),\ \mathbb{F} \leftarrow \mathbb{F} \cup r;$
  end for
  for all $r \in \mathbb{F}$ **do** ExpandTreeWithNegativeLabels$(r);$
  return $\mathbb{F};$

---

the feature files in one of the two presented data-mining **Models** to generate 4 training files (each one for a different class of events). Given the posts to predict, we classify them in one of the 4 classes and then use the corresponding training files in a simulation process to obtain the **Prediction Results**.

3 Simulation

In this section we propose a novel simulation framework for the reposting prediction problem based on a probabilistic Monte-Carlo approach. Zhu et al [5] also introduced a Monte-Carlo approach for reposting simulation, however the presented method did not take the best properties for our semantic-class modeling and the naive implementation proved not to be fitted to our specific requirements. On the simulation process we are given an initial seed of data (test data) with: (i) an original post message (we will call it *root*), which we desire to predict the reposts activity, and probably (ii) some of its reposted messages. This initial test data will feed the simulation process and, given a probabilistic trained model, we are required to predict the metrics M1 and M2.

Our framework is based on a priority queue of events ordered by increasing value of timestamps, which we process from the *root* seed posted time. We dispose of three types of events on the queue: **Original Messages**, messages related to *root*; **Simulation Messages**, messages generated along simulation; and **Stalled Events**: an event which stalls a user from posting a message along a defined period of time. The last one was used to avoid a user from flooding lots of messages during short periods of time.

Algorithm 2 synthesizes the overall one-step simulation process and allow us to introduce a specific characteristic of our semantic model. We use a decay time factor function for each of the 6 events. We define a decay time factor function $\mathbb{F} : \mathbb{N} \rightarrow [0, 1]$ as a decreasing function which maps an elapsed time (on seconds) to a probability decay factor, reducing the expectative of reposting a message. To obtain each of the 6 decay time functions, we have analyzed the whole database and retrieved the expected growing pattern function. We observed that most of the growth functions got the best minimization errors on exponential or non-linear regressions. The inverse of these growth functions gives our desired decay time factor function. We particularly use the decay time function to decrease the probability of a user to repost a message along the simulation time slices. On the WISE 2012 Challenge we ran each step-simulation 100 times for each of the 33 original *root* seed messages and got the expectation value for the required metrics M1 and M2. In sequence we present the final results from the simulation framework.

Algorithm 2. Monte-Carlo Simulation Procedure.

Require: $\mathbb{S}$: Set of seed messages; $\mathbb{M}$: Probabilistic trained model; $\mathbb{F}$: Decay time function.
Ensure: M1 and M2 metrics as described in Section 1.
  Q: Priority queue ordered by events' timestamps;
  B: Set of stalled (blocked) users;
  M1 ← M2 ← 0;
  for all $(s \in \mathbb{S})$ **do** Q.push(s);
  while not $(Q.\text{empty}())$ **do**
    e ← ExtractMin(Q);
    if (e.type = Stalled_Event) **then** B.remove(e.user);
    else if (e.type = Original_Message) **or** (e.type = Simulation_Message)
      M1 ← M1 + 1, M2 ← M2 + size(e.neighbors);
      for all $(v \in e.\text{neighbors})$ **and** $(v \notin B)$ **do**
        w ← GenerateSimulationMessage(v, e);
        p ← $\mathbb{M}(w)$.probability * $\mathbb{F}(w.\text{timestamp} - root.\text{timestamp})$;
        r ← GenerateRandomNumber(0,1);
        if $(r < p)$ **then** Q.push(w), Q.push(GenerateStallEvent(v, e)), B.insert(v);
      end for
    end if
  end while
  return M1 and M2;

---

4 Results

This section presents our final results for repost prediction and a brief analysis of them. The results are presented in Figures 2 and 3 corresponding to the values M1 and M2, for each post and for some of required events, respectively. Figures 2(a) and 2(b) contain the results for each of the 33 original messages obtained through simulation. For each message there are three columns, each representing the results obtained by a different approach. Similar to the message results, Figure 3 contains the individual results for 3 of the 6 events.

The first obtained result is that all approaches generated similar outcomes. This was expected, and can been seen in Figures 2(a) and 2(b). This means that (*i*) all approaches are equally capable and (*ii*) the dataset is rich enough for the results to be consistent. Also, a direct relationship between the M1 and M2 predictions can be observed. This is due to the dependency that view prediction has of repost prediction. The second result obtained is that the Naive approach (each message has 50% probability to be reposted) generated higher prediction values and the CRF approach obtained lower prediction values. This is the result of the repost probability function considered for the Naive Approach. Also, results obtained from the CRF and LR approaches were closer to each other than to the Naive approach. The results obtained from the Naive Approach are considered as baseline for other approaches and are not statistically more significant. Lastly, when analyzing individual results, we found that even though

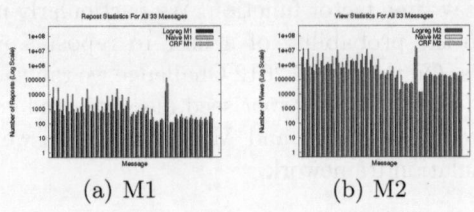

(a) M1   (b) M2

Fig. 2. Simulation results for each message

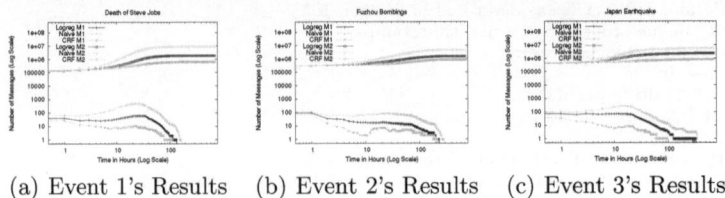

(a) Event 1's Results   (b) Event 2's Results   (c) Event 3's Results

Fig. 3. Simulation results for 3 of the 6 required events

the averages were quite similar, there were internal discrepancies. These are the result of the different mechanisms used by each approach and demonstrates the different capabilities of each one.

5   Conclusions

In this work, two prediction data-mining models were presented and analyzed. It was suggested that these models would be used to predict a series of events based on data provided on the Chinese social network Weibo. Initially, the dataset provided was optimized and it was extracted a representative subset of messages based on a specific semantic classification of the events within the dataset. Further, the semantic classification provided an efficient management and acquisition of our selected set of features used on the models' construction. The 33 target-post messages and its related models were then inserted on our proposed Monte-Carlo framework resulting in the required reposting metrics M1 and M2. The results obtained through simulation were promising and statistically relevant, obtaining consistency and, therefore, stating that the approaches presented were valid and effective. However, there is always room for improvement and we consider that by using characteristics from both approaches it is possible to obtain even better results.

References

1. Sutton, C., McCallum, A.: An introduction to conditional random fields for relational learning. Introduction to statistical relational learning. MIT Press (2006)
2. Fan, R., Chang, K., Hsieh, C., Wang, X., Lin, C.: Liblinear: A library for large linear classification. The Journal of Machine Learning Research, 1871–1874 (2008)

3. Lafferty, J.: CRF++: Yet another crf toolkit,
   `http://crfpp.googlecode.com/svn/trunk/doc/index.html` (accessed in June 16, 2012)
4. Peng, H., Zhu, J., Piao, D., Yan, R., Zhang, Y.: Retweet modeling using conditional random fields. In: 2011 IEEE 11th International Conference on Data Mining Workshops (ICDMW), pp. 336–343. IEEE (2011)
5. Zhu, J., Xiong, F., Piao, D., Liu, Y., Zhang, Y.: Statistically modeling the effectiveness of disaster information in social media. In: 2011 IEEE Global Humanitarian Technology Conference (GHTC), pp. 431–436. IEEE (2011)
6. Sakaki, T., Okazaki, M., Matsuo, Y.: Earthquake shakes twitter users: real-time event detection by social sensors. In: Proceedings of the 19th International Conference on World Wide Web, pp. 851–860. ACM (2010)

Logical Model of Relationship
for Online Social Networks
and Performance Optimizing of Queries[*]
WISE 2012 Challenge -
T1: Performance Track Scalability Winner

Edans F.O. De Sandes, Li Weigang, and Alba Cristina M.A. de Melo

University of Brasilia, Brasilia, Brazil
{edans,weigang,albamm}@cic.unb.br

Abstract. Sina Weibo is currently the microblogging web service with the highest number of registered users in China. As in any large social network, the relationship representation is so huge that executing queries over the network is a very challenging problem. The WISE 2012 conference proposed a challenge based on Sina Weibo with two tracks: performance testing and repost prediction. This paper focuses on the first track challenge, which goal is to implement 19 queries with the highest throughput and the lowest latency, using a scalable parallel paradigm. In the input database, there are 265 millions of relations among more than 60 millions of users and more than 400 millions sent messages. This paper formalizes the logical model of the relationship in order to present the queries in precise and simple manner. Some optimization techniques were also proposed, such as the *aggregate-rank-delete* procedures, which can be applied to some of the queries for improving the performance. The proposed model and optimizations were implemented in a very scalable parallel system and the experimental results show that our solution can obtain high throughput and low latency for most of the queries.

Keywords: Data Mining, Microblog, Performance optimization, Weibo.

1 Introduction

The Sina Weibo microblogging service[1] is the current most popular online social network in China[2]. Some researches have been recently made about this microblog in order to analyze or predict the user behavior[3-7]. Since Sina Weibo has a very huge number of users and connections[8], the analysis and predictions made over this network usually consider only a subset of the network. So, there is an emerging need for fast algorithms that may be able to execute basic queries in a very fast manner. Also, the microblogging systems may use those algorithms to present more detailed information to their own users.

---

[*] This research has been partially supported by the CNPq and FAPDF grants.

X.S. Wang et al. (Eds.): WISE 2012, LNCS 7651, pp. 726–736, 2012.
© Springer-Verlag Berlin Heidelberg 2012

In order to promote the microblogging service research area, the 13th International Conference on Web Information System Engineering (WISE 2012) has presented a challenge[9] based on a dataset collected from Sina Weibo. The challenge is composed of two tracks, which may be attended separately. The first track (T1) is the performance track, aiming the optimization of 19 queries over the microblog dataset. The second track (T2) is the mining track, aiming the prediction of microblog activity based on previous events.

The present paper focuses on the performance track, where the target is to achieve high scalability, low response time and high throughput reported by the BMSA (Benchmark for Social Media Analysis) performance testing tool[10]. The paper firstly formalizes the Weibo user relationship in mathematic model in order to present the queries in precision and simple logic manner. This definition can also be used in Twitter or any other micro-blogging systems.

The paper also proposes the optimization procedure named as *aggregate-rank-delete* to increase the performance of the queries. The queries are separated in 4 classes based on the property of the queries to facilitate the explanation of the proposed algorithms, and each one of these classes is implemented over a very scalable architecture called *query engine*.

The remainder of this paper is organized as follows. Section 2 analyses the dataset files. After that, a formal description of the followship of Weibo is given in Section 3. Indexation and *aggregate-rank-delete* algorithms are proposed in Section 4 and the queries are described in 4 classes in Section 5. Section 6 describes the architecture of the query engine and Section 7 presents the summary of the benchmarks. The conclusion and future work are given in Section 8.

2 Dataset Analysis

The first challenge to be faced in the performance track is the large amount of data extracted from the Weibo microblog. The dataset was previously prepared by the WISE 2012 Challenge committee. After decompression, there are almost 12.9GB of followship network data and 61.8GB of message activities.

The dataset for the followship network consists of many lines, where each line represents that an user A follows an user B. The dataset for the message activities consists of one line for each message, where each line contains properties such as the message ID, sent time, author ID, mentioned users, related events and the reposted message. There are 44 defined events such as natural disasters and social phenomena. Each message can be associated with one or more events.

In the social network files, there are 58,655,849 users and 265,108,370 followship relations and 62,255,232 mentioning relations. There are 2,819,324(4.8%) users that are followed by someone and 58,478,875(99.7%) users that follows someone. There are 7,601,842 bidirectional followship relations, i.e. pairs of users that follow each other (also called as *r*-friends).

Analyzing the message datafile we can find 401,327,390 sent messages. There are 45,722,228 (11,4%) messages that were reposted by 190,926,515 (47,6%) other messages. There are a total of 14,369,994 distinct users being referenced in the message file. There are 8,821,801 (2.2%) messages tagged with some events.

3 Formal Description of Data

In order to supply a theoretical analysis of the data, this section formalizes the data for the followship network, the events and the message relations.

The social network is properly described as a directed graph $G = (V, E)$ where the vertices set V represents the users and the directed edges $E : V \times V$ represents the followship relations among the users, where $(a, b) \in E$ means that user a follows user b.

There are some terms derived from the above definitions that describes the relations among the users. If user a follows b, than we say that a is a *follower* of b and that b is a *followee* of a. If a follows b and b follows a, we say that a and b are *r-friends*.

These terms are formally described by the functions $f_{in}, f_{out}, f_r \colon V \to V^*$.

- $f_{out}(a) = \{b | (a, b) \in E\}$: is the set of all followees of user a.
- $f_{in}(a) = \{b | (b, a) \in E\}$ is the set of all followers of user a.
- $f_r(a) = f_{out}(a) \cap f_{in}(a)$ is the set of all r-friends of user a.

For each of the above function, we define its reverse function. Let $f \in \{f_{in}, f_{out}, f_r\}$, so the reverse function f' is defined in the following way:

$$f' = \begin{cases} f_{in} & \text{if } f = f_{out} \\ f_{out} & \text{if } f = f_{in} \\ f_r & \text{if } f = f_r \end{cases} \tag{1}$$

We can join all the functions in the following way. Let $f_1, f_2 \in \{f_{in}, f_{out}, f_r\}$, so we define $f_1.f_2 : V \to V^*$ in the following way.

$$f_1.f_2(a) = \bigcup_{c \in f_2(a)} f_1(c) \tag{2}$$

If both functions f_1 and f_2 are the same, we can use the notation $f.f = f^2$. Also, this can be generalized like $f.f^{n-1} = f^n$. Joining functions allow us to create many other relationship functions. For example: $f_{in}.f_{out}(a)$ represents the followers of followees of a; $f_{in}^2(a)$ represents the followers of followers of a; $f_{out}^2(a)$ represents the followees of followees of a; and f_r^2 represents the r-friends of r-friends of a. All these concepts can also be applied to other relations, for example the mentioners and mentionees of a users and the reposting users.

4 Proposed Algorithms

In this section we propose some algorithms that are used to optimize the performance of the queries.

4.1 Sorted Intersection

Suppose that two sets $A = \{a_1, a_2, ..., a_n\}$ and $B = \{b_1, b_2, ..., a_m\}$ needs to be intersected. An efficient way to implement this intersection is to maintain

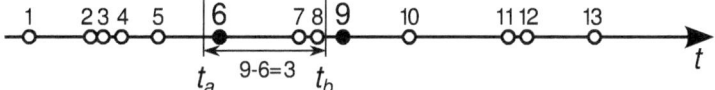

Fig. 1. Indexation Procedure

both sets in an ordered way $a_1 \leq a_2 \leq ... \leq a_n$ and $b_1 \leq b_2 \leq ... \leq b_m$. So, the intersection can be done in $O(m + n)$ time by simply merging both sets maintaining the ordering criteria. During the merge procedure, the algorithm maintains only the elements found on both sets.

4.2   Sorted Counting

Suppose there are m sets $A_1 = \{a_{11}, a_{12}, ..., a_{1n_1}\}$, $A_2 = \{a_{21}, a_{22}, ..., a_{2n_2}\}$, ..., $A_m = \{a_{m1}, a_{m2}, ..., a_{mn_m}\}$ and we want to find the most repeating elements among all these m sets. An efficient way to implement this counting is to sort the elements in each set and merge all the sets maintaining the ordering criteria. During the merge procedure, it is possible to verify how many repeating elements exist in each set. So, the algorithm can sort all the elements ordered by the number of repetitions. This procedure is very efficient if it uses a heap tree during the merge procedure.

4.3   Indexation

Let the set $S = \{(e_1, t_1), (e_2, t_2), ..., (e_n, t_n)\}$ be a series of occurences (e_k, t_k), where $e_k \in \Sigma$ is an element belonged to a generic set Σ and $t_k \in \mathbb{N}$ is the time of occurrence of that element. Let $C(e_k) = \{t_1, t_2, ..., t_m\}$ be the set containing the time of occurrences of element e_k in set S, where $m = |C(e_k)|$ is the number of occurrences. Also, let $S_{[t_i..t_j]}$ be the subset with all occurrences from S that happened in time range $[t_i..t_j]$.

For instance, suppose that user $a \in V$ was reposted at times t_1 and t_2 and $b \in V$ was reposted at times t_3. So we can represent these occurrences as $S = \{(a, t_1), (a, t_2), (b, t_3)\}$ and $C(a) = \{t_1, t_2\}$ and $C(b) = \{t_3\}$.

Suppose that we want to find out the number of occurrences of e_k in $S_{[t_i..t_j]}$. It is not efficient to iterate over all the elements in the set S, as it may have a large number of occurrences. The better way is to maintain the set $C(e_k) = \{t_1, t_2, ..., t_m\}$ ordered by time $(t_1 < t_2 < ... < t_m)$. Using a binary indexed search we can quickly find the first element $t_x \geq t_a$ and the first element $t_y \geq t_b$ in $O(log(n))$. So, the number of occurrences of e in time range $[t_a..t_b]$ is simply the difference $y - x$ between the index of each occurrence.

Figure 1 illustrates an example of indexation with 13 occurrences of the element $e_k \in \Sigma$. The number of occurrences of e_k between times t_a and t_b are the index 9 of the first occurrence after t_b minus the index 6 of the first occurrence after t_a, resulting in $9 - 6 = 3$ elements in the given time range.

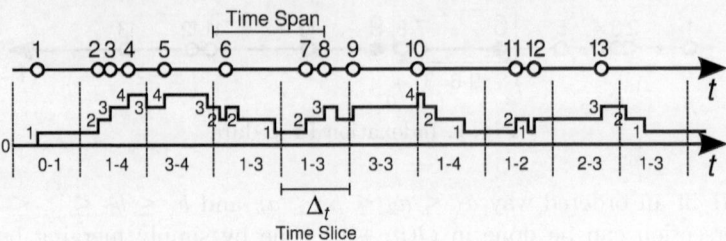

Fig. 2. Aggregation Procedure

4.4   Aggregate-Rank-Delete

Suppose that we want to find the top-x elements that occur in $S_{[t_a..t_b]}$, where the length of the interval $t_b - t_a$ is called *time span*. We can order the elements $e_k \in \Sigma$ by the number of occurrences in the time range $[t_a..t_b]$ using the index search explained in Sect. 4.3. Since this search is executed for every element, this procedure will only be efficient if we have a few indexed elements. If the set of elements is very big, the indexed search will not be sufficient to make the top-x computation efficient. So, three procedures were proposed to make the queries faster: aggregation, ranking and deletion.

– **Aggregation:** The aggregation procedure splits the timeline in regular time slices at every Δ_t second. For each time slice $[t_s..t_s + \Delta_t]$, we compute two values for each element e_k: $min_k(t_s)$ and $max_k(t_s)$. The value $min_k(t_s)$ is the minimum number of occurrences of e_k in any interval $[t_i..t_j]$ such that t_j ends in the given time slice (i.e. $t_j \in [t_s..t_s + \Delta_t]$) and the range $[t_i..t_j]$ has the same size of the time span (i.e. $t_j - t_i = t_b - t_a$). Analogously, $max_k(t_s)$ is the maximum number of occurrences of e_k in the same time slice. With this information, we can find the top-x repetitions in the range $[t_a..t_b]$ by using the indexation method applied only to the elements that were aggregated in the same time slice as t_b. Figure 2 shows an example of aggregation for 13 occurrences of the same element. The timeline in the bottom of the figure shows the number of elements for every possible interval with the size equal to the time span. Suppose that the time span in the figure is equal to an hour. So, by the time of the eighth occurrence, there are 3 occurrences in the previous hour. Then, the timeline is split in many Δ_t time slices and we compute the minimum and the maximum occurrences of elements in time ranges that end in the given time slice. In Fig. 2, the fifth time slice may have from 1 up to 3 occurrences in any interval that ends into it.

– **Ranking:** Using the aggregation method, many information are stored for each element in every time slice. This consumes a lot of memory space and, for this reason, the aggregation method by itself is considered yet a very expensive technique. In order to reduce the used space, we must remove all the aggregated elements that will never be a top-100 element in the given time range (in case of top-$x \leq 100$). For each time slice, we sort the elements ordered by the $min_k(t_s)$ value and maintain the ordered set indexed as $(e_1, e_2, ..., e_{99}, e_{100}, e_{101}, ...e_n)$. We should always keep the first

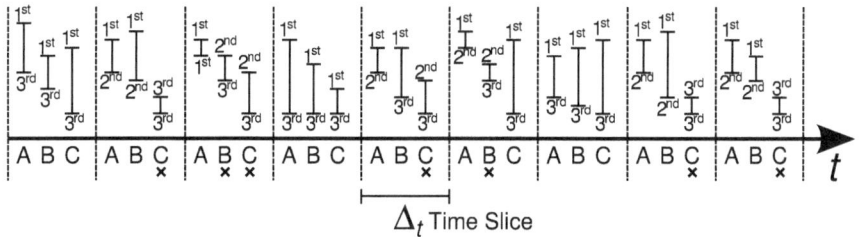

Fig. 3. Ranking Procedure

100 elements. Nevertheless, the remaining elements $e_{i>100}$ must also be kept if $max_i(t_s) \geq min_{100}(t_s)$. This technique significantly reduces the disk space used for the aggregated elements, while keeping the correctness of the results. Figure 3 presents a ranking example for the aggregation of 3 hypothetical elements A, B and C. For each timeslice, every element present a range with its aggregated values $min_k(t_s)$ and $max_k(t_s)$ and each element is labeled with its maximum and minimum rank position. If we would like to maintain only the elements that can be the first element (top-$x = 1$), then we should remove elements with maximum rank position less than 2. The removable elements in Fig. 3 was marked with an "×".

- **Delete:** This procedure deletes all the occurrences $(e_k, t_k) \in S$ that will never contribute to the top-x repetitions in any time range. With the delete procedure, we can reduce the disk space used to store the S set. Also, many iterations of the *aggregate-rank-delete* procedures may be executed in order to create aggregations with smaller Δ_t time slices.

5   Queries

There are 19 queries to be implemented in the Performance Track. When a query requests a top-x operation, this x may be $10, 50$ or 100. Also, when a time range $t \in [t_a..t_b]$ is informed, the time span $(t_b - t_a)$ may be equal to 1 hour (3600s), 1 day (86400s), 1 week (604800s) or 1 year (31536000s).

We grouped 11 of these queries in 4 classes in order to make the comprehension and the implementation easier.

5.1   Class A - Queries Based on Intersection between Users a and b

These queries are supposed to find all users that are related to user a and user b. The queries are described below.

- **Query 4** Find users who are followees of users a and followees of user b.
- **Query 5** Find users who are followees of users a and followers of user b.

For users a and b, these queries are supposed to find all users c that satisfy both conditions $c \in f_1(a)$ and $c \in f_2(b)$, where $f_1, f_2 \in \{f_{in}, f_{out}, f_r\}$. The return set containing all the selected c users is simply obtained by the intersection defined in (3).

$$I(a, b) = f_1(a) \cap f_2(b) \tag{3}$$

The Sorted Intersection algorithm presented in Sect. 4.1 is used to compute the intersection $I(a, b)$ in an efficient way.

5.2   Class B - Queries Based on Common Relations

These queries are supposed to find the top-x users that have common relations with user a. The queries are described below:

– **Query 1** Find top-x r-friends of many of a's r-friends.
– **Query 2** Find top-x followers of many of a's followees.
– **Query 3** Find top-x followees of many of a's followees.

Each query is associated with two functions $f_1, f_2 \in \{f_{in}, f_{out}, f_r\}$ and the order criteria is the number of elements in the same intersection $I(a, b)$ defined in (3). Nevertheless, it is unfeasible to calculate $I(a, b)$ for every user $b \in U$. In order to optimize these queries, we use the Sorted Counting algorithm presented in Sect. 4.2 applied to the reverse function $f_2'(c)$ for each $c \in f_1(a)$. Note that if $I(a, b) = \{c_1, c_2, ..., c_n\}$, then user b belongs to each set $f_2'(c_1), f_2'(c_2), ..., f_2'(c_n)$. So, we simply need to verify the number of duplicated users during the union procedure, without the need to calculate the intersection $I(a, b)$. Then, we return the top-x users (excluding user a and users in $f_1(a)$) accordingly to the functions f_1 and f_2 chosen by each query.

5.3   Class C - Queries Based on Occurrences of Elements in a Time Range

The next queries are supposed to find the top-x occurrences in time range $[t_a..t_b]$. The occurrences is a query specific property over users or events. The queries are described below.

– **Query 7** Find top-x mentioned users in time range $[t_a..t_b]$.
– **Query 10** Find top-x reposted users in time range $[t_a..t_b]$.
– **Query 14** Find top-x tagged events in time range $[t_a..t_b]$.

The goal of the above queries is to find the top-x repeating elements e_k in $S_{[t_a..t_b]}$. The time interval $t_b - t_a$ may be an hour, a day, a week or a year. These intervals may contain more than million occurrences, so the computation of the repeating elements must be done with the indexation and *aggregate-rank-delete* procedures defined in sections 4.3 and 4.4, respectively.

5.4   Class D - Queries Based on Occurrences of Elements in a Time Range Filtered by a Property

The next queries are supposed to find the top-x occurrences in time range $[t_a..t_b]$, such that the occurrences contain a property p.

- **Query 11** Find top-x users that reposted user a in time range $[t_a..t_b]$.
- **Query 15** Find top-x users that tagged event e in time range $[t_a..t_b]$.
- **Query 18** Find top-x users that follows and mentions user a in $[t_0..t_1]$.

Differently from the Class C queries, the Class D queries maintain, for each element occurrence, an additional property p. Then, the set S is redefined as $S = \{(e_1, p_1, t_1), (e_2, p_2, t_2), ..., (e_n, p_n, t_n)\}$ representing a series of occurrences (e_i, p_i, t_i), where e_i is an element belonged to a set of users or events (depending on the query), p_i is the property to be filtered, and $t_i \in \mathbb{N}$ is the time of occurrence of that element. For instance, in Query 11, if user u reposted user a at time t, then $(u, a, t) \in S$.

Let $C(e, p) = \{t_1, t_2, ..., t_m\}$ be the set containing the time of occurrences of element e with property p in set S, where m is the number of occurrences. Also, let $S(p)$ be the set of all occurrences from S with property p and let $S(p)_{[t_x..t_y]}$ be the set with all occurrences from $S(p)$ that happened in time range $[t_x..t_y]$.

The goal of the above queries is to find the top-x repeating elements e_i in $S(p)_{[t_a..t_b]}$. If there are only a few distinct properties p (as in query 15, that p can be one of 44 events), the indexation method and the *aggregate-rank-delete* procedures in the previous subsection can be executed separately for each set $S(p)$. Nevertheless, if p has many different possibilities (as in query 11, that p can be any of the millions users), then the *aggregate-rank-delete* procedures must be executed with very long time slices. In the worst cases, there are so few repeating elements that it is better to use the indexation method without the *aggregate-rank-delete* procedures.

6   Design of Solution

The objective of the present work is to optimize the throughput and latency of the queries, with as much as possible scalability made by parallel requesting threads.

The extract, transform and load (ETL) process was made as follows. The extraction and transformation of the original data was made using the `coreutils` package in Linux and the preprocessed data was generated by a series of scripts in C/C++ and perl. The preprocessed data was then loaded to the SQLite3, an embedded relational database management system. The queries were implemented in C/C++ and the module that executes the queries is called *query engine*.

For getting a proper integration between the Java BSMA benchmark system and the query engine, a Transmission Control Protocol (TCP) socket interface was used to exchange data between the two modules. The BSMA serializes all

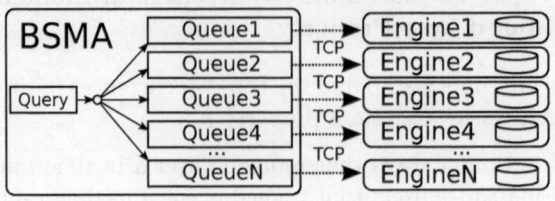

Fig. 4. Architecture of the solution

Table 1. Queries statistics. The best results are presented for each query

Query	Class	DB Size	Throughput	Latency	Threads	Top-x	Time range
Q1	Class B	12GB	390.0 op/s	428.6ms	$\alpha = 200$	$\beta = 10$	$\times$
Q2	Class B	12GB	17.8 op/s	5814.9ms	$\alpha = 125$	$\beta = 10$	$\times$
Q3	Class B	12GB	194.1 op/s	194.1ms	$\alpha = 50$	$\beta = 10$	$\times$
Q4	Class A	12GB	3793.6 op/s	38.5ms	$\alpha = 175$	$\times$	$\times$
Q5	Class A	12GB	2549.7 op/s	65.0ms	$\alpha = 200$	$\times$	$\times$
Q7	Class C	7.1GB	1284.4 op/s	114.9ms	$\alpha = 175$	$\beta = 10$	$\gamma =$w
Q10	Class C	6.4GB	1262.6 op/s	79.4ms	$\alpha = 125$	$\beta = 10$	$\gamma =$w
Q11	Class D	18GB	2314.8 op/s	52.9ms	$\alpha = 200$	$\beta = 10$	$\gamma =$h
Q14	Class C	1.1GB	2539.9 op/s	60.1ms	$\alpha = 175$	$\beta = 10$	$\gamma = $d
Q15	Class D	3.7GB	1460.4 op/s	72.3ms	$\alpha = 150$	$\beta = 10$	$\gamma = $d
Q18	Class D	320MB	4905.6 op/s	36.9ms	$\alpha = 200$	$\beta = 100$	$\gamma =$d

thread queries in a queue and the elements of this queue are sent to the query engine through the socket.

We can use many queues and many query engines to parallelize the computation among different processes. The parallelization is done by a round robin algorithm in the BSMA implementation. Since the query engines use TCP sockets, the query engines can be easily distributed among many different machines.

Figure 4 shows the architecture of the solution. The Java BSMA system connects to many query engines, where each connection have an specific queue.

7   Queries Benchmark

Each query is associated with one of the 4 classes of queries: Class A, Class B, Class C and Class D. Each class has its own algorithm and each query has its own database. Table 1 presents the benchmark statistics with the algorithm association for each query and the size of the input database prepared for each query.

There are three parameters required for each query: α is the number of threads; β is the number of returned records; and γ is the time range analyzed for some queries. The description of the challenge determines that each query must be benchmarked with a different number of times, in order to fulfill all possible parameters. For each query, 8 different numbers of threads must be selected, including a test with a single thread and a test with the maximum supported number of threads. We selected the following number of threads: $\alpha \in \{1, 25, 50, 100, 125, 150, 175, 200\}$. The top-$x$ queries must be run once for each possible return count: $\beta \in \{10, 50, 100\}$. The time range queries must be run once for each possible time range: $\gamma \in \{y,w,d,h\}$. The parameters β and γ are not needed in some queries.

The tests were executed in an AMD Athlon 64 X2 6000+ processor, with 2 CPU cores at 3GHz and 2GB of RAM. The operating system is the Ubuntu 11.10 with Linux 3.0.0 kernel. There were 5 query engines running in parallel.

Table 1 presents the results for each query. As a matter of space, it was not possible to present all the benchmarks in this paper, so only the best results were presented for each query. The best results are considered the ones with better combination of throughput and latency.

8 Conclusion and Future Works

The WISE 2012 Challenge proposed a task to obtain the best throughput and latency of 19 predefined queries with a big dataset. In this paper, a logical model of online network was formally described and 11 queries were grouped in 4 classes of algorithms. The benchmark results of each query were presented. The system architecture is very scalable and 5 query engines were used to execute the queries. We could obtain throughput above 2000 operations per second and latency of less than 40ms in some queries, which is a very satisfactory result. These results were obtained using a platform with low resources and, since the architecture is distributable, the system may be scaled to clusters and grids to obtain even more operations per seconds.

As part of the future work, we intend to implement the remaining queries and to extend the methods of this paper to other social networks such as Twitter. Also, we want to execute the query engines in separate machines in order to obtain higher throughputs.

References

1. Sina Weibo. SINA Corporation (2012), http://www.weibo.com
2. Wu, X., Wang, J.: How about micro-blogging service in China: analysis and mining on sina micro-blog. In: Proceedings of 1st International Symposium on From Digital Footprints to Social and Community Intelligence, SCI 2011, New York, pp. 37–42 (2011)
3. Liu, Z., Chen, X., Sun, M.: Mining the interests of Chinese microbloggers via keyword extraction. Front. Comput. Sci China 6(1), 76–87 (2012)
4. Zhao, B., Zhang, Z., Gu, Y., Gong, X., Qian, W., Zhou, A.: Discovering Collective Viewpoints on Micro-blogging Events Based on Community and Temporal Aspects. In: Tang, J., King, I., Chen, L., Wang, J. (eds.) ADMA 2011, Part I. LNCS, vol. 7120, pp. 270–284. Springer, Heidelberg (2011)
5. Wang, D., Li, Z., Salamatian, K., Xie, G.: The pattern of information diffusion in microblog. In: Proceedings of The ACM CoNEXT Student Workshop, CoNEXT 2011 Student, pp. 3:1–3:2. ACM, New York (2011)

6. Wang, R., Jin, Y.: An Empirical Study on the Relationship between the Followers' Number and Influence of Microblogging. In: Proceedings of the 2010 International Conference on E-Business and E-Government, ICEE 2010, pp. 2014–2017. IEEE Computer Society, Washington, DC (2010)

7. Qu, Y., Huang, C., Zhang, P., Zhang, J.: Microblogging after a major disaster in China: a case study of the 2010 Yushu earthquake. In: Proceedings of the ACM 2011 Conference on Computer Supported Cooperative Work, CSCW 2011, pp. 25–34. ACM, New York (2011)

8. Guo, Z., Li, Z., Tu, H.: Sina Microblog: An Information-Driven Online Social Network. In: Proceedings of the 2011 International Conference on Cyberworlds, CW 2011, pp. 160–167. IEEE Computer Society, Washington, DC (2011)

9. WISE 2012 Challenge, Paphos, Cyprus (2012), http://www.wise2012.cs.ucy.ac.cy/challenge.html

10. IMC, ECNU: BSMA Performance Testing Tool Manual, WISE 2012 Challenge, Paphos, Cyprus (2012), http://www.wise2012.cs.ucy.ac.cy/challenge.html

Predicting Retweet Behavior in Weibo Social Network

Hongbo Zhang[1,2], Qun Zhao[1,2], Hongyan Liu[3,*],
Ke xiao[2], Jun He[1,2,*], Xiaoyong Du[1,2], and Hong Chen[1,2]

[1] Key Labs of Data Engineering and Knowledge Engineering, Ministry of Education, China
[2] School of Information, Renmin University of China
[3] School of Economics and Management, Tsinghua University
{zhanghongbo,mblank.zq,jiaozida,hejun,duyong,chong}@ruc.edu.cn,
liuhy@sem.tsinghua.edu.cn

Abstract. Retweeting ensures the information diffusion in micro-blog services. By this simple way, it is convenient for a user to share and spread interesting information in the whole network. In this paper, we consider many features to compute the probability that a user retweets a tweet. With the probability, we build a retweet model to predict the number of possible-views of a tweet. The model is based on the theory of random walks. Experiments conducted on real dataset show that the proposed method has a good performance than the traditional prediction methods.

Keywords: social network, Weibo, retweet, random walks.

1   Introduction

Online social networks have made a great change in web. One type of online social networks is micro-blog. Weibo (short for Sina Weibo) is a popular micro-blog in China and most of its users are Chinese. In the Weibo, a user can publish a *tweet* (or status) no longer than 140 characters. Besides, a user can choose any other user to *follow*. When a user a follows another user b, a is b's follower and b is a's friend. After a follows b, user a will receive the tweets written by b in a's timeline. To describe the structure of the Weibo, we use a graph $G(V, E)$, where each vertex $v_i \in V$ represents a user, each edge $v_i v_j (v_i \in V$ and $v_j \in V)$ represents a *follow link* between users v_i and v_j, i.e. v_j follows v_i. We call this kind of graph *followship network* or *network* for simplicity. There is a special mechanism called *retweet* in the micro-blog service. Retweet shows the situation that a user a rebroadcasts a tweet t written by another user b. Though b is unnecessary to be a's friend, retweet usually happens between follower and friend.

Weibo is a good place for users to obtain and share information. In fact, it is the mechanism that a user can follow any other users without permission that ensures users to get information they are interested in and retweet it to their followers. The advertisers are interested in it because they can broadcast an advertisement to make

---

[*] Corresponding authors.

X.S. Wang et al. (Eds.): WISE 2012, LNCS 7651, pp. 737–743, 2012.

many users notice and retweet the information. In this paper, two basic tasks in Weibo deserve our attention. One is to predict how many times a tweet will be retweeted and the other is how many users will view the tweet.

In this paper, we give solutions to the two tasks and compare the effect of different solutions. More importantly, we propose a new model to describe the information spread pattern in Weibo and it can extend to other social network like Twitter. The model is based on the idea of random walks and it could simulate the retweeting behavior dynamically as the time goes by. Experiments show that the result of the model is better than other traditional prediction methods.

2   Dataset Description and Preprocessing

In this paper, we choose the dataset from WISE challenge (http://www.wise2012. cs.ucy.ac.cy/challenge.html). The dataset contain two parts: 1) Tweet. It includes information about tweets (time, user ID, message ID, user ID, mentioned, links or not). 2) Followship network. It includes the following network of users (based on user IDs). The dataset is a sample of the whole data in the Weibo. In addition, a small testing dataset is provided. The testing dataset contains 33 tweets and each has been retweeted 100 times. Based on these datasets, we want to predict two figures for each tweet including the number that each tweet is retweeted and the number of possible-view of the original tweet. The number of possible-view of one retweet is defined as the number of followers of the user who conducts the retweet action. Table 1 show some symbols used in the paper. Because of the incompleteness of the dataset, it is important to analyze and preprocess the dataset first.

Table 1. Symbols

Symbol	Explanation
D	Training dataset
TID	Tweet ID
UID	User ID
N_{fr}	The number of friends a user has
N_{fo}	The number of followersa user has
N_{fo2}	The number of followers of a user's followers
IS_l	Contains link or not
IS_{ht}	Contains hash tag or not
IS_m	Contains mention or not
N_{bf}	The number of big followers who have more than 1000 followers
N_r	The number of retweets of a tweet
N_{pv}	The number of possible-views of a tweet
E_t	Event tag of the tweet
N_{cfr}	The number of common friends of two users
N_{rt}	The number of retweet of a test tweet
N_{fot}	The number of followers of a test tweet
N_{fo2t}	The number of followers of a test tweet's user's followers

2.1   Followship Network Expansion

The left part of Figure 1 shows the followers distribution in the followship network. This dataset does not fit the power law distribution very well because N_{fr} and N_{fo} of a user may be smaller than the real value. So we try to expand the followship network. There are two principles: when user A retweets user B, A is very likely to be B's follower; when user A mentions user B, A is always B's follower. The right part of Figure 1 shows the followers distribution of the expanded followship network. Based on the benchmark of power law distribution of the social network, it can be concluded that the expanded followship network is more close to the real situation than the original network.

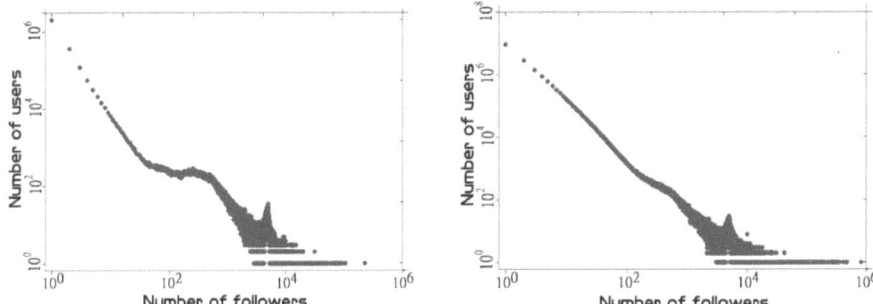

Fig. 1. Comparison of the followers distribution for original network and the expanded network

2.2   Retweet Number Correction

The simplest way to count how many retweets a tweet has is to add up all the retweet paths in the dataset. However, the retweet number of a tweet got from the method is less than the real situation because the incompleteness of the dataset. For example, there is a record "*C B A T*" implying that user A post a tweet T then B and C retweet it sequentially. If the dataset is complete, there should have another record "*B A T*". However, our dataset may miss these records. So in this situation, we should add a record "*B A T*" if the dataset doesn't contain the record. After the correction, we add 80,112,992 retweets accounting for 41.96% of the original retweets.

2.3   Training Dataset

The whole dataset is too huge for some models. So we select a small part of the tweets which are similar with the 33 tweets from the whole dataset as our training data. As all of the 33 tweets have been retweeted 100 times, we choose the tweet whose N_r is equal to or larger than 100 and the author of the tweet has some friends and followers in the original dataset. At last, we get a smaller dataset D which contains about 150,000 tweets. Most models in this paper are trained on this dataset.

3   Predicting the Number of Retweets

3.1   Naïve Methods

In Weibo, we observe one important and basic property that if one has more followers; his tweets will be retweeted more times. Based on this property, we design a simple model M_1 which only depends on N_{fo} and N_{fo2}.

- M_1: basic model based on N_{fo}

$$N_{rt} = \frac{\sum_{t \in D} (N_r/N_{fo})}{n} \times N_{fot} \tag{1}$$

Where n is the number of tweets in the training set, i.e., n=150,000 in our dataset.

- M_2: model based on separation of users

We divide the training set D into two set T_{fm} and T_{fl}. T_{fm} contains tweets the users of which have more followers than the average number of followers in the training set, and users of tweets in T_{fl} have less followers than the average number of followers. Then for a test tweet, if its user has more followers, Equation (1) becomes following Equation (2)

$$N_{rt} = \frac{\sum_{t \in T_{fm}} (N_r/N_{fo})}{n} \times N_{fot} \tag{2}$$

3.2   Regression Model

Another basic model to predict N_r of tweets is regression. On the training data set we try different regression models such as linear regression and regression tree.

3.3   KNN

K-Nearest Neighbor algorithm is one of effective algorithms for classifying objects.

- KNN_1: the basic KNN model

We use all the attributes except UID, TID as the tweet vector to calculate Cosine similarity between two tweets.

- KNN_2: extended KNN model with average retweet number per follower

For model KNN_1, we find the result relies on the similarity function, so we try another method to predict the number of retweets.

Let t be the most similar tweet to the test tweet, N_r be the number of t's retweets, N_{fo} be the number of followers of t's author. Then, the number of retweets of the test tweet is calculated by Equation (3).

$$N_{rt} = \frac{N_r}{N_{fo}} \times N_{fot} \tag{3}$$

3.4   Model Ensemble

We use a linear combination to do model ensemble and use simulated annealing to train the combination parameters. All the 150,000 tweets are used to train the models.

4   Predicting the Number of Possible-Views

The number of retweets of a tweet keeps increasing for a period until it reaches the maximum. We found that more than 90 percent of the retweets happened in the first day after the tweet is posted. Obviously the rate that a tweet is retweeted decreases as time goes by. Based on this point, we apply random walk in the graph. For each tweet, we assume that it is retweeted along all possible paths from its author u. We adopt the symbols such as $followers^1(u)$, $followers^2(u)$ and $followers^n(u)$ to denote the length of the shortest path from author u to the follower. Step 1 represents the time span that extends from the time the tweet is published by user u to the time the tweet is first retweeted by a user $i \in followers^1(u)$. And step n represents the time span extending from the time the tweet is first retweeted by a user $i \in followers^{n-1}(u)$ to the time it is first retweeted by a user $i \in followers^n(u)$. In each step, every user has a probability to retweet the specific tweet. The probability got after k steps is mainly computed by adding the probability after $k-1$ steps to the probability that the user will retweet the tweet in the k_{th} step. After a long time, we regard that the probability for every user is stable. Then the N_r of tweet t could be computed as below:

$$N_r = \sum_{i \in V} \lim_{k \to \infty} P_k(i) \tag{4}$$

In (4), $p_k(i)$ represents the probability that the tweet t is retweeted by user i in step k. According to the above discussion, $p_k(i)$ could be computed as follows.

$$P_K(i) = P_{k-1}(i) + (1 - P_{k-1}(i)) \times (1 - \prod_{j \in friends(i)}(1 - P_{k-1}(j) \times W_k(j,i))) \tag{5}$$

Among Equation (5), $w_k(j, i)$ means the probability that user i will retweet the tweet in step k under the condition that user j retweet the tweet in step $k-1$ and user i is a follower of user j. Further, Equation (6) gives the method to compute $w_k(j, i)$.

$$W_k(j,i) = \alpha_j^k \times w(j,i) \tag{6}$$

$w(j, i)$ is the weight based on which user j will retweet i's tweet. We add a parameter α_j here in order to reflect the fact that the retweet rate for a tweet decreases sharply along the time or step and the rate varies for different users.

Based on the equations given above, given an initial value for $p_0(u)=1$ and $p_0(i)=0$, we can get the retweet probability for every user and thus the total number of retweets of a tweet. Besides in this model, the existence of the limit of $p_k(i)$ should be considered and the different values of a should be tested in the experiment. In our experiment, we add up the followers of the users whose retweet probability is larger than a specific threshold to predict the possible-views.

5   Experiments

For task 1, we try different prediction models as discussed in Section 3. Table 2 lists the model type and error rate. We use relative error(|true-prediction|/true) as our error rate. Results show that model ensemble has the lowest error rate.

Table 2. Error Rate of Models in task 1

Model Type	Error Rate (%)
M_1	467.3989
M_2	65.3655
Liner Regression Model	99.8851
KNN_1	100.9437
KNN_2	68.2651
Model Ensemble	40.9380

For task 2, we compare *KNN model* with the *Random Walk Model* as discussed in Section 4.   The *Random Walk Model* is better than the *KNN* on average. At last, we build model ensemble which combines the two models. The result is listed below.

Table 3. Error Rate of Models in task 2

Model Type	*Error Rate (%)*
The basic *KNN*	183.6743
Random Walk Model	175.9574
Model Ensemble	136.7219

6   Conclusion

In this paper, we focus on the factors that may lead the specific tweets to be retweeted and propose a model to simulate the situation information spreads in Weibo and predict how many users the tweet could influence. The model has a better performance than traditional methods. Besides, the application of the model can be easily extended to other online social networks as long as there exists information spreading behavior.

Acknowledgements. This work was supported by the National Natural Science Foundation of China under Grant No. 70871068, 71110107027 and 61033010.

References

1. Yang, J., Counts, S.: Predicting the Speed, Scale, and Range of Information Diffusion in Twitter. In: ICWSM (2010)
2. Bongwon, S., Lichan, H., Pirolli, P., Chi, E.H.: Want to be Retweeted? Large Scale Analytics on Factors Impacting Retweet in Twitter Network. Social Computing (2010)
3. Wang, W., Wu, B.: Comparing Twitter and Chinese Native Microblog. In: EWI (2011)
4. Yang, Z., Guo, J., Cai, K., Tang, J., Li, J., Zhang, L., Su, Z.: Understanding retweeting behaviors in social networks. In: CIKM, pp. 1633-1636 (2010)

On the Prediction of Re-tweeting Activities
in Social Networks –
A Report on WISE 2012 Challenge

Sayan Unankard[1], Ling Chen[1], Peng Li[1,2], Sen Wang[1], Zi Huang[1],
Mohamed A. Sharaf[1], and Xue Li[1]

[1] School of Information Technology and Electrical Engineering,
The University of Queensland, Brisbane QLD 4072, Australia
[2] College of Computer Science and Technology,
Chongqing University, Chongqing, China 400030
{uqsunank,sen.wang,m.sharaf}@uq.edu.au, ling.chen@uqconnect.edu.au,
pengli@cqu.edu.cn, {huang,xueli}@itee.uq.edu.au

Abstract. This paper reports on our participation in the Data Mining
track of the WISE 2012 Challenge. The challenge is to predict the volume of future re-tweets and possible views for 33 given original short
messages (tweets). Towards this, we compare and contrast four different
methods and highlight our methods of choice for accomplishing this challenge. The first method is a naïve approach that discovers a regression
function based on the popularity of messages and network connectivity.
The second approach is to build a classifier that learns a classification
model based on the user's preferences in different categories of topics.
The third approach focuses on a network simulation that leverages a
Monte Carlo method to simulate re-tweeting paths starting from a root
message. The fourth approach uses collaborative filtering to build a recommendation model. The results of these four methods are compared in
terms of their effectiveness and efficiency. Finally, insights into predicting
message spreading in social networks are also given.

Keywords: social networks, tweet, re-tweet, prediction.

1 Introduction

The prediction of message propagation is one of the major challenges in understanding the behaviors of social networks. In this work, we study that challenge
in the context of the Twitter social network. In particular, our goal is to predict
the propagation behavior of any given short message (i.e., tweet) within a period
of 30 days. This is captured by measuring and predicting two metrics, namely:
1) the number of re-tweets, and 2) the number of possible views.

To model the re-tweeting activities, we use the datasets crawled by the WISE
2012 Challenge [1] from Sina Weibo [2], which is a popular Chinese microbloging

---

[1] http://www.wise2012.cs.ucy.ac.cy/challenge.html
[2] http://weibo.com

X.S. Wang et al. (Eds.): WISE 2012, LNCS 7651, pp. 744–754, 2012.

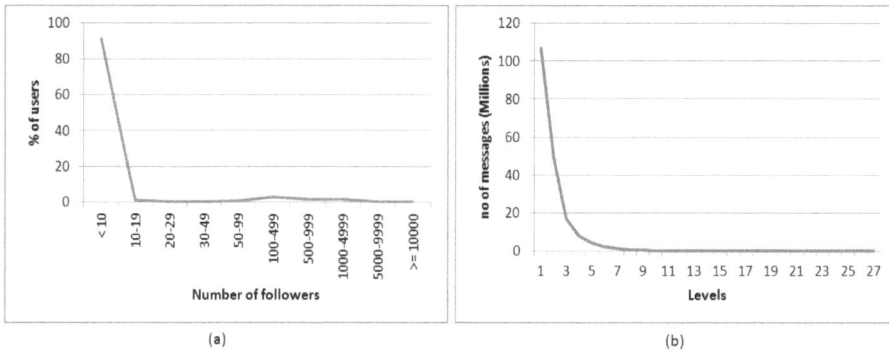

Fig. 1. (a) User distribution based on numbers of followers (b) Number of re-tweets at each level

Table 1. Number of original tweets re-tweeted in 30 days

Number of re-tweets	Original tweets		Tweets with events	
	#msgs	%	#msgs	%
< 10	42,551,891	94.749	882,191	2.073
10 - 99	2,171,214	4.834	65,809	3.031
100 - 499	173,803	0.387	5,464	3.144
500 - 999	10,283	0.023	400	3.890
1,000 - 4,999	2,838	0.006	158	5.567
5,000 - 9,999	26	0.00006	2	7.692
≥10,000	11	0.00002	1	9.091
Total	44,910,066	100.00	954,025	2.124

Table 2. Number of re-tweets in 10 levels

Level	#of re-tweets	%
1	107,025,967	56.056
2	49,401,724	25.874
3	16,934,845	8.869
4	8,045,285	4.213
5	4,196,992	2.198
6	2,315,732	1.212
7	1,294,638	0.678
8	746,494	0.390
9	428,158	0.224
10	240,606	0.126

site similar to Twitter. For the purpose of this challenge, 369 million messages and 68 million user profiles were extracted. The sizes of the followship dataset and the microblog dataset are 12.8 GB and 64.8 GB, respectively.

In preparation for the challenge, we further collected some statistical information for a better understanding of the the available datasets. In particular, for the followship dataset (i.e., the who is following whom relationship), we found that the majority of users have less that 10 followers (approximately 91%) as shown in Figure 1 (a). Additionally, for the microblog dataset (i.e., whose tweets are re-tweeted by whom), we ranked the distribution of the original tweets based on how many re-tweets they received within 30 days as shown in Table 1. The table also shows the subsets of tweets that have been annotated with events. As the table shows, approximately 95% of the original tweets were re-tweeted less than 10 times, of which approximately 2% were annotated with events. In addition, most original tweets were re-tweeted in 3 levels within 30 days (approximately 91%) as shown in Table 2 and Figure 1 (b).

2   Related Work

Microblogging activities in social networks have been attracting growing attentions from researchers in Data Mining and Information Retrieval. One interesting problem is the study on re-tweeting behaviors from an information diffusion perspective. Most works had focused on Twitter, a popular microblogging site. Insightful studies on re-tweeting behaviors can be seen from [1,2,3,4]. Zaman *et al.* [5] adapted a probabilistic collaborative filtering model called Matchbox [6] to predict information spreading on Twitter. Yang *et al.* [7] proposed a factor graph model based on users' re-tweeting history.

Recently, Petrovic *et al.* [8] built a time-sensitive model to automatically predict re-tweets activities. Hong *et al.* [9] trained a multi-class classifier based on logistic regression to predict the volume of re-tweets for a given tweet. Peng *et al.* [10] modelled the re-tweeting activities by using conditional random fields. Naveed *et al.* [11] used logistic regression to compute re-tweet likelihood based on various interesting content features. In our work, the content of tweets has been removed from Sina Weibo microblog dataset pre-processed by WISE 2012 Challenge due to Sina Weibo's Term of Services.

3   Prediction Methods

Based on the given datasets, together with our statistical information presented in Sec. 1, we make the following assumptions:

- The given 33 tweets to be predicted are original tweets.
- The given 6 events associated with the given 33 tweets are popular events.
- An event category is a group of similar events (manually grouped).
- The more popular the event category is, the more likely the tweet will be re-tweeted by a user.
- Similar events have similar re-tweet patterns.
- A user who has re-tweeted frequently in the past is likely to re-tweet in the future.
- Most users are only interested in tweets under certain event categories. Most followers are users who have similar interests.
- Users' interests and preferences are assumed to be stable.

Event Category: In WISE 2012 Challenge, the given original tweets are annotated with some social events together with their corresponding keyword lists. It is difficult to automatically group events into different categories and it is neither in our focus in this report because some events are simply labelled by personal names or by location names. Moreover, their relevant keyword lists are arbitrary and do not show clear contextual information between the keyword list and the event title. To solve this problem, we manually divide the WISE 2012 provided 46 events that have links to Wikipedia pages into 12 categories such as Natural Disaster, Celebrities, Product Release, Sports, and etc.

3.1 Approach 1: Regression Based on Popularity and Connectivity

We develop a model to predict re-tweet activities based on event popularity and user connectivity by using a naïve approach. The intuition is that a tweet is more likely to be re-tweeted if it is about a popular event and its author is highly connected with others.

Prediction of Re-tweets: To compute the connectivity of a user, we design a function $C(uid)$ to find how many re-tweets a uid (user id) may have based on the number of followers she has. We divide users into 10 groups according to the number of followers and randomly pick 10 percent of users from each group to calculate the average number of re-tweets as the group statistics.

To compute event category popularity, we design a function $P(uid, category)$ to predict how the event category popularity influences a tweet being re-tweeted. For a given user, the average number of re-tweets is computed for each category in 3 levels within 30 days. If a given user has never posted any tweet belonging to the event category, we use the average number of re-tweets of the "No Category" instead. The formula for re-tweet prediction is shown as Eq. 1.

$$NumberRetweets = s(\alpha C(uid) + (1 - \alpha)P(uid, category)) \qquad (1)$$

where s is a scaling factor, as Functions C and P only consider up to 3 levels and α is a scaling of weights between Function C and P. Parameter α has been learned from statistics computed from the microblog dataset. We found that the ratio of re-tweeting by followers to non-followers is 3,352,996:261,811,184 (0.024:0.976). Therefore, we set α as 0.024. The scaling factor s has been derived from re-tweet records in the training set. We randomly pick 1000 tweets and calculate the average of s. This process is to be repeated 10 times and the average s over all is 19.950. Here, our equation is shown below:

$$NumberRetweets = 19.950(0.024C(uid) + 0.976P(uid, category)) \qquad (2)$$

Prediction of Possible Views: According to statistics, the percentage of re-tweets at first three levels are 61.73%, 28.50%, and 9.77% respectively. So, we assume that the number of re-tweets r computed from Eq. 2 is distributed accordingly. For each level, we randomly pick up l users who have a history of re-tweeting in the same event category. If the number of users in the category is less than l, we randomly pick up users from "No category". In this case, it is possible that the randomly picked users have no followers. Therefore, we repeat the process for 10 times and compute the average as the resulting prediction. If the number of followers of a given user is zero, the equation is

$$PViews = rtUsers + FollowerRTs \qquad (3)$$

Otherwise, the equation is

$$PViews = FollowerU + 0.976(rtUsers) + FollowerRTs \qquad (4)$$

where $FollowerU$ is the number of followers for a given user, $rtUsers$ is the number of re-tweeters, and $FollowerRTs$ is sum of the number of followers of re-tweeters.

3.2   Approach 2: Classification Based on User Preferences

As people's interests may differ, their interests in types of tweets would also differ. We refer this phenomenon as "user preference". User preferences are used to train a classifier to predict the possible number of re-tweets and the possible number of views in 30 days for a given original tweet. We firstly preprocess the re-tweeting dataset to assign users into certain event categories according to their re-tweeting activities. One of our event category is called "No Category" to indicate the interests of user is unknown. In our method, the process includes three steps as follows:

Computing Interestingness: Given an original tweet, we need to compute how possible a user will re-tweet the original tweet in the category. The candidate users are extracted from re-tweet history in a form of "who-retweet-who". We use $P(r, u, c)$ to denote the interestingness of candidate re-tweet user r to original user u on category c. The function is defined as Eq. 5.

$$P(r, u, c) = \sum RT(r, u, c) / \sum T(u, c) \qquad (5)$$

$RT(r, u, c)$ returns the number of re-tweets by user r from user u on category c. $T(u, c)$ returns the total number of u's tweets on category c.

Classifier Training: We build a classifier using user interestingness scores. During training, the classifier classifies every candidate user to "re-tweet" or "no-retweet", labelled as 1 or 0. For a candidate user, if the user has a high interestingness score on a category, the candidate user is likely to re-tweet the original user's tweet in the future. We use a threshold value λ to build the classifier, where $\lambda \in [0,1]$. The classifier $Q(r, u, c)$ is defined as:

$$Q(r, u, c) = \begin{cases} 1, \ if \ P(r, u, c) \geq \lambda \\ 0, \ if \ P(r, u, c) < \lambda \end{cases} \qquad (6)$$

In order to find the most suitable value for threshold λ, we carry out predictions on tweets from the training dataset with different λ values. For the space limit we omitted the details of this process. Our tests show that when $\lambda = 0.6$ it renders the best performance.

Prediction: we use the classifier described above (Eq. 6) and the function given in Eq. 5 to predict the possible re-tweets and views for each given tweet. The details are explained as follows:

1. Given an original tweet, get its current re-tweets and re-tweeters.
2. For each current re-tweeters, get its followers from the followship dataset as the next-level re-tweet candidates.
3. For each candidate re-tweeter r, compute its interestingness score and use classifier to classify if r will re-tweet the original tweet.
4. Accumulate the predicted number of re-tweets.
5. Repeat 1-4 until no more re-tweets and return the total number of re-tweets.

The number of possible views is computed based on the numbers of followers at every level. For a given tweet at the current level, if the current candidate re-tweeter r's re-tweet possibility is p and the number of r's followers is n, the current number of possible views is $p * n$.

3.3   Approach 3: Network Simulation Based on Re-tweeting Behaviors

The network simulation approach attempts to simulate the re-tweeting propagation starting from a root user who has posted an original tweet. Every user in the network is viewed as a node in a graph of users simulating the re-tweeting behaviors. This probabilistic approach is equivalent to a spanning tree from the root with probabilities predicting at a current user node, 1) if the tweet will be re-tweeted, 2) how many re-tweets will be received by the user, and 3) who will be the further re-tweeters. To address these three aspects, a model is built based on the following three factors.

First-Level Re-tweet: A first-level re-tweet is a re-tweet made directly from an original tweet. We found that the number of first-level re-tweets accounts for more than 50% of the total re-tweets an original tweet had received. For this reason, the average first-level re-tweets of the author of an original tweet is used to predict the number her tweet will be re-tweeted directly.

First-Level User Group and Event Trend: Given an event, users in the simulated network are partitioned into groups based on their average first-level re-tweets in the past. A trend of an event within a group is the average re-tweets on the event from levels 1 to 20. The ratio between two subsequent levels is used as the probability that a current tweet will be re-tweeted at the next level. The average number for the next level is also used as the upper bound for randomly selecting the number of re-tweets to be received at the next level based on a probability distribution. This trend model is built based on the re-tweet history of all events in the same event category.

Re-tweet User Group: We found that only 2% of re-tweets were made by followers and thus probability of using followship network is 2%. If a re-tweet is made by a follower, a follower is randomly selected according to a probability distribution based on total number of re-tweets posted by followers. Otherwise, a user is randomly selected based on re-tweet group probability distribution. Users are categorized into re-tweet groups based on the total number of re-tweets they posted in the past. The average number of re-tweets posted by the group users is used to build a probability distribution over all groups. The implementation of this approach follows a Monte Carlos method. The detailed algorithm is not presented here due to the limitation of the space.

3.4   Approach 4: Collaborative Filtering Based on Tweet Similarity

Collaborative filtering has been widely applied for prediction of interest ratings for a given item by using existing user profiles [5,6]. In our approach, each tweet is regarded as an item and is associated with an event category.

Data Structure: Given M items and K users, the user profiles can be represented by a $K \times M$ matrix, called user-item matrix X. A prediction on a test item m by a test user k is represented by $\hat{x}_{k,m}$. Accordingly, the prediction can be formulated as follows:

$$\hat{x}_{k,m} = PI_{k,m}(x_{k,b}), x_{k,b} \in S_i(i_m), x_{k,b} \neq \emptyset \tag{7}$$

$PI(\cdot)$ in Eq. 7 is a prediction function based on item similarity. $S_i(i_m)$ is a set of similar items to item i_m for user u_k . The prediction function can be constructed in different ways. We have compared a factor graph-based approach [7] with our approach that is designed to count the average of re-tweets per event. We find that our approach performs much faster with similar rating values because tweets in our approach are featured in terms of 12 event categories.

Item Similarity Measurement: In our case, each individual message tweeted by user is viewed as an item and the value of recurrences is regarded as a rating value discovered by out counting function. The more re-tweets an original tweet receives, the higher rating the item receives. We assign "0" or "1" to each attribute field for each event and we then have 12 dimensional binary feature vector for each event for 12 predefined event categories. For two arbitrary tweets Tw_A and Tw_B, we apply cosine similarity metric upon their dimensional event feature vectors as follows:

$$Similarity(Tw_A, Tw_B) = \frac{\sum_{i=1}^{n} A_i \times B_i}{\sqrt{\sum_{i=1}^{n} (A_i)^2} \times \sqrt{\sum_{i=1}^{n} (B_i)^2}} \tag{8}$$

We assume that tweets posted within a time frame contain similar concepts or close topics. When an event happens, a variety of tweets would be posted. Along with time, much newer discussions in tweets are likely to start over about the same topic. Moreover, the average duration starting from the time of posting to the time when it is re-tweeted can reflect the popularity of this tweet. We term it as the average response time and use it as a feature in our method for prediction. The numbers of prediction are formulated as follows:

$$\#retweet_p = \frac{1}{N} \sum_{i=1}^{N} \sum_{j=1}^{K} r \times x_{i,j}, x_{i,j} \neq 0, x_{i,j} \in S_i(retweet_p) \tag{9}$$

$$\#views_p = \frac{1}{N} \sum_{i=1}^{N} \sum_{j=1}^{K} f(x_{i,j}), x_{i,j} \neq 0, x_{i,j} \in S_i(retweet_p) \tag{10}$$

Note that $\|S_i(\cdot)\| = N, N \ll M$, r is a parameter that controls the ration of popularity and $f(\cdot)$ returns the number of followers of a user.

Table 3. Prediction time (second)

Approach 1	Approach 2	Approach 3	Approach 4
4	70	172	15

4   Experiments and Evaluation

4.1   Datasets

In training dataset, we have removed all the tweets belonging to the test events of WISE 2012 Challenge. For the classification approach (i.e., Approach 2), two datasets are used. For training, we extracted 970,125 original tweets and 5,690,837 re-tweet records by 330,386 re-tweet users. For testing, we extracted 11,718,465 related records from the followship dataset. The computations of four methods are all conducted on the PC computers with Core(TM) i7 vPro 2.93 GHz Intel processor and 4 GB of RAM. The average predicting times of four proposed approaches are shown in Table 3.

4.2   Evaluation

We repeat 10 times to randomly select 50 tweets to compute the prediction averages. The evaluation formulas are follows:

$$Prediction\_error = |Actual - Predict|/Actual \qquad (11)$$
$$Accuracy = Num\_CorrectMsgs/50 \qquad (12)$$
$$Average\_Accuracy = \sum_{i=1}^{10} Accuracy_i/10 \qquad (13)$$

where *Prediction_error* is a ratio for the correctly predicted tweets. We define the tweet that has *Prediction_error* $\leq Threshold$ as a correctly predicted tweet and count it as *Num_CorrectMsgs*, where *Threshold* is given as a range of prediction errors. *Accuracy* is the percentage of tweets correctly predicted in each round, and *Average_Accuracy* is the average *Accuracy* of 10 rounds.

Table 4 shows the performance of four approaches. Table 5 lists the predictions for the given 33 original tweets over 6 given events. For the submission of the final predicted results, we remove the smallest and the highest values from the outputs of our four proposed approaches. If any value is less than 100, we round it up as 100 as a minimum predicting value. This is sensible based on our understanding on the general popular events in Sina Weibo.

In Table 4, Approach 2 shows a better performance than others. We consider to give it the highest weight in computing a weighted average of all four approaches. The other three approaches can be ranked as Approach 4, Approach 3,

Table 4. Prediction accuracy with different error threshold of four approaches

Error thrshld	Number of re-tweets				Number of Possible views			
	1	2	3	4	1	2	3	4
0.05	0.02	**0.14**	0.05	0.09	0.05	**0.06**	0.03	0.04
0.10	0.05	**0.30**	0.09	0.19	0.11	**0.14**	0.03	0.09
0.20	0.12	**0.52**	0.14	0.32	0.20	**0.20**	0.05	0.11
0.30	0.22	**0.60**	0.22	0.47	0.26	**0.29**	0.09	0.19
0.40	0.23	**0.70**	0.31	0.62	0.35	**0.39**	0.15	0.24

Table 5. The 33 predicted re-tweets and views

Mid	Approach 1		Approach 2		Approach 3		Approach 4		Weighted average	
	#RT	#Views	#RT	#Views	#RT	#Views	#RT	#Views	#RT	#Views
Death of Steve Jobs										
8872263516485596	228	143678	127	173801	752	1322457	111	10919	147	160891
8872961090747701	135	176704	128	65379	126	360019	111	10919	127	113090
8872983825828431	184	82909	137	425005	19	43450	111	10919	126	73044
8872990233170214	126	66744	140	295061	154	189061	111	10919	137	97323
Fuzhou bombings										
2700059958269443492	476	263420	152	460423	472	584714	196	605635	306	485281
2700117991448817596	93	118802	132	320805	37	105144	195	605603	126	234232
2700176673306864228	223	127116	140	187647	378	518800	196	605635	203	253878
2701374467440601577	418	273773	222	96753	40	50416	195	605603	211	172619
2701431322360449433	10	4718	148	117670	138	250639	195	605603	145	144264
Japan earthquake										
51000180083282169	68	16705	157	224830	266	496369	131	209654	146	219771
51000180083492814	46	92660	142	309648	108	212937	131	209654	122	210748
51000180091104384	46	99404	172	331401	148	291499	131	209654	138	236936
55000180091534860	43	39049	147	141977	42	61202	131	209654	123	125822
55000180527027036	5	29301	134	93188	525	921005	132	209658	133	132011
58000180083553705	30	42697	740	536580	124	266343	131	209654	128	228550
Li Na win French Open tennis										
2709258383303085289	3	4751	260	16983	158	236517	115	131496	132	55154
2709864654666932643	33	6259	117	355694	101	149066	116	131512	110	137363
2709870697693881414	25	5676	114	342211	181	270933	116	131512	115	177986
2709871713230486085	53	73094	132	30334	97	125710	116	131512	110	86248
2709893077170155796	33	2530	130	393903	131	216153	116	131512	124	159726
Xiaomi release										
8896800636296312	20	4804	119	19834	44	39755	154	98647	113	23818
8896822338137478	95	22558	257	1441721	48	71543	154	98647	141	89612
8896858839607761	23	26385	136	15342	58	74233	154	98647	124	45524
8896889634186199	4	9345	178	2256	67	134972	154	98647	132	45066
8896952812610010	12	9142	129	377189	74	92723	155	98652	119	30937
Yao Jiaxin murder case										
2243526721410152330	232	117001	160	377189	447	863848	347	33628	318	265680
2243578214587694822	142	71686	142	145539	70	171629	347	33628	142	113888
51000185683084239O	170	57470	182	193383	305	500303	359	33710	223	135135
510001856834367317	39	7700	298	153050	63	221869	358	33701	232	113267
510001904903643837	946	353153	143	136935	767	2033042	350	33655	517	516156
510001908564754698	9	7022	616	247863	146	407791	350	33655	268	176460
5100019107401880	609	213670	170	290137	2054	3879120	357	33683	420	257365
55000190687383839O	31	65051	184	215313	838	2168249	348	33616	255	150915

and Approach 1 according to their overall accuracy regarding the given thresholds. The overall weighted averages are computed as follows:

$$\#NoRT = \frac{(4 \times rtA2) + (3 \times rtA4) + (2 \times rtA3) + rtA1}{\sum_i weight_i} \tag{14}$$

$$\#NoViews = \frac{(4 \times vA2) + (3 \times vA1) + (2 \times vA4) + vA3}{\sum_i weight_i} \qquad (15)$$

where $rtA1 - rtA4$ are the number of re-tweets from Approaches 1 to 4, $vA1 - vA4$ are the number of possible views from Approaches 1 to 4, and $weight_i$ is the weight of the approaches after the removal of the smallest and highest values.

5   Conclusions

Our final submission to the challenge is presented in the final column of Table 5 as the weighted averages. We choose this way because we find that each approach has its own advantages and disadvantages. Predictions of the first approach (i.e., regression) are based on user connectivity and event popularity of the previous similar event category. The limitation of this method is that the predictions can be the same when giving the same user id and the same event but different tweet ids. The second approach (i.e., classification) makes predictions based on users' re-tweet preferences in different tweet categories. The prediction accuracy is dependent on the partitioning of categories and the stability of user preferences. Predictions made by the third approach (i.e., network simulation) are highly relied on the average first level re-tweets of the authors whose original tweet is to be predicted. Hence, the performance is highly dependent on authors' history. The fourth approach (i.e., collaborative filtering) obtains predictions by considering contributions from top N similar tweets which need less tunes of parameters. However, both predicted re-tweets and possible views are heavily dependent on the item similarity metric. Moreover, prediction of possible views has only used followship hierarchical structure of social networks leading to an ignorance of tweet-retweet network structure. In our future work, we will retrospectively study the assumptions that we have made on the given datasets and develop a hybrid approach to integrate the proposed methods.

References

1. Boyd, D., Golder, S., Lotan, G.: Tweet, Tweet, Retweet: Conversational Aspects of Retweeting on Twitter, pp. 1–10. IEEE (2010)
2. Letierce, J., Passant, A., Breslin, J., Decker, S.: Understanding how twitter is used to spread scientific messages. In: ACM WebSci Conference 2011 (2010)
3. Galuba, W., Aberer, K., Chakraborty, D., et al.: Outtweeting the twitterers-predicting information cascades in microblogs. In: Proc. the 3rd Int. Conf. on Online Social Networks, p. 3. USENIX Association (2010)
4. Suh, B., Hong, L., Pirolli, P., Chi, E.H.: Want to be retweeted? large scale analytics on factors impacting retweet in twitter network. In: 2010 IEEE 2nd Int. Conf. on Social Computing, pp. 177–184 (2010)
5. Zaman, T.R., Herbrich, R., Stern, D.: Predicting information spreading in twitter. Computational Social Science and the Wisdom of Crowds 55, 1–4 (2010)
6. Stern, D., Herbrich, R., Graepel, T.: Matchbox: large scale online bayesian recommendations. In: Proceedings of WWW 2009, pp. 111–120. ACM (2009)

7. Yang, Z., Guo, J., Cai, K., Tang, J., Li, J., Zhang, L., Su, Z.: Understanding retweeting behaviors in social networks. In: CIKM, pp. 1633–1636 (2010)

8. Osborne, M., Lavrenko, V.: Rt to win! predicting message propagation in twitter. Artificial Intelligence 13, 586–589 (2011)

9. Hong, L., Dan, O., Davison, B.D.: Predicting popular messages in twitter. In: Proc. 20th Int. Conf. on Companion on World Wide Web, WWW 2011, pp. 57–58. ACM, New York (2011)

10. Peng, H., Zhu, J., Piao, D., Yan, R., Zhang, Y.: Retweet modeling using conditional random fields. In: ICDM 2011 Workshops, pp. 336–343. IEEE (2011)

11. Naveed, N., Gottron, T., Kunegis, J., et al.: Bad news travel fast: A content-based analysis of interestingness on twitter. In: ACM WebSci Conference, pp. 1–7 (2011)

Acolyte:
An In-Memory Social Network Query System

Ze Tang, Heng Lin, Kaiwei Li, Wentao Han, and Wenguang Chen

Department of Computer Science and Technology, Tsinghua University,
Beijing 100084, China
{tangz10,linheng11,lkw10,hwt04}@mails.tsinghua.edu.cn,
cwg@tsinghua.edu.cn

Abstract. WISE 2012 Challenge provides a data set crawled from Sina Weibo, which consists of hundreds of millions of anonymized following relationships and microblogs. In the performance track, 19 typical queries on the data set are given. Most of the queries are to get the top n users or microblogs which have the largest statistical numbers of some features during a specific period. The solution should focus on both throughput and latency in the execution of these queries. In this report, we present Acolyte, which is an in-memory query system that can solve the performance track efficiently.

1 Introduction

Social networking service grows rapidly in recent years. Popular SNS websites like Facebook and Twitter have more than one hundred million active users. The huge amount of user data and contents bring a lot of new challenges to the storage and query system.

The performance track of WISE 2012 Challenge provides a data set crawled from Sina Weibo, a popular microblog service website in China. The data set has $265,580,802$ pairs of following relationships and $369,797,719$ microblogs. The participants are required to build a system which can deal with 19 given queries. Most of the queries are to get the top n users or microblogs which have the largest statistical numbers of some feature during a specific period. The system should have good performance in both throughput and latency.

To achieve high performance, Acolyte, the system we built, loads all the data *in memory*. We use C++ programming language to write Acolyte for both fine-grained control on data structures and efficient execution. The data structures used in Acolyte were carefully designed to get small footprint, while not affect the performance of query. We also optimize several kinds of slow queries.

The rest of this report is organized as follows. Section 2 describes the design of our system, as well as implementation details. Section 3 elaborates optimization we take on several non-trivial queries. Section 4 presents and analyzes the experimental results. And finally, Section 5 discusses our experiences in doing this track, giving some related works.

X.S. Wang et al. (Eds.): WISE 2012, LNCS 7651, pp. 755–763, 2012.

2   System Design and Implementation

Our goal is to build a system which can process the 19 typical queries as quickly as possible on the given data set. Since the original data we need to query from is very large, commodity database systems such as MySQL[2] and MongoDB[1] do not work very efficiently. To achieve low latency and high throughput, our system remaps, reschedules all the data in the original data set and loads all the data in memory. We also choose a proper selection order and algorithm for each query.

In our design, the server side runs on one machine with comparably large memory (64 GB), and clients run on another machine which connect to server to get the query result.

As Figure 1 shows, we have 4 steps for each query. On the first step, the client transports the query information to the server. On the second step, the server parses the query and transforms all parameters of the query to integers. On the third step, the server creates a new thread to compute the result data. On the last step, the server transforms all the result indices to original strings and sends them back to the client.

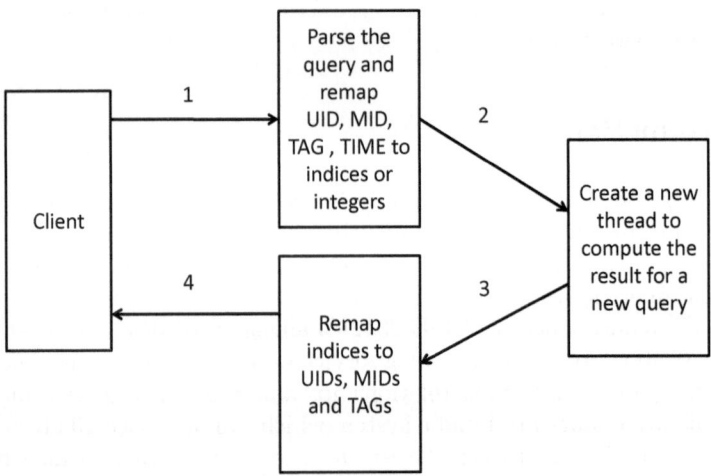

Fig. 1. The architecture of Acolyte

Acolyte does the following things when initializing.

1. Parsing the original data file, and generating the tables defined in the official documentation.
2. Remapping all the parsed tables, and generating new mapped data files.
3. Sorting and rescheduling all the mapped files to get the final version of data files Acolyte needs.
4. Reading the final data files into memory before any queries, and getting ready for processing queries from clients.

2.1   Parsing the Original Data File

We parse the original data file and generate 5 table files: *FRIENDLIST-TABLE*, *MICROBLOG-TABLE*, *EVENT-TABLE*, *MENTION-TABLE*, and *RETWEET-TABLE*. Each table file represents a table layout defined in the official documentation.

2.2   Remapping the Table Files

All the elements in the table files such as UID, MID, TAG, TIME are strings. If we simply load these strings into memory, it will cost a lot of space and a lot of time to locate or compare.

To reduce the response time as well as save memory, we remap UID, MID and TAG to a 32-bit integer. We collect all the UIDs, MIDs and TAGs appeared in the table-files, sort and count these strings, then map the original strings to 32-bit integers.

Take the *FRIENDLIST-TABLE* in Figure 2 as an example. The *UID-MAP-LIST* and the *FRIENDLIST-MAPPED-TABLE* corresponding to it are shown on the right.

For the same reason, we transform a TIME format string to a 32-bit integer representing the number of seconds from 1970-01-01 00:00:00 UTC (the epoch time).

After remapping all the table files, we generated 5 mapped files: *FRIENDLIST-MAPPED-TABLE*, *MICROBLOG-MAPPED-TABLE*, *EVENT-MAPPED-TABLE*, *MENTION-MAPPED-TABLE*, *RETWEET-MAPPED-TABLE*, and 3 map list files: *UID-MAP-LIST*, *MID-MAP-LIST*, and *TAG-MAP-LIST*.

Each of the mapped table has the same data as the original table, besides all the strings are transformed to integers. And the map list files tell us how to transform a UID, MID, or TAG to an integer, and vice versa.

2.3   Rescheduling the Table-Mapped Index Files

To reduce the latency of all queries, we have to reschedule the remapped table files and choose proper data structures to store all these files.

The Relationship Data Structure. The *FRIENDLIST-MAPPED-TABLE* file describes all the following relations in key-value pairs. This structure doesn't

FRIENDLIST-TABLE		UID-MAP-LIST		FRIENDLIST-TABLE-MAPPED	
UID	FRIENDID	UID	UID-index	UID-index	FRIENDID-index
HELEN	JOHN	HELEN	0	0	1
HELEN	TOM	JOHN	1	0	2
TOM	JERRY	TOM	2	2	3
		JERRY	3		

Fig. 2. Remapping example, transforming UID from strings to integers

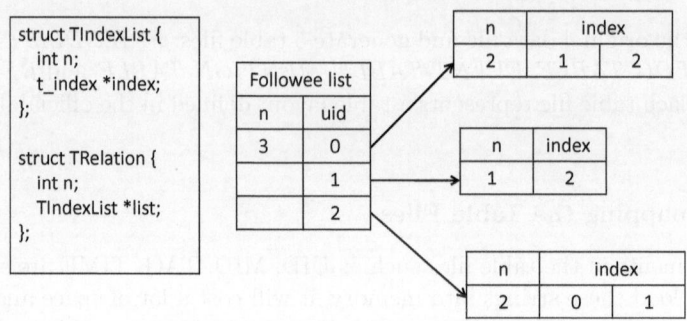

Fig. 3. Data structure for relationship

contain the graph features, so it is not very efficient for some types of queries. For example, when we ask all the followees of a given user U, we may need to scan all the rows.

We use adjacency lists to store the relations among users. We have three kinds of adjacency list: followee adjacency list, follower adjacency list, and R-friend adjacency list.

Figure 3 shows the adjacency list of a given example. In the followee adjacency list, each UID points to an array which represents the followees of it.

The Microblog Related Data Structure. The Microblog Table, Event Table, Mention Table and Retweet Table all have a primary field named MID. So we create a structure named *TMicroblog_all* combining these four tables in one list. *TMicroblog_all* has 6 attributes: *UID, MTIME, REMID, BE_RET_MID, MENTION_ID, TAG*. Each of the attributes is either an integer or an array. For the array attribute, we sort the elements of it to reduce the data locate time.

The C++ struct of *TMicroblog_all* is shown in Figure 4.

The Re-schedule List. The structure in Section 2.3 can be used to easily get all the information relative to a given microblog, but sometimes we need to get information relative to a given user or a given tag. So we need to generate some reschedule tables to achieve these goals. Figure 5 shows the extra lists that we generate.

```
struct TMicroblog_all {
        int n;
        int *uid;
        int *mtime;
        int *remid;
        TIndexList   **be_ret_mid;
        TIndexList   **mention_uid;
        TIndexList   **tag,
};
```

Fig. 4. Microblog related data structure

As shown in Figure 5, *microblog_time* stores all the MID indices and sorts these indices by the microblog times. We can use binary search to get all the MIDs in a given timespan very quickly.

The *miroblog_uid, mention_uid, event_tag* represent microblogs created by a user, mentioned by a user or attached a given tag. The MID indices in *microblog_uid.list[UID_i], mention_uid.list[UID_i], event_tag.list[TAG_i]* are sorted by microblog timestamps.

```
struct TMicroblog_time {
    int n;
    int *mid;
};

struct TList {
    int n;
    TIndexList *list;
};

Tmicroblog_time  microblog_time;
TList  microblog_uid;
TList  mention_uid;
TList  event_tag;
```

```
struct  TUID_desc {
    int n;
    int *uid;
};

TUID_Desc   Mention_uid_desc;
TUID_Desc   Retweet_uid_desc;
```

Fig. 5. Reschedule list data structure   **Fig. 6.** Two special lists used for Query 7 and Query 10

Figure 6 shows another 2 special lists used for optimizing Query 7 and Query 10.

Mention_uid_desc stores all UIDs which are mentioned by at least one microblog. And the UIDs are in decreasing order of the times being mentioned.

Retweet_uid_desc is similar to *Mention_uid_desc*, it stores all UIDs whose microblogs has retweeted by other microblogs at least once. The UIDs in *Retweet_uid_desc* are in decreasing order of the times being retweeted.

2.4   Initialization

On initialization step, we load all the data file described in Section 2.3 into memory. This needs about 50 GB memory. Then the server listens to incoming requests from clients. We use C++ to write the server side code and use Pthreads[5] to do multi-threading tasks.

3   Query Optimization

Using the data structures described above, we can get the data from memory very quickly. For each query, we can divide it into several steps. On each step, we read data from one list or calculate the data from previous steps. In this section, we emphasize the optimization taken to some queries. The other queries are straight-forward.

3.1   Query 6

Steps to get the answer:

- Use *Microblog_uid.list[UID_i]* and *Microblog_all.mention_uid[UID_i]* to get all UIDs which mentioned by A's microblogs
- Use *Mention_uid.list[UID_i]* to get all MIDs which mentioned these UIDs
- Use *Microblog_all.uid[MID_i]* to get all UIDs belong to the MIDs we get from the last step.

3.2   Query 7

We consider the UIDs in *Mention_uid_desc* one by one. For each *UID_i*, use binary search to count the MIDs which mentioned *UID_i* and in the given timespan. We use a heap to stores these candidate UIDs and the count of MIDs. The top of the heap stores the candidate UID which has the minimal count.

If the number of candidates in the heap is not larger than β (returncount), we can just insert the new UID and count number into the heap.

If the heap is full (number of candidates equals β), we compare the number of MIDs which mentioned *UID_i* (not consider the timespan) to the count number of heap-top candidate, if the total number is not larger than the count number of the heap-top candidate, we need not scan all the remaining UIDs in *Mention_uid_desc* from now on, because the total number of mentioned MIDs in *Mention_uid_desc* is in decreasing order. Each of the remaining UIDs cannot be insert into the heap later. We can break the scan loop to save time. Otherwise, we compare the MID count number of new UID to the heap-top one and replace the heap top by the new UID if it is better, then re-organize the heap.

We can easily find that the time complexity of these algorithm is $O(n_{\mathrm{UID}} \cdot (\log \beta + \log n_{\mathrm{mention\_MID}})$. While n_{UID} is the number of UIDs which be mentioned at least one microblog, and $n_{\mathrm{mention\_MID}}$ is the maximal number in all of *Mention_uid_desc.list[UID_i].n*, for any *UID_i*. As described above, for β is always much smaller than n_{UID}, so in most cases, we don't need to scan all the UIDs in *Mention_uid_desc.list[UID_i]*. Thus, the actual time complexity always much smaller than theoretical value.

The pseudocode of the algorithm is shown below:

```
initialize candidate_heap;
for (int i=0; i<Mention_uid_desc.n; i++) {
    UID_i = Mention_uid_desc.uid[i];
    Use binary search to count the number of MIDs which mentioned UID_i and
    in given timespan, name the count number c_MID.
    if (candidate_heap.size() < x) {
        candidate_heap.insert(pair<UID_i, c_MID>);
    } else
    if (Mention_uid.list[UID_i].n <= candidate_heap.top().c_MID) {
        break;
    } else
    if (c_MID > candidate_heap.top().c_MID) {
        candidate_heap.remove(candidate_heap.top());
```

```
        candidate_heap.insert(pair<UID_i, c_MID>);
    }
}
return candidate_heap;
```

3.3   Query 8

In this query, we first read the followee list to get all of A's two-level followees. For each user UID_i in A's two level followees, we can read $Microblog\_uid.list[UID\_i]$ to get all microblogs belong to UID_i. Since the microblogs in $Microblog\_uid.list[UID\_i]$ have been sorted by the microblog time, we just need to consider the last β ones.

We use a heap to store the candidate MIDs, the number of the candidates in the heap is not larger than β and the top of the heap stores the candidate MID which has the minimal microblog time in the heap. If the new MID can't insert to the candidate heap, we don't need to consider the remaining MIDs belong to this followee, because they are earlier than this one.

The pseudocode of the algorithm are shown below:

```
Select all of A's two level follwees
initialize candidate_heap;
For (UID in A's two level followees)
    For (in decreasing order, get last x MID_i in microblog_uid.list[UID]) {
        mtime = microblog_all.mtime[MID_i];
        If (candidate_heap.n < x) {
            candidate_heap.insert(pair<MID_i, mtime>)
            continue;
        }
        If (candidate_heap.top().mtime < microblog_all.mtime[MID_i]) {
            candidate_heap.remove(candidate_heap.top());
            candidate_heap.insert(pair<MID_i, mtime>);
        } else break;
    }
return candidate_heap;
```

4   Evaluation

We ran our experiment on a server with 2 Intel Xeon E5-2680 processors (4 physical cores and 8 logical cores each) and 64 GB main memory.

We choose $\beta = 50$, $\gamma = w$, and choose Query 7, Query 17, and Query 19 to observe their throughputs when α increases. As shown in Figure 7, the throughputs get increasing with α. When α reaches 25, the throughputs begin to drop. Since the server has just 16 physical cores, larger α means more costs in scheduling and shared processor units.

To evaluate the performance of each query, we calculate the average latency of all property setting cases for each query and each of the 19 latencies are shown in Figure 8. We can find that most of the average latency is less than 100 ms. All of the queries except Query 6 and Query 13 have an average latency less than 1000 ms.

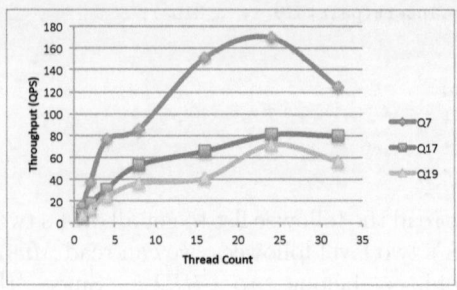

Fig. 7. Throughput under different thread count

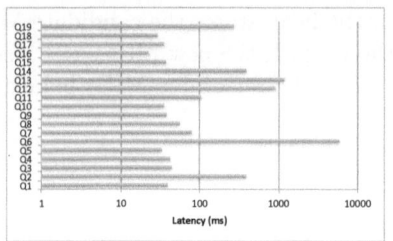 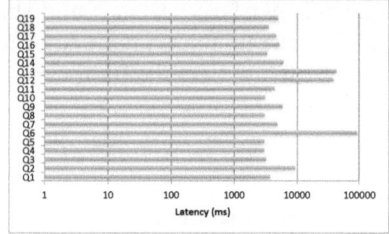

Fig. 8. Average latency of all the queries **Fig. 9.** Maximum latency of all the queries

Besides the average latency and throughputs, we also consider the maximum latency of all queries. Figure 9 shows the maximum latency of all property setting cases for each query. In the worst case, most of the queries can be finished within 5000 ms. Unfortunately, for some individual case, such as Query 6, the worst case needs the client to wait more than one minute. We have tried some simple parallel programming technology such as OpenMP to optimize these bad queries, but the speed-up isn't good enough and the optimizing can also reduce the throughput when more than one client are connecting to the server.

We also compare Acolyte to the solutions based on commodity databases MySQL and MongoDB. Figure 10 and Figure 11 show the latency and throughput on Query 5 when running Acolyte, MongoDB and MySQL. We can see

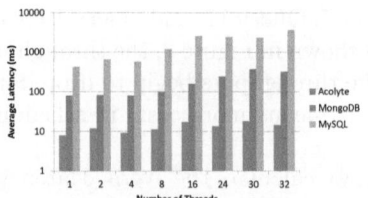

Fig. 10. Average latency of Query 5 using Acolyte, MongoDB and MySQL

Fig. 11. Throughput of Query 5 using Acolyte, MongoDB and MySQL

that Acolyte is about 10 times faster than MongoDB, and 40 times faster than MySQL. On other more complicated queries such as Query 7, Acolyte has even more speed-up. Although the solutions based on MySQL and MongoDB are quite preliminary, the results here indicate the performance of Acolyte is rather good.

5   Discussion

In this report, we present Acolyte, the system for fast social network query. The original data are transformed to get small memory footprint and fast access. All the data reside in memory, so it's much faster than disk-based solutions, which are the common cases for commodity databases. We use C++ programming language to write server side code to get more control over memory management and high performance. Some slow queries are further optimized to improve the overall performance.

There is some opportunity to leverage caching to get even better performance. To do so, the data of following relations and microblogs need to be rearranged by the graph structure. If the total amount of data goes beyond the main memory, flash disks and high-speed local area networks can be used.

Many recent works focus on large-scale graph related processing. Pregel[4] is a system built by Google to process large graphs in a distributed environment. Grace[6] is a graph-aware system which optimizes for fast computation based on graph traverse. WebGraph[3] is a high-ratio compression method to store graph structure and data. We believe that this topic of research is still growing in following years.

References

1. MongoDB, http://www.mongodb.org/
2. MySQL, http://www.mysql.com/
3. Boldi, P., Vigna, S.: The webgraph framework i: compression techniques. In: Proceedings of the 13th International Conference on World Wide Web, pp. 595–602. ACM (2004)
4. Malewicz, G., Austern, M.H., Bik, A.J.C., Dehnert, J.C., Horn, I., Leiser, N., Czajkowski, G.: Pregel: a system for large-scale graph processing. In: Proceedings of the 2010 International Conference on Management of Data, pp. 135–146. ACM (2010)
5. Nichols, B., Buttlar, D., Farrell, J.P.: Pthreads programming. O'Reilly & Associates, Inc., Sebastopol (1996)
6. Prabhakaran, V., Wu, M., Weng, X., McSherry, F., Zhou, L., Haridasan, M.: Managing large graphs on multi-cores with graph awareness. In: Proceedings of the 2012 USENIX Conference on USENIX Annual Technical Conference, pp. 182–193. USENIX Association (2012)

Accelerating Queries over Microblog Dataset via Grouping and Indexing Techniques

Lizhou Zheng, Xiaofeng Zhou, Zhenwen Lin, and Peiquan Jin

School of Computer Science and Technology,
University of Science and Technology of China, 230027, Hefei, China
jpq@ustc.edu.cn

Abstract. We present the ideas and methodologies that we used to address the WISE Challenge 2012 on accelerating different queries over microblog dataset. We employ a grouping and indexing mechanism as our main technique to handle this large dataset, through which the original dataset is partitioned and indexed according to selected attributes. Our experimental results show that the proposed methods are effective and efficient to answer complex queries over microblog dataset.

Keywords: Microblog, Query performance, Grouping, Indexing.

1 Introduction and Task Description

The WISE Challenge 2012 is based on a dataset collected from Sina Microblog [1]. The dataset contains over 200 million microblogs and 70 million users. This report is focused on the performance track. This track requires attendees to build a system for evaluating nineteen typical queries (denoted as Q1, Q2, ..., Q19, see http:// www.wise2012.cs. ucy.ac.cy/challenge.html) over the dataset and to achieve low response time and high throughput.

This report describes the query system we designed and implemented. We build our system in three steps: *Query Grouping*, *Data Preprocessing* and *Indexing*. In the *Query Grouping* step, we group the nineteen queries into five categories according to their query execution styles. In the *Data Preprocessing* step, we design new data formats for each group of queries to accelerate queries. In the *Indexing* step, we build indexes over the dataset produced by the data preprocessing step. In this stage, we focus on a modified B-tree index as well as the inverted index structure.

Our experimental results show that the proposed methods are effective and efficient to answer complex queries over microblog dataset.

2 The General Framework of Our Method

The framework of our method is shown in Fig.1. We first group the original dataset according to different attributes. Then we perform preprocessing on each partitioned dataset and generate files with specific formats. After that, we build indexes for the files of each small dataset.

X.S. Wang et al. (Eds.): WISE 2012, LNCS 7651, pp. 764–770, 2012.
© Springer-Verlag Berlin Heidelberg 2012

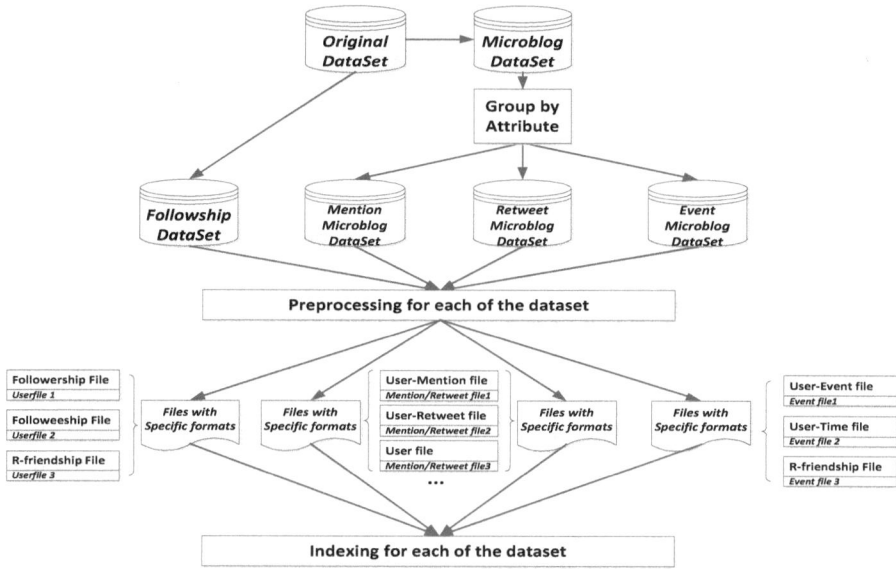

Fig. 1. The framework of our method

3   Query Grouping, Data Preprocessing and Indexing

We analyze the original dataset and queries carefully and group the queries into several categories according to the specific attributes needed when processing the queries. As a result, all the queries are classified into the following categories: *User-based queries, Mention-based queries, Retweet-based queries, Event-based queries* and *All Microblog-based Queries.*

Some of the original data formats provided are too simple and some are redundant. So we first introduce a preprocessing step for each group of queries. Our preprocessing mainly contains two parts: *Dataset-based preprocessing* and *Query-based preprocessing.*

(1) Dataset-Based Preprocessing. This type of preprocessing is applied in the whole dataset. Table 1 summarizes the dataset-based preprocessing rules.

Table 1. Dataset-based Preprocessing Rules

Rule	Description
1	Spilt the original dataset into small files by one day granularity
2	Extract tweets that contain event tag, split by one day granularity
3	Extract tweets that contain mention tag, split by one day granularity
4	Extract tweets that contain retweet tag, split by one day granularity
5	Remove useless information(tags and tag name)

(2) Query-Based Preprocessing. This type of preprocessing is designed specifically for different queries. The basic method is to create files with specific format according to different queries.

For *User-based Queries*, according to the three targets we need to achieve: finding a user's followee, follower and r-friend, we first statistic a user's follower, followee and r-friend information and create three files with specific formats, shown in Table 2.

Table 2. User-based Query Preprocessing Rules

Rule	Data Format (each line)	File Type
1	UserID%Follower count%follower1# follower2#...	UserFile1
2	UserID%Followee count% followee1# followee2#...	UserFile2
3	UserID%R-friend count% r-friend1# r-friend 2#...	UserFile3

For *Event-based Queries*, we first extract all microblogs with event tags then group them by tag name and split by one day granularity. After that we build files with specific formats. Event-user file contains information about a user mentioning an event in a day. User-event file is creating to find out what events a user mentioned in a day. Since searching large amount of users may cost much time, we create a User-Time file for each event which only contains the timestamp and user id of each microblog mentioning the event. The data formats of the three files are shown in Table 3.

Table 3. Event-based Query Preprocessing Rules

Rule	Data Format (each line)	File Type
1	UserID%mention count%time1# time2 #time3#...	Event-User File
2	UserID \t time \t Event tag	User-Event File
3	UserID \t time	User-Time File

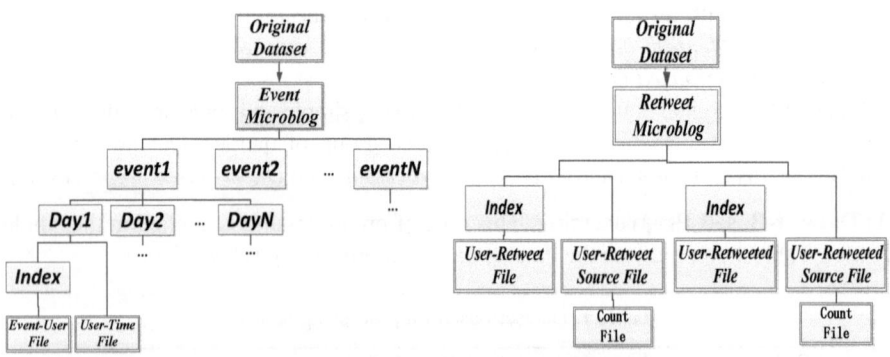

Fig. 2. The structure of event-based and retweet-based query preprocessing

For *Retweet/ Mention-based Queries*, since the two groups are similar, we take retweet-based group as representative. User-retweet file contains information about whose tweets a specific user retweets in a day. User-retweeted file is about whom and when a specific user's tweets are retweeted by. Source file contains only userID and time, and the total count of a user appearing in the file is stored in the count file (one day/ month granularity). The data formats of those files are listed in Table 4. The structure of event-based and retweet/mention-based preprocessing is shown in Fig 2.

Table 4. Mention-based Query Preprocessing Rules

Rule	Data Format (each line)	File Type
1	UserID%Mention count|uid1#time1|uid2#time2|...	User-Mention-File
2	UserID%Mentioned count|uid1#time1|uid2#time2|...	User-Mentioned-File
3	UserID \t time	Source-File
4	UserID \t count	Count-File

For *All Microblog-based Queries*, as there is only one query in this group (Q8), we just create a time-sorted file containing user id and microblog id for each microblog.

We build various indexes for each group of queries and choose the best index for each query based on experiment. Our main goal is to reduce the I/O time on physical disk. Since the physical disk sector size is 4KB as well as the RAM page size, it may cost extra time when reading different 4KB size of data, so in the indexing step we build an index for files in case of $\lceil file\_size/4096 \rceil$ > the index I/O count. Besides, in the experiment we found that the maximum layer count of our B-tree index is 3, so the maximum I/O count is 4 (one for root node, one for layer node, two for leaf node). Therefore we build an index for files which are larger than 16KB.

For user-based queries, we built a modified B-tree index which based on the lexicographic order of user ids and compared it with the inverted index. For event-based queries, we build a three-layer index (event - time - user). Here the events are indexed by their names, while time is indexed with a day granularity. For retweet/mention-based queries, a two layer index (time - user) is constructed.

In user layer we build a modified B-tree index that mainly modifies the leaf nodes of a traditional B-tree index. All non-leaf nodes in our index are organized as 4KB files. To minimize I/O count, we put data into the leaf node. The format of a leaf node file is shown in Fig.3. The first 4KB of a file indicates the offset and size of a data block, here offset represents the i-th 4KB in the file, and size represents how many 4KBs a data block contains. The non-leaf nodes could be loaded into RAM. For the purpose of comparison, we also use Lucene 3.5 to build a inverted index.

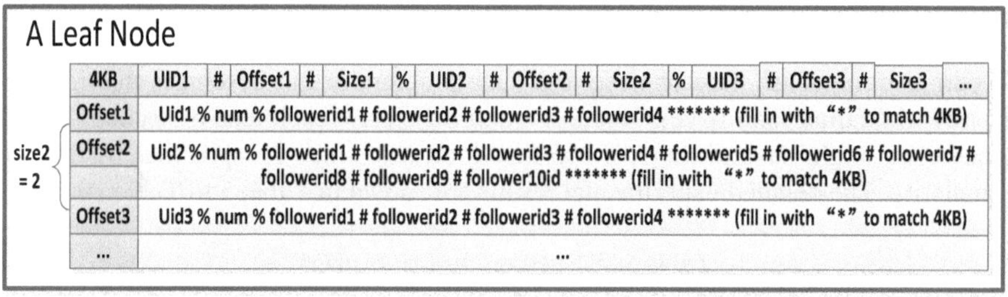

Fig. 3. The leaf node in the modified B-tree Index

4   Query Processing

n this section we describe the query processing of each query. Limited by the length of article, here we just briefly describe the process of the above-mentioned five types of queries. Note the retweet-based and mention-based queries are executed using the same mechanism.

(1) User-Based Queries. We select the modified B-tree index as the indexing mechanism of this group. We just use the index for querying.

(2) Event-Based Queries. In this group combine query of user with query of other attributes, as well as in Retweet/ Mention query group. Since the size of the user result set may be quite large, there are two ways of querying other attributes: using user index or user-time file (source file for retweet/ mention group). Using user index may be time-consuming for large user set, but for a small one, it is a good choice, so the choice is depend on specific query.

(3) Retweet-Based Queries. Those queries are performed using we the user-retweeted/ user-retweet file as well as the source file and count file. The choice is depend on specific query.

(4) Mention-Based Queries. Those queries are treated in a similar way with (3), except that we now use the user-mentioned file.

(5) All Microblog Queries. Since there's only one query in this group, we just traverse the time-sorted file until reaching the return count.

5   Results

We evaluate the performance of our system on a machine with configuration of Intel(R) Core(TM) i3-2120 CPU @ 3.30GHz and 4GB RAM. The operation system is Windows7 Ultimate and the maximum heap space of Java virtual machine is 2.5GB (JDK1.7). The methods we proposed perform well in the given nineteen queries and the modified B-tree index performed the best. Most of the user-based queries and queries with a time span of one hour or one day only need milliseconds of time to finish the execution. Most of the other queries (time span is a week or a year) have a second-level response time or even less. One problem of our system is that in some queries when time span is year, the execution time may reach tens of second. That is because we might have to traverse 365 files for a user since we split files by one day granularity. The detailed experimental results are shown in Table 5 to Table 10.

Table 5. The best results for Q1 to Q5

	Q1	Q2	Q3	Q4	Q5
Best response time	**20 ms** 44 ops/s	**1.4 s** 1.34 ops/s	**160 ms** 6 ops/s	**1.56 ms** 305 ops/s	**32 ms** 28.7 ops/s
Best throughput	**71 ms** **67 ops/s**	**3.28 s** **1.5 ops/s**	**647 ms** **11 ops/s**	**1.56 ms** **305 ops/s**	**43 ms** **39 ops/s**
Operation count = 80 or 100					

Table 6. Response time of Event-based Queries

		Q9	Q13	Q14	Q15	Q19
Best Response Time	Hour	151 ms / 6.4 ops/s	3.7 ms / 176 ops/s	4 ms / 138 ops/s	1.36 ms / 233 ops/s	149 ms / 6.5 ops/s
	Day	140 ms / 7 ops/s	7.8 ms / 93 ops/s	10 ms / 73 ops/s	5 ms / 125 ops/s	163 ms / 6 ops/s
	Week	167 ms / 10 ops/s	56 ms / 29 ops/s	28 ms / 31 ops/s	75 ms / 13 ops/s	396 ms / 4.75 ops/s
	year	308 ms / 3.2 ops/s	4.6 s / 0.2 ops/s	959 ms / 1 ops/s	240 ms / 4.1 ops/s	6.1 s / 0.16 ops/s
Operation count = 80 or 100						

Table 7. Throughput of Event-based Queries

		Q9	Q13	Q14	Q15	Q19
Best Throughput	Hour	487 ms / 11 ops/s	3.7 ms / 176 ops/s	4 ms / 138 ops/s	1.36 ms / 233 ops/s	254 ms / 12 ops/s
	Day	397 ms / 12 ops/s	7.8 ms / 93 ops/s	12 ms / 95 ops/s	5.4 ms / 150 ops/s	300 ms / 8.9 ops/s
	Week	167 ms / 10 ops/s	56 ms / 29 ops/s	61 ms / 43 ops/s	82 ms / 22 ops/s	704 ms / 5 ops/s
	year	315 ms / 6.2 ops/s	24 s / 0.43 ops/s	3.11 s / 2.3 ops/s	533 ms / 6.3 ops/s	19 s / 0.36 ops/s
Operation count = 80 or 100						

Table 8. Response time of Retweet & Mention-based Queries

		Q6	Q10	Q11	Q17
Best Response Time	Hour	4 ms / 85 ops/s	276 ms / 3.5 ops/s	3.9 ms / 190 ops/s	79 ms / 12.3 ops/s
	Day	11 ms / 100 ops/s	557 ms / 1.79 ops/s	7.6 ms / 102 ops/s	145 ms / 13.2 ops/s
	Week	246 ms / 4 ops/s	1.97 s / 0.52 ops/s	38 ms / 24 ops/s	708 ms / 1.41 ops/s
	year	43 s / 0.03 ops/s	5.1 s / 0.3 ops/s	2.7 s / 4.06 ops/s	11 s / 0.48 ops/s
Operation count = 24 (for year) or 80					

Table 9. Throughput of Retweet & Mention-based Queries

		Q6	Q10	Q11	Q17
Best Throughput	Hour	4 ms / 85 ops/s	1.3 s / 7.6 ops/s	3.9 ms / 190 ops/s	112 ms / 26 ops/s
	Day	11 ms / 100 ops/s	4.4 s / 3.12 ops/s	8.8 ms / 134 ops/s	464 ms / 13.5 ops/s
	Week	257 ms / 16 ops/s	8.6 s / 0.96 ops/s	44 ms / 46 ops/s	3.7 s / 3.5 ops/s
	year	78 s / 0.06 ops/s	5.3 s / 0.62 ops/s	2.7 s / 4.06 ops/s	11 s / 0.48 ops/s
Operation count = 24 (for year) or 80					

Table 10. The best result for Q8

Q8		
Return Count	**Best Response Time**	**Best Throughput**
10	**468 ms**, 3.33 ops/s	1.11 s, **8.67 ops/s**
50	**500 ms**, 11 ops/s	500 ms, **11 ops/s**
100	**162 ms**, 6 ops/s	1.23 s, **8.2 ops/s**
Operation count = 80		

Acknowledgements. This work is supported by the National Science Foundation of Anhui Province (NO. 1208085MG117), the National Science Foundation of China (no. 71273010), and the USTC Youth Innovation Foundation.

Classification-Based Prediction
on the Retweet Actions over Microblog Dataset

Lianshuai Zhang, Zequn Zhang, and Peiquan Jin

School of Computer Science and Technology,
University of Science and Technology of China, 230027, Hefei, China
jpq@ustc.edu.cn

Abstract. We present the ideas and methodologies that we used to address the
WISE Challenge 2012 on predicting the retweet actions over microblog dataset.
We employ classification and simulation mechanisms as our main technique, in
which the original event dataset is classified into specific categories and then
the similar curves are simulated to complete the task.

Keywords: Retweet actions, Microblog, Classification, Prediction.

1 Introduction and Task Description

The task of the Mining Track of WISE Challenge 2012 is to predict the re-tweeting
activities of thirty-three tweets involving six events. For each of those six events, only
tweets (and re-tweets) before a given timestamp are given in the original dataset. It is
required to predict two measurements at the time that the original tweet is published
30 days. These two measurements are:

(1) M1: The number of times that the original tweet is retweeted. If a user retweet (or
called re-post, or forward) a tweet twice at different timestamps, it should be counted
two times.

(2) M2: The number of times of possible-view of the original tweet. The number of
possible-view of one re-tweet activity is defined as the number of followers of the
user who conduct the re-tweet action. The number of times of possible-view of a
tweet is defined as the sum of all possible-view numbers of re-tweet actions.

Existing prediction methods are mainly focused on content [1, 2] and link structure
[3]. But based on the characteristics of the data, existing methods are not suitable for
this task. So we use a new method—a classification-simulation method to predict the
retweet actions of microblog users. The motivation comes from an observation of
microblog users' behaviors, which indicates that users usually show different interests
in different types of microblog events. Therefore, we try to first classify the original
events into several pre-defined groups, and conduct different techniques to deal with
those events. In particular, we will find similar events in the provided training dataset,
and construct a similarity-based prediction model to address the trend of the specific
type of events, which will be used to predict the retweet actions.

X.S. Wang et al. (Eds.): WISE 2012, LNCS 7651, pp. 771–776, 2012.
© Springer-Verlag Berlin Heidelberg 2012

In the rest of this report, we first present the general framework of our method in Section 2. After that we describe the details of our method in Section 3 and 4. Finally, the results are discussed in Section 5.

2   The General Framework of Our Method

In this report, we employ classification and simulation methods to complete the mining task. The general framework is shown in Fig.1.

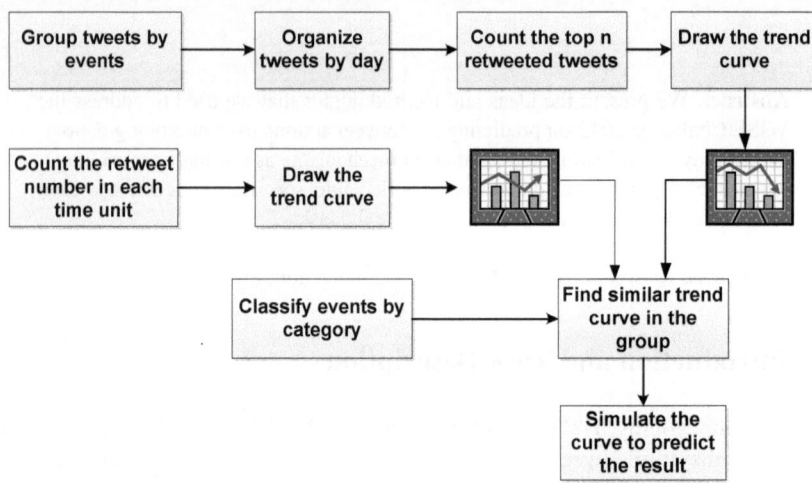

Fig. 1. The framework of our method

The training data set consists of all events except the test events in the given data which size is about 60G. Firstly, we collect the tweets reporting a specific event and generate event-oriented sets of tweets. For example, the tweets reporting "Windows Phone release" will be put into a bucket. Then for each event set, the collected tweets are organized by day. After that, the training events are classified into groups according to the similarity between events. The categories we adopt are shown in Table 1. For each group we deeply analyze the data and select the top n (n ≥ 3, and depends on the number of retweets) most retweeted tweets that were tweeted when the event happened according to the retweet message id (*rtmid*). Then, for each retweet message id got in the previous step, the number of times in a time unit that the original tweet is retweeted is counted. Based on the data we counted, we draw the trend curve in which the X axis represents time and the Y axis represents the retweet number. Fig.2 shows an example of the trend curve constructed on the training dataset.

For the retweet message ids in the test events, we also count the number of retweets in the same time unit and draw the trend curve (an example is given in Fig.2). Then we try to find the similar trend curves in the group that the event belonged to in the training events. Finally we simulate the trend curves of test message ids according to the curve(s) that is (are) similar to the test curves and calculate the values of M1 and M2.

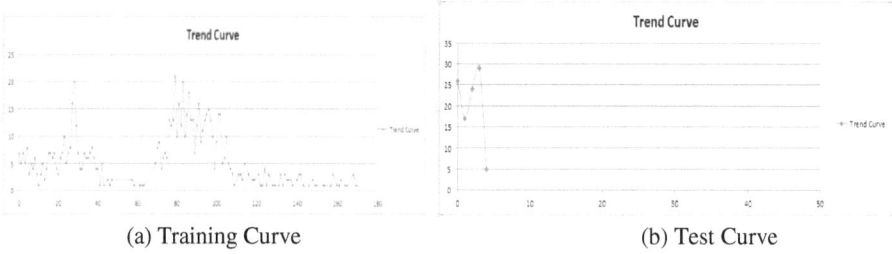

(a) Training Curve   (b) Test Curve

Fig. 2. Examples of the Trend Curve

3   Data Analysis and Preprocessing

In this section, we mainly introduce how we process the data and the basis we adopt to classify events into group.

3.1   Data Process

As the given data contains all the tweets, we have to process the whole data to find out the tweets about the events in the event list. When we do this job, we mainly base on the tags of event list and retweeted event list to classify a tweet into the event it belongs to. After we get every tweet about one event, we sort the tweets by day. Then, we find out the time when each event happened through Wikipedia and count the retweeted number of times of each original tweet after the event happened. We select top n most retweeted message ids from each event. It should be noted that for each event, n is a different value and depends on the popularity of the event.

3.2   Event Classification

The purpose of event classification is to find the events that are similar to test event. The similarity here means the two events have something in common. For example the event *Xiaomi release* and the event Windows Phone release are both about mobile phone release, the event Li Na win French Open in tennis and the event *Yao Ming retire* are both about sports, what's more, Li Na and Yao Ming are both famous sports player in China. All the test events, their similar events and the reasons why they are similar are listed in Table 1.

4   The Prediction Method

The task of the mining track is to predict the values of M1 and M2 at the time that the original tweet is published 30 days. But in the process of analyzing the data, we find that 30 days is a too long time, most of the retweet action disappeared far before that time. So at the time of prediction, we ignore the time limit condition.

Table 1. Similar Events and Categories

Test Event Name	Similar Event Name	Category
Death of Steve Jobs	Windows Phone release	About IT Industry
	Iphone 4s release	About Apple company
	Motorola was purchased by Google	About IT Industry
Fuzhou bombings	Wenzhou train collision	Public Traffic Security
	Foxconn bombing in Chengdu	Public Security
	Gansu school bus crash	Public Traffic Security
Japan Earthquake	The death of Osama Bin Laden	Important Event In Foreign
	Death of Muammar Gaddafi	Important Event In Foreign
	The death of Kim Jongil	Important Event In Foreign
	Yushu earthquake	Natural disasters
	Zhouqu landslide	Natural disasters
Li Na win French Open in tennis	Yao Ming retire	Famous sports player Important event in sports
Xiaomi release	Windows Phone release	Mobile phone release
	Iphone 4s release	Mobile phone release
	Motorola was purchased by Google	IT events
Yao Jiaxin murder case	Deng Yujiao incident	Both reflect social issues
	GuoMeimei	Issues about 2nd Official Generation and 2nd Rich Generation
	Case of running fast car in Hebei University	Issues about 2nd Official Generation and 2nd Rich Generation
	QianYunhui	Both reflect social issues

Out method is mainly based on classification and simulation. Classification has been done in the chapter of Event Classification.

In the process of simulation, we firstly draw the trend curves of the selected training *rtmids* and the test *rtmids* separately. Then we compare the trend curves between each test *rtmid* and each selected *rtmid* in its similar events. If the beginnings of two curves have the similar trend, we regard this *rtmid* in the similar events as one of the candidates. The work of deciding if two trends are similar is done manually. The last step is to calculate the values of M1 and M2 with the candidates. We use a simulation-based method to calculate the two values. The method is based on the hypothesis that if two trends are similar at the beginning, they are similar in the whole process of the development. Based on this hypothesis we need to shrink or magnify the ratio of the number of times that an original tweet is retweeted in a time unit in the train events, so that we can predict the number of times that an original tweet is retweeted in a time unit in the test events. The ratio for each message id in the test events is computed by formula (1).

$$ratio_{ij} = \frac{\sum_{k=1}^{n_i} \frac{NumOfTest_k}{NumOfTrain_k^j}}{n_i} \tag{1}$$

In formula (1), i represents in the ith message id in the test events and j represents the jth similar message id in the similar train events of i, $NumOfTest_k$ represents the number of times that the original message id is retweeted in the kth time unit in the test data, $NumOfTrain_k^j$ represents the number of times that the original message id is retweeted in the kth time unit of the jth similar message id in the similar train events of i, n_i is the number of points in the ith test trend curve.

Then, the M1 value of the ith message id in the test data is computed by formula (2).

$$M1_i = \frac{\sum_{j=1}^{m} M1_{ij}}{m} \tag{2}$$

$M1_{ij}$ is the M1 value of the ith original message id calculated by the jth message id in the train data (as shown in formula (3)).

$$M1_{ij} = ratio_{ij} \times M1_j \tag{3}$$

In formula (3), $M1_j$ is the M1 value of the jth message id in the train data.

The M2 value of the ith message id in the test data is computed by formula(4).

$$M2_i = \frac{\sum_{j=1}^{m} M2_{ij}}{m} \tag{4}$$

$M2_{ij}$ is the M2 value of the ith original message id calculated by the jth message id in the train data (as shown in formula (5)).

$$M2_{ij} = ratio_{ij} \times M2_j \tag{5}$$

In formula (5), $M2_j$ is the M2 value of the jth message id in the train data.

5   Results and Discussions

There are totally 6 events and 33 original message ids in the test events. We choose 16 events, 139 original message ids in the train data as the candidates. Finally we select 74 original message ids to predict the value of M1 and M2. The predicted result is shown in Table2.

Table 1. Predicted Results

Message Id	M1	M2	Message Id	M1	M2
8872983825828431	128	85053	2709870697693881414	3997	3346607
8872990233170214	55	68516	2709864654666932643	7994	6693214
8872961090747701	924	1337848	2709258383303085289	203	49516
8872263516485596	50	64308	8896800636296312	192	68202
2701374467440601577	655	787930	8896858839607761	80	176933
2701431322360449433	1001	883709	8896822338137478	42	9979
2700117991448817596	313	378010	8896952812610010	61	291391
2700176673306864228	270	284742	8896889634186199	9	37388
2700059958269443492	4044	2402834	510001856830842390	317	283474
58000180083553705	1635	1259864	510001856834367317	117	232493
51000180083282169	4223	2043272	5100019107401880	1236	814628
51000180083492814	1251	1511024	510001908564754698	340	470951
51000180091104384	1085	1203673	550001906873838396	1260	833126
55000180091534860	1668	896253	510001904903643837	6553	5145240
55000180527027036	988	758806	2243526721410152330	9575	7050972
2709893077170155796	612	458274	2243578214587694822	572	527390
2709871713230486085	1795	1212483			

Acknowledgements. This work is supported by the National Science Foundation of Anhui Province (NO. 1208085MG117), the National Science Foundation of China (no. 71273010), and the USTC Youth Innovation Foundation.

References

1. Tsur, O.: Ari Rappoport: What's in a hashtag?: content based prediction of the spread of ideas in microblogging communities. In: Proc. of WSDM, Seattle, WA, USA, pp. 643–652 (2012)
2. Macropol, K., Singh, A.K.: Content-based Modeling and Prediction of Information Dissemination. In: Proc. of ASONAM, Kaohsiung, Taiwan, pp. 21–28 (2011)
3. Lichtenwalter, R., Lussier, J.T., Chawla, N.V.: New perspectives and methods in link prediction. In: Proc. of SIGKDD, Washington, DC, USA, pp. 243–252 (2010)

Predicting Retweeting Behavior
Based on Autoregressive Moving Average Model

Zhilin Luo[1,2], Yue Wang[2], and Xintao Wu[2]

[1] Northwestern Polytechnic University, China
[2] University of North Carolina at Charlotte, USA
{zluo5,ywang91,xwu}@uncc.edu

Abstract. In this paper, we consider a fundamental social network issue that illustrates how information dynamically flows through a social media network. Inferring the number of times that a particular message posted by some specific user will be retweeted by his followers and predicting the number of readings of the posted message via various retweeting chains are central to understanding the underlying mechanism of the retweeting behaviors. Specifically we work on the Task 2 of the WISE 2012 Challenge, i.e., predicting retweet behaviors in the Sina Webo data set. We develop an approach based on the Autoregressive-Moving-Average (ARMA). In the approach, we treat retweeting activities of each original tweet as a time series where each value corresponds to the number of times that the original tweet is tweeted or the number of times of possible-view of the original tweet during that particular time period. For each tweet in the test data, our approach first identifies the most similar message from the training data based on the similarity between their time series values in the same length period as provided in the test tweet, fits the ARMA models over the whole time series of the identified message, and then applies the fitted model over the time series of the test tweet to predict future values. We report our prediction results and findings in this paper.

1   Introduction

Retweeting is one of the most important features in microblogging sites such as Twitter and Sina Weibo and examining retweeting behavior has been an active research area recently [5–8]. The retweeting mechanism empowers users to spread their ideas beyond the reach of the original tweet's followers. The process can be regarded as information diffusion in social media.

In this paper, we consider a fundamental social network issue that illustrates how information dynamically flows through a social media network. Inferring the number of times that a particular message posted by some specific user will be retweeted by his followers and predicting the number of readings of the posted message via various retweeting chains are central to understanding the underlying mechanism of the retweeting behaviors. In Microblogging, some particular messages posted by particular users are often retweeted widely and promptly

X.S. Wang et al. (Eds.): WISE 2012, LNCS 7651, pp. 777–782, 2012.

while others attract little attention from other users. Various factors including the message content, associate event, the timing, and the local structure of the followship network may influence the message propagation.

Specifically we work on the Task 2 of the WISE 2012 Challenge [1]. WISE 2012 Challenge is based on a data set collected from Sina Weibo, one of the most popular Microblogging service. The followship network is also provided. In the test data set, a small part of retweeting activities of thirty-three tweets of six events are given. In the challenge, we are required to predict the retweeting activities of those thirty-three tweets. Specifically, we are required to predict two measurements of the original tweet after 30 days:

1. M1: the number of times that the original tweet is retweeted.
2. M2: the number of times of possible-view of the original tweet. The number of possible-view of a tweet is defined as the sum of all possible-view numbers of retweet actions.

In this paper, we study the use of Autoregressive-Moving-Average(ARMA) models to predict retweeting behaviors. ARMA models are mathematical models of the persistence, or autocorrelation, in a time series. ARMA modeling is effective to understanding the physical system by revealing the physical process that builds persistence into the series and predicting the behavior of a time series from past values alone. We treat retweeting activities of each original tweet as a time series where each value corresponds to the number of times that the original tweet is tweeted (for M1) or the number of times of possible-view of the original tweet (for M2) during that particular time period. For each tweet in the test data, we first identify the most similar message from the training data based on the similarity between their time series values in the same length period as provided in the test tweet, fit the ARMA models over the whole time series of the identified message, and then apply the fitted models over the (short) time series of the test tweet to predict future values.

2   Predicting Retweeting Behavior

Autoregressive-moving-average(ARMA) models, also called Box-Jenkins models, are mathematical models of the persistence, or autocorrelation, in a time series. We will work with the mean-adjusted series

$$y_t = Y_t - \overline{Y}, t = 1, ..., N \tag{1}$$

where Y_t is the original time series, $\overline{Y}$ is the sample mean, and y_t is the mean-adjusted series.

The autoregressive model includes lagged terms on the time series itself, and the moving average model includes lagged terms on the residual. We acquire the ARMA model by including both types of lagged terms.

---

[1] http://www.wise2012.cs.ucy.ac.cy/challenge.html

Definition 1. *(Autoregressive-moving-average model [2]) The $ARMA(p,q)$ refers to the model with p autoregressive terms and q moving-average terms. This model contains the $AR(p)$ and $MA(q)$ models,*

$$y_t + \sum_{i=1..p} a_i y_{t-i} = e_t + \sum_{i=1..q} c_i e_{t-i} \tag{2}$$

where $a_1, ..., a_p$ are the autoregressive coefficients; $c_1, ..., c_q$ are the first-order,..., qth-order moving average coefficients; and $e_t, ..., e_{t-q}$ are the regression residuals at times $t, ..., t-q$.

Recall we have two prediction measurements: M1 (the number of times that the original tweet is retweeted) and M2 (the number of times of possible view of the original tweet). Retweeting activities of each original tweet are modeled as a time series. Specifically, for M1, we treat the number of retweeting times of a message as $\mathbf{y} = \{y_1, y_2, \cdots, y_N\}$, where $y_t (t = 1, \cdots, N)$ denotes the number of times that the original tweet is tweeted at the time stamp t. $\mathbf{y}$ can be extracted from the retweet traces. We leave detailed discussions on how to transform tweet traces to time series in Section 3.1. We use $\mathcal{Y}$ to denote the set of time series from all original tweets in the training data. Similarly we use $\mathcal{Z}$ to denote the set of time series from all original tweets in the test data. The retweeting activities of each tweet in the test data are modeled as $\mathbf{z} = \{z_1, \cdots, z_s, z_{s+1}, \cdots, z_N\}$ where $z_t (t = 1, \cdots, s)$ denotes the observed number of retweet times at the time stamp t and $z_t (t = s+1, \cdots, N)$ corresponds to the unknown number of future retweet times. Similarly, for M2, the time series can be extracted from the retweet traces and the followship network. The number of possible-view activity is defined as the number of followers of the user who conduct the retweet action. Each value in the time series denotes the number of possible-view at a particular time stamp.

Algorithm 1. ARMA based Prediction

Input: Training data $\mathcal{Y} = \{\mathbf{y_1}, \cdots, \mathbf{y_n}\}$
Tweet under test $\mathbf{z} = \{z_1, \cdots, z_s, z_{s+1}, \cdots, z_N\}$ where $z_{s+1}, \cdots, z_N$ are unknown
Output: $\sum_{t=1}^{N} z_t$
  1: Identify $\mathbf{y} \in \mathcal{Y}$ such that $\mathbf{y}$ and $\mathbf{z}$ are from the same user and $\sum_{t=1}^{s} (y_t - z_t)^2$ is minimized;
  2: Fit ARMA(p,q) over $\mathbf{y} = \{y_1, \cdots, y_N\}$;
  3: Apply the fitted ARMA(p,q) over $\mathbf{z}$ to generate z_t $(t = s+1, \cdots, N)$;
  4: Output $\sum_{t=1}^{N} z_t$.

Algorithm 1 shows our approach. For each testing message, we firstly identify the most similar message in the training data set by minimizing the Euclidean distance of the partial time series among candidate messages that are authored by the same user. Secondly, we use the whole time series of the identified message to build the ARMA model. Thirdly, we apply the fitted ARMA model on the test tweet to generate future values of the time series. Finally we acquire the prediction results.

ARMA modeling proceeds by a series of well-defined steps. The first step is to identify the model, which consists of determining the structure(AR,MA or ARMA) and the order of the model(p and q). The second step is to estimate the coefficients of the model [3]. The third step is to check the model, which ensures the residuals of the model are random and the estimated parameters are statistically significant. In the paper, we use the $'Forecasting'$ module from SPSS package for ARMA model fitting and forecasting. We use R-square and root-mean-square error (RMSE) to choose parameters p and q.

3   Evaluation

3.1   Data Preprocessing

The data set contains two types of data: tweets and followship network. Tweets includes basic information about tweets (time, user ID, message ID etc.), mentions (user IDs appearing in tweets), retweet paths, and whether containing links. In our prediction, we do not use information on whether containing links or topic information. The test data file contains retweets of thirty-three original messages composed by twenty-seven users. The thirty-three messages are from six events. The number of retweets for each original message in the test data is around 100. The retweet time period T for each original message can be easily derived from the field $rtTime$ and the field $time$ of the message's last retweet. We observe that the retweet periods of some messages are very short (less than 1 minute) whereas the retweet periods of other messages are quite long (more than 1 day). We then divide each time period T to s bins, and derive the time series $\mathbf{z}$ where each value z_t $(t = 1, \cdots, s)$ corresponds to the number of retweets at the time point t. We search the training data set to get candidate messages authored by the same user. For each candidate message, we follow the same strategy to derive its time series. We then identify the candidate message with time series $\mathbf{y}$ for ARMA model fitting such that $\sum_{t=1}^{s}(y_t - z_t)^2$ is minimized. We then run the fitted ARMA model over the test data $\mathbf{z}$ to derive future prediction values z_t $(t = s + 1, \cdots, N)$ where N corresponds to the time period of 30 days. Finally, we output $\sum_{t=1}^{N} z_t$ as the predicted number of times that the message is retweeted in 30 days.

To predict the number of possible-view (for M2), we need to incorporate the number of followers for users who retweet the message. We can easily derive this information from the followship network. The thirty-three messages are composed by twenty-seven authors. There are four users who are not contained the followship network provided in the challenge. We show the distribution of the number of tweets authored by these twenty-seven users in Figure 1(a) and the distribution of the number of total retweets of messages authored by each of these twenty-seven users in Figure 1(b). From the figures, we can observe the distribution of the number of composed tweets and the distribution of the number of times of posts being retweeted are unevenly distributed, which causes the difficulty of prediction tasks.

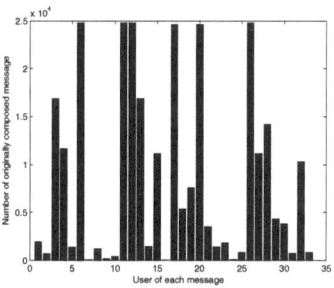

(a) Number of composed tweets

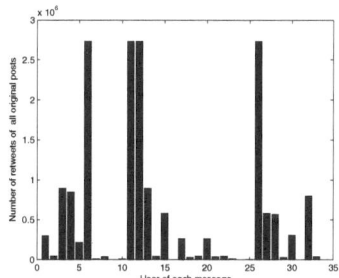

(b) Total number of times of posts being retweeted

Fig. 1. Statistics of twenty-seven users in the test data

3.2 Prediction Results

Figure 2 shows the prediction result respectively for M1: the number of retweeted times of each message after 30 days of original post, and M2: the number of covered users of each message after 30 days of original post. For M1, the thirty-three test tweets received on average 487 times of retweets in the period of 30 days. The second message from the event of *Death of Steve Jobs* received the largest number of retweets (3862) whereas the second message from the event of *Xiaomi Release* received the smallest number of retweets (107). For M2, the thirty-three test tweets received on average 147,831 times of possible-view in the period of 30 days. The first message from the event of *Yao Jiaxin Murder Case* received the largest number of possible-view (1,428,387) whereas the last message from the event of *Xiaomi Release* received the smallest number of possible-view (4,200).

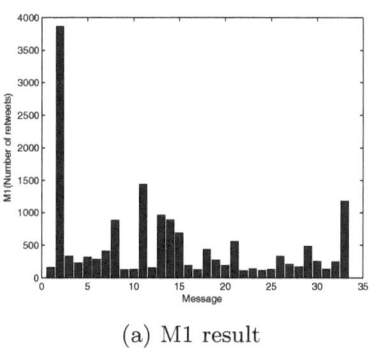

(a) M1 result

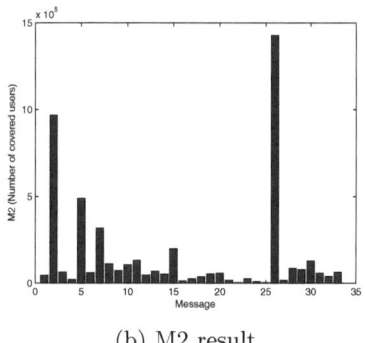

(b) M2 result

Fig. 2. Prediction results

4   Conclusion and Future Work

In this paper, we introduced our approach based on ARMA models to predict retweeting behaviors in WISE 2012 Challenge. We expect that ARMA models can be used to help understand the underlying mechanism of retweeting behaviors. There are some other aspects of this work that merit further research. Among them, we will continue the line of this research by incorporating various factors (e.g., event information associated with a tweet and the fine-grained local topological structure of the followship network) in the modeling process. We will also explore advanced fitting strategies of ARMA models. In analyzing ARMA time series, it is typically assumed that only one realization is available for model fitting. In our current approach, for a given test tweet, we identify the most similar message to build the ARMA models. It is preferable to use multiple similar messages (from the same author or with similar topics) for model fitting. One strategy is that the elements of each time series are averaged across each time point to produce one series for analysis. In [1], the authors show that multiple independent time series from the same ARMA process can be represented by a single univariate ARMA time series through an interleaving of the original series. We will explore these strategies in our future work.

Acknowledgments. This work was supported in part by U.S. National Science Foundation IIS-0546027, CNS-0831204, CCF-0915059, and CCF-1047621. We would like to thank Sina Weibo to provide the retweet data set to the research community and thank the WISE organizers to provide this opportunity for us to participate this challenge. For detailed results, please refer to [4].

References

1. Bowden, R.S., Clarke, B.R.: A single series representation of multiple independent arma processes. Journal of Time Series Analysis 33(2), 304–311 (2012)
2. Brockwell, P., Davis, R.: Time series: theory and methods. Springer (2009)
3. Chatfield, C.: The analysis of time series: an introduction, vol. 59. CRC Press (2004)
4. Luo, Z., Wang, Y., Wu, X.: Predicting retweeting behavior based on autogressive moving average model. Technical Report, UNC Charlotte (2012)
5. Luo, Z., Wu, X., Cai, W., Peng, D.: Examining multi-factor interactions in microblogging based on log-linear modeling. In: ASONAM (2012)
6. Macskassy, S.A., Michelson, M.: Why do people retweet? anti-homophily wins the day!. In: ICWSM (2011)
7. Romero, D.M., Meeder, B., Kleinberg, J.: Differences in the mechanics of information diffusion across topics: idioms, political hashtags, and complex contagion on twitter. In: WWW (2011)
8. Yang, Z., Guo, J., Cai, K., Tang, J., Li, J., Zhang, L., Su, Z.: Understanding retweeting behaviors in social networks. In: CIKM (2010)

A Fast and High Throughput
SQL Query System for Big Data

Feng Zhu, Jie Liu, and Lijie Xu

Technology Center of Software Engineering,
Institute of Software, Chinese Academy of Sciences, Beijing, China 100190
{zhufeng10,ljie,xulijie09}@otcaix.iscas.ac.cn

Abstract. Relational data query always plays an important role in data analysis. But how to scale out the traditional SQL query system is a challenging problem. In this paper, we introduce a fast, high throughput and scalable system to perform read-only SQL well with the advantage of NoSQL's distributed architecture. We adopt HBase as the storage layer and design a distributed query engine (DQE) collaborating with it to perform SQL queries. Our system also contains distinctive index and cache mechanisms to accelerate query processing. Finally, we evaluate our system with real-world big data crawled from Sina Weibo and it achieves good performance under nineteen representative SQL queries.

Keywords: Big Data, Query Processing, NoSQL, HBase, MapReduce.

1 Introduction

Many analytical tasks perform on ever growing massive data. As before, structured data storage and query still play an important role in the big data analysis. But for plenty of online Internet services, relational data queries require high performance, including scalability, low-latency and high-throughput. However, commonly used software is facing too many difficulties to store, manage and process big data, not to mention achieving these three criteria simultaneously.

We analyze the Internet service SQL logic and find that most SQL are read-only (i.e., no insert, update and delete). At the same time, NoSQL systems provide horizontally scalable storage and high-performance *"get(key)"* operations even under heavy read/write workloads. So our idea is to store the structured data in terms of **Key-Value** and convert SQL into a series of imperative operations written against distributed Key-Value stores. This is called **denormalization**. In this way, queries which need to scan large ranges of records or entire tables can be divided into lightweight distributed operations.

Based on the above discussion, we finally choose HBase [1] as our storage layer. There are some reasons: first, it is a distributed architecture and we only need to add servers to increase its scalability; second, the large table is automatically split into small regions. HBase itself can manage the metadata of these regions; third, its data model supports a flexible schema design. We can conveniently add some attributes to

X.S. Wang et al. (Eds.): WISE 2012, LNCS 7651, pp. 783–788, 2012.

the schema; the most important and the last one, HBase is built on top of Hadoop and provides a simple interface. We can process the data very expediently. However, HBase's simple key-value data model cannot satisfy various types of complex queries. So we add a distributed query engine integrated with HBase. The query engine is responsible for collecting and merging query results to clients.

This rest of the paper is organized as follows. Section 2 introduces our data modeling and denormalization method. Section 3 describes our detailed architecture and key technologies. The experiment is illustrated in section 4. Section 5 concludes the paper.

2   Data Modeling

Since NoSQL's data model is flexible and related to specific queries, we should first investigate the contest given SQLs. We find three significant features from them that strengthen our belief on NoSQL solution: (1) read-only queries, no write or update (2) fixed structured data without modification (3) query type is known beforehand with limited variable parameters. Based on these features, we can convert the relational data into Key-Value records and modify SQL into a series of "get(key)" operations. The main problem is that NoSQL does not support JOIN operation natively. We need to handle it at design time by applying "JOIN-free" technique. In other words, we create and partition joined table in advance, so subsequent queries can be done by just looking up the result table. This idea is simple but can dramatically reduce the query latency. Other complex and time-cost queries can be done in the same way.

Formally, this method can be defined as "Denormalization". In opposite to normalization in relational algebra, "Denormalization" encourages to store data in a query-friendly form to simplify query processing. Here, we not only tackle with JOIN, but also let structured data fit Key-Value pattern for scalability and high performance. Note that denormalization may increase our total data volume because of the different forms of duplicated data.

Now, we take the first contest query as an example to illustrate how denormalization works. The first query is to find the Top N suggested followees for user A. User A's followees are users who were followed by A. User A's r-friends are users who followed A and were followed by A. The recommendation algorithm is that: ***get all r-friends of A's r-friends, filter A's r-friends, order them by the number of people in A's r-friends list connecting to them, and then select the top N of them.***

The most time-consuming part arises from the table JOIN operation. To avoid it, we need to pre-create and partition A's r-friend table. If A's r-friend table is available, our query engine selects out all r-friends of A's r-friends, filtering A's r-friends and count the number of each user's connection with them; then the query engine returns the top-x user ids to the client.

Furthermore, the nineteen queries can be divided into two types: (1) Queries need no denormalization. These queries do no need join operation or index. (2) Queries need denormalization. These queries always contain complex operations. Several approaches such as building a secondary index, pre-joining tables and so on can be used to reduce their time cost.

3   System Architecture

Figure 1 shows the main components of our query system. The underlying file system is Hadoop Distributed File System (HDFS) [2, 3]. HDFS creates multiple replicas of data blocks and distributes them to different nodes. Based on HDFS, HBase plays the role of the database. We store all the preprocessed tables into HBase. Apart from the operations of "Query by key" and "Filter" provided by HBase, we also use coprocessor in HBase for complex queries.

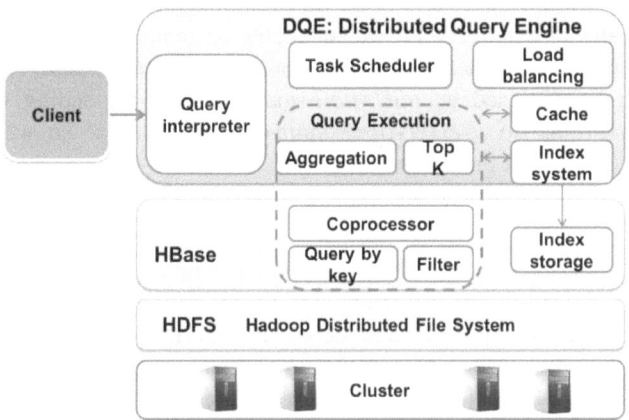

Fig. 1. System architecture

The upper layer is the distributed query engine (DQE). It is a logical processing component of this query system. To achieve high concurrency, we hold the principle of distributing the workloads to multiple nodes. Therefore, the DQE, as a middleware between the client and storage layer, adopts the master-slave structure. The master node is responsible for load balancing, aggregation, top-k selection and so on. Task Scheduler is responsible for distributing the query requests to various slave nodes.

Our system also contains distinctive index and cache mechanisms to accelerate query processing. Because the variables of nineteen queries in this contest are selected from specific collections, the results of these queries are limited. The cache may achieve 100% hit rate. **To test the real performance of our underlying system, we get the experiment data without cache system.**

Next, we list some key technologies while implementing the 19 query interfaces.

- **Data Load**

How to rapidly load the Big Data to the storage system is a basic problem before data query. With the help of Hadoop interfaces, we first upload all the data files to HDFS. Then, we use MapReduce [4] jobs to generate desirable tables. According to the logic of queries, we decide whether to store the generated tables in HDFS for other MapReduce jobs or in HBase for direct queries.

- **Row key design**

As a column-oriented database, HBase stores the data with row key in a lexicographic ascending order. Therefore, it supports not only single key query but also range query. To make the best use of this feature, designing a well-suited row key is very important. Like multi-dimension index, we sometimes combine several attributes as a composite row key. With the composite key, the value can be fetched effectively. Here we take several typical queries as examples to show this skill.

Among these nineteen queries, it is common to get the data in a time range. To avoid scanning a whole table, we put the time attribute into the row key. For example, after de-normalization in query 11, we get a table containing three items: *(uid1, time, uid2)*, which means user *uid1* is re-tweeted by the user *uid2* at *time*. In this case, the row key can be designed as *uid1+time+uid2*. The corresponding range is *userID+timestamp+uidx* to *userID+(timestamp+timespan)+uidy*. So we just need to scan this range to extract the total *uid2* set.

- **Key-list data model**

Most data relationship can be modeled as key-list model. For example, a tag may correspond to a large number of microblog and even more users.

Formally, we define key-list data model as "k-$v_1(attri_1)$, $v_2(attri_2)$, $v_3(attri_3)$...".Item $v_j(attri_j)$ implies that the j-th value belongs to attribute $attri_j$. Then the problem can be defined as: given an attribute set SA and a key k, we need to retrieve the corresponding value set SV from k's values. SV= {v_j | $attri_j \in SA$ && $v_j \in$ value list (k)}.

To make the best use of HBase's data model feature. We first consider and evaluate several table schemas below.

— Link the values together as a single value V, V= v_1 ($attri_1$) +v_2 ($attri_2$) +v_3 ($attri_3$)... and the row key is designed as k.

Row key	Column family
k	v_1 ($attri_1$) +v_2 ($attri_2$) +v_3 ($attri_3$)...

— Consider various values as different columns, then the schema can be designed as:

Row key	Column family			
k	v1 (attri1)	v2 (attri2)	v3 (attri3)	...

— Huge number of columns has negative affection on HBase. Considering putting the values into row key and the corresponding column is the attribute. Then the schema can be designed as:

Row key	Column family
k+ v_i	$attri_i$

— With the previous three designed schemas, we need to scan the whole value set for selecting SV. However, when the attribute set SA satisfies the sort condition, for example the *datetime*, row key can be designed with the attribute.

Row key	Column family
k+ attri$_i$+ v$_i$	...

The different schemas design fit for different cases, a comprehensive consideration is necessary.

- **Coprocessor Endpoint**

For those queries need to scan a wide range of the table or even the whole table, HBase coprocessor is a good choice. Resembling stored procedures in RDBS, coprocessor endpoint is powerful. A table stored in HBase is split into various regions. In the DQE, we invoke the coprocessor endpoint implementation for a table. It will execute in parallel on each region and returns the partial results to the DQE.

For coprocessor endpoint, there is also a tradeoff between the degree of parallelism and network traffic. More regions means less time for scanning a region, but it will cost more memory or even may block the network IO. Therefore, we should carefully set a proper region number.

4   Experimental Evaluation

This contest released a sample of structured big data from microblog. The dataset contains two parts: the first one is "Followship network" of 12.8GB and the second one is "Tweets" of 61.7GB. Through denormalization, we create about 30 tables and the total volume of data is over 300GB.   Our query system is deployed on a cluster with 10 nodes. The configuration of each node is: CPU: Intel(R) Core™ i7-2600 3.4GHz; CPU cores: 4; Memory: 4 * 4GB DDR3; Hard Disk Capacity: 2 * 1TB; OS: Ubuntu-11.04 x86_64; Network: 1Gb/s.

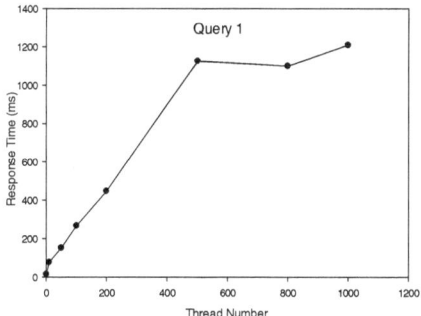
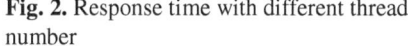

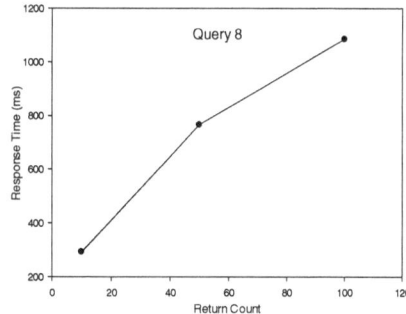

Fig. 2. Response time with different thread number

Fig. 3. Response time with different return count

Thread number, return count and time range are three main factors that affect the performance dramatically. Because there are so many experiments, here we just present a sample of the results and analyze the influence of each factor on response time. (1) Figure-2 shows the relation between response time and thread number of query 1. The return count is 50. (2) Figure-3 shows the influence of return count.

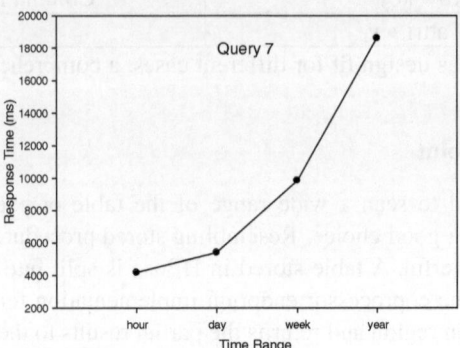

Fig. 4. Response time with different time range

The thread number of query 8 is 10. (3) Figure-4 shows the time range's influence on the scan rate. The thread number is of query 7 is 1.

Above figures just describe three main factors' general influence on performance. Our system actually performs well under different experiment configurations. The results show it is scalable. However, there may be data skew problem. For example, the number of user's followees may vary from one to thousands, which leads to the unpredictable results. Detailed results are available in separate files generated by BSMA tool.

5   Conclusion

In this paper, we first propose a denormalization method. This method is to store the structured data in terms of *Key-Value* and convert SQL into a series of operations written against distributed Key-Value stores. Then we present a query system based on HBase. Experiment shows the system can support high throughput query request with low latency. In the future, we will research how to support complex queries.

Acknowledgements. This work was partially supported by the National Natural Science Foundation of China (61170074), the National Grand Fundamental Research 973 Program of China (2009CB320704), the National Key Technology R&D Program (2012BAH05F02), National Core-High-Base Major Project of China (2010ZX01042-001-001-05).

References

1. Apache HBase, http://hbase.apache.org/
2. Apache Hadoop, http://hadoop.apache.org/
3. HDFS, http://hadoop.apache.org/hdfs/
4. Dean, J., Ghemawat, S.: MapReduce: Simplified Data Processing on Large Clusters. In: OSDI 2004 (2004)

Do-It-Yourself Content Delivery Network Orchestrator

Rajiv Ranjan, Karan Mitra, Suhit Saha,
Dimitrios Georgakopoulos, and Arkady Zaslavsky

CSIRO ICT Center, Acton
Australia, ACT 2601
{ran12a,mit20c,sah007,geo064,zas005}@csiro.au

Content delivery networks (CDNs) [1] provide fast and reliable content access to the end-users. CDN providers (e.g., Akamai [2]), either own the entire infrastructure or it is outsourced to a single Cloud provider. Content owners (e.g., clients and end-users) need to establish expensive contracts with third party ISPs or CDN providers. Hence, existing CDN services are out of reach for all but large enterprises. Current CDNs do not provide services that allow an end-user to create dynamic content such as combining music videos from an existing content source on the Internet. Finally, the content owners do not have low-level control over the orchestration operations such as, multiple Cloud provider selection and resource management for hosting content. Hence, the content owners are dependent on their CDN providers to perform these operations behind the scene.

In this paper, we present MediaWise Cloud Content Orchestrator (MCCO) — a novel system that facilitates do-it-yourself CDN orchestration for simplifying the management of media content (e.g., video) using Cloud services. Unlike existing commercial CDN providers such as Limelight Networks [3] and Akamai [2], MCCO eliminates the need to own and manage expensive infrastructure while facilitating content owner requirements pertaining to price, SLA, privacy and QoS. It offers enhanced flexibility and elasticity as it supports pay-as-you-go model. MCCO content orchestration operations include: (i) production: create and edit; (ii) storage: uploading and scaling of storage space; (iii) keyword-based content tagging and searching and (iv) distribution: streaming and downloading. MCCO capabilities span across a range of operations such as selection, assembly, deployment of services and monitoring their run-time performance (e.g., load, availability, throughput, etc.). MCCO is developed using Java, Cloud services, and open-source Cloud APIs. It supports deployment, configuration and monitoring of content and Cloud services using Web-based widgets. These widgets hide the underlying complexity related to Cloud services and provide *an easy do-it-yourself* interface for content management.

The MCCO architecture shown in Fig. 1(a), consists of two layers. Widget Layer encapsulates user interface components in the form of six principle widgets including Appliance, Instance, Storage, Monitor, Content, and Security. Programming Layer implements the logic for the interface exposed by these widgets. For example, the Media Appliance Manager implements Cloud service API that allows Appliance Widget to list the set of media appliances (e.g., streaming, indexing and editing servers) associated with owner's account. Programming Layer is also designed to

X.S. Wang et al. (Eds.): WISE 2012, LNCS 7651, pp. 789–791, 2012.
© Springer-Verlag Berlin Heidelberg 2012

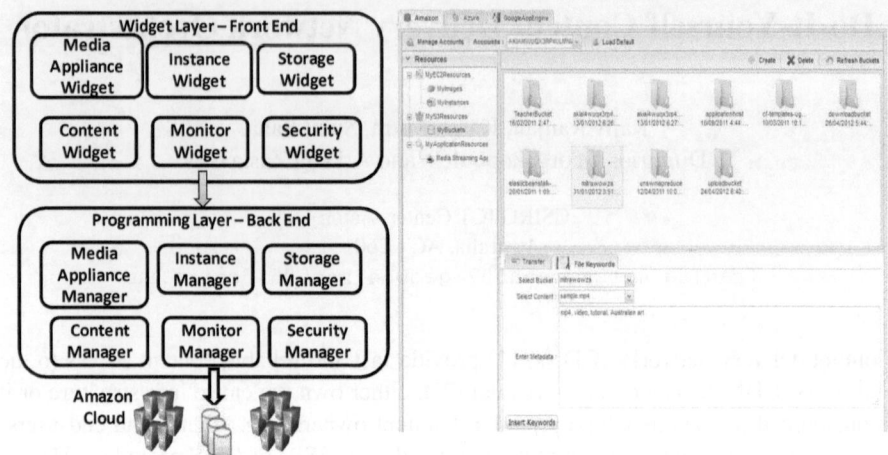

(a) MCCO Architecture.   (b) A screenshot of Content Widget.

Fig. 1.

allow engineers to plug-in different Cloud service APIs. Currently, our implementation works with Amazon Web Service (AWS) [4] and is being extended to support other Cloud providers. The Security Widget manages the authentication and authorization for content orchestration. The basic configuration of media appliances including their architecture, image id, state and virtualization platform is managed via Appliance Widget. The Storage Widget allows owners to upload content and media appliances to Amazon S3. Instance Widget enables the control for deployment (e.g. start, stop and termination) of media appliances. It is also used to select the media appliance deployment configuration parameters including number of instances, their types and SSH key pairs. Content Widget as shown in Fig. 1(b) enables the functionality for tagging content with metadata and its deployment to a media appliance (e.g., streaming server). Monitor Widget is used to monitor the status of media appliance instances, network and storage services for example, CPU load and memory utilization on per media appliance basis.

To the best of our knowledge, no existing CDN providers support: (i) do-it-yourself content and Cloud service orchestration; (ii) drag and drop based deployment of content to media appliance; and (iii) controlling lifecycle activities of both content (upload, edit, delete, stream, etc.) and cloud services (e.g. start, stop, refresh, and undeploy) via shared widgets. This demonstration highlights the effectiveness of MCCO in simplifying the process of end-to-end CDN orchestration operations. We also demonstrate that prior knowledge of existing Cloud service orchestration tools and concepts is not mandatory for content owners. The demonstration will utilize multi-tier streaming consisting of content streaming, indexing, and editing appliance hosted over AWS EC2 and S3 services. A detailed screenshot of the MCCO can be found at: http://rranjans.wordpress.com/MCCO-tool.

References

[1] Buyya, R., Pathan, M., Vakali, A.: Content Delivery Networks. LNEE, vol. 9. Springer, Germany (2008) ISBN: 978-3-540-77886-8
[2] Akamai, http://www.akamai.com/html/solutions/sola-solutions. html (access date: June 07, 2012)
[3] Limelight Networks, http://www.limelight.com/ (access date: June 07, 2012)
[4] Amazon Web Services, http://aws.amazon.com/ (access date: June 07, 2012)

The Semantic Web Linker: A Multilingual and Multisource Framework

Mariantonietta Noemi La Polla, Angelica Lo Duca, and Andrea Marchetti

Institute of Informatics and Telematics
National Research Council (CNR), Pisa, Italy
firstname.lastname@iit.cnr.it

Abstract. In this demonstration we present the Semantic Web Linker (SWL), a framework for helping Name Entity Recognition (NER) procedures. The strength of the SWL is the integration of data coming from different Web sources, such as Wikipedia and DBpedia. The SWL also provides a multilingual repository, in the sense that every entity is associated to its synonyms and translations in many languages. Furthermore, the SWL manages a classification of entities through their hierarchical categorization. The SWL can be browsed through a Web interface.

1 Introduction

Nowadays, the Web is exploited by organized crime to communicate, work or expand their influence. The CAPER European project [1] aims at building a common collaborative and information sharing platform for the detection and prevention of organized crime on the Web. One of the CAPER's issues is the recognizing Named Entities (NEs) in texts collected from the Web (e.g. advertisements for sales of products). NEs can be related to different categories, such as drugs, cars, places and so on. In order to deal with this problem, specific linguistic modules for Named Entities Recognition (NER) are developed within the project. In this demonstration, we present the Semantic Web linker (SWL), which is a framework developed in CAPER to help NER modules during the process of recognition and disambiguation of NEs. In practice, the SWL acts as an oracle for the NER module: if the NER module needs to recognize an entity (e.g. cocaine), it asks the SWL for the information associated to that entity (e.g. illicit drug, alkaloid). The SWL integrates data coming from different Web sources and provides useful information about NEs, like synonyms, translations, slang, URIs, and descriptions. We focus our attention on a scenario in which NEs related to illicit drugs need to be recognized. The SWL belongs to the field of Web data extraction [2] and Web data integration [3,4] systems.

The remainder of the paper is organized as follows: in Section 2 we describe the CAPER project, while in Sections 3 and 4 we illustrate the Semantic Web Linker and the use-case scenario, respectively. Finally in Section 5 we give conclusions and future work.

X.S. Wang et al. (Eds.): WISE 2012, LNCS 7651, pp. 792–795, 2012.

2   Background

This work it has been developed during the CAPER project[1]supported by the European Commission through the Seventh Framework Programme for Research and Technological Development. The main goal of the project is the creation of a common platform for the prevention of organised crime through sharing, exploitation and analysis of Open and private information sources. In particular, the objective of the project is the detection of organised crime that exploits the Web, such as cybercrime, sales of counterfeit products and so on. This is achieved through the automatic analysis of texts, videos and pictures present on the Web in order to mark them as suspicious, if some specific elements (e.g. illicit drugs, weapons) are found within them. From a data point of view we can resume the actions performed by the platform in three different classes: the first one is represented by the gathering of information present on the Web; the second one, that is the most important one, is the analysis of gathered data and the third one is the presentation of the results to the end users. Project's partner coming from both academic and industrial world and Law Enforcement Agencies (LEAs) are involved as end users. Standardisation for interchange of data and tools, integration with Large scale Systems, secure knowledge sharing and collaboration and the study and recommendations about legal, ethical and societal issues are some CAPER's features.

3   The SWL Architecture

Figure 1 shows the SWL framework, which includes three components: the *Scraper*, the *Multilingual NE Repository* (M-NER) and the *Web browsing Interface* (WBI). The Scraper collects as much data as possible concerning the NEs

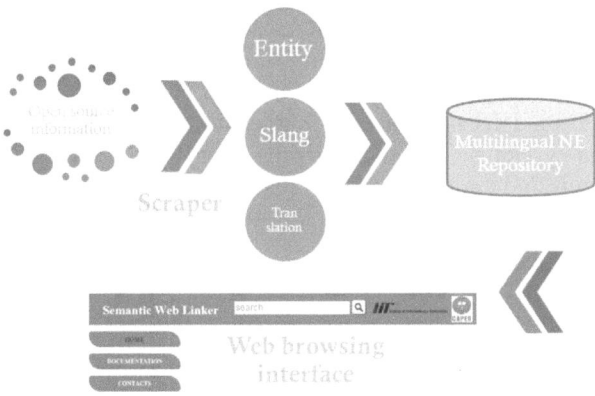

Fig. 1. Semantic Web Linker Framework

---

[1] http://www.fp7-caper.eu/

and their categories from open sources. The information is retrieved through ad-hoc PHP scripts and stored in the M-NER,which is implemented as a MySQL database. The WBI accesses the M-NER to retrieve stored data and shows them to the user. The scraping is performed through two procedures. The first is a scraper that analyzes the HTML code of a page starting from its URL and col-lects NEs using XPath expressions. The second is an *API-based* scraper which uses Wikipedia APIs to recover translations and categories of each NE. Informa-tion retrieved by the Scraper are stored in the M-NER, whose structure derives from the WordNet structure.

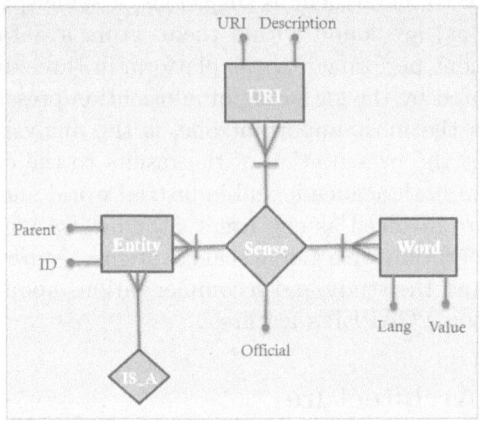

Fig. 2. Multilingual Named Entity Repository Schema

Figure 2 shows the M-NER entity-relationship diagram: an *Entity* is identified by a unique identifier (ID) and it is associated to two relationships: a) *is-a*, b) *sense*. The *is-a* relationship defines the hierarchical nexus among entities, while the *sense* relationship sets a link among the entity, one or many words and one or many URIs. A Word specifies the value (e.g. cocaine) and its language (e.g. English), while an URI specifies the entity description and the web resource where that entity can be found. Generally every entity is associated to many words (synonyms and translations) and many URIs, deriving from many sources (e.g. Wikipedia and DBpedia).

4   Scenario

In this demonstration we focus on a scenario where information about illicit drugs are collected, stored and visualized.

Table 1 shows the Web data sources from which we have collected information. They have been chosen manually. For each source, its Web URI and supported languages are shown. Wikpedia is the only ananlyzed source providing more than one language (i.e. it manages 146 languages). Furthermore, the table high-lights whether a Web source is specific for the domain of illicit drugs or not. The

Table 1. Web data sources used by the SWL

Source	languages	specific	Web URI
Dbpedia	English	NO	http://dbpedia.org
Encyclopedia Britannica	English	NO	http://www.britannica.com
Erowid	English	YES	http://www.erowid.org
FRANK	English	YES	http://www.talktofrank.com
Treccani	Italian	NO	http://www.treccani.it
Sapere	Italian	NO	http://www.sapere.it
Wikipedia	146 lang.	NO	http://www.wikipedia.org

only specific Web sources are FRANK and Erowid. FRANK is a national drug education service developed by the Department of Health and Home Office of the British government in 2003, while Erowid is a member-supported organization providing access to reliable, non-judgmental information about psychoactive drugs, The other Web sources are online encyclopedias publicly available on the Web.

5 Conclusions and Future Work

In this paper we have described the SWL, a framework which extracts, integrates and visualizes information from Web sources. The SWL has been deployed within the context of the CAPER projects. As future work, we will transform it into a SPARQL node, in order to be RDF compliant.

Acknowledgments. This work has been supported by EU FP7 Project CA-PER (Grant Agreement no. FP7-261712).

References

1. The CAPER project official Web site, http://www.fp7-caper.eu
2. Ferrara, E., De Meo, P., Fiumara, G., Baumgartner, R.: Web data extraction, applications and techniques: A survey. ACM Computing Surveys (to appear, 2012)
3. DeRose, P., Shen, W., Chen, F., Lee, Y., Burdick, D., Doan, A., Ramakrishnan, R.: Dblife: A community information management platform for the database research community (demo). In: CIDR, pp. 169–172 (2007)
4. Szomszor, M., Cattuto, C., Alani, H., O'Hara, K., Baldassarri, A., Loreto, V., Servedio, V.: Folksonomies, the semantic web, and movie recommendation. In: 4th European Semantic Web Conference, Bridging the Gap between Semantic Web and Web 2.0., June 3-7 (2007)

SMART: Supporting the Design and Execution of User-Centric Service-Based Applications*

Piergiorgio Bertoli, Raman Kazhamiakin, Michele Nori, and Marco Pistore

SAYservice srl, via alla Cascata 56/C, 38123, Trento, Italy
{bertoli,raman,nori,pistore}@sayservice.it
http://www.sayservice.it

Abstract. This demo showcases the SMART approach, which allows the effective provision of user-centric service-based applications where specific user needs are tackled by customizable combinations of services and functionalities, by encompassing the integration, combination, monitoring, presentation and customization of heterogeneous services. The demo shows how a personal activity support application, easing fruition of territorial events, services and opportunities, can be effectively built and managed at runtime using SMART.

1 Motivation and Relevance

The growth in the end-user availability of IT services paves the way to a range of service-based applications (SBAs) which address user-tailored needs by allowing a strong degree of flexibility in the selection, integration, provision and presentation of services. In this demo, we showcase the SMART approach and platform, an all-round contribution enabling the effective design and realization of user-centric service based applications (SBAs), based on the key design concept of *domain objects* to bridge the technological service view with the user view, focused on perceived user *assets* such as time, location or social network. The SMART platform stands on the conceptual architecture represented in Fig. 1, which is structured into a service integration, a service model, a user interaction and a context awareness components.

In particular, the connection of SMART to 3rd party services is realized by the abstraction layer provided by its *Service Integration* subsystem. Its task is to compensate the technological and conceptual heterogeneity of the services, providing mechanisms for service invocation as well as for monitoring, contextualizing, filtering and proactively proposing extracted information. This is achieved thanks to two design support elements. First, a pattern-oriented language, named XS*, specifically designed to extract data from XML/HTML formats, which integrates and enriches features from the XSLT and XML Schema languages. Second, a simple yet expressive flow-based language for combining and exposing bidirectionally communicating services.

Vice versa, the connection of SMART with the user goes through the *Metaphor-based User Interaction* subsystem, providing the necessary mechanisms to enable user operativity on services, and to notify the user on relevant applicative/domain events.

* This work is supported by project "SMART4U: User-Centric Platform for the Internet of Services", funded by the Operational Programme "FESR 2007-2013" of the Province of Trento.

X.S. Wang et al. (Eds.): WISE 2012, LNCS 7651, pp. 796–799, 2012.

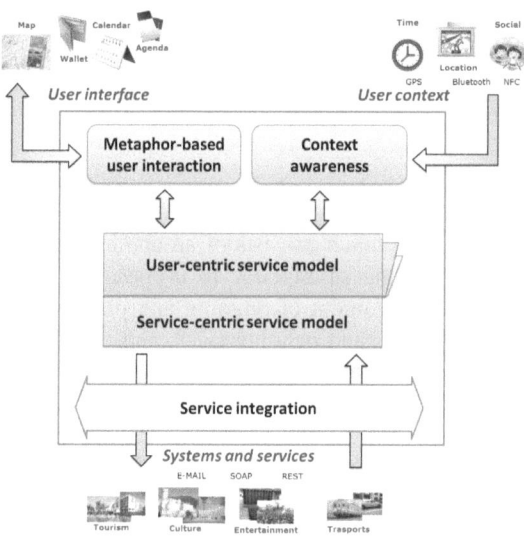

Fig. 1. Conceptual architecture

These mechanisms are based on interactive metaphors associated with key user assets such as time (calendar), location (map) and social network. This is complemented by a *Context Awareness* subsystem that, via appropriate technologies, observes the user context to analyze the impact of context changes on user activities.

Central to the platform stands the *Service model* subsystem, which provides the model and run-time environment for service integration, monitoring, and processing, and enables different forms of service compositions: information integration, events aggregation, orchestration of transactional services. This core level enacts the user-centric perspective, enabling personalization of services to represent domain concepts and functionalities related to the user needs, through the key conceptual Domain Object element. Domain Objects are stateful, event-driven entities that form a dynamic network and represent (instances of) relevant domain concepts for a user-centric SBA, defining their runtime evolution logics and the way they relate to other elements of the framework. Domain Objects in SMART are described via a rich Domain Object Specification Language (DOSL), that allows for specifying event-driven evolution and propagating events down from services up to high-level entities directly associated to user activities such (e.g. ToDo lists, reservations...), and is connected with the XS* and the flow-based Service Integration languages via the definition of common events. Remarkably, Domain Object instances also used to represent the evolution of essential resources for specific users (e.g. time, geo-location, social relations), and as such allow defining a bridge between the technological and user-related levels, enabling the customization of (integrated) objects and services in terms of resources and user ownership.

Technologically, SMART is a Java-based platform organized around SOA principles, adopting the OSGi layering to support dynamic evolution of its modules. It is a platform that embeds significant research results, integrating different paradigms in a synergic, holistic and pragmatic way, hence providing a relevant and important contribution for the web information system engineering audience, and witnessing a novel approach for dynamic web information integration. We now discuss the demonstration setting for such platform.

2   Demo

The demo of the SMART approach consists of an application that supports the user while visiting a town, using the smartphone to propose events and services in the area, and to support the management of ongoing and planned activities, taking into account the user's profile, plans and pending tasks, and monitoring and notifying divergences in the progress of activities against the plan. In this way, the demo, whose working is more comprehensively explained by a presentation which is available at http://www.sayservice.it/smart-cockpit/wise12.pdf, is capable to cover both the design and runtime support given by SMART. Specifically, it is organized as follows:

Service integration: first, DFlow and XS* code snippets show how the SMART toolset is used to wrap and integrate SBA-relevant services. This also clarifies how even passive services that provide information in a static way can be associated to dynamic events monitoring information changes.

Composing and Monitoring Domain Objects: the demo then shows how DOSL is used to define a network of concepts that originates user-level events, via DOSL code snippets. We will show the dynamics of such concepts and events, such as the expiration of booking places for an event in the user agenda or the delay of a booked train, via a technical control console like shown in Fig. 2, where a semaphore-light visualization is used to highlight different statuses of monitored objects.

Fig. 2. A snapshot of the objects monitoring console

Runtime of the example application: finally, the demo shows the user run-time perception of our sample SBA. This takes place on a scenario concerning a person who visits Trento (Italy) and adopts the app to support his/her activities. First, the demo shows how SMART supports the user by continuously monitoring his context and events from relevant services, proactively proposing information via the interactive metaphors. E.g. this may be triggered by some incoming event recommendation by some other user via a Facebook interface. Upon adopting the recommendation (see Fig. 3, left), the event is linked to relevant services: e.g., for a concert event, transportation to the venue and monitoring of remaining places for the sake of timely booking. Then, the demo shows how context monitoring identifies services relevant for the user's interests, activities and tasks, considering situations where the user encounters geo-located services while visiting the city, e.g. museums or pharmacies (see Fig. 3, center). Further, the demo shows how commitments that conflict over resources are catched and handled by the SMART platform. E.g. this can be triggered by a chosen recommendations that adds a conflicting event in the user agenda; upon this, conflicts are discovered and alerted to the user. Finally, the demo highlights how the continuous validation of the consistency of the actual user context vs. the one foreseen from his/her planned activities results in notifications, in case of mismatches. In our scenario, this can be triggered by some "expiration" on a personal user event, e.g. free places for a concert event in agenda running below a threshold (see Fig. 3, right).

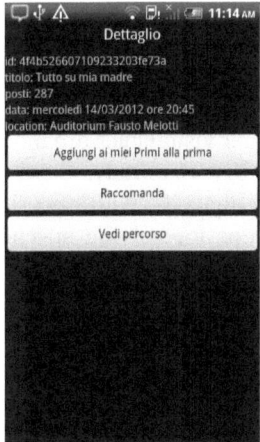  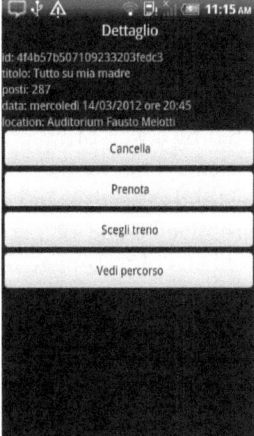

Fig. 3. Snapshots of the runtime

WPPS: A Framework for Web Page Processing*

Ruslan R. Fayzrakhmanov

Institute of Information Systems, TU Vienna,
Favoritenstraße 9, A-1040 Vienna, Austria
fayzrakh@dbai.tuwien.ac.at

Abstract. In this paper, we present WPPS, a new configurable Java-based framework for developing web page processing methods. The key innovations of WPPS are 1) a unified ontological model which describes the visual representation of web pages; 2) an API and abstractions which allow the application of both declarative and object-oriented mechanisms to develop new methods and approaches.

Keywords: web information extraction, web page understanding, ontological models, object-oriented paradigm, declarative approach.

1   Introduction

Web page understanding (WPU) and web information extraction (WIE) play an important role in information search, information integration, and competitive intelligence. Current methods of WPU and WIE (hereinafter referred to as *methods*) analyse different web page representations: source code (X/HTML, XML), DOM tree, and visual representation—particularly CSS attributes computed by a web browser. Methods that additionally consider visual cues are more efficient. This fact is experimentally confirmed in [4,2]. Unfortunately, there is no standardised visual model suitable for the automatic analysis. And also the unresolved issue of determining features and relationships for solving specific problems of web page object identification still exists.

In this paper, we present a Web Page Processing System (WPPS) that gives the developer a possibility to investigate different aspects of web page representations modelled by a unified ontological model (UOM) [3,1] in order to find more appropriate objects, features and relations for developing an efficient method using an API. The API of WPPS allows the application of declarative methodologies and object-oriented paradigms.

2   WPPS Architecture and API

The WPPS framework is an Eclipse RCP-based application implemented in Java. It provides a convenient means for developing various methods utilising the abundance of different web page representations, features and relationships. The main functional components are (cf. Fig. 1):

---

* This work is funded by the Austrian Forschungsförderungsgesellschaft FFG under grant 829614 (TAMCROW).

X.S. Wang et al. (Eds.): WISE 2012, LNCS 7651, pp. 800–803, 2012.

1. UOM manager is an implementation of the UOM [3,1] with reasoners applied and realised using Jena[1]. The UOM consists of two main sub-models: *physical* and *logical*. The *physical model* describes web page's aspects, such as layout (spatial configuration of visualised elements), interface, (e.g. web forms, links, images), perceptible visual features (e.g. saliency as a degree of uniqueness, emphasis, such as bold, italic, underlined, etc.), and DOM trees. The *logical model* provides an interpretation of the subset of individuals from the physical model as a result of WPU and WIE. The logical model is a contribution to the development of Semantic Web technology, providing information about a web page's content that is accessible to computers for further automatic analysis and reasoning. **2. Configuration manager** provides control over configuration parameters of the UOM as well as modes of computing features and relations. **3. Core** provides main functionality to interact with the UOM. It allows the application of SPARQL queries and logical inference rules. Core is responsible for handling different inaccuracies (fuzziness) while computing qualitative features and relations. **4. Adapter layer** provides object-oriented abstraction for concepts and individuals of the UOM according to their configuration, which allows the application of heuristics over the ontologies. **5. Physical model generator** is responsible for generating the physical model according to the web page and configuration provided. WPPS has Mozilla (XULRunner v. 1.9.2) web browser integrated to retrieve all necessary data. **6. API** is based on the adapter layer and provides all functions necessary for developing methods. It is designed for querying the UOM and processing the information acquired, building the logical model and its integration. It uses R-tree for indexing two-dimensional objects and performing efficient spatial querying. The API makes it possible to use an object-oriented approach provided by the adapter layer and declarative mechanism realised by SPARQL queries and logical inference rules. A method can be implemented as a JAR file or Eclipse's plug-in fragment. **7. GUI** is used for applying developed methods over web pages to check their performance as well as to see results of their work by visual demonstration.

3   Web Page Processing

Concerning the UOM, we present WPU and IE processes consisting of three main phases (cf. Fig. 2): 1) web page analysis, 2) web page understanding and direct information extraction, 3) information transformation.

By "web page analysis" we mean physical model generation. The process is performed by WPPS based on the data (DOM tree, computed CSS attributes) provided by the web browser and configuration of the UOM. Model generation occurs during the depth-first pass through the tree of web pages related by means of frames and corresponding DOM trees.

Web page understanding and direct information extraction are processes performed using the physical model. The aim of these processes is to provide an

---

[1] http://jena.apache.org/

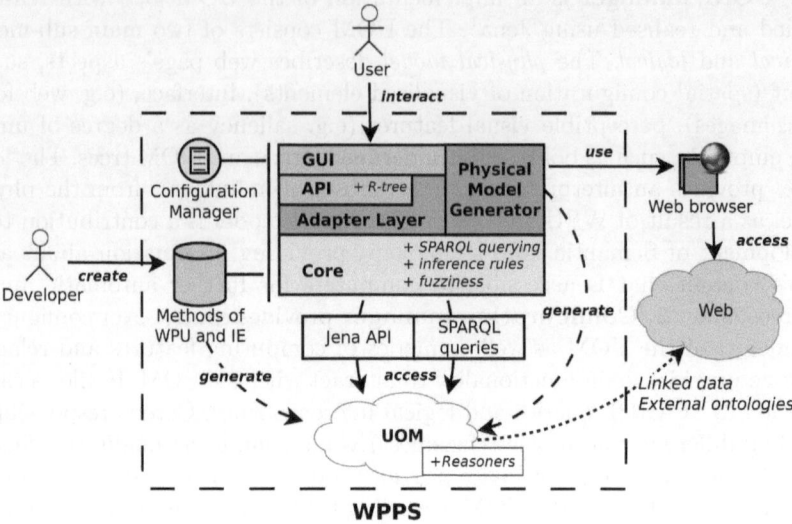

Fig. 1. Architecture of the WPPS framework

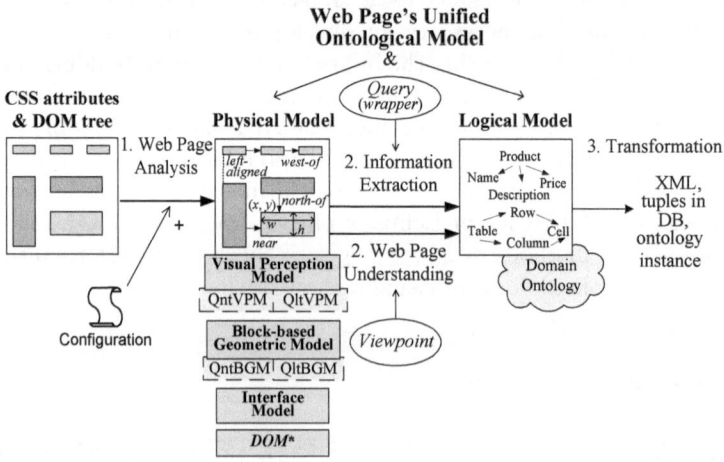

Fig. 2. The web page's unified ontological model in the processes of web page understanding and web information extraction

interpretation in the form of a logical model for the concepts of the physical model by means of domain ontologies and linked data.

Information transformation is a process aimed at providing information in the form appropriate for external applications. Information can be represented as an ontology, XML document, or tuples in a database.

4   WPPS API

The WPPS API provides all necessary functionality for developing methods. It is implemented on top of the *adapter layer*, treating ontological concepts as Java objects. The functional can be split into 3 groups: 1) selectors, 2) processing functions, 3) statistical functions.

1. Selectors are those functions that allow the selection of a specified subset of objects from the UOM. Selection can be performed based on the type of object (e.g. "image", "box", "html link"), based on the specified predicate (function that returns boolean value) or SPARQL query. Furthermore, the object contained or intersecting specified area can be selected; in this case R-tree is involved for efficiency.

2. Processing functions were designed to process wrapped objects acquired from the UOM. Treating a collection of objects as a set, the following functions are available: intersection, union, selection. Moreover, the functions make it possible to order elements and group elements of a set into subsets, sequences, trees and grids using the predicates provided.

3. Statistical functions provide means for computing aggregated values, such as mean, median, variance, minimal and maximal values, over a set of objects and pairs of objects. The latter is useful for computing some characteristics regarding the relations between objects in some sets, such as average spatial distance between neighbouring elements.

By leveraging the API and adapters one can perform queries on the UOM and manipulate individuals as Java objects abstracting form the ontologies and reasoners.

References

1. Fayzrakhmanov, R.R.: Information Extraction from Web Pages Based on Their Visual Representation. In: Harth, A., Koch, N. (eds.) ICWE 2011. LNCS, vol. 7059, pp. 342–346. Springer, Heidelberg (2012)
2. Hiremath, P.S., Algur, S.P.: Extraction of flat and nested data records from web pages. In: SivaKumar, K., Selvi, A. (eds.) IJCSE, vol. 2, pp. 36–45. SIPS Tech. (2010)
3. Krüpl-Sypien, B., Fayzrakhmanov, R.R., Holzinger, W., Panzenböck, M., Baumgartner, R.: A versatile model for web page representation, information extraction and content re-packaging. In: Proc. of DocEng 2011, pp. 129–138. ACM (2011)
4. Zhai, Y., Liu, B.: Web data extraction based on partial tree alignment. In: Proc. of WWW 2005, pp. 76–85 (2005)

Information Systems for Knowledge Workers:
The Kpeople Enterprise 2.0 Tool

Ugo Barchetti, Antonio Capodieci, Anna Lisa Guido, and Luca Mainetti

Dip. Ingegneria dell'Innovazione, University of Salento, Via Monteroni, 73100 Lecce, Italy
{ugo.barchetti,antonio.capodieci,
annalisa.guido,luca.mainetti}@unisalento.it

Abstract. Every day companies deal with internal problems to manage human resources during the execution of knowledge processes. In these situations, the ability to quickly identify and rapidly apply effective business practices for recurring problems becomes crucial in order to improve the efficiency of the organization. To address this problem, we demonstrate the Kpeople tool that enables organizations configuring a set of process patterns to codify business practices. Kpeople allows tracking unstructured knowledge, improving knowledge management and fostering collaboration.

Keywords: Business Practices, Collaboration, Coordination, Enterprise 2.0.

1 Introduction

In the past 50 years, a new form of worker – the "knowledge worker" [1] – is becoming more and more important for companies. The knowledge worker is "one who works primarily with information or one who develops and uses knowledge in the workplace". Typically, knowledge workers operate on multiple tasks at the same time. They are involved in many parallel "knowledge processes" [2] that, often, are not codified in formal procedures, are unstructured, collaborative and continuously changing. The advent of Web 2.0 tools in the workplace is dramatically amplifying the need to keep coherent knowledge processes (unstructured) and business processes (structured ones) [3], moving from tacit to explicit knowledge, and shaping a new kind of information system known as Enterprise 2.0.

Here we demonstrate the Kpeople tool that extends traditional information systems, giving flexible support to networked human processes, and allowing codifying them as business practices. The tool has been experimented by the Italian Association for Computing (AICA) to trace informal activities for new product budgeting, by an Italian large-scale hospital to manage the deployment lifecycle of IT products, by a Brazilian IT company (Elogroup) specialized in BPM tools, by a Hungarian company (John Von Neumann Computer Society) and a Korean company (KPC) that work in the field of computer driving licensing, and by an Italian ICT company (Webscience s.r.l.) to support the project proposal process.

X.S. Wang et al. (Eds.): WISE 2012, LNCS 7651, pp. 804–807, 2012.

2   Related Work

This section briefly reports on existing tools that support knowledge workers executing collaborative and cooperative processes [4], and enable companies to manage and preserve social capital. As the author observes in [5], if properly deployed, new technologies allow companies to cost-effectively increase their productivity and their competitive advantage.

In this context, the COIN project [6] proposes a framework for web service composition and knowledge extraction from process execution based on process description. COIN allows Collaborative Concurrent Competitive Enterprises integrating knowledge processes in their information systems.

The Active Project [7] addresses a similar challenge through an integrated knowledge management workspace that reduces information overload by significantly improving the mechanisms for creating, managing, and using information. The project's approach addresses three topics: (i) sharing information through tagging, wikis, and ontologies; (ii) prioritizing information delivery by understanding users' current-task context; and (iii) leveraging informal processes that are learned from user behavior.

Finally, we observe that more than 40 IT companies provide Enterprise 2.0 tools, [7], ranging from global players, that are turning their traditional products into social software for enterprises (IBM Lotus Connections and Oracle Beehive), to innovative small and medium enterprises offering software-as-a-service (SaaS) solutions with blogs, wikis, file sharing, RSS, community and group management, and instant messaging functionalities.

3   System Architecture

The Kpeople tool demonstrator was built adopting a web service oriented and an event-driven architecture [8], which – thanks to custom adapters – is able to trace and store events generated by traditional enterprise information systems (CMS, BI, CRM, ERP, etc.), communication tools (e-mail), unified communications & collaboration tools (UCC), and Web 2.0 facilities.

The system architecture is shaped to give flexible support to unstructured and complex processes within a networked enterprise environment (such as decisional, collaborative, and creative contexts), to improve the management of information and communication, and to optimize workspace, recovering the time spent in low-value activities, in particular to find relevant information to execute knowledge tasks and to recover tasks' context.

A lot of open source frameworks have been employed: WSO2/WSAS at the business logic and data access layer, Pentaho to implement business intelligence services, MySQL for the persistence of knowledge process events, Sesame OpenRDF to represent the Enterprise Ontology, SOLR as search engine, and Alfresco for documents storage.

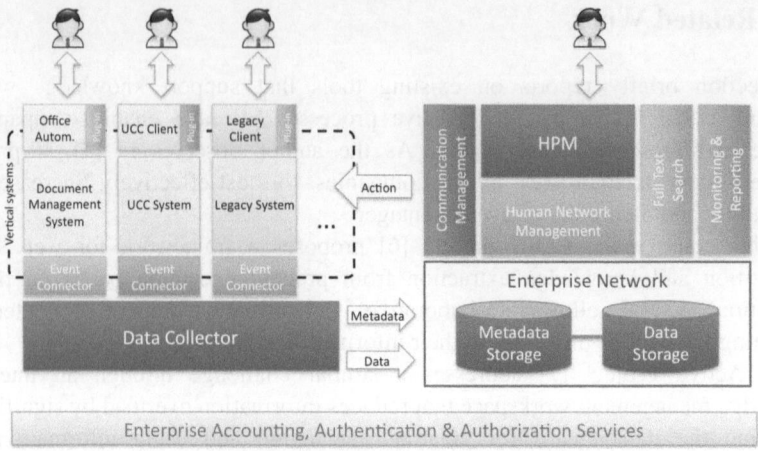

Fig. 1. The Kpeople system architecture

As Fig. 1 draws, executing Kpeople the Knowledge workers can collaborate with colleagues by exchanging information, files and tasks through the HPM (Human Process Management) tool that allows users to apply patterns and examine the progress of the processes, the activities to be completed, the flow of communication, exchanged documents and e-mails, and to examine a set of indicators useful to evaluate performances and to identify bottlenecks. All data, information and documents are collected in a common database (Data Storage) enabling easy data retrieval (through Metadata) for knowledge workers and improving their efficiency. Events are tagged and clustered using a domain ontology. Event streams may be analyzed by social networks analysis tools (during the case study execution the Cytoscape open source platform for complex network analysis and visualization has been exploited).

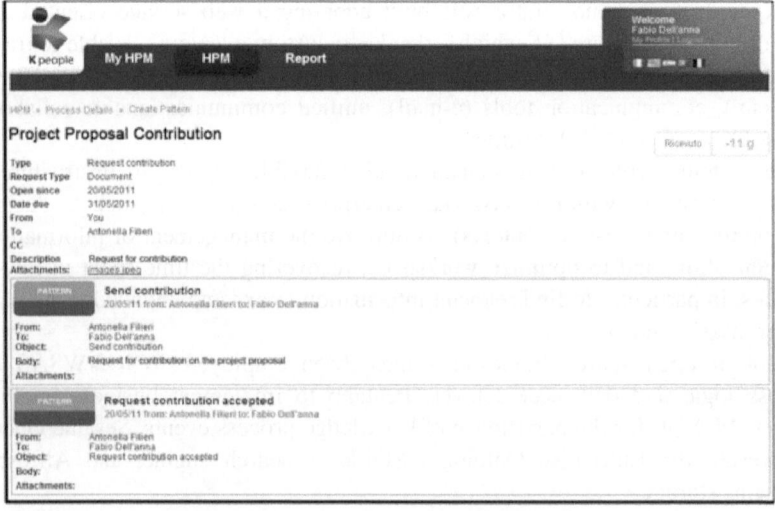

Fig. 2. A screenshot of the Kpeople system

The Kpeople tool enables organizations configuring a set of business patterns. Patterns can be modeled and dynamically injected in the Kpeople system using the Activiti BPM platform. Fig. 2 shows a screenshot of the application of the Retrieve Contribution pattern during the configuration phase of Kpeople. The system also provides an open-source execution environment, deployed as a Java process engine, and a monitoring interface (the Activiti Explorer).

4   Demo Scenario

We demonstrate Kpeople, showing how the tool supports knowledge workers during collaborative activities, reducing their efforts. We show how business practices can be modeled as process patterns and how the Kpeople tool enacts their workflows connecting information systems (a company's intranet) and knowledge productivity tools (office automation and e-mails). The demo's case study we work on is the "project proposal drafting" process, which is made up of two sub-processes: the "proposal writing" and the "budget creation".

A prototype of the Kpeople tool is available at http://kpeople.webscience.it and on the Source Forge open source community http://sourceforge.net/projects/K-people.

Acknowledgments. This research has been supported by the FESR PO Italian funded 4RCPCJ0 Kpeople project and Webscience s.r.l.

References

1. Davenport, T.H.: Thinking for a Living: How to Get Better Performances and Results from Knowledge Workers. Harvard Business Press, Boston (2005)
2. Yongchareon, S., Liu, C., Zhao, X.: An Artifact-Centric View-Based Approach to Modeling Inter-organizational Business Processes. In: Bouguettaya, A., Hauswirth, M., Liu, L. (eds.) WISE 2011. LNCS, vol. 6997, pp. 273–281. Springer, Heidelberg (2011)
3. Distante, D., Rossi, G., Canfora, G., Tilley, S.: A comprehensive design model for integrating business processes in web applications. Int. J. Web Eng. Technol. 3(1), 43–72 (2007), doi:10.1504/IJWET.2007.011527
4. Hall, H., Goody, M.: KM, culture and compromise: interventions to promote knowledge sharing supported by technology in corporate environments. Journal of Information Science 33(2), 181–188 (2007), doi:10.1177/0165551506070708
5. Andriole, S.J.: Business impact of Web 2.0 technologies. Communications of the ACM 53(12), 67–79 (2010)
6. Moisescu, M.A., Sacala, I.S., Stanescu, A.M., Serbanescu, C.: Towards Integration of Knowledge Extraction from Process Interoperability in Future Internet Enterprise Systems. In: 14th IFAC Symposium on Information Control Problems in Manufacturing Bucharest, Romania (2012) ISBN: 978-3-902661-98-2
7. Simperl, E., Thurlow, I., Warren, P., Dengler, F., Davies, J., Grobelnik, M., Mladenić, D., Gomez-Perez, J.M., Ruiz Moreno, C.: Overcoming information overload in the enterprise: the ACTIVE approach. IEEE Internet Computing 14, 39–46 (2010)
8. Achilleos, A., Kapitsaki, G.M., Papadopoulos, G.A.: A Model-Driven Framework for Developing Web Service Oriented Applications. In: Harth, A., Koch, N. (eds.) ICWE 2011 Workshops. LNCS, vol. 7059, pp. 181–195. Springer, Heidelberg (2012)

Intelligent and Semantic Real-Time Process of the Greek LOD for Enhancing Citizen Awareness in Public Expenditures

M. Vafopoulos[1], M. Meimaris[1], A. Papantoniou[1], I. Anagnostopoulos[2], I. Xidias[1], G. Alexiou[1], G. Vafeiadis[1], and V. Loumos[1]

[1] School of Electrical & Computer Engineering, National Technical University of Athens
[2] Computer Science and Biomedical Informatics Dpt., University of Central Greece
{vaf,apapant}@medialab.ntua.gr, janag@ucg.gr

Abstract. The provision of publicly available Open Data targets to provide transparency in several public sector decisions and actions. However, this information is served massively and in different forms - mostly due to different bureaucratic procedures- while its diffusion does not occur at regular or at least generally predictable time intervals. Thus, even though the information is available by the involved public sectors and enterprises. citizens are overwhelmed from the size/inconsistency of the information they deal with. In this paper, we present a publicly accessible Web point (publicspending.gr/), which aims to promote transparency and enhance citizen awareness regarding public spending in Greece. The information is based on intelligent and semantic real-time processed open data that are derived from the open API of "Diavgeia" (Clarity) Program (diavgeia.gov.gr/en/) and TAXIS - Taxation Information System (gsis.gr/), while it is served through easily consumed visualization charts and diagrams.

Keywords: Public Spending, Linked Open Data, Transparency, Intelligent Information Processing, Semantics.

1   Background

Technologies of the Semantic Web or the Web of Data has significantly expanded and are now applied to bridge previously autonomous business domains and to concatenate independent government activities [1], [2], [3]. A consequence of the Web of Data underlying technologies brought up lately, the term "Linked Open Data" (LOD) or "Big Data". LOD now form a quite extraordinary cloud, but more than that its datasets involve provenance and governmental data (e.g. openspending.org). In addition, Linked Data enable the creation of better and massive services for use and reuse for many of these data, driving existing infrastructure in its full potential [4]. For government bodies, Linked Data adoption is focused on open, transparent, collaborative and more efficient governance. Related approaches with the one proposed in this paper are "Where does my money go: Showing you where your taxes get spent" (wheredoesmymoneygo.org), "Open Public Procurement Project"

X.S. Wang et al. (Eds.): WISE 2012, LNCS 7651, pp. 808–811, 2012.

(tender.sme.sk), "Accountability Initiative" (accountabilityindia.in), "Pera Natin 'to! (It's Our Money!)" (transparencyreporting.net/), and "Texas Transparency" (texastransparency.org/moneygoes) in U.S.

2   System Description and Data Visualization

Our system is mainly based on data feeds provided by "Diavgeia", the first Greek Government Open Data API (opendata.diavgeia.gov.gr/). "Diavgeia", offers the possibility for publicly accessing the full set spending decisions of the Greek Public Sector Organizations in XML files. Metadata include information concerning the names and the VAT registration numbers of payers and payees, the types of government expenditures in a single classification system for public procurement, aimed at standardizing the references used by contracting authorities and entities to describe the subject of procurement contracts (CPVs)[1]. Tax Information System (TAXIS, gsis.gr), which is the official web portal where citizens and legal entities submit taxation-related information and documents, is also employed. Through TAXIS, queries are performed in the form of SOAP calls with the VAT registration number of the entity as the reference key. The response contains metadata about the legal entity, including contact details, activity descriptions, registration dates and current operational status. Within the scope of this project, the web service is used for querying legal entities on their first appearance as payment agents, while the response data are "RDFized" and stored as payment agent metadata. The Web point is functional from the beginning of June 2012 under the hosting domain of publicspending.gr. Currently project deliverables run on two "sandbox" domains, one hosting the demo site and one hosting the SPARQL API endpoint of the project (publicspending.medialab.ntua.gr/en/data.php). Through this API one can run SPARQL queries against the dataset and output the results in various formats, such as HTML, spreadsheet, XML, JSON, JavaScript, N-triples, RDF/XML and CSV. A free version of the Highcharts visualization API[2] was used in order to visualize the resulting JSON data. Diagrams are provided in terms of daily, weekly, monthly and yearly time intervals, as well as since the beginning of the "Diavgeia" project (overall time period). It involves visual information regarding public expenditure according to the respective CPV codes, information about payers, payees, as well as their one-to-one payment connections. Some diagram examples are illustrated in the Appendix (Figure 1). Moreover, Figure 1a presents the total public expenditure on a weekly basis, while Figure 1b depicts the top-10 awardees of public contracts. For both diagrams, the amounts (in MEuros) are calculated from the beginning of "Diavgeia". Furthermore, Figure 1c illustrates the top-20 expenditure categories according to the respective CPV classification used, for the period between 17/06/2012 and 24/06/2012 (weekly-based information). Finally, Figure 1d highlights through a bubble depiction, the top-10 payees for the payer "Ministry of Finance". This is an important type of diagrams because it demonstrates the binary relation between the payer and its top-payees. Similarly, in this figure, the amounts are calculated from the inception of "Diavgeia" up to the very last day of submitting this work, thus covering a time period where Greece suffers from the so-called economic "crisis".

---

[1] http://simap.europa.eu/codes-and-nomenclatures/codes-cpv/
codes-cpv_en.htm
[2] http://www.highcharts.com

3   Research Issues – Future Work

Having been inspired from a widely used initiative, the British "Opening up government" project (http://data.gov.uk/) and the corresponding "payments" ontology [5] , our ontology was developed from scratch. Through the publicspending.gr ontology we are initiating cross-governmental interlinking between Greece and UK public spending. However, the expansion of such functions demands to resolve compatibility issues concerning existing "Big Data" datasets. This subject is related to the linkage of our dataset to openspending.org, as well as to opencorporates.com and the core person, business and location vocabularies[3]. Parallel to that, we plan to link with the Greek Public Contracts Registry[4] and the Greek DBpedia (el.wikipedia.org). Apart from the typical graph-based approach described, there will be further improvements/enhancements involving Open GeoSpatial APIs (e.g. OpenStreetMap[5]) for comparing public expenditures in respect to demographic data. Future work also involves the connection to the National Typography Service (et.gr) and to other similar Open Data portals that are under development in local and global scale. Most importantly, we plan to launch a powerful and elastic dashboard, which can be used by citizens and data journalists to perform comparative queries, to identify irregularities among national and European governmental expenditures (e.g. comparing per unit costs for public spending in health, education, etc.).

References

1. Passant, A., Laublet, P., Breslin, J.G., Decker, S.: Enhancing Enterprise 2.0 Ecosystems Using Semantic Web and Linked Data Technologies: The SemSLATES Approach. In: Linking Enterprise Data, pp. 79–102. Springer US (2010), doi:10.1007/978-1-4419-7665-9_5
2. Kobilarov, G., Scott, T., Raimond, Y., Oliver, S., Sizemore, C., Smethurst, M., Bizer, C., Lee, R.: Media Meets Semantic Web – How the BBC Uses DBpedia and Linked Data to Make Connections. In: Aroyo, L., Traverso, P., Ciravegna, F., Cimiano, P., Heath, T., Hyvönen, E., Mizoguchi, R., Oren, E., Sabou, M., Simperl, E. (eds.) ESWC 2009. LNCS, vol. 5554, pp. 723–737. Springer, Heidelberg (2009)
3. Siegel, D.: Pull—the Power of the Semantic Web to Transform your Business. Portfolio—Penguin Publishing Group, New York (2010) ISBN: 9781591842774
4. Vafopoulos, M.: The Web economy: goods, users, models and policies. Foundations and Trends® in Web Science 3(1-2), 1–136 (2012), http://dx.doi.org/10.1561/1800000015
5. http://data.gov.uk/resources/payments

---

[3] http://ec.europa.eu/isa/actions/01-trusted-information-exchange/1-1action_en.htm
[4] dev.opengov.gr/d/agora/?page_id=322
[5] http://www.openstreetmap.org/

Appendix

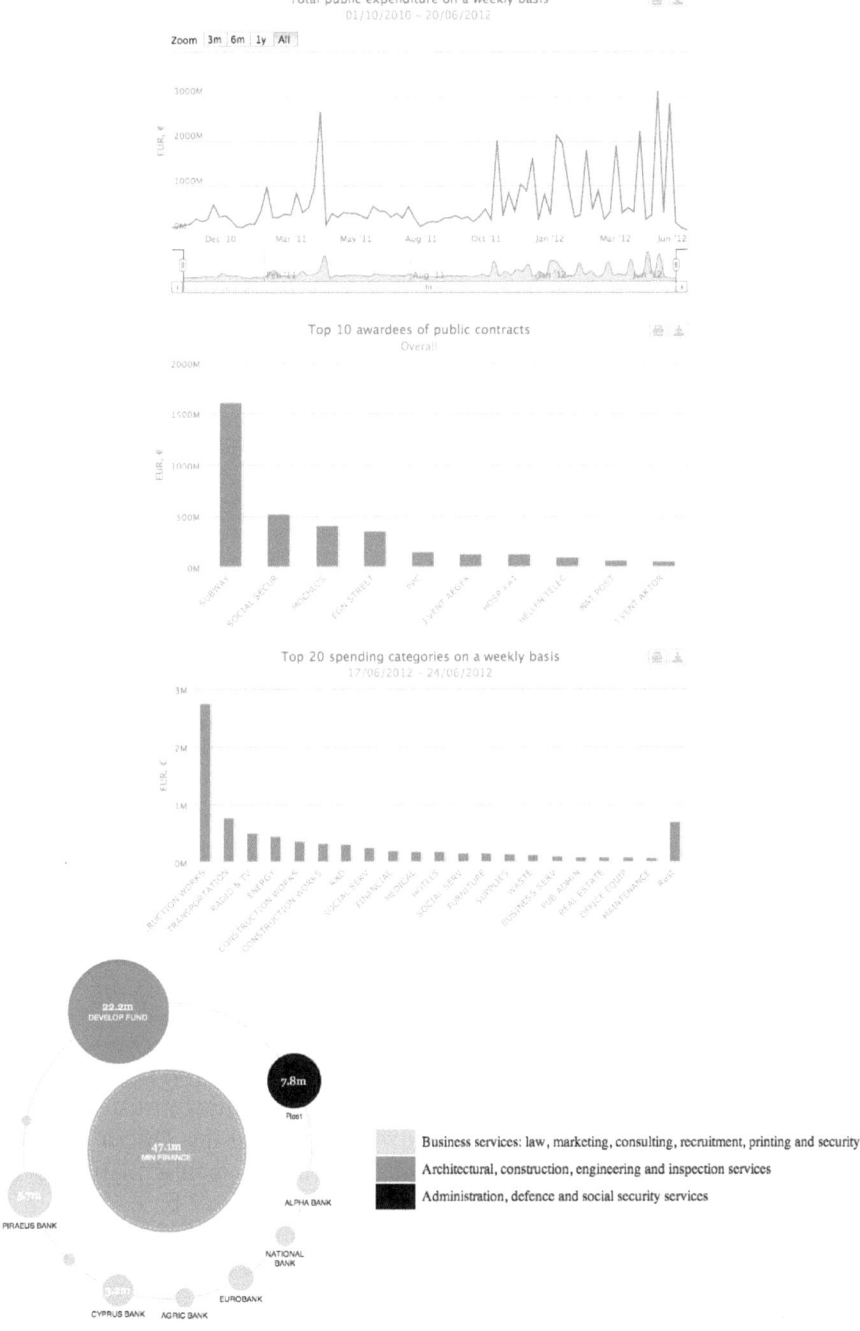

Fig. 1. Indicative visualizations derived from publicspending.gr

PoliMaR-Web: Multi-source Semantic Matchmaking of Web APIs

Luca Panziera, Marco Comerio, Matteo Palmonari,
Carlo Batini, and Flavio De Paoli

University of Milano-Bicocca, Milan, Italy
{panziera,comerio,palmonari,batini,depaoli}@disco.unimib.it

Abstract. PoliMaR-Web provides experts and ordinary Web users with a tool to discover suitable Web APIs among the ones published in repositories. Given a set of constraints, either soft or hard, semantic descriptions are extracted from repositories and heterogeneous sources available on the Web, and then matchmade to deliver a personalized ranked list of APIs. PoliMaR-Web performs at run time to ensure up-to-date answers.

1 Introduction

Public repositories like *Programmable Web*[1], and *Webmashup*[2] publish a growing number of functional-equivalent Web APIs, characterized by different properties such as service availability and licensing. The API descriptions are mainly keyword-based, which means that names and values are simple tags with no explicit meaning associated with them. Moreover different and possibly incomplete descriptions of a same API can appear in different sites and repositories. Therefore, service discovery cannot be automated with a satisfactory level of effectiveness. As a consequence, the process of selecting APIs from the ones available on the Web is mainly a manual activity, where users have to deal with disperse and heterogeneous sources to collect enough information to make conscious choices.

PoliMaR-Web is an automated discovery engine that, extending the functionalities of the semantic *policy* matcher PoliMaR [1], supports the discovery of APIs described in existent Web sources. PoliMaR-Web is novel w.r.t. other semantic and non semantic service discovery engines proposed so far [3,2]: despite we use RDF and semantics for describing services and user preferences, these descriptions are built at run-time by extracting and fusing descriptions from heterogeneous, non semantic, Web sources. Therefore, semantic matching techniques, which support the evaluation of expressive requests based on soft and hard constraints on non-functional properties, can be applied to descriptions that are always as up-to-date as the source descriptions are.

---

[1] http://www.programmableweb.com/
[2] http://www.webmashup.com

X.S. Wang et al. (Eds.): WISE 2012, LNCS 7651, pp. 812–814, 2012.

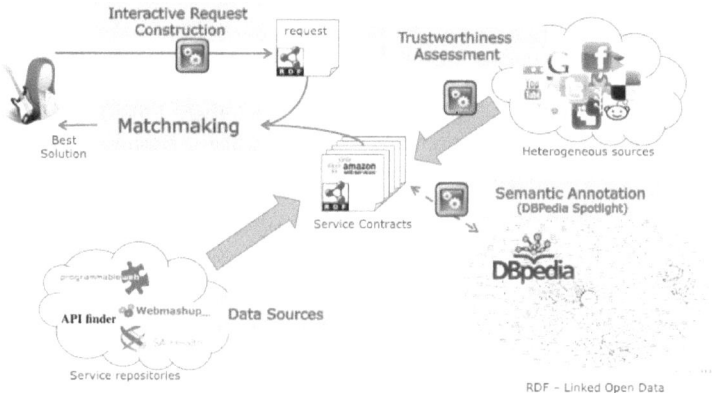

Fig. 1. The logical behavior of PoliMaR-Web

2   PoliMaR-Web

PoliMaR-Web[3] has been designed as a discovery service; the logical behaviour of
the service is sketched in Figure 1. APIs are represented as *service contracts*, to
state that the descriptions represent the conditions under which services can be
executed. Through an initial Google-like interface the user types in a set of tags
to immediately perform discovery to get a list of functional-equivalent APIs, or
can refine the request by adding a set of constraints to get a customized ranked
list. Constraints are in the form of expressions associated with indicators to state
their relevance (from low/optional to high/mandatory), which will be used to
weight the computed matching scores. Users are provided with autocompletion
and assisted interface to define properties and constraints.

PoliMaR-Web dynamically selects a set of services matching the request by
collecting information from the heterogeneous sources and building the associated contracts. Such contracts are made semantic by mapping terms to linked
data. During the extraction process, we take into account that collected information has different levels of accuracy, currency, and trustworthiness (computed
using information from the social Web) to state a confidence level for the results. Contracts of a same API built from different sources are fused by evaluating the quality of the sources, the extracted values, and the overall descriptions
[2]. After matchmaking, a ranked list of services is returned based on the degree of matching and the quality of the fused descriptions. The user can browse
the list by looking at the matching details of each constraint, and the RDF
descriptions.

---

[3] Demo available at
   `http://jeeg.siti.disco.unimib.it:8080/polimar/discovery.jsp`

References

1. Palmonari, M., Comerio, M., De Paoli, F.: Effective and Flexible NFP-Based Ranking of Web Services. In: Baresi, L., Chi, C.-H., Suzuki, J. (eds.) ICSOC-ServiceWave 2009. LNCS, vol. 5900, pp. 546–560. Springer, Heidelberg (2009)
2. Panziera, L., Comerio, M., Palmonari, M., De Paoli, F., Batini, C.: Quality-driven extraction, fusion and matchmaking of semantic web api descriptions. J. Web Eng. 11(3), 247–268 (2012)
3. Panziera, L., Comerio, M., Palmonari, M., De Paoli, F.: Distributed Matchmaking and Ranking of Web APIs Exploiting Descriptions from Web Sources. In: Proc. of the IEEE International Conference on Service-Oriented Computing and Applications, SOCA 2011, pp. 1–8 (2011)

ODCleanStore: A Framework for Managing and Providing Integrated Linked Data on the Web⋆

Tomáš Knap, Jan Michelfeit, Jakub Daniel, Petr Jerman, Dušan Rychnovský,
Tomáš Soukup, and Martin Nečaský

Charles University in Prague,
Faculty of Mathematics and Physics, Dept. of Software Engineering
Malostranské nám. 25, 118 00 Prague, Czech Republic
`tomas.knap@mff.cuni.cz`

Abstract. We present an ODCleanStore framework (1) enabling management of Linked Data on the Web and (2) providing web applications with a possibility to consume cleaned and integrated Linked Data according to their needs.

The advent of Linked Data [1] accelerates the evolution of the Web into an exponentially growing information space (see the linked open data cloud[1]) where the unprecedented volume of data will offer information consumers a level of information integration and aggregation agility that has up to now not been possible. Consumers – linked data applications – can now "mashup" and readily integrate information for use in a myriad of alternative end uses. Indiscriminate addition of information can, however, come with inherent problems, such as the provision of poor quality, inaccurate, irrelevant or fraudulent information. All will come with an associate cost of the data integration which will ultimately affect data consumer's benefit and linked data applications usage and uptake.

To overcome these issues, as part of the *OpenData.cz initiative* and *LOD2 project*[2], we are developing the *ODCleanStore (ODCS) framework* (1) enabling management of Linked Data – data cleaning, linking, transformation, and quality assessment – and (2) providing web applications with a possibility to consume cleaned and integrated data, which reduces the the costs of the web application development.

The overall picture of ODCS is depicted in Figure 1; information about ODCS and the source code can be obtained from `http://sourceforge.net/p/odcleanstore`. ODCS processes data *feeds* (collections of RDF triples) in the *staging area*; feeds can be uploaded to the staging area by any third-party application registered to ODCS. Based on the identifier of the feed, the appropriate

⋆ The work presented in this article has been funded in part by EU ICT FP7 under No.257943 (LOD2 project), the Technology Agency of the Czech Republic (TAČR, project TA02010182), the Czech Science Foundation (GAČR, grant number 201/09/H057), and GAUK 3110.

[1] `http://richard.cyganiak.de/2007/10/lod/`
[2] `http://opendata.cz`, `http://lod2.eu`

X.S. Wang et al. (Eds.): WISE 2012, LNCS 7651, pp. 815–816, 2012.
© Springer-Verlag Berlin Heidelberg 2012

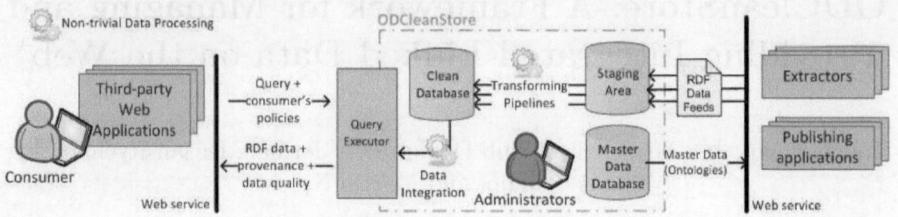

Fig. 1. ODCleanStore Framework

transforming pipeline is launched; the pipeline successively executes a defined (and customizable) set of transformers ensuring that data in the processed feed is cleaned, resources deduplicated and linked to already existing resources in the *clean database* or in the linked open data cloud, data is enriched with new one, arbitrarily transformed, and the quality score of the feed is assessed. When the pipeline finishes, the augmented RDF feed is populated to the clean database together with any auxiliary data and metadata created during the pipeline execution, such as links to other resources or metadata about the feed's quality.

Consumers can query the clean database to obtain data about the certain resource (e.g. business entity, such as "IBM"). Since the same resource can be described by various sources (feeds), conflicts may arise when integrating data about that business entity. To solve this, ODCS applies certain conflict handling strategies which may resolve, ignore, or avoid the data conflicts in the resulting RDF data. Furthermore, the resulting RDF data is supplemented with provenance metadata (data origin) and quality influenced by the quality of the feed the data originates from and by the applied conflict resolution policy [2]. The consumer can customize the query execution by specifying: (1) the preferred conflict handling strategy, (2) any restriction on the provenance of the consumed data (e.g. preferring data originating from the given source or being processed by the pipeline created by the given agent), and (3) quality assessment policies (e.g. to filter out data coming from feeds with low quality).

The practical demonstration at the conference will show (on the set of scenarios) the ODCleanStore framework in action – how the transforming pipelines can be configured, the conflict handling works, and how the web application can query the framework, customize the results of the query, and benefit from that.

References

1. Bizer, C., Heath, T., Berners-Lee, T.: Linked Data - The Story So Far. International Journal on Semantic Web and Information Systems 5(3), 1–22 (2009)
2. Knap, T., Michelfeit, J., Nečaský, M.: Linked Open Data Aggregation: Conflict Resolution and Aggregate Quality. In: METHOD 2012: The 1st IEEE International Workshop on Methods for Establishing Trust with Open Data, COMPSAC, Turkey (2012), http://www.ksi.mff.cuni.cz/~knap/files/method.pdf

An Interactive Exploratory System with Real-Time Preference Elicitation

Panagiotis Papadakos, Yannis Tzitzikas, and Dafni Zafeiri

Institute of Computer Science, FORTH-ICS, Greece, and
Computer Science Department, University of Crete, Greece
{papadako,tzitzik,dzafiri}@ics.forth.gr

Keywords: Exploratory Search, Multi-Dimensional Information Spaces, Preferences, Decision Making.

1 Introduction

Current proposals for preference-based information access [4] seem to ignore that users should be acquainted with the information space and the available choices for describing effectively their preferences. Furthermore, users rarely formulate complex (preference or plain) queries, because it is a laborious and difficult task for them. We will demonstrate a system for interactive exploration of multi-dimensional and hierarchical information spaces, enriched with actions that allow users to dynamically express their *preferences*, based on the preference framework described in [5]. Specifically, the system supports progressive preference elicitation, inherited preferences with scope-based resolution of conflicts, and preference composition over multi-dimensional and hierarchical information spaces. We argue that such functionality can ease the interaction and speed up the restriction of the focus to those parts of the information space that the user is interested in.

2 Preference Enabled Exploration

The proposed interaction model can be implemented over a variety of exploratory methods like the interaction paradigm of *Faceted and Dynamic Taxonomies (FDT)* [2,3]. We will demonstrate an implementation over a system for browsing and exploring RDF sources, described in detail in [1] (the fuzzy aspect is ignored). This system uses Jena[1], which is a Java framework for building Semantic Web applications. Our information base holds about 4036 cars and trucks, which are described by classes like `Manufacturer` and `Drive_System` (hierarchically organized), `Vehicle_Type`, and `Transmission`, as shown in Fig. 1. In this figure, continuous arrows denote *subClassOf* relationships while dashed arrows denote *typeOf* relationships.

The architecture of the system and its components is given in Figure 2. The preference actions are offered through HTML 5 *context menus*[2] and AJAX, which are

---

[1] http://jena.apache.org/
[2] Available only to firefox 8 and up.

X.S. Wang et al. (Eds.): WISE 2012, LNCS 7651, pp. 817–820, 2012.
© Springer-Verlag Berlin Heidelberg 2012

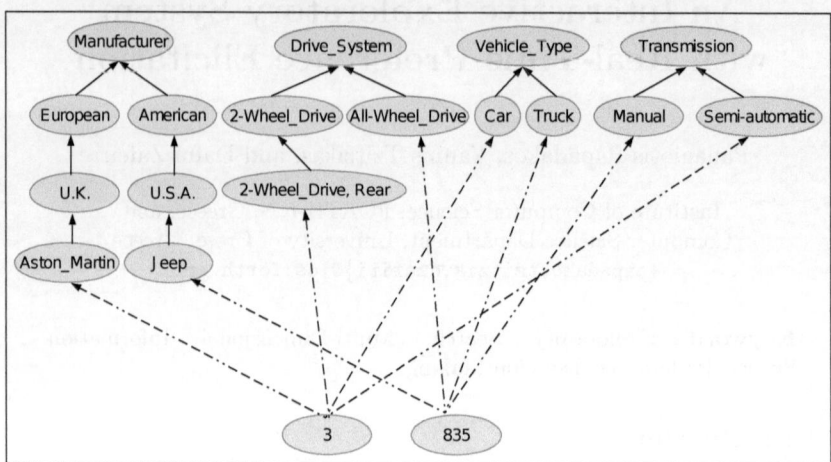

Fig. 1. The RDF Knowledge Base

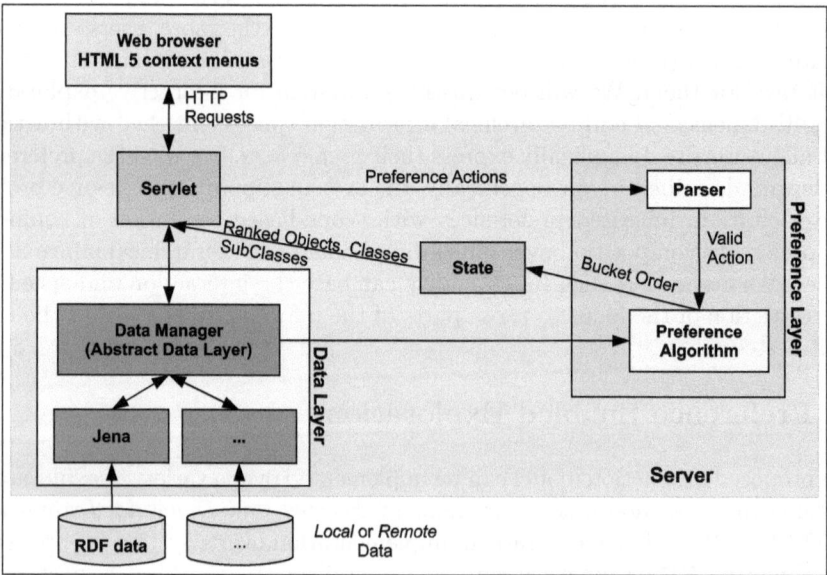

Fig. 2. The architecture of the system

enacted by right clicking in the browser window. The user is able to order classes, subclasses and objects using *best*, *worst* and *prefer to* actions (i.e. relative preferences), or actions that order them lexicographically or based on their count values. Regarding objects, since their number can be very large, the user is able to define a threshold, so that preferences are applied only when the number of objects is reduced under this specific threshold[3]. Furthermore, he can compose object scoped

---

[3] The user can reduce the number of objects by navigating over the classes, subclasses, and objects and restricting his focus.

preferences anchored to classes, using *Priority*, *Pareto* and *Pareto Optimal* compositions. Composition is offered by defining the appropriate composition mode and selecting classes through the appropriate classes' context menus. Finally, the user is able to store and load his preferences, since exploration is a time depth process.

A number of screenshots of the system with the available options and indicative preference actions is shown below. Figure 3 depicts the available system options (i.e. composition, threshold, session store and load) and the expression of a simple preference action. Figure 4 shows how we can define a relative preference action anchored to terms that affects the order of objects. Figure 5 shows how to order all objects lexicographically, and finally Figure 6 shows how the user can easily express more complicated actions, like *Pareto* composition.

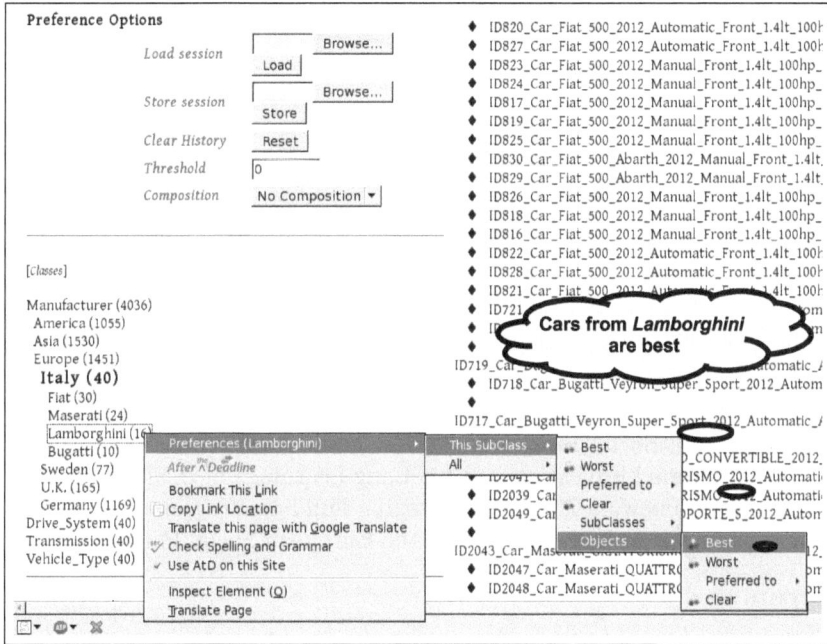

Fig. 3. Overview of the system options and a simple preference action (i.e. Cars from Lamborghini are best)

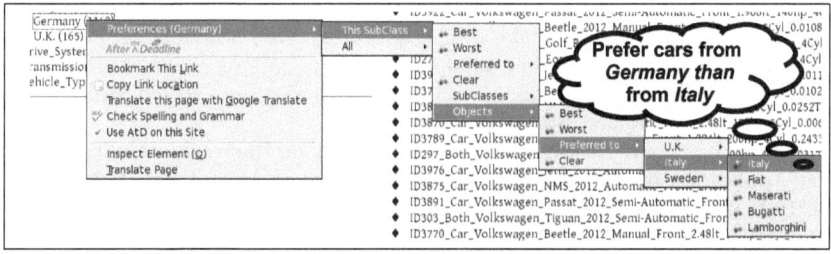

Fig. 4. Action to Prefer German to Italian cars

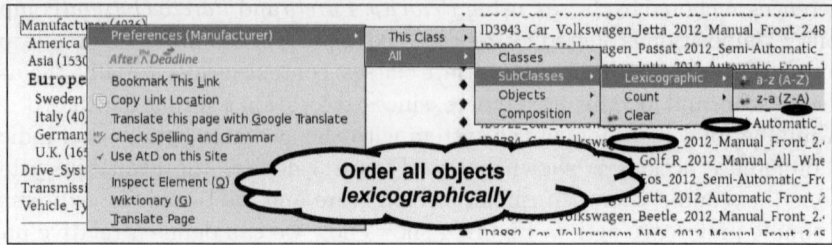

Fig. 5. Order all objects lexicographically in ascending order

Fig. 6. Add *Manufacturer* class to Pareto composition

This demonstration focuses on the flexibility of the provided preference actions and how with a few actions the user can select the desired car from an information base of 4036 cars.

Acknowledgements. This research has been co-financed by the European Union (European Social Fund - ESF) and Greek national funds through the Operation Program "Education and LifeLong Learning" of the National Strategic Reference Framework (NSRF) - Research Funding Program: Herakleitus II. Investing in knowledge society through the European Social Fund.

References

1. Manolis, N., Tzitzikas, Y.: Interactive Exploration of Fuzzy RDF Knowledge Bases. In: Antoniou, G., Grobelnik, M., Simperl, E., Parsia, B., Plexousakis, D., De Leenheer, P., Pan, J. (eds.) ESWC 2011, Part I. LNCS, vol. 6643, pp. 1–16. Springer, Heidelberg (2011)
2. Papadakos, P., Armenatzoglou, N., Kopidaki, S., Tzitzikas, Y.: On Exploiting Static and Dynamically Mined Metadata for Exploratory Web Searching. Knowledge and Information Systems 30(3), 493–525 (2012)
3. Sacco, G.M., Tzitzikas, Y. (eds.): Dynamic Taxonomies and Faceted Search: Theory, Practise and Experience. Springer (2009)
4. Stefanidis, K., Koutrika, G., Pitoura, E.: A Survey on Representation, Composition and Application of Preferences in Database Systems. ACM Trans. Database Syst. 36, 19:1–19:45 (2011)
5. Tzitzikas, Y., Papadakos, P.: Interactive Exploration of Multi-Dimensional and Hierarchical Information Spaces with Real-time Preference Elicitation. Fundamenta Informaticae (accepted for publication in 2012)

Author Index